SYSTEM IDENTIFICATION
(SYSID 2000)

A Proceedings volume from the 12th IFAC Symposium on System Identification, Santa Barbara, California, USA, 21 – 23 June 2000

Edited by

R. SMITH

Department of Electrical and Computer Engineerin
University of California, Santa Barbara, CA, USA

(In three volumes)

Volume 1

Published for the

INTERNATIONAL FEDERATION OF AUTOMATIC CONTROL

by

PERGAMON

An Imprint of Elsevier Science

UK Elsevier Science Ltd, The Boulevard, Langford Lane, Kidlington, Oxford, OX5 1GB, UK
USA Elsevier Science Inc., 660 White Plains Road, Tarrytown, New York 10591-5153, USA
JAPAN Elsevier Science Japan, Tsunashima Building Annex, 3-20-12 Yushima, Bunkyo-ku, Tokyo 113, Japan

First edition 2001

Library of Congress Cataloging in Publication Data

A catalogue record for this book is available from the Library of Congress

British Library Cataloguing in Publication Data

A catalogue record for this book is available from the British Library

ISBN 0-08-043545 9

Printed in Great Britain

12th IFAC SYMPOSIUM ON SYSTEM IDENTIFICATION (SYSID 2000)

Sponsored by
International Federation of Automatic Control (IFAC)
IFAC Technical Committee on Modeling, Identification and Signal Processing
The American Automatic Control Council
Department of Electrical and Computer Engineering at the University of California, Santa Barbara

International Programme Committee (IPC)

Michèle Basseville,	France	Ioan Landau,	France
David Bayard,	U.S.A.	Wally Larimore,	U.S.A.
Robert Bitmead,	Australia	Lennart Ljung,	Sweden
Mogens Blanke,	Denmark	Pertti Mäkilä,	Finland
Jie Chen,	U.S.A.	Iven Mareels,	Australia
Han-Fu Chen,	China	Tomas McKelvey,	Sweden
Manfred Deistler,	Austria	Mario Milanese,	Italy
Bart De Moor,	Belgium	Brett Ninness,	Australia
Guy Dumont,	Canada	Kameshwar Poolla,	U.S.A.
Peter Gawthrop,	U.K.	Boris Polyak,	Russia
Laszlo Gerencsèr,	Hungary	Phillip Regalia,	France
Michel Gevers,	Belgium	Dan Rivera,	U.S.A.
Laura Giarré,	Italy	Johan Schoukens,	Belgium
Mike Grimble,	U.K.	Roy Smith (Chair),	U.S.A.
Lei Guo,	China	Torsten Söderström,	Sweden
Fred Hadaegh,	U.S.A.	Roberto Tempo,	Italy
Shinji Hara,	Japan	Heinz Unbehauen,	Germany
Håkan Hjalmarsson,	Sweden	Paul Van den Hof,	The Netherlands
Rolf Isermann,	Germany	Bo Wahlberg,	Sweden
Hidenori Kimura,	Japan	Eric Walter,	France
Vikram Krishnamurthy,	Australia		
Kousuke Kumamaru,	Japan		

National Organizing Committee (NOC)

Co-chair
Dale Seborg
Dept. of Chemical Engineering,
University of California,
Santa Barbara, CA 93106, U.S.A.
+1 (805) 893-3352,
+1 (805) 893-4731 (fax)
seborg@engineering.ucsb.edu

Co-chair
Roy Smith
Dept. of Electrical
& Computer Engineering,
University of California,
Santa Barbara, CA 93106, U.S.A.
+1 (805) 893-2967,
+1 (805) 893-3262 (fax)
roy@ece.ucsb.edu

Finance
Duncan Mellichamp
Dept. of Chemical Engineering,
University of California,
Santa Barbara, CA 93106, U.S.A.
+1 (805) 893-2821,
+1 (805) 893-4731 (fax)
dmell@engineering.ucsb.edu

Facilities
Kameshwar Poolla
Dept. of Mechanical Engineering,
University of California,
Berkeley, CA 94720, U.S.A.
+1 (510) 642-4642,
+1 (510) 643-5599 (fax)
poolla@jagger.me.berkeley.edu

Registration
Dan Rivera
Dept. of Chemical, Bio.
& Materials Engineering,
Mail stop 876006
Arizona State University
Tempe, AZ 85287-6006, U.S.A.
+1 (480) 965-9476,
+1 (480) 965-2910 (fax)
daniel.rivera@asu.edu

Publications
Mario Sznaier
Dept. of Electrical Engineering
227E Electrical Engineering West
The Pennsylvania State University
University Park, PA 16802, U.S.A.
+1 (814) 865-0196,
+1 (814) 865-7065 (fax)
msznaier@frodo.ee.psu.edu

Foreword

Santa Barbara was the host city for the IFAC Symposium on System Identification (SYSID), which was held from the 21st to the 23rd of June, 2000 at the Doubletree Resort. Sponsorship for this symposium was provided by: the IFAC technical committee on Modeling, Identification and Signal Processing; the American Automatic Control Council; and the Department of Electrical and Computer Engineering at the University of California, Santa Barbara.

The Symposium on System Identification is IFAC's longest running, and largest, symposium. It is without doubt the premier forum for the presentation of research and application results in our field. SYSID is held every three years and the previous U.S. location was Baltimore in 1982. Santa Barbara was the ideal location for SYSIDs return to the U.S., being centrally located in California with close connections to a large number of active research centers; some local, some further afield.

A total of 291 manuscripts were submitted to SYSID 2000. Each was sent to at least two members of the International Program Committee who were responsible for obtaining reviews. The final program contained 226 papers. Of these 183 were scheduled to be presented in a lecture format, 28 were to be given in poster sessions and 15 were presented as software demonstrations. A novel scheduling arrangement was adopted for the poster sessions. One time slot—Thursday midday—was dedicated solely to poster presentations.

This proceedings volume contains 205 papers including the papers that accompany the plenaries given by Jorma Rissanen, *"Complexity and Information in Data"*, and Tomas McKelvey, *"Frequency Domain Identification."* Unfortunately there is no paper accompanying David Luenberger's plenary, *"System Identification in Modern Finance."* You had to be there for that one.

A symposium of this size relies on the work of a large number of dedicated people. I would personally like to thank both the National Organizing Committee and the International Program Committee for their significant effort in ensuring that the symposium was a success.

Roy Smith

Chair, International Program Committee

CONTENTS

VOLUME 1

PLENARY PAPERS

IDENTIFICATION FOR ROBUST CONTROL

NEURAL NETWORKS

SUBSPACE METHODS I

BIOLOGICAL APPLICATIONS

MODEL ERROR MODELING AND MODEL VALIDATION

CONTINUOUS TIME METHODS

SUBSPACE METHODS II

ENVIRONMENTAL MODELING

TIME-VARYING SYSTEMS

IDENTIFICATION FOR CONTROL

BLIND IDENTIFICATION, EQUALIZATION AND SOURCE LOCALIZATION

MODEL SELECTION

ADAPTIVE CONTROL I

VOLUME 2

PROCESS APPLICATIONS I

MODELING AND IDENTIFICATION WITH ORTHOGONAL BASIS FUNCTIONS

FAULT DETECTION AND MONITORING

SOFTWARE DEMONSTRATION I (GENERAL)

ADAPTIVE CONTROL II

PROCESS APPLICATIONS II

AUTOMOTIVE AND MECHANICAL APPLICATIONS

IDENTIFICATION APPLICATIONS

SOFTWARE DEMONSTRATION II (MATLAB)

CLOSED-LOOP IDENTIFICATION

RECURSIVE IDENTIFICATION

BASES FOR NONLINEAR SYSTEM IDENTIFICATION

EDUCATION AND TRAINING IN SYSTEM IDENTIFICATION

VOLUME 3

SOFTWARE DEMONSTRATION III (MATLAB)

BOUNDED-ERROR PARAMETER AND STATE ESTIMATION I

PARAMETER ESTIMATION

NONLINEAR SYSTEM IDENTIFICATION I

DISTRIBUTED PARAMETER SYSTEMS

BOUNDED-ERROR PARAMETER AND STATE ESTIMATION II

ESTIMATION AND FILTERING

NONLINEAR SYSTEM IDENTIFICATION II

INPUT SEQUENCES IN LINEAR AND NONLINEAR IDENTIFICATION

SET MEMBERSHIP IDENTIFICATION

STOCHASTIC ESTIMATION AND CONTROL

NONLINEAR SYSTEM IDENTIFICATION III

EXPERIMENT DESIGN

COMPLEXITY AND INFORMATION IN DATA

J. Rissanen
IBM Research Division
Almaden Research Center, DPE-B2/802,
and
University of London, Royal Holloway
Egham, Surrey TW20 OEX, UK
jormarissanen@msn.com

Abstract: In this paper we describe recent results in model building and selection based on a formal definition of complexity and information in a given data sequence. This permits a decomposition of data into the information bearing part and noise and a similar parameter free sufficient statistics factorization of a model which is universal for a parametric class. *Copyright ©2000 IFAC*

Keywords: Models, Estimation, Inference, Universal data compression

1 INTRODUCTION

It seems tautological to state that the objective of statistical model building is to obtain information from data. In the absence of a clear definition of what information is, or how to measure it, the common substitute is some probabilistic machinery that we assume to have generated the data, and which when 'estimated' will provide us with the sought-for information. Although 'information' certainly has to do with some such machinery, the procedure proposed is far too simplistic to provide any coherent and solid foundation for a theory of model building. The main difficulty, of course, is that since we cannot generally touch and measure the machinery that could have generated the data we have to leave its creation to imagination. A fundamental difficulty in sorting out the plausible ones from the rest is the fact that a complex model fits the data better than a simple one, which seems to lead to the inevitable and often disastreous conclusion that the more complex model we fit to the data the more we learn. Rather than dwelling further in the confusion that the absence of modeling theory has created we move on to outline a formal definition of information on which a sound modeling theory can be erected.

It stands to reason that we cannot expect to define information without some type of language in which to express it. The language for us will be a set $\mathcal{M}_\gamma$ of probability models of type $f(y^n;\theta,\gamma)$, or $f(y^n|x^n;\theta,\gamma)$ expressing a statistical relationship between the data strings $y^n = y_1,\ldots,y_n$ and $x^n = x_1,\ldots,x_n$, where for simplicity $\theta = \theta_1,\ldots,\theta_k$ is a collection of real-valued parameters, γ a structure index such as the pair of ARMA orders p,q, and k depends on γ. We further may consider larger model classes such as $\mathcal{M} = \cup_\gamma \mathcal{M}_\gamma$. An example is the class of all ARMA models with some probability distribution modeled for the input.

In broad terms, we want to separate the observed data sequence y^n into two parts $y^n = \hat{y}^n + e^n$, such that $\hat{y}^n$ has all the useful information that can be extracted from the data with the model class, and e^n is just 'noise' having nothing to contribute to $\hat{y}^n$. These intuitive properties are formalized by the requirement that e^n cannot be compressed with the model class selected, and that the same holds about $\hat{y}^n$. Also, by knowing e^n we cannot shorten the code length or the description of $\hat{y}^n$. Hence, in this sense both parts are independent, and the entire code length $L(y^n;\mathcal{M}_\gamma)$ (or $L(y^n;\mathcal{M})$ depending on the class of models) of the data sequence y^n is minimum. We call that length the *stochastic complexity* of the sequence, given the class of models considered. The amount of the (useful) *information* in the data is the code length $L(\hat{y}^n;\mathcal{M}_\gamma)$ (or $L(\hat{y}^n;\mathcal{M})$) of the sequence $\hat{y}^n$.

Such properties can also be expressed as follows: The stochastic complexity defines a *universal* model for the model class $\mathcal{M}_\gamma = \{f(y^n;\theta,\gamma)\}$ by

$$\hat{f}(y^n;\mathcal{M}_\gamma) = e^{-L(y^n;\mathcal{M}_\gamma)}. \quad (1)$$

It factors as

$$\hat{f}(y^n;\mathcal{M}_\gamma) = p(\hat{\theta}(y^n))q(e^n), \tag{2}$$

where $\hat{\theta}(y^n)$ is the maximum likelihood estimate of θ. This is a generalization of the usual sufficient statistics decomposition into a *universal* sufficient statistics decomposition, which has no parameters.

We can then conclude that the objective in modeling is to achieve the decomposition (2) rather than minimize this or that criterion, and if we succeed in this we have extracted all the useful information in the data that on the whole can be extracted with the models proposed. For instance, we cannot predict the data better than by use of the universal model, and so on. It is no accident that the criteria that tend to give best results, regardless of what reasoning has led to them, are those that happen to achieve the decomposition above to some approximation. Also, principles like Occam's razor and its variants express only the obvious: redundancy cannot add to the useful information, and since such a model cannot achieve the stochastic complexity the redundancy should be removed.

Notice that we have made no assumption about the data generating mechanism. The model classes selected are just our way of being able to formalize the crucial notions of 'useful information' and 'complexity'. Their selection cannot be optimized and must be left to ingenuity and experience. In brief, we have formalized the part in modeling process that can be formalized and left open the part that must be left open. Another important point to realize is that we cannot meaningfully ask for more than what lies in the data given. If the data sequence is short, it is completely meaningless to ask for models that are too complex, regardless of how complex the 'true' data generating mechanism may be. The arguments sometimes raised that complex models are required for complex 'truth' are just foolish.

In the body of this paper we discuss in some detail how to achieve the desired decomposition to an approximation that improves with the growing amount of data. As can be expected, it can be achieved to a near perfection for the Gaussian model classes and, more generally, for the exponential class.

2 CONSTRUCTION OF UNIVERSAL MODELS

As indicated in Introduction the key to the desired factorization and decomposition is the shortest (ideal) code length with which the data sequence x^n can be encoded, when the codes are somehow designed with the models in the classes $\mathcal{M}_\gamma = \{f(x^n;\gamma,\theta)\}$ and $\mathcal{M} = \bigcup_\gamma \mathcal{M}_\gamma$, respectively. The word 'ideal' refers to the convenient habit of dropping the requirement that a real code length must be integer-valued, and hence the negative logarithm of any probability or density is regarded as an ideal code length. By a harmless abuse of notation even the word 'ideal' is often dropped. For those unfamiliar with coding we add that all these conventions are justified, because we can design codes for a large set of objects such that the real integer-length code lengths differ from the ideals by a negligible amount. And since our objects are typically the sequences x^n and real-valued parameters, their sets are certainly 'large'. Whenever we need to consider code lengths for small sets of objects, such as structure indices, we use the real code lengths or their accurate estimates. To summarize, a probability distribution, a model, and a code can be identified, and the negative logarithm of a probability is taken as a code length. With these agreements the shortest code length, relative to the class $\mathcal{M}_\gamma$, we are searching for will be of the form $-\ln f(x^n;\gamma)$, which means that we are looking for a model that is universal for the class $\mathcal{M}_\gamma$ in question. The desired decomposition then is a matter of factoring it.

If we go back to Shannon's measure of the mean information, the entropy, of a given (discrete) random variable with distribution $P(x)$, it can be defined as the solution to a minimization problem, namely,

$$\min_Q \sum_x P(x) \ln \frac{1}{Q(x)},$$

where $Q(x)$ ranges over all discrete distributions. The solution is $Q(x) = P(x)$ for all x, and the shortest mean ideal code length $E_P \ln(1/Q(X))$ is the entropy $E_P \ln(1/P(X))$. Shannon's problem can also be stated in another form, which suggests generalizations

$$\min_Q \sum_x P(x) \ln \frac{P(x)}{Q(x)}. \tag{3}$$

The expression $D(P\|Q) = E_P \ln(P(x)/Q(x))$ to be minimized, is the important Kullback-Leibler distance (KL-distance) between the two distributions.

We need to generalize this process for our case, where instead of a fixed distribution we have a class of them. We also discuss the general case of density function models. The first and the traditional way in universal coding theory to generalize (3) is the following minmax problem

$$R_n(k) = \min_q \max_\theta E_\theta \ln \frac{f(X^n;\theta)}{q(X^n)}, \tag{4}$$

where E_θ denotes the expectation with respect to the model $f(x^n;\theta)$ in the class $\mathcal{M}_k$. We drop the structure index γ and let $\theta = \theta_1, \ldots, \theta_k$ range over some subset Ω of the $k-$dimensional Euclidean space. We also replace the data string symbol y^n by x^n, for we consider the case where only one data sequence is given; the other case is handled in the same manner. In words, we are looking for a density function which is closest to the worst case model in the KL-distance $D(f(X^n;\theta)\|q(X^n))$. The minmax KL-distance is called the *redundancy*.

We give the quite elegant and well known arguments, [4], to describe the solution. The first step is to widen the problem by letting the maximization be done over all 'priors' w in the mixture densities

$$f_w(x^n) = \int w(\theta) f(x^n;\theta) d\theta \tag{5}$$

thus

$$\min_q \max_w \int w(\theta) E_\theta \ln \frac{f(X^n;\theta)}{q(X^n)} d\theta. \tag{6}$$

Since singular priors, which put all the probability to a single point θ, are allowed, the new minmax is at least as large as the original. Suppose the minimization can be done first. By integrating over θ first and then over x^n in (6), we see by Shannon's theorem that for any w the minimizing density is given by the mixture (5) (because the expectation over the logarithm of the numerator does not depend on q). We are left with finding the maximizing prior $\bar{w}$ of

$$\max_w \int w(\theta) E_\theta \ln \frac{f(X^n;\theta)}{f_w(X^n)} d\theta. \tag{7}$$

Let $\bar{q}$ denote the minimizing distribution in (4). Then

$$\int \bar{w}(\theta) E_\theta \ln \frac{f(X^n;\theta)}{f_{\bar{w}}(X^n)} d\theta \leq \int \bar{w}(\theta) E_\theta \ln \frac{f(X^n;\theta)}{\bar{q}(X^n)} d\theta. \tag{8}$$

Because the mean does not exceed the maximum the right hand side is less than or equal to $R_n(k)$, and the two minmax problems are equal. The solution to (4) is then the mixture with the prior maximizing (7). For many processes, above all when they are iid processes, [3], the maximizing mixture is obtained with the so-called Jeffreys' prior,

$$\bar{w}(\theta) = \frac{\sqrt{|I(\theta)|}}{\int_\Omega \sqrt{|I(\theta)|} d\theta}, \tag{9}$$

where $I(\theta)$ is the Fisher information matrix

$$I(\theta) = \{-E_\theta \frac{\partial^2 \ln f(X;\theta)}{\partial\theta_i \partial\theta_j}\}.$$

The redundancy then is given by

$$R_n(k) = \frac{k}{2} \ln \frac{n}{2\pi e} + \ln \int_\Omega \sqrt{|I(\theta)|} d\theta + o(1). \tag{10}$$

Since $R_n(k)$ does not depend on θ the optimal mixture is seen to be at equal distance from all the models $f(x^n;\theta)$ in $\mathcal{M}_k$.

It is clear that the models in the class $\mathcal{M}_k$ cannot express all the statistical properties in real world data sequences x^n, no matter how large n is. For instance, if we generate the data with a density function $g(x^n)$ which is outside the class, the data will have statistical properties different from those expressible with the models in the class. Notice, that we do not want to make the claim that the data are a sample from any distribution. Rather, we are simply using density functions to describe statistical properties in the data. This suggests the following generalization of the minmax problem (4), as studied in [12]: First, one defines the parameter θ_g that minimizes the KL-distance $E_g \ln g(X^n)/f(X^n;\theta)$, where g is a distribution outside the class $\mathcal{M}_k$, and then (4) is replaced by

$$\min_q \max_g E_g \ln \frac{f(X^n;\theta_g)}{q(X^n)}, \tag{11}$$

where g runs over a large class that includes $\mathcal{M}_k$ as a proper subclass. The new minmax value, the *relative redundancy*, was found in [12] to be asymptotically the same $R_n(k)$ resulting from a modified Jeffreys' mixture.

Before discussing further the universal model defined by Jeffreys' mixture we consider another related minmax problem. If we consider the model $f(x^n;\theta_g)$ as the best in the given class $\mathcal{M}_k$, it has the drawback that we cannot compute it, because we do not have the distribution g. What we instead can compute is $f(x^n;\hat{\theta}(x^n))$, which certainly gives an ideal code length which no model in the class can beat. However, this is not a model at all, because it does not integrate to unity. This then suggests the following minmax problem, [2],

$$\min_q \max_\theta E_\theta \ln \frac{f(X^n;\hat{\theta}(X^n))}{q(X^n)}, \tag{12}$$

where $q(x^n)$ is any density function and the expectation is with respect to $f(x^n;\theta)$. Its generalization is

$$\min_q \max_{g \in G} E_g \ln \frac{f(X^n;\hat{\theta}(X^n))}{q(X^n)}, \tag{13}$$

where G is a class larger than $\mathcal{M}_k$. In fact, we can let G consist of all distributions such that $G = \{g : |E_g \ln(g(X^n)/f(X^n;\hat{\theta}(X^n))| < \infty\}$. This excludes the singular distributions, which clearly do not restrict the data in any manner and hence do not specify any properties in them. Also, both the minimum and the maximum will be reached.

Theorem 1 *The solution to the minmax problem (13) is the universal* NML *(normalized maximum likelihood) model*

$$\hat{f}(x^n) = \frac{f(x^n;\hat{\theta}(x^n))}{C_n(k)}, \tag{14}$$

where

$$C_n(k) = \int_{\hat{\theta}(y^n)\in\Omega} f(y^n;\hat{\theta}(y^n)) dy^n, \tag{15}$$

and Ω is such that the integral is finite. Clearly,

$$E_g \ln \frac{f(X^n;\hat{\theta}(X^n))}{\hat{f}(X^n} = \ln C_n(k) \tag{16}$$

for all g.

The proof is given in [10].

The solution $\hat{f}(x^n)$ also solves Shtarkov's minmax problem, [11],

$$\min_q \max_{x^n} \ln \frac{f(x^n;\hat{\theta}(x^n))}{q(x^n)} = \ln C_n(k). \tag{17}$$

Notice the important fact that the best model $\hat{f}$ involves only the models in the class $\mathcal{M}_k$. Under certain conditions, satisfied for the classes of exponential distributions, [8],

$$\ln C_n(k) = \frac{k}{2} \ln \frac{n}{2\pi} + \ln \int_\Omega \sqrt{|I(\theta)|} d\theta + o(1), \tag{18}$$

where $I(\theta)$, is the Fisher information matrix

$$I(\theta) = \{-E(\theta \frac{\partial^2 \ln f(X|\theta)}{\partial\theta_i \partial\theta_j}\}.$$

By comparing $R_n(k)$ and $C_n(k)$ we see that the mean code length difference $E_g \ln(\hat{f}(X^n)/f_{\bar{w}}(X^n))$ amounts to the difference $E_g \ln(f(X^n;\hat{\theta}(X^n))/f(X^n;\theta_g)) + o(1)$. We know that when g is restricted to $\mathcal{M}_k$ the difference iid processes goes to zero as n grows.

It was shown in [7] that the mean of the universal code length given by (9) as well as that of the *NML* model

$$-\ln \hat{f}(x^n) = -\ln f(x^n;\hat{\theta}(x^n)) + \ln C_n(k), \qquad (19)$$

the mean taken with respect to models $f(x^n;\theta)$, is in essence a lower bound that no code can beat, except when θ ranges over a set whose volume goes to zero as n grows to infinity. A generalization of this is given in [10]. Accordingly we are justified to define the ideal code length (19) to be the *stochastic complexity* of the data string x^n, relative to the model class $\mathcal{M}_k$. We can rewrite the *NML* model as

$$\hat{f}(x^n) = f(x^n|\hat{\theta}(x^n))\pi(\hat{\theta}(x^n)) \qquad (20)$$

$$f(x^n|\hat{\theta}(x^n)) = \frac{f(x^n;\hat{\theta}(x^n))}{g(\hat{\theta}(x^n))} \qquad (21)$$

$$\pi(\hat{\theta}(x^n)) = \frac{g(\hat{\theta}(x^n))}{C_n(k)}, \qquad (22)$$

where $g(\hat{\theta}(x^n))$ is a density induced by $f(x^n;\hat{\theta}(x^n))$.

For sequences y^n such that $\hat{\theta}(y^n) = \hat{\theta}(x^n)$ the code lengths $-\ln f(y^n|\hat{\theta}(y^n))$, divided by n, are virtually equal for large n, which means that we cannot compress them below this length with the given model class; it is the length of a code defining just incompressible noise having virtually no information to add to the regular learnable features that are in the optimal model $f(y^n;\hat{\theta}(x^n))$. We call $-\ln \pi(\hat{\theta}(x^n))$ the amount of *information* in the data sequence x^n that can be extracted with the given model class.

For exponential families of models the factorization of $\hat{f}(x^n)$, (20), generalizes the usual parametric sufficient statistics decomposition to a parameter free *universal* sufficient statistics decomposition. For many other families the decomposition has the desired properties only asymptotically. For the special case of linear regression with normal models such a decomposition is given in an exact form in [9]. The paper also includes a numerical example of denoising speech data with results superior to anything we are aware of.

It was shown in [6] by quite elegant arguments of differential geometry without any appeal to code lengths that one can derive a continuum of distinguishable distributions such that their equal distance lattice in the natural metric, defined by the quadratic form $\theta' I(\theta)\theta$, for data sequences of length n has the number of elements given by $C_n(k)$, (18). Its logarithm is called there the *geometric complexity*. Also the *NML* density function itself $\hat{f}(x^n)$ has an interpretation in terms of differential geometry. One could also regard $\ln C_n(k)$ as the information in the data sequence. The difference is that $-\ln \pi(\hat{\theta}(x^n))$ is an ideal code length while $\lceil \ln C_n(k) \rceil$ is a real code length. If we convert the former into a real code length by optimal quantization of the parameter $\hat{\theta}(x^n)$, the probability mass induced by $g(\hat{\theta}(x^n))$ in the so-defined equivalence classes converges to unity, and we get the real code length as $\lceil \ln C_n(k) \rceil$.

We mention to this end that the mixture universal model has the advantage that it defines a random process while the *NML* model does not. However, since generally $\hat{\theta}(x^n)$ converges we may approximate $\hat{f}(x^{n+1})/\hat{f}(x^n)$ by the predictive conditional density $f(x_{n+1}|x^n,\hat{\theta}(x^n))$, which define a random process and a closely related universal model.

3 AKAIKE'S QUEST

In the early seventies, [1], Akaike set out to obtain a model selection criterion by asking for the model in a parametric class which is closest in KL-distance to a data generating model lying outside of the parametric class. This then by dubious arguments to estimate this distance led into his criterion *AIC*. That there is a problem in the arguments is best illustrated by the fact that if we consider a class of nested subclasses, such as the set of polynomials of all orders k, where $k > m$ implies that $f(x^n;\hat{\theta}^k) \geq f(x^n;\hat{\theta}^m)$, and let the data generating model be a nonparametric one, then the nearest model is either the one with the maximum number of parameters or it does not exist at all. In fact, for $k > m$, $D(g(X^n)\|f(X^n;\hat{\theta}^k)) < D(g(X^n)\|f(X^n;\hat{\theta}^m))$, and the search for the nearest model in $\mathcal{M} = \bigcup_k \mathcal{M}_k$ does not give a meaningful model selection criterion in the ideal case where we know the data generating model! Also *AIC* fails to recover the data generating model in the case where it lies within the model class.

The difficulty can be avoided if we consider the problem of finding the nearest *universal* model, written now as $\hat{f}(X^n;k)$,

$$\min_k D(g(X^n)\|\hat{f}(X^n;k)) \qquad (23)$$

rather than the parametric model. Minimization of the natural data dependent estimate of the KL-distance results in the *MDL* criterion, which seeks to minimize the stochastic complexity (19). Although the *MDL* estimates of k and more generally γ have been shown to be strongly consistent only in the case where the data generating model is within the family $\mathcal{M}$, in view of the convergenge of the estimates $\hat{\theta}(x^n)$ to θ_g and $-(1/n)\ln f(x^n;\hat{\theta}(x^n))$ to $-E_g \ln f(x^n;\theta_g))$ for iid processes, [5], [13], it is quite plausible that this criterion gives consistent estimates of the number of parameters in θ_g as well.

References

[1] Akaike, H. (1974), 'Information Theory and an Extension of the Maximum Likelihood Principle,' *Second international symposium on information theory, ed. B.N. Petrov and F. Csaki*, Akademia Kiedo, Budapest, 267-81

[2] Barron, A.R., Rissanen, J., and Yu, B. (1998) 'The MDL Principle in Modeling and Coding', special issue of *IEEE Trans. Information Theory* to commemorate 50 years of information theory, Vol. **IT-44**, No. 6, October 1998, pp 2743-2760

[3] Clarke, B. S. and Barron, A. R. (1990), 'Information-Theoretic Asymptotics of Bayes Methods', *IEEE Trans. Information Theory*, Vol. **IT-36**, No. 3, 453-471, May 1990.

[4] Davisson, L.D., 'Minimax Noiseless Universal Coding for Markov Sources', *IEEE Trans. Information Theory*, Vol. **IT-29**, No. 2, 211-215, 1983

[5] Grünwald, P.D. (1998), *The Minimum Description Length Principle and reasoning under Uncertainty*, PhD thesis, Institute for Logic, Language and Computation, Universiteit van Amsterdam, 296 pages

[6] Myung, I.J., Balasubramanian, Vijay, and Pitt, Mark, A. (1999), 'Counting Probability Distributions: Differential Geometry and Model Selection', submitted to Proc. of National Academy of Science; see also http://arxiv.org/list/nlin/9601 for Balasubramanian, V. ,'Statistical Inference, Occam's Razor and Statistical Mechanics on the Space of Probability Distributions', *Neural Computation*, **9**, No. 2, 349-268, 1997

[7] Rissanen, J. (1986), 'Stochastic Complexity and Modeling', *Annals of Statistics*, Vol **14**, 1080-1100

[8] Rissanen, J. (1996), 'Fisher Information and Stochastic Complexity', *IEEE Trans. Information Theory*, Vol. **IT-42**, No. 1, pp 40-47

[9] Rissanen, J. (1999), 'MDL Denoising', (submitted to IEEE Trans. on Information Theory), http://www.cs.tut.fi/ rissanen/

[10] Rissanen, J. (2000), 'Strong Optimality of Normalized ML models as Ideal Codes', (submitted to IEEE Trans. on Information Theory), http://www.cs.tut.fi/ rissanen/

[11] Shtarkov, Yu. M. (1987), "Universal Sequential Coding of Single Messages", Translated from Problems of Information Transmission, Vol. 23, No. 3, 3-17, July-September 1987.

[12] Takeuchi, Jun-ichi and Barron, Andrew R. (1998), 'Robustly Minimax Codes for Universal Data Compression', *The 21'st Symposium on Information Theory and Its Applications*, Gifu, Japan, December 2-5, 1998.

[13] White, H. (1994), *Estimation, inference, and specification analysis*, Cambridge University press, Cambridge, UK, 349 pages

FREQUENCY DOMAIN IDENTIFICATION

T. McKelvey *,[1]

* *Dept. of Electrical Engineering, Linköping University, SE-581 83 Linköping, Sweden.*
tomas@isy.liu.se

Abstract: Techniques to identify parametric transfer functions from noisy frequency domain data are considered. A maximum-likelihood estimation method is presented which in parallel with the system transfer function also estimates a parametric noise transfer function. This leads to a consistent and efficient estimator. It is shown how the discrete Fourier transform can be applied to generate frequency domain data from sampled time domain data. For the finite data case the exact frequency domain expressions are derived relating the transfer function with the discret Fourier transformed data for both continuous and discrete time systems.

Keywords: Identification, spectral estimation, discrete Fourier transform, maximum-likelihood estimation

1. INTRODUCTION

Building mathematical models based on measured input and output signals of a dynamical system is known as system identification. Such models based on empirical information are important if the dynamical system is unknown or is only partially known and when it is in-feasible to derive a theoretical model from first principles. The availability of accurate models is important in order to derive high performance solutions, e.g., for model based control design or model based signal processing.

Almost all measurements originating from real world devices intrinsically belong to the time domain, i.e. are samples of continuous time signals. Consequently most system identification methods and the theory developed around them deals with how to determine parametric models from such time domain measurements (Ljung 1999, Söderström and Stoica 1989). The very basic and old technique of frequency analysis departs from the time domain technique.

In frequency analysis the system is assumed to be excited by a (sum of) pure sinusoidal signal(s). When the output has settled to a (sum of) stationary sinusoidal signal(s), the complex value of the transfer function at the specific excitation frequencies is determined by the discrete Fourier transform (DFT).

So called frequency analyzers are dedicated pieces of equipment which performs such experiments and deliver the result as a non-parametric transfer function. In a second step a parametrized transfer function model can be fitted to the transfer function data using some complex curve fitting technique (Levy 1959, Sanathanan and Koerner 1963, Whitfield 1986, Whitfield 1987, Pintelon *et al.* 1994). A discussion of techniques to fit parametric models to noisy frequency domain data is the scope of this paper. An early reference for the related time series problem is (Whittle 1951) wherein classical inferential procedures, e.g. maximum-likelihood estimation, were combined with spectral theory. See also (Hannan 1970, Robinson 1976). The book (Schoukens and Pintelon 1991) covers many aspects of frequency domain identification techniques and presents

[1] Supported by the Swedish Foundation for International Cooperation in Research and Higher Education (STINT). This work was completed while the author was visiting Dept. of Electrical and Computer Engineering, University of Newcastle, Australia.

a method for the frequency domain errors-in-variables problem.

Since the transformation of a signal from time to frequency domain using the discrete Fourier transform is nothing but a unitary transformation it might appear, at first sight, that nothing is gained.

However, an important difference arises when colored noises are present. Specifically, for such a colored signal $v(t)$ which have a covariance function $c(\tau) = \mathbf{E}\{v(t+\tau)v(t)\}$ (decaying to zero at an exponential rate), the vector of all time domain measurements have a full covariance matrix. The optimal estimation procedure, i.e. the maximum-likelihood method, requires the use the full covariance matrix since all time domain signals are statistically dependent (Hannan and Deistler 1988). The unitary transformation, represented by the DFT, however decouples this statistical dependence. That is, in the frequency domain each frequency point is asymptotically (as the number of time domain data points tends to infinity) independent of the others (Brillinger 1981).

Furthermore under very mild assumptions regarding the probability distribution of the time domain noise signal, the frequency domain data are all (asymptotically) normally distributed. Optimal estimators therefore only include a diagonal covariance matrix which simplifies the actual computations.

Prefiltering with the inverse of the filter describing the noise color will also make the time domain covariance matrix diagonal. This is also known as pre-whitening since the resulting noise is white. However this requires the knowledge of the noise color or it needs to be jointly estimated together with the rest of the unknown parameters. So-called prediction error methods (Åström 1980, Ljung 1999, Söderström and Stoica 1989) are based on prefiltering with a parametric dependent prefilter and are asymptotically equivalent to the method of maximum-likelihood. In case only non-parametric noise information is at hand, the design of whitening-filter is more demanding. Finally it is important to recognize that the unitary transformation does not change any asymptotic properties if all frequency domain data points are retained. The time-domain maximum-likelihood methods and the frequency domain counterpart both have equal asymptotic properties as pointed out by (Hannan 1970, Ljung and Glover 1981, Schoukens *et al.* 1997*a*) and others. The actual choice of one over the other must therefore be based on the finite data properties of each estimator and also on computational complexity.

A distinctive feature of frequency domain techniques is that the modeling of continuous time systems from sampled data can be done in a straightforward fashion if a certain class of band-limited excitation signals is employed (Robinson 1976, Schoukens *et al.* 1994). This is a great advantage in contrast with the rather involved time domain techniques which, even in the noise free case, are only approximate if a finite set of sampled data is available. See (Young 1981, Unbehauen and Rau 1990). A continuous time system with a time delay is rather difficult to model in the time domain since it cannot be described by a finite system of ordinary differential equations. In the frequency domain however a nice finite dimensional parametric description exists which lends itself to identification using parametric methods.

Frequency domain identification of nonlinear systems of Wiener-Hammerstein type have been reported in (Vandersteen *et al.* 1997, Vandersteen and Schoukens 1999). These techniques rely on a convergent Volterra representation of the nonlinear system.

Techniques based on higher order spectra (polyspectra) can be found in (Giannakis 1995, Tugnait 1998) which by assuming particular noise and input distributions generate consistent estimates.

1.1 *Problem formulation*

Let us assume that we are interested in obtaining a model of a system that can be described by the following linear time-invariant form

$$y(t) = G_0(q)u(t) + H_0(q)e(t) \tag{1}$$

where $y(t)$, $u(t)$ and $e(t)$ are the real valued output, input and innovation signals, respectively. The operators $G_0(q)$ and $H_0(q)$ represent the discrete time linear transfer functions. We assume that H_0 is a stable monic filter. For a continuous time representation, exchange the delay operator q with the differentiation operator p. The noise signal $e(t)$ is assumed to be i.i.d. and zero mean with variance λ_0 and independent of the input signal $u(t)$. We will later in Section 3.5 discuss the closed loop case where u and e are dependent through a feedback controller.

Let us postulate the following frequency domain system equation

$$Y(\omega) = G_0(e^{i\omega})U(\omega) + H_0(e^{i\omega})E_0(\omega). \tag{2}$$

where Y and U are the (weak) limits of the Fourier transform of y and u respectively and E_0 is the frequency domain innovation which is a zero mean complex normal random variable with frequency invariant variance λ_0. The complex functions $G_0(e^{i\omega})$ and $H_0(e^{i\omega})$ are the frequency functions of the linear operators G and H respectively. In Section 2 we will more formally justify (2). For a continuous time system just exchange the argument $e^{i\omega}$ for $i\omega$.

Assume the relation (2) can be sampled at a sequence of arbitrary frequencies in the set $\Omega_N = \{\omega_k\}_{k=1}^N$ yielding the set of measurements $Z^N = \{Y_k, U_k | k = 1, \ldots, N\}$ where $U_k = U(\omega_k)$ and

$Y_k = Y(\omega_k)$. The aim is to find a model of (1) and to do so we construct a model set by using a parametrized model structure

$$Y(\omega) = G_\theta(e^{i\omega})U(\omega) + H_\theta(e^{i\omega})E(\omega), \qquad (3)$$

where $G_\theta(z)$ and $H_\theta(z)$ are respectively the rational transfer functions of the system and noise which are parametrized by a real valued vector θ. Let $D_{\mathcal{M}}$ denote the set of valid parameters. We assume $H_\theta(z)$ is stable and inversely stable monic transfer function for all $\theta \in D_{\mathcal{M}}$. We impose no particular structure on how the parameters enter into $G_\theta(z)$ or $H_\theta(z)$ and this enables the use of various parametrizations such as fraction of polynomials or state-space models. Hence, parametrized *gray-box models* which are only partially unknown can also be used.

Given the parametrized model class and data Z^N, the estimate of the system is found by parametric optimization of some criterion function

$$\hat{\theta} = \arg\min_{\theta} V_N(\theta, Z^N) \qquad (4)$$

where V_N is a function which provides a metric on how to optimally fit the model (3) to the given data Z^N.

1.2 *Outline*

The paper is organized as follows. In the next section we provide the background to the frequency domain equation (2) for both the discrete time case as well as the continuous time case, with a key feature being the provision of insight into how the noise properties carry over to the frequency domain by using a probabilistic setup. In Section 3 an identification criterion V_N is derived which is based on the technique of maximum-likelihood. The asymptotic properties of such a criterion are analyzed and their relation with the time domain counterpart is discussed.

The aim of the paper is to provide a basic understanding for the frequency domain identification problem. In order to accommodate this we will restrict ourself to consider the case when the system and noise models G_0 and H_0 are finite dimensional and scalar. Extensions to the multivariable case are straightforward.

2. FROM TIME TO FREQUENCY DOMAIN

The exact noise free relationships between the discrete Fourier transform (DFT) of the input and outputs and the frequency response function of the linear system will be derived under rather general excitation conditions. It is well known that the Fourier transform of the noise free output is exactly the frequency response function multiplied by the transform of the input signal if the transform make use of data sequences of infinite length (Ljung 1999). In the finite length case an extra term appears which accounts for the history of the system prior to the measurement interval. As the number of data tends to infinity the extra term tends to zero at a rate proportional to $\frac{1}{\sqrt{N}}$. In the following analysis we will use state-space models to describe the finite-dimensional linear systems. A similar derivation using transfer function models can also be found in (Pintelon *et al.* 1997).

2.1 *Discrete Fourier Transform*

Assume a signal $s(t)$ is sampled at N equidistant time instances $t = kh, k = 0, 1, \dots, N-1$ where h is the sampling interval. The N-point discrete Fourier transform (DFT) of the set $\{s(kh)\}_{k=0}^{N-1}$ is then defined as

$$S_N(\omega) \triangleq \frac{1}{\sqrt{N}} \sum_{k=0}^{N-1} s(kh)e^{-i\omega k} \qquad (5)$$

where $\omega \in [-\pi, \pi]$ is the normalized angular frequency in radians per second. Hence ω/h is the unnormalized angular frequency. Unless h is explicitly included we assume $h = 1$ in the sequel. Predominately we will focus on the N distinct values of S_N given by the argument $\omega_k \triangleq \frac{2\pi k}{N}$ for $k = 0, 1, \dots, N-1$. Notice that $e^{i\omega_k} = e^{-i\omega_{N-k}}$ and $S_N(\omega_k) = S_N^*(\omega_{N-k}) = S_N^*(-\omega_k)$ for $k = 1, \dots, N-1$. Here X^* denotes the complex conjugate of X.

2.2 *Discrete time systems*

In this section a derivation of the DFT relation for a discrete time system is presented. The discussion of the influence of noise is deferred to Section 2.5. A noise free discrete time system of finite order admits a state-space realization

$$\begin{aligned} x(t+1) &= Ax(t) + Bu(t) \\ y(t) &= Cx(t) \end{aligned} \qquad (6)$$

where $x(t)$ is a state-vector of length n, A is a $n \times n$ real matrix, B and C^T are vectors of size n. If the realization is of minimal order, n is the McMillan degree of the linear system (Kailath 1980). The transfer function is given by

$$G(z) = C(zI - A)^{-1}B \qquad (7)$$

which is a fraction between two polynomials. The *frequency response function* at frequency ω is defined as the transfer function evaluated on the unit circle $G(e^{i\omega})$.

Consider the system described by (6) and assume N points of the input and output signals are available, i.e. $\{y(t), u(t)\}_{t=0}^{N-1}$. The history of the input up to time $t < 0$ is unknown but its impact on the future is captured by the state at time zero, $x(0) = x_0$. Assume $\det(e^{i\omega_k}I - A)$ is nonzero for all $k = 0, 1, \dots, N-1$ which is equivalent to requiring the frequency response function to be finite for all ω_k or equivalently no poles of the system to be located at $e^{i\omega_k}$.

Let $Y_N(\omega)$ and $U_N(\omega)$ denote the N-point DFT of the output and input signals respectively. Then for $\omega_k = \frac{2\pi k}{N}$, $k = 0, 1 \ldots, N-1$ the following equation holds

$$Y_N(\omega_k) = G(e^{i\omega_k})U_N(\omega_k) + T(e^{i\omega_k})\frac{1}{\sqrt{N}} \quad (8)$$

where

$$\begin{aligned} T(z) &= zC(zI - A)^{-1}(I - A^N)(x_0 - x_p) \\ x_p &= (I - A^N)^{-1}\sum_{t=0}^{N-1} A^t Bu(N-1-t). \end{aligned} \quad (9)$$

A proof is detailed in Appendix A.1.

The transient term $T(z)\frac{1}{\sqrt{N}}$, which picks up the transient effects of the unmatched initial condition, decays as $\frac{1}{\sqrt{N}}$ if the system is stable, i.e., the eigenvalues of the A matrix have a modulus strictly less than one or, equivalently, all poles lie strictly inside the unit circle. If the input signal is non-periodic then $U_N(\omega_k)$ will stay bounded as N increases and hence $G(e^{i\omega_k})U_N(\omega_k)$ will be the dominating term for large N. If the input is periodic with period time P and $N = PM$, i.e. the measurements are collected over M full periods, then $U_N(\omega)$ will at most have P non-zero values at $\omega = \frac{2\pi k}{P}$, $k = 0, \ldots, P-1$. The size of the non-zero values will grow at a rate of $\sqrt{M}$. In this case we obtain

$$\frac{|T(e^{i\omega_k})|\frac{1}{\sqrt{N}}}{|G(e^{i\omega_k})U_N(\omega_k)|} \sim O(\frac{1}{N})$$

The explicit form of $T(z)$ enables the possibility of estimating it along with the transfer function $G(z)$. This could be beneficial when the data record is short and when a non-periodic excitation signal is used. Furthermore note that $G(z)$ and $T(z)$ share the same denominator polynomial and only the n parameters of the numerator polynomial need to be determined. If the input excitation is such that $U_N(\omega_k)$ is constant for all frequencies it is however not possible to distinguish between $G(z)$ and $T(z)$.

Note that the relation (8) holds also for unstable systems. In such a case the size of transient term will grow exponentially as N increases and it is essential to include it in any open loop estimation scheme.

2.3 *Continuous time systems*

The output of a finite dimensional continuous time system can be described as the solution to a system of first order differential equations

$$\begin{aligned} \dot{x}(t) &= A_c x(t) + B_c u(t) \\ y(t) &= C_c x(t) \end{aligned} \quad (10)$$

where $x(t)$ is the size n state vector. The transfer function is

$$G_c(s) = C_c(sI - A_c)^{-1}B_c \quad (11)$$

and the continuous time *frequency response function* at frequency ω is defined as the transfer function evaluated along the imaginary axis, i.e. $G_c(i\omega)$.

To successfully identify a continuous time system from sampled data it is important to consider how the input signal excites the continuous time system. If the input is piecewise constant between the sampling instances then the continuous system has a discrete time representation which exactly describes the output signal at the sample points and hence the expressions (8) and (9) hold. The mapping which takes a continuous time system into a discrete one is called zero order hold (ZOH) sampling (Åström and Wittenmark 1984). A restriction is that this mapping is not bijective. Several continuous time systems are represented by the same discrete time one. Furthermore the inverse mapping is not defined for certain discrete time systems (Åström and Wittenmark 1984). The method of first identifying a discrete time model and then employ the inverse ZOH mapping might consequently fail. The correct approach is to parameterize a continuous time model and then via the ZOH mapping derive the discrete time model which is matched with the sampled data (Ljung 1999).

A different approach which is well suited for the frequency domain, is to excite the continuous time system using a band limited input with a zero spectrum for all frequencies on and above the Nyquist frequency ($\frac{\pi}{h}$). Furthermore, assume that the input signal is such that it admitts a band-limited Nh-periodic continuation outside the measurement interval, i.e. $u(t) = u(t + Nh)$ for all t. Fourier analysis then tells us that all such signals can be described as (for N odd)

$$u(t) = \sum_{k=-\frac{N-1}{2}}^{\frac{N-1}{2}} f_k e^{i\frac{2\pi k}{Nh}t} \quad (12)$$

where $u(t)$ is real if $f_k = f^*_{-k}$. Clearly all such real valued signals are composed of a sum of sinusoids with (normalized) frequencies constrained to belong to the finite set $\{\omega_k \,|\, \omega_k = \frac{2\pi k}{N} < \pi, k = 0, 1, \ldots\}$. This type of excitation signal is known as multi-sine excitation (Schoukens and Pintelon 1991).

Assume a continuous time system (10) from $t = 0$ and onwards is excited by an input signal which is band limited and has a band limited Nh-periodic extension outside the measurement interval. The input and output signals are sampled at N points with sampling interval h. No assumptions are made about the character of the input to the system prior to to tome $t = 0$. The input history is concisely represented by the initial state x_0. Also assume $\det(i\frac{\omega_k}{h}I - A_c) \neq 0$ for all ω_k. Then the following equation holds.

$$Y_N(\omega_k) = G_c(i\frac{\omega_k}{h})U_N(\omega_k) + T_c(e^{i\omega_k})\frac{1}{\sqrt{N}} \quad (13)$$

where

$$G_c(s) = C_c(sI - A_c)^{-1}B_c$$
$$T_c(z) = zC_c(zI - e^{A_c h})^{-1}(I - e^{A_c hN})(x_0 - x_p)$$
$$x_p = (I - e^{A_c hN})^{-1} \int_{\tau=0}^{Nh} e^{A_c \tau} B_c u(Nh - \tau)\, \mathrm{d}\tau \quad (14)$$

where $e^{A_c\tau}$ is the matrix exponential associated with the matrix A_c (Kailath 1980). The proof of (13) can be found in Appendix A.2. The relation (13) is quite similar to the discrete time version (8) with one important exception. Here $T_c(z)$ is a discrete time transfer function. However just as in the discrete time case $T_c(z)$ is completely defined by the system, the input and the initial condition at time $t = 0$. In (Pintelon *et al.* 1997) an approximate expression was derived for the transient term which made use of a continuous time transfer function.

2.4 *Discussion*

In both domains we have left out systems which have a direct feed-through term, which exists if the input instantaneously can influence the output. Most physical systems do not have such properties. However the inclusion of a direct feed-through term is straightforward and only involves an additional notational effort. By comparing (8) with (13) we conclude that the only differences between the frequency response function for discrete time systems and continuous time system is the argument,(i.e. $e^{i\omega}$ or $i\omega$) and the transient terms. The extension to the multivariable case is straightforward since the derivations made use of a state-space representation. In the general case when the system is infinite dimensional and stable, the equations (8) and (13) still hold. In this case the transient transfer function $T(z)$ does not admit any finite representation but the frequency response can be upper bounded (Ljung 1999).

2.5 *The noise*

Recall from (1) that we assume the noise to be described as $v(t) = H_0(q)e(t)$ where $H_0(q)$ is a stable linear filter and $e(t)$ is zero mean i.i.d. with variance λ_0. Denote by $E_N(\omega_k)$ the DFT of the noise signal $\{e(t)\}_{t=0}^{N-1}$. It is easy to establish that $E_N(\omega_k)$ is a zero mean random variable with the second moment properties

$$\mathbf{E}\{E_N(\omega_k)E_N^*(\omega_s)\} = \begin{cases} \lambda_0, & \omega_k = \omega_s \\ 0, & \omega_k \neq \omega_s \end{cases} \quad (15)$$

and $\mathbf{E}\{E_N(\omega_k)E_N(\omega_s)\} = 0$ for all $\omega_k, \omega_k > 0$ and $\omega_k = \omega_s \neq \pi$. If $e(t)$ is normally distributed or if $N \to \infty$ then $E_N(\omega_k)$ has a complex normal distribution (Brillinger 1981)

$$E_N(\omega_k) \in N^c(0, \lambda_0) \quad (16)$$

for $\omega_k \in \{\frac{2\pi k}{N}, k = 1 \ldots, N-1\}$ and $\omega_k \neq \pi$. For $\omega_k = 0$ and $\omega_k = \pi$, $E_N(\omega_k)$ is real, zero mean, with variance λ_0 and normal distributed.

Using equation (8) the DFT of $v(t) = H_0(q)e(t)$ is conveniently described by

$$V_N(\omega_k) = H_0(e^{i\omega_k})E_N(\omega_k) + T_{H_0}(e^{i\omega_k})\frac{1}{\sqrt{N}} \quad (17)$$

where the last frequency function $T_{H_0}(z)$ is due to the "unmatched" initial condition of the noise filter and is a linear function of the of the innovations $e(t)$ for $t < N$. This implies it has zero mean for all frequencies. When $H_0(z)$ is finite dimensional it is also possible to derive an exact expression for the covariance of T_H. We refrain from doing so here and just note that for our purposes it suffices to use the uniform bound

$$C_{H_0} = \max_k |T_{H_0}(e^{i\omega_k})|^2 \quad (18)$$

which exists since $H_0(z)$ is assumed strictly stable. Since (by definition) $e(t)$ is a non-periodic signal $E_N(\omega)$ will remain bounded and $H_0(e^{i\omega_k})E_N(\omega_k)$ will be the dominating term in $V_N(\omega_k)$ when N is large. All together we obtain for the first and second moments:

$$\mathbf{E}\{V_N(\omega_k)\} = 0, \quad \forall \omega_k \quad (19)$$

$$\mathbf{E}\{V_N(\omega_k)V_N^*(\omega_s)\} = \\ = \begin{cases} |H_0(e^{i\omega_k})|^2\lambda_0 + \dfrac{\xi_1(\omega_k)}{N}, & \omega_k = \omega_s \\ \dfrac{\xi_2(\omega_k)}{N}, & \omega_k \neq \omega_s \end{cases} \quad (20)$$

where $|\xi_2(\omega_k)| \leq |\xi_1(\omega_k)| \leq C_H$. Asymptotically, as the number of data points tends to infinity, the term $T_{H_0}(e^{i\omega_k})\frac{1}{\sqrt{N}}$ tends to zero. Then $V_N(\omega_k)$ becomes complex normal distributed with zero mean and variance $|H_0(e^{i\omega_k})|^2\lambda_0$ and independent between frequencies. It is now clear that the DFT of the output signal asymptotically has the properties as postulated in (2). For a much more thorough treatment and relaxed assumptions we refer to (Brillinger 1981, Ljung 1999).

We make no distinction between the continuous time case and the discrete time case regarding the noise signal $v(t)$. For both cases we use a *discrete time model* to describe the noise properties for the sampled data. The validity of this approach can be argued as follows. Consider a continuous time noise signal described by a stochastic differential equation which is equidistantly sampled in time. The first and second order moments of the signal at the sampling instances can equally be described by a discrete time stochastic model of the same order as the continuous time stochastic model (Åström 1970).

2.6 *Periodic input and averaging techniques*

The expressions derived above hold for arbitrary excitation signals and we made no assumption about the history of the input signal prior to time $t = 0$. If on the other hand a stable system is excited with an P-periodic signal which has been applied prior to taking the measurements, say at time $t = -s$ for some integer s and the measurement record is a multiple of the period length, then the initial condition x_0 will approach x_p as s increases. The size of the transfer function $T(z)$ will therefore also decrease, see equations (9) and (14).

If measurements are collected over M full periods then $U_N(\omega)$ will at most have P non-zero values. However the size of the non-zero values will grow at a rate of $\sqrt{M}$. Consequently the noise free counterpart of $Y_N(\omega)$ at the non-zero excitation frequencies will also grow at the same rate. Since the noise signal is non-periodic, $V_N(\omega)$ will remain bounded for all frequencies and the signal to noise ratio at the excitation frequencies will increase with a factor M. For a given size of the measurement window $N = MP$ the choice between period length P and number of periods M involves a fundamental tradeoff between frequency resolution and the signal to noise ratio.

An alternative view is given by averaging the measurements over one period. Assume the input is periodic with period time P. A natural estimate of the noise free output is then

$$\hat{y}_M(t) = \frac{1}{M}\sum_{k=0}^{M-1} y_m(t + kP), \quad 0 \leq t \leq P - 1 \tag{21}$$

and periodically continued for larger t. This gives a noise estimate

$$\hat{v}(t) = y_m(t) - \hat{y}(t)$$

from which the noise properties can be estimated. Furthermore we can use the averaged period as data $\{u(t), \hat{y}_M(t)\}_{t=0}^{N-1}$ in the model building session. This gives both a data size reduction as well as a noise level reduction. In the paper (Schoukens *et al.* 1997*b*) it is shown that such a non-parametric noise estimate based on averaging over only four periods can be used instead of the true ones and still preserve asymptotic optimality of a maximum-likelihood type estimator. See also Section 3.6. Finally it is worth mentioning that $\hat{y}_M(t)$ from (21) is a consistent (non-parametric) estimate as $M \to \infty$ of the noise free periodic output under very mild assumptions on the noise properties. Consequently $\hat{y}_M(t)$ and $u(t)$ can in a second step be used to fit a parametric model in either domain. See (McKelvey and Akçay 1995, McKelvey 1996, Forssell *et al.* 1999).

3. IDENTIFYING A PARAMETRIC MODEL

After providing the necessary information about the measurement signals we are now ready to use the frequency domain data in order to find a model of the underlying system. Certainly the estimate

$$\hat{\hat{G}}(\omega_k) = \frac{Y_N(\omega_k)}{U_N(\omega_k)}$$

which is known as the empirical transfer function estimate (Ljung 1999) would be one possible alternative. Its simplicity is appealing but the variance at each frequency is asymptotically, c.f. (20), $|H_0(e^{j\omega_k})|^2\lambda_0/|U(\omega)|^2$ which does not decrease with an increasing N unless $u(t)$ is periodic. To reduce the variance the transforms can be smoothed which decreases the variance but introduces a bias in the estimate. Here we will approach the problem by instead fitting a parametrized finite order transfer function (3) to the noisy frequency domain data. If the model set is large enough such that the true system is included, the estimate will asymptotically be unbiased.

3.1 *Method of maximum-likelihood*

The method of maximum-likelihood is frequently used for many estimation problems (Kendall and Stuart 1967, Åström 1980). Most frequency domain identification techniques do not explicitly model the noise properties with a parametric model. Instead consistent estimates are obtained by output-error type algorithms (Tugnait 1998) or IV-type methods (McKelvey 1997) or by using a non-parametric noise model which is either known or estimated a-priori (Schoukens and Pintelon 1991, Schoukens *et al.* 1997*b*). Disregarding the correct noise properties leads to an increased variance of the estimated system. Here we will present the frequency domain maximum-likelihood estimator which explicitly models the unknown noise transfer function. To simplify notation in what follows let $G_{0,\omega} \triangleq G_0(e^{i\omega})$ and $G_{\theta,\omega} \triangleq G_\theta(e^{i\omega})$ and similarly for H. Notice that in this section the frequencies ω_k in the set Ω_N can be arbitrary and not constrained to the frequency grid implied by a DFT.

Recall the postulated identification setup given by equation (2) which was motivated in Section 2. Accordingly the samples Y_k of the output Fourier transform have a complex normal distribution with mean value $G_{0,\omega_k}U_k$ and variance $|H_{0,\omega_k}|^2\lambda_0$. The probability density function for each measurement Y_k is thus [2] (Brillinger 1981)

[2] Here we assume that the frequencies $\omega = 0$ and $\omega = \pi$ is not part of our set Ω_N since at these frequencies $Y(\omega)$ is real valued and has a slightly different probability density function. Including them will however not change any asymptotic properties.

$$\frac{1}{\pi |H_{0,\omega_k}|^2 \lambda_0} \exp\left(-\frac{|Y_k - G_{0,\omega_k} U_k|^2}{|H_{0,\omega_k}|^2 \lambda_0} \right) \quad (22)$$

Since the measurements are independent of each other the joint probability density function for Z^N is the product of the individual functions (22).

If instead of the true (but unknown) transfer functions G_0, H_0 and variance λ_0 we insert the parametrized counterparts G_θ, H_θ and λ of our model and consider the measurements Z^N fixed we obtain a parametrized likelihood function. By taking the negative logarithm of the parametrized likelihood function and removing terms which do not depend on the parameters (θ and λ) we obtain (Ljung 1994, McKelvey and Ljung 1997)

$$V_N(\theta, \lambda) = \frac{1}{N} \sum_{k=1}^{N} \left[\log\left(|H_{\theta,\omega_k}|^2 \lambda\right) + \frac{|Y_k - G_{\theta,\omega_k} U_k|^2}{|H_{\theta,\omega_k}|^2 \lambda} \right] \quad (23)$$

which for given values of θ and λ can be calculated. Minimizing V_N with respect to θ and the noise variance λ yields the *maximum-likelihood estimate*,

$$\hat{\theta}_N, \hat{\lambda}_N = \arg\min_{\theta,\lambda} V_N(\theta). \quad (24)$$

To reduce the number of free parameters, the variance λ can be eliminated from the criterion by analytical minimization, see (McKelvey and Ljung 1997).

If the data record is short and it is expected that the transient terms in (8) or (13) cannot be neglected then $G_{\theta,\omega} U_k$ should be exchanged for $G_{\theta,\omega} U_k + T_{\theta,\omega} \frac{1}{\sqrt{N}}$ in order to capture the transient effects. Note that the n parameters in T_θ which depend on the initial condition $x_0 - x_p$ cannot consistently be estimated since the effect of $T(z)$ on Y_k decays as $\frac{1}{\sqrt{N}}$.

In general the minimization of (23) can not be performed analytically and an iterative optimization strategy needs to be employed. Often Newton type methods (Dennis and Schnabel 1983) perform well for this class of problems.

3.2 *Asymptotic properties*

As the number of frequency points increases to infinity the ML criterion (23) converges to a limit function which can be described by an integral.

Denote by Ω the interval of the real line to which the set of sample frequencies belongs and let Ω_N (as before) denote the set of sample frequencies. Let us define

$$W_N(\omega) = \frac{|\{k \,|\, \omega_k < \omega, \omega_k \in \Omega_N\}|}{N} \quad (25)$$

where $|S|$ denotes the cardinality of the set S. In probability theory $W_N(\omega)$ corresponds to a distribution function. We assume that the sequence of sample frequencies is such that $W_N(\omega)$ converges to a function $W(\omega)$ in all points of continuity of $W(\omega)$. By using a Stieltjes integral notation (Rudin 1987), an infinite sum can be written as an integral. In our case we have that as $N \to \infty$ (see (McKelvey and Ljung 1997) for the details)

$$V_N(\theta,\lambda) \to \bar{V}(\theta,\lambda) \triangleq \int_\Omega \frac{|G_{0,\omega} - G_{\theta,\omega}|^2 \Phi_u(\omega) + \Phi_v(\omega)}{|H_{\theta,\omega}|^2 \lambda} + \log |H_{\theta,\omega}|^2 \lambda \, dW(\omega), \quad (26)$$

where $\Phi_u(\omega) = |U(\omega)|^2$ and $\Phi_v(\omega) = |H_{0,\omega}|^2 \lambda_0$. Under some regularity conditions on G_θ and H_θ the convergence in (26) is uniform with probability one. It then follows that the estimate $\hat{\theta}_N, \hat{\lambda}_N$ converges to values which minimize $\bar{V}(\theta, \lambda)$.

Assume the model is sufficiently flexible such that there exists a non-empty set Θ_* such that for all $\theta_* \in \Theta_*$

$$\int_\Omega |G_{0,\omega} - G_{\theta_*,\omega}|^2 \Phi_u(\omega) \, dW(\omega) = 0 \quad (27)$$

$$\int_\Omega \left| |H_{0,\omega}|^2 - |H_{\theta_*,\omega}|^2 \right|^2 dW(\omega) = 0 \quad (28)$$

Then it can be shown that $\bar{V}(\theta, \lambda)$ is minimized by all $\theta \in \Theta_*$ and $\lambda = \lambda_0$. If the model structure is restricted such that the limiting set Θ_* is a singleton then $\hat{\theta}_N \to \theta_*$ with probability one as $N \to \infty$ which means that the estimator is consistent (in the sense of satisfying (27) and (28)).

By using a fixed noise model which does not depend on the parameters θ the limiting estimate is the minimizer of

$$\int_\Omega \frac{|G_{0,\omega} - G_{\theta,\omega}|^2 \Phi_u(\omega)}{|H_\omega|^2} dW(\omega) \quad (29)$$

If the true system G_0 is not part of the model class then an approximate model will result. In this case the estimated model will be the model which in a weighted mean square sense best approximates the transfer function of the system. As shown in (29) the weight is dependent on the spectrum of the excitation signal, the inverse of the assumed noise model and the distribution function of the frequency samples. Note that the true noise spectrum does not influence the limiting estimate.

If the frequency distribution function $W(\omega)$ is differentiable in the interior of the interval Ω the Stieltjes integrals simplifies into $\int_\Omega (\cdot) w(\omega) \, d\omega$ where $w(\omega) = \frac{d}{d\omega} W(\omega)$. In this case $w(\omega)$ acts as a standard weighting function.

3.3 *Asymptotic variance*

Consider the case when a fixed noise model H_ω is used in the criterion (23) and assume the limiting set Θ_*, defined by (27) is a singleton

θ_*. Furthermore assume that $G'_{\omega,\theta}$ and $G''_{\omega,\theta}$ are Lipschitz continuous. Define

$$Q = \int_{\Omega} \frac{\Phi_v(\omega)\Phi_u(\omega)2\text{Re}\{G'_{\omega,\theta_*}(G'_{\omega,\theta_*})^*\}}{|H_\omega|^4} \, dW(\omega) \tag{30}$$

and

$$R = \int_{\Omega} \frac{\Phi_u(\omega)2\text{Re}\{G'_{\omega,\theta_*}(G'_{\omega,\theta_*})^*\}}{|H_\omega|^2} \, dW(\omega) \tag{31}$$

and assume $R > \delta I$ for some $\delta > 0$. Then the estimate given by minimizing (23) is asymptotically normally distributed (McKelvey and Ljung 1997)

$$\sqrt{N}(\hat{\theta}_N - \theta_*) \in AsN(0, P_\theta) \tag{32}$$

with covariance matrix $P_\theta = R^{-1}QR^{-1}$. If the noise model is chosen equal to the true one i.e.

$$\int_{\Omega} \left||H_{0,\omega}|^2 - |H_\omega|^2\right|^2 dW(\omega) = 0$$

then the size of the covariance is minimized[3] and is equal to $P_\theta^{opt} = \lambda_0 R^{-1}$. The estimator is then asymptotically *efficient.* If an independently parametrized noise model H_ξ is estimated together with G_θ then the variance result (32) still holds and the optimal variance P_θ^{opt} is obtained if the estimated noise model satisfies (28).

3.4 *Discussion*

It is interesting to compare the frequency domain ML-estimator (23) with its time-domain counterpart as described in (Ljung 1999, Söderström and Stoica 1989). Applying Parsevals formula to the time-domain equation reveals that the time domain ML estimator minimizes the function

$$\int_{-\pi}^{\pi} \frac{|Y_N(\omega_k) - G_{\theta,\omega_k}U_N(\omega_k)|^2}{|H_{\theta,\omega_k}|^2} \, d\omega. \tag{33}$$

A few points are worth noticing. For a fixed known noise model $H_\theta(z) = H(z)$ the frequency domain ML-estimate (23) and the time domain estimate (33) are essentially the same. Whenever the noise model is estimated, the additional term

$$\frac{1}{N}\sum_{k=1}^{N} \log(|H_{\theta,\omega_k}|^2\lambda) \tag{34}$$

occurs in the criterion. In fact (34) is the determinant of the transformation which change variables from Y to E (output to innovations). In the time domain this transformation is triangular with 1's along the diagonal. Hence, this transformation has a determinant equal to 1, so it does not affect the ML criterion.

However if the frequencies ω_k are equidistantly distributed between $-\pi$ and π this additional term again becomes of no importance since, independently of $\theta \in D_{\mathcal{M}}$,

$$\frac{1}{2\pi}\int_{-\pi}^{\pi} \log(|H_{\theta,\omega}|^2\lambda) \, d\omega = \log\lambda$$

for any stable and inversely stable monic transfer function. The sum in (34) is also $\log\lambda$ if $\omega_k = \frac{2\pi(k-1)}{N}$, $k = 1, \ldots, N$. This implies that the time-domain ML estimator and the frequency domain estimator are (asymptotically) identical if the frequency domain estimator obtains the data by DFT and all frequencies are used (see also (Schoukens *et al.* 1997*a*)). If the frequency domain estimator uses only a subset of frequencies the limiting criterion function will be weighted by the distribution function of the frequencies and it is vital to use the full ML-criterion (23) to guarantee consistency. If a good fit is required at a certain frequency band the input spectrum should be large for these frequencies or the input and output data could be bandpass filtered prior to estimation. For example if the model is intended for control design it is desired to find a low complexity model with a good fit around the desired crossover frequency. In the time domain it does not quite make sense to prefilter the data prior to the identification if a noise model is also estimated since the noise model will then undo the effect of the prefilter (Ljung 1999). In the frequency domain a prefiltering effect is obtained by only including a subset of frequencies in Ω_N corresponding to the frequency band where a good fit is desired. Here an estimate of a parametric noise model (to improve the variance properties) still makes sense if it can be expected that G_θ is flexible enough to provide an unbiased estimate (in the sense of (27)). It could also be expected that it suffices with a less complex noise model since only a part of the true noise transfer function needs to be accounted for.

3.5 *Closed loop*

Consider the case when the system operates in closed loop with a linear controller K. The input is then described by

$$u(t) = r(t) - K(q)y(t)$$

where $r(t)$ is an external reference signal. The output signal is just as before $y(t) = G_0(q)u(t) + H_0(q)e(t)$. Assume that U_k and Y_k are used in the criterion function (23). The dependence between U_k and Y_k make the criterion function converge to a limiting function different from (26). Straightforward calculations then reveal that

$$V_N(\theta,\lambda) \to \bar{V}(\theta,\lambda) \triangleq \int_{\Omega} \frac{|G_{0,\omega} - G_{\theta,\omega}|^2|R(\omega)|^2 + |1 + K_\omega G_{\theta,\omega}|^2\Phi_v(\omega)}{|1 + K_\omega G_{0,\omega}|^2|H_{\theta,\omega}|^2\lambda} + \log|H_{\theta,\omega}|^2\lambda \, dW(\omega). \tag{35}$$

where $R(\omega)$ is the Fourier transform of $r(t)$. First consider the case when the frequencies ω_k are

[3] Minimized in the sense that $P_\theta - P_\theta^{opt}$ is a positive semidefinite matrix for all choices of fixed noise models.

uniform in $[-\pi, \pi]$. Hence the integral of the log term in (35) is zero. Also assume that the product $K(q)G_0(q)$ has no direct feed-trough term. Then $1 + K(q)G_0(q)$ is a monic transfer function which in turn implies that

$$\int_\Omega \frac{|1 + K_\omega G_{\theta,\omega}|^2 \Phi_v(\omega)}{|1 + K_\omega G_{0,\omega}|^2 |H_{\theta,\omega}|^2 \lambda} d\omega$$

is minimized when the model equals the true system. Hence, whenever the correct noise model is used consistency is preserved also for the closed loop case. This property is analogous to the time domain case, see (Ljung 1999). If frequencies are drawn from a non-uniform distribution a slight modification is needed to the criterion function. By subtracting the term $\log |1 + K_{\omega_k} G_{\theta,\omega_k}|^2$ from the summand in (23) the estimator again become consistent. Note however that in this case we need to know the controller K and could thus have reformulated the closed loop problem to an open loop one, so called indirect identification.

3.6 *Some variants of the criterion*

Consider a system of the form

$$y(t) = \frac{b(q)}{a(q)} u(t) + \frac{1}{a(q)} e(t)$$

where $a(q)$ and $b(q)$ are polynomials. By disregarding λ and the log term the ML-criterion (23) simplifies into

$$\frac{1}{N} \sum_{k=1}^{N} |a(e^{i\omega_k}, \theta) Y_k - b(e^{i\omega_k}, \theta) U_k|^2 \qquad (36)$$

The criterion (36) is quadratic in the coefficients of the a and b polynomials and the minimizing parameters can be found by simple linear regression. Asymptotically this estimator equals the time domain ARX method (Ljung 1999). For continuous time models this estimator was proposed in (Levy 1959). If on the other hand

$$y(t) = \frac{b(q)}{a(q)} u(t) + e(t)$$

the use of the simple criterion (36) will lead to bias and an undesired weighting $|a(e^{i\omega})|^2$. A remedy to this problem was given in (Sanathanan and Koerner 1963) where an iterative procedure was suggested where at iteration m a weighted criterion was minimized

$$\frac{1}{N} \sum_{k=1}^{N} |a(e^{i\omega_k}, \theta) Y_k - b(e^{i\omega_k}, \theta) U_k|^2 W_k^{(m)} \qquad (37)$$

where $W_k^{(m)} = |\hat{a}^{(m-1)}(e^{i\omega_k})|^{-2}$ and $\hat{a}^{(m-1)}(z)$ is the estimated a polynomial from step $m-1$. The scheme is not guaranteed to converge to the true system (Whitfield 1987). A multivariable version is described in (de Callafon *et al.* 1996).

A number of subspace based methods have recently been developed for the frequency domain multivariable problem. A key feature is that they provide accurate state-space models for high order systems (McKelvey *et al.* 1996*a*, McKelvey *et al.* 1996*b*, Jacques *et al.* 1996, Liu *et al.* 1996, McKelvey 1997, Van Overschee and De Moor 1996). An accurate method for high order systems based on more standard curve fitting and polynomial matrix descriptions is given in (Bayard 1994). These simplified procedures are well suited to provide the optimal ML-estimator with good staring values for the iterative nonlinear optimization.

If the input signal also is corrupted with noise we obtain a dynamical errors-in-variables problem. Assuming the true system is given by $G_0(q) = b(q)/a(q)$, the input and output power spectrums of the noise, denoted by $\sigma_U^2(\omega)$ and $\sigma_Y^2(\omega)$ are known, the maximum-likelihood estimate is given by minimizing (Schoukens and Pintelon 1991)

$$\sum_{k=1}^{N} \frac{|b(e^{\omega_k}, \theta) U(\omega_k) - a(e^{\omega_k}, \theta) Y(\omega_k)|^2}{\sigma_U^2(\omega_k) |b(e^{\omega_k}, \theta)|^2 + \sigma_Y^2(\omega_k) |a(e^{\omega_k}, \theta)|^2} \qquad (38)$$

if the input and output noise sources are uncorrelated. For correlated noise a correction is subtracted from the denominator in (38). Recent extensions in (Schoukens *et al.* 1997*b*, Schoukens *et al.* 1999) show that the true input and output noise spectra can be exchanged by non-parametric estimates derived from four repeated experiments while still providing consistent estimates.

3.7 *Numerical issues for continuous time modeling*

It is well known that the parameter estimation problem is better conditioned for discrete time transfer functions than continuous time ones, since powers of $e^{i\omega}$ form a natural orthogonal basis (Bayard 1993). This property is amplified if the model order is high or if a large frequency band is used. By use of the bilinear transformation the continuous-time identification problem can be solved in the discrete domain without introducing any approximation errors but drastically improve the numerical conditioning (McKelvey *et al.* 1996*b*).

The bilinear transformation maps the complex values in the s domain to the z domain as

$$s \to \frac{2(z-1)}{T(z+1)}, \text{ with inverse } z \to \frac{2+sT}{2-sT}$$

The parameter T is a parameter which the user is free to specify under constraint that $2/T$ is not a pole of the continuous-time system (Al-Saggaf and Franklin 1988), and can be seen as a sort of sampling period.

If the continuous-time transfer function is given by $G_c(s)$ the bilinear transformation gives the discrete time transfer function

$$G(z) = G_c\left(\frac{2(z-1)}{T(z+1)}\right).$$

Stability and minimum-phase properties are preserved by the transformation.

An important feature of the bilinear transformation is that the frequency response is invariant if we pre-warp the frequency scale. Let the continuous-time transfer function be evaluated at $i\omega_k$ and let the bilinearly transformed discrete time transfer function be evaluated at $e^{i\omega_k^d}$. Then it holds that (Åström and Wittenmark 1984)

$$G_c(i\omega_k) = G_c\left(\frac{2(e^{i\omega_k^d}-1)}{T(e^{i\omega_k^d}+1)}\right) = G(e^{i\omega_k^d})$$

if $\tan(\omega_k^d/2) = \omega_k T/2$. Assume we have samples U_k and Y_k from a continuous time system at frequencies ω_k. Then they also are samples of the corresponding bilinearly transformed discrete time system at new frequencies ω_k^d

$$\omega_k^d = 2\,\mathrm{atan}(\frac{\omega_k T}{2}), \quad k = 1, \dots, M, \qquad (39)$$

where atan denotes the inverse of tan. After estimation of a discrete time transfer function $\hat{G}(z)$, the sought continuous time transfer function is obtained through the inverse map

$$\hat{G}_c(s) = \hat{G}\left(\frac{2+sT}{2-sT}\right).$$

4. CONCLUSIONS

It has been demonstrated that the frequency domain ML method asymptotically equals the time domain version in the case where the frequency data is obtained by DFT. An important question to pose then is in which cases can it be advantageous to use the frequency domain method? The following points can shed some light on the answer.

Partial modeling Often it is enough to find a model which accurately describes the true system in a limited frequency band. A low order model could thus be sufficient. In the frequency domain this is simply accomplished by only including the desired frequencies in the set Ω_N. In the time domain the raw data needs to be prefiltered and the influence of the initial conditions of the filters might distort the end results. Also it makes no sense to estimate a noise model in the time domain if a prefilter has been applied.

Continuous time systems If the experimental conditions are such that a band limited input can be used then the modeling in the frequency domain is straightforward as demonstrated in Section 2. Direct identification of continuous time models in the time domain requires difficult user choices of how to approximate higher order derivatives from sampled data. Systems with time delay are also straightforward to handle by using $G_c(s)e^{-\tau s}$ as the model structure where τ is the time delay.

Periodic input When a system is excited with a periodic input applied for a sufficiently long time prior to the actual measurement interval, then the effect of the initial condition is diminished even for a finite measurement interval. Hence the transient effect need not be included in the model. Periodic excitation also make it possible to first estimate a non-parametric noise model which can be used as fixed noise model in the ML-criterion, see (Schoukens *et al.* 1997*b*).

Merging Data If data is obtained by different experiments all frequency data can be merged into one estimation data set. By performing several experiments using different sampling rates a single wide band frequency domain data set can efficiently be assembled in order to estimate a continuous time model valid over a large frequency band (Schoukens *et al.* 1994).

ACKNOWLEDGMENT

I would like to thank Lennart Ljung, Johan Schoukens, Rik Pintelon and Petre Stoica who at various times have generously shared with me their expertise on this subject. Thanks also to Brett Ninness and Will Heath for providing valuable comments during the preparation of this manuscript.

5. REFERENCES

Al-Saggaf, U. M. and G. F. Franklin (1988). Model reduction via balanced realizations: An extension and frequency weighting techniques. *IEEE Trans. on Automatic Control* **33**(7), 687–692.

Åström, K. J. (1970). *Introduction to Stochastic Control Theory.* Academic Press. New York.

Åström, K. J. (1980). Maximum likelihood and prediction error methods. *Automatica* **16**, 551–574.

Åström, K. J. and B. Wittenmark (1984). *Computer Controlled Systems.* Prentice-Hall. Englewood Cliffs, New Jersey.

Bayard, D. S. (1993). Multivariable state-space identification in the delta and shift operators: Algorithms and experimental results. In: *Proc. American Control Conference, San Francisco, CA, June.* pp. 3038–3042.

Bayard, D. S. (1994). High-order multivariable transfer function curve fitting: Algorithms, sparse matrix methods and experimental results. *Automatica* **30**(9), 1439–1444.

Brillinger, D. R. (1981). *Time Series: Data Analysis and Theory.* McGraw-Hill Inc., New York.

de Callafon, R. A., D. de Roover and P. M. J. Van den Hof (1996). Multivariable least squares frequency domain identification using polynomial matrix fraction descriptions. In: *Proc. 35th IEEE Conference on Decision and Control.* Kobe, Japan.

Dennis, J. E. and R. B. Schnabel (1983). *Numerical Methods for Unconstrained Optimization and Nonlinear Equations.* Prentice-Hall. Englewood Cliffs, New Jersey.

Forssell, U., F. Gustafsson and T. McKelvey (1999). Time-domain identification of dynamic errors-in-variables systems using periodic excitation signals. In: *Preprints of the 14th World Congress of IFAC, Beijing, P. R. China* (H.F. Chen and B. Wahlberg, Eds.). Vol. H. Elsevier Science. pp. 421–426.

Giannakis, G. B. (1995). Polyspectral and cyclostationary approaches for identification of closed-loop systems. *IEEE Trans. on Automatic Control* **AC-40**, 882–885.

Hannan, E. G. (1970). *Multiple Time Series.* John Wiley and Sons, Inc.

Hannan, E. J. and M. Deistler (1988). *The Statistical Theory of Linear Systems.* Wiley.

Jacques, R. N., K. Liu and D. W. Miller (1996). Identification of highly accurate low order statespace models in the frequency domain. *Signal Processing, EURASIP* **52**(2), 195–208.

Kailath, T. (1980). *Linear Systems.* Prentice-Hall. Englewood Cliffs, New Jersey.

Kendall, M. G. and A. Stuart (1967). *The Advanced Theory of Statistics.* Vol. 2. second ed.. Griffin, London.

Levy, E. C. (1959). Complex curve fitting. *IRE Trans. on Automatic Control* **AC-4**, 37–44.

Liu, K., R. N. Jacques and D. W. Miller (1996). Frequency domain structural system identification by observability range space extraction. *ASME J. Dyn. Syst., Meas. and Control.*

Ljung, L. (1994). Building models from frequency domain data. In: *IMA Workshop on Adaptive Control and Signal Processing.* Minneapolis, Minnesota.

Ljung, L. (1999). *System Identification: Theory for the User.* second ed.. Prentice-Hall. Englewood Cliffs, New Jersey.

Ljung, L. and K. Glover (1981). Frequency domain versus time domain methods in system identification. *Automatica* **17**(1), 71–86.

McKelvey, T. (1996). Periodic excitation for identification of dynamic errors-in-variables systems operating in closed loop. In: *Proc. 13th World Congress of International Federation of Automatic Control.* Vol. J. San Francisco, California. pp. 155–160.

McKelvey, T. (1997). Frequency domain system identification with instrumental variable based subspace algorithm. In: *Proc. 16th Biennial Conference on Mechanical Vibration and Noise, DETC'97, ASME.* Sacramento, California, USA. pp. VIB-4252.

McKelvey, T. and H. Akçay (1995). Subspace based system identification with periodic excitation signals. *Systems & Control Letters* **26**(5), 349–361.

McKelvey, T. and L. Ljung (1997). Frequency domain maximum likelihood identification. In: *Proc. of the 11th IFAC Symposium on System Identification.* Vol. 4. Fukuoka, Japan. pp. 1741–1746.

McKelvey, T., H. Akçay and L. Ljung (1996*a*). Subspace-based identification of infinite-dimensional multivariable systems from frequency-response data. *Automatica* **32**(6), 885–902.

McKelvey, T., H. Akçay and L. Ljung (1996*b*). Subspace-based multivariable system identification from frequency response data. *IEEE Trans. on Automatic Control* **41**(7), 960–979.

Pintelon, R., J. Schoukens and G. Vandersteen (1997). Frequency domain system identification using arbitrary signals. *IEEE Trans. on Automatic Control* **42**(12), 1717–1720.

Pintelon, R., P. Guillaume, Y. Rolain, J. Schoukens and H. Van Hamme (1994). Parametric identification of transfer functions in the frequency domain – A survey. *IEEE Trans. of Automatic Control* **94**(11), 2245–2260.

Robinson, P. M. (1976). The estimation of linear differential equations with constant coefficients. *Econometrica* **44**(4), 751–764.

Rudin, W. (1987). *Principles of Mathematical Analysis.* third ed.. McGraw-Hill.

Sanathanan, C. K. and J. Koerner (1963). Transfer function synthesis as a ratio of two complex polynomials. *IEEE Trans. on Automatic Control* **8**, 56–58.

Schoukens, J. and R. Pintelon (1991). *Identification of Linear Systems: a Practical Guideline to Accurate Modeling.* Pergamon Press, London, UK.

Schoukens, J., G. Vandersteen, R. Pintelon and P. Guillaume (1999). Frequency-domain identification of linear systems using arbitrary excitations and a nonparametric noise model. *IEEE Trans. on Automatic Control* **AC-44**(2), 343–347.

Schoukens, J., R. Pintelon and H. Van Hamme (1994). Identification of linear dynamic systems using piecewise constant excitations: Use, misuse and alternatives. *Automatica* **30**(7), 1153–1169.

Schoukens, J., R. Pintelon and Y. Rolain (1997*a*). Study of conditional ML estimators in time and frequency domain system identification. In: *Proc. European Control Conference.* Brussels, Belgium.

Schoukens, J., R. Pintelon, G. Vandersteen and P. Guillaume (1997*b*). Frequency domain identification using non-parametric noise models estimated from a small number of data sets. *Automatica* **33**(6), 1073–1086.

Söderström, T. and P. Stoica (1989). *System Identification.* Prentice-Hall International. Hemel Hempstead, Hertfordshire.

Tugnait, J. K. (1998). Identification of multivariable stochastic linear systems via polyspectral analysis given noisy input-output time-

domain data. *IEEE Trans. on Automatic Control* **43**(8), 1084–1100.
Unbehauen, H. and G. P. Rau (1990). Continuous-time approaches to system identification – A survey. *Automatica* **26**(1), 23–35.
Van Overschee, P. and B. De Moor (1996). Continuous-time frequency domain subspace system identification. *Signal Processing, EURASIP* **52**(2), 179–194.
Vandersteen, G. and J. Schoukens (1999). Measurement and identification of nonlinear systems consisting of linear dynamic blocks and one static nonlinearity. *IEEE Trans. on Automatic Control* **44**(6), 1266–1271.
Vandersteen, G., Y. Rolain and J. Schoukens (1997). Non-parametric estimation of the frequency-response functions of the linear blocks of a Wiener-Hammerstein model. *Automatica* **33**(7), 1351–1355.
Whitfield, A. H. (1986). Transfer function synthesis using frequency response data. *Int. J. Control* **43**(5), 1413–1426.
Whitfield, A. H. (1987). Asymptotic behaviour of transfer function synthesis methods. *Int. J. Control* **45**(3), 1083–1092.
Whittle, P. (1951). *Hypothesis testing in time series analysis.* Thesis. Uppsala University, Almqvist and Wiksell, Uppsala. Hafner, New York.
Young, P. (1981). Parameter estimation for continous-time models – A survey. *Automatica* **17**(1), 23–39.

Appendix A. PROOFS

A.1 *Proof of* (8)-(9)

Consider the state space equations (6). The state at time $t = N$ is given by

$$x(N) = A^N x(0) + \sum_{k=0}^{N-1} A^k Bu(N-1-k)$$

Now with the initial condition $x(0) = x_p$ for (6) where x_p is given by (9) then $x(N) = x_p$. Let $x^{\text{per}}(t)$ denote the state evolution of (6) when $x(0) = x_p$ and let $y^{\text{per}}(t) = Cx^{\text{per}}(t)$ denote the corresponding output. Define the transient output as $y^{\text{tra}}(t) = CA^t(x_0 - x_p)$. It is straightforward to verify that the original output $y(t)$ from the state-space system (6) with initial condition $x(0) = x_0$ can be decomposed as $y(t) = y^{\text{per}}(t) + y^{\text{tra}}(t)$. The DFT of the transient output is simply a finite geometric sum

$$\begin{aligned} Y_N^{\text{tra}}(\omega_k) &= \frac{1}{\sqrt{N}} \sum_{t=0}^{N-1} CA^t(x_0 - x_p)e^{-i\omega_k t} \\ &= \frac{1}{\sqrt{N}} e^{i\omega_k} C(e^{i\omega_k}I - A)^{-1}(I - A^N)(x_0 - x_p) \end{aligned}$$

where we used the property that $e^{i\omega_k N} = 1$. Denote the DFT of $x^{\text{per}}(t)$ as $X_N^{\text{per}}(\omega)$. The DFT of $x^{\text{per}}(t+1)$ at $\omega_k = \frac{2\pi k}{N}$ is given by

$$\frac{1}{\sqrt{N}} \sum_{t=0}^{N-1} x^{\text{per}}(t+1)e^{-i\omega_k t} =$$
$$\frac{1}{\sqrt{N}} e^{i\omega_k} \sum_{t=0}^{N-1} x^{\text{per}}(t+1)e^{-i\omega_k (t+1)} = e^{i\omega_k} X_N^{\text{per}}(\omega_k)$$

where the last equality follows from $x^{\text{per}}(N) = x^{\text{per}}(0)$. By taking the DFT of both sides of the state-space equations governing $x^{\text{per}}(t)$ we have

$$e^{i\omega_k} X_N^{\text{per}}(\omega_k) = AX_N^{\text{per}}(\omega_k) + BU_N(\omega_k)$$

By eliminating the state $X_N^{\text{per}}(\omega_k)$ we finally obtain for the DFT of $y^{\text{per}}(t)$

$$Y_N^{\text{per}}(\omega_k) = C(e^{i\omega_k}I - A)^{-1}BU_N(\omega_k)$$

which concludes the proof. □

A.2 *Proof of* (13)-(14)

This proof parallels the discrete time case with some obvious modifications. In the proof we assume $h = 1$, N is an odd integer and define $n_f \triangleq \frac{N-1}{2}$. The proof for the general case is analogous.

The evolution of the continuous time state variable is given by (Kailath 1980)

$$x(t) = e^{A_c t}x(0) + \int_{\tau=0}^{t} e^{A_c \tau} B_c u(t-\tau) \qquad \text{(A.1)}$$

Now if we assume that $x(0)$ in (10) is given by x_p defined by (14), straight forward calculations yields $x(N) = x(0) = x_p$. Define $x^{\text{per}}(t)$ to be the state response when $x(0)$ is given by x_p and similarly define $y^{\text{per}}(t) = C_c x^{\text{per}}(t)$. By introducing $y^{\text{tra}}(t) \triangleq C_c e^{A_c t}(x_0 - x_p)$ it is easy to verify that the measured output for $t = [0, N]$ is given by $y(t) = y^{\text{per}}(t) + y^{\text{tra}}(t)$ Obviously, the state x^{per} is also an N-periodic signal. By explicitly using the structure of the input given by (12) we obtain

$$x_p = \sum_{k=-n_f}^{n_f} (i\omega_k I - A_c)^{-1} B_c f_k$$

Straightforward calculations using the expression (A.1) with $x(0) = x_p$ yields

$$x^{\text{per}}(t) = \sum_{k=-n_f}^{n_f} (i\omega_k I - A_c)^{-1} B_c f_k e^{i\omega_k t}$$

and the DFT of the state is simply

$$X_N^{\text{per}}(\omega_k) = (i\omega_k I - A_c)^{-1} B_c U_N(\omega_k)$$

For the output we simply have $Y_N^{\text{per}}(\omega_k) = C_c X^{\text{per}}(\omega_k)$. As in the discrete time case the DFT of the transient output is a finite geometric sum

$$Y_N^{\text{tra}}(\omega_k) = \frac{1}{\sqrt{N}} \sum_{t=0}^{N-1} C_c e^{A_c t}(x_0 - x_p)e^{-i\omega_k t} =$$
$$\frac{1}{\sqrt{N}} e^{i\omega_k} C_c(e^{i\omega_k}I - e^{A_c})^{-1}(I - e^{A_c N})(x_0 - x_p)$$

□

MODEL VALIDATION FOR ROBUST CONTROL AND CONTROLLER VALIDATION IN A PREDICTION ERROR FRAMEWORK

Michel Gevers *, Xavier Bombois *, Benoît Codrons *, Franky De Bruyne ** and Gérard Scorletti ***

* *CESAME, Université Catholique de Louvain*
Bâtiment EULER, 4 av. Georges Lemaitre, B-1348 Louvain-la-Neuve, Belgium *E-mail:* Gevers@csam.ucl.ac.be
** *Siemens, Service EIT ES5, Demeurslaan 132, Building 15/0+, B-1654 Huizingen, Belgium* *E-mail:* Franky.De-Bruyne@Siemens.be
*** *LAP ISMRA, 6 boulevard du Maréchal Juin, F-14050 Caen Cedex, France* *Email:* gerard.scorletti@greyc.ismra.fr

Abstract:
This paper presents a coherent framework for model validation for control and for controller validation (for stability and for performance) in the context where the validated model uncertainty sets are obtained by prediction error identification methods. Thus, these uncertainty sets are parametrized transfer function sets, with parameters lying in ellipsoidal regions in parameter space. Our results cover two distinct aspects: (1) *Control-oriented model validation results,* where we show that a measure of size of the validated model set is connected to the size of the model-based controller set that robustly stabilizes the model set, leading to validation design guidelines. (2) *Controller validation results,* where we present necessary and sufficient conditions for a controller to stabilize all models, or to achieve a given level of performance for all models, in such uncertainty set. *Copyright © 2000 IFAC*

Keywords: identification, validation, robust stability, robust performance

1. INTRODUCTION

This paper presents in summary form the key ideas and results that we have developed over the last few years on *model validation for control* and *controller validation*, in the prediction error identification framework. The full presentation of these results, as well as the proofs, can be found in the comprehensive paper (Gevers *et al.*, 2000), and in the more technical supporting papers (Gevers *et al.*, 1999*a*), (Bombois *et al.*, 1999*b*), (Bombois *et al.*, 1999*c*), (Gevers *et al.*, 1999*b*), (Bombois *et al.*, 1999*a*), (Bombois *et al.*, 2000*b*), (Bombois *et al.*, 2000*a*). Our results on validation establish a robust stability and robust performance theory for model uncertainty sets produced by prediction error identification and validation, which we call PE uncertainty sets for ease of reference. In the case of full order models, these validated PE model uncertainty sets are defined by parametrized transfer function sets whose parameter vectors lie in ellipsoids in parameter space. In the case of restricted complexity models, they are obtained by a stochastic embedding technique (see (Goodwin *et al.*, 1992)) and are made up of ellipsoids at every frequency in the space of transfer functions (Bombois *et al.*, 2000*a*). These

[1] The authors acknowledge the Belgian Programme on Inter-university Poles of Attraction, initiated by the Belgian State, Prime Minister's Office for Science, Technology and Culture. The scientific responsibility rests with its authors.

PE uncertainty sets are not the classical uncertainty sets used in robust control analysis and design, such as additive, multiplicative, coprime factor uncertainty sets, etc. Our robust stability and robust performance results take the form of necessary and sufficient conditions for a controller to stabilize all models in such PE sets, or to achieve some given level of performance with all models in such PE sets, without the need to first approximate or overbound these sets by one of the more standard uncertainty sets of robust control theory.

In addition to these *controller validation* results, we also present results on *model validation for control* that impact on validation design. To put it simply, we define a 'control-oriented measure of size' for the validated PE uncertainty sets mentioned above, that is directly connected, via a robust stability result, to the size of the sets of model-based controllers that stabilize all models in the validated PE model sets. This measure of size is called the worst case ν-gap between the validated model and the validated model set. We show how this worst case gap is affected by the design of the validation experiment that delivers the validated PE set. Thus, we now have a complete connection all the way from the validation experiment that defines a PE uncertainty set to the set of controllers that are guaranteed to stabilize all models in this PE uncertainty set. This allows one to formulate 'validation for control design guidelines'.

Because of the space limitations imposed by the format of a Conference paper, we shall not attempt here to describe the state of the art in model validation, model validation for control, and controller validation. [1] We shall also not review the abundant literature that already exists on these very broad and important questions. In the full paper (Gevers *et al.*, 2000) we have attempted to perform this review of the literature and we have also put our results in perspective with respect to the existing literature.

This paper is organized as follows. In Section 2 we introduce the prediction error framework for validation, and we present our results on model set validation for robust control, while Section 3 covers controller validation in this prediction error framework, both for stability and for performance. For reasons of space, our results are presented here without proof; we refer the reader to the appropriate references for the proofs and for further details.

[1] We just want to repeat, as we have done in our previous work, that these are three different issues, each requiring a different formulation and hence different solutions.

2. MODEL SET VALIDATION FOR CONTROL

We consider that input-output data y and u are generated from a Linear Time Invariant "*true system*":

$$\mathcal{S} : y(t) = G_0(z)u(t) + v(t), \qquad (1)$$

where $G_0(z)$ is a linear time-invariant causal operator, u is a measured input signal, and v is zero mean stationary noise.

We consider the situation where one is allowed to perform experiments on the true system in order to contruct a model set to which the true system is guaranteed to belong, at some a priori fixed probability level α, say 95%. The experiment delivers a data record of N input-output data: $Z^N = \{y(1), u(1), \ldots, y(N), u(N)\}$. The construction of the model set is achieved, in our framework, by a Prediction Error (PE) identification experiment. For the sake of simplicity and brevity, we shall consider here that this PE identification is performed using a model set of the same structure as that of the true system, leading to an unbiased estimate $G(z, \hat{\delta})$ of $G_0(z)$. This may appear like a severe restriction. However, as shown in (Gevers *et al.*, 2000), the results that follow apply without any modification to the case where a low order model is used, and where the set validation is performed by a prediction error identification step applied to an unbiased model of the Model Error Model (see e.g. (Ljung, 1997)). One can even entirely remove the need for the identification of an unbiased model (of the true system or of the Model Error Model) by the use of stochastic embedding techniques à la Goodwin et al. (Goodwin *et al.*, 1992), as we have recently shown in (Bombois *et al.*, 2000*a*). In such case, the parametrization of the validated PE uncertainty set is somewhat different from the one we present below, but the methodology presented here remains unchanged and all the results and conclusions of this paper apply.

Thus, using N input-output data, collected in open or in closed loop, and a model set $\mathcal{M}$ containing $\mathcal{S}$, we estimate a model $G(z, \hat{\delta})$ and a covariance matrix P_{δ} of $\hat{\delta}$ using classical PE identification (Ljung, 1999). We have shown in (Bombois *et al.*, 1999*d*) and (Bombois *et al.*, 2000*b*) that, whether the validation is performed in open loop or in closed loop, by identification of a full order model for G_0 or by identification of an unbiased estimate of a Model Error Model, the validated model sets can all be described in the following generic structure that we call the **generic PE model uncertainty set**.

Proposition 2.1. The model sets resulting from prediction error validation, which contain the true system $G_0 = G(z, \delta_0)$ at a prescribed probability level, can all be described in the following generic form called the generic PE model uncertainty set:

$$\mathcal{D} = \Big\{G(z,\delta) \mid G(z,\delta) = \frac{e + Z_N\delta}{1 + Z_D\delta} \text{ and } \delta \in U = \{\delta \mid (\delta - \hat{\delta})^T R(\delta - \hat{\delta}) < 1\}\Big\} \quad (2)$$

where
- $\delta \in \mathbf{R}^{k\times 1}$ is a real parameter vector, and $\hat{\delta}$ is the parameter estimate resulting from the identification/validation step.
- $R \in \mathbf{R}^{k\times k}$ is a symmetric positive definite matrix, proportional to the inverse of the covariance matrix P_δ of $\hat{\delta}$.
- $Z_N(z)$ and $Z_D(z)$ are row vectors of size k of known transfer functions.
- $e(z)$ is a known transfer function with a delay equal to the delay of G_0.

In the special case where the validation is performed in open loop, or in closed loop using a direct identification method, the PE uncertainty set $\mathcal{D}$ reduces to the more familiar set:

$$\Big\{G(z,\delta) \mid G(z,\delta) = \frac{b_1 z^{-1} + \ldots + b_m z^{-m}}{1 + a_1 z^{-1} + \ldots + a_n z^{-n}} = \frac{Z_2\delta}{1 + Z_1\delta} \; and \; \delta \in U\Big\} \quad (3)$$

with

- $\delta^T = [a_1 \ldots a_n \; b_1 \ldots b_m] \in \mathbf{R}^{k\times 1}$, $k \triangleq n + m$
- $Z_1(z) = [z^{-1} \; z^{-2} \ldots z^{-n} \; 0 \ldots 0] \in \mathbf{C}^{1\times k}$
- $Z_2(z) = [0 \ldots 0 \; z^{-1} \; z^{-2} \ldots z^{-m}] \in \mathbf{C}^{1\times k}$

Comments

- Observe that the PE model uncertainty sets are very different from the classical uncertainty sets that are used in mainstream robust control design procedures, such as additive, multiplicative, feedback, coprime factor uncertainty sets, etc.
- The validated model set $\mathcal{D}$ depends very much on the experimental conditions under which the validation has been performed. This is perhaps not so apparent in the definition (2) of $\mathcal{D}$ via the parameter covariance matrix P_δ which defines R. However, let us recall that the covariance of the transfer function estimate $G(z, \hat{\delta})$ is a function of the input spectrum and, in a closed loop experiment, of the controller and the reference spectrum; see e.g. (Ljung, 1999). Thus, two different validation data sets will yield two different validated regions $\mathcal{D}^{(1)}$ and $\mathcal{D}^{(2)}$, both of which contain the true G_0 with probability α.

In the study of *model validation for control* it is therefore important to examine whether one PE validated region, $\mathcal{D}^{(1)}$, is better tuned for robust control design than another one, $\mathcal{D}^{(2)}$.

We shall from now on assume that the control design is based on some nominal model G_{mod}, and that this nominal model is any validated model, i.e. any model in $\mathcal{D}$. An obvious option would be to use the center of the validated set, i.e. $G(z, \hat{\delta})$, as the model used for control design. However, one may often prefer to use a low order model for control design, since this results in a lower order controller. For the theory that follows, any validated model can be used for control design. [2]

In order to address the question of whether a PE uncertainty set is tuned for robust control design, we have defined a 'measure of size' of a validated PE set $\mathcal{D}$ that can be connected, via a robust stability theorem, to the size of the model-based controller set that robustly stabilizes all models in $\mathcal{D}$. This measure is called the *worst-case ν-gap* (and its frequency by frequency version, the worst-case chordal distance) between some model G_{mod} and all models in a validated PE model set $\mathcal{D}$. In order to define these worst-case measures, we first recall the definitions of chordal distance and of ν-gap between two transfer functions, introduced by Vinnicombe (Vinnicombe, 1993). To keep things simple we consider scalar transfer functions only.

Definition 2.1. (Vinnicombe, 1993) The ν-gap metric between two transfer functions G_1 and G_2 is defined as

$$\delta_\nu(G_1, G_2) = \max_\omega \kappa\left(G_1(e^{j\omega}), G_2(e^{j\omega})\right) \quad (4)$$

if $W(G_1, G_2) = 0$, and 1 otherwise, where $\kappa\left(G_1(e^{j\omega}), G_2(e^{j\omega})\right)$

$$\triangleq \frac{|G_1(e^{j\omega}) - G_2(e^{j\omega})|}{\sqrt{1 + |G_1(e^{j\omega})|^2}\sqrt{1 + |G_2(e^{j\omega})|^2}} \quad (5)$$

and where $W(G_1, G_2) = wno(1 + G_1^* G_2) + \eta(G_2) - \tilde{\eta}(G_1)$.

Here $\eta(G)$ (resp. $\tilde{\eta}(G)$) denotes the number of poles of G in the complement of the closed (resp. open) unit disc, while $wno(G)$ denotes the winding number about the origin of $G(z)$ as z follows the unit circle indented into the exterior of the unit disc around any unit circle pole and zero of $G(z)$. The function $\kappa(G_1(e^{j\omega}), G_2(e^{j\omega}))$ is the chordal distance between the projections of $G_1(e^{j\omega})$ and $G_2(e^{j\omega})$ onto the Riemann sphere of unit diameter with South Pole at the origin of the complex plane.

---

[2] In fact, our theory applies even to the case when the model used for control design lies outside the PE validated region $\mathcal{D}$.

Consider now a closed loop system with transfer function matrix

$$T(G,C) = \begin{pmatrix} T_{11} & T_{12} \\ T_{21} & T_{22} \end{pmatrix} = \begin{pmatrix} \frac{GC}{1+GC} & \frac{G}{1+GC} \\ \frac{C}{1+GC} & \frac{1}{1+GC} \end{pmatrix} \quad (6)$$

Definition 2.2. (Vinnicombe, 1993) The **generalized stability margin** of this closed loop system is defined as

$$b_{GC} = \|T(G,C)\|_\infty^{-1} = \min_\omega \kappa(G(e^{j\omega}), -\frac{1}{C(e^{j\omega})}) \quad (7)$$

if (G, C) is stable, and 0 otherwise.

Thus, the generalized stability margin of a closed loop system $[G\ C]$ is measured by the least chordal distance between the projections on the Riemann sphere of G and of the inverse of $-C$.

The main interest of the ν-gap metric and the chordal distance is their use in robust stability. Here we recall a result that is most useful for our 'validation for control' analysis.

Proposition 2.2. (Vinnicombe, 1993) Let C stabilize a model G_{mod}. Then C stabilizes all G such that $\delta_\nu(G_{mod}, G) < 1$ and such that $\forall\omega$

$$\kappa\left(G_{mod}(e^{j\omega}), G(e^{j\omega})\right) < \kappa(G_{mod}(e^{j\omega}), -\frac{1}{C(e^{j\omega})}).$$

In particular, C stabilizes all models in the set :

$$\mathcal{G} = \{G : \delta_\nu(G, G_{mod}) < b_{G_{mod}C}\} \quad (8)$$

We now build on these robust stability results to connect validated PE uncertainty sets to sets of robustly stabilizing controllers. To do this, we have introduced two new notions that extend definitions of distance between two models to definitions of **worst case distance** between a model and all models in a PE model set.

Definition 2.3. Consider a PE uncertainty set $\mathcal{D}$ of the form (2) and a model G_{mod}. The 'worst case chordal distance' at frequency ω between G_{mod} and $\mathcal{D}$ is defined as

$$\kappa_{WC}(G_{mod}(e^{j\omega}), \mathcal{D}) = \sup_{G_\mathcal{D} \in \mathcal{D}} \kappa(G_{mod}(e^{j\omega}), G_\mathcal{D}(e^{j\omega}))$$

The 'worst case ν-gap' between G_{mod} and $\mathcal{D}$ is defined as

$$\delta_{WC}(G_{mod}, \mathcal{D}) = \sup_{G_\mathcal{D} \in \mathcal{D}.} \delta_\nu(G_{mod}, G_\mathcal{D}) \quad (9)$$

An alternative characterization of the worst case ν-gap is as follows: see (Bombois *et al.*, 1999*c*).

Lemma 2.1. If $W(G_{mod}, G_\mathcal{D}) = 0$ for one plant $G_\mathcal{D} \in \mathcal{D}$, then $\delta_{WC}(G_{mod}, \mathcal{D})$ can also be expressed as

$$\delta_{WC}(G_{mod}, \mathcal{D}) = \sup_\omega \kappa_{WC}\ (G_{mod}(e^{j\omega}), \mathcal{D})).\Box$$

We have shown in (Bombois *et al.*, 1999*a*) that the worst-case chordal distance can be computed as the optimal value of a convex optimization problem involving Linear Matrix Inequality (LMI) constraints.

Having extended the concepts of chordal distance and of ν-gap between plants to those of worst-case chordal distance and worst-case ν-gap between a model and a validated PE set, we can now also extend the stability results of Proposition 2.2 to the context of our validated PE sets.

Theorem 2.1. Consider a PE uncertainty set $\mathcal{D}$ containing the true G_0 with probability α, a model G_{mod} with $\delta_{WC}(G_{mod}, \mathcal{D}) < 1$, and let C be a controller that stabilizes G_{mod}. Then C stabilizes all models in $\mathcal{D}$ if $\forall\omega$

$$\kappa_{WC}(G_{mod}(e^{j\omega}), \mathcal{D}) < \kappa(G_{mod}(e^{j\omega}), -\frac{1}{C(e^{j\omega})})$$

and, a fortiori, if $\delta_{WC}(G_{mod}, \mathcal{D}) < b_{G_{mod}C}$. □

Observe that the left hand side depends on the validated set, while the right hand side depends on the controller; both quantities are known. Both of these stability conditions are sufficient conditions only, and that the latter is a conservative (Min-Max) version of the former. Necessary and sufficient conditions for the stabilization of all models in a validated PE model set by a controller C will be given in Section 3. The main use of Theorem 2.1 is therefore not so much for checking the stability of a particular controller, but rather it allows us to link the validation experiment (through the measure of size $\kappa_{WC}(G_{mod}(e^{j\omega}), \mathcal{D})$ of the validated set) with the set of robustly stabilizing controllers for $\mathcal{D}$. Indeed, for each validated PE uncertainty set $\mathcal{D}$ we can now define, by the use of Theorem 2.1, the corresponding set of G_{mod}-based controllers that robustly stabilize all models in the set $\mathcal{D}$ as[3] :

$$\mathcal{C}(G_{mod}, \mathcal{D}) = \Big\{C \mid \kappa_{WC}(G_{mod}(e^{j\omega}), \mathcal{D}) < \kappa\left(G_{mod}(e^{j\omega}), -\frac{1}{C(e^{j\omega})}\right)\ \forall\omega\Big\} \quad (10)$$

We then have the following result (Gevers *et al.*, 2000).

Theorem 2.2. Consider two different validated PE sets $\mathcal{D}^{(1)}$ and $\mathcal{D}^{(2)}$, obtained from two different validation experiments, both containing the model G_{mod}. If for all ω

$$\kappa_{WC}(G_{mod}(e^{j\omega}), \mathcal{D}^{(1)}) < \kappa_{WC}(G_{mod}(e^{j\omega}), \mathcal{D}^{(2)}),$$

then $\mathcal{C}(G_{mod}, \mathcal{D}^{(2)}) \subset \mathcal{C}(G_{mod}, \mathcal{D}^{(1)})$. □

[3] Observe that there may be other controllers, outside the set $\mathcal{C}(G_{mod}, \mathcal{D})$, that stabilize all models in $\mathcal{D}$, because the stability condition used to define this set is only a sufficient condition.

Theorem 2.2 establishes a clear link between a measure of size defined on validated PE model sets and the sets of stabilizing controllers for all models in these validated model sets. A more compact measure of size is $\delta_{WC}(G_{mod}, \mathcal{D})$ (i.e. just one number to characterize the size of $\mathcal{D}$) for which a corresponding but more conservative link can be established with a set of stabilizing controllers: see (Bombois *et al.*, 1999*c*), (Gevers *et al.*, 2000). Thus, $\kappa_{WC}(G_{mod}(e^{j\omega}), \mathcal{D})$ and $\delta_{WC}(G_{mod}, \mathcal{D})$ are two measures of the ability of a validated PE model set to be robustly stabilized by a large set of controllers. It makes sense therefore, in identification for robust control, to choose those experimental conditions for validation that lead to a validated set with the smallest possible worst-case chordal distance, or the smallest possible worst-case ν-gap. Thus, our results give the designer a handle on what to aim for in terms of "control-oriented validated sets". In a Prediction Error context, we believe that these results are the first that give substance to the concept of *control-oriented model validation.*

3. CONTROLLER VALIDATION FOR STABILITY AND PERFORMANCE

In this section, we present *necessary and sufficient conditions* for some given controller $C(z)$ to stabilize all models in a PE uncertainty set, as well as necessary and sufficient conditions for this controller to achieve a specified level of performance with all models in such PE uncertainty set. In other words, we develop a robust stability analysis (for both stability and performance) for uncertainty sets of the generic PE form described in (2).

Robust stability

For our PE uncertainty sets, necessary and sufficient conditions for robust stability have been obtained by showing that the set of feedback connections of C with all models in $\mathcal{D}$ can be reformulated, using an LFT framework, as a set of feedback connections $[M_{\mathcal{D}}(z) \;\; \phi]$, where $M_{\mathcal{D}}(z)$ is fixed and where the uncertainty part ϕ is a real vector, linearly related to the parameter vector δ that defines $\mathcal{D}$: see (2). We have then shown that the (real) stability radius linked with the set of loops $[M_{\mathcal{D}}(z) \;\; \phi]$ can be computed exactly. The main robust stability result is as follows (Bombois *et al.*, 2000*b*).

Theorem 3.1. Consider a generic PE uncertainty set of the form (2) and a controller $C(z) = X(z)/Y(z)$ that stabilizes $G(\hat{\delta})$. Then all models in $\mathcal{D}$ are stabilized by $C(z)$ if and only if $\max_{\Omega} \mu_{\phi}(M_{\mathcal{D}}(e^{j\omega})) \leq 1$, where

- $M_{\mathcal{D}}(z) = \frac{-(Z_D + \frac{X(Z_N - eZ_D)}{Y+eX})T^{-1}}{1+(Z_D + \frac{X(Z_N - eZ_D)}{Y+eX})\hat{\delta}}$
- T is defined as $\mathcal{D} : R = T^T T$.
- $\phi = T(\delta - \hat{\delta})$, whereby $\delta \in U \Leftrightarrow |\phi|_2 < 1$
- $\mu_{\phi}(M_{\mathcal{D}}(e^{j\omega}))$ is the *stability radius* of the loop $[M_{\mathcal{D}}(z) \;\; \phi]$. For a real vector ϕ, μ_{ϕ} is computed as:

$$\sqrt{|Re(M)|_2^2 - \frac{(Re(M)Im(M)^T)^2}{|Im(M)|_2^2}} \quad \text{if } Im(M) \neq 0$$

$$|M|_2 \quad \text{if } Im(M) = 0.$$

The necessary and sufficient conditions for controller validation depend critically on the uncertainty set $\mathcal{D}$, hence the importance of the validation design. Of course, this observation applies to all robust control methodologies. What distinguishes our approach from most others is that, in mainstream robust control theory, the uncertainty sets are assumed a priori, while here they are the result of experiments, and our analysis of the previous sections has given us at least some handle on how we can shape these uncertainty sets towards robust control design.

Robust performance

Most commonly used performance criteria are derived from some norm of a frequency weighted version of the transfer matrix $T(G, C)$ of the closed-loop system $[G \;\; C]$ defined in (6). Thus we shall start from the following definition.

Definition 3.1. The performance of a closed loop system (G, C) is defined by the following frequency function:

$$J(G, C, W_l, W_r, \omega) = \sigma_{\max}\left(W_l T(G(e^{j\omega}), C(e^{j\omega}))W_r\right), \quad (11)$$

where $W_l(z)$ and $W_r(z)$ are diagonal frequency weights that allow one to define specific performance levels, and where $\sigma_{\max}(A)$ denotes the largest singular value of A. □

The frequency function J defines a template. Most commonly used performance functions are functions of J, or special cases of J. The worst case performance over a validated PE set is then defined as follows.

Definition 3.2. The worst case performance achieved by a controller $C(z)$ at a frequency ω over all models in a validated PE uncertainty set $\mathcal{D}$ is defined as

$$J_{WC}(\mathcal{D}, C, W_l, W_r, \omega) \quad (12)$$
$$= \max_{G(z,\delta)\in\mathcal{D}} \sigma_{\max}\left(W_l T(G(e^{j\omega}, \delta), C(e^{j\omega}))W_r\right).$$

The following theorem, established in (Bombois *et al.*, 2000*b*), gives a procedure for the computation of the criterion $J_{WC}(\mathcal{D}, C, W_l, W_r, \omega)$ at the frequency ω. A crucial feature that makes this computation possible is the rank one property of the matrix $T(G, C)$.

Theorem 3.2. Consider a PE uncertainty region $\mathcal{D}$ defined by (2) and a robustly stabilizing controller $C = \frac{X}{Y}$. Then, at frequency ω, the criterion function $J_{WC}(\mathcal{D}, C, W_l, W_r, \omega)$ is obtained as

$$J_{WC}(\mathcal{D}, C, W_l, W_r, \omega) = \sqrt{\gamma_{opt}}, \qquad (13)$$

where γ_{opt} is the optimal value of γ for the following standard convex optimization problem involving LMI constraints:

$$\begin{array}{ll} \textit{minimize} & \gamma \\ \textit{over} & \gamma,\ \tau \\ \textit{subject to} & \tau \geq 0 \ \ \textit{and} \end{array}$$

$$\begin{pmatrix} Re(a_{11}) & Re(a_{12}) \\ Re(a_{12}^*) & Re(a_{22}) \end{pmatrix} - \tau \begin{pmatrix} R & -R\hat{\delta} \\ (-R\hat{\delta})^T & \hat{\delta}^T R\hat{\delta} - 1 \end{pmatrix} < 0$$

where

$$a_{11} = (Z_N^* W_{l1}^* W_{l1} Z_N + Z_D^* W_{l2}^* W_{l2} Z_D) - \gamma(Q Z_1^* Z_1);$$
$$a_{12} = Z_N^* W_{l1}^* W_{l1} e + W_{l2}^* W_{l2} Z_D^* - \gamma(Q Z_1^* (Y + eX));$$
$$a_{22} = e^* W_{l1}^* W_{l1} e + W_{l2}^* W_{l2} - \gamma(Q(Y + eX)^* (Y + eX));$$
$$Q = 1/(X^* W_{r1}^* W_{r1} X + Y^* W_{r2}^* W_{r2} Y);$$
$$Z_1 = X Z_N + Y Z_D.$$

4. CONCLUSIONS

Using Prediction Error identification methods, we have developed a coherent framework for *model validation for robust control* and for *controller validation.* Our model validation procedure is nothing but an identification experiment with a full order model, [4] which leads to a validated PE model set made up of parametrized transfer functions, whose parameters lie in ellipsoidal regions. The true system belongs to this set with some probability, say α. The sets resulting from such PE validation method are non-standard in mainstream robust control theory. We have therefore developed robust analysis tools, that have led to necessary and sufficient conditions for controller validation with PE uncertainty sets, both for stability and for worst-case performance. This is one major contribution of our work.

Another major focus has been to highlight the design aspects of our model set validation procedure, and how they impact on the set of robustly stabilizing controllers. In order to establish a connection all the way from the design of the validation step to the specification of a set of robustly stabilizing controllers, we have defined a measure of the size of the PE validated sets, and shown that it is connected to the size of the controller sets that are guaranteed to robustly stabilize all models in these PE validated sets. Even though these results are based on sufficient conditions for stability only, they provide us with guidelines for *model validation for robust control.*

In the full paper (Gevers *et al.*, 2000), our results and techniques are illustrated on two real-life applications: a flexible transmission system that has been used as benchmark for control design, and a ferrosilicon production process.

[4] We have recently shown in (Bombois *et al.*, 2000*a*) that biased models can also be handled.

5. REFERENCES

Bombois, X., M. Gevers and G. Scorletti (1999*a*). Controller validation based on an identified model. In: *38th IEEE Conference on Decision and Control.* Phoenix, Arizona. pp. 2816–2821.

Bombois, X., M. Gevers and G. Scorletti (1999*b*). Controller validation for a validated model set. In: *CD-ROM Proc. European Control Conference 99, paper 869.* Karlsruhe, Germany.

Bombois, X., M. Gevers and G. Scorletti (1999*c*). A measure of robust stability for an identified set of parametrized transfer functions. *submitted to IEEE Transactions on Automatic Control.*

Bombois, X., M. Gevers and G. Scorletti (2000*a*). Controller validation for stability and performance based on a frequency domain uncertainty region obtained by stochastic embedding. In: *39th IEEE Conference on Decision and Control.* Sydney, Australia.

Bombois, X., M. Gevers, G. Scorletti and B.D.O. Anderson (1999*d*). Controller validation for stability and performance based on an uncertainty region designed from an identified model. *CESAME Tech Rept 99.36, submitted to Automatica.*

Bombois, X., M. Gevers, G. Scorletti and B.D.O. Anderson (2000*b*). Controller validation for stability and performance based on an uncertainty region designed from an identified model. In: *IFAC Symposium on System Identification, SYSID 2000.* Santa Barbara, California.

Gevers, M., B. Codrons and F. De Bruyne (1999*a*). Model validation in closed-loop. In: *Proc. American Control Conference 99.* San Diego, USA. pp. 326–330.

Gevers, M., X. Bombois, B. Codrons, F. De Bruyne and G. Scorletti (1999*b*). The role of experimental conditions in model validation for control. In: *Robustness in Identification and Control - Proc. of Siena Workshop, July 1998* (A. Garulli, A. Tesi and A. Vicino, Eds.). Vol. 245 of *Lecture Notes in Control and Information Sciences.* Springer-Verlag. pp. 72–86.

Gevers, M., X. Bombois, B. Codrons, G. Scorletti and B.D.O. Anderson (2000). Model validation for control and controller validation: a prediction error identification approach. *submitted for publication to Automatica.*

Goodwin, G.C., M. Gevers and B. Ninness (1992). Quantifying the error in estimated transfer functions with application to model order selection. *IEEE Trans. Automatic Control* **37**, 913–928.

Ljung, L. (1997). Identification, model validation and control. *36th IEEE Conf. on Decision and Control, plenary lecture.*

Ljung, L. (1999). *System Identification: Theory for the User, 2nd Edition.* Prentice-Hall. Englewood Cliffs, NJ.

Vinnicombe, G. (1993). Frequency domain uncertainty and the graph topology. *IEEE Trans Automatic Control* **AC-38**, 1371–1383.

UNCERTAINTY MODEL UNFALSIFICATION

Robert L. Kosut

SC Solutions, 3211 Scott Blvd., Santa Clara, CA 95054, USA
kosut@scsolutions.com

Abstract An *uncertainty model* embeds the behavior of a system in a set whose elements consist of uncertain parameters, signals, and dynamics. By uncertain is meant "unknown, but bounded." In this paper the bounds are determined from finite data for specified types of sets useful for robust control design.

1 Introduction

Although most attempts at defining adaptive control fail, the working definition here is this: *adaptive control means to extract from available data the knowledge necessary to design the "best" controller to try on the system.* How do we quantify "best?" Put another way, how certain is the knowledge gained from the data?

In this paper we examine how *unfalsification* can estimate an uncertainty model set from finite data which includes both disturbance and dynamic uncertainty as well as the usual parameters in the standard prediction error models, *e.g.*, [12],[14]. Unfalsification has been proposed as a replacement for the system identification step in adaptive control [6], and as a means for direct controller falsification, *i.e.*, using available data to falsify a candidate controller without implementation [13], [7]. Unfalsification also helps quantify the connections between adaptation and learning as espoused in the dual control concept [1, Ch. 7], and in the windsurfer approach to adaptation [9]. Future applications may involve direct tuning methods, *e.g.*, [8], where, as in [13], no plant model is needed, The plant model is replaced by the closed-loop performance specification.

In this paper we estimate parameters in the sets describing a stochastic disturbance and a deterministic dynamic uncertainty. We will draw some connections to the "stochastic embedding" ideas from [4] and the "model error modeling" approach from [11].

Notation An ℓ-length sequence $x = \{x_1, \ldots, x_\ell\}$ will be equivalently viewed as a vector and we will sometimes write "the sequence $x \in \mathbf{R}^\ell$." The norm of the sequence $x \in \mathbf{R}^\ell$ is defined as $\|x\|^2 = \sum_{t=1}^{\ell} x_t^2$, and the RMS-norm as $\|x\|_{\text{rms}} = \|x\| / \sqrt{\ell}$ The dynamic systems considered are scalar, causal, map sequences in $\mathbf{R}^\ell$ into sequences in $\mathbf{R}^\ell$, and are initially at rest, *i.e.*, they have no memory of input sequences for time $t < 1$. For the system G, the value of the mapping Gu at time t is denoted by $(Gu)_t$ and a subsequence of the mapping by $(Gu)_{[1,t]}$. Depending on the context, z denotes either the unit forward shift operator or the z-transform variable; n-forward shifts are denoted by z^n and n-backward shifts (delays) by z^{-n}.

2 Unfalsification

In this paper we will focus on the *prediction error uncertainty model*,

$$y = G(\theta)u + H(\theta)\left(w + \Delta(\delta)u\right) \quad \begin{cases} \theta \in \boldsymbol{\Theta} \\ w \in \mathbf{W} \\ \delta \in \boldsymbol{\Delta} \end{cases} \tag{1}$$

where $G(\theta), H(\theta)$ are LTI and represent the nominal system model with parameter $\theta \in \mathbf{R}^p$, u and y are the applied input and sensed output, respectively, w is an exogenous disturbance, and $\Delta(\delta)$ is an LTI system with uncertain parameter $\delta \in \mathbf{R}^k$. The sets $\boldsymbol{\Theta}, \mathbf{W}, \boldsymbol{\Delta}$ are assembled from our prior assumptions about the uncertain character of θ, w, δ and may depend on some unknown bounding parameters. The uncertainty model (1) is of course not the most general form. It is, however, sufficiently flexible for most purposes and is useful to illustrate the main ideas. A more general form is the *linear fractional uncertainty model*, *e.g.*, [14].

The *prediction error* associated with the θ-dependent part of the model is given in standard form by [10]

$$e(\theta) := H(\theta)^{-1}(y - G(\theta)u) \tag{2}$$

By combining (1) with (2) we can express the uncertainty model in error form by:

$$e(\theta) = w + \Delta(\delta)u \quad \begin{cases} \theta \in \boldsymbol{\Theta} \\ w \in \mathbf{W} \\ \delta \in \boldsymbol{\Delta} \end{cases} \tag{3}$$

The problem considered here is the following:

> *Given finite data $y, u \in \mathbf{R}^\ell$, does there exist $\theta \in \boldsymbol{\Theta}$, $w \in \mathbf{W}$, and $\delta \in \boldsymbol{\Delta}$ consistent with (3),* i.e.,
>
> $$e(\theta) = w + \Delta(\delta)u$$

If the answer is yes, then the uncertainty model (3) is said to be **unfalsified** *with respect to the data. If the answer is no, then the uncertainty model is said to be* **falsified**.

We also consider a more relaxed problem where the uncertainty sets have free parameters which are selected to unfalsifiy the model.

An uncertainty model with deterministic uncertainty sets, once falsified, can never be unfalsified by another data set. If, however, the disturbance or dynamic uncertainty is probabilistic, then unfalsification can be stated *with probability*, *e.g.*, the uncertainty model in unfalsified with probability 99%. In this case, new data can unfalsifiy the same uncertainty model with a different probability. This aspect will not be pursued here.

3 ARX Uncertainty Model

We will restrict attention to the ARX uncertainty model,

$$A(\theta)y = B(\theta)u + \Delta(\delta)u + w \quad \left\{ \begin{array}{l} \theta \in \mathbf{R}^p \\ w \in \mathbf{W} \\ \delta \in \mathbf{\Delta} \end{array} \right. \tag{4}$$

with

$$\begin{array}{rcl} A(\theta) & = & 1 + a_1 z^{-1} + \cdots + a_n z^{-n} \\ B(\theta) & = & b_1 z^{-1} + \cdots + b_m z^{-m} \\ \theta & = & [a_1 \cdots a_n \; b_1 \cdots b_m]^T \in \mathbf{R}^p, \; p = n + m \\ \Delta(\delta) & = & \delta_1 z^{-1} + \cdots + \delta_k z^{-k} \end{array} \tag{5}$$

An equivalent expression for the ARX uncertainty model is:

$$y = \Phi\,\theta + \Psi\,\delta + w \quad \left\{ \begin{array}{l} \theta \in \mathbf{R}^p \\ w \in \mathbf{W} \\ \delta \in \mathbf{\Delta} \end{array} \right. \tag{6}$$

with

$$\begin{array}{l} \delta = [\delta_1 \cdots \delta_k]^T \in \mathbf{R}^k \\ \Phi = \left[-z^{-1}y \cdots - z^{-n}y \;\; z^{-1}u \cdots - z^{-m}u \right] \in \mathbf{R}^{\ell \times p} \\ \Psi = \left[\; z^{-1}u \cdots - z^{-k}u \right] \in \mathbf{R}^{\ell \times k} \end{array} \tag{7}$$

The data dependent variables are $y \in \mathbf{R}^\ell$, $\Phi \in \mathbf{R}^{\ell \times p}$, and $\Psi \in \mathbf{R}^{\ell \times k}$. The uncertainty parameters are $\theta \in \mathbf{R}^p$, $\delta \in \mathbf{R}^k$, and $w \in \mathbf{R}^\ell$. We will see later (section 6) that for control design other choices for constructing Ψ are more appropriate.

4 Uncertainty Sets

Disturbance Uncertainty

There are many ways to characterize the disturbance set $\mathbf{W}$. The following define finite length sequence sets.

RMS-Bounded Disturbance

$$\mathbf{W} = \left\{ w \in \mathbf{R}^\ell \;\middle|\; \frac{1}{\ell} \|w\|^2 \leq \lambda \right\} \tag{8}$$

Here $\sqrt{\lambda}$ is the maximum RMS of the disturbance over ℓ samples. Norm bounded sets such as this are the simplest of choices for deterministically characterizing "noise." The main advantage is that it is a convex set and therefor easy to handle in optimization. However, there are no restrictions preventing correlation of w with sensed inputs, which would disqualify w as "noise."

White Noise Disturbance A white noise disturbance is characterized as

$$\mathbf{W} = \left\{ w \in \mathbf{R}^\ell \mid \mathbf{E}w = 0, \, \mathbf{E}ww^T = \lambda I_\ell \right\} \tag{9}$$

where λ is the variance. A further refinement, actually a more restrictive set, is Gaussian white noise:

$$\mathbf{W} = \left\{ w \in \mathbf{R}^\ell \mid w \in \mathcal{N}(0, \lambda I_\ell) \right\} \tag{10}$$

This additional restriction also brings more detail that could be utilized, *e.g.*, the density function is known.

Dynamic Uncertainty

Uncertain dynamics can also be characterized in a number of ways. Consider the following gain-bounded, time-invariant (TI), dynamic uncertainty set:

$$\mathbf{\Delta} = \left\{ \Delta \in \mathrm{TI} \;\middle|\; \sup_{\|u\|_{[1,\ell]} \neq 0} \frac{\|\Delta(\delta)u\|_{[1,\ell]}}{\|u\|_{[1,\ell]}} \leq \gamma, \;\; \forall \ell \right\} \tag{11}$$

Linear-Time-Invariant (LTI) If $\Delta \in \mathrm{LTI}$, it follows that Δ has the stable z-transform $\Delta(z)$, and additionally, the gain bounded set is equivalent to the set of $\mathbf{H}_\infty$ bounded transfer functions,

$$\mathbf{\Delta} = \left\{ \Delta(z) \;\middle|\; \|\Delta\|_{\mathbf{H}_\infty} = \sup_{0 \leq \omega \leq \pi} |\Delta(e^{j\omega})| \leq \gamma \right\} \tag{12}$$

Since Δ is stable, it has the z-transform,

$$\Delta(z) = \sum_{k=1}^{\infty} \delta_k z^{-k} \tag{13}$$

where $\{\delta_1, \delta_2, \ldots\}$ are the impulse response coefficients. Let $v \in \mathbf{R}^\ell$ denote the ℓ-length output sequence of the LTI system Δ with ℓ-length input sequence $u \in \mathbf{R}^\ell$. Hence, we can write the error equation (3) in vector form up to length (response time) ℓ as

$$\begin{array}{rcl} e(\theta) & = & w + v \\ v & = & \mathcal{T}\{\delta\}\,u = \mathcal{T}\{u\}\,\delta \end{array} \tag{14}$$

where $\delta \in \mathbf{R}^\ell$ is the vector of the first ℓ impulse response coefficients of Δ, and $\mathcal{T}\{u\}, \mathcal{T}\{\delta\}$ are the $\ell \times \ell$ Toeplitz matrices formed from $u \in \mathbf{R}^\ell$, $\delta \in \mathbf{R}^\ell$, respectively, *e.g.*,

$$\mathcal{T}\{\delta\} = \begin{bmatrix} \delta_1 & 0 & \cdots & 0 \\ \delta_2 & \delta_1 & \cdots & 0 \\ \vdots & \vdots & \ddots & \vdots \\ \delta_\ell & \delta_{\ell-1} & \cdots & \delta_1 \end{bmatrix} \tag{15}$$

The following theorem is established in [5]:

Theorem *Let* $\Delta(z) = \sum_{n=1}^{\infty} \delta_n z^{-n} \in \mathbf{H}_\infty$, *and let* $\delta = [\delta_1, \ldots, \delta_\ell]^T \in \mathbf{R}^\ell$. *Then, as* $\ell \to \infty$,

$$\|\mathcal{T}\{\delta\}\| \nearrow \|\Delta\|_{\mathbf{H}_\infty}$$

The symbol $\nearrow$ means "approaches from below." Hence, for finite ℓ, $\|\mathcal{T}\{\delta\}\| < \gamma$, unless $\Delta(z)$ is FIR of order ℓ, in which case $\|\mathcal{T}\{\delta\}\| = \gamma$. This could mean that without prior information, it is not possible from finite data to overbound model error. *Hence, with finite data alone, we can never precisely estimate this error.*

The inequality $\|\mathcal{T}\{\delta\}\| < \gamma$ is equivalent to the matrix inequality,

$$\mathcal{T}\{\delta\}^T \mathcal{T}\{\delta\} - \gamma^2 I_\ell < 0 \tag{16}$$

Observe that this is a *linear* matrix inequality (LMI) in δ and γ, *i.e.*, equivalently [2],

$$\begin{bmatrix} \gamma I_\ell & \mathcal{T}\{\delta\}^T \\ \mathcal{T}\{\delta\} & \gamma I_\ell \end{bmatrix} > 0 \tag{17}$$

This LMI can also be expressed in terms of the sequences $e(\theta)$ and w. Applying the Toeplitz operation (15) to both sides of (14) gives,

$$\mathcal{T}\{e(\theta)\} = \mathcal{T}\{w\} + \mathcal{T}\{\delta\}\mathcal{T}\{u\}$$

After some linear algebra, it follows that (16) holds if and ony if,

$$\mathcal{T}\{e(\theta) - w\}^T \mathcal{T}\{e(\theta) - w\} - \gamma^2 \mathcal{T}\{u\}^T \mathcal{T}\{u\} < 0 \tag{18}$$

Equvalently,

$$\begin{bmatrix} \gamma \mathcal{T}\{u\} & \mathcal{T}\{e(\theta) - w\}^T \\ \mathcal{T}\{e(\theta) - w\} & \gamma \mathcal{T}\{u\} \end{bmatrix} > 0 \tag{19}$$

This is an LMI in γ, w and $e(\theta)$. In the case of the ARX model, $e(\theta)$ is affine in θ, and hence, is also an LMI in θ.

5 Uncertainty Tradeoff

In this section we show how a tradeoff between disturbance and dynamic uncertainty can be obtained from a finite data set for the ARX uncertainty model (4)-(7), *i.e.*,

$$y = \Phi\,\theta + \Psi\,\delta + w \quad \left\{ \begin{array}{l} w \in \mathbf{W} \\ \delta \in \mathbf{\Delta} \end{array} \right. \tag{20}$$

with $\Phi \in \mathbf{R}^{\ell \times p}$ and $\Psi \in \mathbf{R}^{\ell \times k}$ as defined in (7). Observe that if $k = \ell$, then from the definition (15), $\Psi = \mathcal{T}\{u\} \in \mathbf{R}^{\ell \times \ell}$. Again, for control Ψ is constructed with filters (section 6).

The physics of many problems motivates the combination of deterministic dynamic uncertainty with stochastic disturbance uncertainty. Assume that the uncertainty sets are $\mathbf{H}_\infty$ bounded dynamics with Gaussian white disturbance:

$$\begin{aligned} \mathbf{W} &= \{\, w \in \mathbf{R}^\ell \mid w \in \mathcal{N}(0, \lambda I_\ell) \,\} \\ \mathbf{\Delta} &= \{\, \delta \in \mathbf{R}^k \mid \|\mathcal{T}\{\delta\}\| \le \gamma \,\} \end{aligned} \tag{21}$$

Adding the Gaussian assumption may be too restrictive in some cases, despite the central limit theorem. Computations, however, are often simpler.

Under these assumptions, those values of (θ, δ, w) which are unfalsified by the data $(y, u) \in \mathbf{R}^\ell$ are in the set:

$$\begin{aligned} & y = w + \Phi\theta + \Psi\delta \\ & w \in \mathcal{N}(0, \lambda I_\ell) \\ & \|\mathcal{T}\{\delta\}\| \le \gamma \end{aligned} \tag{22}$$

To test for mewmbership in this set appears to require defining a "measure of Gaussianity," or more generally, a measure of whiteness. We will return to this issue after a brief aside.

RMS-bounded disturbance In [6, 7], we formulated a definitive test under the assumption that the disturbance set is RMS-bounded (8). In this case we posed an optimization problem to find the smallest value of λ for a given γ, thereby resulting in a γ-family of unfalsified uncertainty models specified by the parameters $(\hat{\lambda}_\gamma, \hat{w}_\gamma, \hat{\delta}_\gamma, \hat{\theta}_\gamma)$. Specifically, seeking the smallest RMS-bound for a given γ reduces to solving the *convex* optimization problem:

$$\begin{aligned} \hat{\delta}_\gamma &= \arg \min_{\substack{\delta \\ \|\mathcal{T}\{\delta\}\| \le \gamma}} \left\| M^{1/2}(y - \Psi\delta) \right\|^2 \\ \hat{\theta}_\gamma &= (\Phi\Phi^T)^{-1}\Phi^T \left(y - \Psi\hat{\delta}_\gamma \right) \\ \hat{w}_\gamma &= M \left(y - \Psi\hat{\delta}_\gamma \right) \\ \hat{\lambda}_\gamma &= \frac{1}{\ell} \|\hat{w}_\gamma\|^2 \\ M &= I_\ell - \Phi(\Phi\Phi^T)^{-1}\Phi^T \end{aligned} \tag{23}$$

For each value of γ, the graph of $\hat{\sigma}_\gamma$ versus γ ($\hat{\sigma}_\gamma^2 = \hat{\lambda}_\gamma$) establishes a tradeoff between model uncertainty, γ, and the minimum possible corresponding RMS disturbance uncertainty bound $\hat{\sigma}_\gamma$. With the ARX model, the tradeoff function, $\hat{\sigma}_\gamma$, is convex in γ (because $e(\theta)$ is affine in θ) and this "curve" separates falsified from unfalsified uncertainty models.

Along with the nominal parameter $\hat{\theta}_\gamma \in \mathbf{R}^p$, the uncertainty parameter $\hat{\delta}_\gamma \in \mathbf{R}^k$ has to be explicitly determined in this approach. Typically, $k >> p$, so in effect this is, a high order model of the dynamic model error constrained only by the $\mathbf{H}_\infty$ bound, γ, represented by (16). Consequently, the γ-family of frequency responses,

$$\hat{\Delta}_\gamma(e^{j\omega}) = \sum_{k=1}^{\ell} \hat{\delta}_\gamma e^{-j\omega k}, \quad \omega \in [0, \pi] \tag{24}$$

provides very detailed information about modeling uncertainty. Surely this is very closely related (probably equivalent, in principle) to the *model-error-modeling* approach described in [11]. The latter approach produces a frequency response in the output error whereas here we produce an ARX error.

To every point on the tradeoff curve there is a different parameter $\hat{\theta}_\gamma$, and hence, a different set of nominal transfer functions, $G(\hat{\theta}_\gamma), H(\hat{\theta}_\gamma)$. Many simulations reveal that $\hat{\theta}_\gamma$ does not vary much with γ and is often very close to the ordinary least-squares solution,

$$\hat{\theta}_0 = (\Phi\Phi^T)^{-1}\Phi^T y \tag{25}$$

This is that value of $\hat{\theta}_\gamma$ for which $\gamma = 0$ (and $\delta = 0$). The corresponding transfer function estimates $G(\hat{\theta}_0), H(\hat{\theta}_0)$ may thus be quite good, particularly after appropriate data filtering and model order selection, which is the normal course in system identification. Unfalsification can then be used after identification to quantify model quality. (More on this in section 6 when we discuss control design.) If the model parameters are "pre-selected" either by ordinary least-squares or any other means, the uncertainty tradeoff is now obtained by solving the following simplified version of (23):

$$\begin{aligned} \hat{\delta}_\gamma &= \arg \min_{\substack{\delta \\ \|\mathcal{T}\{\delta\}\| \leq \gamma}} \left\| y - \Phi\hat{\theta}_0 - \Psi\delta \right\|^2 \\ \hat{w}_\gamma &= y - \Phi\hat{\theta}_0 - \Psi\hat{\delta}_\gamma \\ \hat{\lambda}_\gamma &= \frac{1}{\ell}\|\hat{w}_\gamma\|^2 \end{aligned} \tag{26}$$

Back to stochastic uncertainty ... There seems to be no good way (as yet) to deal *directly and completely* with this type of uncertainty, *i.e.*, no precise mathematical formalism. There are, however, solutions that work well. For example, assuming a Gaussian distribution for the noise leads naturally to a Maximum-Likelihood (ML) approach. Here we seek to minimize the *negative log-likelihood function* [3], which to within a constant is,

$$L = \frac{1}{\lambda}\|w\|^2 + \ell \log \lambda \tag{27}$$

subject to the constraints in (22). It is easy to establish that the resulting optimization is *identical* to the convex optimization posed in (23). Therefore, the resulting optimal L depends only on γ,

$$L_\gamma = \ell + \ell \log \frac{1}{\ell}\|\hat{w}_\gamma\|^2$$

That value of γ which produces a minimum, equivalent to minimizing $\|\hat{w}_\gamma\|^2$, is always going to be the *largest* γ. This choice, however, seems to makes no sense, and *may* be exposing a limitation of this approach to representing the uncertain dynamics. However, there are other "knobs" buried within the variables for which L may have a sharply defined minimum. Examples include the ARX model order(s), data filters used during the identification, and additional filters used to help shape the uncertain dynamics. The latter is interesting particularly when the intended use of the model is control design. (see next section).

Stochastic embedding Another approach is to replace the $\mathbf{H}_\infty$ bound assumption, $\|\mathcal{T}\{\delta\}\| \leq \gamma$ with the assumption that δ is a random variable. Now both the disturbance and dynamic uncertainty are stochastic. This view is articulated in [4] and referred to as *stochastic embedding.* Maximum likelihood techniques are again invoked, and although the details differ from what is proposed here, and from the model-error-modeling approach in [11], the practical consequences appear not to be so different. Perhaps in the end, it's all more or less "least-squares."

Goodness-of-Fit An as yet untested approach, in the context of this paper, which seems to directly address the issue of knowledge about the distribution, is the classical "goodness-of-fit" criterion, *e.g.*, [3]. In terms of the problem formulation here, the idea is to test the hypothesis that the disturbance uncertainty sequence, $\{w_1, \ldots, w_\ell\}$ are samples of a random variable with known distribution, *e.g.*, $w \in \mathcal{N}(0, \lambda I_\ell)$. The test involves comparing the *cumulative distribution function* associated with the assumed distribution (*e.g.*, the Gaussian assumption) with that of the *empirical distribution function* engendered by the data. Two common measures of goodness-of-fit are the Kolomogorov-Smirnov statistic, and the Pearson statistic. The interesting feature of both of these statistics, and several variants, is that as the data length increases, the statistics have asymptotic distributions that are *independent* of the cumulative distribution. Thus a confidence level can be associated with the criteria which is independent of the distribution assumption. This still remains an idea for unfalsification, or for that matter, identification.

6 Control Design

In this section we utilize identification followed by uncertainty model unfalsification, both for control design, and then present a simulation example. A typical identification session involves two finite data sets: $(y_{\text{id}}, u_{\text{id}})$ for identification and $(y_{\text{val}}, u_{\text{val}})$ for validation/unfalsification. The sets are first used to find a good nominal model by following standard procedures of least-squares prediction error estimation:

$$\begin{aligned} \hat{\theta} &= \arg\min_{\theta} \|F_{\text{id}} e_{\text{id}}(\theta)\|^2 \\ e_{\text{id}}(\theta) &= H(\theta)^{-1}(y_{\text{id}} - G(\theta)u_{\text{id}}) \\ \hat{G} &= G(\hat{\theta}), \quad \hat{H} = H(\hat{\theta}) \end{aligned} \tag{28}$$

Here F_{id} is a filter which is used to emphasize important frequencies, *e.g.*, a lowpass filter whose bandwidth is the same as that desired for closed-loop control. It is often best to repeat the process for varying filter bandwidths and model orders and then compare the RMS values of the prediction errors on the two sets, *i.e.*, $\left\|F_{\text{id}} e_{\text{id}}(\hat{\theta})\right\|_{\text{rms}}$ and $\left\|F_{\text{id}} e_{\text{val}}(\hat{\theta})\right\|_{\text{rms}}$ where

$$e_{\text{val}}(\hat{\theta}) = \hat{H}^{-1}(y_{\text{val}} - \hat{G}u_{\text{val}}) \tag{29}$$

As the model order increases, the RMS level on the identification data set will continually decrease (assuming persistent excitation). On the validation data set, the RMS level will at first decrease, but then will begin to increase with model order. The inflection value (usually the minimum) is most often the best choice for model order. (See row 2 of figure 1). This simple test for selecting model order can be refined and made more precise, *e.g.*, [10, Ch.16], [4].

After this procedure, we posit the following uncertainty model for control design:

$$\hat{H}^{-1}(y - \hat{G}u) = w + \Delta \hat{H}^{-1}\hat{G}F_{\Delta}u \tag{30}$$

with w an uncertain stochastic disturbance with variance λ and Δ an uncertain system with $\mathbf{H}_{\infty}$ bound γ. Equivalently, the control design uncertainty model is:

$$y = (1 + \Delta F_{\Delta})\hat{G}u + \hat{H}w, \quad \mathbf{E}w^2 = \lambda, \quad \|\Delta\|_{\mathbf{H}_{\infty}} \leq \gamma \tag{31}$$

Hence, Δ is a *multiplicative model error*. The closed-loop system output response with controller

$$u = -\hat{K}y \tag{32}$$

is

$$y = \frac{\hat{S}\hat{H}}{1 + \Delta F_{\Delta}\hat{T}} w, \quad \hat{S} = 1 - \hat{T} = \frac{1}{1 + \hat{G}\hat{K}} \tag{33}$$

Suppose $\hat{K}$ is robustly stabilizing against all $\|\Delta\|_{\mathbf{H}_{\infty}} \leq \gamma$, *i.e.*,

$$\hat{S}, \hat{T} \text{ are stable and } \gamma \left\|F_{\Delta}\hat{T}\right\|_{\mathbf{H}_{\infty}} < 1 \tag{34}$$

Then the output variance of the uncertainty model is bounded by,

$$\mathbf{E}y^2 \leq \lambda \left\| \left(1 - \gamma |F_{\Delta}\hat{T}|\right)^{-1} \hat{S}\hat{H} \right\|_{\mathbf{H}_2}^2 \tag{35}$$

The above inequalities form further constraints along the (γ, λ)-uncertainty tradeoff.

Simulation Example The system used to generate the data is

$$y = Gu + v, \quad G = \mathbf{zoh}\left\{ \frac{\omega_0 \omega_1^2}{(s + \omega_0)(s^2 + 2\zeta\omega_1 s + \omega_1^2)} \right\}$$

with sampling frequency of 2 hz, $\omega_0 = 2\pi(.01), \omega_1 = 2\pi(.1), \zeta = .15$. The disturbance v is Gaussian white with variance $(.1)^2$. The input sequence u is a sequence of four pulses each 64 samples long resulting in the responses shown in the first row of figure 1 . The noise to signal RMS-ratio at the output is 0.20. The identification data is the first 128 samples and the validation/unfalsification data are the remaining 128 samples.

Identification was performed using an ARX model with no data filter, *i.e.*, $F_{\text{id}} = 1$. Row 2 of figure 1 shows how $\|e_{\text{id}}\|_{\text{rms}}$ and $\|e_{\text{val}}\|_{\text{rms}}$ vary with ARX model order. The minimum value of $\|e_{\text{val}}\|_{\text{rms}}$ occurs at $n = 17$ (filled $\triangle$) which is chosen as the nominal model, *i.e.*, $\hat{G} = \hat{B}/\hat{A}$, $\hat{H} = 1/\hat{A}$ with $(\hat{A}, \hat{B})$ as defined in (4) with $n = m = 17$. Unfalsification was performed using the uncertainty model:

$$\begin{aligned} e_{\text{val}} &= w + \Psi\delta \\ \Psi &= \mathcal{T}\left\{\hat{H}^{-1}\hat{G}F_{\Delta}u_{\text{val}}\right\} = \mathcal{T}\left\{\hat{B}F_{\Delta}u_{\text{val}}\right\} \end{aligned} \tag{36}$$

The model error filter was chosen as a 4th order highpass butterworth with passband at 0.1 hz, low frequency gain of 0.01, and a high frequency gain of 10. Row 3 of figure 1 shows 5 points on the (γ, σ) uncertainty tradeoff $(\sigma^2 = \lambda)$. The true values (at X) are $\sigma_0 = \left\|\hat{A}v\right\|_{\text{rms}} = 0.11$ and $\gamma_0 = \left\|(G/\hat{G} - 1)/F_{\Delta}\right\|_{\mathbf{H}_{\infty}} = 6.6$. Rows 4 and 5 show the unfalsifed model error in the frequency domain for these γ values compared to the true error shown in the heavy black line. Row 4 shows $\gamma|F_{\Delta}(e^{j\omega})|$ for the selected γ and row 5 shows $|\hat{\Delta}_{\gamma}(e^{j\omega})F_{\Delta}(e^{j\omega})|$ for the selected γ. These frequency response plots are very similar, both following $|F_{\Delta}(e^{j\omega})|$ as might be expected, but with the row 4 plots being smoother. Once these plots are available a controller can be easily designed to achieve robust stability and performance *of the uncertainty model* as indicated in

(34)-(35). It is also possible to replace (35) with

$$\mathbf{E}\,y^2 \le \lambda \left\| \left(1 - |\hat{\Delta}_\gamma F_\Delta \hat{T}|\right)^{-1} \hat{S}\hat{H} \right\|_{\mathbf{H}_2}^2 \qquad (37)$$

The heavy dashed lines in rows 4 and 5 is $1/|\hat{T}(e^{j\omega})|$ for a design with closed-loop bandwidth of 0.01 hz and clearly provides robust performance of both the model and the true system. The latter would not be confirmed until the control was implemented.

7 Closing Remarks

Unfalsification of an uncertainty model consisting of a Gaussian white disturbance and an $\mathbf{H}_\infty$ bounded model error leads to conclusions, in the control design context, that are not all that different from other methods, *e.g.*, ***stochastic embedding*** [4] and ***model-error-modeling*** [11]. As already stated, all these methods are utilizing "least-squares" in some form, and perhaps that is the connecting piece, despite the fact that the interpretations seem different. Although often extremely effective, I view all these methods as somewhat piece-meal, and not satisfactorily elegant. Perhaps this is due to fundamental limitations, *e.g.*, it is not possible to uniquely separate dynamic and disturbance uncertainty from finite data. For control design, however, this limitation may have no consequence. No model is needed in direct controller falsification [13] or direct tuning [8] - the data "speaks" to the closed-loop specification.

The answer will come, I believe, by resolving what I consider a fundamental challenge: ***to develop a self consistent theory of identification and validation from a single set of assumptions.***

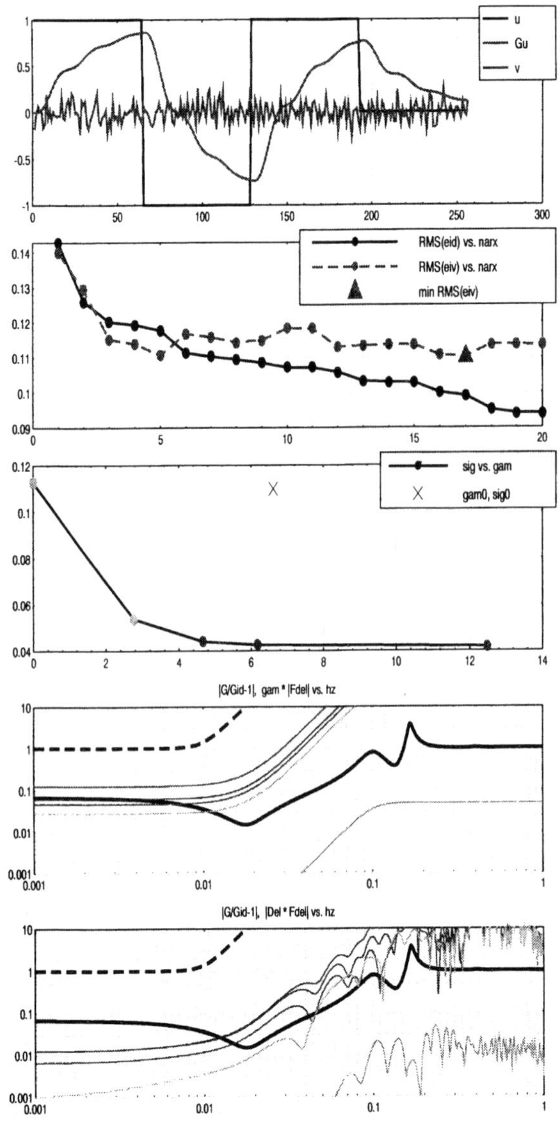

Figure 1: Simulation example: **row 1**: u, Gu, v vs. sample; **row 2**: $\|e_{\text{id}}\|_{\text{rms}}, \|e_{\text{val}}\|_{\text{rms}}$ vs. ARX model order; **row 3**: (γ, λ) uncertainty tradeoff; **row 4**: $\gamma|F_\Delta(e^{j\omega})|$ compared to true error; **row 5**: $\hat{\Delta}(e^{j\omega})|F_\Delta(e^{j\omega})|$ compared to true error.

References

[1] K. J. Åström and B. Wittenmark, *Adaptive Control*, Addison-Wesley, 1995.

[2] S. Boyd, L. El Ghaoui, E. Feron, and V. Balakrishnan, *Linear Matrix Inequalities in System and Control Theory*, SIAM Studies in Applied Mathematics, 1994.

[3] H. Cramer, *Mathematical Methods of Statistics*, Almqvist & Wiksells, 1945, Princeton Univ.Press, 1974.

[4] G.C. Goodwin, M. Gevers, and B. Ninness, "Quantifying the error in estimated transfer functions with applications to model order selection," *IEEE Trans. on Automatic Control*, vol. 37, no.7, pp. 913-928, July 1992.

[5] U. Grenander and G. Szego, *Toeplitz Forms and Their Applications*, Chelsea Publishing Co., New York, 1958.

[6] R.L. Kosut, "Uncertainty model unfalsification: a system identification paradigm compatible with robust control design," *Proc. 1995 CDC*, New Orleans, LA, Dec. 1995.

[7] R. L. Kosut, "Iterative adaptive robust control via uncertainty model unfalsification," *Proc. 1996 IFAC World Congress*, June, 1996, San Francisco, CA

[8] L. Kammer, R. Bitmead, P. Bartlett, "Direct iterative tuning via spectral analysis," *Proc. 1998 CDC*, pp.2874-2879, Tampa, FL, Dec. 1998.

[9] W.S. Lee, B.D.O. Anderson, I.M.Y. Mareels, and R.L. Kosut, "On some key issues in the windsurfer approach to robust adaptive control", *Automatica*, Vol. 31, No. 11, pp. 1619-1636, 1995.

[10] L. Ljung, *System Identification: Theory for the User*, Prentice-Hall, 1987.

[11] L. Ljung, "Model validation and model error modeling," *The Åström Symposium on Control*, pp.15-42, Lund, Sweden, August 27-28, 1999.

[12] K. Poolla, P. Khargonnekar, A. Tikku, J. Krause, and K. Nagpal, "A time-domain approach to model validation", *Proc. 1992 ACC*, Chicago, IL, June 1992, and *IEEE Trans. Aut. Contr.*, vol. 39, no. 5, pp. 951-959, May 1994.

[13] M. G. Safonov and T. C. Tsao, "The unfalsified control concept and learning", *Proc. IEEE CDC*, pp.2819-2824, Lake Buena Vista, FL, Dec.1994, and *IEEE Trans. Aut. Contr.*, vol. 42, no. 6, pp. 843-847, June 1997.

[14] R. S. Smith and J. C. Doyle, "Model validation - a connection between robust control and identification," *Proc. 1989 ACC*, pp. 1435-1440, June 1989, and *IEEE Trans. Aut. Contr.*, vol. 37, pp. 942-952, July 1992.

MODEL ERROR MODELING AND CONTROL DESIGN

Lennart Ljung *

* *Division of Automatic Control, Linköping University, SE-58183, Linköping, Sweden, email:* ljung@isy.liu.se

Abstract: Model validation and estimating the size of a possible model error is a central aspect of System Identification. In this contribution we discuss the model error concepts and model error modeling for control design. Of special interest is how to make use of periodic inputs, and how to deal with non-linear error models. The discussion is limited to SISO models and stability robustness issues.

1. INTRODUCTION

In this contribution we shall consider the estimation of linear models, of the type

$$y(t) = G(q,\theta)u(t) + H(q,\theta)e(t) \tag{1}$$

This is of course a standard topic in the literature, e.g., (Ljung 1999*a*). The parameter vector θ will be estimated using a prediction error method, leading to an estimate $\hat{\theta}_N$ with corresponding transfer function estimate

$$\hat{G}_N(q) = G(q,\hat{\theta}_N) \tag{2}$$

This estimate will then be used to design a regulator

$$u(t) = K(q)(r(t) - y(t)) \tag{3}$$

so that the closed loop system

$$y(t) = \frac{K(q)\hat{G}_N(g)}{1 + K(q)\hat{G}_N(q)} \tag{4}$$

behaves well.

The problem is that $\hat{G}_N$ will only be an approximate description of the true system. The extensive literature on robust control design, e.g., (Zhou *et al.* 1996), (Skogestad and Postlethwaite 1996) deals with the question to ensure good behavior of the closed loop system, despite inaccuracies in the model $\hat{G}_N$.

When the model has been identified from data, there is the additional question of how to estimate the uncertainty of the model. The interplay between the estimation and the design processes has received substantial attention over the past year. e.g., (Smith and Doyle 1992), (Kosut *et al.* 1992*a*), (Rangan and Poolla 1996), (Skelton 1989),(Gevers 1993), (Kosut *et al.* 1992*b*), (Smith and Dullerud 1996), (Poolla *et al.* 1994).

Part of this problem is to estimate, in reliable way, the model error or model uncertainty, not assuming that the true system can be described within the chosen structure (1). See, e.g., (Ninness and Goodwin 1995), (Ljung 1999*b*) Then it is mostly assumed that the true system is linear, and given by a transfer function $G_0(q)$, and that the model error is captured by estimates of the size of the frequency function

$$\tilde{G}_N(e^{i\omega}) = \hat{G}_N(e^{i\omega}) - G_0(e^{i\omega}) \tag{5}$$

In this contribution we shall consider the more general situation of possibly non-linear model errors, and discuss the roles that an explicit model error model will play in such a set-up.

The discussion is confined to single-input-single-output systems/models and to stability robustness issues only. This *is* a simplification and does mean a loss of generality.

2. MODEL VALIDATION AND MODEL ERROR MODELING

To validate a model is to confront it with facts about the system. An ideal situation is that there is prior knowledge about the system which can be compared with the model's properties. Often, however, the only "facts" that are at hand are the data themselves. It is desirable that the data used for validation are not the same as for estimation, a separate "validation data set" should be used. It then boils dow to comparing what the model thinks the output should be with what the actual measured output turned out ot be. That is, the *residuals*, the model left-overs, should be examined:

$$\varepsilon(t) = y(t) - \hat{y}_N(t) \tag{6}$$

Here $\hat{y}$ is the simulated output

$$\hat{y}_N(t) = \hat{G}_N(q)u(t) \tag{7}$$

$\hat{G}_N$ is the *nominal model*, which may be the one estimated in (2), be may also be chosen in other ways.

If a noise model $\hat{H}$ has been estimated it is natural to consider a filtered version

$$\varepsilon_F(t) = \hat{H}_N^{-1}(q)\varepsilon(t) \tag{8}$$

and subject that to the analysis below.

2.1 *Classical Correlation Analysis of Residuals*

One of the most basic tests, (Draper and Smith 1981), is to compute the correlation between the regressors, in our case the past inputs, and the residuals:

$$\hat{r}^N(\tau) = \frac{1}{N}\sum_{t=1}^{N} u(t-\tau)\varepsilon(t) \tag{9}$$

It is customary to plot these estimates as a function of τ and compare with their standard deviations to check if they are significantly different from zero. If not, we have not traced any significant influence of u in ε, so we cannot say that the model $\hat{G}$ has not picked up all the influence of u on y. (Note the double negation: we are not saying that "$\hat{G}$ has picked up all ..."). It is convenient to form

$$\varphi(t) = \left[u(t-1) \ \dots \ u(t-M)\right]^T \tag{10}$$

$$h_N^M = \begin{bmatrix} \hat{r}^N(1) \\ \vdots \\ \hat{r}^N(M) \end{bmatrix} = \frac{1}{N}\sum_{t=1}^{N}\varphi(t)\varepsilon(t) \tag{11}$$

Under the assumption that ε is white noise with variance λ, h has-a normal distribution with zero mean and variance λ/NR_N, where

$$R_N = \frac{1}{N}\sum_{t=1}^{N}\varphi(t)\varphi^T(t) \tag{12}$$

so

$$\xi_N^M = \frac{1}{\lambda N}\|\sum_{t=1}^{N}\varphi(t)\varepsilon(t)\|_{R_N^{-1}} \tag{13}$$

will in this case have a χ^2 distribution, and the familiar χ^2-test

$$\xi_N^M < \alpha \tag{14}$$

is based on this. The result is typically presented as a plot of the autocorrelation of the residuals and a plot of (9). See Figure 1.

Note that other kind of dependences can be tested quite analogously by letting $\varphi(t)$ be other, nonlinear, functions of past inputs:

$$\varphi(t) = f(u^t) \tag{15}$$

2.2 *Correlation Analysis as Model Error Modeling*

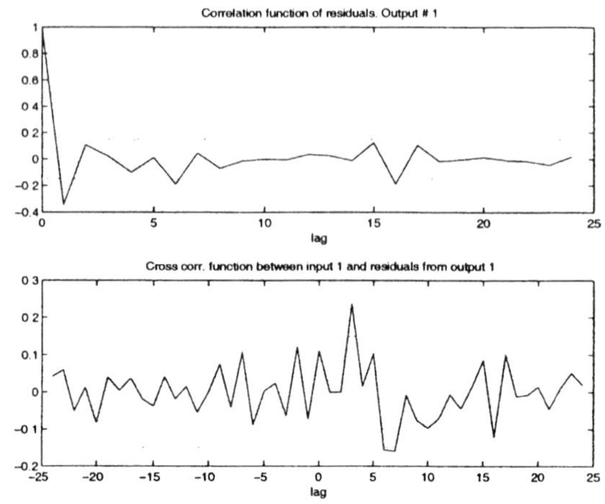

Fig. 1. Traditional residual analysis: Auto- and cross-correlation functions with uncertainty regions.

The information from the cross correlation analysis between ε and u, can also be interpreted as an implicit FIR model for the transfer function $\tilde{G}$ in

$$\varepsilon(t) = \tilde{G}(q)u(t) + w(t) \tag{16}$$

from u to ε. For control purposes, it is much more effective to present the (amplitude) frequency function of this *model error model*, with uncertainty bounds as in Figures 2 – 4. The data used in these figures are simulated from a second order ARMAX model. It is clear that conventional model validation corresponds to increasing the model complexity until the model error model has uncertainty bounds that include zero (as in Figure 4), since then there is no clear evidence that $\tilde{G}$ is not zero – the estimated model is then not falsified. But it is also clear that the two plots together; the model and its "sidekick", the model error model, can be used for control design, even if the model is falsified. Look at Figure 3. According to the model error model there is significant, but rather small errors in the mid frequency range.

The model is thus falsified, but could still very well be used for control design if the information in the lower plot is taken into account.

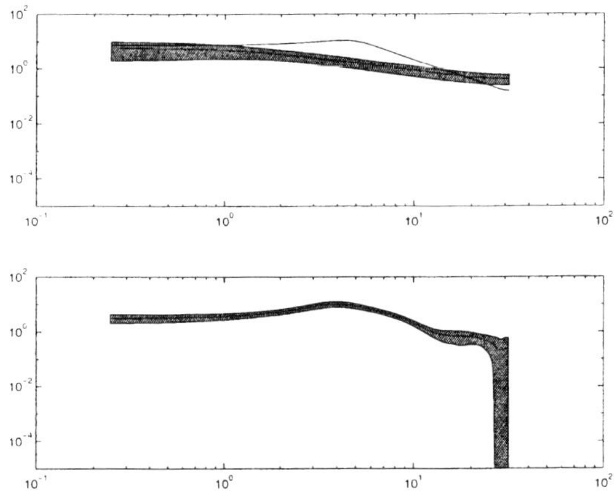

Fig. 2. Upper plot: Amplitude Bode plot of a first order model with estimated uncertainty bounds. The true system is also plotted. Lower plot: The model error model computed as a 20:th order ARX model from u to $y - \hat{G}u$

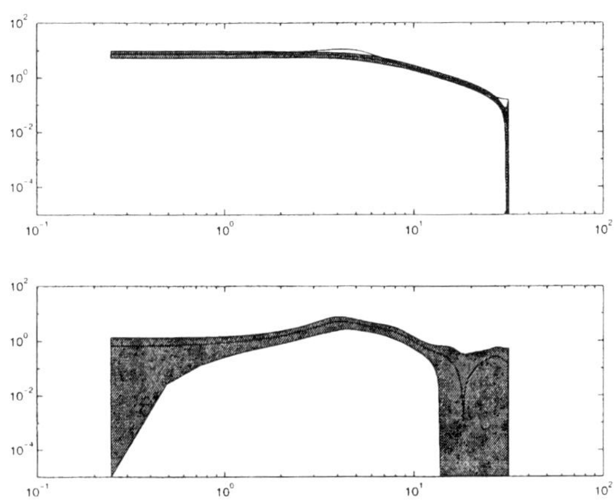

Fig. 3. As in previous figure, but second order ARX model

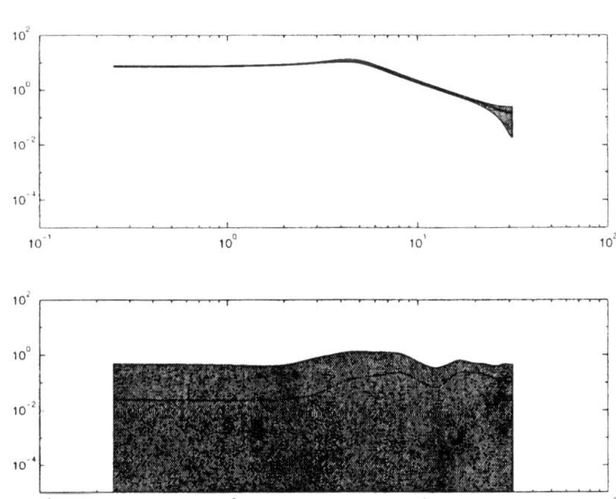

Fig. 4. As in previous figure, but second order ARMAX model. Here the model error model contains zero in its uncertainty region, which means that the top model is not falsified.

3. MODEL ERROR MODELS AND THEIR GAINS

3.1 *The General Form*

A model error model is a description of how u affects ε. We will typically not be interested in a detailed such description – that should rather be done as a more detailed nominal model. Primarily we focus on bounds on the gain of such a model. To go well together with linear control design, we shall work with combinations of frequency weightings and unstructured, unknown models with bounded (estimated) gain. That means that the general structure of the model from u to ε can be depicted as in Figure 5. Formally we have

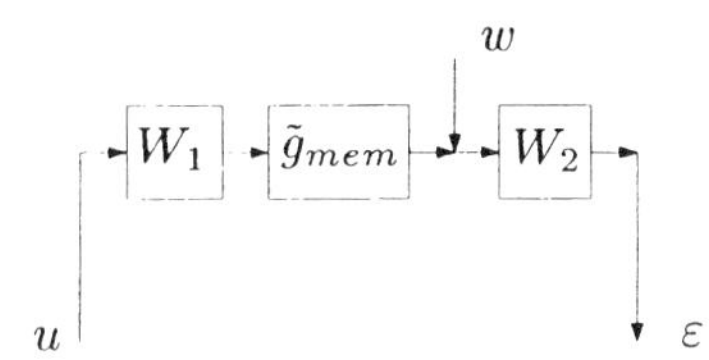

Fig. 5. The model error model

$$\varepsilon_F(t) = W_2^{-1}(q)\varepsilon(t), \quad u_F(t) = W_1(q)u(t) \quad (17)$$

$$\varepsilon_F(t) = \tilde{g}_{mem}(u_F^{t-1}) + w(t) \quad (18)$$

Here, W_1 and W_2 are given linear filters, superscript $t-1$ denotes all the signal values up to time $t-1$, and $\tilde{g}_{mem}$ is an uncertain model. The bottom line is that, with some amount of confidence, based on the measured data, and possibly on prior information, we should be able to say that its gain is bounded by 1:

$$||\tilde{g}_{mem}||_\infty < 1 \quad (19)$$

Normalization can always be achieved using W_1 and W_2. We shall shortly discuss how to estimate such a bound.

The model $\tilde{g}_{mem}$ can come in different shapes. It could be

- Linear Time Invariant. In this case we can take either W_1 or W_2 to be unity, since all the blocks in in Figure 5 commute.
- Linear Time Varying.
- General Non-Linear.

3.2 *Estimating the Gain of the Model Error Model: Distinguishing Model Error from Noise*

The estimation task is to determine the filters W_i and estimate the gain of $\tilde{g}_{mem}$ in

$$\varepsilon_F(t) = \tilde{g}_{mem}(u_F^{t-1}) + w(t)$$

i.e. estimate

$$\sup_t \sup_{u^t} \frac{||\tilde{g}_{mem}(u^t)||}{||u^t||}$$

Now, let us say right away that it is *impossible* to determine this gain without prior knowledge. Even in the linear case, an arbitrarily thin resonance peak will require an arbitrarily long sequence of data, due to the uncertainty principle. After having considered a finite data record, no guarantee can be given for the "true" gain. The estimate of the gain will always reflect the data that have been measured, possibly in conjunction with prior knowledge.

A first fundamental problem is to distinguish the model error $\tilde{g}_{mem}$ from the noise contribution w. Loosely speaking, the noise w would be that contribution to ε that would not change if u is changed. This problem is addressed in (Smith and Dullerud 1996) by looking at trade-offs between model error size and hard amplitude bounds on w. From a statistical perspective, it is more natural to try and "correlate out" w. There are two basic methods for that:

- Use a periodic input, and eliminate/reduce w by averaging over the periods.
- Build a (parametric) model of $\tilde{g}_{mem}$ in (18) assuming u and w to be independent.

3.3 *Using Periodic Inputs*

Suppose the noise contribution is additive as in (18) and that u is periodic with period P. Then, after a transient has died out, also $\hat{y}_N(t)$ (see (7)) will be periodic with the same period. Moreover, the part of y that originates from the input will, under weak assumptions, also be periodic with period P. Let $\bar{y}$ denote the output over one period, obtained by averaging over all the periods, and similarly for $\bar{\varepsilon}_F(t)$:

$$\bar{\varepsilon}_F(t) = W_2^{-1}(q)(\bar{y}(t) - \hat{y}_N(t)), \quad t = 1, ..., P$$

From (18) we now have

$$\bar{\varepsilon}_F(t) = \tilde{g}_{mem}(u_F^{t-1}) + \bar{w}(t)$$

where $\bar{w}$ will tend to zero as N tends to infinity (more and more periods are averaged over).

A simple under-bounding estimate of the gain of $\tilde{g}_{mem}$ is

$$\widehat{\|\tilde{g}_{mem}\|^2} = \frac{\sum_{t=1}^{P} \bar{\varepsilon}_F^2(t)}{\sum_{t=1}^{P} u_F^2(t)} \tag{20}$$

A natural choice of filter is $W_1(q) = \hat{G}(q)$ (the nominal model). This makes

$$u_F(t) = \hat{y}_N(t) \tag{21}$$

Moreover, W_2 can be chosen as $W_2(q) = \tilde{H}(q)$, a "noise model" for $\bar{y}(t) - \hat{y}_N(t)$,

$$\bar{y}(t) - \hat{y}_N(t) = \tilde{H}(q)\nu(t) \tag{22}$$

with $\nu(t)$ being white noise. That will make the numerator in (20) as small as possible. Note that this makes

$$\widehat{\|\tilde{g}_{mem}\|^2} = \frac{\sum_{t=1}^{P}[W_2^{-1}(q)(\bar{y}(t) - \hat{y}_N(t))]^2}{\sum_{t=1}^{P} \hat{y}_N^2(t)} \tag{23}$$

a measure of the proportion of "unexplained" output variation. This has a close relationship to the *multiple correlation coefficient* used in residual analysis, page 33 in (Draper and Smith 1981).

3.4 *General Linear Black Box Model Error Models*

The model error model concept gives us more freedom in investigating the residuals than the classical residual correlation test, which essentially employs a FIR model. A more general linear model

$$\varepsilon(t) = \tilde{G}(q)u(t) + \tilde{H}(q)w(t) \tag{24}$$

could improve the estimate of the error model, since a model of the disturbance is used.

Instead of parametric linear models, we may apply spectral analysis to try and extract any linear influence of u on ε, (Kosut 1986), (Stenman and Tjärnström 2000). In any case, there is a close relationship between the Blackman-Tukey spectral analysis estimate of this transfer function and the one obtained by a FIR-model.

Once we have a specific, linear model of the influence from u to ε, along with an uncertainty (confidence) region, the error model in Figure 5 is directly defined: Take W_2=1 and W_1 as $\tilde{G}_N + \Delta$. Here Δ is the estimated variance error (at a certain confidence level), so that the amplitude curve of W_1 corresponds to the upper curves in (the lower plots of) Figures 2-4. If no nonlinearity has been detected in the error model, this choice of W_i in Figure 5 gives a correct description of the error and uncertainty associated with the nominal model.

3.5 *Non-linear Model Error Models*

In the literature, most model error discussions as well as the *identification-for-control* approaches are dealt with in a setting where "the true system" is a high order linear model, and the models are of lower order. In practical use, it is of course more common that the model errors are ignored non-linearities rather than unmodeled linear dynamics. From a model error perspective, this simply means that we should test non-linear models:

$$\varepsilon_F(t) = \tilde{g}_{mem}(u_F^{t-1}) + w(t) \tag{25}$$

In the absence of specific, suspected non-linearities, it is reasonable to test non-linear black boxes like neural network NNFIR model, cf (Sjöberg *et al.* 1995):

$$\varepsilon_F(t) = \tilde{g}_{mem}(u_F(t-1), \ldots, u_F(t-m), \eta) + w(t) \quad (26)$$

The number of lagged inputs can be chosen relatively small here, like $m = 5$ or so. To appreciate the size (gain) of any estimated non-linearity (in particular for control applications) it is natural to use the sup-norm

$$\|\tilde{g}_{mem}\|_\infty = \sup_{u_1,\ldots,u_m} \frac{|\tilde{g}_{mem}(u_1, u_2, \ldots, u_m, \hat{\eta})|^2}{u_1^2 + \ldots + u_m^2} \quad (27)$$

Then also determine the worst case value of this gain in a properly chosen confidence region for the estimate $\hat{\eta}$:

$$\|\tilde{g}_{mem}\| = \sup_{\hat{\eta}\in\Theta} \|\tilde{g}_{mem}(u(t-1), \ldots, u(t-m), \hat{\eta})\|_\infty \quad (28)$$

4. ROBUST CONTROL DESIGN FOR UNCERTAIN SYSTEMS

4.1 *Linear Model Errors*

Once a model with its model error uncertainty is delivered, the question is how to design a controller that will stabilize the system robustly. By this we would mean that the chosen controller should *stabilize all models in the "region" defined by the nominal model and the model error model.*

In case we have used a linear model error model, this region is easily depicted in the frequency domain. It will look like a strip in the Bode, or Nyquist plot. See Figures 2-4, i.e.

$$G \in \mathcal{G} = \{G| \quad |G(e^{i\omega}) - G_{nom}(e^{i\omega})| < \Delta(\omega)\} \quad (29)$$

How to achieve robust stability for such a set of models is well known: Choose a regulator K, such that the complementary sensitivity function

$$T = \frac{G_{nom}K}{1 + G_{nom}K} \quad (30)$$

is less than the inverse relative model error bound:

$$|T(e^{i\omega})| < \frac{G_{nom}(e^{i\omega})}{\Delta(\omega)}, \quad \forall\omega \quad (31)$$

H_∞ techniques can be used to determine if such a K exists, for given G_{nom} and Δ. See, e.g. (Zhou et al. 1996).

4.2 Frequency Weighted Non-linear Model Error Model

The error model (18) corresponds to a closed loop block diagram as in Figure 6. This can be

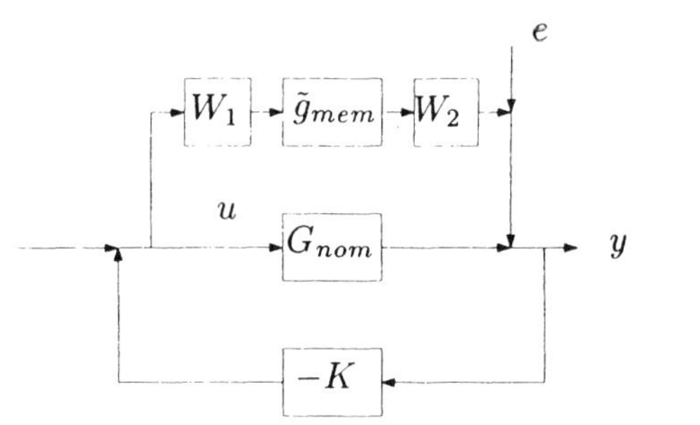

Fig. 6. Block diagram of the feedback loop with model error

rearranged to be seen as feedback between the non-linear part of the error model $\tilde{g}_{mem}$ and $KW_1W_2/(1 + KG_{nom})$ (keeping in mind that we only consider SISO models here). Suppose that we normalize W_1W_2 so that the gain of the non-linear part is subject to

$$\|\tilde{g}_{mem}\|_\infty < 1 \quad (32)$$

The small gain theorem tells us that stability is assured if

$$\left|\frac{W_1(e^{i\omega})W_2(e^{i\omega})K(e^{i\omega})}{1 + K(e^{i\omega})G_{nom}(e^{i\omega})}\right| < 1 \quad \forall\omega \quad (33)$$

Comparing with (31) we realize that we just can consider the set of possible system descriptions to be linear and given by

$$G \in \mathcal{G} = \{G| \quad |G(e^{i\omega}) - G_{nom}(e^{i\omega})| < \ldots < W_1(e^{i\omega})W_2(e^{i\omega})\} \quad (34)$$

By stabilizing any linear model in this set, i.e., achieving (31) for $\Delta = W_1W_2$, we have also made the linear control design robust against non-linear model errors of the type (18). (Considering only stability robustness.)

5. ESTIMATING NON-LINEAR FREQUENCY WEIGHTED ERROR MODELS

When dealing with non-linear model error models for control design, the option of frequency weighting is essential. As an example, suppose the true system is

$$\tilde{y}(t) = \frac{1}{(q-1)(q-\alpha)}u(t) \quad (35)$$

$$y(t) = f(\tilde{y}(t)) \quad (36)$$

for some static, linearly bounded, nonlinearity f. Suppose, for simplicity, that the nominal model is $G_{nom}(q) = 1/(q-1)(q-\alpha)$ so the model error is this nominal system followed by the static nonlinearity $f(y) - 1$. Suppose that $\|f(y) - 1\| < 1$. The gain of this model error model is infinite, due to the infinite gain of the linear part. So, this error model does not do much good for control design.

If we now pick the weightings $W_1(q) = 1/(1-q)$, $W_2(q) = 1$ the error model $\tilde{g}_{mem}$ in (18) has a maximum gain of 1. Handling the infinite gain weighting W_2 in (33) is no major problem, since the denominator also has infinite gain at zero.

Note that it is only the product $W_1 W_2$ that affects the robustness test, while the split into W_1 and W_2 may have a substantial influence on the norm of $\tilde{g}_{mem}$. This gives an extra dimension to the model error modeling. In loose terms the weightings should be chosen so as to minimize $\|\tilde{g}_{mem}\|_\infty$ while letting $W_1 W_2$ have a shape that suits (33). A more comprehensive discussion of how to deal with this freedom has to be deferred to another occasion. Some hints are, however, as follows:

- Using $W_2 = \hat{H}$ (the nominal noise model) reduces the power of the output of the non-linear block as much as possible, giving the possibility of a smaller gain of the non-linear part.
- Using $W_1 = \hat{G}$ (the nominal model), will focus the error model on the relative error.

6. CONCLUSIONS

Building linear models may be deceptive. Given any data set, of arbitrary length, we can always build a linear model, that will pass standard residual analysis validation tests. The reason is that any (non-linear, time-varying) system from which quasistationary input-output sequences are collected, has a linear second order equivalent. We cannot distinguish the true system from this linear model, using just second order tests. This means that the standard linear identification setup will deliver a model with uncertainty

$$\hat{G} \pm \Delta$$

where the (variance) error Δ will tend to zero as more data are processed. Subjecting the data to the model-error-model treatment described in this contribution will give a model

$$\hat{G} \pm \Delta, \quad \Delta = W_1 W_2$$

where Δ will tend to zero, only if non-linearity tests show no sign of non-linearities. Otherwise Δ will reflect a remaining uncertainty/model error. However, if we just consider stability robustness, it will be safe to treat the system as a linear one, with the given uncertainty, even if non-linearities have been detected.

7. REFERENCES

Draper, N.R. and H. Smith (1981). Applied Regression Analysis, 2nd ed.. Wiley, New York.

Gevers, Michel (1993). Towards a joint design of identification and control?. In: Essays on control: Perspectives in the theory and its applications (H L Trentelman and J C Willems, Eds.). ECC '93 Groningen.

Kosut, R., M. K. Lau and S. P. Boyd (1992a). Set-membership identification of systems with parametric and nonparametric uncertainty. IEEE Trans. Automatic Control **AC-37**, 929–941.

Kosut, R.L. (1986). Adaptive calibration: An approach to uncertainty modeling and on-line robust control design. In: Proc. 25th IEEE Conference on Decision an Control. Vol. 1. Athens, Greece. pp. 455–461.

Kosut, R.L., G. C. Goodwin and M. P. Polis (Eds) (1992b). Special Issue on System Identification for Robust Control Design, IEEE Trans. Automatic Control, Vol 37.

Ljung, L. (1999a). System Identification - Theory for the User. 2nd ed.. Prentice-Hall. Upper Saddle River, N.J.

Ljung, Lennart (1999b). Model validation and model error modeling. In: The Åström Symposiium on Control (B. Wittenmark and A. Rantzer, Eds.). Studentlitteratur. Lund, Sweden. pp. 15 –42.

Ninness, B. and G. C. Goodwin (1995). Estimation of model quality. Automatica **31**(12), 1771–1797.

Poolla, K., P. P. Khargonekar, A. Tikku, J. Krause and K.Nagpal (1994). A time-domain approach to model validation. IEEE Trans. on Automatic Control **AC-39**, 951–059.

Rangan, S. and K. Poolla (1996). Time-domain validation for sample-data uncertainty models. IEEE Trans. Automatic Control **AC-41**, 980–991.

Sjöberg, J., Q. Zhang, L. Ljung, A. Benveniste, B. Delyon, P.Y. Glorennec, H. Hjalmarsson and A. Juditsky (1995). Nonlinear black-box modeling in system identification: A unified overview. Automatica **31**(12), 1691–1724.

Skelton, R. E. (1989). Model error concepts in control design. Int. J. Control **49**, 1725–1753.

Skogestad, S. and I. Postlethwaite (1996). Multivariable Feedback Control. John Wiley. New York.

Smith, R. and G.E. Dullerud (1996). Continuous-time control model validation using finite experimental data. IEEE Trans. Automatic Control **AC-41**, 1094–1105.

Smith, R. and J. C. Doyle (1992). Model validation: a connection between robust control and identification. IEEE Trans. Automatic Control **AC-37**, 942–952.

Stenman, Anders and Fredrik Tjärnström (2000). A nonparametric approach to model error modeling. In: Preprints of the 12th IFAC Symposium on System Identification. To appear.

Zhou, K., J. C. Doyle and K. Glover (1996). Robust and Optimal Control. Prentice-Hall. Englewood Cliffs, NJ.

SET MEMBERSHIP IDENTIFICATION FOR H_∞ ROBUST CONTROL DESIGN

Mario Milanese and Michele Taragna

Dipartimento di Automatica e Informatica, Politecnico di Torino
Corso Duca degli Abruzzi 24 - 10129 Torino - Italy
e-mail: milanese@polito.it, taragna@polito.it

Abstract: The Set Membership H_∞ identification of LTI discrete-time exponentially stable SISO systems, from noise corrupted measurements in the time and/or the frequency domain, is considered. The assumptions on the noise can account for information on its maximal magnitude and deterministic uncorrelation properties. A test is given for validating the assumptions on system and noise. The main focus of the paper is on the trade-off between identified model set complexity and optimality properties, and on the non conservativeness in the identification error evaluation. By using a suitably weighted H_∞ norm, the identification error measures the degradation of the closed loop performance that is underlying the control design, due to the mismatch between the actual plant and the model. Thus, optimality properties of the identified model set reflect in minimality of conservativeness of guaranteed closed loop performances. Identification of the optimal model set is a NP-hard problem and alternative algorithms are investigated, simpler to be computed, at the expense of identification accuracy degradation. This is measured by the suboptimality level of the identified model set, i.e. the ratio between the achieved identification error and the minimal one. A method is given to evaluate upper and lower bounds on this suboptimality level. Reduced order model sets that tightly include the optimal one are derived. Then, model set order selection can be performed by trading between model set complexity and suboptimality level degradation.

Keywords: Set Membership identification, identification for control, H_∞ identification

1. INTRODUCTION

The basic issue in most practical control problems is to design a control law achieving given performance specifications for a plant, whose behavior is not exactly known. The classical approach consists in designing a control that meets the desired performance specifications for a model identified on the basis of available information on the plant and on measured data. However, in this way it is not taken into account that any identified model is only an approximation of the actual system. Indeed, according to the size of the modeling error, the performances actually achievable by the designed control when applied to the plant may be very poor, and even the closed loop stability may be missed. In order to face these problems, robustness had become in past years a central issue in system and control theory, focusing the attention of researchers from the study of a single model to the investigation of a set of models. Such a set, often indicated as *uncertainty model set* or model set for short, has to be identified from available information and data in a suitable way to be used for analysis and design purposes. Due to space limitations, it is impossible to reference the main contributions in the literature, which can be found in surveys and books, see e.g. (Milanese and Vicino, 1993; Ninness and Goodwin, 1995; Mäkilä *et al.*, 1995; Milanese *et al.*, 1996; Partington, 1997; Garulli *et al.*, 1999; Chen and Gu, 2000).

In this paper, identification of model sets for linear, time-invariant, discrete-time, BIBO stable SISO systems is considered. A bound on the system impulse response decay rate is assumed to be known. Experimental information consists of measurements in the time and/or the frequency domain, corrupted by an output additive noise. The assumptions on the noise can account for information on its maximal magnitude and possible uncorrelation properties in deterministic sense. The aim is to deliver not only a nominal model, but also a set of model perturbations, whose size in a given norm measures the uncertainty about the system to be identified. If prior assumptions on system and noise are "true", the identified model set includes the actual plant, an important property in view of subsequent use for robust control design. As required in any identification theory, the problem of checking the "truth" of priors arises. Indeed, as it is always the case in validation problems, the only thing that can be actually done is to check if measured data falsify

This research was supported in part by funds of Ministero dell'Università e della Ricerca Scientifica e Tecnologica under the Project "Algorithms and architectures for the identification and control of industrial systems".

the priors. Necessary and sufficient conditions for such a check are presented in section 3.

The identification of optimal model sets in the setting considered here is a NP-hard problem and methods for their computation are not known, at the best of the authors' knowledge. This motivated the investigation of alternative algorithms, like linear, two-stage or interpolatory ones. These algorithms are simpler to be computed, at the expense of some degradation in the identification accuracy. Such a degradation is typically measured by the suboptimality level, defined as the ratio between the achieved identification error and the minimal error of the optimal model set. The suboptimality level of model sets identified by linear and two-stage algorithms is not known. On the contrary, interpolatory algorithms are known to identify model sets having suboptimality level at most 2. Indeed, interpolatory algorithms can be found, which identify model sets having suboptimality level quite close to 1, (Milanese and Taragna, 2000). Such model sets may have high order nominal model, and in section 4 a method is given for deriving reduced order model sets that tightly include in the H_∞ norm the optimal one. The evaluation of the suboptimality level of model sets of different order gives then a sensible method for selecting among them, by suitably trading between model set complexity and achievable suboptimality level.

2. SET MEMBERSHIP H_∞ IDENTIFICATION

In this section, the Set Membership (SM) identification problem studied in the paper is defined.

Let $\mathcal{S}$ be the Banach space of SISO, causal, linear time-invariant, discrete-time, BIBO stable dynamical systems and $h^S = \{h_0^S, h_1^S, \ldots\}$ be the impulse response of $S \in \mathcal{S}$. Let $S^o \in \mathcal{S}$ be the system to be identified on the basis of noise corrupted measurements and prior information on the system and the noise.

The measurements can be in the time, frequency or mixed time and frequency domain. Such experimental information can be expressed in the form:

$$y^N = F_N h^S + e^N$$

where $y^N = [y_0 \ldots y_{N-1}]^T \in \Re^N$ is a known vector depending on the actual measurements, F_N is a known semi-infinite matrix indicating how the measurements depend on S^o, and $e^N \in \Re^N$ is an unknown vector representing the measurement noise. Explicit expressions of matrix F_N for time or frequency domain experiments can be found in (Milanese and Taragna, 1999).

The prior information on S^o and e^N consist in the following assumptions.

SM assumptions:

1) $S^o \in K = \{S \in \mathcal{S} : |h_\ell^S| \leq L\rho^\ell, \ \forall \ell \geq 0\}$ with $L>0$ and $0<\rho<1$ known constants

2) $e^N \in \mathcal{B}_e = \{\tilde{e}^N = [\tilde{e}_0 \ldots \tilde{e}_{N-1}]^T \in \Re^N : \|A\tilde{e}^N\|_\infty^{W_e} \leq 1\}$ where $A \in \Re^{l \times N}$ is a known matrix, $W_e = \text{diag}(w_{e,1}, \ldots, w_{e,l}) \in \Re^{l \times l}$ is a known weighting matrix with $w_{e,k} > 0 \, \forall k$, and $\|A\tilde{e}^N\|_\infty^{W_e} = \|W_e^{-1} A\tilde{e}^N\|_\infty = \max_{1 \leq k \leq l} w_{e,k}^{-1} \left|\left(A\tilde{e}^N\right)_k\right|$ ■

Note that assumption 2) can accommodate for information on maximal noise magnitude, i.e. $\|\tilde{e}^N\|_\infty^{W_e} = \max_{1 \leq k \leq N} w_{e,k}^{-1} |\tilde{e}_k| \leq 1$, as typically done in most of the literature, but can also account for possible information on deterministic uncorrelation properties of the noise, see e.g. (Paganini, 1996).

A key role is played by the Feasible Systems Set, often called "unfalsified systems set", i.e. the set of systems consistent with prior information and measured data.

Definition 1. *Feasible Systems Set:*

$$FSS = \{S \in K : \|A \cdot (y^N - F_N h^S)\|_\infty^{W_e} \leq 1\}$$

■

If prior assumptions on system $S^o \in K$ and on noise $e^N \in \mathcal{B}_e$ are "true", this set includes S^o, an important property in view of subsequent use for robust control design. As required in any identification theory, the problem of checking the validity of priors arises. Indeed, the only thing that can be actually done is to *check if measured data falsify the priors*. This is equivalent to check if FSS is empty, being the FSS the set of unfalsified systems. Necessary and sufficient conditions for such a check are given in section 3. The FSS can be considered an uncertainty model set for S^o and, in the line with the robustness paradigm, control should be designed to be robust versus such an uncertainty model set. However, the FSS is not represented in a suitable form to be used by robust control design techniques, and model sets with such a property have to be looked for. To be consistent with robust control design philosophy, *model sets including the set of unfalsified systems* have to be looked for. This is formalized by the following definition.

Definition 2. A set of models $\mathcal{M} \subseteq \mathcal{S}$ is called a *model set for* S^o if:

$$\mathcal{M} \supseteq FSS$$

■

In this paper, additive frequency shaped model sets are considered, of the form:

$$\hat{\mathcal{M}}(W) = \left\{\hat{M} + \Delta : |\Delta(\omega)| \leq W(\omega), \forall \omega \in [0, 2\pi]\right\} \quad (1)$$

The nominal model $\hat{M}$ has to be obtained by some identification algorithm, i.e. an operator ϕ mapping the available information, represented by the triple $(K, \mathcal{B}_e, y^N)$, into a model $\hat{M} \in \mathcal{S}$:

$$\phi\left(K, \mathcal{B}_e, y^N\right) = \hat{M}$$

Moreover, for given $\hat{M}$, $W(\omega)$ has to be chosen such that $\hat{\mathcal{M}}(W)$ is a model set for S^o. A necessary and sufficient condition for this choice is provided by the following theorem.

Theorem 1. For given nominal model $\hat{M}$, then:

i) $\hat{\mathcal{M}}(W)$ is a model set for S^o if and only if:

$$W(\omega) \geq \sup_{S \in FSS} |S(\omega) - \hat{M}(\omega)| = W^*(\omega), \forall \omega \in [0, 2\pi]$$

ii) $\hat{\mathcal{M}}(W^*)$ is the smallest model set for S^o of the form (1)

Proof. See (Milanese and Taragna, 2000). ■

Due to this result, model sets $\hat{\mathcal{M}}(W)$ of the form (1) with $W(\omega) = W^*(\omega)$, $\forall \omega \in [0, 2\pi]$, will be considered in the sequel, and the dependence on W will be omitted when unnecessary.

The error between S^o and the identified nominal model $\hat{M}$ is measured by a weighted H_∞ norm:

$$\|S^o - \hat{M}\|_\infty^{W_C} = \sup_{0\le\omega\le 2\pi} \left| W_C^{-1}(\omega)\left[S^o(\omega) - \hat{M}(\omega)\right]\right|$$

where $W_C(\omega)$ is a known weighting function, suitably selected according to the intended use of the identified model. For example, if identification is aimed to derive an uncertainty model set for H_∞ robust control design, $W_C(\omega)$ can be computed so that the identified model optimally approximates the closed loop performance that is underlying the control design. As shown in (Milanese *et al.*, 1997), this quantity turns out to be equivalent to a weighted H_∞ norm of the dual Youla parametrization of the plant and of the model.

Clearly the error $\|S^o - \hat{M}\|_\infty^{W_C}$ cannot be known exactly since, under the assumptions of the paper, S^o is only known to belong to FSS. Thus, only the following bound on this error can be given:

$$\|S^o - \hat{M}\|_\infty^{W_C} \le \sup_{S\in FSS} \|S - \hat{M}\|_\infty^{W_C}$$

and no tighter bound can be obtained.

These considerations motivate the following definition of the identification error of the identification algorithm ϕ, and of the corresponding smallest additive model set $\hat{\mathcal{M}}(W^*)$.

Definition 3.
Algorithm and *model set identification error:*

$$E_y(\phi) = E(\hat{\mathcal{M}}(W^*)) = \sup_{S\in FSS} \|S - \phi(y^N)\|_\infty^{W_C}$$

■

In the literature, other worst-case errors are often defined and investigated, see e.g. the surveys (Milanese and Vicino, 1993; Ninness and Goodwin, 1995; Mäkilä *et al.*, 1995). The error $E_y(\phi)$ considered here is the minimal identification error that can be guaranteed by using the model $\hat{M}$, identified by algorithm ϕ operating on available information.

The model set uncertainty can be minimized by appropriately choosing the nominal model $\hat{M}$, thus leading to the following optimality concepts.

Definition 4. *Optimal algorithm* and *model set, radius of information.*
An algorithm ϕ^* is called optimal if:

$$E_y(\phi^*) = \inf_\phi E_y(\phi) = r_y$$

The corresponding model set $\mathcal{M}^*(W^*)$ is called optimal and r_y is called radius of information. ■

In the H_∞ case considered in this paper, the problem of computing optimal model sets is NP-hard and methods for their computation are not known. This motivates the interest in deriving algorithms having lower complexity, at the expense of some degradation in the accuracy of the identified model set. The following definition is introduced to give a measure of such a degradation.

Definition 5. *Model set suboptimality level.*
Given a model set $\hat{\mathcal{M}}$, its suboptimality level $\alpha(\hat{\mathcal{M}})$ is defined by:

$$\alpha(\hat{\mathcal{M}}) = E(\hat{\mathcal{M}})/r_y$$

■

Note that $\alpha(\hat{\mathcal{M}}) \ge 1$, since $\inf_{\hat{\mathcal{M}}} E(\hat{\mathcal{M}}) = r_y$.

3. VALIDATION OF SM ASSUMPTIONS

The problem of checking the validity of priors is now considered. We say that SM assumptions are invalidated by data if no unfalsified system exists, i.e. if FSS is empty. We say that *SM assumptions are validated* if not invalidated, i.e. if FSS is non-empty. Note that the fact that the priors are validated by using the present data does not exclude that they may be invalidated by future data, as it is always the case in validation problems.

For a given system $S \in \mathcal{S}$ with impulse response $h^S = \{h_0^S, h_1^S, \ldots\}$, let S^ν be the FIR_ν system having the same first ν impulse response samples of S, i.e. $h^{S^\nu} = \{h_0^S, h_1^S, \ldots, h_{\nu-1}^S, 0, 0, \ldots\}$. Then, we have the following validation result.

Theorem 2. For given positive integer ν, let γ^* be the solution of the linear programming problem:

$$\gamma^* = \min_{\gamma, S^\nu : \left\{\begin{array}{l} S^\nu \in K \\ \|A\cdot(y^N - F_N h^{S^\nu})\|_\infty^{W_e} \le \gamma \end{array}\right.} \gamma$$

i) A sufficient condition for SM assumptions being validated by data is:

$$\gamma^* \le 1$$

ii) A necessary condition for SM assumptions being validated by data is:

$$\gamma^* \le 1 + \varepsilon_\nu \|W_e^{-1}A\|_{\infty,\infty}$$

where $\left\|W_e^{-1}A\right\|_{\infty,\infty} = \max_{1\le i\le l} \sum_{j=1}^{N} \left|\left(W_e^{-1}A\right)_{ij}\right|$

and $\varepsilon_\nu = \frac{L\rho^\nu}{1-\rho}$

iii) In the case of time domain data, choosing $\nu \ge N$, a necessary and sufficient condition for SM assumptions being validated by data is:

$$\gamma^* \le 1$$

Proof. See (Milanese and Taragna, 2000). ■

Note that the gap between conditions i) and ii) can be made as small as desired by increasing ν. Indeed, in case of time domain measurements, no gap exists just for $\nu = N$.

4. LOW ORDER MODEL SET IDENTIFICATION

The identification of the optimal model set is a NP-hard problem and methods for computing such a model set are not known, at the best of the authors' knowledge. Indeed, an interpolatory algorithm which may give quite good approximation of the optimal model set can be obtained as follows, (Milanese and Taragna, 2000).

For given integer ν, let us define:

$$\overline{FSS}^\nu = \left\{S^\nu \in K : \|A\cdot(y^N - F_N h^{S^\nu})\|_\infty^{W_e + \widetilde{W}} \le 1\right\}$$

where $\widetilde{W} = \varepsilon_\nu \cdot \mathrm{diag}(\|a_1\|_1, \ldots, \|a_l\|_1) \in \Re^{l\times l}$ with a_k the k-th row of A and $\varepsilon_\nu = \frac{L\rho^\nu}{1-\rho}$.

For given $\omega \in [0, 2\pi]$ and positive integer m, compute in complex plane the points $\overline{t_k}$, $k = 1, \ldots, m$, by solving the following m linear programming problems:

$$\overline{t_k}(\omega) = \Omega(\omega)\cdot \operatorname*{argmin}_{S^\nu \in \overline{FSS}^\nu} \left[-\Omega_1(\omega)\cdot\sin(s_k) + \Omega_2(\omega)\cdot\cos(s_k)\right]\cdot h^{S^\nu}$$

where $s_k = 2\pi k/m$ and:

$$\Omega(\omega)=\begin{bmatrix}\Omega_1(\omega)\\ \Omega_2(\omega)\end{bmatrix}=\begin{bmatrix}\Re e(\Psi(\omega))\\ \Im m(\Psi(\omega))\end{bmatrix}\in\Re^{2\times\infty} \quad (2)$$

$$\Psi(\omega)=\begin{bmatrix}1\; e^{-j\omega}\; e^{-j2\omega}\cdots\end{bmatrix}\in\mathcal{C}^{1\times\infty}$$

For $k = 1,\dots,m$, consider the line l_k in complex plane with slope $s_k = 2\pi k/m$ and containing point $\overline{t_k}(\omega)$. Compute the intersection points $\overline{v_k}(\omega)$, $k = 1,\dots,m$ of the supporting lines l_k and $l_{\text{mod}(k,m)+1}$ of V^ν, with $\text{mod}(k,m) = k$ if $k < m$ and 0 otherwise. Let $\overline{VO}^\nu_m(\omega)$ be the polytope having vertices $\overline{v_k}(\omega)$, $k = 1,\dots,m$.

Consider the FIR_ν model $\hat{M}^{no}_\nu \in FSS$, whose impulse response is obtained as solution of the following problem:

$$h^{\hat{M}^{no}_\nu} = \arg\min_{S^\nu\in FSS} \|s^* - \Omega^*\cdot h^{S^\nu}\|^{W_C}_\infty$$

where q is a given integer and:

$$s^*=\begin{bmatrix}s^*(\omega_1)\\ \vdots\\ s^*(\omega_q)\end{bmatrix}\in\Re^{4q},\; s^*(\omega)=\begin{bmatrix}\max\limits_{s=\left[\begin{smallmatrix}s_1\\ s_2\end{smallmatrix}\right]\in\overline{VO}^\nu_m(\omega)} s_1\\ \max\limits_{s=\left[\begin{smallmatrix}s_1\\ s_2\end{smallmatrix}\right]\in\overline{VO}^\nu_m(\omega)} s_2\\ \min\limits_{s=\left[\begin{smallmatrix}s_1\\ s_2\end{smallmatrix}\right]\in\overline{VO}^\nu_m(\omega)} s_1\\ \min\limits_{s=\left[\begin{smallmatrix}s_1\\ s_2\end{smallmatrix}\right]\in\overline{VO}^\nu_m(\omega)} s_2\end{bmatrix}\in\Re^4$$

$$\Omega^*=\begin{bmatrix}\Omega^*(\omega_1)\\ \vdots\\ \Omega^*(\omega_q)\end{bmatrix}\in\Re^{4q\times\nu},\; \Omega^*(\omega)=\begin{bmatrix}\Omega^\nu(\omega)\\ \Omega^\nu(\omega)\end{bmatrix}\in\Re^{4\times\nu}$$

with $\Omega^\nu(\omega)$ given by the first ν columns of the matrix $\Omega(\omega)$ defined in (2).

Note that this problem can be easily solved by linear programming techniques.

By choosing ν, m and q sufficiently large, the model set $\hat{\mathcal{M}}^{no}_\nu$ has suboptimality level typically very close to 1, as shown in the example section, but its nominal model $\hat{M}^{no}_\nu$ has transfer function of order ν, which may be large. In order to obtain model sets suitable for robust control design, it may be of interest to identify model sets with low order nominal models. This can be obtained by looking for model sets of the form:

$$\mathcal{M}_n(W,p) = \{M_n(p)+\Delta : |\Delta(\omega)|\le W(\omega), \forall\omega\in[0,2\pi]\}$$

where $M_n(p)$ is a class of parametric models with transfer function of order n. The best model set of this form is:

$$\mathcal{M}_n(W^*,\hat{p}) = \{M_n(\hat{p})+\Delta : |\Delta(\omega)|\le W^*(\omega), \forall\omega\in[0,2\pi]\}$$

where:

$$\hat{p} = \arg\min_p E\left(\mathcal{M}_n(W^*,p)\right) \quad (3)$$

In order to solve (3) by means of iterative nonlinear optimization algorithms, the evaluation of $E(\mathcal{M}_n(W^*,p))$ for given p is needed. Moreover, in order to describe the model set $\mathcal{M}_n(W^*,p)$, the frequency error bound $W^*(\omega) = \sup_{S\in FSS}|S(\omega) - M_n(\omega,p)|$ has to be computed. Exact computation of such quantities is hard, but convergent upper and lower bounds are derived in the next result.

For given integer ν, let us define:

$$\underline{FSS}^\nu=\left\{S^\nu\in K : \|A\cdot(y^N - F_N h^{S^\nu})\|^{W_e}_\infty\le 1\right\}$$

For given $\omega\in[0,2\pi]$ and positive integer m, compute in complex plane the points $\underline{t_k}$, $k = 1,\dots,m$, by solving the following m linear programming problems:

$$\underline{t_k}(\omega) = \Omega(\omega)\cdot\operatorname*{argmin}_{S^\nu\in\underline{FSS}^\nu}\left[-\Omega_1(\omega)\cdot\sin(s_k)+\Omega_2(\omega)\cdot\cos(s_k)\right]\cdot h^{S^\nu}$$

with $s_k = 2\pi k/m$ and $\Omega(\omega)=\begin{bmatrix}\Omega_1(\omega)\\ \Omega_2(\omega)\end{bmatrix}$ defined in (2).

Let $\underline{VI}^\nu_m(\omega)$ be the polytope having vertices $\underline{t_k}(\omega)$ $k = 1,\dots,m$ and $r_2[\underline{VI}^\nu_m(\omega)]$ be its radius in the euclidean norm. This radius can be computed by using any of the methods available for computing the minimum circle covering a given set of points in $\Re^2$, see e.g. (Preparata and Shamos, 1985).

Theorem 3.

i) For given model $\hat{M}$, $\omega\in[0,2\pi]$ and positive integers ν and m, $W^*(\omega)$ is bounded as:

$$\underline{W}^\nu_m(\omega)\le W^*(\omega)\le\overline{W}^\nu_m(\omega)$$

where:

$$\underline{W}^\nu_m(\omega) = \max_{k=1,\dots,m}\left\|\hat{M}(\omega)-\underline{t_k}(\omega)\right\|_2$$

$$\overline{W}^\nu_m(\omega) = \max_{k=1,\dots,m}\left\|\hat{M}(\omega)-\overline{v_k}(\omega)\right\|_2+\varepsilon_\nu$$

Moreover:

$$\lim_{\nu,m\to\infty}\overline{W}^\nu_m(\omega) = \lim_{\nu,m\to\infty}\underline{W}^\nu_m(\omega) = W^*(\omega)$$

ii) For given nominal model $\hat{M}$, the identification error of the corresponding model set $\hat{\mathcal{M}}$ is bounded as:

$$\underline{E}^\nu_m(\hat{\mathcal{M}})\le E(\hat{\mathcal{M}})\le\overline{E}^\nu_m(\hat{\mathcal{M}})$$

where:

$$\underline{E}^\nu_m(\hat{\mathcal{M}}) = \sup_{0\le\omega\le2\pi}\underline{W}^\nu_m(\omega)\,W_C^{-1}(\omega)$$

$$\overline{E}^\nu_m(\hat{\mathcal{M}}) = \sup_{0\le\omega\le2\pi}\overline{W}^\nu_m(\omega)\,W_C^{-1}(\omega)$$

Proof. See (Milanese and Taragna, 2000). ■

In the literature only upper bounds on the identification error are available, whose degree of conservativeness is not known, see e.g. (Chen and Gu, 2000) and the references therein. Indeed, as shown in the numerical examples reported in section 5 and in (Milanese and Taragna, 2000), the actual overall conservativeness of the bounds in the literature may be relevant. On the contrary, Theorem 3 allows to compute the identification error to the desired accuracy by increasing ν and m, or, if this is not possible due to computational problems, to evaluate the achieved accuracy.

Note that $\underline{VI}^\nu_m(\omega)$ and $\overline{VO}^\nu_m(\omega)$ provide convergent inner and outer approximations of the value set $V(\omega)$, i.e. the set of transfer function values at frequency ω of all the feasible systems in FSS, see (Milanese and Taragna, 1999).

The model $M_n(\hat{p})$, where $\hat{p}$ is solution of (3), is the so called "conditional center" of FSS (Kacewicz *et al.*, 1988). Finding conditional centers is even harder than finding non-conditional ones, see e.g. (Kacewicz *et al.*, 1988; Kacewicz, 1999) for some results in the case that the class of models $M_n(p)$ is linear in the parameters, e.g. represented by FIR, Laguerre, Kautz or orthonormal basis functions. These models are essentially with fixed poles and the problem arises of choosing "good" basis, i.e. "good" model poles, since this choice affects, for given model order, the identification accuracy. On the other hand, models nonlinear in

the parameters, able to arrange also their poles, may improve the identification accuracy, but make problem (3) non-convex, and trapping in local minima may take place. In such a case, the choice of a good starting point to be used by iterative nonlinear optimization algorithms for solving (3) is very important. Indeed, very good results may be achieved by using as starting point reduced order models $\hat{M}_n^r$ of $\hat{M}_\nu^{no}$, obtained by Hankel norm approximation or balanced realization and SVD truncation. This is due to the well-known relationship between Hankel and H_∞ norms, and to the fact that $\hat{M}_n^r$ is a "nearly central" projection model in the Hankel norm. Indeed, central projection models are known to provide good approximations of conditionally central models (Kacewicz, 1999).
The selection of the nominal model order n can be made by suitably trading between model complexity and achievable suboptimality level. Such a level can be evaluated from the following result, reported with slight modifications from (Milanese and Taragna, 1999).

Theorem 4. For given nominal model $\hat{M}$ and positive integers ν and m, the suboptimality level of the corresponding model set $\hat{\mathcal{M}}$ is bounded as:

$$\underline{\alpha}(\hat{\mathcal{M}}) \leq \alpha(\hat{\mathcal{M}}) \leq \overline{\alpha}(\hat{\mathcal{M}})$$

where:

$$\underline{\alpha}(\hat{\mathcal{M}}) = \max\left\{1,\ \underline{E}_m^\nu(\hat{\mathcal{M}})/\,\overline{r}_y\right\}$$

$$\overline{\alpha}(\hat{\mathcal{M}}) = \overline{E}_m^\nu(\hat{\mathcal{M}})/\,\underline{r}_y$$

$$\overline{r}_y = \sup_{0\leq\omega\leq 2\pi}\left[\max_{k=1,\ldots,m}\left\|\hat{M}^{no}(\omega)-\overline{v_k}(\omega)\right\|_2 + \varepsilon_\nu\right] W_C^{-1}(\omega)$$

$$\underline{r}_y = \sup_{0\leq\omega\leq 2\pi} r_2\left[\underline{VI}_m^\nu(\omega)\right] W_C^{-1}(\omega)$$ ■

5. EXAMPLE

Identification of the following system proposed in (Ljung, 1999) is considered:

$$S^o(z) = \frac{z^3+0.5z^2}{z^4-2.2z^3+2.42z^2-1.87z+0.7225}$$

The available experimental information is made up by $N = 1100$ samples of the system output to a PRBS input of unitary amplitude. These samples are corrupted by a pointwise bounded noise e^N with $|e_i| \leq 4$, $i = 0, 1, \ldots, 1099$. In figure 1 the system input and output are reported.
The prior information assumed on S^o is $L = 6$ and $\rho = 0.93$, which has been validated using Theorem 2. This is a quite "safe" prior assumption, since stronger assumptions are still not invalidated by data, e.g. $L = 6$ and $\rho = 0.91$, or $L = 5$ and $\rho = 0.93$.
The identification error is measured by unweighted H_∞ norm, i.e. $W_C(\omega) = 1\ \forall\omega \in [0, 2\pi]$.
Different nominal models $\hat{M}$ have been identified, with their corresponding model sets $\hat{\mathcal{M}}$:
- a 150-th order FIR model $\hat{M}_{150}^{no}$ provided by the "nearly optimal" algorithm described in section 4, considering $q = 500$ equispaced frequencies in the interval $[0, \pi]$;
- approximations of $\hat{M}_{150}^{no}$ of order $2 \div 6$, denoted by $\hat{M}_2 \div \hat{M}_6$, provided by the procedure described in section 4, using as starting point for the optimization problem (3) the transfer function coefficients of models $\hat{M}_2^r \div \hat{M}_6^r$ of corresponding order obtained by applying model reduction algorithms available under MATLAB to $\hat{M}_{150}^{no}$.

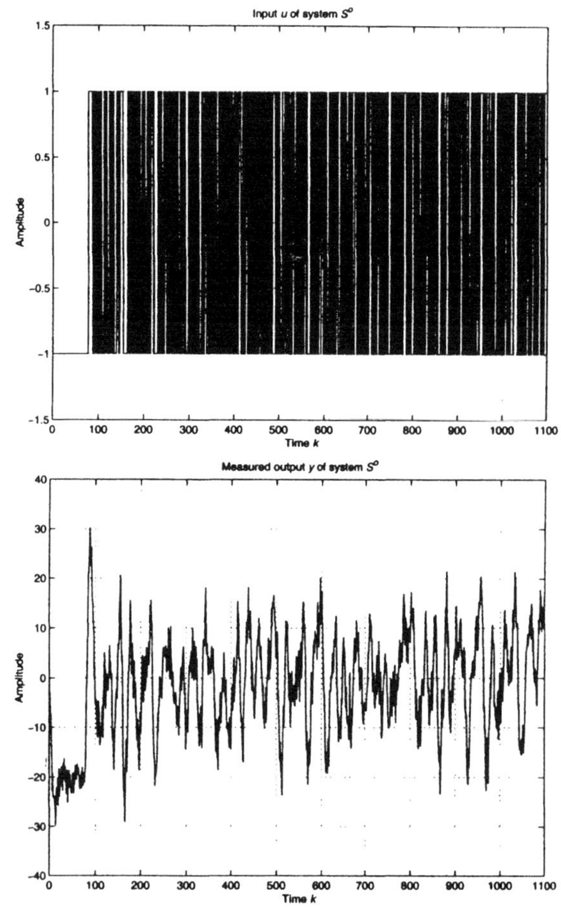

Fig. 1. System input u and measured output y.

The computations of the $\sup_{0\leq\omega\leq 2\pi}$ required by Theorems 3 and 4 have been carried on by gridding ω on a sufficiently coarse set of frequencies. Indeed, a set T_ω of 500 equispaced frequencies in the interval $[0, \pi]$ has been chosen. The Nyquist plots of system S^o and the value sets $V(\omega)$ computed on the frequency set T_ω are reported in figure 2.
The values $\nu = 150$ and $m = 16$ have been chosen, giving $\frac{\overline{E}_{16}^{150}(\hat{\mathcal{M}})-\underline{E}_{16}^{150}(\hat{\mathcal{M}})}{\underline{E}_{16}^{150}(\hat{\mathcal{M}})} \leq 1.3\%$ for any identified model set $\hat{\mathcal{M}}$.
In the next table, it is possible to compare the obtained values of the lower and upper bounds on the identification error $E(\hat{\mathcal{M}})$ and on the suboptimality level $\alpha(\hat{\mathcal{M}})$ for the identified model sets $\hat{\mathcal{M}}_{150}^{no}$ and $\hat{\mathcal{M}}_n$, with $n = 2, \cdots, 6$.

$\hat{\mathcal{M}}$	$\underline{E}_{16}^{150}(\hat{\mathcal{M}})$	$\overline{E}_{16}^{150}(\hat{\mathcal{M}})$	$\underline{\alpha}(\hat{\mathcal{M}})$	$\overline{\alpha}(\hat{\mathcal{M}})$
$\hat{\mathcal{M}}_{150}^{no}$	3.57	3.61	1.00	1.04
$\hat{\mathcal{M}}_2$	6.49	6.52	1.77	1.87
$\hat{\mathcal{M}}_3$	5.31	5.34	1.45	1.53
$\hat{\mathcal{M}}_4$	3.48	3.49	1.00	1.01
$\hat{\mathcal{M}}_5$	3.48	3.49	1.00	1.01
$\hat{\mathcal{M}}_6$	3.48	3.49	1.00	1.01

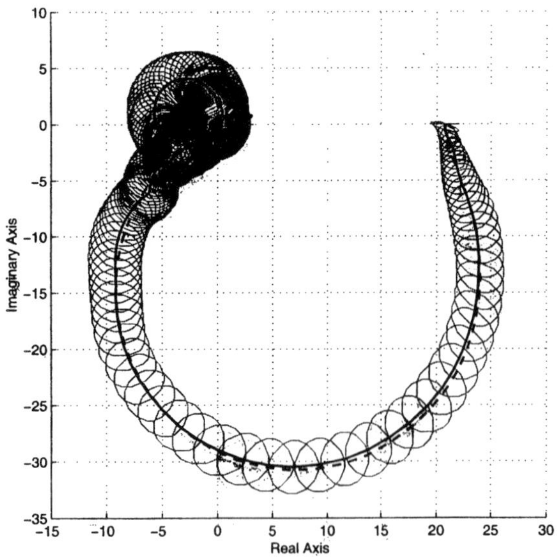

Fig. 2. Nyquist plots of S^o (thick dashed line), "nearly optimal" model $\hat{M}^{no}_{150}$ (thick solid line) and value sets $V(\omega)$ (thin solid lines).

The magnitude frequency response of the identified model set $\hat{\mathcal{M}}_4$ is reported in figure 3 and compared with the actual system S^o.

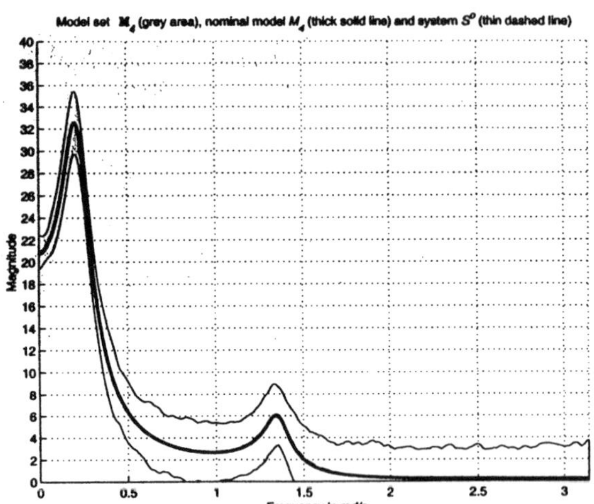

Fig. 3. Magnitude frequency response of $\hat{\mathcal{M}}_4$.

The above results may be compared with the model sets identified by the interpolatory algorithm proposed in (Chen and Nett, 1995). This algorithm is convergent and "almost optimal", i.e. the identified model sets is guaranteed to have suboptimality level not greater than 2. Note that the prior information used in (Chen and Nett, 1995) is: $S^o \in \overline{B}\mathcal{H}_{\infty,\rho}(L) = \{S \in \mathcal{S} : \|S(\rho \cdot z)\|_\infty \leq L\}$. It is well known that $\overline{B}\mathcal{H}_{\infty,\rho}(L) \subset K(L,\rho)$. Thus the comparison is fair, since the algorithm presented in the present paper uses weaker prior assumptions. Algorithm in (Chen and Nett, 1995) typically identifies nominal models of order equal to the number of data, in the present example 1100. However, for such number of data, this algorithm is affected by numerical and computational problems, that prevented us from obtaining the nominal model. Nevertheless, upper bounds on the identification error provided by Theorems 4.2–4.3 of (Chen and Nett, 1995) can be easily computed, and are indeed valid for any interpolatory algorithm. The obtained values for this example are reported in the following table.

	Error bounds
Th. 4.2 of (Chen and Nett, 1995)	167.43
Th. 4.3 of (Chen and Nett, 1995)	16.99

It can be seen that these bounds, though based on stronger assumptions than those made in the present paper, are very conservative, while the method here presented allows to derive very tight bounds on the identification error.
A numerical example for the case of frequency domain data can be found in (Milanese and Taragna, 1999). It has to be remarked that the bounds available in the literature for frequency domain measurements appears to be not so largely conservative (Milanese and Taragna, 2000).

6. REFERENCES

Chen, J. and C. N. Nett (1995). The Carathéodory-Fejér problem and H_∞/ℓ_1 identification: a time domain approach. *IEEE Transactions on Automatic Control* **AC-40**(4), 729–735.

Chen, J. and G. Gu (2000). *Control-Oriented System Identification: An H_∞ Approach.* John Wiley & Sons, Inc.. New York.

Garulli, A., Tesi, A. and Vicino, A., (Eds.) (1999). *Robustness in Identification and Control.* Vol. 245 of *Lecture Notes in Control and Information Sciences.* Springer-Verlag. Godalming, UK.

Kacewicz, B. (1999). Worst-case conditional system identification in a general class of norms. *Automatica* **35**(6), 1049–1058.

Kacewicz, B. Z., M. Milanese and A. Vicino (1988). Conditionally optimal algorithms and estimation of reduced order models. *Journal of Complexity* **4**, 73–85.

Ljung, L. (1999). Model validation and model error modeling. In: *Proc. of the Åström Symposium on Control.* Lund, Sweden. To be published by Studentliteratur, Lund, Sweden.

Mäkilä, P. M., J. R. Partington and T. K. Gustafsson (1995). Worst-case control-relevant identification. *Automatica* **31**(12), 1799–1819.

Milanese, M. and A. Vicino (1993). Information-based complexity and nonparametric worst-case system identification. *Journal of Complexity* **9**, 427–446.

Milanese, M. and M. Taragna (1999). Suboptimality evaluation of approximated models in H_∞ identification. In: *Proc. of the 38th IEEE Conference on Decision and Control.* Phoenix, AZ. pp. 1494–1499.

Milanese, M. and M. Taragna (2000). Optimality, approximation and complexity in Set Membership H_∞ identification. Technical Report MiTa00. DAUIN - Politecnico di Torino.

Milanese, M., M. Taragna and P. M. J. Van den Hof (1997). Closed-loop identification of uncertainty models for robust control design: a set membership approach. In: *Proc. of the 36th IEEE Conference on Decision and Control.* San Diego, CA. pp. 2447–2452.

Milanese, M., Norton, J., Piet-Lahanier, H. and Walter, É., (Eds.) (1996). *Bounding Approaches to System Identification.* Plenum Press. New York.

Ninness, B. and G. C. Goodwin (1995). Estimation of model quality. *Automatica* **31**(12), 1771–1797.

Paganini, F. (1996). A set-based approach for white noise modeling. *IEEE Transactions on Automatic Control* **AC-41**(10), 1453–1465.

Partington, J. R. (1997). *Interpolation, Identification, and Sampling.* Vol. 17 of *London Mathematical Society Monographs New Series.* Clarendon Press - Oxford. New York.

Preparata, F. P. and M. I. Shamos (1985). *Computational Geometry: An Introduction.* Springer-Verlag. New York.

CLOSED-LOOP MODEL VALIDATION USING COPRIME FACTOR UNCERTAINTY MODELS

Raymond A. de Callafon *,1 and **Paul M.J. van den Hof** **

* *University of California, San Diego*
Dept. of Mechanical and Aerospace Engineering
9500 Gilman Drive
La Jolla, CA 92093-0411, U.S.A.
** *Department of Applied Physics*
Delft University of Technology
Lorentzweg 1, 2628 CJ Delft
The Netherlands

Abstract: Model (in)validation techniques are used to bridge the gap between models used in robust control synthesis and uncertainty models obtained from identification experiments. In most applications the aim is to design a robust controller and therefore it is valuable to validate or invalidate an uncertainty model in view of this application by considering a closed-loop model validation technique. In this paper a model validation approach is presented that generalizes the (in)validation of possibly unstable models on the basis of closed-loop experiments with a stabilizing, but possibly unstable, controller. The approach is presented in a robust control framework with an uncertainty model described with coprime factor perturbations. It is shown that this approach yields an affine expression of the uncertainty model in all possible transfer functions that can be measured via a closed-loop experiments, which facilitates the optimization involved with a model invalidation.

Keywords: identification; model invalidation; coprime factors; closed-loop

1. INTRODUCTION

1.1 *Model invalidation*

In trying to model a plant, a distinction can be made between exact and approximate modeling. In case of exact modeling, the attention is focused on trying to model a plant meticulously by trying to capture the dynamical behaviour of plant exactly. Approximate identification is concerned with system identification problems in which the identification technique is used to find approximate models of the plant. Especially if models have to be used for control design, deliberate undermodeling is often required as model-based control design procedures tend to yield controllers that have the same complexity as the model used to compute the controller (Boyd and Barrat, 1991; Zhou *et al.*, 1996). By limiting the order of the model, the order of the controller can be limited during the control design.

In the event of a (deliberate) undermodeling, i.e. models that are too simple to describe the plant completely, a mismatch between the plant and the model is unavoidable. Such a mismatch can be captured in an uncertainty model that consists of a nominal model and an additional model uncertainty or allowable model perturbation. Even if exact modeling is desired on the basis of experimental data, such a model uncertainty

[1] Author to whom all correspondence can be addressed, email: `callafon@ucsd.edu`, phone:+1.858.5343166

in inevitable due to incomplete knowledge of the plant or the finite data sequence used during the modeling.

For (robust) control purposes, it is important to verify the validity of the uncertainty model. Without such a verification, the robustness of a controller can not be guaranteed fully. Considering the uncertainty model to be given, the validity of the model can be verified on the basis of an (additional) experiment by a so-called model validation procedure (Smith and Doyle, 1992). The model validation procedure is used to determine whether or not the experimental data could have been produced by the uncertainty model. In general, this is done by determining the level of the smallest model perturbation and the smallest noise signal that could have produced the experimental data. In case these levels exceed the assumed bounds of the uncertainty model, the model is said to be invalidated by the experiment. Although these techniques allow only ***invalidation*** of uncertainty models, the term model validation is being used extensively.

Several techniques have been reported in the literature that are either based on time domain (Poolla *et al.*, 1994; Rangan and Poolla, 1996) or frequency domain (Smith, 1995; Smith *et al.*, 1997; Token and Chen, 1998) characterizations of the model (in)validation problem. In the time domain approaches, a distinction can be made between a strictly discrete time setting (Poolla *et al.*, 1994) and the more practical framework of continuous time sampled data systems (Rangan and Poolla, 1996; Smith and Dullerud, 1996). In general, these techniques exploit a formulation for the model validation problem that allows an affine optimization to determine the level of the smallest model perturbation and the smallest noise signal that could have produced the experimental data.

Unfortunately, most of these existing techniques focus on an open-loop type of model validation. In the open-loop type of model validation, an uncertainty model is being validated on the basis of open-loop experiments. The open-loop type of model validation techniques exploit the linear appearance of an additive or multiplicative model uncertainty description to solve an affine optimization that addresses the model (in)validation problem (Smith and Doyle, 1992).

1.2 *Closed-loop model invalidation*

Obviously, in the case where the actual plant and/or the uncertainty model are unstable, an open-loop experiment and an open-loop model validation are not desirable. In most situations, open-loop experiments can not be performed as normal operating conditions require a feedback system to comply with safety or performance requirements.

Additionally, in most situations one is specifically interested in validating a model under closed-loop or controlled conditions on the basis of data obtained with a (stabilizing) feedback controller. The reason for this lies in the possible approximations being made during the formulation of the uncertainty model:

- An approximate model with model uncertainty has been developed that specifically aims at modeling the closed-loop behavior of the plant.
- A validation of the uncertainty model under operating conditions (with a feedback controller) is more realistic and desirable.

However, in case of closed-loop experiments, the uncertainty model may appear in a non-affine way in any of the closed-loop transfer functions needed for model validation (Dullerud and Smith, 1999). As a result, the resulting affine optimization used in the open-loop model validation might not be convex when applied to closed-loop systems.

The necessity and the problems associated to closed-loop model validation have been recognized in Dullerud and Smith (1999) and applied to a closed-loop experimental situation in Chen and Smith (1997) and Chen and Smith (1998). It has been observed in these references that the convexity of the model invalidation problem can be preserved by using the knowledge of the controller used in the closed-loop experiments. Furthermore, it has been recognized that a closed-loop configuration has a positive effect on the model validation. This effect is caused by the closed-loop configuration that weighs the effects of the model uncertainties via a specific closed-loop transfer function and addresses the fact that a validation of the uncertainty model under operating conditions (with a feedback controller) is more realistic and desirable.

In the line of the requirements on closed-loop model validation, this paper will generalizes the results on the use of controller information to facilitate closed-loop model validation techniques. The results in this paper focus on the development of model validation techniques that can be applied to data obtained under closed-loop or feedback controlled conditions. Furthermore, in order to present a framework that is applicable to the validation of stable and unstable models using closed-loop experiments, a framework of fractional model representations is being used (de Callafon and Van den Hof, 1997; Van den Hof and de Callafon, 1999). The model validation will be posed as closed-loop validation criterion that involves a closed-loop control objective. In this way, the model is being (in)validated towards its intended model application: robust control design.

2. THE MODEL VALIDATION PROBLEM

2.1 *Problem definition*

In order to provide a formal definition of the model validation problem, it is necessary to provide more details on the uncertainty model, denoted by $\mathcal{P}$, that consists of a nominal model $\hat{P}$ and an additional model uncertainty or allowable model perturbation Δ. A general description of the uncertainty model $\mathcal{P}$ can be given by using an LFT framework (Boyd and Barrat, 1991) that can characterize the uncertainty model $\mathcal{P}$ via an upper fractional transformation $\mathcal{F}_u(Q, \Delta)$ as follows

$$\mathcal{P} = \{P \mid P = \mathcal{F}_u(Q, \Delta)\} \text{ with} \\ \mathcal{F}_u(Q, \Delta) := Q_{22} + Q_{21}\Delta(I - Q_{11}\Delta)^{-1}Q_{12} \quad (1)$$

where the entries of Q contain the information on the nominal model $\hat{P}$ and how the uncertainty Δ is characterized around the nominal model. Typically, $Q_{22} = \hat{P}$ as for $\Delta = 0$ the nominal model should be obtained. For an (unweighted) output multiplicative uncertainty description it can be verified that $Q_{11} = Q_{21} = 0$ and $Q_{12} = I$.

Given the LFT representation of the uncertainty set $\mathcal{P}$, the entries of Q are assumed to be LTI discrete-time systems. Furthermore, Δ is assumed to be a member of a class of perturbations with $\Delta \in \mathcal{R}H_\infty$ and $||\Delta||_\infty < 1$. Given these assumption, the discrete time experimental data $\{u(t), y(t)\}$, coming from the plant, consists of N observations and is described by

$$y(t) = \mathcal{F}_u(Q, \Delta)u(t) + Hd(t) \quad (2)$$

where d denotes an unknown but bounded noise $d \in l_2$ with $||d||_2 < 1$ and H is a stable and stably invertible noise filter.

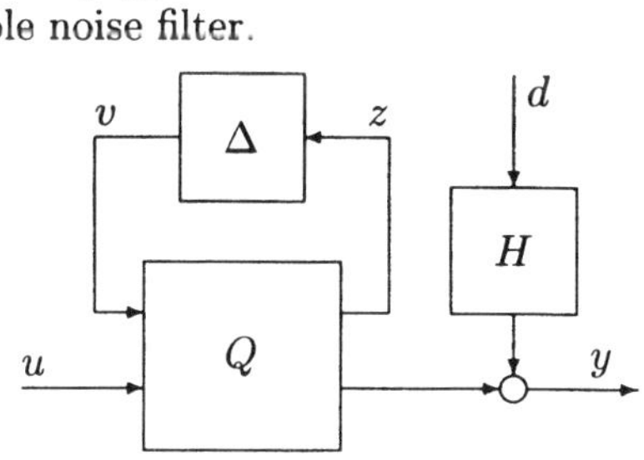

Fig. 1. Uncertainty and noise model

A block diagram of the uncertainty model and the noise model that are subjected to model (in)validation is depicted in Figure 1. The model validation problem can be formalized as follows (Smith and Doyle, 1992).

Problem 1. Given the Q of the uncertainty model $\mathcal{P}$ in (1), the noise model H and observations $\{u(t), y(t)\}$ for $t = 0, 1, \ldots, N-1$, does there exists a discrete time signal d with $||d||_2 < 1$ and an uncertainty Δ with $\Delta \in \mathcal{R}H_\infty$ and $||\Delta|| < 1$ such that (2) holds.

2.2 *Model validation via optimization*

The general approach taken in model validation is an optimization where the value of the smallest model perturbation Δ and the noise signal $d(t)$ is found that could have produced the experimental data in (2). In case the smallest model perturbation or noise signal exceeds the assumed bounds, the model is invalidated by the experimental data.

Posing the condition $\Delta \in \mathcal{R}H_\infty$ and $||\Delta|| < \alpha$ is a key step in solving the model validation problem by optimizing the value of the smallest model perturbation Δ. This condition has been solved in Poolla *et al.* (1994) for the LTI case and can be summarized as follows.

Lemma 2. Given the signals $v(t)$ and $z(t)$ for $t = 0, 1, \ldots, N-1$ with $v = \Delta z$, as indicated in Figure 1. Then there exists a $\Delta \in \mathcal{R}H_\infty$ with $||\Delta||_\infty < \alpha$ if and only if

$$V^T V < \alpha^2 Z^T Z$$

where V and Z are block Toeplitz matrices derived from $v(t)$ and $z(t)$.

With this result, the model validation can be solved via a convex optimization. Crucial in this convex optimization is the fact that the uncertainty Δ appears linearly in (2). With the general LFT form of the uncertainty model in (1), it can be seen that this is the case for $Q_{11} = 0$ and holds for example for an uncertainty model with a multiplicative uncertainty description.

With an the uncertainty Δ appearing linearly in (2), finding the smallest model perturbation Δ and the noise signal $d(t)$ is simply an additional linear constraint added to the convex minimization. With $Q_{11} = 0$ and $z(t) = Q_{12}u(t)$, the following convex optimization needs to be solved.

$$\begin{gathered} \min_{\alpha}, \text{ subjected to} \\ V^T V < \alpha^2 Z^T Z \\ d^T d < 1 \\ z(t) = Q_{12}u(t) \\ y(t) - Q_{22}u(t) = Q_{21}v(t) + Hd(t) \end{gathered} \quad (3)$$

In case $\alpha \geq 1$, the model is invalidated by the experimental data $\{u(t), y(t)\}$

Although the optimization mentioned above addresses the model validation problem, the question arises whether or not there are alternative uncertainty descriptions that still allow the use of a convex optimization. Moreover, it is preferable to perform the model validation in closed-loop, as mentioned in the introduction. In case a feedback connection is created around the signals $u(t)$ and $y(t)$, the resulting optimization involved with the model validation may not be convex.

The rest of the paper is devoted to the choice of an uncertainty model that allows for a model validation in closed-loop. The uncertainty structure and the resulting uncertainty model Q will be based on perturbation on coprime factorizations, which generalizes the approach to the validation of stable and unstable models.

3. MODEL STRUCTURE FOR CLOSED-LOOP MODEL INVALIDATION

3.1 *Uncertainty model based on coprime factor perturbations*

For the purpose of the closed-loop model validation using coprime factor uncertainty models, the nominal model $\hat{P}$ is represented in a coprime factor representation

$$\hat{P} = \hat{N}\hat{D}^{-1}$$

where $(\hat{N}, \hat{D})$ denotes a right coprime factorization (*rcf*) of the model $\hat{P}$. The accompanying uncertainty on the model is assumed to be modeled as a perturbations in a dual-Youla parametrization (de Callafon and Van den Hof, 1997). This perturbation uses the knowledge of the feedback controller to characterize the model uncertainty and yields a uncertainty model $\mathcal{P}$ that can be characterized as follows.

Proposition 3. Let a nominal model $\hat{P}$ with a *rcf* $(\hat{N}, \hat{D})$ and a controller C with a *rcf* (N_c, D_c) form an internally stable feedback connection. Then an uncertainty model $\mathcal{P}$ is constructed by

$$\begin{aligned} &\mathcal{P}(\hat{N}, \hat{D}, N_c, D_c, \hat{V}, \hat{W}) := \\ &\{P \mid P = (\hat{N} + D_c\Delta_R)(\hat{D} - N_c\Delta_R)^{-1} \text{with} \\ &\Delta_R \in \mathcal{R}H_\infty,\ \Delta := \hat{V}\Delta_R\hat{W} \text{ and } \|\Delta\|_\infty < 1\} \end{aligned} \quad (4)$$

and $\hat{V}$, $\hat{W}$ are stable and stably invertible weighting functions.

As indicated in the proposition, the uncertainty model $\mathcal{P}$ essentially depends on the factorization $(\hat{N}, \hat{D})$ of the nominal model $\hat{P}$, the factorization (N_c, D_c) of the known controller C and the weighting functions $\hat{V}$, $\hat{W}$ that take into account the shape and scaling of the uncertainty. It can be observed that the uncertainty Δ_R acts in a non-trival way on the (coprime factors of the) nominal model $\hat{P}$. It can be seen in Figure 2 that the weighted uncertainty $\Delta_R = \hat{W}^{-1}\Delta\hat{V}^{-1}$ appears in a similar way as a Youla parametrization. To distinguish between the familiar Youla parametrization, the perturbation in Figure 2 is indicated by a dual-Youla parametrization (de Callafon and Van den Hof, 1997).

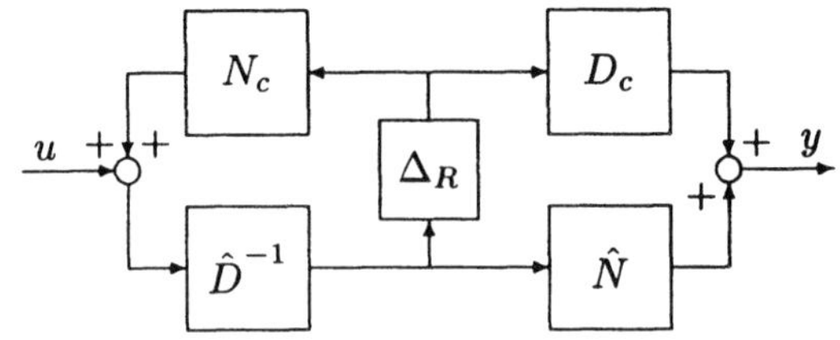

Fig. 2. Representation of $\mathcal{P}$ given in (4) with $\Delta_R = \hat{W}^{-1}\Delta\hat{V}^{-1}$.

With little effort, the nominal model and the perturbation Δ_R in the form of a dual-Youla parametrization can be rewritten in a standard LFT form of (1) with a norm bounded uncertainty $\Delta \in \mathcal{R}H_\infty$. On the basis of Figure 2, a characterization of the coefficient matrix Q in (1) can be given and an alternative representation of the uncertainty model $\mathcal{P}$ in (4) can be obtained.

Corollary 4. The uncertainty model $\mathcal{P}$ given in (4) can be written as

$$\mathcal{P} = \{P \mid P = \mathcal{F}_u(Q, \Delta),\ \|\Delta\|_\infty < 1 \text{ and}$$

$$Q = \left[\begin{array}{c|c} \hat{W}^{-1}\hat{D}^{-1}N_c\hat{V}^{-1} & \hat{W}^{-1}\hat{D}^{-1} \\ \hline (D_c + \hat{P}N_c)\hat{V}^{-1} & \hat{P} \end{array}\right] \quad (5)$$

As mentioned in Section 2.2, it is crucial that $Q_{11} = 0$ in order to exploit a convex optimization for the *open-loop* model validation problem. It can be seen that this model structure does not exhibit this property and is not suitable for an open-loop model validation. However, in the next section it will be illustrated that this model structure is beneficial in a closed-loop model validation.

Although the coefficient matrix Q in (5) looks complicated, it can be written as as simple multiplication

$$Q = \begin{bmatrix} \hat{W}^{-1} & 0 \\ 0 & I \end{bmatrix} \begin{bmatrix} \bar{Q}_{11} & \bar{Q}_{12} \\ \bar{Q}_{21} & \bar{Q}_{22} \end{bmatrix} \begin{bmatrix} \hat{V}^{-1} & 0 \\ 0 & I \end{bmatrix}$$

where the entries of the (unweighted) coefficient matrix $\bar{Q}$ are given by

$$\begin{bmatrix} \hat{D}^{-1} & 0 \\ 0 & I \end{bmatrix} \begin{bmatrix} I & 0 \\ \hat{P} & I \end{bmatrix} \begin{bmatrix} C & I \\ I & 0 \end{bmatrix} \begin{bmatrix} D_c & 0 \\ 0 & I \end{bmatrix}. \quad (6)$$

The multiplication of the transfer functions in (6) indicates the construction of the (unweighted) coefficient matrix $\bar{Q}$. Furthermore, it can be observed that $\bar{Q}$ is invertible, as all the matrices in (6) are known to be invertible, whereas

$$\begin{bmatrix} I & 0 \\ \hat{P} & I \end{bmatrix} \text{ and } \begin{bmatrix} C & I \\ I & 0 \end{bmatrix}$$

are unimodular. From the multiplication in (6) it can be observed that Q will have a McMillan degree equal to the sum of the McMillan degree of the (nominal) model $\hat{P}$ and the controller C.

3.2 *Favorable properties of coprime factor based uncertainty model*

The appearance of the (unstructured) model uncertainty Δ_R in the uncertainty model $\mathcal{P}$ in (5) is more complicated than a standard additive or multiplicative uncertainty. However, the only knowledge needed to construct the uncertainty model $\mathcal{P}$ is a coprime factorization of the nominal model $\hat{P}$ and the controller C. With this information, the construction of the model uncertainty Δ_R is not more complicated than a standard uncertainty description. Due to the specific appearance of the model uncertainty in the uncertainty model of (5), the following favorable properties can be summarized.

- All models in the uncertainty model $\mathcal{P}$ are stabilized by the know feedback controller C, irrespective of the size or shape of Δ_R.
- The uncertainty Δ_R in (5) will appear in an affine way in any closed-loop transfer function.

The first property indicates the usefulness of the uncertainty set $\mathcal{P}$ of (5). Basically, only those models P are captured in the set $\mathcal{P}$ that are stabilized by the controller C. This is an appealing and intuitive result as the plant, currently operating in a feedback connection, is know to be stabilized by the feedback controller C. Hence, only those models can be validated that are known to be stabilized by the controller C.

The second property will enable the application of an affine optimization, similar to (3) to address the closed-loop model validation problem. Clearly, this generalizes the application of model validation techniques to closed-loop systems. Furthermore, the use of stable coprime factorizations in the uncertainty model $\mathcal{P}$ of (5) generalizes the approach to stable and unstable plants and controllers.

For an explanation of these properties, consider the transfer function matrix

$$T(P,C) := \begin{bmatrix} P \\ I \end{bmatrix} (I+CP)^{-1} \begin{bmatrix} C & I \end{bmatrix} \quad (7)$$

which captures all possible closed-loop transfer functions. With the general closed-loop transfer function matrix $T(P,C)$ given in (7) and the uncertainty model given in (5), the following result can be given.

Lemma 5. Consider the uncertainty model $\mathcal{P}$ of (5) and the controller C used in the construction of $\mathcal{P}$. With the definition of the weighted closed-loop transfer function matrix $T(P,C)$ given in (7), the uncertainty model $\mathcal{P}$ satisfies

$$\mathcal{P} = \{P \mid T(P,C) = \mathcal{F}_u(M,\Delta) \text{ with } \Delta \in \mathcal{RH}_\infty,\ \|\Delta\|_\infty < 1\}$$

where the entries of M are given by

$$\begin{aligned} M_{11} &= 0 \\ M_{12} &= \hat{W}^{-1}(\hat{D}+C\hat{N})^{-1}\begin{bmatrix} C & I \end{bmatrix} \\ M_{21} &= -\begin{bmatrix} I \\ -C \end{bmatrix} D_c \hat{V}^{-1} \\ M_{22} &= T(\hat{P},C) \end{aligned} \quad (8)$$

The entries of M in (8) are all known quantities. It can be verified that all these entries are stable *if and only if* the controller C internally stabilizes the nominal model $\hat{P}$, as mentioned in the construction in the set $\mathcal{P}$ in Proposition 3. With $\hat{V}$, $\hat{W}$ stable and stably invertible and $\Delta \in \mathcal{RH}_\infty$, all models $P \in \mathcal{P}$ are stabilized by the know feedback controller C, irrespective of the size or shape of Δ_R.

It can also be observed that the uncertainty Δ appears affinely in all possible (weighted) closed-loop transfer functions of $T(P,C)$. The affine representation of the dual-Youla based model uncertainty Δ can be exploited to formulate a framework to (in)validate uncertainty models on the basis of closed-loop data.

4. APPLICATION TO CLOSED-LOOP MODEL VALIDATION

With the LFT representation of the uncertainty model $\mathcal{P}$ in (5) and the application of the feedback controller C to the models in $\mathcal{P}$, a closed-loop uncertainty model depicted in is obtained.

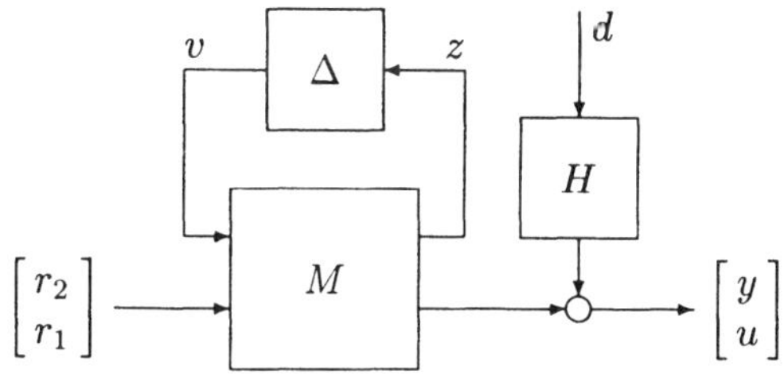

Fig. 3. Closed-loop uncertainty model

In Figure 3 the signals r_2 and r_1 indicate external closed-loop reference signals, whereas u and y denote the input and output of the plant. On the basis of this closed-loop representation, a closed-loop model validation problem can be formulated where the affine appearance of the model uncertainty Δ in (8) is used. The discrete time experimental data $\{u(t), y(t)\}$, coming from the closed-loop plant, consists of N observations and is described by

$$\begin{bmatrix} y(t) \\ u(t) \end{bmatrix} = \mathcal{F}_u(M, \Delta) \begin{bmatrix} r_2(t) \\ r_1(t) \end{bmatrix} + Hd(t) \quad (9)$$

where d denotes an unknown but bounded noise $d \in l_2$ with $||d||_2 < 1$ and H is a stable and stably invertible noise filter. The closed-loop model validation problem can now be formalized as follows.

Problem 6. Given the M of the uncertainty model $\mathcal{P}$ in (8), the noise model H and observations $\{r_2(t), r_1(t), y(t), u(t)\}$ for $t = 0, 1, \ldots, N-1$, does there exists a discrete time signal d with $||d||_2 < 1$ and an uncertainty Δ with $\Delta \in \mathcal{R}H_\infty$ and $||\Delta|| < 1$ such that (9) holds.

Along the lines of Section 2.2, the closed-loop model validation problem can be solved via a convex optimization, as $M_{11} = 0$. With the uncertainty Δ appearing linearly in (8), finding the smallest model perturbation Δ and the noise signal $d(t)$ is simply an additional linear constraint added to the convex minimization:

$$\min_{\alpha}, \text{ subjected to}$$
$$V^T V < \alpha^2 Z^T Z$$
$$d^T d < 1$$
$$z(t) = M_{12} \begin{bmatrix} r_2(t) \\ r_1(t) \end{bmatrix}$$
$$\begin{bmatrix} y(t) \\ u(t) \end{bmatrix} - M_{22} \begin{bmatrix} r_2(t) \\ r_1(t) \end{bmatrix} = M_{21} v(t) + Hd(t)$$

In case $\alpha \geq 1$, the model is closed-loop invalidated by the close-loop experimental data $\{r_2(t), r_1(t), y(t), u(t)\}$.

5. CONCLUSIONS

In this paper a model validation approach is presented that generalizes the (in)validation of possibly unstable models on the basis of closed-loop experiments with a stabilizing, but possibly unstable, controller. The approach is presented in a robust control framework with an uncertainty model described with coprime factor perturbations. It is shown that this approach yields an affine expression of the uncertainty in all possible transfer functions that can be measured via a closed-loop experiments. This allows the model validation problem to be solved via standard convex optimization techniques.

6.- REFERENCES

Boyd, S.P. and C.H. Barrat (1991). *Linear Controller design – Limits of Performance.* Prentice–Hall, Englewood Cliffs, New Jersey, USA.

Chen, L. and R. Smith (1997). Closed–loop model validation for an inverted pendulum via a linear matrix inequality approach. In: *Proc. IEEE Conference on Decision and Control.* San Diego, USA. pp. 2565–2566.

Chen, L. and R. Smith (1998). Closed–loop model validation: Application to an unstable experimental system. In: *Proc. American Control Conference.* Philadelphia, USA. pp. 618–622.

de Callafon, R.A. and P.M.J. Van den Hof (1997). Connecting system identification and robust control by a factorization approach. In: *Prepr. 11th IFAC Symp. on System Identification.* Kitakyushu, Japan. pp. 117–122.

Dullerud, G. and R. Smith (1999). The validation of model sets on the basis of closed–loop feedback system generated data. In: *Proc. Int. Symposium on Computer Aided Control System Design.* Hawaii, USA. pp. 28–33.

Poolla, K., P. Khargonekar and A. Tikku (1994). A time–domain approach to model validation. *IEEE Trans. on Automatic Control, AC–39* pp. 951–959.

Rangan, S. and K. Poolla (1996). Time–domain validation for sampled–data uncertainty models. *IEEE Trans. on Automatic Control, AC–41* pp. 980–991.

Smith, R. and G. Dullerud (1996). Validation of continuous time control models by finite experimental data. *IEEE Trans. on Automatic Control, AC–41* pp. 1094–1105.

Smith, R., G. Dullerud, S. Rangan and K. Poola (1997). Model validation for dynamically uncertain systems. *Mathematical Modelling of Systems,* Vol. *3* pp. 43–58.

Smith, R.S. (1995). Model validation for robust control: an experimental process control application. *Automatica,* Vol. *31* pp. 1637–1648. (reviewed for Automatica, TS–072).

Smith, R.S. and J.C. Doyle (1992). Model validation: a connection between robust control and identification. *IEEE Trans. on Automatic Control, AC–37* pp. 942–952.

Token, O. and J. Chen (1998). On validation of structured uncertainty model setsback systems. *Systems & Control Letters,* Vol. *33,* No. *3* pp. 748–760.

Van den Hof, P.M.J. and R.A. de Callafon (1999). Process control–relevant and closed–loop identification. In: *Prepr. 14th IFAC World Congress.* Beijing, China. pp. 7a–03–5.

Zhou, K., J.C. Doyle and K. Glover (1996). *Robust and Optimal Control.* Prentice–Hall, Upper Saddle River, New Jersey, USA.

Studies on Initialization for Multilayer Networks

HIROSHI SHIRATSUCHI†, HIROMU GOTANDA††,
KATUHIRO INOUE††† AND KOUSUKE KUMAMARU†††

Email: *sira@ie.u-ryukyu.ac.jp*

†Faculty of Engineering, Univ. of the Ryukyus
Nishihara, Okinawa 903-0213, Japan
††Faculty of Engineering, Kinki Univ. in Kyushu
Iizuka, Fukuoka 820-8555, Japan
†††Faculty of Computer Science and Systems Engineering, Kyushu Institute of Technology
Iizuka, Fukuoka 820-8502, Japan

Abstract: This paper proposes an initialization of back propagation (BP) networks for pattern classification problems; the weights of hidden units are initialized so that hyperplanes should pass through the center of input pattern set, and those of the output layer are initialized to zero. Several simulation results confirm that the proposed initialization gives better convergence than the ordinary initialization that all the weights are initialized by uniform random values with zero mean. *Copyright © 2000 IFAC*

Keywords: Multilayer net, initialization, pattern classification

1 Introduction

The BP networks with single hidden layer are generally applied in various fields such as pattern recognition and control(Haykin 1994, Luo & Unbehauen 1997). But they have convergence problems such as falling into local minima and slow learning speeds(Bigus 1996). This is mainly due to the fact that the convergence of BP algorithm depends on the initial weight values of the network because the algorithm is based on the gradient principle.

Jia et al. (1990) have suggested a weight initializing guideline that matches active regions to an input domain. The guideline is based on the fact that, if the common region between the input domain and the active region is moderately wide, the differential value of sigmoid function is non-zero and the weights are updated smoothly. The networks of equal structure allow the same cardinality of admissible solutions for the learning tasks of which input patterns are related by affine transform, even if their sigmoid functions are different in polarity and range(Gotanda et al. 1996). Nevertheless bipolar networks with their sigmoid functions ranging -0.5 to 0.5 show better convergence of BP learning algorithm than unipolar networks with those ranging 0 to 1 under the initialization that all the initial weights are given by uniform random values with zero mean(Gotanda et al. 1995). For convenience, we call this initialization *an ordinary random initialization*. This can be explained from the fact that the common region of two distribution of normal vectors, one at the ordinary random initialization and the other when separation hyperplanes pass through the input domain, is wider for bipolar networks than for unipolar ones(Shiratsuchii et al. 1998). However, the input set is a subset of the input domain. Thus the separation hyperplane passing through the input domain do not necessarily pass through the input set.

In the present paper, we propose an initialization of bipolar networks with one hidden layer trained for pattern classification problems: weights of hidden units are initialized so that the hyperplanes should pass through the center of input pattern set, and those of output units are initialized to zero. At the same time, we suggest that if there exist two or more hidden layers, their initial bias weights except for the first hidden layer are given by zero so that the hyperplanes should pass through the center of input domain. From simulation results for MONK's problem, Iris classification problem and Sonar target identification problem, it is verified that the proposed initialization is effective.

2 Input set and domain

Consider a three-layer-network $(m - n - l)$ of bipolar type with m input units, n hidden units and l output units trained for classification problems of which input patterns and targets are given by $\boldsymbol{x}_s = [x_1^s, x_2^s, \cdots, x_m^s]^T$

and $\boldsymbol{d}_s = [d_1^s, d_2^s, \cdots, d_l^s]^T$ $(s = 1, 2, \cdots, S)$, respectively; $d_k^s = -0.5$ or 0.5 $(k = 1, 2, \cdots, l)$.

For the input pattern $\boldsymbol{x}_s$, the net-input u_j^s and the output y_j^s of the j-th hidden unit $(j = 1, 2, \cdots, n)$ are calculated as follows:

$$u_j^s = \sum_{i=1}^{m} x_i^s w_{ji} + b_j \quad (1)$$

$$y_j^s = f(u_j^s) = \frac{1}{2}\tanh(\frac{u_j^s}{2}) \quad (2)$$

where $f(\cdot)$ is a bipolar type sigmoid function, w_{ji} the connection weight from the i-th input unit$(i = 1, 2, \cdots, m)$, and b_j the bias weight. Of the output units, the connection weights are similarly denoted by v_{kj}, the bias weights by c_k, the net-inputs by q_k^s and the output by z_k^s.

Let H_j be a separation hyperplane $\{\boldsymbol{x} \in R^m | \boldsymbol{w}_j^T \boldsymbol{x} + b_j = 0\}$ due to the j-th hidden unit where R^m denotes an m-dimensional Euclid space. Then R^m is separated into high and low output regions at the boundary of H_j, and each region is further divided into an active region $\mathcal{A}_j = \{\boldsymbol{x} \mid -0.5 + \rho < f(\boldsymbol{w}_j^T \boldsymbol{x} + b_j) < 0.5 - \rho\}$ and a saturation region $\mathcal{S}_j = \{\boldsymbol{x} \mid f(\boldsymbol{w}_j^T \boldsymbol{x} + b_j) < -0.5 + \rho,\ f(\boldsymbol{w}_j^T \boldsymbol{x} + b_j) > 0.5 - \rho\}$ where $\|\cdot\|$ denotes a norm and ρ is a small positive number around 0.1 that defines the active region. The width of the active region is given by

$$g_j = \frac{2\ln\{(1-\rho)/\rho\}}{\|\boldsymbol{w}_j\|}. \quad (3)$$

The separation hyperplane can be specified by *a normal vector*

$$\boldsymbol{h}_j = -\frac{b_j}{\|\boldsymbol{w}_j\|^2}\boldsymbol{w}_j \quad (4)$$

which is the vector extending from the origin of R^m and intersecting orthogonally at H_j. Of the output units, the separation hyperplane, the normal vector, the active region and its width are similarly defined by H_k, $\boldsymbol{h}_k$, $\mathcal{A}_k$ and g_k, respectively.

Consider the input pattern set $\mathcal{X} = \{\boldsymbol{x}_s\}$, the input pattern domain $\boldsymbol{X}$ (a parallelepiped hyper-rectangular circumscribing $\mathcal{X}$) and the input set for the output units $\mathcal{Y} = \{\boldsymbol{y}_s = [y_1^s, y_2^s, \cdots, y_n^s]^T\}$. Then it is found that there exist the following differences among them. $\mathcal{X}$ is invariant in course of learning although the range of input patterns depends on learning tasks. The inputs $\boldsymbol{y}_s$ for the output units exist in an n-dimensional hyper-cube $\boldsymbol{Y} = [-0.5, 0.5]^n$ prescribed by the sigmoid function and then $\mathcal{Y} \subset \boldsymbol{Y}$. $\mathcal{Y}$ varies in course of learning because the weights in hidden units are updated. Therefore the input set $\mathcal{X}$ for the hidden layer is invariant in course of learning and then can be treated as the same input set settled finally at convergence; while the input set $\mathcal{Y}$ for the output layer is variant and then differs from a input set settled finally at convergence.

3 Initialization

From the above reason, the initialization for the hidden layer is discussed separately from that for the output layer.

3.1 Initialization for hidden layer

It is considered that some separation hyperplanes must pass through the input set if the network has a solution for pattern classification problems. From this perspective, by a set of normal vectors Ω we have formulated such a separability condition that the hyperplanes pass through the parallelepiped hyper-rectangular circumscribing the input set. Then we have revealed that the common region to two sets, *i.e.*, the set Ω satisfying the separability condition and the initial normal vector set Ψ_0 resulting from the ordinary random initialization, is wider in the bipolar case than in the unipolar case(Shiratsuchi et al. 1998). Finally, we have theoretically supported the superiority of bipolar networks to unipolar ones in convergence of the BP learning under the ordinary random initialization.

In the case of the ordinary random initialization, however, not all the separation hyperplanes pass through the input pattern set $\mathcal{X}$ as shown in Fig. 1. Therefore the separation hyperplane H_j is translated to generate a new separation hyperplane H_j^* passing through $\mathcal{X}$.

In the net-input set $\{u_j^s | s = 1, 2, \cdots, S\}$ for the input pattern set $\mathcal{X}$, there exist the maximum u_j^{max} and the minimum u_j^{min} for each hidden unit. The corresponding input patterns to u_j^{max} and u_j^{min} are represented by $\boldsymbol{x}_j^{max}$ and $\boldsymbol{x}_j^{min}$, respectively.

$$u_j^{max} = \boldsymbol{w}_j^T \boldsymbol{x}_j^{max} + b_j \quad (5)$$

$$u_j^{min} = \boldsymbol{w}_j^T \boldsymbol{x}_j^{min} + b_j \quad (6)$$

The new separation hyperplane H_j^* is defined so that it should be parallel to H_j and pass at a point dividing the line between $\boldsymbol{x}_j^{min}$ and $\boldsymbol{x}_j^{max}$ internally in the

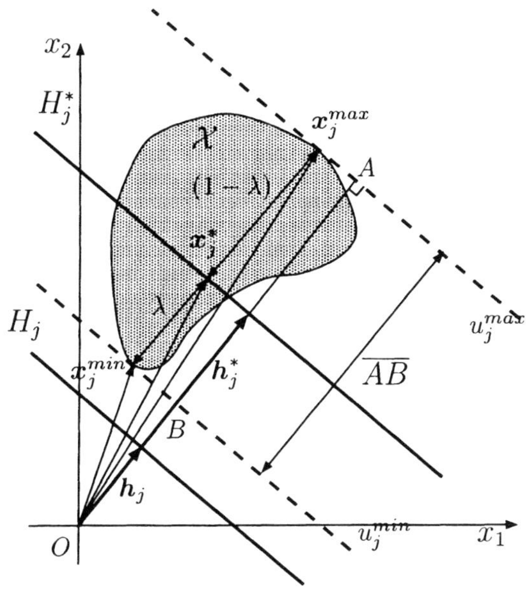

Figure 1: Separation hyperplane and input pattern set

proportion of λ to $(1-\lambda)$:

$$x_j^* = (1-\lambda)x_j^{min} + \lambda x_j^{max} \tag{7}$$

Hereafter, we call λ *a dividing ratio.* In fact, H_j^* surely passes through $\mathcal{X}$ if $0 \le \lambda \le 1$. The normal vector h_j^* accompanying to H_j^* is derived as follows:

$$h_j^* = \frac{h_j^T\{(1-\lambda)x_j^{min} + \lambda x_j^{max}\}}{\|h_j\|^2} h_j. \tag{8}$$

Substituting Eq. (8) to Eq. (4) with keeping g_j the same, the weights w_j^* and b_j^* resulting in H_j^* are obtained as

$$w_j^* = w_j \tag{9}$$
$$b_j^* = -w_j^T\{(1-\lambda)x_j^{min} + \lambda x_j^{max}\}. \tag{10}$$

Further, it is found from Eqs. (5) and (6) that the bias weights b_j^* are easily calculated in terms of the maximum net-input u_j^{max} and the minimum net-input u_j^{min} as follows:

$$b_j^* = b_j - \{\lambda u_j^{max} + (1-\lambda)u_j^{min}\} \tag{11}$$

It has been confirmed by simulation that if the common region between the input pattern set $\mathcal{X}$ and the active region $\mathcal{A}_j$ is wide, the converging performance increases(Shiratsuchi et al. 1998). Viewing the dividing ratio from this perspective, it is found that the common region becomes the widest at $\lambda = 0.5$ when the separation hyperplanes pass through the center of $\mathcal{X}$. On the other hand, when $\lambda = 0.0$ and 1.0, the separation hyperplanes pass through both ends of $\mathcal{X}$, with the result that the common region is smaller than that at $\lambda = 0.5$. Therefore $\lambda = 0.5$ is adopted. This adoption will be verified by several simulation results in the following section.

Of the hidden units, therefore, we employ the initial connection weights w_j^* being kept the same to w_j as Eq.(9), and employ the initial bias weights b_j^* set over again as Eq. (11) under the condition of $\lambda = 0.5$.

When the weights in the first hidden layer are initialized so that H_j should pass through the center of $\mathcal{X}$ as described above, at least the outset of learning, the input set $\mathcal{Y}$ for the next layer is considered to distribute around the origin of R^m, *i.e.*, the center of the input domain $\boldsymbol{Y}$. Thus, if there exist two or more hidden layers, the bias weights are initialized to zero except for the first hidden layer.

3.2 Initialization for output layer

The input domain for the output layer is divided into high and low output regions. It is uniquely determined by the bias sign which region produces high outputs(Gotanda et al. 1997): in the case of $v_k \ne \mathbf{0}$ and $c_k \ne 0$, the region including the origin of R^m becomes high (low) output region if c_k is positive (negative), as seen from

$$h_k^T(y - h_k) = -\frac{c_k q_k}{\|v_k\|^2} < 0. \tag{12}$$

However, in the case of $v_k = \mathbf{0}$ and $c_k = 0$, the input set is entirely covered in the separation hyperplane as $\boldsymbol{Y} \subset H$. This means that the network outputs become zero for all the input patterns and then the network produces a flat and even output surface.

Figure 2 conceptually shows the network input/output relation when the training data, 4 pairs of input pattern and target, are given by $(x_s,\ d_s)$ $(s = 1,\ 2,\ 3,\ 4)$. In the figure, the horizontal axis represents any input patterns $x \in R^m$, and the vertical axis the network outputs $z(x,\ W)$ for $\{x\}$ where W stands for all the weights in the network. Note that $z(x,\ W)$ is a hypersurface, $z_s = z(x_s,\ W)$ and $e_s = d_s - z_s$. W^* represents a converged solution, and W_1 and W_2 represent two kinds of initial weights due to the ordinary random initialization. It is considered that the learning works

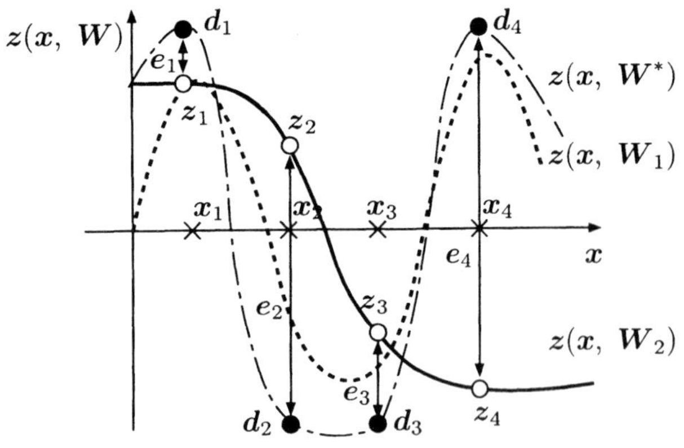

Figure 2: Network output surface

well to converge when the network outputs $z(\boldsymbol{x}_s, \boldsymbol{W}_1)$ match with their targets $\boldsymbol{d}_s$ at the outset of learning. In this case, $\boldsymbol{W}_1$ is considered to properly reflect *a priori* information about the training data. However, there is no methods for extracting such *a priori* information properly.

In the case of the ordinary random initialization, the width of the active region g_j defined by Eq. (3) reduces with increasing the number of hidden (output) units $m(n)$ while the size of the input domain $\boldsymbol{X}$ ($\boldsymbol{Y}$) for the hidden (output) layer increases in proportion to $\sqrt{m}$ ($\sqrt{n}$)(Gotanda et al. 1995). In this case, thus, the number of inputs belonging to the saturation region increases much more. The solid line indicated by $z(\boldsymbol{x}, \boldsymbol{W}_2)$ in Figure 2 shows such an example where the inputs in the saturation region increase. In this example, presentation of $\boldsymbol{x}_1$ and $\boldsymbol{x}_4$ to the network does not lead to update the weights because the corresponding $\boldsymbol{y}_1$ and $\boldsymbol{y}_4$ exist in the saturation region with $f'(\cdot)$ being almost zero, while presentation of $\boldsymbol{x}_2$ and $\boldsymbol{x}_3$ lead to update the weights because the corresponding $\boldsymbol{y}_2$ and $\boldsymbol{y}_3$ exist in the active region with relatively large $f'(\cdot)$ where $f'(\cdot)$ denotes the differentiation of $f(\cdot)$. For a successful learning, it is necessary to decrease such errors that are very large at the outset like $\boldsymbol{e}_4$ as well as the other errors that are small at the outset like $\boldsymbol{e}_1$, $\boldsymbol{e}_2$ and $\boldsymbol{e}_3$. Therefore, since the training data resulting in large errors at the outset can not update the weights by themselves, these large errors must be decreased through the weight update by the other remaining training data.

However, the weight update due to the other remaining training data does not necessarily guarantee to decease those large errors. Therefore, it is considered that the larger the network becomes, the harder the learning started with the ordinary random initialization converge. When the weights of output units are initialized to zero, every input $\boldsymbol{y}_s$ for the output layer exists in the active region because $\boldsymbol{Y} \subset H$ as described before. Then all the network outputs become zero and the absolute values of the corresponding errors become unbiasedly 0.5. Thus the weights of the output units are updated as

$$\Delta v_{kj} = \eta f'(q_k^s) e_k^s y_j^s \tag{13}$$

for presentation of all the input patterns except for those with $\boldsymbol{y}_s = \boldsymbol{0}$, where η is a learning coefficient. Note that, at the outset of learning, the weight update is greatly influenced by the target since $e_k^s = d_k^s - z_k^s = d_k^s$. The weights of hidden units are updated as follows.

$$\Delta w_{ji} = \eta f'(u_j^s) \sum_{k=1}^{l} v_{kj} f'(q_k^s) e_k^s y_j^s \tag{14}$$

Note that, even at the outset of learning, w_{ji} is updated since v_{kj} is not equal to zero but to $\Delta v_{kj} \neq 0$ after finishing the weight update of the output units.

From the above, the connection and bias weights of the output units are initialized as follows:

$$\boldsymbol{v}_k = \boldsymbol{0} \tag{15}$$

$$c_k = 0 \tag{16}$$

4 Simulation Results

First, preliminary experiments were performed to show the validity of setting the dividing ratio to 0.5 by comparing the following cases; $\lambda = 0.0$, 0.5, and 1.0 in addition to the case where λ is given by a uniform random value [0, 1]. The BP learning was run at the learning coefficients $\eta = 0.01$, 0.05, 0.1 and the momentum coefficient $\alpha = 0.9$ with the maximum iteration 100,000 under the convergence criterion that the absolute error for every training datum is less than 0.1. Converging performance was evaluated under fifty trials in terms of convergence rates (convergent trials to 50 trials) and averaged epochs (averaged iterations in the convergent trials divided by the number of training patterns). Experiments were performed on Iris classification problem(4–3–3, continuous inputs, #30 training data), Sonar target identification problem(60–4–1, continuous inputs, #104 training data) and MONK's problem(6–4–1, discrete inputs, #124 training data). Figure 3 shows the learning results of the Iris classification problem(4–3–3). In the figures, the horizontal axis represents a magnitude γ of uniform random values $[-\gamma, \gamma]$ ($\gamma = 0.01, 0.1, 0.5, 1.0, 1.5, \cdots, 6.0$) and the vertical axis represents the convergence rate and the averaged epoch. From this results, it is verified that the converging performance is improved by setting $\lambda = 0.5$ so that the hyperplanes should pass through the center of the input set. The validity of $\lambda = 0.5$ was also confirmed for the other problems.

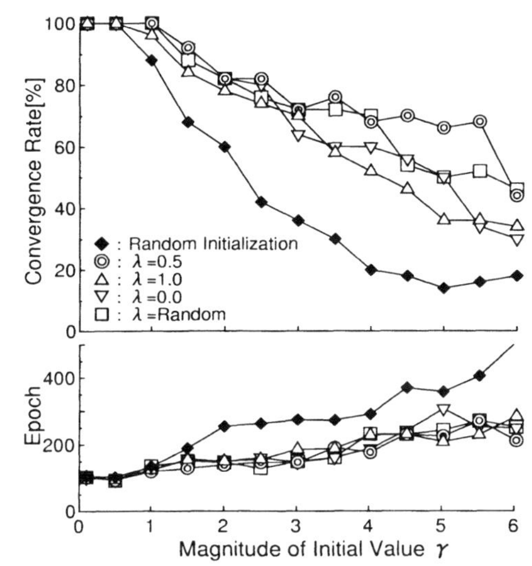

Figure 3: Validity of setting λ to 0.5

Next, in order to verify the proposed initialization, experiments were performed on the above problems. The convergence criterion for these problem is that the mean square error is less than 0.01. Table 1 shows the initial weights of hidden and output units; the mark ○ indicates results for the proposed initialization, ◆ indicates those for the ordinary random initialization, and "Random" means that the weights are initialized by uniform random values with zero mean.

The results are shown in Figure 4 to 6. From the results marked by ●, ◎ and ○, it is also found that the converging performance is improved by setting $\lambda = 0.5$. The proposed initialization gives much better converging performance for a wider range of γ than the other initializations. This fact was also true at $\eta = 0.01$ and 0.1. In Iris classification problem(Figure 4), the

Table 1: Initial weights of hidden and output units

	w_j and b_j	v_k and c_k
◆	Random	Random
●	Eqs.(9)(11)	Eqs.(9)(11)
◎	Eqs.(9)(11)	Random
○	Eqs.(9)(11)	$v_k = 0,\ c_k = 0$

convergence rates about 100% for every γ. This fact implies that, at least for Iris classification problem, dependence of the BP learning convergence on initial values is improved. From these experimental results, BP learning is found to converges well when the initial weights of output units are zero, even if the initial weights of hidden units are given by large values to some extent.

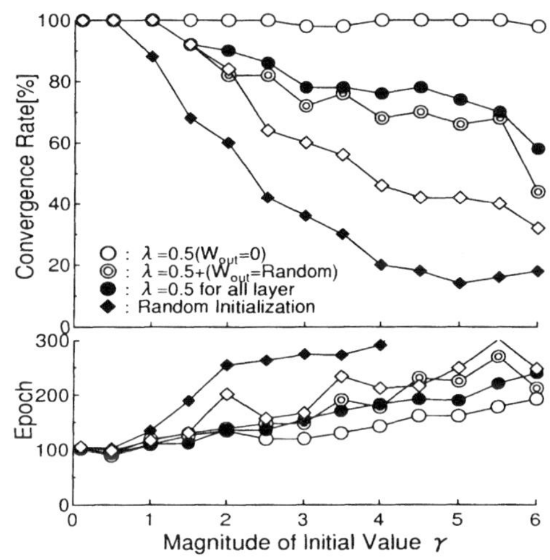

Figure 4: Learning results for Iris problem

5 Conclusions

In the present paper, we have proposed the initialization of bipolar networks trained for pattern classification problems. The weights of the first hidden layer are initialized so that the hyperplanes should pass through the center of the input pattern set. Of the output units, all the weights are initialized to zero. From the simulation results for MONK's problem, Iris classification problem and Sonar target identification problem, it has been confirmed that the proposed initialization gives better convergence.

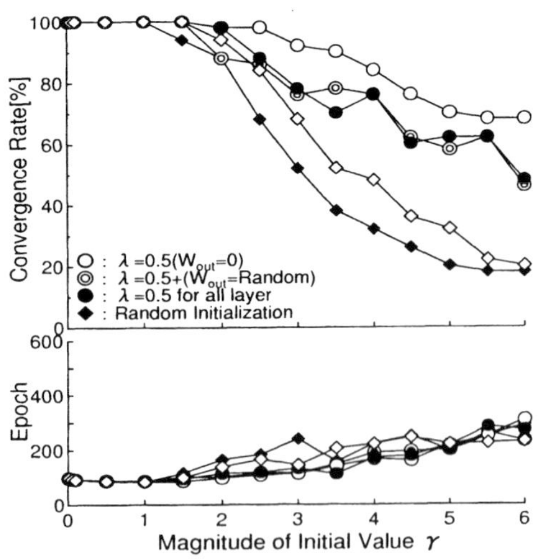

Figure 5: Learning results for MONK'S problem

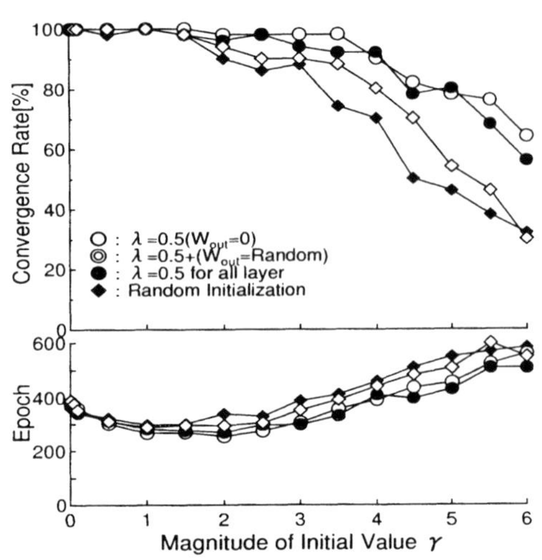

Figure 6: Learning results for Sonar problem

Thus far, even if the input patterns could be approximately clustered and the clustering results could be reflected on the initial weights of hidden units, it could not be judged whether or not the clustering effect contributes to improve the convergence of BP learning because the weights of output units are initialized by random values. From the experimental results, BP learning is found to converges well when the initial weights of output units are zero, even if the initial weights of hidden units are given by large values to some extent. This implies the possibility that the initialization problem of the networks with one hidden layer can be reduced to an initialization problem only of the hidden layer. From this viewpoint, therefore, we will examine an initialization of the hidden layer, taking into account some existing clustering methods.

References

Bigus, J. P. (1996), *Data Mining with Neural Networks*, The McGraw-Hill Companies.

Gotanda, H., Shiratsuchi, H., Inoue, K. & Kumamaru, K. (1996), 'Studies on cardinality of solutions for multilayer nets and a scaling method in hardware implementations(in japanese)', *IPSJ Trans.* **37**(8), 1535–1542.

Gotanda, H., Shiratsuchi, H., Inoue, K. & Kumamaru, K. (1997), 'Difference of solution regions due to net polarity(in japanese)', *IEICE Trans.* **J80-D-II**(2), 696–699.

Gotanda, H., Ueda, Y. & Kawasaki, T. (1995), 'Effects on back propagation learning by unit's polarity(in japanese)', **J78-D-II**(9), 1372–1382.

Haykin, S. (1994), *Neural Networks – A Comprehensive Foundation*, Macmillan Publishing Company.

Jia, A., Toda, N. & Usui, S. (1990), 'A study on initial value setting of the back-propagation learning algorithm(in japanese)', *IEICE Trans.* **J73–D-II**(8), 1179–1185.

Luo, F. & Unbehauen, R. (1997), *Applied Neural Networks for Signal Processing*, Cambridge University Press.

Shiratsuchi, H., Gotanda, H., Inoue, K. & Kumamaru, K. (1998), 'Convergent superiority of bipolar nets from the viewpoint of input set separability(in japanese)', *ISCIE Trans.* **11**(4), 190–197.

ON THE DYNAMICS OF LOCAL LINEAR MODEL NETWORKS WITH ORTHONORMAL BASIS FUNCTIONS

Oliver Nelles, Masayoshi Tomizuka

Mechanical Systems Control Laboratory, UC Berkeley
6189 Etcheverry Hall, Berkeley, CA 94720-1740, USA
E-mail: nelles@mechatro2.me.berkeley.edu

Abstract: This paper deals with local linear model networks for nonlinear system identification. It compares the standard nonlinear autoregressive with exogenous input (NARX) model structure with the new nonlinear orthonormal basis functions (NOBF) model structure. In particular, the dynamics of interpolated local ARX and OBF models are studied and significant advantages for the NOBF approach are pointed out. Furthermore, the dynamic effect of scheduling variables in local linear model networks based on OBFs is investigated. It is concluded that the NOBF approach is a promising alternative for nonlinear system identification and deserves more attention in the future.

Keywords: nonlinear system identification, local linear model network, orthonormal basis functions, NARX, interpolation

1. INTRODUCTION TO ORTHONORMAL BASIS FUNCTIONS

Most nonlinear dynamic model architectures are an extension of the linear model structures such as ARX (autoregressive with exogenous input), ARMAX (autoregressive moving average with exogenous input), and OE (output error), etc. Tapped delay lines are used to generate previous inputs and outputs which are fed to a nonlinear static approximator (e.g. a neural network, fuzzy system, polynomial, etc.). In the case of a NARX model, the following model structure arises for a system with a single input u, a single output y, and the dynamic order m

$$\hat{y}(k) = f(u(k-1), \ldots, u(k-m), y(k-1), \ldots, y(k-m)). \quad (1)$$

Recently, as an alternative to these models with output feedback, orthonormal basis functions gained significant interest and have been studied thoroughly for linear system identification (Van den Hof *et al.*, 1995), (Wahlberg, 1991). Relatively few work has been done on their extension for nonlinear system identification, see (Back and Tsoi, 1996), (Nelles, 1997), (Sbarbaro, 1997), (Schram *et al.*, 1996). The approach pursued in this paper is based on the following model structure (q denotes the time forward shift operator)

$$\hat{y}(k) = f\left(L_1(q)u(k), \ldots, L_n(q)u(k)\right) \quad (2)$$

where the $L_i(q)$ are linear filters that are chosen as orthonormal[1] basis functions (OBFs). A general construction mechanism for OBFs is given in (Heuberger *et al.*, 1995). In the most simple case, the filters can be chosen as time delays $L_i(q) = q^{-i}$. Then, (2) represents an NFIR (nonlinear finite impulse response) model. The disadvantage of NFIR models is that the number of filters n must be chosen extremely high if the process possesses slow dynamics (compared to the sampling time). If the major time constant of the process to be

[1] The impulse responses of these filters are given by orthonormal Laguerre polynomials.

modeled is roughly known, Laguerre functions are a much better choice

$$L_i(q) = \frac{\sqrt{1-p^2}}{q-p}\left(\frac{1-pq}{q-p}\right)^{i-1} \tag{3}$$

where p denotes the pole of the Laguerre filter. The closer the choice of this pole p reflects the process dynamics, the smaller the order n can be chosen for a good approximation. Note that for $p = 0$ the FIR model is recovered. Laguerre models only poorly describe weakly damped oscillatory systems, i.e., the required n would be very large. In these cases, the Kautz basis can be used which allows the incorporation of prior knowledge on a conjugate complex pole pair. An arbitrary number of poles can be incorporated into the filters with the generalized OBF construction introduced in (Heuberger *et al.*, 1995). The Kautz, Laguerre, and FIR models are special cases of this generalized basis. For the sake of simplicity, the remaining part of this paper focuses on Laguerre models.

Comparing the NOBF model in (2) with the NARX model in (1), the following observations can be made (assuming that the models shall be used for simulation not for one-step prediction):

(1) Both model structures have an infinite impulse response (IIR). The NARX model possesses external feedback which is influenced by the determination of $f(\cdot)$, while the NOBF model has internal feedback which is fixed by the user a priori when choosing the characteristics of the filter $L_i(q)$.

(2) As a consequence, dependent on $f(\cdot)$ the NARX model can become unstable, while the NOBF model is guaranteed to be stable for all possible $f(\cdot)$ if the pole is chosen properly.

(3) The NOBF model can only approximate the process dynamics (the higher the order n is chosen the better the approximation), whereas the NARX model can exactly capture the process dynamics if the model order is chosen equivalent to the process order. The importance of this issue fades if one realizes that, in practice, models will be usually low order approximations of the process and that further (presumably more significant) errors are introduced by the approximation of $f(\cdot)$.

(4) The NOBF model requires rough prior knowledge about the process dynamics[2]. As the quality of the knowledge decreases the number of filters required for the same accuracy increases, and as a consequence, the input space dimensionality of $f(\cdot)$ increases. Due to the curse of dimensionality, too inaccurate prior knowledge may let the NOBF approach fail. Furthermore, with a NOBF model, an unstable system cannot be modeled at all and processes with weakly damped or dispersed poles require Kautz functions respectively generalized OBFs with an increasing demand on prior knowledge.

(5) Both model structures are linear in their parameters if a linearly parameterized function approximator for $f(\cdot)$ is selected.

(6) The NOBF model structure represents an output error model, the NARX model structure represents an equation error model. Therefore, the estimates of the NARX model's parameters are biased, while the NOBF model's parameters are estimated unbiased in the presence of uncorrelated additive output noise. Furthermore, an exact expression for the variance error of NOBF models can be derived if the approximator of $f(\cdot)$ is linearly parameterized because the regression matrix is deterministic. This also allows the calculation of the NOBF model's confidence intervals (errorbars), whereas this is not possible for NARX models.

(7) The data distribution in the input space of $f(\cdot)$ is different for NOBF and NARX approaches. This can have a significant influence on the achievable approximation quality of $f(\cdot)$, in particular if axis-orthogonal partitioning strategies are applied.

Note that in a comparison between NOBF and NOE (instead of NARX) model structures, NOE can break even with NOBF at the sixth property since both are output error models. However, NOE has significant drawbacks at the fifth property because it is nonlinear in its parameters and thus requires much higher computational effort with the risk of convergence to local optima.

The appealing features of NOBF models clearly motivate further research in this area. The remainder of this paper is organized as follows. First, the architecture of local linear model networks is introduced. They shall be used for approximation of the function $f(\cdot)$ in (2). Then, the dynamics of interpolated local OBF models are analyzed and compared with local ARX/OE models. Next, the dynamic effects of scheduling variables that govern the nonlinear model behavior are discussed and the differences between local OBF and ARX/OE models are pointed out. Finally, the results achieved in this study are summarized.

[2] Approaches that are less sensitive to prior knowledge have been proposed. For example, the Laguerre pole can be estimated by nonlinear optimization or from an initial (N)OBF model based on prior knowledge a better choice for the pole can be found by model reduction techniques (this idea can be applied in an iterative manner). These ideas clearly require further investigation and are not pursued in this paper because they are computationally more demanding than the calculation of a least squares solution.

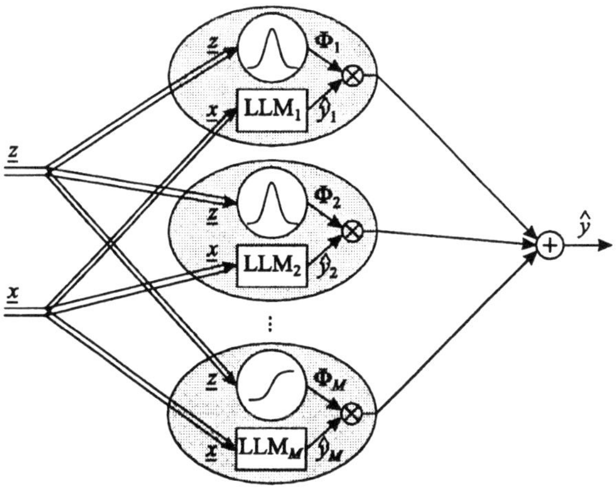

Fig. 1. Architecture of a local linear model network with M neurons for nx LLM inputs $\underline{x}$ and nz validity function inputs $\underline{z}$.

2. LOCAL LINEAR MODEL NETWORKS

The architecture of a local linear model network is depicted in Fig. 1 (Johansen and Foss, 1993), (Murray-Smith, 1994), (Nelles, 1999). Each neuron realizes a local linear model (LLM) and an associated validity function that determines the region of validity of the LLM. The validity functions form a partition of unity, i.e., they are normalized such that

$$\sum_{i=1}^{M} \Phi_i(\underline{z}) = 1 \tag{4}$$

for any model input $\underline{z}$. The output of a local linear model network is calculated as

$$\hat{y} = \sum_{i=1}^{M} (w_{i,0} + w_{i,1}x_1 + \ldots + w_{i,nx}x_{nx})\, \Phi_i(\underline{z}) \tag{5}$$

where the local linear models depend on $\underline{x} = [x_1\ x_2\ \cdots\ x_{nx}]^T$ and the validity functions depend on $\underline{z} = [z_1\ z_2\ \cdots\ z_{nz}]^T$. Thus, the network output is calculated as a weighted sum of the outputs of the local linear models where the $\Phi_i(\cdot)$ are interpreted as operating point dependent weighting factors. The network interpolates between different LLMs with the validity functions. The choice of the inputs $\underline{x}$ and $\underline{z}$ depends on the specific problem. In the most general case, $\underline{x} = \underline{z}$ and both contain all relevant model inputs. In particular for dynamic systems, in many applications, however, the variables that mainly cause the nonlinear effects are known a priori. Then, only these *scheduling variables* have to be included in $\underline{z}$ so that its dimension can be kept low and (in some cases) independent on the dynamic order of the system.

The validity functions are typically chosen as normalized Gaussians. If these Gaussians are furthermore axis-orthogonal (i.e. have diagonal covariance matrices), the validity functions become

$$\Phi_i(\underline{z}) = \frac{\mu_i(\underline{z})}{\sum_{j=1}^{M} \mu_j(\underline{z})} \tag{6}$$

with $\mu_i(\underline{z}) =$

$$\exp\left(-\frac{1}{2}\left(\frac{(z_1 - c_{i,1})^2}{\sigma_{i,1}^2} + \ldots + \frac{(z_{nz} - c_{i,nz})^2}{\sigma_{i,nz}^2}\right)\right).$$

Note that the network output is ***linear*** in the weights $w_{i,j}$ but is ***nonlinear*** in the centers $c_{i,l}$ and standard deviations $\sigma_{i,l}$ ($i = 1, \ldots, M$, $j = 1, \ldots, nx$, $l = 1, \ldots, nz$).

The local linear model network possesses an interpretation in terms of fuzzy logic in which each neuron represents one rule (Hunt *et al.*, 1996). The validity functions represent the normalized multi-dimensional membership functions of the rule premise and the LLMs represent the rule consequents. If the input space partitioning is axis-orthogonal, one-dimensional Gaussian membership functions

$$\mu_{i,j}(z_j) = \exp\left(-\frac{1}{2}\frac{(z_j - c_{i,j})^2}{\sigma_{i,j}^2}\right) \tag{7}$$

can be combined by a t-norm (conjunction) realized with the product operator to form the multi-dimensional membership functions in (2). One of the major strengths of these model architectures is as described above that premises and consequents do not have to depend on identical variables, i.e., $\underline{z}$ and $\underline{x}$ can be chosen independently.

For construction of local linear model networks from data various algorithms exist. The arguably two most common alternatives are input-output space clustering (Babuška and Verbruggen, 1996) and incremental tree-construction strategies that partition the input space by axis-orthogonal splits like the local linear model tree (LOLIMOT) algorithm in (Nelles, 1999).

3. DYNAMICS OF INTERPOLATED LOCAL ARX AND OBF MODELS

Local linear model networks have many appealing features that make them well suited for approximation of $f(\cdot)$ in (1) or (2), see (Babuška and Verbruggen, 1996), (Johansen and Foss, 1993), (Murray-Smith, 1994), (Nelles, 1999). An mth order local linear ARX model network can also be written as

$$\begin{aligned}\hat{y}(k) = {} & b_1(\underline{z})u(k-1) + \ldots + b_m(\underline{z})u(k-m) \\ & -a_1(\underline{z})y(k-1) - \ldots - a_m(\underline{z})y(k-m) + \xi(\underline{z})\end{aligned} \tag{8}$$

with operating point dependent parameters $b_i(\underline{z})$, $a_i(\underline{z})$ and the offset $\xi(\underline{z})$ (not denoted as c to avoid confusion with the validity functions' centers). When used for ***simulation***, the measured process outputs $y(k-i)$ on the right hand side are replaced by the model outputs $\hat{y}(k-i)$ introducing *feedback* [3]. In contrast, an nth order local linear OBF model network possesses no such output feedback

$$\hat{y}(k) = b_1(\underline{z})\,[L_1(q)u(k)] + \ldots + b_n(\underline{z})\,[L_n(q)u(k)] + \xi(\underline{z}) \qquad (9)$$

where the brackets $[\cdot]$ emphasize that the filtered versions of the input can be calculated independently of the local model network.

For both model structures, the operating point dependent parameters are obtained by an interpolation of the LLM parameters. For the NOBF model, this is equivalent to an interpolation of the time responses of the LLMs, i.e., the output of each LLM for a given input signal can be calculated separately and the overall model output is a weighted average of these. This is not the case for the NARX model because of the output feedback. This observation has important consequences for the dynamics of the nonlinear dynamic model. These consequences shall be illustrated in the following for the simple case of two LLMs that are interpolated by $\Phi_1(z)$ and $\Phi_2(z)$ where (for the sake of simplicity) the scalar z is an external scheduling variable which does not enter $\underline{x}$. Note that a stability analysis becomes more complex if validity functions depend on model states. Then, linear matrix inequalities may be use for checking stability (Tanaka and Sugeno, 1992).

3.1 *First Order Systems*

For first order NARX models, the interpolation of the denominator parameters of the LLMs is equivalent to the interpolation of their poles since the parameters are just the negative poles. Thus, the operating point dependent time constant of the NARX model is guaranteed to lie in between the time constants of the two LLMs and the NARX model yields reasonable results. However, Fig. 2 demonstrates that the operating point influences the gain and the time constant in a nonlinear manner. For example, at $\Phi_1 = \Phi_2 = 1/2$ the gain is not the mean of the the two LLM gains ($K_1 = 1$, $K_2 = 2$) but rather $K_{\text{int}} = 1.33$. Such a behavior, although undesirable, may be tolerated in practice. In contrast, the NOBF model behaves exactly as intuitively expected. Note that for this example, the order of the NOBF model is chosen as $n = 6$ and the Laguerre pole is chosen as $p = 0.7$. Worse choices for the Laguerre pole or lower NOBF model orders can lead to

[3] Note that for ***simulation*** it makes no difference whether the model structure is (N)ARX, (N)ARMAX, (N)OE, etc.

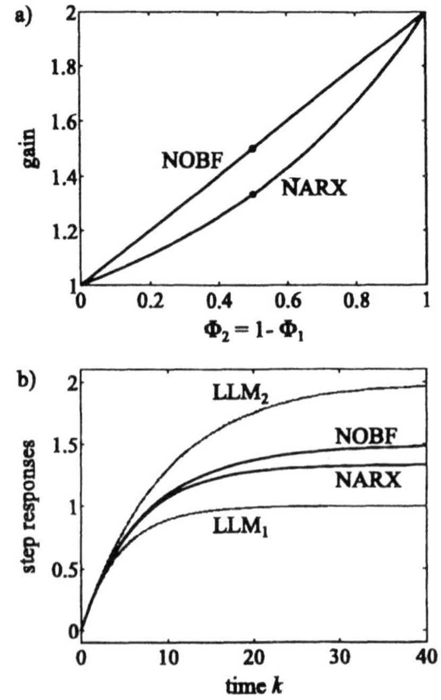

Fig. 2. a) Gain of the overall model dependent on the operating point. b) Step responses of two LLMs and the interpolation of the corresponding ARX and OBF models with $\Phi_1 = \Phi_2 = 1/2$. LLM$_1$: $G_1(q) = 0.2/(q-0.8)$, LLM$_2$: $G_2(q) = 0.2/(q-0.9)$.

visible deviations between the LLM dynamics and their OBF approximations. However, the major *characteristics* of the interpolation behavior is not affected by this.

3.2 *Second Order Systems*

For second order NARX models, the situation becomes worse. Figure 3 illustrates that the interpolation of two second order ARX models with real poles may yield (dependent on the operating point) an overall model with a conjugate complex pole pair. Note that this effect only occurs if both poles of one LLM are slower/faster than both poles of the other LLM. Although the imaginary part of this pole pair is the smaller the closer the LLM poles are to each other, such a change in the character of the dynamics is highly undesirable. It is caused by the fact that the parameters are interpolated and not the poles. In (Chow *et al.*, 1995) it is proposed to overcome this problem by interpolating the poles directly instead of the parameters. However, it is unclear which poles of each LLM correspond to each other and thus should be chosen for mutual interpolation, in particular for higher order models. Different choices yield different results. For NOBF models all these difficulties do not exist since parameter interpolation is equivalent to an interpolation in the time domain.

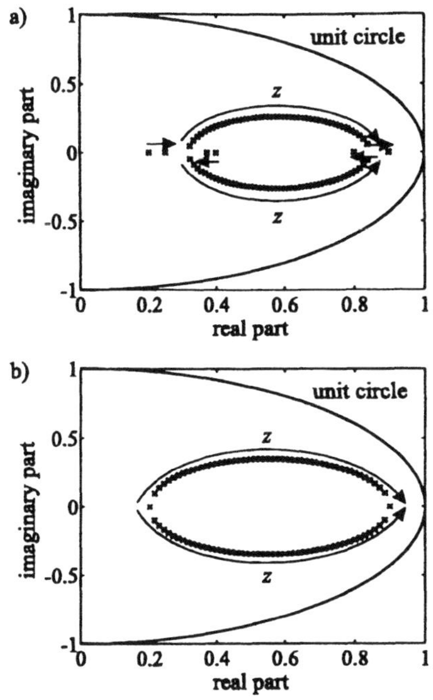

Fig. 3. Poles of the overall NARX model dependent on the operating point. a) LLM_1: $p_{1/2} = 0.2/0.4$, LLM_2: $p_{1/2} = 0.7/0.9$. b) LLM_1: $p_{1/2} = 0.2$, LLM_2: $p_{1/2} = 0.9$.

3.3 *Higher Order Systems*

For NARX models with the order higher than two, the situation becomes even worse. While for second order NARX models stability is guaranteed if both LLMs are stable (and the scheduling variable z is an external signal, i.e. no state of the model), this is not the case for higher order NARX models. Figure 4 explains the reason for this by illustrating that the stability region for second order models is convex but for third order models it becomes nonconvex. These stability regions in the parameter space can be easily calculated by the Schur-Cohn-Jury criterion (Isermann, 1991). Again, NOBF models do not face this problem.

4. DYNAMIC EFFECT OF THE SCHEDULING VARIABLES

Local linear model networks offer the important feature of distinct input spaces for the local linear models (given by $\underline{x}$) and for the validity functions that define the operating point (given by $\underline{z}$). Thus, these networks can be seen as an extension of the gain or parameter scheduling approach. In fact, if $\underline{z}$ does not contain any states of the model then it is equivalent to gain or parameter scheduling. As in the previous section, such a simple case shall be considered in the sequel.

For the NARX model, the scheduling variable defining the operating point can be directly used as the input for the validity functions. Assume the

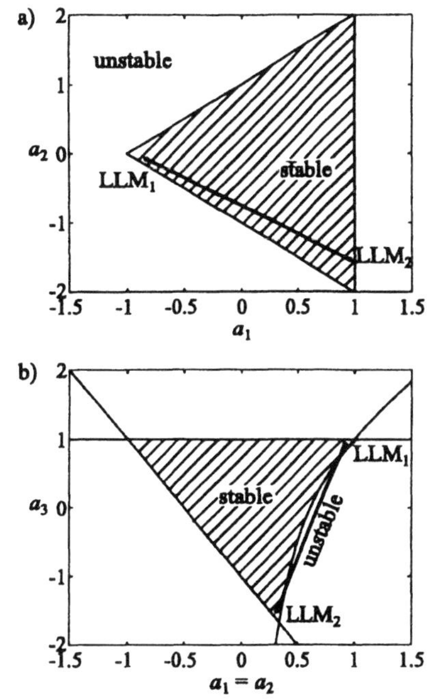

Fig. 4. Stability regions for the NARX model in the parameter space. The thick lines represent the interpolated parameters for $\Phi_1 = 0 \ldots 1$. a) For a second order model the interpolated denominator parameters stay in the stability region. b) For a third order model they can lie outside the stability region (here shown for $a_1 = a_2$).

following scenario. The process and the model are in steady state. Then, the operating point changes suddenly to a different value, i.e., $\underline{z}$ changes stepwise activating other LLMs in the model. As a consequence, the NARX model performs a transient until it settles to its new steady-state, in spite of the fact that the operating point dependent parameters $b_i(\underline{z})$, $a_i(\underline{z})$, $\xi(\underline{z})$ in (9) change suddenly. The reason for this transient is the output feedback within the NARX model. The dynamics of the transient depend on the parameters of activated LLMs.

Now, consider the same scenario for an NOBF model. Obviously, a sudden change in the scheduling variable causes an equivalently fast change in the model output because no model output feedback exists. This can cause unrealistic spikes in the model output. In order to achieve a similar transient as in the NARX model case, it is necessary to filter the scheduling variable. The choice of this filter is an open issue. Preliminary simulation studies indicate that it might be in many situations sufficiently accurate to filter with the first OBF $L_1(q)$ which is assumed to roughly capture the process dynamics.

The effect of scheduling variables on the dynamics is an important issue and requires further investigation in the future. It also arises in the identi-

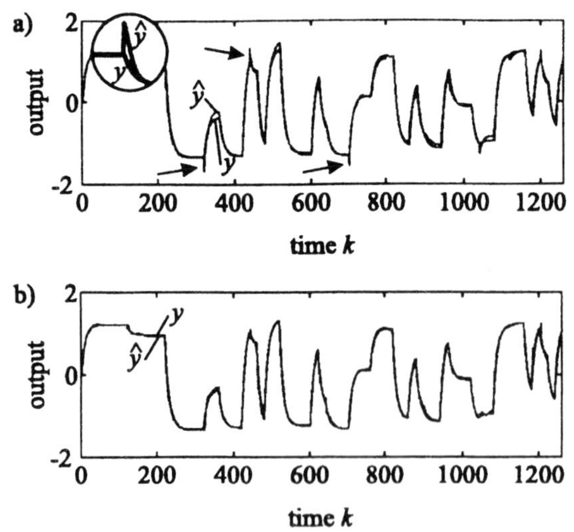

Fig. 5. Local linear Laguerre model network for a Hammerstein process with unfiltered (a) and filtered (b) scheduling variable (which for the Hammerstein structure is equivalent to the input). The arrows indicate peaks due to the sudden parameter change.

fication of a process with the common Hammerstein structure consisting of a static nonlinearity followed by a linear dynamic system. For such processes, the nonlinearity depends only on the process input u. Identification of a NARX model e.g. with the LOLIMOT algorithm indeed verifies that it is sufficient to choose the input as scheduling variable, see (Nelles, 1999). Nevertheless, for a NOBF model the input must not be used as a scheduling variable without pre-filtering, see Fig 5.

5. SUMMARY

It has been pointed out why local linear model networks based on orthonormal basis functions (OBFs) are a promising new approach for nonlinear system identification. They represent linearly parameterized output error models with inherent stability. They require and exploit prior knowledge about the approximate process dynamics for the design of the OBFs. The dynamics of interpolated local linear ARX/OE and OBF models has been studied and some serious shortcomings for NARX/NOE compared to NOBF models have been demonstrated. Finally, the different effects that scheduling variables have in the NARX/NOE and NOBF models' dynamics respectively have been investigated.

6. REFERENCES

Babuška, R. and H.B. Verbruggen (1996). An overview of fuzzy modeling for control. *Control Engineering Practice* **4**(11), 1593–1606.

Back, A.D. and A.C. Tsoi (1996). Nonlinear system identification using discrete laguerre functions. *Journal of Systems Engineering* **6**(3), 194–207.

Chow, C.-M., D.W. Clarke and A.G. Kuznetsov (1995). Using multiple models in predictive control. In: *European Control Conference.* Rome, Italy. pp. 1732–1737.

Heuberger, P.S.C., P.M.J. Van den Hof and O.H. Bosgra (1995). A generalized orthonormal basis for linear dynamical systems. *IEEE Transactions on Automatic Control* **40**(3), 451–465.

Hunt, K.J., R. Haas and R. Murray-Smith (1996). Extending the functional equivalence of radial basis functions networks and fuzzy inference systems. *IEEE Transactions on Neural Networks* **7**(3), 776–781.

Isermann, R. (1991). *Digital Control Systems.* Springer. Berlin.

Johansen, T.A. and B.A. Foss (1993). Constructing NARMAX models using ARMAX models. *International Journal of Control* **58**(5), 1125–1153.

Murray-Smith, R. (1994). A Local Model Network Approach to Nonlinear Modeling. PhD thesis. University of Strathclyde. Strathclyde, UK.

Nelles, O. (1997). Orthonormal basis functions for nonlinear system identification with local linear model trees (LOLIMOT). In: *IFAC Symposium on System Identification.* Kitakyushu, Fukuoka, Japan. pp. 667–672.

Nelles, O. (1999). *Nonlinear System Identification with Local Linear Neuro-Fuzzy Models.* Automatisierungstechnik series. Shaker Verlag, ISBN: 3-8265-4880-9. Aachen, Germany.

Sbarbaro, D. (1997). Local Laguerre models. In: *Multiple Model Approaches to Modelling and Control* (R. Murray-Smith and T.A. Johansen, Eds.). Chap. 10, pp. 257–268. Taylor and Francis, London.

Schram, G., A. Krijgsman and M.H.G. Verhaegen (1996). System identification with orthogonal basis functions and neural networks. In: *IFAC World Congress.* San Francisco, USA. pp. 221–226.

Tanaka, K. and M. Sugeno (1992). Stability analysis and design of fuzzy control systems. *Fuzzy Sets & Systems* **45**(2), 135–156.

Van den Hof, P.M.J., P.S.C. Heuberger and J. Bokor (1995). System identification with generalized orthonormal basis functions. *Automatica* **31**(12), 1821–1834.

Wahlberg, B. (1991). System identification using Laguerre models. *IEEE Transactions on Automatic Control* **36**(5), 551–562.

BACKWARD CENTRE SELECTION METHOD FOR ON-LINE ADAPTATION OF RBF NETWORK MODELS

J. B. Gomm and D. L. Yu

Control Systems Research Group, School of Engineering,
Liverpool John Moores University, Byrom Street, Liverpool L3 3AF, U.K.
E-mail j.b.gomm@livjm.ac.uk

Abstract: An approach to tracking process time variations using an adaptive radial basis function network model is described. The method is based on a numerically robust recursive algorithm for updating the network output layer weights. It is shown how network centres contributing least to the network output can be found and removed from model calculations. Thus, both the structure and weights of the network are adaptive. An illustrative example is given, to demonstrate the effectiveness of the algorithm and illustrate its performance, in an application to modelling a real chemical process. Results show more accurate model predictions compared to using a network with only weight updating. *Copyright © 2000 IFAC.*

Keywords: neural-network models, process identification, radial basis function networks, recursive least squares, self-adapting algorithms, time-varying plants

1. INTRODUCTION

In recent years, applications of neural networks to non-linear process modelling and control have been intensively investigated. Several neural control strategies that have been studied, including model reference, inverse and internal model control, are reviewed in Hunt, *et al.* (1992). Some of these control techniques rely on a non-linear process model (usually represented by a neural network) within the scheme. Only a fixed neural network model is typically considered. However, it is realised that, based on a fixed neural model, robust control is difficult to maintain for some systems due to their time-varying nature. One strategy for adaptive neural control (to cope with process time-variations) with these model based schemes, is to develop an adaptive neural network model that can be updated on-line.

Radial basis function neural networks (RBFNs) are particularly suited to on-line adaptation because their linear feature of the weights relating to the network output allows linear optimisation methods, e.g. least squares, to be applied. Algorithms for recursive adaptation of the parameters (centres and weights) of an RBFN with a fixed structure have been developed for process identification (Chen, *et al.*, 1992). However, only updating the weights and/or centre positions restricts the full utilisation of RBFNs. Adaptation of both the structure and weights would enable RBFNs to model time-varying and severely non-linear processes with reasonable network size. Some off-line adaptation techniques have been proposed for RBFNs to achieve an optimal structure (number and location of centres) for a certain data set (e.g. Chen, *et al.*, 1991; Karayiannis and Mi, 1997). However, the structure will not be optimal when the process or environment changes. Therefore, on-line adaptation of both structure and parameters of an RBFN for use with control is needed.

Orthogonal decomposition is a well known numerically robust technique to eliminate ill-conditioning effects in solving least squares problems. Chen, *et al.* (1991) showed how batch orthogonal least squares (OLS) could be used to develop RBFNs by sequentially selecting centres based on evaluating an error reduction ratio in a forward selection algorithm. Yu, *et al.* (1997a) developed a recursive OLS (ROLS) algorithm for training the output layer weights in multi-input, multi-output (MIMO) RBFNs. Results showed that the ROLS algorithm can be applied to a large training data set with a low requirement on computer memory, and can eliminate ill-conditioning in the data set, so that the accuracy of an RBFN model is greatly improved. The batch RBFN training method of Chen, *et al.* (1991) was extended by Luo, *et al.* (1996), where a forward centre selection algorithm, using the same error reduction ratio, based on ROLS was used for on-line structure and parameter adaptation of RBFNs. More recently, Gomm and Yu (2000) derived both forward and backward centre selection algorithms for the MIMO case within the ROLS framework. The algorithms were applied in off-line network training and, in the applications reported, smaller RBFNs with acceptable accuracy were obtained with backward selection.

This paper describes a recursive implementation of the backward selection algorithm based on ROLS for on-line adaptation of the structure and weights of an RBFN. Operation and performance of the algorithm is demonstrated in an application to training an RBFN sub-system model of a real MIMO chemical reactor.

2. RBFN ADAPTATION ALGORITHM

The RBFN adaptation algorithm is described in two parts. ROLS updating of the network output weights is first described, followed by the backward centre selection method. Finally, a summary of the adaptation algorithm is given.

2.1 Recursive Orthogonal Least Squares Training of RBFNs

A standard RBFN performs a non-linear mapping $x \in \Re^m \rightarrow \hat{y} \in \Re^p$ via the linear output transformation,

$$\hat{y}^T(t) = \phi^T(t,x)W \tag{1}$$

where $\hat{y}(t)$ and $x(t)$ are the network output and input vectors at sample time t respectively; $W \in \Re^{n_h \times p}$ is the weighting matrix with element w_{ij} denoting the weight connecting the ith hidden node output to the jth network output; $\phi(t) \in \Re^{n_h}$ is the output vector of the non-linear basis functions in the hidden layer with the ith element $\phi_i(t)$. In this work, the weighting matrix, W, is computed to minimise the error, $\varepsilon(t) = y(t) - \hat{y}(t)$ where y(t) is the target network output, using an ROLS algorithm. The form of the ROLS algorithm used here is applied to determining the full weighting matrix W, and not the increment in the weighting matrix as reported in Yu, *et al.* (1997a). The algorithm is therefore given below.

Considering (1) for a set of N input-output training data, we have

$$Y = \hat{Y} + \mathrm{E} = \Phi W + \mathrm{E} \tag{2}$$

where $Y \in \Re^{N\times p}$ is the desired output matrix; $\hat{Y} \in \Re^{N\times p}$ is the neural network output matrix; $\Phi \in \Re^{N\times n_h}$ is the hidden layer output matrix; $\mathrm{E} \in \Re^{N\times p}$ is the error matrix and

$$Y^T = [y(1),\cdots,y(N)],\ \hat{Y}^T = [\hat{y}(1),\cdots,\hat{y}(N)],$$
$$\Phi^T = [\phi(1),\cdots,\phi(N)],\ \mathrm{E}^T = [\varepsilon(1),\cdots,\varepsilon(N)].$$

Now a recursive MIMO least squares problem can be formulated to solve W(t) at each stage t such that the following time-varying cost function is minimised

$$J(t) = \left\| \begin{bmatrix} \lambda Y(t-1) \\ \cdots \\ y^T(t) \end{bmatrix} - \begin{bmatrix} \lambda\Phi(t-1) \\ \cdots \\ \phi^T(t) \end{bmatrix} W(t) \right\|_F \tag{3}$$

where the *F-norm* of a matrix is defined as $\|A\|_F^2 = trace(A^T A)$. Eqn (3) includes exponential forgetting when the forgetting factor λ<1. It can be shown (Bobrow and Murray, 1993; Yu, *et al.*, 1997a) that minimising (3) is equivalent to minimising the following cost function

$$J(t) = \left\| \begin{bmatrix} \lambda \hat{Y}(t-1) \\ \cdots \\ y^T(t) \end{bmatrix} - \begin{bmatrix} \lambda R(t-1) \\ \cdots \\ \phi^T(t) \end{bmatrix} W(t) \right\|_F \tag{4}$$

where R is an $n_h \times n_h$ upper triangular matrix. R and $\hat{Y}$ are computed by an orthogonal decomposition as follows

$$\begin{bmatrix} \lambda R(t-1) \\ \cdots \\ \phi^T(t) \end{bmatrix} = Q(t) \begin{bmatrix} R(t) \\ \cdots \\ 0 \end{bmatrix},$$
$$\begin{bmatrix} \hat{Y}(t) \\ \cdots \\ \eta^T(t) \end{bmatrix} = Q^T(t) \begin{bmatrix} \lambda \hat{Y}(t-1) \\ \cdots \\ y^T(t) \end{bmatrix} \tag{5}$$

where Q is an orthogonal matrix. This gives, from (4) and (5), the revised cost function

$$J(t) = \left\| Q(t) \begin{bmatrix} \hat{Y}(t) \\ \cdots \\ \eta^T(t) \end{bmatrix} - Q(t) \begin{bmatrix} R(t) \\ \cdots \\ 0 \end{bmatrix} W(t) \right\|_F = \left\| \begin{bmatrix} \hat{Y}(t) - R(t)W(t) \\ \cdots \\ \eta^T(t) \end{bmatrix} \right\|_F \quad (6)$$

since the *F-norm* of a matrix is preserved by an orthogonal transformation. Hence, the optimal $W(t)$ in (6) can be solved from

$$R(t)W(t) = \hat{Y}(t) \quad (7)$$

which leaves the residual at stage t as $\left\| \eta^T(t) \right\|_F$.

Recall that $R(t)$ is an upper triangular matrix and therefore $W(t)$ can be easily solved from (7) by backward substitution. The decomposition in (5) can be achieved efficiently by applying Givens rotations to obtain the transformation (e.g. Bobrow and Murray, 1993)

$$\begin{bmatrix} \lambda R(t-1) & \lambda \hat{Y}(t-1) \\ \phi^T(t) & y^T(t) \end{bmatrix} \rightarrow \begin{bmatrix} R(t) & \hat{Y}(t) \\ 0 & \eta^T(t) \end{bmatrix} \quad (8)$$

The procedure of the ROLS algorithm is therefore the following: at stage t, calculate $R(t)$ and $\hat{Y}(t)$ according to (8), then solve $W(t)$ in (7). Initial values for R and $\hat{Y}$ can be assigned as $R(0) = \alpha I$ and $\hat{Y}(0) = 0$, where α is a small positive number, e.g. 0.01.

2.2 *Backward Selection of Network Centres*

The triangular system of (7) contains important information with regard to the learned network operation. In particular, the increase in the residual for the whole training data set up to time t caused by removing a particular network centre (or, equivalently, hidden node), j, can be calculated efficiently as follows (Gomm and Yu, 2000). Remove the jth column from R(t) then re-triangularise the system by the orthogonal decomposition

$$R' = Q_j \begin{bmatrix} R_j \\ \cdots \\ 0 \end{bmatrix}, \quad \begin{bmatrix} \hat{Y}_j \\ \cdots \\ \eta_j^T \end{bmatrix} = Q_j^T \hat{Y}(t) \quad (9)$$

where R' is R(t) with the jth column removed. This system then minimises the following cost function

$$J_j = \left\| Q_j \begin{bmatrix} \hat{Y}_j \\ \cdots \\ \eta_j^T \end{bmatrix} - Q_j \begin{bmatrix} R_j \\ \cdots \\ 0 \end{bmatrix} W_j \right\|_F = \left\| \begin{bmatrix} \hat{Y}_j - R_j W_j \\ \cdots \\ \eta_j^T \end{bmatrix} \right\|_F \quad (10)$$

Hence, the increase in the residual caused by removing the jth centre is $\left\| \eta_j^T \right\|_F$. Similar to the ROLS algorithm, the orthogonal decomposition in (9) is computed efficiently by applying Givens rotations to obtain the transformation

$$\begin{bmatrix} R' & \hat{Y}(t) \end{bmatrix} \rightarrow \begin{bmatrix} R_j & \hat{Y}_j \\ 0 & \eta_j^T \end{bmatrix} \quad (11)$$

The above analysis provides a method for optimally pruning an RBFN by removing one centre at each stage so as to retain the highest achievable accuracy with the remaining centres. Equation (11) are repeated for each centre in turn and the centre contributing the least to the output accuracy is removed from the network. The procedure is repeated at each stage of the pruning process. Note that the method does not require re-training of the pruned network, since the remaining triangular system after pruning, $R_j W_j = \hat{Y}_j$ from (10), provides the weight solution to (4) with the jth centre removed, and is identical to that which would be obtained if re-training up to time t using the ROLS algorithm were done.

The final element of the backward centre selection is a method to determine when to stop the algorithm. The approach used in this work is based on the parsimony principle of finding a trade-off between model complexity and accuracy. Several metrics to measure this balance for linear models have been applied to RBFNs (Pottmann and Seborg, 1992). In this research Akaike's Final Prediction Error (FPE) criterion is used because it is simple to calculate compared to alternative metrics,

$$FPE = \frac{N + n_p}{N - n_p} V(t) \quad (12)$$

where the loss function, V(t), is the mean squared error between the actual and model outputs, n_p is the total number of network output layer weights and N is the number of data points used. With exponential forgetting, N was approximated in this research by N=1/(1-λ), the asymptotic memory length of the ROLS estimator. An estimate of V(t) for the network is easily obtained during training with the ROLS algorithm from (6) as:

$$V(t) = \lambda V(t-1) + \left\| \eta^T(t) \right\|_F / N \qquad (13)$$

with the initial value $V(0) = 0$. Thus, before the removal of each network centre, the pruned network FPE can be calculated from the previous network loss function, the increase in the loss function caused by removing the hidden node from (13) and the number of weights in the pruned network. Network pruning is then stopped when the FPE increases indicating a balance between complexity and accuracy has been met.

The above selection algorithm is implemented at each sampling instant thus, adapting the network structure. In this mode, centres are not actually removed from the network but re-ordered, because a centre which is not significant at one point in time may be significant later. After centre selection at each iteration, the matrix R(t) is rearranged so that the first columns correspond to the final centres chosen by the selection procedure.

3. APPLICATION TO MODELLING A CHEMICAL PROCESS

The methods described in Section 2 are applied to RBFN modelling of a chemical process. The performance of the adaptive RBFN model is investigated. The process is first described followed by the RBFN modelling.

3.1 *The Chemical Process*

The process used in this research is a pilot, chemical reactor system established in the laboratory to generally represent the dynamic behaviour of real chemical processes in industry. A schematic of the chemical reactor is shown in Fig.1. It consists of a continuously stirred tank to which the chemical solutions, ammonium hydroxide (NH_4OH), acetic acid (CH_3COOH) and sodium sulphite ($N_{a2}SO_3$), and air are added. The volume of liquid in the reactor is maintained at a pre-specified constant level by an outflow pump system. The concentrations and flow rates of CH_3COOH and $N_{a2}SO_3$ are constant, while the flow rates of NH_4OH and air are adjustable to control the pH and the dissolved oxygen in the

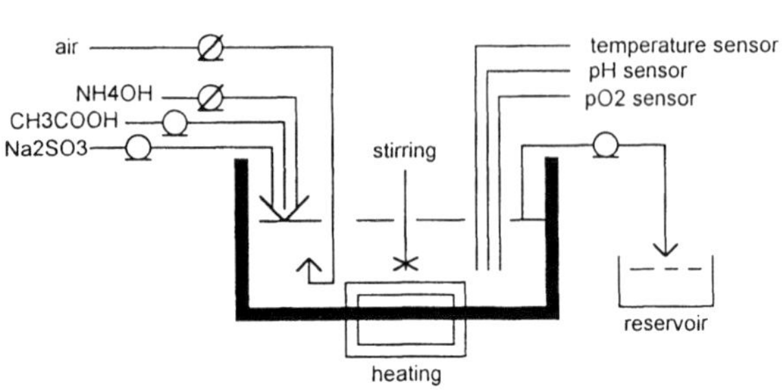

Fig. 1. The chemical reactor process.

reactor. The liquid temperature is controlled by a heating system. With the three inputs, the heating power (Q), the flow rate of NH_4OH (f_b) and the flow rate of air (f_a), and the three outputs, liquid temperature (T), pH and percentage of dissolved oxygen (pO_2), the process is a MIMO, non-linear system with complex, interactive dynamics.

3.2 *Adaptive RBFN Modelling*

In previous research a fixed neural network model of the chemical process studied has been developed. It was observed that the dissolved oxygen exhibited time-varying behaviour and the accuracy of the fixed model deteriorated during the experiments. Hence, the modelling of this process variable is considered in this section to illustrate the adaptive RBFN.

For neural network modelling of a non-linear, dynamical system, a widely adopted approach is to train the network to predict the system output at sample t based on measurements obtained at previous samples (e.g. Chen, *et al.*, 1992; Pottmann and Seborg, 1992; Gomm, *et al.*, 1996). Following some further investigations on the chemical reactor process, the system sample interval was chosen to be 10 seconds based on examining the rise times for the output variables. The RBF network inputs were chosen as

$$x(t) = [T(t-1), pH(t-1), pO_2(t-1), \\ pO_2(t-2), f_a(t-1)]^T$$

using the model order and time-delay selection method of Yu, *et al.* (1997b). The network output corresponded to process output $pO_2(t)$.

There are several possible choices for the basis function, $\phi_i(t)$, in the RBFN, such as Gaussian or multiquadric. The thin-plate-spline was chosen in this work because, unlike other possible basis functions including the Gaussian, this basis function does not require the choice of an additional width parameter. The thin-plate-spline function is defined as

$$\phi_i(t) = d_i^2(t) \log_e[d_i(t)] \quad i = 1, \ldots, n_h \qquad (14)$$

$$d_i(t) = \|x(t) - c_i\|_2 \quad i = 1, \ldots, n_h \qquad (15)$$

where $c_i \in \Re^m$ is the ith centre vector; n_h is the number of nodes in the hidden layer; $d_i(t)$ is the Euclidean distance of the input vector to the ith hidden node centre.

The adaptive RBFN was tested with 1000 input-output data points collected from the process, which were normalised to have zero mean and unit variance. The RBFN model had 5 inputs and 1 output, using the input-output structure described above, and the

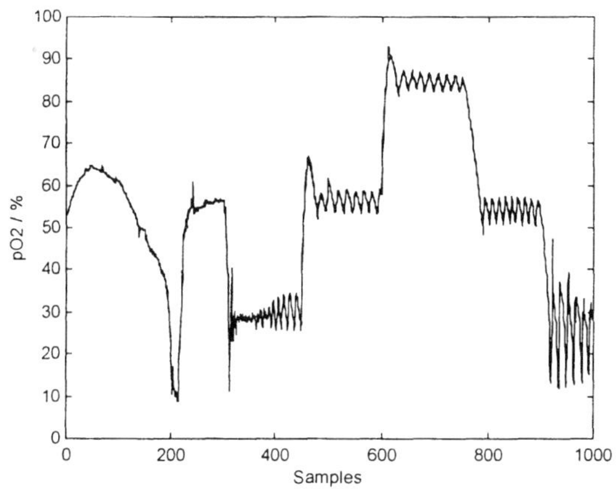

Fig. 2. Adaptive RBFN model and process outputs.

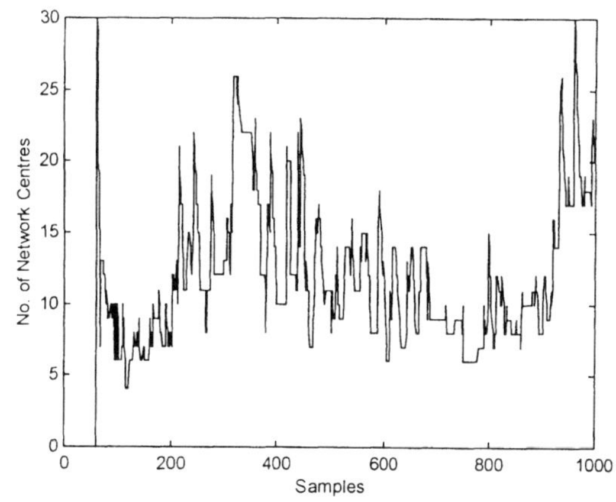

Fig. 4. Variation of the number of centres chosen in the adaptive RBFN model

number of hidden layer nodes was chosen to be 30. The network centres were chosen to randomly cover the input space in the range [-2,2]. A forgetting factor of λ=0.99 was used in the ROLS algorithm.

The one-step-ahead predicted output of the adaptive RBFN model, $\hat{y}(t) = \phi^T(t)W_s(t-1)$ where W_s are the weights corresponding to the chosen network centres, was in good agreement with the process output as shown in Fig.2. Fig.3 shows the one-step-ahead prediction errors for the adaptive structure RBFN and the RBFN using all 30 centres, i.e. only the output layer weights are adapted. It can be seen that the adaptive structure errors are generally smaller than using the full structure, which demonstrates the benefit of adapting both the structure and weights of the network. This is also confirmed in the variances of the prediction errors which were 6.36 for the full structure and 1.93 for the adaptive structure. Fig.4 shows the number of centres chosen in the network by backward selection. The number of centres used in the adaptive RBFN are often less than half of the total number of centres available (thirty).

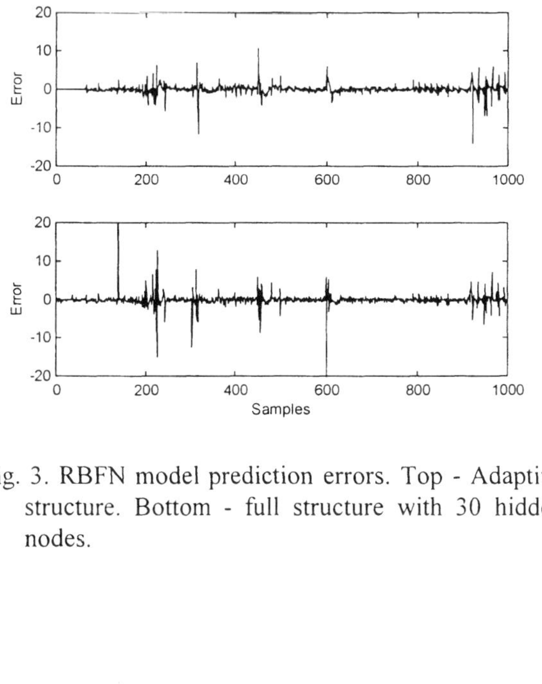

Fig. 3. RBFN model prediction errors. Top - Adaptive structure. Bottom - full structure with 30 hidden nodes.

4. CONCLUSIONS

A backward selection method based on an ROLS weight update algorithm was described for recursively selecting the number of centres in RBFN models. An adaptive RBFN was applied to modelling a real chemical reactor process using the developed algorithm. The adaptive model achieved good accuracy in predicting the process output, with reduced numbers of network centres. The prediction accuracy was better than an RBFN using the full set of candidate centres, i.e. only weight updating. Thus, an improved performance with both structure and weight adaptation was demonstrated.

Further work is to develop techniques for automatically adding and removing network centres. Also, to investigate the performance of a model based control scheme incorporating the adaptive RBFN.

REFERENCES

Bobrow, J.E. and W. Murray (1993). An algorithm for RLS identification of parameters that vary quickly with time. *IEEE Trans on Automatic Control*, **38**(2), 351-354.

Chen, S., C.F.N. Cowan and P.M. Grant (1991). Orthogonal least squares learning algorithm for radial basis function networks. *IEEE Trans on Neural Networks*, **2**, 302-309.

Chen, S., S.A. Billings and P.M. Grant (1992). Recursive hybrid algorithm for non-linear system identification using radial basis function networks. *Int. J. Control*, **55**, 1051-1070.

Gomm, J.B., D. Williams, J.T. Evans, S.K. Doherty, and P.J.G. Lisboa (1996). Enhancing the non-linear modelling capabilities of MLP neural networks using spread encoding. *Fuzzy Sets and Systems*, **79**(1), 113-126.

Gomm, J.B. and D.L. Yu (2000) Selecting radial basis function network centres with recursive orthogonal least squares training. *IEEE Trans on Neural Networks*, In press.

Hunt, K.J., D. Sbarbaro, R. Zbikowsky and P.J. Gawthrop (1992). Neural networks for control systems - a survey. *Automatica*, **28**(6), 1083-1112.

Karayiannis, N.B. and G.W. Mi (1997). Growing radial basis neural networks: merging supervised and unsupervised learning with network growth techniques. *IEEE Trans on Neural Networks*, **8**(6), 1492-1506.

Luo, W., M.N. Karim, A.J. Morris and E.B. Martin (1996). Control relevant identification of a pH waste water neutralisation process using adaptive radial basis function networks. *Computers Chem. Engng*, **20**, Suppl., S1017-S1022.

Pottmann, M. and D.E. Seborg (1992). Identification of non-linear processes using reciprocal multiquadric functions. *J. Proc. Cont.*, **2**(4), 189-203.

Yu, D.L., J.B. Gomm and D. Williams (1997a). A recursive orthogonal least squares algorithm for training RBF networks. *Neural Processing Letters*, **5**(3), 167-176.

Yu, D.L., J.B. Gomm and D. Williams (1997b). Selection of neural model order and time-delay for MIMO non-linear systems and a case study on a CSTR process. *Proc. IFAC Symposium SYSID'97*, Fukuoka, Japan, **1**, 213-218.

COMPARISON OF A HIERARCHICALLY CONSTRUCTED NEURAL NETWORK AND A HIERARCHICAL LOOK-UP TABLE

Susanne Töpfer, Oliver Nelles, Rolf Isermann

Darmstadt University of Technology
Institute of Automatic Control, Landgraf-Georg-Strasse 4
D-64283 Darmstadt, Germany
Phone: (+49 6151) 16-4524
E-Mail: stoepfer@iat.tu-darmstadt.de

Abstract: Two black-box models, LOLIMOT and hierarchical look-up tables, are presented which are well suited to deal with identification of complex nonlinear systems. LOLIMOT is a local linear neuro-fuzzy model which preferably can be used for off-line simulation. Hierarchical look-up tables represent an extension of classical look-up tables and allow the on-line simulation of the identified nonlinear systems under time-critical conditions. Both models share a recursive partitioning approach for solving the structure identification problem. *Copyright ©2000 IFAC*

Keywords: identification, hierarchy, recursive partitioning, neuro-fuzzy, look-up tables

1. INTRODUCTION

Static and dynamic nonlinear models of real processes are required in many disciplines, i.e. for system analysis, controller design, model-based control, process optimization, supervision and fault diagnosis. According to the considered application, models have to predict or simulate nonlinear characteristics of processes or control strategies in off-line or real-time (on-line) operation. Real-time applications are characterized by increasing demands due to the model implementation under limited storage capacity and computational resources.

Three different modeling approaches can be distinguished: theoretical modeling (white-box models), experimental modeling (black-box models) and a combination of these (grey-box models). In this paper two black-box models will be presented. That means their parameter and structure identification is solely based on measurement data. Generally, parameters and structures of black-box models have no direct relationship to the underlying first principles. Such model architectures need a carefully chosen complexity to reasonably fit the complexity of the nonlinear process behaviour. On one hand, the model quality with respect to the training data increases with a growing model size. However, generalization is poor in case of too many model parameters (overfitting). On the other hand, model structures with too few parameters are unable to capture the whole nonlinearity of the process to be identified (underfitting). The problem of balancing over- and underfitting is known as *bias-variance trade-off* and can be approached through the optimization of the model complexity. In order to accomplish this task, different structure optimization and regularization techniques can be distinguished. One elegant way is the usage of *constructive* structure optimization approaches for local model architectures. *Local model approaches* are based on the decomposition of the input/output data into smaller subsets. These homogeneous subsets are used for the development of simple local submodels where each submodel is valid in a specified part of the operating range. The before mentioned constructive techniques increase the number of submodels iteratively. Simultaneously, the input space is recursively partitioned into a hierarchically ordered manner as long as a suitable complexity is reached. This procedure leads to effective identification techniques which are superior to basis-function expansions with respect to training speed, minimum model structure and parameter interpretability. Well known hierarchically constructed local model approaches are for instance CART (Breiman *et al.*, 1984), MARS (Friedman, 1991), Basis Function Trees (Sanger, 1991) and k-d Trees (Bently, 1975).

In the next two sections two hierarchically constructed black-box models will be presented. Section 2 describes the LOLIMOT algorithm. This hierarchical optimization algorithm constructs local linear neuro-fuzzy models. In Section 3 the concept of hierarchical look-up tables is presented. While LOLIMOT constructs a neuro-fuzzy model whose evaluation is rather computationally expensive since it requires exponential functions the hierarchical look-up table allows the implementation on low-cost hardware. Both similar approaches will be compared in Section 4. Section 5 contains identification results with respect to an artificial nonlinear function.

2. LOLIMOT

The network structure of a local linear neuro-fuzzy model is as follows. Each neuron realizes a local linear model (LM) and an associated validity function that determines the region of validity of the LM. The validity functions form a partition of unity, i.e., they are normalized such that

$$\sum_{i=1}^{M} \Phi_i(\underline{u}) = 1 \qquad (1)$$

for any model input $\underline{u} = [u_1 \; u_2 \; \ldots u_p]^T$. The output of a local linear neuro-fuzzy model is calculated as

$$\hat{y} = \sum_{i=1}^{M} (w_{i,0} + w_{i,1}u_1 + \ldots + w_{i,p}u_p)\, \Phi_i(\underline{u}) \quad (2)$$

Thus, the network output is calculated as a weighted sum of the outputs of the local linear models where the $\Phi_i(\cdot)$ are interpreted as the operating point dependent weighting factors. The network interpolates between different LMs with the validity functions. The weights $w_{i,j}$ are *linear* network parameters.

The validity functions are typically chosen as normalized Gaussians. If these Gaussians are furthermore axis-orthogonal the validity functions are

$$\Phi_i(\underline{u}) = \frac{\mu_i(\underline{u})}{\sum\limits_{j=1}^{M} \mu_j(\underline{u})} \qquad (3)$$

with

$$\mu_i(\underline{u}) = \prod_{j=1}^{p} \exp\left(-\frac{1}{2}\left(\frac{(u_j - c_{i,j})^2}{\sigma_{i,j}^2}\right)\right) \qquad (4)$$

The centers and standard deviations are *nonlinear* network parameters.

The LOLIMOT algorithm consists of an outer loop in which the input space is decomposed by determining the parameters of the validity functions and a nested inner loop in which the parameters of the local linear models are optimized by local weighted least squares estimation (Nelles, 1999), (Nelles and Isermann, 1996).

The input space is decomposed in an axis-orthogonal manner yielding (hyper-)rectangles in whose center the Gaussian validity functions $\mu_i(u)$ are placed. The standard deviations of these Gaussians are chosen proportionally to the extensions of (hyper-) rectangles to account for the varying granularity. Thus, the nonlinear parameters $c_{i,j}$ and $\sigma_{i,j}$ are determined by a heuristic avoiding explicit nonlinear optimization. LOLIMOT starts with a single linear model that is valid in the whole input space. In each iteration, it splits one local linear model into two new submodels. Only the (locally) worst performing LM is considered for further refinement. Splits along all input axes are compared and the best performing alternative is carried out.

The idea behind the local estimation approach is to consider the optimization of the local model parameters as individual problems. The parameters for each local linear model are estimated *separately* neglecting the interactions between the local models (Murray-Smith and Johansen, 1995). This reduces the variance error and increases the bias error of the model and thus can be seen as a regularization technique which makes the estimation robust against noise, for more details refer to (Murray-Smith and Johansen, 1995), (Nelles, 1999). The motivation for this approach stems from the case $\sigma_{i,j} \to 0$ in which no interaction between the different local models takes place. The output of the model for a given input is solely determined by a single local linear model while all others are inactive. In this case the local estimation approach is exactly equivalent to the global one. As the standard deviations are increased the interaction between neighbored local

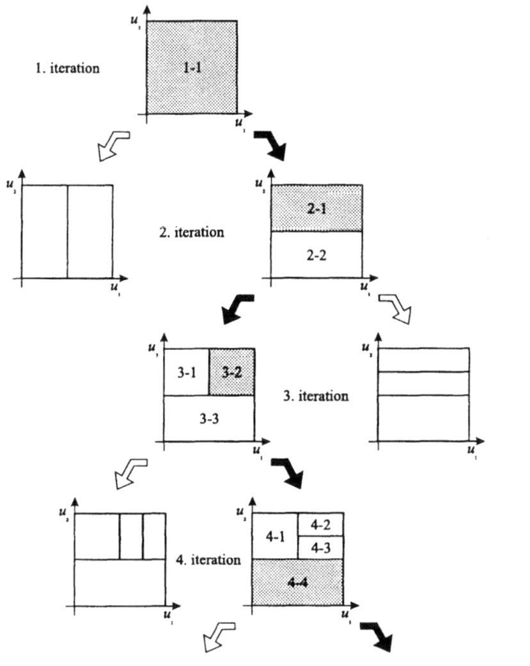

Fig. 1. Tree Construction of LOLIMOT-Algorithm

models grows and the error caused by their neglect increases.

Instead of estimating all $M(p+1)$ parameters simultaneously as it is done in the global estimation approach, M separate local estimations are carried out for the $p+1$ parameters of each local linear model. The parameter vector for each of these $i = 1, \ldots, M$ estimations is

$$\underline{w}_i = [w_{i,0} \; w_{i,1} \; \cdots \; w_{i,p}]^T . \qquad (5)$$

The corresponding regression matrices are (N denotes the number of data samples)

$$\underline{X}_i = \begin{bmatrix} 1 & u_1(1) & u_2(1) & \cdots & u_p(1) \\ 1 & u_1(2) & u_2(2) & & u_p(2) \\ \vdots & \vdots & \vdots & & \vdots \\ 1 & u_1(N) & u_2(N) & \cdots & u_p(N) \end{bmatrix} . \qquad (6)$$

Note that the regression matrices of all local linear models $i = 1, \ldots, M$ are identical since the elements of $\underline{X}_i$ do not depend on i. A local linear model with the output ($\underline{\hat{y}}_i = [\hat{y}_i(1) \; \hat{y}_i(2) \; \cdots \; \hat{y}_i(N)]^T$)

$$\underline{\hat{y}}_i = \underline{X}_i \, \underline{w}_i \qquad (7)$$

is valid only in the region where the associated validity function $\Phi_i(\cdot)$ is close to one. This will be the case close to the center of $\Phi_i(\cdot)$. Data in this region is highly relevant for the estimation of $\underline{w}_i$. As the validity function decreases, the data becomes less relevant for the estimation of $\underline{w}_i$ and more relevant for the estimation of the neighboring models. Consequently, it is straightforward to apply a weighted least squares optimization where the weighting factors are given by the validity function values, i.e.

$$J_i = \sum_{j=1}^{N} \Phi_i(\underline{u}(j)) \, e^2(j) \longrightarrow \min_{\underline{w}_i} \qquad (8)$$

where $e(j) = y(j) - \hat{y}(j)$ represent the model errors for data sample. For the extreme case, $\sigma_{i,j} \to 0$, $\Phi_i(\underline{u}(j))$ is either equal to 0 or 1, that is only a subset of the data is used for the estimation of $\underline{w}_i$. For $\sigma_{i,j} > 0$ all data samples are exploited for estimation but those whose validity value is close to 0 are virtually ignored.

3. HIERARCHICAL LOOK-UP TABLES

The concept of hierarchical decomposition of the input space can be extended to look-up table models too. That leads to the *hierarchical look-up table approach* which is comprehensively presented in (Ernst(Töpfer), 1999).

Classical grid-based look-up tables are the most common type of nonlinear static models for real-time applications. Especially in automotive engineering due to cost reasons computational resources are strongly restricted and therefore the very simple look-up table models are applied. Nonlinear modeling of these grid-based models is based on the interpolation between a number of data points or nodes positioned on a multi-dimensional lattice structure (Fig. 2 (a)). Each node comprises two components: the *data point position* and the corresponding scalar *data point height* or *weight* being an estimate of the approximated function at the data point position. Because all these nodes are positioned on an axis-orthogonal lattice structure this approach suffers from the curse of dimensionality problem. Therefore application of grid-based look-up tables is restricted to one- or two-dimensional input spaces. Additional input variables usually have to be incorporated in an additive or multiplicative manner.

Equivalent to these grid-based look-up tables, nonlinear modeling of hierarchical look-up tables is based on the interpolation between a set of data points too. However, hierarchical look-up tables allow a much more flexible data point distribution according to the complexity of nonlinearity in different regions of the operating range. The data points of the model are estimated into a hierarchically ordered manner, by a heuristic strategy similar to LOLIMOT. That means, in each iteration step one operating regime is axis-orthogonally partitioned into two hypercubes. New data points are positioned on the vertices of these resulting hypercubes (see Fig. 2 (b)).

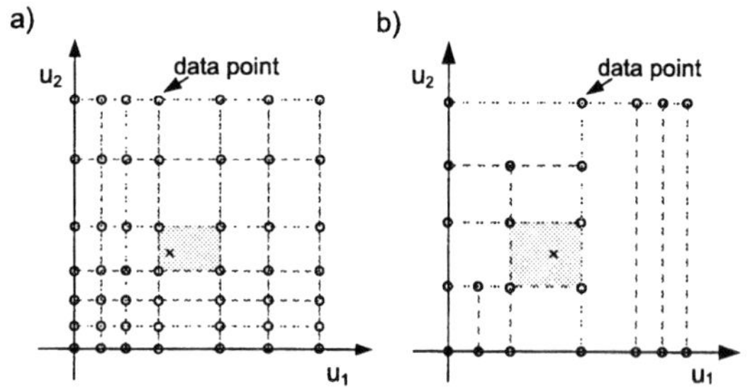

Fig. 2. Data point positioning (a) for grid-based look-up tables (b) for hierarchical look-up tables

The recursive partitioning is performed in each iteration step in the hypercube with the worst local error measure. From a candidate set of p possible cuts orthogonal to the p input space directions, the cut mostly decreasing the loss function value is selected. The corresponding data point weights can be found by linear optimization. However, since data point weights serve as model parameters within different hypercubes, these parameters are strongly coupled and can not be estimated locally. Therefore, after adding new nodes all weights $\underline{w}$ have to be reestimated.

$$J = \sum_{j=1}^{N} e^2(j) \longrightarrow \min_{\underline{w}} \tag{9}$$

Like in the grid-based look-up table model case, model output at data point positions corresponds exactly to their data point heights. For input vectors inside of a hypercube bilinear interpolation between all 2^p nodes at the vertices enclosing the actual hypercube is utilized.

Hierarchical look-up tables constructed with this algorithm are binary tree structured. Each branch describes one input space decomposition and the terminal nodes represent the distinct hypercubes resulting from the recursive partitioning. Accordingly, simulation or access to hierarchical look-up tables is a two-stage process.

In the first step, a very fast *binary tree search* can be performed searching the hypercube or operating regime enclosing the actual input vector (Fig. 3).

In the second step, equivalent to classical grid-based look-up tables nodes on the vertices of the actual hypercube are used for *local output interpolation.* The applied hypercube interpolation rules can be rewritten as polynomials of order p where all terms containing one input variable of an order higher than one are discarded. In two-dimensional case the area interpolation can be formulated as

$$\hat{y}(\underline{u}) = \theta_0 + \theta_1 \cdot u_1 + \theta_2 \cdot u_2 + \theta_{12} \cdot u_1 \cdot u_2 \tag{10}$$

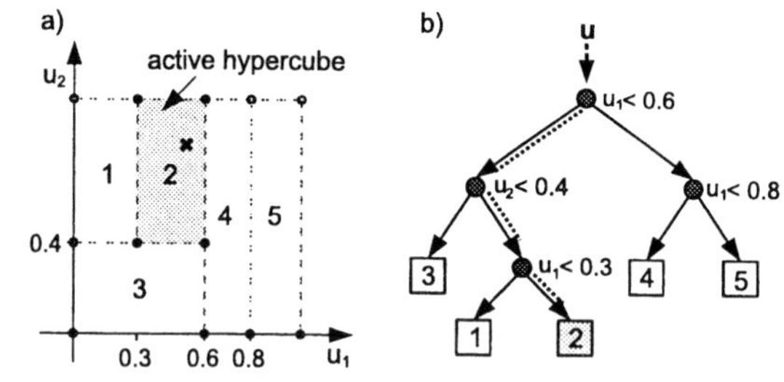

Fig. 3. Tree search within the hierarchical model structure

where the parameters θ_i are determined by the enclosing data point heights.

Hierarchical look-up tables permit the adjustment of their data points to the local complexity represented by the measurement data. Due to the parsimonious placement of data points, hierarchical look-up tables do not suffer from the curse of dimensionality and can be applied to the representation of higher-dimensional nonlinear relationships. The binary tree structure and the simple output calculation allow the implementation of hierarchical look-up tables for information processing in electronic control units.

4. COMPARISON

LOLIMOT models belong to the class of local model approaches. Hierarchical look-up tables can be considered as local model architectures too. With respect to (Murray-Smith and Johanson, 1997) these model structures can be characterized by the following four features:

(1) Decomposition algorithm
(2) Structure of local submodel
(3) Parameter estimation algorithm
(4) Combination of local submodels

Both considered model structures use the same hierarchical decomposition scheme for recursive partitioning of the input space. The proposed hierarchical algorithms perform axis-orthogonal cuts within the respective operating regimes dividing them into two new smaller subregions of same size. However, local model structures valid within these hypercube regions are different. While LOLIMOT employs linear submodels with $p+1$ parameters, hierarchical look-up tables use polynomial submodels with 2^p parameters given by the corresponding data point weights (10). In both cases, the parameter optimization problem is linear. LOLIMOT estimates local model parameters independently with a local weighted least squares method. Since data point weights of hierarchical look-up tables influence parameter determination in several submodels, the weights have to be estimated all at once by least squares method for the global model. LOLIMOT implements a smooth deterministic transition from one submodel to neighboring ones by weighting each submodel with its corresponding validity function. Hierarchical look-up tables use discrete partition of the operating range. That means at each operating point exactly one local submodel is valid. Both models can be described by the unifying equation:

$$\hat{y}(\underline{u}) = \sum_{i=1}^{M} \Phi_i(\underline{u}) \cdot \hat{y}_i(\underline{u}, \underline{\theta}) \qquad (11)$$

The M linear or polynomial submodels $\hat{y}_i$ are weighted with their respective validity functions Φ_i. The LOLIMOT procedure applies normalized Gaussian validity functions. For hierarchical look-up tables (3 and 4), "indicator functions" can be introduced as validity functions with:

$$\Phi_i(\underline{u}) = I_i(\underline{u}) = \begin{cases} 1 & \underline{u} \text{ belongs to the the } i\text{th submodel} \\ 0 & \text{otherwise} \end{cases} \qquad (12)$$

These discussed features are summarized in the Table 1. Although similar decomposition methods are applied for LOLIMOT and hierarchical look-up tables, final model structures and simulation of these models differ significantly. The final neuro-fuzzy model of LOLIMOT algorithm is "flat". Information gathered in the hierarchical decomposition process is reflected by the parameters of the Gaussian validity functions. All estimated local models contribute for each operating point to the model output. That means, for simulation all local model outputs have to be calculated and weighted with their corresponding validity functions (2). This procedure yield desirable smooth model transitions without discontinuities. Contrary to this, hierarchical look-up tables possess a binary tree structure allowing fast access to the solely active local submodel by a discrete logic. For each operating point, model output is equivalent to the active local submodel output. Consequently, hierarchical look-up tables can exhibit discontinuities at model transitions. Under condition of restricting some data point heights, discontinuities can be avoided. However, non-differentiable submodel transitions are structure inherent. The main intention of hierarchical look-up table usage is simulation under real-time conditions.

5. EXAMPLE

The identification capabilities of both models are shown by means of an illustrative example, the approximation of the artificial nonlinear function with exponential behaviour shown in Fig. 4.

Table 1. Comparison

	LOLIMOT	**Hierarchical Look-up Tables**
1.	Hierarchical decomposition	
2.	Linear	Polynomial
3.	Local Weighted Least Squares for each submodel	Global Least Squares for all parameters
4.	Smooth partition	Hard partition

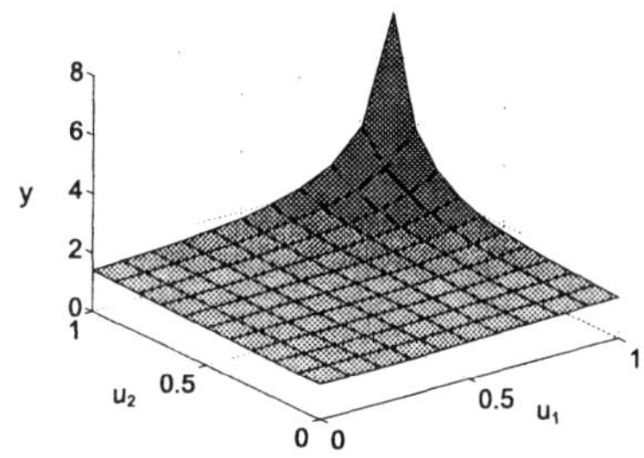

Fig. 4. Nonlinear function

LOLIMOT and hierarchical look-up tables were trained for 20 local models (LM). Figure 5 illustrates axis-orthogonal partitioning and distribution of local models. In both cases, the operating range is densely subdivided in the region of strong increasing function values. In other regions only few local models are valid within large operating regimes. Both investigated models solve the structure identification problem in a similar way. Model outputs are shown in Fig. 6.

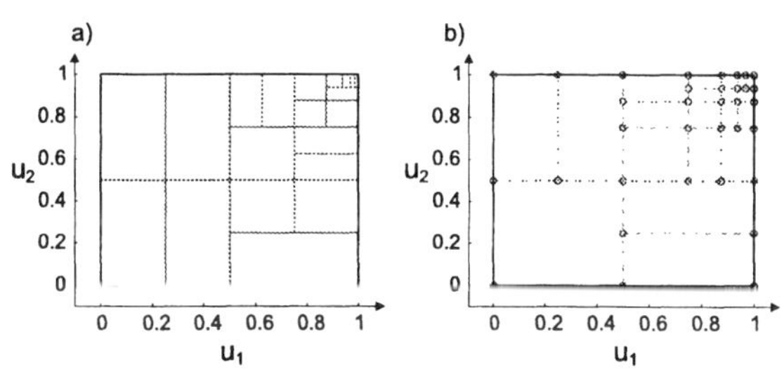

Fig. 5. Partition of two-dimensional input spaces for a) LOLIMOT and b) hierarchical look-up tables

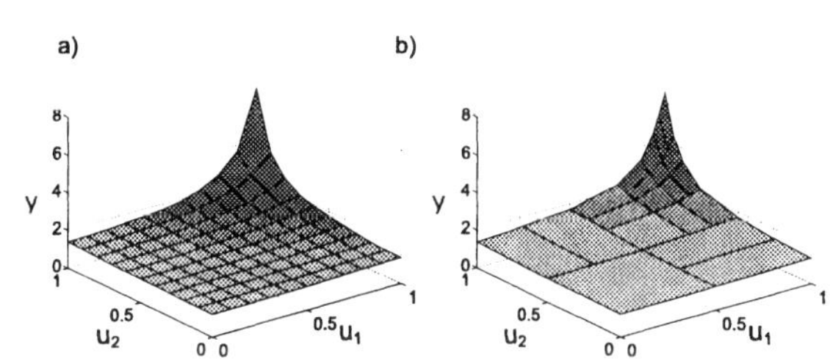

Fig. 6. Model output for a) LOLIMOT and b) hierarchical look-up tables

Table 2 summarizes identification results. Final model quality is measured with respect to generalization data with root mean square error (RMSE). For LOLIMOT the nominal and effective number of local model parameters have to be distinguished. While the nominal number is 60, the effective number is smaller, around 40 for the chosen validity function overlap, since local estimation has a regularization effect. Following from the same hierarchical decomposition scheme, the number of structure parameters is the same for both models. Yet, LOLIMOT structure parameters are reflected in parameters of Gaussian validity functions while hierarchical look-up table structure parameters describe the data point positions. Hierarchical look-up tables have an average response time of about $90\mu sec$ on 16-bit microcontroller hardware for the example presented here. This allows their application for instance in automotive control units with typical sampling times of few milliseconds.

Table 2. Identification results

	LOLIMOT	**Hierarchical Look-up Tables**
no. of LMs	20	20
RMSE	0.011	0.017
no. of LM parameters	nominal: 60 effective: 40	34
response time	–	$90\mu sec$

6. CONCLUSIONS

In this paper, two black-box models for identification of nonlinear systems are compared. The LOLIMOT algorithm builds local linear neuro-fuzzy models. Hierarchical look-up tables extend the approach of grid-based look-up tables for real-time applications. As presented in this paper, hierarchical look-up tables can be interpreted as local models too. Under this unifying point of view both approaches pursue the same strategy to solve the structure identification problem and similar model structures result. Both hierarchical construction algorithms lead to recursive partitioning of the input space in accordance with the complexity of the underlying nonlinear process. Because of this inherent structure identification feature the *curse of dimensionality* can be avoided. Advantages of the divide-and-conquer principle are especially short training times and parsimonious models. Training times are often orders of magnitude faster than nonlinear optimization algorithms optimizing all parameters.

7. REFERENCES

Bently, J. L. (1975). 'Multidimensional binary search trees used for associative searching'. *Communications of the ACM* **18**(9), 509–517.

Breiman, L., J. Friedman, R. Olshen and C. Stone (1984). *Classification and Regression Trees.* Chapman & Hall. New York.

Ernst(Töpfer), S. (1999). Hierarchical look-up tables. In 'European Congress on Intelligent Techniques and Soft Computing (EUFIT)'. Aachen, Germany.

Friedman, J. H. (1991). 'Multivariate adaptive regression splines (with discussion)'. *The Annals of Statistics* **19**, 1–141.

Murray-Smith, R. and T. A. Johansen (1995). Local learning in local model networks. In 'IEE International Conference on Artificial Neural Networks'. pp. 40–46.

Murray-Smith, R. and T. A. Johanson (1997). *Multiple Model Approaches to Modelling and Control.* Tayler & Francis. London.

Nelles, O. (1999). *Nonlinear System Identification with Local Linear Neuro-Fuzzy Models.* Shaker Verlag. Darmstadt, Germany.

Nelles, O. and R. Isermann (1996). Basis function networks for interpolation of local linear models. In 'IEEE Conference on Decision and Control (CDC)'. Kobe, Japan. pp. 470–475.

Sanger, T. D. (1991). 'A tree-structured algorithm for reducing computation in networks with separable basis functions'. *Neural Computation* **3**, 67–78.

AN INTERPOLATION TECHNIQUE FOR LEARNING WITH SPARSE DATA

Jürgen Van Gorp and Yves Rolain

Vrije Universiteit Brussel, Dept. ELEC, Pleinlaan 2, B-1050 Brussel, Belgium
Email: Jurgen.Van.Gorp@vub.ac.be, Yves.Rolain@vub.ac.bc

Abstract — The occurence of sparse data is a stringent problem for Black Box system identification. The lack of measurement data leads to a high modeling error in the resulting gaps. There are a number of techniques described in the literature to fill in these gaps. Some of these techniques are only useable for a low input count (splines). Others suffer from noisy data (interpolation), severe parameter biasing (adding input noise) or time-consuming recursive calculations (Errors-In-Variables). This paper describes a linear feedforward interpolation technique that uses probability techniques, based on 95% or 99% certainty bounds. A smoothing factor can be used to trade off bias and variance, based upon probability theory. *Copyright ©2000 IFAC*

Keywords — System Identification, Sparse Data, Interpolation, Modeling, Piecewise Linear

1.INTRODUCTION

Consider a general Multi-Input Multi-Output sampled data system F with m inputs and n outputs. Consider also that the m inputs and n outputs are sampled N times simultaneously at a sampling rate T_s. At the k-th sampling instant the vectors

$$\mathbf{u}_k = [u_{1,k}, u_{2,k}, \ldots, u_{m,k}] \in \mathbb{R}^m \qquad (1)$$

and

$$\mathbf{y}_k = [y_{1,k}, y_{2,k}, \ldots, y_{m,k}] \in \mathbb{R}^n \qquad (2)$$

represent the sampled data. In estimation, the input-output pairs $(\mathbf{u}_k, \mathbf{y}_k)$ are used to identify the parameters θ of a linear or nonlinear model $f_{bb}(\theta)$. In this paper a Black Box (BB) model

$$f_{BB}: \mathbb{R}^m \to \mathbb{R}^n: y = f_{BB}(\mathbf{u}, \theta) \qquad (3)$$

is chosen. Consider the case where a plant is persistently excited but not regularly measured, i.e. gaps remain in between the measurements. In measurement based black box (BB) identification, where f_{bb} can be a Neural Network (NN), a Radial Basis Function Network (RBFN) or a Fuzzy Logic System (FLS), it is a well known problem that unpredictable behaviour (high variability) exists in the measurement gaps. The goal of this paper is to circumvent this problem through the synthesis of additional measurement data to span the gaps.

Definition 1: In the sequel of this paper, the following definitions are used to describe the intersample behaviour of the nonlinear mapping f_{bb}. BB mappings are said to have a large **variability** if different BB mappings tend to have a large variance when compared to each other. The **variance** of a BB model is used when a BB mapping has large variations compared to the true function, e.g. caused by overfitting on noisy data. A mapping is called **biased** when the mean output of an infinite number of BB mappings does not map the true function.

□

The most direct way to prevent a high variability of the BB models, is to perform better experiments and fill in the gaps with additional measurements. When this is not possible, synthetic data can be used instead of measurements, at the cost of some assumptions concerning the plant or controller. In most cases a smoothness condition is put upon the system characteristics, usually called "generalization". The most commonly used techniques for generalization, are

1) The use of a **validation set**, which typically needs large and representative data sets (An [2]) Gaps in the measurements are worsened, since the data has to be split in multiple data sets.
2) **Model selection methods** The idea is that an optimal choice is made for the number of parameters in the model. Although this seems a trivial problem, model order selection is an open problem for nonlinear models. E.g. in case of Neural Networks, typically a large number of neurons is chosen to improve the possibility to fit details in the data. The result is an increased danger for high variability.
3) **Regularization** adds a penalty term in the cost

function to reduce variability and relates to the Akaike criterion in linear system theory. E. g. Sjöberg [8] adds an increasing penalty for an increasing number of parameters. The way how regularization is reached heavily depends on the specific properties of the penalty term (An [2]). The property that is usually asked from the penalty term, is a low variance of the BB model. Translating this into a regularization term is not a trivial task.

4) **Curvature-driven smoothing** (Bishop [3]) is based on a penalty term that uses the second order derivatives of the model. It is also called **Bias correction** and is described for linear systems (Amemiya [1]) and nonlinear systems (Santhanan, Vandersteen et al. [9]). It is our observation that bias correction leads to a vast number of local minima during the minimization of the cost function.
5) **Nearest-neighbor regression** (Geman et al. [4]) calculates a response output vector y for any input vector u, based on the average of the k measurements that are the closest to u. The value of k should be somewhere in between one and the total number of measurements to obtain a "reasonable" amount of smoothing. In this paper, a variant of the nearest-neighbor driven smoothing and a rule of thumb to select k based on the variance of the measurements is derived.
6) **Parzen-window regression** resembles the the nearest-neighbor regression, but the neighbor u_i of an input u is multiplied with a weight, based on a Parzen-window

$$W(u) = \left(\frac{1}{\sqrt{2\pi}\sigma}\right)^d \exp\left(-\frac{1}{2}\frac{|u-u_i|^2}{\sigma^2}\right) \tag{4}$$

with d a smoothing factor. The choice of σ determines the allowable model variance (Bishop [3]). This paper also uses Parzen windows for the recombination of local-linear systems. The width of the windows will be based on 95% or 99% certainty bounds.
7) **Adding parameter noise** adds random noise to the weights of a BB model, i.e. its parameter space θ. The inevitable result of this is that the obtained parameters differ from the true minimizer of the cost function. It is shown that this results in regularization (An [2]), but it is highly doubtful that the optimal parameters θ^* that describe the true plant, are reached.
8) **Injecting data noise** (An [2]), also called jitter. The results of the doubtfull, but much used, technique of adding data noise inevitably leads to biasing of the NN parameters (Bishop [3]). The technique of adding noisy input samples, which has become popular within NN, is actually a form of zero-order interpolation, and was proven to be of low effect on generalization (Van Gorp et al. [10]).

All of the above techniques demand a "smooth" behaviour of the resulting models in the case of sparse or noisy data. In practise, it is our finding that the only good technique to fill in gaps, is to provide extra "measurement" samples that fill the gap, based on a smoothing demand. Regardless of the choice of a too high number of BB model parameters, the model is forced to fit the data, thus decreasing variability. A well known technique for such interpolation is the use of splines. The drawback of using splines, however, is the lack of a physical meaning of the smoothing factor, and the fact that it is not defined for high-order dimensions of the input. Before introducing an interpolation method, let us therefore first define general demands for an interpolation technique:

1) The user must be able to impose a smoothing factor. This smoothing factor must preferably have a physical meaning, such that the user gains clear insight upon the limitations put on the model.
2) The interpolation technique must be useful for noisy and noiseless measurement data.
3) The interpolation should be applicable for models with multiple inputs (MIMO models).
4) The variance of the interpolated data points must be lower than, or equal to, the variance of the measurement data.
5) It is preferred that the interpolation technique is useable with non-equidistant grids. The backlaying idea is that certain regions can be fitted in more detail.
6) It is prefered that the technique is feedforward, i.e. that an absolute minimum of iterative steps is needed to obtain the desired interpolation. Consecutively with this demand, the interpolation should be fast.
7) When used for extrapolation (instead of interpolation), the method must not "explode" the data, i.e. by resorting to a linear extrapolation. While the risk for extrapolation is low for SISO systems, this demand is certainly necessary for high-dimensional multi-input systems that are not persistently measured.
8) It is prefered that the interpolation technique has low memory and calculation needs. Since the goal is interpolation, an increase in complexity of $\mathcal{O}(n^2)$ with n the input dimension, is acceptable.

2. INTERPOLATION OF SPARSE DATA FOR SISO SYSTEMS

To simplify the explanation of the method used, consider a single-input single output (SISO) system with N input and output measurements

$$\mathbf{u} = [u_1, u_2, ..., u_N]^T \in \mathbb{R}^N \tag{5}$$

and

$$\mathbf{y} = [y_1, y_2, ..., y_N]^T \in \mathbb{R}^N. \tag{6}$$

Assumption 1: The input and output measurements are contaminated by additive noise

$$\begin{aligned} u_i &= u_{i,0} + \Delta u_i & \Delta u_i \in \mathbb{R} \\ y_i &= y_{i,0} + \Delta y_i & \Delta y_i \in \mathbb{R} \end{aligned} \tag{7}$$

in which the $u_{i,0}$ and $y_{i,0}$ are the true, but unknown input and output values and Δu_i and Δy_i are independent, mutually uncorrelated, zero mean Gaussian distributed random variables with standard deviations $\sigma_{u,i}$, $\sigma_{y,i}$ and $\sigma_{uy,i} = 0; \forall i$. □

Assumption 2: The input measurements are ordered. This assumption can easily be met for single-input systems by sorting $\mathbf{u}$. For multi-input systems the data is expected to be sorted per dimension, i.e. that the measurements are performed over one dimension while the values for other dimensions are kept constant. If a multiple-input system cannot be measured following a strict grid, it is assumed that a (non-equidistant) grid can be given that encloses the data with a sufficient resolution. □

The general idea behind the interpolation scheme is that the ordered measurements, described in assumption 2, are resampled using a fine grid. There is no need that the fine grid uses equidistant points. For the SISO case, the sorted measurement points are used as the grid points. The interpolation scheme starts with the calculation of the local linear sections. Since the inputs are sorted, it is very well possible that measurements skip the boundaries of a local linear section due to noise displacements. For that reason, all measurements that lay "in a neighborhood" of the calculated section are used for the calculation of the section parameters. This procedure is also used when an arbitrary coarse grid has been chosen. The selection of the points that lay within a "neighborhood" is based upon the 95% ($\Delta_{95} = 1.955$) or 99% ($\Delta_{99} = 2.577$) confidence intervals of the measurement points. A measurement (u_i, y_i) participates to the calculation of the a local linear section if its confidence interval $\Delta_{u,i} = [u_i - \sigma_{u,i}\Delta_C, u_i + \sigma_{u,i}\Delta_C]$ with $\Delta_C = \Delta_{95}$ or $\Delta_C = \Delta_{99}$, overlaps with the segment $[u_k, u_{k+1}]$ to be interpolated. Define $\hat{\mathbf{u}}_k$ as the vector of all input measurements that lie in this segment and define $\hat{\mathbf{y}}_k$ as the corresponding vector of output measurements. The number of rows in the vector $\hat{\mathbf{u}}_k$ is denoted as P_k. The calculation of the parameters of the local linear section depends on the values of σ. In the case that $\sigma_{u,k} = 0$ and $\sigma_{y,k} = 0$, the calculation of the local-linear sections is based on a Least Squares (LS) cost function and the parameters of the linear system can easily be determined as

$$A_k = [\hat{\mathbf{u}}_k \,|\, 1_{P_k}] \backslash [\hat{\mathbf{y}}_k] \tag{8}$$

in which the operator "\" solves the LS problem using QR decomposition and 1_{P_k} is a unity vector with the same number of rows as $\hat{\mathbf{u}}_k$. The $N-1$ vectors $A_k = [a_k, b_k]^T$, $k = 1, 2, ..., N-1$, form the coefficients of a first order interpolation between the noiseless data points. In the case of exact input measurements and noisy output measurements ($\sigma_{u,k} = 0$ and $\sigma_{y,k} \neq 0$), the A_k are calculated using a Weighted Least Squares (WLS) cost function, such that (8) becomes

$$A_k = (\Sigma_y [\hat{\mathbf{u}}_k \,|\, 1_{P_k}]) \backslash (\Sigma_y [\hat{\mathbf{y}}_k]) \tag{9}$$

in which the matrix of output variances Σ_y is calculated as

$$\Sigma_y = \mathrm{diag}([\sigma_{y,k}, \sigma_{y,k+1}]). \tag{10}$$

If both the inputs and outputs are noisy (case $\sigma_{u,k} \neq 0$ and $\sigma_{y,k} \neq 0$) the Errors-In-Variables (EIV) cost function is used. This results in an iterative scheme

$$A_{k,t+1} = (\Sigma_{uy,t} [\hat{\mathbf{u}}_k \,|\, 1_{P_k}]) \backslash (\Sigma_{uy,t} [\hat{\mathbf{y}}_k]) \tag{11}$$

with

$$\Sigma_{uy,t} = \mathrm{diag}\Big(\Big[\sqrt{\sigma_{y,1}^2 + a_{k,t}^2 \sigma_{u,1}^2}, \sqrt{\sigma_{y,2}^2 + a_{k,t}^2 \sigma_{u,2}^2}, \ldots, \sqrt{\sigma_{y,P_k}^2 + a_{k,t}^2 \sigma_{u,P_k}^2}\Big]\Big). \tag{12}$$

The scheme is repeated for t until convergence is reached on the $a_{i,t}$ parameter. The initial parameter $a_{k,0}$ can be found by solving (9). Equation (11) has only one global minimum and can be solved in polynomial time.

Lemma 1: The local-linear parameters A_k that are found using (8), (9) and (11) are locally unbiased, i.e. are unbiased with respect to the $\hat{\mathbf{u}}_k$ and $\hat{\mathbf{y}}_k$ vectors.

Proof: see Ljung [5]. □

The recombination of the $N-1$ sections is done using a Parzen-window, as defined by Nelles et al. [7]. The width of each section is defined as

$$\Delta_i = u_{i+1} - u_i \tag{13}$$

and the centers of the sections are calculated as

$$c_i = u_i + \Delta_i / 2 \tag{14}$$

with $i = 1, 2, ..., N-1$ and $\mathbf{c} = [c_1, c_2, ..., c_{N-1}]^T$. The fine grid with the interpolated inputs is denoted as $\mathbf{x} = [x_1, x_2, ..., x_K]^T$. Define the smoothing variance

$$\sigma_{sm,i}^2 = \frac{f_s \Delta_i}{\Delta_{99}} \tag{15}$$

with $f_s \in [0, \infty]$ the smoothing factor. To avoid calculation precision problems, it is recommended to avoid smoothing values lower than $f_s < 0.01$. Typical values are $f_s = 1$ for regular interpolation, while it doesn't make much sense to choose $f_s > N$. A rule of thumb is given in section 4.

The recombination of the sections is done as follows. Define

$$\varphi_{i,j} = \exp\left(-\frac{(c_j - x_i)^2}{2\sigma_{sm,i}^2}\right) \text{ and } \phi_i = \sum_{j=1}^{N-1} \varphi_{i,j} \tag{16}$$

with $i = 1, ..., K$ and $j = 1, ..., N-1$ The recombined response is

$$y_i = \frac{1}{\phi_i}\sum_{j=1}^{N-1} \varphi_{i,j}(a_j x_i + b_j)\ ;\ i = 1, 2, ..., K. \quad (17)$$

Illustrative example

To fix the ideas, consider the regression example of An [2], where an arbitrary functional

$$y = \sin(3(u+0.8)^2) \text{ with } u \in [-1, 1] \quad (18)$$

is sampled at 16 points. Noise is added on the inputs and outputs with standard deviations $\sigma_u = 0.05$ and $\sigma_y = 0.05$. The sampling is repeated 10 times, such that the sample variances of the inputs u and outputs y can be calculated. This results in 16 input-output measurements that are sorted for u. These samples are shown in Fig. 2. for all samples and in Fig. 1 for the first five samples. The sample means are shown with circles. The 99% interval based on the sample variance is shown with a horizontal line.

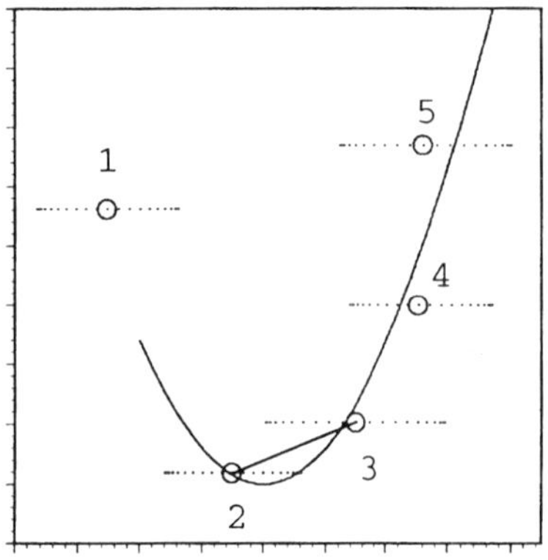

Fig. 1. Selection of measurements for calculation of the local-linear section 2-3. Based upon the 99% boundaries of the sample variances on the input, also samples 4 and 5 are used.

The calculation of the local-linear section 2-3 is shown in detail in Fig. 1. The 99% boundaries of samples 4 and 5 overlap with the section 2-3 and are used for the calculation of the local-linear piece 2-3. The vectors $\hat{\mathbf{u}}_2$ and $\hat{\mathbf{y}}_2$ therefore each contain 4 elements, such that $\hat{\mathbf{u}}_2 = [u_2, u_3, u_4, u_5]^T$ and $\hat{\mathbf{y}}_2 = [y_2, y_3, y_4, y_5]^T$.

Since both $\sigma_{u,k} \neq 0$ and $\sigma_{y,k} \neq 0$, the iterative scheme (11) is used to calculate the local-linear sections. For the given example, the local-linear sections based upon (11) are depicted in Fig. 2.

Remark that the sections don't necessarily form a first-order interpolation between the different measurements, since each section is based upon all measurements that lay within a "neighborhood". For recombination of the different local-linear sections, we choose an arbitrary value for the smoothing factor that equals $f_s = 2$. In the sequel of the paper a more detailed analysis of the smoothing factor is given. The results of the recombination are shown in Fig. 3. The solid line is the true function, while the interpolated points are shown as crosses.

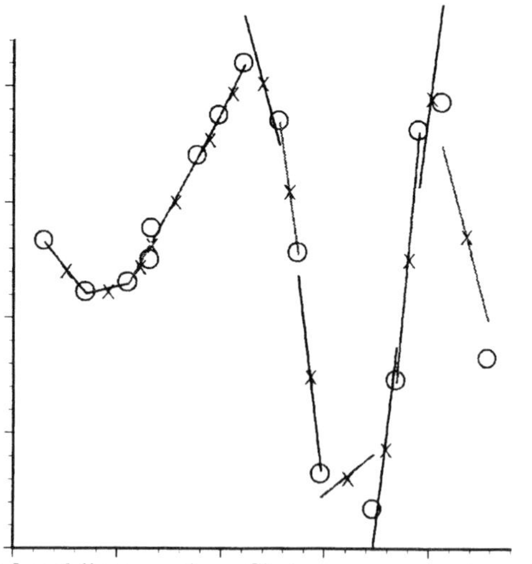
Fig. 2. Local-linear sections. Circles: measurement samples; crosses: section centers, lines: local-linear interpolation.

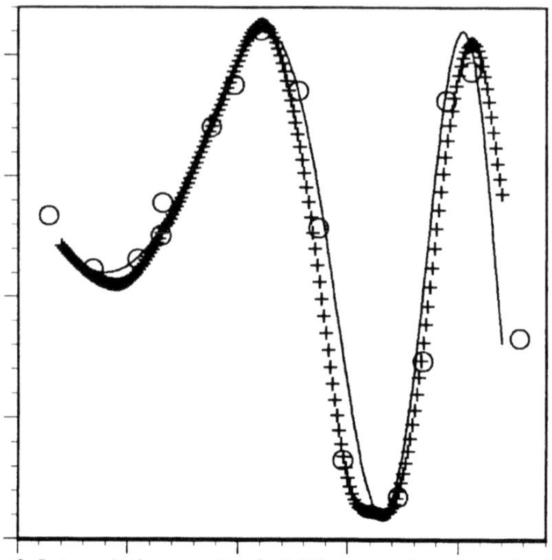
Fig. 3. Interpolation results. Solid line: true function, circles: measurement data, crosses: interpolated samples with $f_s = 2$.

3.INTERPOLATION OF SPARSE DATA FOR MIMO SYSTEMS

Due to restrictions in paper length, the MIMO case is not treated in detail. The general concept is that hypercubes are formed, based on e.g. an equidistant grid. Local linear sections are formed based on all measurements that lay within a region with boundaries formed by the hypercube

$$\frac{(u_{i,1}-x_1)^2}{\sigma^2_{u,i,1}} + \frac{(u_{i,2}-x_2)^2}{\sigma^2_{u,i,2}} + \dots + \frac{(u_{i,m}-x_m)^2}{\sigma^2_{u,i,m}} \le \Delta^2_{99} \quad (19)$$

Next, interpolated measurement points are generated, based on a finer grid, and using the same technique as for the SISO case.

4.ANALYSIS OF THE SMOOTHING FACTOR

Note that in the above interpolation scheme, only one variable is chosen by the user: the smoothing factor. This factor directly influences the width of the Parzen windows, i.e. the region in which each local-linear section is effective. If the smoothing factor is chosen as $f_s = 1$, then each local-linear section width equals the Δ_{99} width, i.e each point of the fine grid has a 99% probability that it falls at least partially within the correct local-linear section. Choosing larger values for f_s means that the region of influence is enlarged accordingly, thus doubling the region of influence if the smoothing factor is chosen as $f_s = 2$. It is possible to choose $f_s < 1$ in order to fully eliminate smoothing. However, in practice care should be taken that $\sigma_{sm,i}$ is chosen large enough (typically $\sigma_{sm,i} > 10^{-3}$ for normalized measurement data) to avoid numerical stability problems. For the same reason, smoothing factors should be chosen that are larger than $f_s > 0.1$. It must be stressed that choosing larger smoothing factors also increases biasing effects.

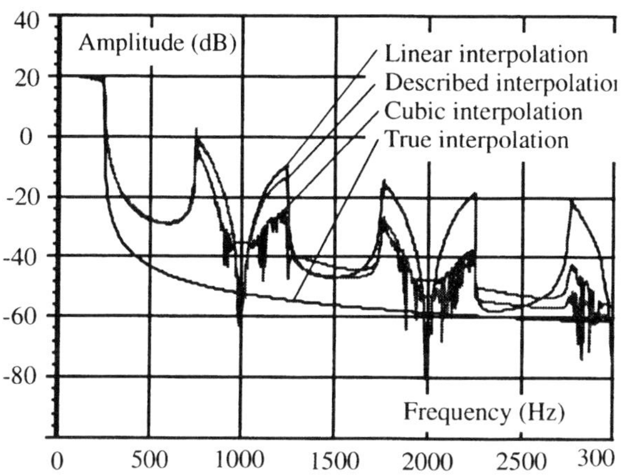

Fig. 4. Frequency analysis of the interpolated signals

The effects of the smoothing factor are better explained using frequency analysis. A multisine signal is sampled 1000 times on an equidistant grid. The highest frequency of the multisine signal is arbitrarily chosen as 250Hz. The samples are interpolated with the described SISO method with $f_s = 1$, a simple linear interpolation scheme and using a cubic interpolation. The grid used for interpolation has a density which is 10 times higher than the original grid, leading to 9991 interpolated points. The frequency analysis of the interpolated signals is shown in Fig. 4. It can be seen that the interpolation scheme given in this paper, has the same distortion as a linear interpolation scheme for in band frequencies, while it outperforms the cubic interpolation in out-of-band rejection. The reconstruction of equation (16) can be seen as a convolution of the local linear sections with the impulse response e^{-x^2}. If the exponential has a constant width (once σ_{sm}^2 is fixed), the interpolation scheme acts as a low pass filter with a cut-off frequency that only depends on the smoothing factor. On an empirical basis we found that the -3dB frequency of the filter (f_{-3dB}) can be calculated as

$$\frac{f_{-3dB}}{N} = \frac{0.78127}{f_s} + 4.06 10^{-3} \tag{20}$$

in the particular case of an equidistant grid such that $\Delta_u = 1/N$, and with values $0.01 < f_{-3dB}/N < 0.4$. The value $f_{-3dB}/N < 0.4$ showed to be a practical limit for the given optimization scheme, which yields a lower limit $f_s \geq 1.97$ for the smoothing factor. If the number of samples of the original grid N, and a certain cut-off frequency f_{-3dB} is given, the proper smoothing factor can be calculated as

$$f_s = 0.78127\left(\frac{f_{-3dB}}{N} - 4.06 10^{-3}\right)^{-1}. \tag{21}$$

Consider again the example of Fig. 4, choosing the highest cut-off frequency $f_{-3dB} = 400\text{Hz}$. With $N = 1000$ the optimal smoothing factor becomes $f_s = 1.97$. The result of this interpolation is shown in Fig. 5. It is clear that the interpolation scheme causes far less distortion than the cubic interpolation scheme, while the original signal is largely preserved.

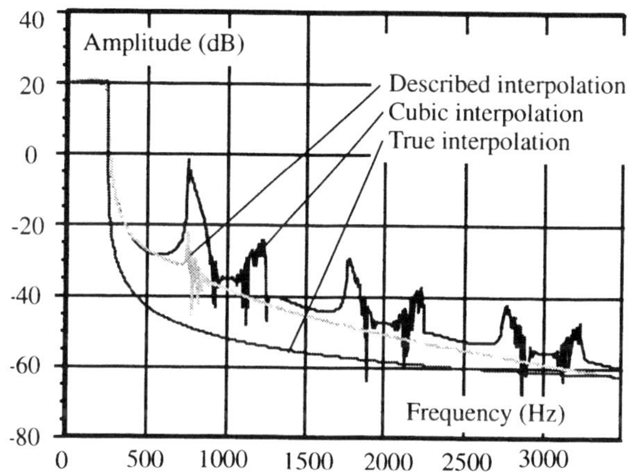

Fig. 5. Frequency analysis of the interpolation scheme with cut-off frequency at 400Hz.

5.SIMULATION RESULTS: OVERFITTING EXAMPLE (MACKAY [6])

A NN is trained to fit 37 input-output measurements. The input-output pairs (u, y) were not measured on an equidistant grid and gaps remain within the measurement points. Mackay noticed that overtraining led to a large variability of the NN in these gaps. To illustrate this Fig. 6. shows the mean result and variance of 100 simulations of a feedforward perceptron NN with one hidden layer and 20 neurons in the hidden layer.

The given solution for this overfitting was to increase the number of "measurements" by taking a large number of copies of the data samples and add Gaussian noise to the inputs of the resulting data set. This results in a "smoothing" of the NN mapping but also creates bias on the NN parameters. Since the final goal of adding noise is to become a "smooth" behaviour of the NN, we used the interpolation scheme to fill in the gaps. The smoothing factor was chosen as $f_s = 1.97$ a the measurements were interpolated with 200 interpolated points in between the measurements. The results of the NN mappings for the same type of NN is shown in Fig. 7. It comes clear that the variability of the NN has decreased significantly, while the original measurements are preserved.

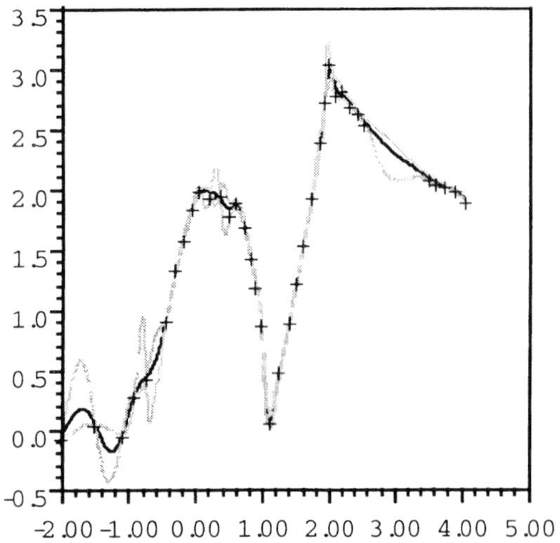

Fig. 6. Variability of NN in the case of sparse data. Solid line: mean NN mapping; gray line: variability on the NN mappings.

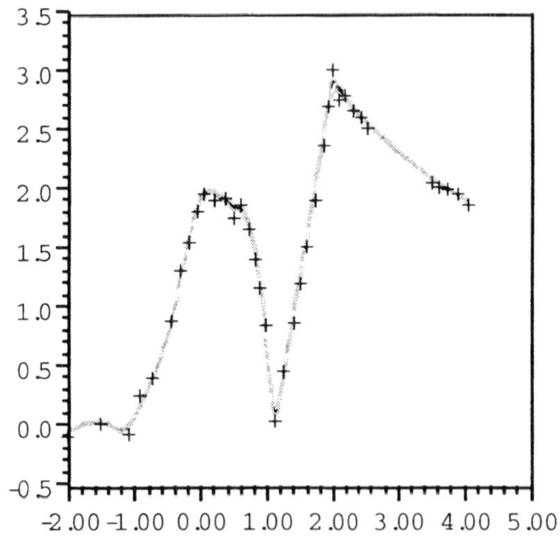

Fig. 7. Variability of the same NN structure with prior interpolation.Solid line: mean NN mapping; gray line: variability on the NN mappings.

6.CONCLUSION

This paper presented an interpolation scheme, based upon a local linearization of the data. The combination of the different local-linear parts leads to a polynomial interpolation of the data, which has a relationship with spline fitting. The advantages of the method are a physical meaning of the smoothing factor, the applicability on higher dimension multi-input systems, the straightforward calculation, low bias, and a near-linear behaviour when the scheme is used for extrapolation. A drawback of the interpolation scheme is that typically a large number of parameters are used to represent the measurement data. The scheme is therefore not recommended for use as a black box model.

Acknowledgments

This paper presents research results of the Belgian Programme on Interuniversity Poles of Attraction (IUAP 4/2), initiated by the Belgian State, Prime Minister's Office for Science, Technology and Culture; and the Flemish Government (GOA, IMMI2). The research is supported by the Fund for Scientific Research - Flanders (FWO).

References

[1] Amemiya Y., "Two-Stage Instrumental Variable Estimators for the Nonlinear Errors-In-Variables model", Journal of Econometrics, No. 44, 1990, pp. 311 - 332.

[2] An G., "The Effects of Adding Noise During Backpropagation Training on a Generalization Performance", Neural Computation, Vol. 8, 1996, pp. 643 - 674.

[3] Bishop C. M., "Neural Networks for Pattern Recognition", Clarendon Press, Oxford, 1995.

[4] Geman S., Bienenstock E., Doursat R., "Neural Networks and the Bias/Variance Dilemmma", Neural Computation, Vol. 4, 1992, pp. 1 - 58.

[5] Ljung L., "System Identification, Theory for the User", Prentice-Hall information and system sciences series, 1994.

[6] MacKay D. J. C., "Bayesian Interpolation", Neural Computation, Vol. 4, 1992, pp. 415 - 447.

[7] Nelles O., Hecker O., Isermann R., "Automatic Model Selection in Local Linear Model Trees (LOLIMOT) for Nonlinear System Identification of a Transport Delay Process", SYSID '97, Japan, July 1997, pp. 727 - 732

[8] Sjöberg J., "Non-Linear System Identification with Neural Networks", Linköping Studies in Science and Technology Dissertation No. 381, Sweden, 1995.

[9] Vandersteen G., Rolain Y., Schoukens J., Pintelon R., "On the Use of System Identification for Accurate Parametric Modeling of Nonlinear Systems Using Noisy Measurements", IEEE Trans. on Instrumentation and Measurement, Vol. 45, No. 2, April 1996, pp. 605 - 609.

[10] Van Gorp J., Schoukens J., Pintelon R., "Adding Input Noise to Increase the Generalization of Neural Networks is a Bad Idea", ANNIE 1998, Volume 8, pp. 127 - 132.

SIMULATION OF CENTRAL PATTERN GENERATORS VIA CHAOTIC NEURON POPULATIONS

P. Arena, L. Fortuna, D. Garofalo

Università degli Studi di Catania
Dipartimento Elettrico Elettronico e Sistemistico
Viale A. Doria, 6, 95125 Catania, ITALY
Tel. +39 095 7382322 - Fax: +39 095 339535
e-mail: parena, lfortuna@dees.unict.it

Abstract: In this paper a simulation environment is presented where biological neural networks are can be efficiently built and simulated, at the aim to study spatio-temporal organization arising from population of chaotic neurons. Some models of Central Pattern Generators are also presented with application to the generation of insect walking patterns, implemented on a robotic structure directly driven by the simulator. *Copyright © 2000 IFAC*

Key words: Simulation, Motor patterns, Neural network models, Neural dynamics, Chaotic behaviour

1. INTRODUCTION

Recently great attention has been paid to the simulation and reproduction of locomotion patterns in animals. This great interest since animals, even the simplest ones, show a great degree of adaptation and succeed in moving in nearly any type of environment. Therefore neurobiological studies and experiments are a great source for engineers to derive efficient locomotion models

Indeed, neurobiologists agree on the fact that the neural architectures for locomotion control must have evolved in a hierarchical way. For this reason the organisation of biological neural networks is commonly functionally studied as a hierarchical structure. In particular, the study is mostly performed by following some stereotypes both for invertebrates and vertebrates. In general it is agreed that the Central Nervous System (CNS) must produce specific patterns of motor neuron impulses during a coordinated motion. The central hypothesis is that there is a Central Pattern Generator (CPG), within the CNS, that produces the basic motor programs. Information derived from sensory inputs may modify the output of the pattern generator so as to adapt locomotion to the environment.

These are the main guidelines that depict the whole system for locomotion generation in animals. Of course the CPG is a very high-level representation of the biological motor system. Indeed, looking at the details of the dynamics of the single neuron, outlined by the Nobel Prizes Hodkin and Huxley in 1952, it clearly emerges that the neuron is, by itself, a complex dynamical system, showing even chaotic motions. Therefore the dynamics of a population of neurons is clearly able to show very complex spatio-temporal dynamics. At this stage, simulation is a powerful technique to investigate on populations of neurons mutually connected that could give rise to self-organized phenomena, generating motor patterns. Indeed, the role of chaos in neural systems is acquiring an increasing interest. In fact, a particular organised pattern of activity could be seen as a particular state deriving from the control of a chaotic spatio-temporal multidimensional system towards a stable oscillation. Under this perspective, the rich variety of dynamics contained in chaotic motions assumes a fundamental role, and could explain how some biological neural networks are able to generate coordinated movements. Referring to terrestrial insects, there are several types of motor patterns. Each species is able to abruptly shift from one pattern to another one according to impelling needs. Therefore, the existence of a central neural network can be supposed, able to manage all the locomotor patterns in animals endowed with six legs. To obtain a general neural structure several models of neural networks will be investigated, able to generate coordinated movements in terrestrial insects; the most significant results obtained will be shown through simulations. First the classical fourth order Hodkin-Huxley (HH) non linear neuron model will be evaluated, and subsequently the simpler (HR) Hindmarsh-Rose model, also drawn experimentally, will be analyzed, able to maintain all the main non linear neural dynamics, including the presence of chaos. Then, some neuron populations will be defined through synaptic links. Finally, some new models of CPGs able to generate three different motor patterns on an hexapod structure will be carried out. To perform the simulations reported in the paper, a simulation environment has been built, where it is possible to describe, from the point of view of a dynamical system, the peculiarities of the neuron model selected to build a neural network, together with the suitable synaptic model. The main aspects of the simulator built are reported in the next Section.

2. BoNNie: A BIOLOGICAL NEURAL NETWORK SIMULATOR

The typical problems that arise when simulating systems of nonlinear differential equations, with particular attention to the interlinks among the equations are to select an efficient and easy strategy to design the whole system. Moreover attention has to be paid to the selection of a suitable integration method. In our case, the characteristic of these systems to swiftly change state can generate some integration problems. Therefore, care must be taken of some serious problems, like accuracy and speed, during the integration routine, and of the important issue to create a simple method to build system network. To solve these problems an new simulator is proposed: BoNNie. This true multi-threaded C++ software implements the embedded Runge-Kutta formula with Cash-Karp coefficients and an adaptive stepsize control. The integration method used is able to assure accuracy and a good speed level. Of course some other integration routines are currently being selected and implemented in the simulator, in order to improve its characteristics.

Actually the models implemented in the simulator are the HH and HR neuron, electrical and chemical synapses, as well as a simple model of the muscle system, in order to generate suitable signals to drive, via the parallel port, an hexapode robot prototype. A simulation performed with this software can be divided into three steps: a) nodes specification; b) simulation setup; c) simulation and results analysis.

a) For the first point, an important feature is the capability to describe relevant neural networks with two very simple text files. The first file describes how many nodes (or systems, that can be a HR or HH neuron, for example) are considered and how these nodes are grouped. The second text file contains the connections description. In the following are reported the two text files which define a neural network consisting of two groups, each one built-up of 10 neurons: each neuron inside a group is connected, via synaptic inhibitory connections to all the neurons of the other group.

Node file

```
BNNNODES
# example of a nodes script
GROUP1 ISTYPE GROUP 10
GROUP2 ISTYPE GROUP 10
```

Connection file

```
BNNCONNECTIONS
#example of a connections script
GROUP1  ISATARGETFOR  GROUP2  BY  INHIBITORY  WEIGHTBASE  0.6
RANDOM 0.3
GROUP2  ISATARGETFOR  GROUP1  BY  INHIBITORY  WEIGHTBASE  1.5
RANDOM 0.5
```

b) To describe the simulation setup, in the following figures are reported some dialog boxes to set other very important parameters, such as time length, accuracy, and other system parameters.

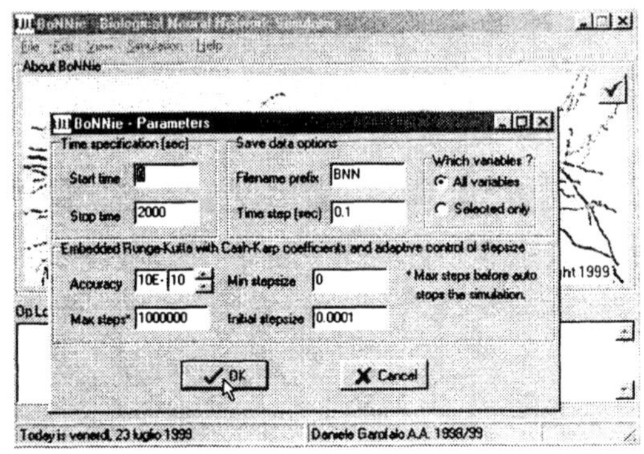

Fig. 1. The BoNNie dialog box

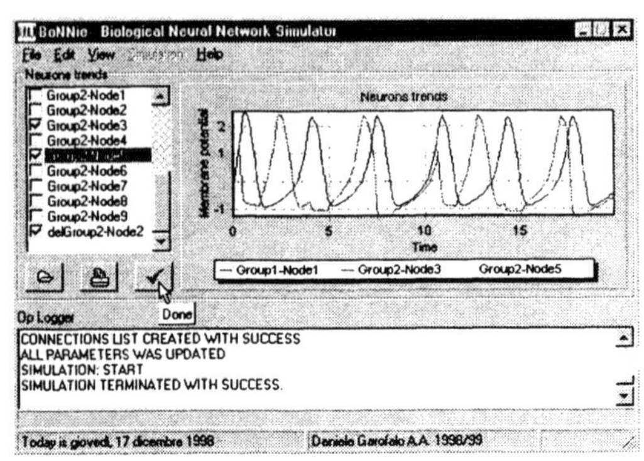

Fig. 2. BoNNie's built-in viewer

c) Finally, the simulation can be analyzed by the built-in viewer or by other software (MATLAB for example). This capability is possible since BoNNie saves all the selected time series in files with an user-defined name at the simulation setup step.

3. THE NEURON MODELS

To simulate neural assemblies able to generate coordinated motions, two models of neuron are considered. The classical HH formulation of the dynamics of ion channels is described by the following fourth-order dynamical nonlinear system:

$$\frac{dv(t)}{dt} = -(G_{Na}m^3(t)h(t)(v(t)-V_{Na}) + G_k n^4(t)(v(t)-V_K) + \\ + G_l(v(t)-V_l) - i_{app})/cm \qquad (1)$$
$$\frac{dn(t)}{dt} = \alpha_n(1-n(t)) - \beta_n n(t)$$
$$\frac{dm(t)}{dt} = \alpha_m(1-m(t)) - \beta_m m(t)$$
$$\frac{dh(t)}{dt} = \alpha_h(1-h(t)) - \beta_h h(t)$$

with:

$$\alpha_n = \frac{0{,}01(10-v)}{e^{\frac{10-v}{10}}-1};\ \beta_n = 0{,}125e^{-\frac{v}{80}}$$
$$\alpha_m = \frac{0{,}1(25-v)}{e^{\frac{25-v}{10}}-1};\ \beta_m = 4e^{-\frac{v}{18}}$$
$$\alpha_b = 0{,}7\left(-\frac{v}{20}\right);\ \beta_h = \frac{1}{e^{\frac{30-v}{10}}+1}$$

In these equations v is the membrane potential (mV), m and n describe respectively Na^+ and K^+ ion channels activation while h represents Na^+ inhibition (Scott, 1995). The parameters used are following: cm=1.0, G_{Na}=120.0, G_K=36.0, G_l=0.3, V_{Na}=50.0, V_K=-77.0, V_l=-54.4. Some examples of an HH typical time series are depicted in Fig. 3.
The main feature of the other model used, an HR neuron, is the ability to capture spiking-bursting chaotic behavior observed in intracellular voltage measurements. The HR model is

$$\frac{dx(t)}{dt} = y(t) + \phi[x(t)] - z(t) + I$$
$$\frac{dy(t)}{dt} = \psi[x(t)] - y(t) \qquad (2)$$
$$\frac{dz(t)}{dt} = -rz(t) + rS[x(t)-c_x]$$

with

$$\phi(x) = ax^3 - x^2$$
$$\psi(x) = 1 - bx^2$$

The three variables are respectively, the membrane potential *x(t)* (dimensionless, associated with the fast subsystem Na^+ K^+), an auxiliary variable called *y(t)* representing a set of fast ion channels and a "slow" variable *z(t)*, which captures the slower dynamics of yet other ion channels (Rabinovich et al., 1997). Typical parameters used for simulation are following: a=3.0, b=5.0, c_x=-1.6, S=4.0, r=0.0021 and I=3.281. In Fig. 3 the time series from an HR neuron with chaotic and regular behavior is shown.

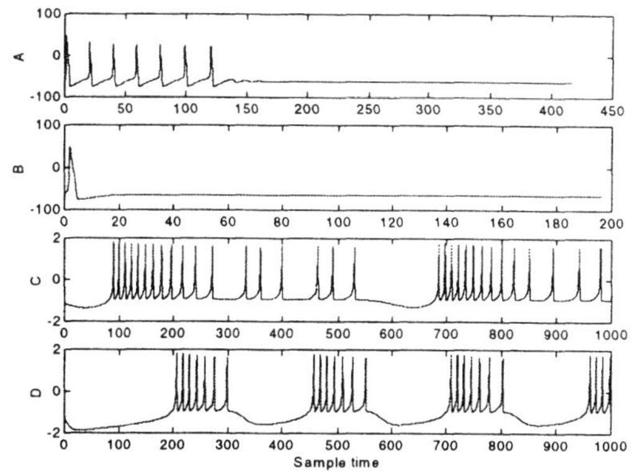

Fig. 3. Time series from HH and HR neurons for different values of the applied current i_{app} to stimulate neural activity.(A) HH i_{app}=6.25 mA; (B) HH i_{app}=0.0 mA; (C) HR I=3.281, chaotic behaviour; (D) HR I=2.5, bursting activity.

Chaos is very important in neural networks since, as outlined earlier, the rich variety of nonlinear dynamics embedded into systems with chaos, could be a key issue to explain the extraordinary ability of neural systems to adapt, make transitions from one activity pattern to another one when the environment is altered. The key aspect of this concept is the ability to create a rich variety of patterns with chaotic neurons (Rabinovich et al., 1997; Rabinovich et al., 1998; Wang, 1993; Freeman, 1992). The HR model, although being a low order model, maintains main features of the neuron, including also the chaotic characteristics. Therefore it will be used in all the following simulations.
Another aspect of neural networks lies in the characteristics of synaptic connections. There are two types of synapses: chemical and electrical. Chemical coupling can define an inhibitory or excitatory synaptic action. The chemical synaptic model is:

$$\varepsilon(x_1(t)+V_c)\Theta(x_2(t-\tau_c)-X) \qquad (3)$$

where ε is the strength of the coupling, V_c is the reverse potential, X is the threshold, τ_c is the synaptic delay and $\Theta(\omega)$ is the Heaviside function (Rabinovich et al., 1997). Electrical coupling is simpler than chemical synaptic action. The model is:

$$\varepsilon(x_1(t)-x_2(t)) \qquad (4)$$

with ε strength of electrical coupling (Rabinovich et al., 1997). The main feature of this type of coupling is that it does not introduce synaptic delay. Generally, electrical synapses are observed in localized areas of the human nervous system, such as heart, reflex arc, or in the neural circuits controlling escape motions in animals. Of course, one or the other synaptic coupling will be added to the first equation in the model, which represents the membrane potential, as it can be seen from equations

(5), defining a neuron connected via one way chemical synapses to another one:

$$\frac{dx_1(t)}{dt} = y_1(t) + \phi[x_1(t)] - z_1(t) + I + \varepsilon(x_1(t) + V_c)\Theta(x_2(t-\tau_c) - X) \quad (5)$$
$$\frac{dy_1(t)}{dt} = \psi[x_1(t)] - y_1(t)$$
$$\frac{dz_1(t)}{dt} = -rz_1(t) + rS[x_1(t) - c_x]$$

4. EXAMPLES OF CPGs

Several models of CPGs have been formulated recently; some of them are depicted in Fig.4. In a CPG called Half-Center-Model (HCM), the driver group controls two populations of interneurons (by chemical inhibitory synapses); these ones control muscular activation (flexor and extensor) with an alternate signal. Another, more complex, model of CPG is the Closed-Loop-Model, where there are several populations of interneurons connected with chemical inhibitory synapses. In this model every group is a part of a ring. There are corresponding pools of motoneurons activated, or inhibited, in sequence. Finally, the Pacemaker Model (PM) of CPG can be considered. For this model an oscillatory clock generator (pacemaker group) can be localized, which directly drives a muscle and its antagonist by a population of interneurons coupled with chemical inhibitory synapses (Sheperd, 1994).
These CPG models have emerged from studies from both vertebrates and invertebrates; in fact, one emerging common opinion in neurobiology is that despite the diversity of locomotion patterns and animal body structures the basic types of neural architectures to generate organized activities are quite limited to some basic schemes, like the ones reported in Fig. 4, and through some slight generalizations the other specific motion generator can be derived.

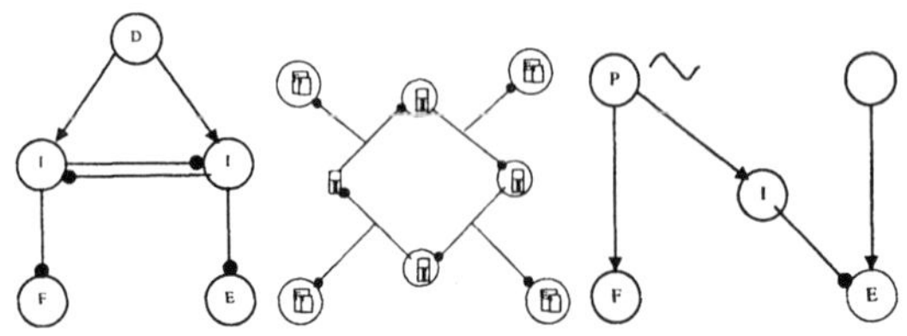

Fig. 4. The HCModel - CLModel - PModel

5. A CPG FOR INSECT LOCOMOTION

In this Section some CPG models for insect locomotion, partially derived from the CPGs previously described, are presented. Moreover ring configurations to model CPGs have already been used (Rabinovich et al., 1997). As previously discussed the ability to obtain several distinct patterns by HR populations of neurons that wonder in a chaotic motion is a target for this work. A structure with two rings of HR groups (Fig. 6a) is first considered. One of two rings (the inner one) drives the other one by chemical inhibitory couplings. The first important result that can be observed (Fig. 5) is that, starting from a chaotic motion for each isolated neuron, the whole dynamics for the structure built lies in a highly organized state. This important result opens the way to the use of different structures to obtain different organized states, deriving from the same chaotic dynamics, but controlled towards a different oscillation state. In other words, an interesting investigation issue regards the role of the neural structure in the type of organized state to be reached. In fact, a first result obtained by a lot of simulations is that only rings connections containing an even number of neurons can reach an organized state. However, in this work, the main interest is to obtain a number of different patterns in the same structure. Indeed, this model is able to show only simple locomotion patterns. In fact, in the outer group, only alternating motions can be derived, while our aim is to obtain general patterns where motions among the animal legs are linked one another with particular phase shifts, which could be realized by suitable synaptic delay modulations. The answer to this problem can be found if the number of synapses is reduced and a distributed CPG is considered, where every peripheral group directly controls the movements of one insect leg. The scheme is reported in Fig.6b, where L1 to L3 represent the drivers for the legs belonging to the left part of the animal, while R1 to R3 drive the right part. The driver is formed by two reciprocally coupled populations of 10 neurons where every neuron of each group is connected to any other neuron in the other group. All synaptic actions are chemical inhibitory. The peripheral groups can be formed even of only one neuron, since here the organization is imposed by the driver and moreover there is no feedback action within the peripheral centers. In this model this fine modulation cannot be realized. This means that ring coupling is not sufficient to establish the correct pattern modulations required. The neural activity pattern can be easily and finely changed with this structure by using synaptic delays. A fundamental issue regards the connection strength ε in (3), which is a key factor to reach global organization in the whole structure. In particular, lots of simulation have shown that a with $\varepsilon<0.6$ no global organization can be obtained. Moreover, it has to be outlined that the ε value was varied by a random factor η in the interval [0;1]. Simulations have shown the robustness of the organized activity also with a disturbance in the spatial synaptic connections. Our aim is to obtain suitable locomotion patterns for the insect motion. In Fig. 8 the slow, medium and fast gait are reported (Arena et al., 1998; Conti et al., 1998). To obtain all the patterns presented earlier the structure in Fig. 6b with synaptic delay modulation is used. Patterns can be varied by changing the delay template. The implementation of the above delay templates allowed to obtain the three locomotion patterns of Fig. 7, as it can be seen in Fig. 8. The role of the driver groups is to generate an oscillatory signal which synchronizes the six peripheral groups. The delay template provides the suitable phase delays for the correct pattern formation. In Fig. 9 the driver trends are depicted.

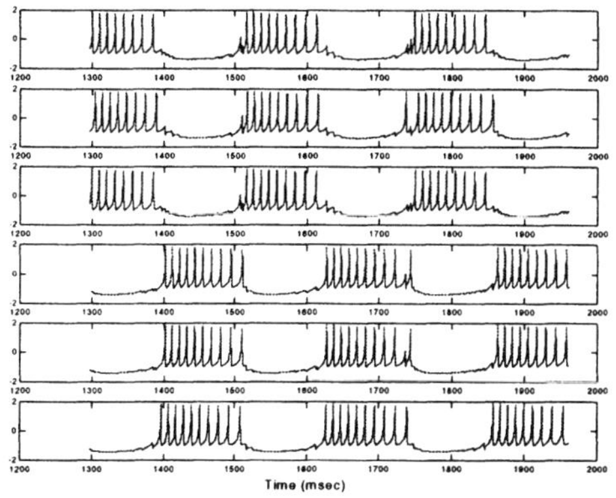

Fig. 5. Double ring time series. The centralized driver isn't able to give a different pattern to the outer ring.

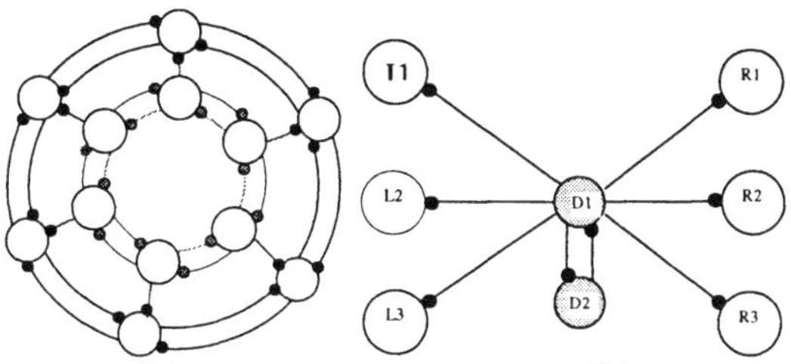

(a) (b)

Fig. 6. (a) double ring with groups mutually inhibited. The inner ring is a driver for the outer one. Only chemical inhibitory synapses are used. Each group inside the rings is formed by 10 chaotic neurons. (b) driver groups (Filled circles) directly drive peripheral centers by chemical inhibitory synapses. Each peripheral neurons directly drives the actuators of each leg.

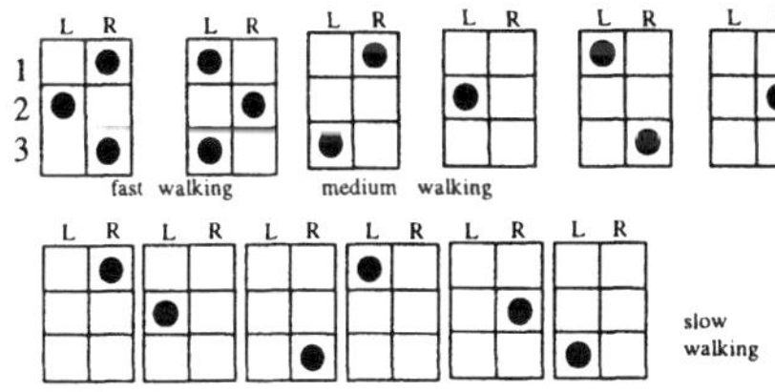

Fig. 7. Insect walking patterns

Table 1 Templates to obtain all pattern discussed earlier. Delays are espressed in computational step

	Delay templates					
Pattern	L1	R2	L3	R1	L2	R3
Slow	200	400	600	800	1000	1200
Medium	2	450	900	900	1350	2
Fast	2	2	2	1200	1200	1200

They are the same for all the simulations performed on this model.

6. LOCOMOTION PATTERS IN A ROBOT PROTOTYPE

The strategy presented in the previous sections was implemented on a robot prototype, presented in Fig.10. It is a robot consisting of six legs, each one possessing two degrees of freedom. The actuation consists of 12 dc servomotors, driven by using a PWM technique.

Signals from the simulator, representing the locomotion pattern for each leg, have now to be suitably manipulated to provide motion commands for the servo-motors. The filter designed represents a rough model of the muscle system, sufficient to obtain reliable results. In particular the muscle was modeled as a first order model with a low time constant. Moreover, a non-linearity is needed to limit high and low contraction activity. This characteristics directly meets the needs for the servomotors to be driven; in fact the whole spiking period is translated into a unique rising wave with an upper and a lower saturation level, while the remaining phase corresponds to a relaxation period for the output signal.

Up to now the strategy runs in open loop, and the locomotion pattern modulation is decided at the "keyboard level". A suitable control strategy is currently under investigation in order to select a suitable sensor network to make this stage autonomous.

7. CONCLUSIONS

In this paper a new simulator has been introduced, where neuron populations can be efficiently and easily described, to reproduce self-organized dynamics, starting from chaotic wanderings for the isolated cells. In particular, populations of HR neurons, organized in groups and mutually connected via inhibitory chemical synapses have been considered. Among the great number of neural geometries simulated, the most significant ones have been presented in the paper. An interesting application of the arising spatio-temporal regularizations regards the reproduction of several locomotion patterns, commonly found in insect, on a biological inspired walking robot. The simulation results appear very interesting to build efficient control strategies for artificial multipodes, able to easily adapt the locomotion pattern to the environment needs. The key features of the simulator are the easiness to build a neural network architecture, based on neuron and synapse models, given as nonlinear dynamical systems, without needing the characterization of all of the biological details required in traditional simulators of this type. In fact it is assumed that the neuron model derives from a previous characterization from a biological point of view, and the model drawn is ready to be implemented in neural architectures. Of course the simulator is at its early stage. Several interesting tools and models are going to be inserted in the simulator. One fundamental issue regards the possibility to derive models of neural cells directly by using experimental data. In fact, powerful identification strategies, based on artificial neural networks as well as fuzzy systems have deeply been investigated and are ready to be included in this environment. This aspect seems really appealing as a useful framework for the common links that nowadays are becoming more and more strict among the experimental neurobiology and the system engineering approach.

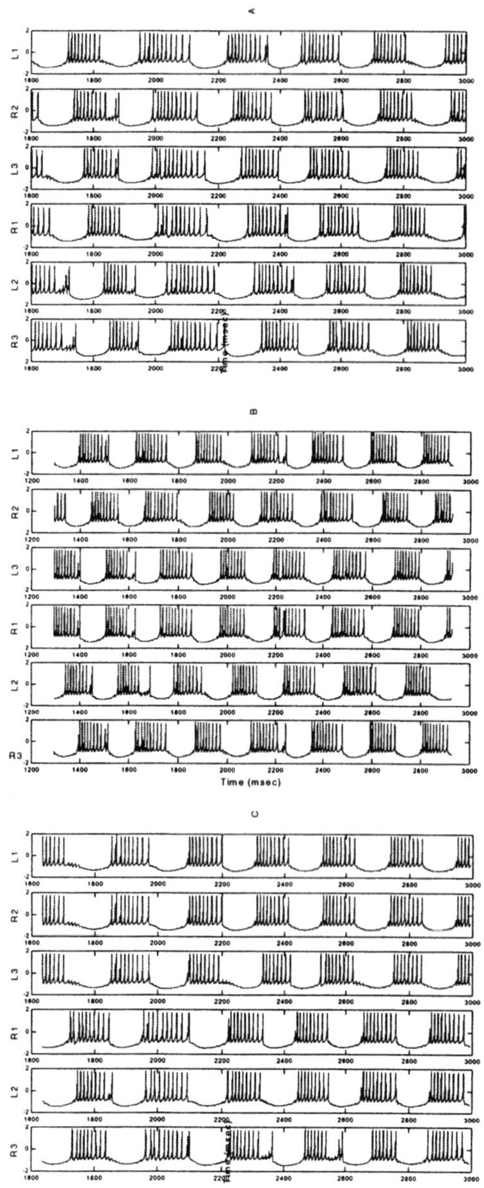

Fig. 8. Trends of neural activity for each pattern. (A) slow walking, (B) medium walking, (C) fast walking

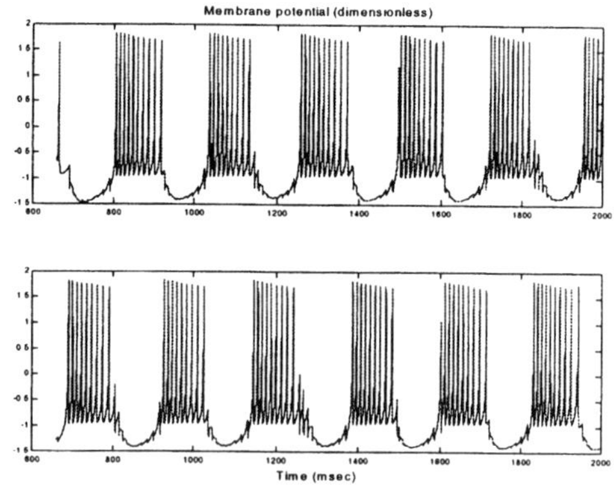

Fig. 9. Time series from driver groups

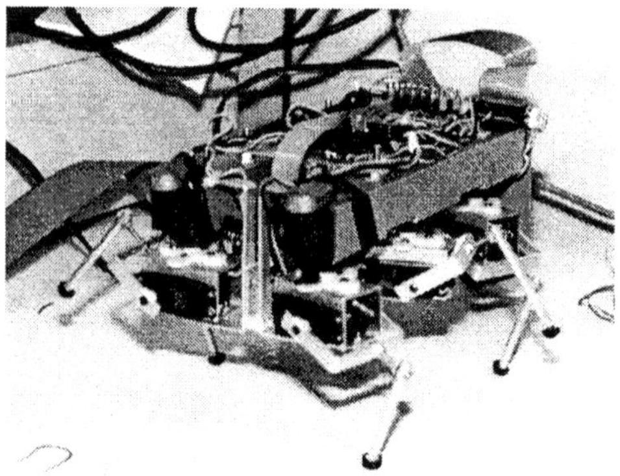

Fig.10. The hexapod robot prototype

REFERENCES

Arena, P., L. Fortuna (1998). Spatio-temporal waves in CNN to induce and control motion in artificial multipodes, *Proc. NOLTA 98, Crans-Montana.*

Conti, E., G. Costa, A. Petralia (1998). Rhythmic leg movements among terrestrial insects: some hypotheses on a unified pattern generator, *Proceedings NOLTA 98, Crans-Montana.*

Freeman, W.J. (1992). Tutorial on neurobiology: from single neuron to brain chaos. *International Journal of Bifurcation and Chaos, Vol. 2 No. 3 451-482*

Rabinovich, M.I., H.D.I. Abarbanel, R. Huerta, R. Elson, A.I. Selverston (1997). Self-Regularization of Chaos in Neural Systems: Experimental and Theoretical Results, *IEEE Trans. on circuits and systems Vol. 44, No 10, 1997*

Rabinovich, M.I., H.D.I. Abarbanel (1998). The role of chaos in neural systems, *Neuroscience Vol. 87 No. 1 pp. 5-14, 1998*

Scott, A. (1995). *Stairway to the mind. The controversial New Science Of Consciousness*, Springer-Verlag, New York

Sheperd, G.M. (1994). *Neurobiology*, Oxford

Wang, X.J. (1993). Genesis of bursting oscillations in the Hindmarsh-Rose model and homoclinicity to a chaotic saddle, *Physica D 62, 263-274*

Error Analysis of Certain Subspace Methods

Alessandro Chiuso * and Giorgio Picci**

* *Dipartimento di Elettronica e Informatica, University of Padova, 35131 Padua, Italy; Email:* chiuso@dei.unipd.it

** *Dipartimento di Elettronica e Informatica, University of Padova and LADSEB-CNR, 35131 Padua, Italy; Email:* picci@dei.unipd.it

Abstract. Some subspace methods make use of an oblique projection to determine the observability matrix. There is experimental evidence (Chiuso and Picci, 1999; Kawauchi et al., 1999) that these methods perform poorly in certain situations. In this paper we present an error analysis of oblique projections and show that in some situations small errors in the subspace along which the projection is done may lead to large errors in the estimate. This may happen for instance in the N4SID algorithm. Sometimes an orthogonal projection should be recommended instead; in fact, it can be shown that the worst case "signal to noise" ratio on the rows of the estimated observability matrix is larger for orthogonal projections than for oblique projections. A variance analysis is presented and simulations are included which compare oblique projection based algorithms with the orthogonal projection. *Copyright ©2000 IFAC*

Key Words. Subspace Identification; Exogenous inputs; Canonical correlation analysis; Oblique projection; Error analysis.

1. Background and notations

Assume we have two time-series,

$$\{u_0, \dots, u_t, \dots, u_N\} \quad \{y_0, \dots, y_t, \dots, y_N\} \quad (1)$$

with $u_t \in \mathbb{R}^p$, $y_t \in \mathbb{R}^m$, generated by a "true system"

$$\begin{cases} \mathbf{x}(t+1) &= A\mathbf{x}(t) + B\mathbf{u}(t) + K\mathbf{e}(t) \\ \mathbf{y}(t) &= C\mathbf{x}(t) + D\mathbf{u}(t) + \mathbf{e}(t) \end{cases} \quad (2)$$

where $\{\mathbf{y}(t)\}$, $\{\mathbf{u}(t)\}$, $\{\mathbf{x}(t)\}$, are zero mean (wide-sense) stationary processes. The system is taken, without loss of generality, in "innovation form", meaning that $\{\mathbf{e}(t)\}$ is a white noise process uncorrelated with the past history of all other variables. We shall assume that the sample size N is very large and that the observed data are second-order ergodic, so that for any linear functional (statistics), say $\xi(u_0, y_0, u_1, y_1 \dots)$, $\eta(u_0, y_0, u_1, y_1 \dots)$ the sum

$$\langle \xi, \eta \rangle := \frac{1}{N+1} \sum_{t=0}^{N} \xi_t \eta_t \quad (3)$$

where $\xi_t := \xi(u_t, y_t, u_{t+1}, y_{t+1} \dots)$ and similarly for η_t, converges as $N \to \infty$ to a finite number which necessarily coincides with the expected value (inner product in L^2) of the corresponding random variable, $\xi(\mathbf{uy})\eta(\mathbf{uy})$.

Fix a "present" time $t = k$ with k larger than n, the model order. In the following we shall assume that the order n of the "true" system generating the data is known, so there will be no need to discuss order estimation.

Define as usual the block Hankel matrix $U_{0|k-1}$ of dimension $pk \times (N+1)$ as:

$$U_{0|k-1} := \begin{bmatrix} u_t & u_{t+1} & \dots & u_{t+N} \\ u_{t+1} & u_{t+2} & \dots & u_{t+N+1} \\ \vdots & \dots & \ddots & \vdots \\ u_{t+k-1} & u_{t+k} & \dots & u_{t+k+N-1} \end{bmatrix}.$$

and, in perfect analogy, define $U_{k|2k-1}$, formed with future input data and $Y_{0|k-1}$, $Y_{k|2k-1}$, both of dimension $mk \times (N+1)$, formed with past and future output data. The rowspaces $\mathcal{U}_{0|k-1}$, $\mathcal{Y}_{0|k-1}$ of the first two matrices are the past input and output spaces generated by the rows of the $p \times (N+1)$ "tail" matrices $u(t) := [u_t, \dots, u_{t+N}]$ and, respectively, by the tail matrices $m \times (N+1)$ matrices $y(t) := [y_t, \dots, y_{t+N}]$ for $0 \le t < k$. Likewise $\mathcal{U}_{k|2k-1}$ and $\mathcal{Y}_{k|2k-1}$ are the future input and output vector spaces spanned by $u(t)$, resp. $y(t)$, for $k \le t < 2k$. A trivial although important observation for what follows is that the tail matrices obey exactly the same dynamical equations (2) as

[1] This work was supported by the TMR network *System Identification* and by the national project *Identification and control of industrial processes* funded by MURST.

the generating stochastic processes, namely

$$\begin{cases} x(t+1) &= Ax(t) + Bu(t) + Ke(t) \\ y(t) &= Cx(t) + Du(t) + e(t) \end{cases} \quad (4)$$

Therefore one can think of the identification problem of the system (2) as a (partial) realization problem starting from the data (1). For infinitely long tail matrices we have $e(t) \perp (x(s), u(s))$, $s \leq t$ exactly. Of course for finite N the orthogonality holds only approximately.

2. Problem statement

The basic object which is first computed in most subspace methods is the predictor matrix based on joint input-output data

$$\begin{aligned} \hat{Y}_{k|2k-1} &:= E[Y_{k|2k-1} \mid \mathcal{Y}_{0|k-1} \vee \mathcal{U}_{0|2k-1}] \\ &= [L_1^k \quad L_2^k \quad L_3^k] \begin{pmatrix} U_{0|k-1} \\ U_{k|2k-1} \\ Y_{0|k-1} \end{pmatrix} \end{aligned}$$

(the wedge denotes vector sum). From this, an estimate $\hat{\Gamma}_k$ of the observability matrix Γ_k of (2) can be obtained by an oblique projection along $\mathcal{U}_{k|2k-1}$ followed by a weighted SVD step (see (van Overschee and De Moor, 1994)). From the SVD factorization of the weighted oblique projection

$$\begin{aligned} USV^T &:= W_1 [L_1^k \quad L_3^k] \begin{pmatrix} U_{0|k-1} \\ Y_{0|k-1} \end{pmatrix} W_2 \\ &= W_1 \Gamma_k \tilde{\mathcal{X}}_k W_2 \end{aligned} \quad (5)$$

one obtains

$$\hat{\Gamma}_k := W_1^{-1} U S^{1/2} \quad , \quad \tilde{X}_k = S^{1/2} V^T W_2^{-1} \quad (6)$$

In this paper we want to examine the effects of noise due to finite data length on the estimate $\hat{\Gamma}_k$. It has been shown (van Overschee and De Moor, 1994), that the estimation of the observability matrix is done essentially by the same procedure in most subspace algorithms, the difference being only in the choice of the two weighting matrices W_1, W_2. Asymptotically (as $N \to \infty$) the estimates of the observability matrix corresponding to different weights are consistent (under a reasonable assumption of "generic data" (Jansson and Wahlberg, 1997)) and differ by a non singular change of basis. In this paper we shall assume that W_1 is fixed equal to the identity matrix and only discuss the choice of the "right" weight, W_2. In fact we restrict to the case where W_2 is either the identity matrix or the projector onto the orthogonal complement of future inputs, $W_2 = \Pi_{U^\perp_{k|2k-1}}$ (orthogonal projection (Verhaegen, 1994)) [2]. Later we shall argue that the choice $W_2 = \Pi_{U^\perp_{k|2k-1}}$ is sometimes (although not always) to be recommended.

Once $\hat{\Gamma}_k$ is computed, the second crucial step of the procedure in (van Overschee and De Moor, 1994) is to solve the overdetermined linear system

$$\begin{bmatrix} \hat{\Gamma}_{k-1}^{-L} \hat{Y}_{k+1|2k-1} \\ y(k) \end{bmatrix} = \begin{bmatrix} A \\ C \end{bmatrix} \hat{\Gamma}_k^{-L} \hat{Y}_{k|2k-1} + \\ + \begin{bmatrix} K_1 \\ K_2 \end{bmatrix} U_{k|2k-1} + W^\perp \quad (7)$$

for the unknown parameters (A, C) and (K_1, K_2), (see eq. (43) in (van Overschee and De Moor, 1994)). If $N \to \infty$ the term $W^\perp$ is orthogonal to the row span of $U_{k|2k-1}$ and $\hat{Y}_{k|2k-1}$, and hence for $N \to \infty$, we can recover exactly the matrices (A, C) and (K_1, K_2) from equation (7). In reality, due to the fact that only a finite amount of data is available, $W^\perp$ is not orthogonal to the regression spaces and an error in the parameter estimates is introduced.

The solution of (7) can be computed by solving a linear least-squares problem. It follows, see e.g. (Katayama and Picci , 1999), that the LS parameter estimates of (A, C) and (K_1, K_2) are the oblique projections of the LHS of (7) onto the rowspace of $\hat{\Gamma}_k^{-L} \hat{Y}_{k|2k-1}$ along the rowspace of $U_{k|2k-1}$ and, respectively, onto $U_{k|2k-1}$ along the rowspace of $\hat{\Gamma}_k^{-L} \hat{Y}_{k|2k-1}$. For nearly parallel row spaces of $U_{k|2k-1}$ and of $\hat{\Gamma}_k^{-L} \hat{Y}_{k|2k-1}$ one expects that the solution of (7) will be ill-conditioned, and the parameter estimates will be affected by large errors. Here "nearly parallel" subspaces means subspaces for which some canonical angles are nearly zero (see (Golub and van Loan, 1989), p. 584). The conditioning of the problem (7) is analyzed in (Chiuso and Picci, 1999). We just note that near parallelism is likely in N4SID-type methods since the rowspace of

$$\Gamma_k^{-L} \hat{Y}_{k|2k-1} = \hat{X}_k + \Gamma_k^{-L} H_k U_{k|2k-1} \quad (8)$$

may be "close" to that of $U_{k|2k-1}$. Here $\hat{X}_k$ is the $n \times N$ (Kalman-filter) state matrix and H_k is a lower triangular block Toeplitz matrix made with the first k Markov parameters $(D, CB, CAB, ..., CA^{k-2}B)$, of the "deterministic" subsystem (A, B, C, D) (see equation (15) in (van Overschee and De Moor, 1994)).

The finite data length, besides causing the estimate $\hat{\Gamma}_k$ to be affected by errors implies that $\hat{Y}_{k|2k-1} \neq \Gamma_k \hat{X}_k + H_k U_{k|2k-1})$, so that the oblique projection for computing (K_1, K_2) is done along a perturbed subspace. Perturbations due to sampling errors may get amplified by the ill-conditioning of (7).

[2] It can be shown that other choices of instruments may be treated in a completely similar way

3. Oblique Projections

It is well known that, unlike orthogonal projections, oblique projections are not contractions. Roughly speaking, this is equivalent to the existence of vectors whose norm is increased by the projection operator; therefore small errors on the data can be amplified into large perturbations in the projections. Consider a vector space $\mathcal{H}$ and a direct sum decomposition

$$\mathcal{H} = \mathcal{A} + \mathcal{B}. \tag{9}$$

Denote by $E(\cdot \mid \mathcal{H})$ the orthogonal projection onto $\mathcal{H}$, and by $E_{\|\mathcal{B}}(\cdot \mid \mathcal{A})$ the oblique projection onto $\mathcal{A}$ along $\mathcal{B}$. The operator norm of a linear operator T on $\mathcal{H}$ is defined by the well-known relation $\|T\|^2 = \sup_{h \in \mathcal{H}} \frac{\|Th\|^2}{\|h\|^2}$. The following Lemma is due to H. Wimmer (unpublished)

Lemma 3.1 Let $d(x, \mathcal{B})$ be the distance, between the vector x and the subspace $\mathcal{B}$; then the following holds:

$$\begin{aligned} \|E_{\|\mathcal{B}}(\cdot \mid \mathcal{A})\| &= \sup_{x \in \mathcal{A},\ \|x\|^2=1} \frac{1}{d(x,\mathcal{B})} \\ &= \frac{1}{\sqrt{1-\sigma_{max}^2}} \end{aligned} \tag{10}$$

where σ_{max} is the maximum canonical correlation coefficient between $\mathcal{A}$ and $\mathcal{B}$.

Let now $z \in \mathcal{H}$ and assume that only an approximation (estimate), say $\mathcal{C}$, of $\mathcal{B}$ is available. We want to compute the norm of the difference $(z - \hat{z})$ between the oblique projections of z onto $\mathcal{A}$ along $\mathcal{B}$ and onto $\mathcal{A}$ along $\mathcal{C}$ respectively.

Suppose that the subspace $\mathcal{B}$ along which we project is perturbed by errors. Denote by $\mathcal{C}$ the perturbed $\mathcal{B}$ subspace. We will still assume that the sum is direct:

$$\mathcal{H} = \mathcal{A} + \mathcal{C}. \tag{11}$$

The situation in which $\mathcal{A} + \mathcal{C} \neq \mathcal{A} + \mathcal{B}$. will be examined below. Clearly the oblique projection on $\mathcal{A}$ along $\mathcal{B}$ and $\mathcal{C}$ are different; we will provide below a formula for the norm of the perturbation on the projections which depends upon the "distance" between $\mathcal{A}$ and $\mathcal{B}$ measured by the smallest canonical angle, and a measure of the perturbation.

Let $\theta_{\mathcal{BC}}^{max}$ be the largest canonical angle between $\mathcal{B}$ and its perturbed version $\mathcal{C}$. We define a "neighborhood" $B(\mathcal{B}, \bar{\theta})$ of $\mathcal{B}$ made by all subspaces $\mathcal{C}$ of the same dimension making an angle which is smaller than a given bound $\bar{\theta}$, namely

$$B(\mathcal{B}, \bar{\theta}) := \{\mathcal{C} \mid \theta_{\mathcal{BZ}}^{max} \leq \bar{\theta}\}$$

The following lemma, whose proof is given in (Chiuso, 2000) provides an estimate for the perturbation on the projections:

Lemma 3.2 Let $\theta_{\mathcal{AB}}^{min}$ be the smallest angle between $\mathcal{A}$ and $\mathcal{B}$. Then the following holds true:

$$\begin{aligned} &\sup_{\mathcal{C} \in B(\mathcal{B},\bar{\theta})} \|E_{\|\mathcal{B}}(\cdot \mid \mathcal{A}) - E_{\|\mathcal{C}}(\cdot \mid \mathcal{A})\|^2 = \\ &= \frac{1-\cos^2(\bar{\theta})}{\left(1-\cos^2\left(\theta_{\mathcal{AB}}^{min}-\bar{\theta}\right)\right)\left(1-\cos^2\left(\theta_{\mathcal{AB}}^{min}\right)\right)} \end{aligned} \tag{12}$$

The bound is sharp, in the sense that it is attained for subspaces such that the singular vector corresponding to the maximum angle $\theta_{\mathcal{BC}}^{max}$ belongs to the plane spanned by the singular vectors in $\mathcal{A}$ and $\mathcal{B}$ corresponding to the angle $\theta_{\mathcal{AB}}^{min}$.

The bound tells us that, for a fixed error on the "direction" along which the projection is performed, the closer two subspaces are, the larger the norm of the error in the projection. This result has been found independently also in (Cantoni and Vinnicombe, 1999) in a similar setup.

In order to discuss the case where $\mathcal{A} + \mathcal{C} \neq \mathcal{A} + \mathcal{B}$ (but $\mathcal{B}$ and $\mathcal{C}$ are of the same dimension) we need some preliminary results. It is easy to see that, without loss of generality, one may restrict the ambient space $\mathcal{H}$ to be $\mathcal{H} = \mathcal{A} \vee \mathcal{B} \vee \mathcal{C}$ since the computation of the norm may be restricted to this subspace. Let us define the subspaces

$$\bar{\mathcal{B}} := \mathcal{B} \oplus (\mathcal{A} + \mathcal{B})^{\perp} \qquad \bar{\mathcal{C}} := \mathcal{C} \oplus (\mathcal{A} + \mathcal{C})^{\perp}$$

Lemma 3.3 Let the symbols have the meaning of above. Then the following holds:

$$E_{\|\bar{\mathcal{B}}}^{\mathcal{A}} = E_{\|\mathcal{B}}^{\mathcal{A}} \quad , \quad E_{\|\bar{\mathcal{C}}}^{\mathcal{A}} = E_{\|\mathcal{C}}^{\mathcal{A}} \tag{13}$$

See (Chiuso, 2000) for a proof.

It follows from the Lemma that

$$\|E_{\|\mathcal{B}}^{\mathcal{A}} - E_{\|\mathcal{C}}^{\mathcal{A}}\| = \|E_{\|\bar{\mathcal{B}}}^{\mathcal{A}} - E_{\|\bar{\mathcal{C}}}^{\mathcal{A}}\|$$

and therefore we can make use of the previous results using $\bar{\mathcal{B}}$ and $\bar{\mathcal{C}}$ in place of $\mathcal{B}$ and $\mathcal{C}$. From (13) the norm of the oblique projections are the same and $\theta_{\bar{\mathcal{B}}\bar{\mathcal{C}}}^{max}$ have to be used in place of $\theta_{\mathcal{BC}}^{max}$.

Example Recall that in N4SID the estimate of $(\mathcal{K}_1, \mathcal{K}_2)$ is obtained by computing the oblique projections of the LHS of (7) onto $U_{k|2k-1}$ along the rowspace of $\hat{\Gamma}_k^{-L}\hat{Y}_{k|2k-1}$. Due to finite sample fluctuations, and uncertainties on the estimated observability matrix only a perturbed version of the regressor space $\hat{\Gamma}_k^{-L}\hat{Y}_{k|2k-1} \neq \hat{X}_k + \Gamma_k^{-L}U_{k|2k-1}$ is available. Using Lemma 3.2 with $\mathcal{A} := \mathcal{U}_{k|2k-1}$, $\mathcal{B} := \Gamma_k^{-L}\hat{Y}_{k|2k-1}$ and $\mathcal{C} := \hat{\Gamma}_k^{-L}\hat{\mathcal{Y}}_{k|2k-1}$ the influence of this perturbation on $(\mathcal{K}_1, \mathcal{K}_2)$ can be estimated.

In particular, it can be seen from formula (12) that if the nominal regressor spaces are "close enough", meaning that the minimum angle between the rowspaces of $U_{k|2k-1}$ and $\Gamma_k^{-L} Y_{k|2k-1}$ is "nearly zero", the perturbations can easily have a dramatic effect. In the next section we shall try to obtain a rough estimate of the errors in $\hat{\Gamma}_k$ which may be used to compute a worst-case distance between the subspaces $\Gamma_k^{-L} \hat{\mathcal{Y}}_{k|2k-1}$ and $\mathcal{C} := \hat{\Gamma}_k^{-L} \hat{\mathcal{Y}}_{k|2k-1}$

4. Error estimates on Γ_k

An asymptotic distribution for estimates of Γ_k has been derived in the literature (Viberg et al., 1997); however this expression is complicated and not easy to use for our purposes. We shall rather look directly for expressions for the error variances. Let us recall that we are operating with fixed observed data $U_{0|2k-1}$, $Y_{0|2k-1}$; denote $P_{0|k-1} := \begin{pmatrix} U_{0|k-1} \\ Y_{0|k-1} \end{pmatrix}$ for short. We may describe the effect of finite sample fluctuations by writing the predictor matrix $\hat{Y}_{k|2k-1} = E\left[Y_{k|2k-1} | P_{0|k-1} \vee U_{k|2k-1}\right]$ as a "true predictor" affected by an additive error term

$$\begin{aligned} \hat{Y}_{k|2k-1} &= E\left[Y_{k|2k-1} | P_{0|k-1} \vee U_{k|2k-1}\right] \\ &= \Gamma_k \hat{X}_k + H_k U_{k|2k-1} + \hat{W}_{k|2k-1} \end{aligned}, \quad (14)$$

where Γ_k, H_k are the *true parameters* (in an appropriate basis) and $\hat{W}_{k|2k-1}$ is a random fluctuation term due to the fact that orthogonality of $W^{\perp}$ to the regression spaces is only approximately true for finite data length N: $\hat{W}_{k|2k-1} = E\left[W_{k|2k-1} | P_{0|k-1} \vee U_{k|2k-1}\right]$ where $W_{k|2k-1} = H_k^s E_{k|2k-1}$ having denoted by H_k^s the Hankel matrix formed with the first k Markov parameters $(I, CK, CAK, \ldots, CA^{k-2}K)$ of the "stochastic" subsystem (A, K, C, I). Clearly for $N \to \infty$, $\hat{W}_{k|2k-1} \to 0$.

In order to obtain an estimate $\hat{\Gamma}_k$ one has to get rid of the term $H_k U_{k|2k-1}$ at the same time reducing the effect of $\hat{W}_{k|2k-1}$. Splitting the contributions of $\hat{X}_k$ and $\hat{W}_{k|2k-1}$ in two components: $\hat{X}_k^p$, $\hat{W}_{k|2k-1}^p \in \mathcal{P}_{0|k-1}$, $\hat{X}_k^u$, $\hat{W}_{k|2k-1}^u \in \mathcal{U}_{k|2k-1}$, which are in direct sum, equation (14) may be rewritten as

$$\begin{aligned} \hat{Y}_{k|2k-1} = {} & \Gamma_k \hat{X}_k^p + \hat{W}_{k|2k-1}^p + \\ & + \Gamma_k \hat{X}_k^u + H_k U_{k|2k-1} + \hat{W}_{k|2k-1}^u . \end{aligned} \quad (15)$$

In order to eliminate the effect of terms depending on future inputs, one has to perform an oblique projection along $\mathcal{U}_{k|2k-1}$ onto some complementary subspace $\mathcal{U}_{k|2k-1}^c$.

Ideally the choice of the space $\mathcal{U}_{k|2k-1}^c$ should be done in order to minimize the effect of the perturbations $\hat{W}_{k|2k-1}$. The choice $\mathcal{U}_{k|2k-1}^c = \mathcal{P}_{0|k-1}$ is done for instance in N4SID (van Overschee and De Moor, 1994) while $U_{k|2k-1}^c = \left(P_{0|k-1} \mid U_{k|2k-1}^{\perp}\right)$, which corresponds to an orthogonal projection, is done, for instance, in (Verhaegen, 1994).

The choice between orthogonal and oblique projections corresponds to different state spaces which may be viewed as finite time oblique and orthogonal splitting subspaces. Denote by $\mathcal{X}_k^{ort}$ the state space corresponding to the orthogonal projection and by $\mathcal{X}_k^{obl}$ the one corresponding to the oblique projection, the following holds true:

$$\mathcal{Y}_{k|2k-1} \perp \mathcal{U}_{k|2k-1}^{\perp} | \mathcal{X}_k^{ort} \quad (16)$$

and

$$\mathcal{Y}_{k|2k-1} \perp (\mathcal{Y}_{0|k-1} \vee \mathcal{U}_{0|k-1}) \mid (\mathcal{X}_k^{obl} \vee \mathcal{U}_{k|2k-1}) \quad (17)$$

where $\mathcal{U}_{k|2k-1}^{\perp} := E\left(\mathcal{P}_{0|k-1} \mid \mathcal{U}_{k|2k-1}^{\perp}\right)$. It is easily seen that, if we denote by $\mathcal{X}_k^{+/-}$ the minimal markovian oblique predictor subspace (Chiuso, 2000), then the minimal finite time orthogonal and oblique splitting subspaces are given by

$$\mathcal{X}_k^{ort} = E\left[\mathcal{X}_k^{+/-} | \mathcal{U}_{k|2k-1}^{\perp}\right] \quad (18)$$

and

$$\mathcal{X}_k^{obl} = E_{\|\mathcal{U}_{k|2k-1}}\left[\mathcal{X}_k^{+/-} | \mathcal{Y}_{0|k-1} \vee \mathcal{U}_{0|k-1}\right] \quad (19)$$

If the basis $\mathbf{x}^{ort}(k)$ of $\mathcal{X}_k^{ort}$ and $\mathbf{x}^{obl}(k)$ of $\mathcal{X}_k^{obl}$ were chosen uniformly, i.e.

$$\mathbf{x}^{ort}(k) = E\left[\mathbf{x}^{obl}(k) | \mathcal{X}_k^{ort}\right] \quad (20)$$

then the corresponding estimates A and C would the same. (Unfortunately, however, $\mathcal{X}_k^{ort}$ and $\mathcal{X}_k^{obl}$ are not markovian and do not give rise to stationary and causal realizations).

Let $E_{0|k-1}$ be an orthonormal basis of $\mathcal{P}_{0|k-1}$ constructed by the following canonical correlation analysis procedure: pick the singular vectors in $\mathcal{X}_k^{obl}$ with respect to $\mathcal{U}_{k|2k-1}$ and complete this set of n vectors to form a basis of $\mathcal{P}_{0|k-1}$. Let $\hat{\sigma}_i$'s, $i = 1, \ldots, n$ be the (sample) canonical correlation coefficients between $\mathcal{X}_k^{obl}$ and $\mathcal{U}_{k|2k-1}$, and let σ_i's be their "true" values. Recall that in the generic case where the consistency condition ((Jansson and Wahlberg, 1997)) holds, all the σ_i's are strictly less than one. Define

$$\hat{\Sigma} = \ diag\{1 - \hat{\sigma}_1^2, 1 - \hat{\sigma}_2^2, \ldots, 1 - \hat{\sigma}_n^2\}$$

Consider a "corresponding" dual orthonormal basis $E^{ort}_{0|k-1}$ in $\mathcal{U}^{\perp}_{k|2k-1}$ formed as follows

$$E^{ort}_{0|k-1} = \begin{bmatrix} \hat{\Sigma}^{-1/2} & 0 \\ * & * \end{bmatrix} E \left[E_{0|k-1} | U^{\perp}_{k|2k-1} \right]$$

where the elements denoted with "*" are chosen in order to make $E^{ort}_{0|k-1}$ orthonormal. Recall that the first n rows of $E_{0|k-1}$ are the singular vectors in $\mathcal{X}^{obl}_k$ with respect to $\mathcal{U}_{k|2k-1}$. Therefore their projections onto $\mathcal{U}^{\perp}_{k|2k-1}$ are still orthogonal but not necesarily normalized. As far as the projection of the last $r - n$ rows we just need to recall that since $P_{0|k-1}$ and $U_{k|2k-1}$ have independent rows, the projection is still full rank (which in particular shows that the choice of projection does not influence consistency). Defining $\tilde{\Gamma}_k E_{0|k-1} := \hat{W}^p_{k|2k-1}$ and $\tilde{\Gamma}^{ort}_k E^{ort}_{0|k-1} := \hat{W}^p_{k|2k-1} \Pi_{\mathcal{U}^{\perp}_{k|2k-1}}$, it may be easily seen that using an orthogonal projection we perform SVD on the matrix $\hat{\Gamma}^{ort}_k = [\Gamma^{ort}_k \;\; 0] + \tilde{\Gamma}^{ort}_k = [\Gamma_k \hat{\Sigma}^{1/2} \;\; 0] + \tilde{\Gamma}^{ort}_k$ while in the oblique projection case on $\hat{\Gamma}_k = [\Gamma_k \;\; 0] + \tilde{\Gamma}_k$. By γ_i we denote the i-th row of the matrix Γ_k and use similar notations for γ^{ort}_i, $\tilde{\gamma}_i$, $\tilde{\gamma}^{ort}_i$ etc. The following result states that the worst case signal to noise ratio is always larger for the orthogonal projection.

Proposition 4.1 The worst case signal to noise ratio satisfies the following inequality

$$\inf_{\hat{w}_i} \frac{\|\gamma^{ort}_i\|^2}{\|\tilde{\gamma}^{ort}_i\|^2} \geq \inf_{\hat{w}_i} \frac{\|\gamma_i\|^2}{\|\tilde{\gamma}_i\|^2} \tag{21}$$

where $\hat{w}_i$ is the i-th row of the perturbation matrix $\hat{W}_{k|2k-1}$.

The next result deals instead with the statistical effect of the perturbations. The proof can be found in (Chiuso, 2000). Let ρ_i, $i = 1, \ldots, r$ be the canonical correlation coefficients between $\mathcal{P}_{0|k-1}$ and $\mathcal{U}_{k|2k-1}$ and $\hat{\rho}_i$ their sample version. Note that $r = k(p + m)$. Let also $\tilde{\phi} := \tilde{\gamma}^T_i \tilde{\gamma}_i$, $\phi := E\tilde{\phi}$. The following proposition quantifies the relative error due to the terms $\tilde{\Gamma}^{ort}_k$ and $\tilde{\Gamma}_k$.

Proposition 4.2 Let $SNR^{ort}_i := \frac{\|\gamma^{ort}_i\|^2}{E\|\tilde{\gamma}^{ort}_i\|^2}$ and

$$SNR^{obl}_i := \frac{\|\gamma_i\|^2}{E\|\tilde{\gamma}_i\|^2} = \frac{\|\gamma_i\|^2}{\sum_{j=1}^{r} \phi^2_{jj}}$$

where γ_{ij} is the ij-th element of the matrix Γ_k.

Then the following holds:

$$\frac{SNR^{ort}_i}{SNR^{obl}_i} = \frac{\left(\sum_{j=1}^{r} \phi_{jj}\right) \left(\sum_{j=1}^{n} \gamma^2_{ij} \left(1 - \sigma^2_j\right)\right)}{\left(\sum_{j=1}^{r} \phi_{jj} \left(1 - \rho^2_j\right)\right) \left(\sum_{j=1}^{n} \gamma^2_{ij}\right)} \tag{22}$$

Since $\mathcal{X}^{obl}_k \subset \mathcal{P}_{0|k-1}$ then $\rho_j \geq \sigma_j \quad j = 1, \ldots, n$; moreover, in general $\rho_j > 0$, $j = 1, \ldots, mk$ and

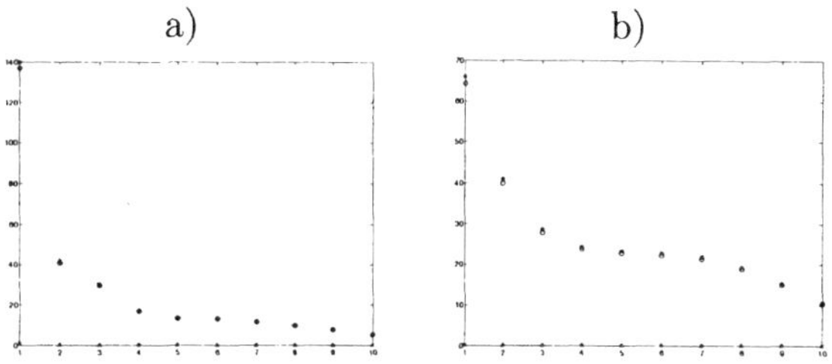

Fig. 1. Signal to noise ratio on the rows of the prediction matrix; a) joint identification b) deterministic identification stars: orthogonal projection, crosses: oblique projection, circles: estimated SNR orthogonal projection with the approximation of the previous section.

$mk \geq n$. This implies that it is likely that the effect of amplified perturbations in oblique projections may worsen the picture more than what is done by the norm reduction on the rows of the predictor matrix caused by the orthogonal projection.

Using an argument similar to the one presented in (Viberg et al., 1997) (theorem 1), one could show that the asymptotic variance on the estimated observability space (see (Viberg et al., 1997) for an explanation) is the same for the two procedures when $\mathcal{X}^{obl}_k$ is spanned by (a subset of) the singular vectors in $\mathcal{P}_{0|k-1}$ w.r.t. $\mathcal{U}_{k|2k-1}$ (implying that the σ_i's are a subset of the ρ_i's). When this is not the case the situation is not clear and further investigation is necessary. However there is experimental evidence that the orthogonal projection approach performs better in a broad range of situations.

4.1. *Experimental Results*

In this section we shall present an experimental study which compares the use of oblique and orthogonal projections. To obtain an estimate of the signal to noise ratio we have run Monte Carlo simulations. Estimates of the observability space (assuming the order to be known) via SVD are computed. A comparison between the true and estimated observability spaces via the canonical angles is performed and an estimate of the system poles is obtained via SVD (Total Least Squares). Both estimation of the observability space associated to the deterministic system and to the joint deterministic-stochastic system have been computed. This corresponds, for the orthogonal projetion case to the well-known PI-MOESP and PO-MOESP schemes (Verhaegen, 1994). The data length was taken to be $N = 500$; every experiment has been repeated 500 times and averages are taken.

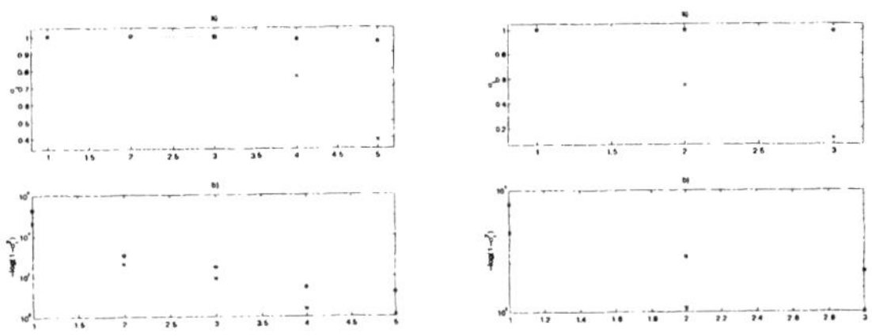

Fig. 2. Canonical correlations between true and estimated extended observability space: left joint identification right deterministic identification a) canonical correlations σ_i b) $-\log(1 - \sigma_i^2)$. stars: orthogonal projection, crosses: oblique projection.

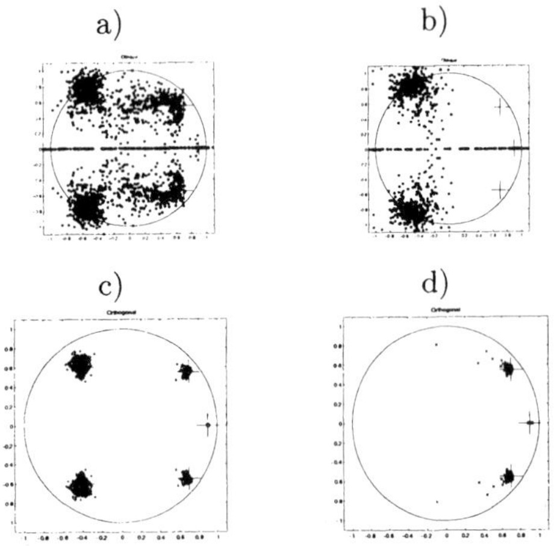

Fig. 3. Estimated poles with shift invariance method: a) and c) joint identification b) and d) deterministic identification a) and b) oblique projection; c) and d) orthogonal projection.

5. Conclusions

In this paper we have presented an error analysis of two most common subspace identification algorithms. This analysis is based on the conditioning of oblique projections. We have shown that generically the determination of the observability matrix should be done using an orthogonal projection instead of oblique projection.

6. REFERENCES

M. Cantoni and G. Vinnicombe (1999). Linear Feedback Systems and the Graph Topology. *submitted to IEEE Transaction of Automatic Control*

Chiuso, A. Geometric methods for Subspace Identification, *Ph.D. Thesis*, Dipartimento di Elettronica ed Informatica, Universitá di Padova, Italy, 2000.

Chiuso, A. and Picci G. (1999). Subspace Identification by Orthogonal Decomposition. In *Proc. 14th IFAC World Congress* , Vol. , 241–246.

Chiuso, A. and Picci G. (2000). Probing inputs for subspace identification, *submitted to CDC2000.*

Bauer, D. and Jansson M. Analysis of the asymptotic properties of the moesp type of subspace algorithms. *submitted to Automatica*, 1998.

Golub, G. H. and van Loan, C. R. (1989). *Matrix Compuations (2nd ed.)*. The Johns Hopkins Univ. Press.

Jansson M. and Wahlberg B. Counterexample to general consistency of subspace system identification methods. In *Proceedings of SYSID97*, Fukuoka, Japan,(1997) 1677–1682.

T. Katayama and G. Picci (1999) Realization of Stochastic Systems with Exogenous Inputs and Subspace Identification Methods, *Automatica*, accepted for pubblication

H. Kawauchi, A. Chiuso, T. Katayama, G.Picci. Comparison of Two Subspace Identification Methods for Combined Deterministic - Stochastic Systems, *Proc. of The 31st ISCIE International Symposium on Stochastic Systems Theory and its Applications*, Yokohama, Japan, 1999.

Larimore, W. E. (1990). Canonical Variate Analysis in Identification, Filtering, and Adaptive Control. In *Proc. 29th IEEE Conf. Decision & Control*, Honolulu, 596–604.

Lindquist, A. and G. Picci. Canonical Correlation Analysis, Approximate Covariance Extension, and Identification of Stationary Time Series. *Automatica*, **32**, 709-733.

M. Moonen, B. De Moor, L. Vanderberghe and J. Vandewalle (1989). On- and off-line identification of linear state-space models. *International Journal of Control* **49**, 219–232.

van Overschee, P. and B. De Moor (1994). N4SID: Subspace algorithms for the identification of combined deterministic– stochastic systems. *Automatica* **30**, 75–93.

van Overschee P. and B. De Moor, Choice of State-space Basis in Combied Deterministic-Stochastic Subspace Identification, *Automatica* **31** (1995), pp. 1877–1883.

van Overschee P. and B. De Moor, A unifying theorem for subspace system identification algorithms and its interpretation. *Proc. 10th IFAC Symposium on System Identification* **2** (1994), pp. 145–156.

Verhaegen, M. (1994). Identification of the deterministic part of mimo state space models given in innovations form from input-output data. *Automatica* **30**, 61–74.

Viberg M., Wahlberg B. and Ottersten B. (1997).Analysis of State Space Identification Methods Based on Instrumental Variables and Subspace Fitting. *Automatica* **33**, 1603–1616.

Wimmer, H. Lipschitz continuity of oblique projections. Working Notes.

ASYMPTOTIC VARIANCE ANALYSIS OF SUBSPACE IDENTIFICATION METHODS *

Magnus Jansson

Dept. of Signals, Sensors and Systems
Signal Processing
Royal Inst. of Technology (KTH), Stockholm, Sweden

Abstract: The class of subspace algorithms for system identification is an interesting complement to the maximum likelihood or prediction error methods, especially for multivariable systems. The statistical analysis of the subspace methods is difficult since the estimates depend on the data in a rather complicated manner. Previous results include proofs of generic consistency and asymptotic normality of the estimates. However, no explicit transparent expression for the covariance matrix of the limiting distribution has so far been reported because of the aforementioned difficulties. The main objective of this paper is to provide a methodology that simplifies the asymptotic analysis of subspace based estimation algorithms. The basic idea is illustrated by deriving the asymptotic covariance matrix corresponding to the estimates of the state-space matrices (or the transfer function estimate) of a quite general subspace algorithm.

Keywords: System identification, subspace methods, asymptotic analysis, performance analysis.

1. INTRODUCTION

The subspace methods for system identification estimates linear state-space models directly from time-discrete observations. The ideas behind these methods go back to classical state-space realization theory. The main observation used in the more recent algorithms (Van Overschee and De Moor, 1996; Verhaegen, 1994; Larimore, 1990; Peternell *et al.*, 1996) is that, under the assumption that there exists a true underlying finite order linear time invariant system, a certain data cross correlation matrix is rank deficient with rank equal to the system order. Moreover, an estimate of the observability matrix or the state-trajectory can be obtained by utilizing the singular value decomposition of the aforementioned cross correlation matrix. In contrast to the more classical approaches of maximum likelihood or prediction error methods (PEM), the subspace methods avoid the use of (multivariable) canonical forms and the system order is the only structural parameter needed. The subspace methods are also computationally attractive since only standard matrix operations are utilized to calculate the estimates and they do not use iterative minimization techniques as, in general, are needed for the computation of the PEM estimates. On the other hand, the theoretical statistical properties of PEM are by now well established (see e.g. (Ljung, 1987)).

Several papers devoted to the statistical analysis of the subspace methods have appeared in recent years. An asymptotic analysis of a frequency domain method was undertaken in (McKelvey, 1995). Consistency of some of the time domain methods for systems with or without observed inputs have been studied in (Deistler *et al.*, 1995; Peternell *et al.*, 1996; Bauer *et al.*, 1999; Jansson and Wahlberg, 1998; Bauer and Jansson, 2000). For systems without observed inputs, (Bauer *et al.*, 1999) prove consistency and derive the lim-

* This work was supported in part by the Swedish Foundation for International Cooperation in Research and Higher Education (STINT).

iting distribution for the estimates. In the current paper we will focus on subspace algorithms that estimate a state-space model of systems with observed inputs. For this case, the asymptotic distribution for the pole estimates was derived in (Viberg *et al.*, 1993; Viberg *et al.*, 1997). A more complete analysis presented in (Bauer and Jansson, 2000) showed that the estimates of the matrices in the state-space model are generically consistent under weak conditions, with an asymptotic normal distribution. The approach taken in (Bauer and Jansson, 2000) (which is similar to that in (Bauer *et al.*, 1999)) consists of a straightforward linearization of the mapping from the data to the estimates. No explicit expression for the covariance matrix of the limiting distribution was given in (Bauer and Jansson, 2000) partly because of the complexity of this mapping. One of the difficulties in the statistical analysis of estimates based on quantities from singular value or eigenvalue decompositions is caused by the possible non-uniqueness of these quantities. For example, there is always a sign ambiguity among the singular vectors. See e.g. (Bauer *et al.*, 1999) for an extensive discussion on this issue.

In this paper we use a transformation of the estimates which makes the derivation of the limiting distribution simpler compared to the previous approaches. The idea is to use the fact that we can apply a similarity transformation to the estimates of the state-space matrices without altering the estimate of the transfer function or any other canonical system quantity. It turns out that for a suitably chosen similarity transformation the above mentioned linearization is significantly simplified. Additionally, the analysis to be presented becomes invariant to the ambiguity in the orientation of the singular vectors.

2. PROBLEM FORMULATION

Consider a linear time-invariant system described by the state-space equation

$$\begin{cases} \mathbf{x}(t+1) = \mathbf{A}\mathbf{x}(t) + \mathbf{B}\mathbf{u}(t) + \mathbf{K}\mathbf{e}(t) \\ \mathbf{y}(t) = \mathbf{C}\mathbf{x}(t) + \mathbf{D}\mathbf{u}(t) + \mathbf{e}(t) \end{cases} \tag{1}$$

where

(i) $\mathbf{u}(t) \in \mathbb{R}^{n_u}$ and $\mathbf{y}(t) \in \mathbb{R}^{n_y}$ denote the input and output signals, respectively. The input $\mathbf{u}(t)$ is assumed to be a quasi-stationary signal (Ljung, 1987).

(ii) $\mathbf{x}(t) \in \mathbb{R}^{n}$ is the state-vector and n is the system order. We assume that the system is minimal in the sense that the system cannot be described by a state-space form of order less than n. Furthermore, the system is assumed to be stable.

(iii) $\mathbf{e}(t) \in \mathbb{R}^{n_y}$ denotes the innovation process and it is assumed to be a zero-mean white Gaussian random sequence. We also assume that $\mathbf{e}(t)$ is independent of the input signal $\mathbf{u}(s)$ for all t and s (the system is assumed to operate in open loop).

In the model (1), the matrices $\mathbf{A} \in \mathbb{R}^{n\times n}$, $\mathbf{B} \in \mathbb{R}^{n\times n_u}$, $\mathbf{C} \in \mathbb{R}^{n_y\times n}$, $\mathbf{D} \in \mathbb{R}^{n_y\times n_u}$, $\mathbf{K} \in \mathbb{R}^{n\times n_y}$ and the covariance matrix of the innovations are unknown and need to be estimated from observations of $\mathbf{u}(t)$ and $\mathbf{y}(t)$ for $t = 1, 2, \ldots, N$. In this paper we will focus on the estimation of the transfer function from $\mathbf{u}(t)$ to $\mathbf{y}(t)$ and, hence, on the estimation of $\{\mathbf{A}, \mathbf{B}, \mathbf{C}, \mathbf{D}\}$. In the next section we introduce the estimation method considered in the analysis, which in turn is presented in Section 4.

3. ESTIMATION METHOD

In the following we give a brief description of the method for estimating $\{\mathbf{A}, \mathbf{B}, \mathbf{C}, \mathbf{D}\}$ that we will be concerned with. However, first we need to introduce some preliminary notation. Define the vectors of stacked inputs, outputs and innovations as

$$\mathbf{y}_f(t) = \left[\mathbf{y}^T(t)\ \mathbf{y}^T(t+1)\ \ldots\ \mathbf{y}^T(t+f-1)\right]^T$$
$$\mathbf{u}_f(t) = \left[\mathbf{u}^T(t)\ \mathbf{u}^T(t+1)\ \ldots\ \mathbf{u}^T(t+f-1)\right]^T$$
$$\mathbf{e}_f(t) = \left[\mathbf{e}^T(t)\ \mathbf{e}^T(t+1)\ \ldots\ \mathbf{e}^T(t+f-1)\right]^T$$

where f is an integer chosen by the user. By employing (1), we can establish the relation

$$\mathbf{y}_f(t) = \mathbf{\Gamma}\mathbf{x}(t) + \mathbf{\Phi}\mathbf{u}_f(t) + \mathbf{v}(t) \tag{2}$$

where $\mathbf{v}(t) = \mathbf{\Psi}\mathbf{e}_f(t)$ and

$$\mathbf{\Gamma} = \begin{bmatrix} \mathbf{C} \\ \mathbf{CA} \\ \vdots \\ \mathbf{CA}^{f-1} \end{bmatrix} \tag{3}$$

$$\mathbf{\Phi} = \begin{bmatrix} \mathbf{D} & \mathbf{0} & \cdots & \mathbf{0} \\ \mathbf{CB} & \mathbf{D} & & \\ \mathbf{CAB} & \mathbf{CB} & \ddots & \vdots \\ \vdots & & \ddots & \mathbf{0} \\ \mathbf{CA}^{f-2}\mathbf{B} & & \mathbf{CB} & \mathbf{D} \end{bmatrix} \tag{4}$$

$$\mathbf{\Psi} = \begin{bmatrix} \mathbf{I} & \mathbf{0} & \cdots & \mathbf{0} \\ \mathbf{CK} & \mathbf{I} & & \\ \mathbf{CAK} & \mathbf{CK} & \ddots & \vdots \\ \vdots & & \ddots & \mathbf{0} \\ \mathbf{CA}^{f-2}\mathbf{K} & & \mathbf{CK} & \mathbf{I} \end{bmatrix}. \tag{5}$$

Since $\mathbf{u}(t)$ is assumed to be a quasi-stationary signal, we will deal with mixed deterministic and random signals. It is then convenient to follow the conventions introduced in (Ljung, 1987) and define the "averaging operator"

$$\bar{\mathrm{E}}\{\cdot\} = \lim_{N\to\infty} \frac{1}{N} \sum_{t=1}^{N} \mathrm{E}\{\cdot\}$$

where $\mathrm{E}\{\cdot\}$ is the statistical expectation operator. In what follows we will make extensive use of the notation

$$\mathbf{R}_{\mathbf{uv}}(\tau) = \bar{\mathrm{E}}\{\mathbf{u}(t)\mathbf{v}(t-\tau)^T\} \tag{6}$$

for the covariance matrix[1] of any (zero-mean) random or quasi-stationary vector processes $\mathbf{u}(t)$ and $\mathbf{v}(t)$. The sample counterpart of (6) will be denoted as

$$\hat{\mathbf{R}}_{\mathbf{uv}}(\tau) = \frac{1}{N}\sum_{t=1}^{N} \mathbf{u}(t)\mathbf{v}^T(t-\tau).$$

If $\tau = 0$ we simply write $\mathbf{R}_{\mathbf{uv}}$ and $\hat{\mathbf{R}}_{\mathbf{uv}}$ in lieu of $\mathbf{R}_{\mathbf{uv}}(0)$ and $\hat{\mathbf{R}}_{\mathbf{uv}}(0)$.

Next, let $\mathbf{z}(t) \in \mathbb{R}^{n_z}$ be a vector valued quasi-stationary signal with the following properties:

$$\hat{\mathbf{R}}_{\mathbf{u}_f\mathbf{z}} \equiv \mathbf{0}, \qquad \text{for all } N \tag{7}$$

$$\mathbf{R}_{\mathbf{vz}}(t-s) = \mathbf{0}, \qquad \text{for } t \geq s \tag{8}$$

$$\text{rank}(\mathbf{R}_{\mathbf{xz}}) = n \tag{9}$$

where rank($\mathbf{A}$) denotes the rank of the matrix $\mathbf{A}$. The above conditions on $\mathbf{z}(t)$ are typical for subspace identification methods. A common choice of $\mathbf{z}(t)$ is

$$\mathbf{z}(t) = \hat{\mathbf{W}}_c\big(\mathbf{p}(t) - \hat{\mathbf{R}}_{\mathbf{pu}_f}\hat{\mathbf{R}}^{-1}_{\mathbf{u}_f\mathbf{u}_f}\mathbf{u}_f(t)\big) \tag{10}$$

where $\hat{\mathbf{W}}_c$ is a data dependent "weighting matrix" and $\mathbf{p}(t)$ is a vector with "past" observations

$$\mathbf{p}(t) = \Big[\mathbf{y}^T(t-p)\ \ldots\ \mathbf{y}^T(t-1)$$
$$\mathbf{u}^T(t-p)\ \ldots \mathbf{u}^T(t-1)\Big]^T.$$

Here, p is an integer chosen by the user. With the choice of $\mathbf{z}(t)$ given in (10) it can be readily verified that the conditions (7) and (8) hold under the assumptions made in Section 2. The condition (9), which is crucial for the consistency of the method, has been extensively analyzed in (Jansson and Wahlberg, 1998) and (Bauer and Jansson, 2000). The conclusion from these papers is that (9) holds generically under weak persistence of excitation conditions on $\mathbf{u}(t)$.

We are now prepared to outline the estimation method which is similar to the MOESP algorithm (Verhaegen, 1994) (see also (Bauer and Jansson, 2000) and (Stoica and Jansson, 1999)). Let $\hat{\mathbf{W}}_r$ be a symmetric positive definite data dependent weighting matrix and compute the singular value decomposition (SVD)

$$\hat{\mathbf{Q}}\hat{\mathbf{\Sigma}}\hat{\mathbf{V}}^T = \hat{\mathbf{W}}_r\hat{\mathbf{R}}_{\mathbf{y}_f\mathbf{z}} \tag{11}$$

where $\hat{\mathbf{Q}} \in \mathbb{R}^{fn_y\times fn_y}$ and $\hat{\mathbf{V}} \in \mathbb{R}^{n_z\times n_z}$ are the matrices made from the left and right singular vectors, respectively, and the elements of the diagonal matrix $\hat{\mathbf{\Sigma}} \in \mathbb{R}^{fn_y\times n_z}$ equal the singular values. Assume that $\hat{\mathbf{W}}_r \to \mathbf{W}_r$ with probability one (wp 1) as $N\to\infty$, and note that $\hat{\mathbf{W}}_r\hat{\mathbf{R}}_{\mathbf{y}_f\mathbf{z}} \to \mathbf{W}_r\mathbf{R}_{\mathbf{y}_f\mathbf{z}} = \mathbf{W}_r\mathbf{\Gamma}\mathbf{R}_{\mathbf{xz}}$ (wp 1) as $N\to\infty$. If f is large enough (e.g., $f > n$) then the rank of $\mathbf{W}_r\mathbf{R}_{\mathbf{y}_f\mathbf{z}}$ will be equal to the system order n and the range space is determined by the columns of $\mathbf{W}_r\mathbf{\Gamma}$. This suggests that we may obtain an estimate of the observability matrix in (3) as

$$\hat{\mathbf{\Gamma}} = \hat{\mathbf{W}}_r^{-1}\hat{\mathbf{Q}}_s\mathbf{T}_1 \tag{12}$$

where $\hat{\mathbf{Q}}_s \in \mathbb{R}^{fn_y\times n}$ is the matrix made of the left singular vectors corresponding to the n dominating singular values in (11) (the first n columns of $\hat{\mathbf{Q}}$). The $n\times n$ matrix $\mathbf{T}_1$ in (12) reflects the fact that we only can estimate $\mathbf{\Gamma}$ up to an arbitrary similarity transformation. Common choices of $\mathbf{T}_1$ in the estimation algorithms are $\mathbf{T}_1 = \mathbf{I}$ or $\mathbf{T}_1 = \hat{\mathbf{\Sigma}}_s^{1/2}$, where $\hat{\mathbf{\Sigma}}_s$ is a diagonal matrix containing the n largest singular values in (11). Next, define the selection matrices[2]

$$\mathbf{J}_1 = \big[\mathbf{I}_{(f-1)n_y}\ \ \mathbf{0}_{(f-1)n_y\times n_y}\big]$$
$$\mathbf{J}_2 = \big[\mathbf{0}_{(f-1)n_y\times n_y}\ \ \mathbf{I}_{(f-1)n_y}\big]$$

and note that $\mathbf{J}_1\mathbf{\Gamma}\mathbf{A} = \mathbf{J}_2\mathbf{\Gamma}$. Given an estimate of $\mathbf{\Gamma}$ we may estimate $\mathbf{A}$ as the least-squares (LS) solution to the over-determined system of equations:

$$\mathbf{J}_1\hat{\mathbf{\Gamma}}\mathbf{A} \approx \mathbf{J}_2\hat{\mathbf{\Gamma}}.$$

The LS estimate of $\mathbf{A}$ can be written as

$$\hat{\mathbf{A}} = (\mathbf{J}_1\hat{\mathbf{\Gamma}})^\dagger\mathbf{J}_2\hat{\mathbf{\Gamma}} \tag{13}$$

where $(\cdot)^\dagger$ denotes the Moore-Penrose pseudo inverse. If $\mathbf{X}$ has full column rank then $\mathbf{X}^\dagger = (\mathbf{X}^T\mathbf{X})^{-1}\mathbf{X}^T$.

The $\mathbf{C}$ matrix can be estimated as the first block-row of $\hat{\mathbf{\Gamma}}$. For simplicity of presentation we introduce the selection matrix

$$\mathbf{J}_3 = \big[\mathbf{I}_{n_y}\ \ \mathbf{0}_{n_y\times(f-1)n_y}\big]$$

and write

$$\hat{\mathbf{C}} = \mathbf{J}_3\hat{\mathbf{\Gamma}}. \tag{14}$$

To estimate $\mathbf{B}$ and $\mathbf{D}$ we use the fact that $\mathbf{\Phi}$ is linear in $\mathbf{B}$ and $\mathbf{D}$. Let $\hat{\mathbf{P}} \in \mathbb{R}^{fn_y\times(fn_y-n)}$ be a full rank matrix such that $\hat{\mathbf{P}}^T\hat{\mathbf{\Gamma}} = \mathbf{0}$. ($\hat{\mathbf{P}}$ is often taken to be $\hat{\mathbf{W}}_r\hat{\mathbf{Q}}_n$ where $\hat{\mathbf{Q}}_n$ is the matrix made of the left singular vectors corresponding to the smallest singular values in (11) such that $\hat{\mathbf{Q}} = \big[\hat{\mathbf{Q}}_s\ \hat{\mathbf{Q}}_n\big]$.) Furthermore, let $\boldsymbol{\xi}(t) \in \mathbb{R}^{n_\xi}$ be a vector valued quasi-stationary signal with the following properties:

$$\mathbf{R}_{\mathbf{v}\boldsymbol{\xi}}(t-s) = \mathbf{0}, \qquad \text{for } t\geq s \tag{15}$$

$$\text{rank}(\mathbf{R}_{\mathbf{u}_f\boldsymbol{\xi}}) = fn_u. \tag{16}$$

Typically, $\boldsymbol{\xi}(t)$ is made from the input sequence, e.g., $\boldsymbol{\xi}(t) = \hat{\mathbf{R}}^{-1}_{\mathbf{u}_f\mathbf{u}_f}\mathbf{u}_f(t)$. By recalling the relation (2) and using the notation introduced above, we have

$$\hat{\mathbf{P}}^T\hat{\mathbf{R}}_{\mathbf{y}_f\boldsymbol{\xi}} \approx \hat{\mathbf{P}}^T\hat{\mathbf{\Phi}}\hat{\mathbf{R}}_{\mathbf{u}_f\boldsymbol{\xi}} \tag{17}$$

where $\hat{\mathbf{\Phi}}$ is the matrix $\mathbf{\Phi}$ in (4) with $\mathbf{A}$ and $\mathbf{C}$ replaced by $\hat{\mathbf{A}}$ and $\hat{\mathbf{C}}$. The approximation in (17)

[1] It is a slight abuse of notation to use the term "covariance matrix" unless both $\mathbf{u}(t)$ and $\mathbf{v}(t)$ are zero-mean random processes.

[2] $\mathbf{I}_n$ denotes an $n\times n$ identity matrix and $\mathbf{0}_{n\times m}$ denotes an $n\times m$ matrix of zeros.

is due to the facts that we use estimates $\hat{\mathbf{A}}, \hat{\mathbf{C}}$ and $\hat{\mathbf{P}}$ in lieu of the true quantities and that $\hat{\mathbf{R}}_{\mathbf{v}\xi}$ is not identically zero in finite samples. Since $\hat{\mathbf{\Phi}}$ is linear in $\mathbf{B}$ and $\mathbf{D}$, there exists a matrix $\hat{\mathbf{L}}$ that depends on $\hat{\mathbf{A}}$ and $\hat{\mathbf{C}}$ such that

$$\text{vec}(\hat{\mathbf{\Phi}}) = \hat{\mathbf{L}}\boldsymbol{\theta} \tag{18}$$

where we have defined

$$\boldsymbol{\theta} = \text{vec}([\mathbf{B}\ \mathbf{D}]). \tag{19}$$

Here, $\text{vec}(\mathbf{A})$ denotes a vector where the columns of $\mathbf{A}$ have been stacked on top of each other. Next we use (18) and the fact that for three matrices of conformable size $\mathbf{X}$, $\mathbf{Y}$ and $\mathbf{Z}$ it holds that $\text{vec}(\mathbf{XYZ}) = \mathbf{Z}^T \otimes \mathbf{X}\,\text{vec}(\mathbf{Y})$, where $\otimes$ denotes Kronecker matrix product, to rewrite (17) as

$$\text{vec}(\hat{\mathbf{P}}^T\hat{\mathbf{R}}_{\mathbf{y}_f\xi}) \approx (\hat{\mathbf{R}}^T_{\mathbf{u}_f\xi} \otimes \hat{\mathbf{P}}^T)\hat{\mathbf{L}}\boldsymbol{\theta}. \tag{20}$$

Hence, we can estimate $\mathbf{B}$ and $\mathbf{D}$ (or $\boldsymbol{\theta}$, cf. (19)) by solving the approximate equation (20) in a least-squares sense

$$\hat{\boldsymbol{\theta}} = (\hat{\mathbf{L}}^T(\hat{\mathbf{R}}_{\mathbf{u}_f\xi}\hat{\mathbf{R}}^T_{\mathbf{u}_f\xi} \otimes \hat{\mathbf{P}}\hat{\mathbf{P}}^T)\hat{\mathbf{L}})^{-1} \times \hat{\mathbf{L}}^T(\hat{\mathbf{R}}_{\mathbf{u}_f\xi} \otimes \hat{\mathbf{P}})\,\text{vec}(\hat{\mathbf{P}}^T\hat{\mathbf{R}}_{\mathbf{y}_f\xi}). \tag{21}$$

4. ASYMPTOTIC ANALYSIS

As is well known, the state-space realization in (1) is unique only up to a similarity transformation. The estimated transfer function

$$\hat{\mathbf{G}}(z) = \hat{\mathbf{C}}(z\mathbf{I} - \hat{\mathbf{A}})^{-1}\hat{\mathbf{B}} + \hat{\mathbf{D}} \tag{22}$$

remains the same if $\{\hat{\mathbf{A}}, \hat{\mathbf{B}}, \hat{\mathbf{C}}, \hat{\mathbf{D}}\}$ is replaced by $\{\mathbf{T}^{-1}\hat{\mathbf{A}}\mathbf{T}, \mathbf{T}^{-1}\hat{\mathbf{B}}, \hat{\mathbf{C}}\mathbf{T}, \hat{\mathbf{D}}\}$ for any non-singular $n\times n$ matrix $\mathbf{T}$. Also, the estimated poles or the eigenvalues of $\hat{\mathbf{A}}$ are identical to the eigenvalues of $\mathbf{T}^{-1}\hat{\mathbf{A}}\mathbf{T}$. A similar result holds for any canonical system quantity related to the transfer function $\hat{\mathbf{G}}(z)$. Hence, to evaluate the accuracy of the estimation method outlined in the previous section, it is natural to study the quantities

$$\begin{aligned}
\tilde{\mathbf{A}}_\mathbf{T} &\triangleq \mathbf{T}^{-1}\hat{\mathbf{A}}\mathbf{T} - \mathbf{A}\\
\tilde{\mathbf{B}}_\mathbf{T} &\triangleq \mathbf{T}^{-1}\hat{\mathbf{B}} - \mathbf{B}\\
\tilde{\mathbf{C}}_\mathbf{T} &\triangleq \hat{\mathbf{C}}\mathbf{T} - \mathbf{C}\\
\tilde{\mathbf{D}} &\triangleq \hat{\mathbf{D}} - \mathbf{D}
\end{aligned}$$

for some, possibly data dependent, matrix $\mathbf{T}$. Henceforth, $\{\mathbf{A}, \mathbf{B}, \mathbf{C}, \mathbf{D}\}$ denote the system matrices of some realization of the true system (1). In the following we will use the symbol $\simeq$ to denote an equality up to first order perturbations in terms of $\tilde{\mathbf{A}}_\mathbf{T}$ etc.. Define the true transfer function as

$$\mathbf{G}(z) = \mathbf{C}[z\mathbf{I} - \mathbf{A}]^{-1}\mathbf{B} + \mathbf{D}.$$

A first order linearization of $\hat{\mathbf{G}}(z)$ in (22) around $\mathbf{G}(z)$ can then readily be obtained:

$$\begin{aligned}
\hat{\mathbf{G}}(z) &= \hat{\mathbf{C}}\mathbf{T}(z\mathbf{I} - \mathbf{T}^{-1}\hat{\mathbf{A}}\mathbf{T})^{-1}\mathbf{T}^{-1}\hat{\mathbf{B}} + \hat{\mathbf{D}}\\
&= (\mathbf{C} + \tilde{\mathbf{C}}_\mathbf{T})[z\mathbf{I} - (\mathbf{A} + \tilde{\mathbf{A}}_\mathbf{T})]^{-1}\\
&\quad \times (\mathbf{B} + \tilde{\mathbf{B}}_\mathbf{T}) + (\mathbf{D} + \tilde{\mathbf{D}})\\
&\simeq \mathbf{C}[z\mathbf{I} - \mathbf{A}]^{-1}\mathbf{B} + \mathbf{D} + \tilde{\mathbf{C}}_\mathbf{T}[z\mathbf{I} - \mathbf{A}]^{-1}\mathbf{B}\\
&\quad + \mathbf{C}[z\mathbf{I} - \mathbf{A}]^{-1}\tilde{\mathbf{A}}_\mathbf{T}[z\mathbf{I} - \mathbf{A}]^{-1}\mathbf{B}\\
&\quad + \mathbf{C}[z\mathbf{I} - \mathbf{A}]^{-1}\tilde{\mathbf{B}}_\mathbf{T} + \tilde{\mathbf{D}}.
\end{aligned}$$

Hence if we, for notational simplicity, define

$$\begin{aligned}
\mathbf{H}_1 &= ((z\mathbf{I} - \mathbf{A})^{-1}\mathbf{B})^T \otimes \mathbf{I}\\
\mathbf{H}_2 &= \mathbf{I} \otimes \mathbf{C}(z\mathbf{I} - \mathbf{A})^{-1}
\end{aligned}$$

then it is clear that

$$\text{vec}(\hat{\mathbf{G}}(z) - \mathbf{G}(z)) \simeq [\mathbf{H}_1\mathbf{H}_2\ \mathbf{H}_1\ \mathbf{H}_2\ \mathbf{I}]\,\text{vec}([\tilde{\mathbf{A}}_\mathbf{T}\ \tilde{\mathbf{C}}_\mathbf{T}\ \tilde{\mathbf{B}}_\mathbf{T}\ \tilde{\mathbf{D}}]). \tag{23}$$

To establish the (asymptotic) statistics of $\text{vec}(\hat{\mathbf{G}}(z) - \mathbf{G}(z))$ we need to derive the statistical properties of $\text{vec}([\tilde{\mathbf{A}}_\mathbf{T}\ \tilde{\mathbf{C}}_\mathbf{T}\ \tilde{\mathbf{B}}_\mathbf{T}\ \tilde{\mathbf{D}}])$. It may be believed that the non-uniqueness of the system matrices will cause difficulties in the analysis. However, as will be shown shortly, the similarity transformation introduces some freedom that actually can be used to simplify the analysis. Indeed, we only need to consider a specific $\mathbf{T}$ and not the whole set of $n \times n$ non-singular matrices.

We will work under the assumption that the quantities $\mathbf{T}^{-1}\hat{\mathbf{A}}\mathbf{T} - \mathbf{A}$, $\mathbf{T}^{-1}\hat{\mathbf{B}} - \mathbf{B}$, $\hat{\mathbf{C}}\mathbf{T} - \mathbf{C}$ and $\hat{\mathbf{D}} - \mathbf{D}$ all converge to zero wp 1 as $N \to \infty$; i.e., that the estimate of e.g. the system transfer function is consistent. As mentioned before, this essentially follows from condition (9) on $\mathbf{z}(t)$. Detailed consistency analyses can be found in (Bauer and Jansson, 2000; Jansson and Wahlberg, 1998; Peternell *et al.*, 1996).

In the following, let

$$\mathbf{T} = \hat{\mathbf{\Gamma}}^\dagger\mathbf{\Gamma} \tag{24}$$

and note that from (13) we have

$$\begin{aligned}
\mathbf{T}^{-1}\hat{\mathbf{A}}\mathbf{T} &= (\mathbf{J}_1\hat{\mathbf{\Gamma}}\mathbf{T})^\dagger\mathbf{J}_2\hat{\mathbf{\Gamma}}\mathbf{T}\\
&= (\mathbf{J}_1\mathbf{\Pi}_{\hat{\mathbf{\Gamma}}}\mathbf{\Gamma})^\dagger\mathbf{J}_2\mathbf{\Pi}_{\hat{\mathbf{\Gamma}}}\mathbf{\Gamma}.
\end{aligned} \tag{25}$$

We will use the notation $\mathbf{\Pi}_\mathbf{X} = \mathbf{X}\mathbf{X}^\dagger$ to denote the orthogonal projection onto the range-space of $\mathbf{X}$, and $\mathbf{\Pi}^\perp_\mathbf{X} = \mathbf{I} - \mathbf{\Pi}_\mathbf{X}$ will denote the orthogonal projection onto the null-space of $\mathbf{X}^T$. By defining $\tilde{\mathbf{\Pi}}_{\hat{\mathbf{\Gamma}}} = \mathbf{\Pi}_{\hat{\mathbf{\Gamma}}} - \mathbf{\Pi}_\mathbf{\Gamma}$ and using (25) we obtain a first order perturbation expression for $\mathbf{T}^{-1}\hat{\mathbf{A}}\mathbf{T} - \mathbf{A}$:

$$\begin{aligned}
&\mathbf{T}^{-1}\hat{\mathbf{A}}\mathbf{T} - \mathbf{A}\\
&= (\mathbf{J}_1\mathbf{\Pi}_{\hat{\mathbf{\Gamma}}}\mathbf{\Gamma})^\dagger(\mathbf{J}_2\mathbf{\Pi}_{\hat{\mathbf{\Gamma}}}\mathbf{\Gamma} - \mathbf{J}_1\mathbf{\Pi}_{\hat{\mathbf{\Gamma}}}\mathbf{\Gamma}\mathbf{A})\\
&= (\mathbf{J}_1\mathbf{\Pi}_{\hat{\mathbf{\Gamma}}}\mathbf{\Gamma})^\dagger[\underbrace{\mathbf{J}_2\mathbf{\Pi}_\mathbf{\Gamma}\mathbf{\Gamma} - \mathbf{J}_1\mathbf{\Pi}_\mathbf{\Gamma}\mathbf{\Gamma}\mathbf{A}}_{=0}\\
&\quad + \mathbf{J}_2\tilde{\mathbf{\Pi}}_{\hat{\mathbf{\Gamma}}}\mathbf{\Gamma} - \mathbf{J}_1\tilde{\mathbf{\Pi}}_{\hat{\mathbf{\Gamma}}}\mathbf{\Gamma}\mathbf{A}]\\
&\simeq (\mathbf{J}_1\underbrace{\mathbf{\Pi}_\mathbf{\Gamma}\mathbf{\Gamma}}_{=\mathbf{\Gamma}})^\dagger(\mathbf{J}_2\tilde{\mathbf{\Pi}}_{\hat{\mathbf{\Gamma}}}\mathbf{\Gamma} - \mathbf{J}_1\tilde{\mathbf{\Pi}}_{\hat{\mathbf{\Gamma}}}\mathbf{\Gamma}\mathbf{A}).
\end{aligned} \tag{26}$$

We see that the error in the estimation of $\mathbf{A}$ to first order is related to $\tilde{\mathbf{\Pi}}_{\hat{\Gamma}}\mathbf{\Gamma}$. This observation is important of two reasons that will be explained next. First, note that the singular vectors in (11) are not uniquely determined. If the singular values are distinct, the vectors in $\hat{\mathbf{Q}}_s$ are only determined up to a sign. If some of the singular values coincide, there is also a rotational ambiguity among the corresponding singular vectors. This introduces some technical difficulties when we want to determine the statistical properties of the random matrix $\hat{\mathbf{Q}}_s$ (or $\hat{\mathbf{\Gamma}}$, cf. (12)). However, by using the $\mathbf{T}$ in (24) we showed that $\mathbf{T}^{-1}\hat{\mathbf{A}}\mathbf{T} - \mathbf{A}$ only depends on the data via $\mathbf{\Pi}_{\hat{\Gamma}}$ which is uniquely defined and converges to $\mathbf{\Pi}_{\Gamma}$ wp 1 as $N \to \infty$.

The second and maybe more important reason as to why it was useful to relate $\mathbf{T}^{-1}\hat{\mathbf{A}}\mathbf{T}-\mathbf{A}$ to $\tilde{\mathbf{\Pi}}_{\hat{\Gamma}}\mathbf{\Gamma}$ stems from the fact that

$$\tilde{\mathbf{\Pi}}_{\hat{\Gamma}}\mathbf{\Gamma} \simeq \mathbf{\Pi}_{\Gamma}^{\perp}\tilde{\mathbf{\Pi}}_{\hat{\Gamma}}\mathbf{\Gamma} \simeq \mathbf{\Pi}_{\Gamma}^{\perp}\hat{\mathbf{R}}_{\mathbf{vz}}\mathbf{R}_{\mathbf{xz}}^{\dagger}. \tag{27}$$

Due to space limitations, a proof of this relation will be reported elsewhere.

Combining (27) and (26) immediately gives

$$\mathrm{vec}(\mathbf{T}^{-1}\hat{\mathbf{A}}\mathbf{T} - \mathbf{A}) \simeq \mathbf{F}_A\,\mathrm{vec}(\hat{\mathbf{R}}_{\mathbf{vz}}) \tag{28}$$

where

$$\mathbf{F}_A = (\mathbf{R}_{\mathbf{xz}}^{\dagger})^T \otimes \left((\mathbf{J}_1\mathbf{\Gamma})^{\dagger}\mathbf{J}_2\mathbf{\Pi}_{\Gamma}^{\perp}\right) - \left(\mathbf{A}^T(\mathbf{R}_{\mathbf{xz}}^{\dagger})^T\right) \otimes \left((\mathbf{J}_1\mathbf{\Gamma})^{\dagger}\mathbf{J}_1\mathbf{\Pi}_{\Gamma}^{\perp}\right)$$

is a deterministic matrix that depends on the true system parameters.

Using (14), (24) and (27) we can write

$$\begin{aligned}\hat{\mathbf{C}}\mathbf{T} - \mathbf{C} &= \mathbf{J}_3\mathbf{\Pi}_{\hat{\Gamma}}\mathbf{\Gamma} - \mathbf{J}_3\mathbf{\Pi}_{\Gamma}\mathbf{\Gamma}\\ &= \mathbf{J}_3\tilde{\mathbf{\Pi}}_{\hat{\Gamma}}\mathbf{\Gamma} \simeq \mathbf{J}_3\mathbf{\Pi}_{\Gamma}^{\perp}\hat{\mathbf{R}}_{\mathbf{vz}}\mathbf{R}_{\mathbf{xz}}^{\dagger}\end{aligned}$$

or, equivalently,

$$\mathrm{vec}(\hat{\mathbf{C}}\mathbf{T} - \mathbf{C}) \simeq \mathbf{F}_C\,\mathrm{vec}(\hat{\mathbf{R}}_{\mathbf{vz}}) \tag{29}$$

with the matrix $\mathbf{F}_C$ defined by

$$\mathbf{F}_C = (\mathbf{R}_{\mathbf{xz}}^{\dagger})^T \otimes \mathbf{J}_3\mathbf{\Pi}_{\Gamma}^{\perp}.$$

Note that, because of the special choice of $\mathbf{T}$ made in (24), the error terms could to first order be related to the data cross covariance matrix $\hat{\mathbf{R}}_{\mathbf{vz}}$, the asymptotic covariance matrix of which is straightforward to calculate (as will be shown shortly). For this particular choice of $\mathbf{T}$ it was also shown that the estimation error only depends on the error in $\hat{\mathbf{\Gamma}}$ that is orthogonal to the true observability space (that is, on $\mathbf{\Pi}_{\Gamma}^{\perp}\hat{\mathbf{\Gamma}}$).

From the expressions above we can also observe that the asymptotic variances corresponding to $\hat{\mathbf{A}}$ and $\hat{\mathbf{C}}$ do not depend on $\mathbf{W}_r$. In fact we can make a more general observation since the quantity $\mathbf{\Pi}_{\Gamma}^{\perp}\mathbf{\Pi}_{\hat{\Gamma}}\mathbf{\Gamma}$, which can be viewed as the "subspace estimation error," does not depend on $\mathbf{W}_r$ (cf. (27)). Previously it was only known that the pole estimation error variance is invariant to $\mathbf{W}_r$ (Jansson and Wahlberg, 1996).

The derivation of similar expressions for the $\mathbf{B}$ and $\mathbf{D}$ estimates is slightly more involved. One fact that complicates matters is that the estimates of $\mathbf{B}$ and $\mathbf{D}$ depend on $\hat{\mathbf{A}}$ and $\hat{\mathbf{C}}$. Define

$$\bar{\mathbf{T}} = \begin{bmatrix}\mathbf{I}\otimes\mathbf{T}^{-1} & \mathbf{0}\\ \mathbf{0} & \mathbf{I}\end{bmatrix}$$

and note that $\bar{\mathbf{T}}\hat{\boldsymbol{\theta}} = \mathrm{vec}([\mathbf{T}^{-1}\hat{\mathbf{B}}\ \hat{\mathbf{D}}])$. By using the facts that $\hat{\mathbf{P}}^T = \hat{\mathbf{P}}^T\mathbf{\Pi}_{\hat{\Gamma}}^{\perp}$ and $\hat{\mathbf{R}}_{\mathbf{y}_f\xi} = \mathbf{\Gamma}\hat{\mathbf{R}}_{\mathbf{x}\xi} + \mathbf{\Phi}\hat{\mathbf{R}}_{\mathbf{u}_f\xi} + \hat{\mathbf{R}}_{\mathbf{v}\xi}$ we obtain from (21):

$$\begin{aligned}\bar{\mathbf{T}}\hat{\boldsymbol{\theta}} &= (\bar{\mathbf{T}}^{-T}\hat{\mathbf{L}}^T(\hat{\mathbf{R}}_{\mathbf{u}_f\xi}\hat{\mathbf{R}}_{\mathbf{u}_f\xi}^T \otimes \hat{\mathbf{P}}\hat{\mathbf{P}}^T)\hat{\mathbf{L}}\bar{\mathbf{T}}^{-1})^{-1}\\ &\quad\times \bar{\mathbf{T}}^{-T}\hat{\mathbf{L}}^T(\hat{\mathbf{R}}_{\mathbf{u}_f\xi} \otimes \hat{\mathbf{P}}\hat{\mathbf{P}}^T)\\ &\quad\times \mathrm{vec}[\mathbf{\Pi}_{\hat{\Gamma}}^{\perp}(\mathbf{\Gamma}\hat{\mathbf{R}}_{\mathbf{x}\xi} + \mathbf{\Phi}\hat{\mathbf{R}}_{\mathbf{u}_f\xi} + \hat{\mathbf{R}}_{\mathbf{v}\xi})]\\ &= (\bar{\mathbf{T}}^{-T}\hat{\mathbf{L}}^T(\hat{\mathbf{R}}_{\mathbf{u}_f\xi}\hat{\mathbf{R}}_{\mathbf{u}_f\xi}^T \otimes \hat{\mathbf{P}}\hat{\mathbf{P}}^T)\hat{\mathbf{L}}\bar{\mathbf{T}}^{-1})^{-1}\\ &\quad\times \bar{\mathbf{T}}^{-T}\hat{\mathbf{L}}^T(\hat{\mathbf{R}}_{\mathbf{u}_f\xi} \otimes \hat{\mathbf{P}}\hat{\mathbf{P}}^T)\\ &\quad\times \left[(\hat{\mathbf{R}}_{\mathbf{u}_f\xi}^T \otimes \mathbf{I})(\mathbf{L}\boldsymbol{\theta} - \hat{\mathbf{L}}\bar{\mathbf{T}}^{-1}\boldsymbol{\theta} + \hat{\mathbf{L}}\bar{\mathbf{T}}^{-1}\boldsymbol{\theta})\right.\\ &\quad\left.+ \mathrm{vec}(\mathbf{\Pi}_{\hat{\Gamma}}^{\perp}(\mathbf{\Gamma}\hat{\mathbf{R}}_{\mathbf{x}\xi} + \hat{\mathbf{R}}_{\mathbf{v}\xi}))\right]\end{aligned} \tag{30}$$

where $\mathbf{L}$ is the matrix $\hat{\mathbf{L}}$ (cf. (18)) constructed from $\mathbf{A}$ and $\mathbf{C}$. Observe that $\hat{\mathbf{L}}\bar{\mathbf{T}}^{-1} \to \mathbf{L}$ as $N \to \infty$ (wp 1) and from the definition of $\tilde{\mathbf{\Pi}}_{\hat{\Gamma}}$ we have $\mathbf{\Pi}_{\hat{\Gamma}}^{\perp}\mathbf{\Gamma} = -\tilde{\mathbf{\Pi}}_{\hat{\Gamma}}\mathbf{\Gamma}$. Making use of these observations in (30) gives

$$\begin{aligned}\bar{\mathbf{T}}\hat{\boldsymbol{\theta}} - \boldsymbol{\theta} &= (\bar{\mathbf{T}}^{-T}\hat{\mathbf{L}}^T(\hat{\mathbf{R}}_{\mathbf{u}_f\xi}\hat{\mathbf{R}}_{\mathbf{u}_f\xi}^T \otimes \hat{\mathbf{P}}\hat{\mathbf{P}}^T)\hat{\mathbf{L}}\bar{\mathbf{T}}^{-1})^{-1}\\ &\quad\times \bar{\mathbf{T}}^{-T}\hat{\mathbf{L}}^T(\hat{\mathbf{R}}_{\mathbf{u}_f\xi} \otimes \hat{\mathbf{P}}\hat{\mathbf{P}}^T)\\ &\quad\times \left[(\hat{\mathbf{R}}_{\mathbf{u}_f\xi}^T \otimes \mathbf{I})(\mathbf{L} - \hat{\mathbf{L}}\bar{\mathbf{T}}^{-1})\boldsymbol{\theta}\right.\\ &\quad\left.+ \mathrm{vec}(\mathbf{\Pi}_{\hat{\Gamma}}^{\perp}(\mathbf{\Gamma}\hat{\mathbf{R}}_{\mathbf{x}\xi} + \hat{\mathbf{R}}_{\mathbf{v}\xi}))\right]\\ &\simeq (\mathbf{L}^T(\mathbf{R}_{\mathbf{u}_f\xi}\mathbf{R}_{\mathbf{u}_f\xi}^T \otimes \mathbf{P}\mathbf{P}^T)\mathbf{L})^{-1}\\ &\quad\times \mathbf{L}^T(\mathbf{R}_{\mathbf{u}_f\xi} \otimes \mathbf{P}\mathbf{P}^T)\\ &\quad\times \left[(\mathbf{R}_{\mathbf{u}_f\xi}^T \otimes \mathbf{I})(\mathbf{L} - \hat{\mathbf{L}}\bar{\mathbf{T}}^{-1})\boldsymbol{\theta}\right.\\ &\quad\left.+ \mathrm{vec}(\hat{\mathbf{R}}_{\mathbf{v}\xi} - \tilde{\mathbf{\Pi}}_{\hat{\Gamma}}\mathbf{\Gamma}\mathbf{R}_{\mathbf{x}\xi})\right].\end{aligned} \tag{31}$$

As expected, the error in the estimation of $\mathbf{B}$ and $\mathbf{D}$ is due to three terms: the first is due to the fact that we use $\hat{\mathbf{A}}$ and $\hat{\mathbf{C}}$ in lieu of $\mathbf{A}$ and $\mathbf{C}$ in $\mathbf{\Phi}$, the second is due to the noise, and the third is due to the error in $\hat{\mathbf{P}}$.

To evaluate the covariance matrix corresponding to $\mathbf{B}$ and $\mathbf{D}$ we first note that the vector $(\hat{\mathbf{L}}\bar{\mathbf{T}}^{-1} - \mathbf{L})\boldsymbol{\theta}$ will contain elements of the form

$$(\hat{\mathbf{C}}\hat{\mathbf{A}}^k\mathbf{T} - \mathbf{C}\mathbf{A}^k)\mathbf{B}. \tag{32}$$

Next observe that we can write

$$\begin{aligned}\hat{\mathbf{C}}\hat{\mathbf{A}}^k\mathbf{T} - \mathbf{C}\mathbf{A}^k &= \hat{\mathbf{C}}\mathbf{T}\mathbf{T}^{-1}\hat{\mathbf{A}}^k\mathbf{T} - \mathbf{C}\mathbf{A}^k\\ &= [\mathbf{C} + \tilde{\mathbf{C}}_{\mathbf{T}}][\mathbf{A} + \tilde{\mathbf{A}}_{\mathbf{T}}]^k - \mathbf{C}\mathbf{A}^k\\ &\simeq \tilde{\mathbf{C}}_{\mathbf{T}}\mathbf{A}^k + \mathbf{C}\sum_{p=0}^{k-1}\mathbf{A}^p\tilde{\mathbf{A}}_{\mathbf{T}}\mathbf{A}^{k-p-1}.\end{aligned} \tag{33}$$

By combining (33) and (32) it is clear that there exists a matrix $\mathbf{F}_L$ that depends on $\mathbf{A}$, $\mathbf{B}$ and $\mathbf{C}$ such that

$$(\hat{\mathbf{L}}\bar{\mathbf{T}}^{-1} - \mathbf{L})\boldsymbol{\theta} = \mathbf{F}_L\,\mathrm{vec}([\tilde{\mathbf{A}}_{\mathbf{T}}\ \tilde{\mathbf{C}}_{\mathbf{T}}]). \tag{34}$$

From (31), (27), (28), (29) and (34) we finally get

$$\bar{\mathbf{T}}\hat{\boldsymbol{\theta}} - \boldsymbol{\theta} = \mathbf{F}_{\boldsymbol{\theta}} \operatorname{vec}([\hat{\mathbf{R}}_{\mathbf{vz}}\ \hat{\mathbf{R}}_{\mathbf{v}\xi}]) \qquad (35)$$

for a deterministic matrix $\mathbf{F}_{\boldsymbol{\theta}}$ given by the above expressions.

Define

$$\zeta(t) = \begin{bmatrix} \mathbf{z}(t) \\ \xi(t) \end{bmatrix}$$

and note that

$$\operatorname{vec}(\hat{\mathbf{R}}_{\mathbf{v}\zeta}) = \operatorname{vec}([\hat{\mathbf{R}}_{\mathbf{vz}}\ \hat{\mathbf{R}}_{\mathbf{v}\xi}]). \qquad (36)$$

We have shown above that $\tilde{\mathbf{A}}_{\mathbf{T}}$, $\tilde{\mathbf{B}}_{\mathbf{T}}$, $\tilde{\mathbf{C}}_{\mathbf{T}}$ and $\tilde{\mathbf{D}}$ to first order can be written as linear functions of $\operatorname{vec}(\hat{\mathbf{R}}_{\mathbf{v}\zeta})$. Since $\operatorname{vec}(\hat{\mathbf{R}}_{\mathbf{v}\zeta})$ can be shown to be asymptotically zero-mean Gaussian by using the central limit theorem (see, e.g., (Ljung, 1987)), it follows that $\operatorname{vec}([\tilde{\mathbf{A}}_{\mathbf{T}}\ \tilde{\mathbf{C}}_{\mathbf{T}}\ \tilde{\mathbf{B}}_{\mathbf{T}}\ \tilde{\mathbf{D}}])$ has a limiting zero-mean Gaussian distribution. Here we assume that the mapping from $\operatorname{vec}(\hat{\mathbf{R}}_{\mathbf{v}\zeta})$ to $\operatorname{vec}([\tilde{\mathbf{A}}_{\mathbf{T}}\ \tilde{\mathbf{C}}_{\mathbf{T}}\ \tilde{\mathbf{B}}_{\mathbf{T}}\ \tilde{\mathbf{D}}])$ exists, but this follows from the consistency assumption (see e.g. (Bauer and Jansson, 2000)).

To derive the covariance matrix in the limiting distribution we first need the following result (which can be proved by straightforward calculations, see e.g. (Viberg *et al.*, 1997))

$$N\bar{\mathrm{E}}\{\operatorname{vec}(\hat{\mathbf{R}}_{\mathbf{v}\zeta}) \operatorname{vec}^T(\hat{\mathbf{R}}_{\mathbf{v}\zeta})\} \to \sum_{|\tau|<f} \mathbf{R}_{\zeta\zeta}(\tau) \otimes \mathbf{R}_{\mathbf{vv}}(\tau) \quad \text{as } N \to \infty.$$

If we partition $\mathbf{F}_{\boldsymbol{\theta}}$ in (35) as $\mathbf{F}_{\boldsymbol{\theta}} = [\mathbf{F}_{\boldsymbol{\theta}_z}\ \mathbf{F}_{\boldsymbol{\theta}_\xi}]$, corresponding to the partitioning in (36), and define

$$\mathbf{F} = \begin{bmatrix} \mathbf{F}_A & \mathbf{0} \\ \mathbf{F}_C & \mathbf{0} \\ \mathbf{F}_{\boldsymbol{\theta}_z} & \mathbf{F}_{\boldsymbol{\theta}_\xi} \end{bmatrix}$$

then

$$\begin{aligned} \lim_{N\to\infty} N\bar{\mathrm{E}}\{&\operatorname{vec}([\tilde{\mathbf{A}}_{\mathbf{T}}\ \tilde{\mathbf{C}}_{\mathbf{T}}\ \tilde{\mathbf{B}}_{\mathbf{T}}\ \tilde{\mathbf{D}}]) \\ &\times \operatorname{vec}^T([\tilde{\mathbf{A}}_{\mathbf{T}}\ \tilde{\mathbf{C}}_{\mathbf{T}}\ \tilde{\mathbf{B}}_{\mathbf{T}}\ \tilde{\mathbf{D}}])\} \\ &= \mathbf{F}[\sum_{|\tau|<f} \mathbf{R}_{\zeta\zeta}(\tau) \otimes \mathbf{R}_{\mathbf{vv}}(\tau)]\mathbf{F}^T. \qquad (37) \end{aligned}$$

By combining (23) and (37), it is straightforward to compute the covariance matrix for the transfer function estimate.

5. CONCLUSIONS

This contribution extends previous analyses of subspace identification methods. We used the fact that different state-space realizations are related by similarity transformations to relate the estimates of the state-space matrices and a given realization of the true system. Under the assumption that the true system belongs to the model class, this observation allowed us to prove that the estimated transfer function (or other canonical system quantities, such as the poles and zeros) has a limiting normal distribution with mean equal to the true transfer function and a certain covariance matrix. We also derived an expression for the covariance matrix that, even though it is quite complicated, is much simpler than those that previous analyses have provided.

6. REFERENCES

Bauer, D. and M. Jansson (2000). Analysis of the asymptotic properties of the MOESP type of subspace algorithms. *Automatica.* to appear.

Bauer, D., M. Deistler and W. Scherrer (1999). Consistency and asymptotic normality of some subspace algorithms for systems without observed inputs. *Automatica* **35**(7), 1243–1254.

Deistler, M., K. Peternell and W. Scherrer (1995). Consistency and relative efficiency of subspace methods. *Automatica* **31**(12), 1865–1875.

Jansson, M. and B. Wahlberg (1996). A linear regression approach to state-space subspace system identification. *Signal Processing* **52**(2), 103–129.

Jansson, M. and B. Wahlberg (1998). On consistency of subspace methods for system identification. *Automatica* **34**(12), 1507–1519.

Larimore, W. E. (1990). Canonical variate analysis in identification, filtering and adaptive control. In: *Proc. 29th CDC.* Honolulu, Hawaii. pp. 596–604.

Ljung, L. (1987). *System Identification: Theory for the User.* Prentice-Hall. Englewood Cliffs, NJ. 2nd ed. 1999.

McKelvey, T. (1995). Identification of State-Space Models from Time and Frequency Data. PhD thesis. Linköping University, Sweden.

Peternell, K., W. Scherrer and M. Deistler (1996). Statistical analysis of novel subspace identification methods. *Signal Processing* **52**(2), 161–177.

Stoica, P. and M. Jansson (1999). Mimo system identification: State-space versus transfer function. Submitted to IEEE Trans. on Signal Processing.

Van Overschee, P. and B. De Moor (1996). *Subspace Identification for Linear Systems: Theory – Implementation – Applications.* Kluwer Academic Publishers.

Verhaegen, M. (1994). Identification of the deterministic part of MIMO state space models given in innovations form from input-output data. *Automatica* **30**(1), 61–74.

Viberg, M., B. Ottersten, B. Wahlberg and L. Ljung (1993). Performance of subspace-based system identification methods. In: *Proc. IFAC 93.* Sydney, Australia.

Viberg, M., B. Wahlberg and B. Ottersten (1997). Analysis of state space system identification methods based on instrumental variables and subspace fitting. *Automatica* **33**(9), 1603–1616.

ON THE IMPACT OF WEIGHTING MATRICES IN SUBSPACE ALGORITHMS

D. Bauer, M. Deistler and W. Scherrer

Institut für Ökonometrie, Operations Research und Systemtheorie, Technische Universität Wien, Argentinierstr. 8, A-1040 Wien, Austria
Wolfgang.Scherrer@tuwien.ac.at

Abstract: In this paper some results concerning the asymptotic distribution of the estimates of the transfer function obtained by using subspace methods are given. The discussion centers on the estimation of the transfer function rather than on the estimation of the system matrices, since the transfer function is a system invariant. Therefore the accuracy of the transfer function estimate can be directly used to evaluate the relative efficiency of different methods. The present paper uses the techniques and the asymptotic expressions derived in (Bauer *et al.*, 1999). In particular the effects of certain weighting matrices used in the algorithms are investigated.

Keywords: subspace methods, asymptotic properties, linear systems, state space models

1. INTRODUCTION

Subspace identification methods are used for the estimation of linear, time invariant, finite dimensional state space systems. Many different algorithms have been proposed in the past, in particular `MOESP` (Verhaegen, 1994), `N4SID` (Van Overschee and DeMoor, 1994) and `CCA` (Larimore, 1983). The asymptotic properties of subspace estimates have been analyzed recently in a couple of papers: (Bauer *et al.*, 1999; Bauer and Jansson, 2000; Deistler *et al.*, 1995; Jansson, 1997; Jansson and Wahlberg, 1998; Peternell *et al.*, 1996; Viberg *et al.*, 1993). In (Jansson, 1997) it is shown, that the asymptotic accuracy of the `MOESP`-estimates of the poles of the system does not depend on a particular weighting matrix used in the algorithms, on the other hand, as is shown in (Bauer and Jansson, 2000), this is not true for the estimates of the transfer function.

In this paper we heavily draw on the results obtained in (Bauer, 1998; Bauer *et al.*, 1999). Here the aim is to analyze the influence of certain weighting matrices on the asymptotic variance of `CCA/N4SID` type estimates. In particular it will be shown, that as far as the transfer function estimates are concerned, the choice of the "past" weighting matrix does not influence the asymptotic distribution. One consequence of this result is, that the numerical computation of the asymptotic variances are simplified.

The paper is organized as follows: In the next section the model set is defined and the algorithms are described in some detail. Also the asymptotic properties are stated. Section 3 then presents the main result of this paper, i.e. the analysis of the asymptotic expressions. In section 4 the results obtained are discussed. In order to simplify the exposition, we will restrict ourselves to the case of systems with no observed inputs. Also we will mainly deal with `CCA/N4SID` type algorithms, i.e.

[1] Support by the Austrian 'Fonds zur Förderung der wissenschaftlichen Forschung', Projekt P-11213-MAT is gratefully acknowledged. Some of this work has been done, while D. Bauer was holding a postdoc position at the Department for Automatic control, Linköping University, Sweden, funded by the EU TMR project 'SI'.

with subspace algorithms, where in a first step the states are estimated, and then these state estimates are used for the estimation of the system matrices.

2. MODEL SET AND ALGORITHMS

In this paper finite dimensional, discrete time, linear state space systems without observed inputs of the form

$$\begin{aligned} x_{t+1} &= Ax_t + K\varepsilon_t \\ y_t &= Cx_t + \varepsilon_t \end{aligned} \tag{1}$$

are considered. Here $(y_t; t \in \mathbb{Z})$ denotes the s-dimensional observed output process and $(\varepsilon_t; t \in \mathbb{Z})$ denotes s-dimensional ergodic white noise of mean zero and positive definite variance Ω. $(x_t; t \in \mathbb{Z})$ denotes the n-dimensional state sequence. It will be assumed throughout that the system is stable, i.e. that $|\lambda_i(A)| < 1$ holds for all eigenvalues λ_i of A, and that the system is strictly minimum-phase, i.e. $|\lambda_i(\bar{A})| < 1$ holds for all eigenvalues of $\bar{A} = (A - KC)$. These assumptions in particular imply that the system is in innovation form and that the innovation variance is nonsingular. In order to guarantee asymptotic normality of the parameter estimates, we also impose some additional conditions:

$$\begin{aligned} \mathbb{E}\{\varepsilon_t|\mathcal{F}_{t-1}\} &= 0 \\ \mathbb{E}\{\varepsilon_t\varepsilon_t'|\mathcal{F}_{t-1}\} &= \Omega = \mathbb{E}\varepsilon_t\varepsilon_t' \\ \mathbb{E}\{\varepsilon_{t,a}\varepsilon_{t,b}\varepsilon_{t,c}|\mathcal{F}_{t-1}\} &= \omega_{a,b,c} \\ \mathbb{E}\{\varepsilon_{t,a}^4\} &< \infty \end{aligned}$$

Here $\mathbb{E}$ denotes expectation, $\mathcal{F}_t$ denotes the σ-algebra spanned by $(y_s, s \leq t)$ and $\varepsilon_{t,a}$ denotes the a-th component of the vector ε_t.

All subspace algorithms, that have been mentioned so far, hinge upon the same basic fact: Let $Y_{t,f}^+ = [y_t', \ldots, y_{t+f-1}']'$ denote the vector of the stacked future of the output process and let $Y_{t,p}^- = [y_{t-1}', \ldots, y_{t-p}']'$ denote the vector of the stacked past of the output process. Define $E_{t,f}^+$ analogously for the noise. Here f and p are integer parameters, which have to be chosen by the user. For the moment we will only impose the condition $f, p \geq n$. For the various procedures different restrictions on f and p have to be imposed, in order to ensure asymptotic normality. It follows easily from the system equations (1) that

$$Y_{t,f}^+ = \mathcal{O}_f\mathcal{K}_pY_{t,p}^- + \mathcal{O}_f\bar{A}^px_{t-p} + \mathcal{E}_fE_{t,f}^+. \tag{2}$$

Here $\mathcal{O}_f = [C', (CA)', \ldots, (CA^{f-1})']'$ and $\mathcal{K}_p = [K, \bar{A}K, \ldots, \bar{A}^{p-1}K]$. Further $\mathcal{E}_f$ denotes the block Toeplitz matrix, whose i-th block row is equal to $[CA^{i-2}K, \ldots, CK, I_s, 0^{s\times(f-i)s}]$, where I_s is the s-dimensional identity and $0^{a\times b}$ is an $a \times b$ nullmatrix. Note that in this equation the future of the output is decomposed into three terms, where the first corresponds to the finite past of the outputs, the second to the far past (i.e. the term involving x_{t-p}), and the third to the future of the noise. The various procedures proposed use this decomposition in different ways, but all procedures rely on the fact, that the term corresponding to the future of the noise is uncorrelated with the remaining terms. In the case of additional observed inputs a similar equation can be derived (cf. e.g. (Verhaegen, 1994)).

CCA/N4SID further use the fact, that $\bar{A}^p \to 0$ for $p \to \infty$. Thus when p tends to infinity this term vanishes and $\mathcal{O}_f\mathcal{K}_p$ can be estimated consistently from a regression of $Y_{t,f}^+$ onto $Y_{t,p}^-$.

The main structure of the CCA/N4SID type algorithms can be summarized as follows:

(1) Obtain an estimate $\hat{\beta}$ of $\beta = \mathcal{O}_f\mathcal{K}_p$ by regressing $Y_{t,f}^+$ on $Y_{t,p}^-$.
(2) Typically $\hat{\beta}$ will be of full rank, whereas $\beta = \mathcal{O}_f\mathcal{K}_p$ is of rank n. Thus find a rank n approximation of $\hat{\beta}$ using a weighted singular value decomposition $(\hat{W}_f^+)^{-1/2}\hat{\beta}(\hat{W}_p^-)^{1/2} = \hat{U}_n\hat{\Sigma}_n\hat{V}_n' + \hat{R}$, where $\hat{U}_n$ denotes the matrix, whose columns are the left singular vectors corresponding to the dominating n singular values, the latter are the diagonal entries in the diagonal matrix $\hat{\Sigma}_n$. $\hat{V}_n$ corresponds to the respective right singular vectors. $\hat{R}$ accounts for the remaining singular values. An estimate of $\mathcal{K}_p$ is defined as $\hat{\mathcal{K}}_p = \hat{T}\hat{V}_n'(\hat{W}_p^-)^{-1/2}$. The weightings $\hat{W}_p^-$, $\hat{W}_f^+$ and the matrix $\hat{T}$ are explained in more detail below. For the moment it suffices to state, that $\hat{T}$ is related to basis change and thus is required to be nonsingular.
(3) Compute an estimate of the state as $\hat{x}_t = \hat{\mathcal{K}}_pY_{t,p}^-$ and then use the system equations to obtain estimates of the various system matrices: Estimate C by regression of y_t onto $\hat{x}_t$. Let $\hat{C}$ denote this estimate, then $\hat{\varepsilon}_t = y_t - \hat{C}\hat{x}_t$ is an estimate of the noise. Finally $[A, K]$ is estimated by regressing $\hat{x}_{t+1}$ on $[\hat{x}_t', \hat{\varepsilon}_t']'$, and the variance Ω is estimated by the sample covariance of $\hat{\varepsilon}_t$.

Some important notes and remarks on these algorithms are given in the following:

- The approximation in step 2 uses two positive definite weighting matrices $\hat{W}_f^+$ and $\hat{W}_p^-$. It is easy to see, that the estimate $\hat{\mathcal{K}}_p$ and thus the estimates $(\hat{A}, \hat{K}, \hat{C}, \hat{\Omega})$ do not depend on the particular square roots $(\hat{W}_f^+)^{1/2}$, $(\hat{W}_p^-)^{1/2}$ of these matrices, where e.g. $(\hat{W}_f^+)^{1/2}(\hat{W}_f^+)^{\prime/2} = \hat{W}_f^+$.

- $\hat{\mathcal{K}}_p$ is obtained as a right factor for the rank n matrix $\hat{U}_n \hat{\Sigma}_n \hat{V}_n'$. Of course some kind of normalization is needed to make this factor unique. In the analysis below it will be convenient to use a normalization such that the first n columns of $\hat{\mathcal{K}}_p$ are equal to the identity matrix. Thus we choose $\hat{T} = [\hat{V}_n'(\hat{W}_p^-)^{-1/2} S_n]^{-1}$, where $S_n = [I_n, (0^{n \times (sp-n)})']'$.

  In many cases a different normalization is chosen, e.g. $\tilde{T}$ is chosen such that $\tilde{\mathcal{K}}_p = \tilde{T} \hat{V}_n' (\hat{W}_p^-)^{-1/2}$ fullfills $\tilde{\mathcal{K}}_p \hat{\Gamma}_p^- \tilde{\mathcal{K}}_p' = I_n$, where $\hat{\Gamma}_p^-$ denotes the sample variance of $Y_{t,p}^-$. It is easy to see from the regression equations defining the estimates, that the estimates $(\tilde{A}, \tilde{K}, \tilde{C}, \tilde{\Omega})$ using this normalization are related to $(\hat{A}, \hat{K}, \hat{C}, \hat{\Omega})$, via $\tilde{A} = T\hat{A}T^{-1}$, $\tilde{K} = T\hat{K}$, $\tilde{C} = \hat{C}T^{-1}$ and $\tilde{\Omega} = \hat{\Omega}$ where $T = \tilde{T}\hat{T}^{-1}$. This means that both estimates represent the same system.
- We will use the same normalization for the *true* matrices $\mathcal{O}_f$ and $\mathcal{K}_p$, i.e. we assume that $\mathcal{K}_p S_n = I_n$. Doing this, we have to assume in addition that the first n columns of the matrix $\beta = \mathcal{O}_f \mathcal{K}_p$ have full rank n. Then by a basis change, we can achieve that $\mathcal{K}_p S_n = I_n$ holds.

  In (Bauer *et al.*, 1999), it is assumed that the diagonal entries of the a.s. limit Σ_n of $\hat{\Sigma}_n$ are distinct. Using the above normalization $\mathcal{K}_p S_n = \hat{\mathcal{K}}_p S_n = I_n$, we can replace this assumption by the somewhat simpler assumption, that the first n columns of $\mathcal{K}_p$ have full rank. It is easy to see that this assumption holds on a generic set of all systems (1) of order n. (For a proof of this statement arguments completely paralelling the ones given in (Bauer *et al.*, 1999) can be used).
- In this paper only the following choices of the weighting matrices are considered:
  - $\hat{W}_f^+ = \hat{\Gamma}_f^+$, $\hat{W}_f^+ = I_{(sf)}$, and $\hat{W}_f^+ = (\gamma_{i-j})_{i,j=1,\dots f}$, where $\gamma_i = \mathbb{E} z_t z_{t-i}'$ denotes the covariances of a process (z_t) with a rational spectral density, which has no spectral zeros.
  - $\hat{W}_p^- = \hat{\Gamma}_p^-$, and $\hat{W}_p^- = I_{(ps)}$.

  Here $\hat{\Gamma}_f^+$ denotes the sample variance of $Y_{t,f}^+$. In all cases the weightings are either deterministic or functions of the sample covariances of the observed process (y_t). In the following W_f^+, W_p^- denote a.s. limits of the weightings, i.e. the sample covariances are replaced by their population counterparts. Note that `CCA` corresponds to the choice $\hat{W}_f^+ = \hat{\Gamma}_f^+$, $\hat{W}_p^- = \hat{\Gamma}_p^-$ and `N4SID` uses $\hat{W}_f^+ = I_{(sf)}$, $\hat{W}_p^- = \hat{\Gamma}_p^-$.
- Note also, that for consistency or asymptotic normality the integer p has to tend to infinity at a certain rate depending on the sample size. The integer $f \geq n$ will be chosen to be fixed and finite for notational simplicity. Similar results would be possible for procedures allowing f to increase to infinity as a function of the sample size, see (Bauer, 1998).

The following result follows from the respective results in (Bauer *et al.*, 1999):

Theorem 1. Let $f \geq n$ be a fixed integer. Further assume that p is chosen as a function of the sample size T such that $p \geq -\frac{d}{2}\frac{\log T}{\log \rho_0}$, where $d > 1$ is some real value and $\rho_0 = \max_i\{|\lambda_i(\bar{A})|\}$, and such that $p/(\log T)^a \to 0$ for some $a > 0$. Then given all the assumptions on the ergodic white noise, the true system and the integer f mentioned above:

$$\sqrt{T}\text{vec}\left[\hat{A} - A, \hat{K} - K, \hat{C} - C, \hat{\Omega} - \Omega\right] \xrightarrow{d} Z$$

where Z is a multivariate Gaussian random variable with mean zero and variance V and $\xrightarrow{d}$ denotes convergence in distribution.

Note, that the theorem prescribes rates for the increase of p, which depends on the true system. It has been proposed in (Peternell *et al.*, 1996) to use $p = 2\hat{p}_{AIC}$, where $\hat{p}_{AIC}$ is the estimate of the order in a long autoregression for modeling y_t using `AIC`. This leads to a consistent estimate in the sense, that the restriction on p is fulfilled a.s. for T large enough.

Also note that similar results can be derived for the case of additional observed inputs. For the corresponding methods and preliminary results see (Bauer, 1998) in the case of `CCA/N4SID` type of procedures or (Bauer and Jansson, 2000) for the `MOESP` type of procedures respectively.

3. THE MAIN RESULT

The result of Theorem 1 is obtained by linearizing the mapping attaching the estimates $(\hat{A}, \hat{K}, \hat{C}, \hat{\Omega})$ to the sample covariances of the observed process, and by using the fact that these sample covariances are asymptotically normal. In this paper the linearization of the SVD-step is analyzed in detail, which will lead to some results on the influence of the weighting matrices on the asymptotic distribution of the estimates.

Lemma 2. (Linearization of weighted SVD). Let

$$\mathcal{O}_f^\dagger = (\mathcal{O}_f'(W_f^+)^{-1}\mathcal{O}_f)^{-1}\mathcal{O}_f'(W_f^+)^{-1}$$

denote a generalized left inverse of $\mathcal{O}_f$, then

$$\begin{aligned}(\hat{\mathcal{K}}_p - \mathcal{K}_p) = \; & \mathcal{O}_f^\dagger(\hat{\beta} - \beta)(I - S_n\mathcal{K}_p) + \\ & O(\|\hat{\beta} - \beta\|\|\hat{W}_p^- - W_p^-\|) + \\ & O(\|\hat{\beta} - \beta\|^2)\end{aligned} \quad (3)$$

PROOF. From the properties of the SVD, it follows that the following orthogonality relations

$$\hat{\mathcal{O}}_f'(\hat{W}_f^+)^{-\prime/2}\hat{R} = 0 \qquad \hat{R}(\hat{W}_p^-)^{\prime/2}\hat{\mathcal{K}}_p' = 0,$$

hold, where $\hat{\mathcal{O}}_f$ is defined such that $\hat{U}_n\hat{\Sigma}_n\hat{V}_n' = (\hat{W}_f^+)^{-1/2}\hat{\mathcal{O}}_f\hat{\mathcal{K}}_p(\hat{W}_p^-)^{1/2}$. Using the fact that $\beta = \mathcal{O}_f\mathcal{K}_p$ has rank n and that $\mathcal{K}_pS_n = \hat{\mathcal{K}}_pS_n = I_n$, (3) can be derived from straightforward calculations.

Since the weighting matrices can only influence the estimates of the transfer function via the estimates of the states $\hat{x}_t = \hat{\mathcal{K}}_pY_{t,p}^-$ we may conclude:

- The linearization does not depend on $(\hat{W}_p^- - W_p^-)$ and $(\hat{W}_f^+ - W_f^+)$. This implies that the asymptotic distribution of the estimates is the same, if the stochastic weightings are replaced by their asymptotic limits.
- The linearization does not depend on $\hat{W}_p^-$. This implies that the asymptotic distribution does not depend on the past weighting matrix.

These observations lead directly to the main result:

Theorem 3. Given the assumptions of Theorem 1, then the asymptotic distribution of the transfer function estimates is unchanged if $\hat{W}_f^+$ and $\hat{W}_p^-$ are replaced by their a.s. limits (for fixed f and p). Also the matrix $\hat{W}_p^-$ does not influence the asymptotic distribution and thus can be set arbitrarily.

The asymptotic distribution of the transfer function estimate does not depend on the distribution of the noise.

PROOF. The first statement follows directly from equation (3). The second statement follows from the fact, that the only significant source of uncertainty in equation (3) is due to $\hat{\beta} - \beta$. This does not depend on the distribution of the noise, but only on the second moments for $p = p(T) \to \infty$ (for a proof see e.g. Lewis and Reinsel, 1985). Furthermore let $\langle a_t, b_t\rangle = \frac{1}{T}\sum_{t=p+1}^{T-p} a_tb_t'$. Then e.g.

$$\begin{aligned}\hat{C} - C &= \langle y_t - \hat{x}_t, \hat{x}_t\rangle\langle\hat{x}_t, \hat{x}_t\rangle^{-1} \\ &= (\langle\varepsilon_t, \hat{x}_t\rangle + \langle x_t - \hat{x}_t, \hat{x}_t\rangle)\langle\hat{x}_t, \hat{x}_t\rangle^{-1} \\ &= \left(\langle\varepsilon_t, \hat{x}_t\rangle + (\mathcal{K}_p - \hat{\mathcal{K}}_p)\langle Y_{t,p}^-, \hat{x}_t\rangle\right)\langle\hat{x}_t, \hat{x}_t\rangle^{-1}\end{aligned}$$

which does not depend on the noise distribution, if $\mathcal{K}_p - \hat{\mathcal{K}}_p$ has the same property. Similar straightforward calculations show the same result for $\hat{A}, \hat{K}$.

$s = 1$	p		
f	16	18	20
16	30.89	24.99	18.84
18	24.34	21.63	17.69
20	17.77	17.15	15.30
$s = 2$	p		
f	16	17	18
16	1.92	1.82	1.67
17	1.80	1.78	1.70
18	1.63	1.68	1.67

Table 1. In this table the percentage of FLOPS as counted by MATLAB needed for the calculation of the asymptotic variance using the fixed weightings as compared to the number of calculations needed for the estimated weightings is shown.The upper part shows $s = 1$, the lower part $s = 2$. In both cases $n = 2$.

4. DISCUSSION OF THE RESULTS

The main result contains a number of interesting implications: First, the result shows, that estimates of certain covariances, which are used for the weighting matrices might be exchanged with their respective expectations without changing the asymptotic distribution. This means that the (numerical) computation of the asymptotic variance can be simplified considerably. In particular there is no need to compute the linearization of the Cholesky factorization of $\hat{W}_f^+$ or $\hat{W}_p^-$ which is computationally very burdensome.

As a small example consider two systems of output dimension one and two respectively with two states. The actual values in the matrices are of no importance for our discussion. In (Bauer *et al.*, 1999) an algorithm for the calculation of the asymptotic variances is outlined. The approximation of the asymptotic variance requires p to be chosen large. When p is large, also the derivative of the Cholesky decomposition of a large matrix has to be calculated. In our implementation of the algorithms for estimating the asymptotic variances, using fixed weighting matrices (and thus omitting the calculation of the derivative of the Cholesky factor) leads to the reduction in calculations as given in Table 1.

As can be seen from the results, the difference gets worse for increasing f and p. For $s = 1$ and for $f = p = 20$ the number of flops required for the calculation assuming the weightings to be estimated is about 6 times the amount when assuming them to be fixed, while the calculated asymptotic variances differ by less than $1.7e-5$. For $s = 2$ the results are even more impressing: Less than two percent of the calculations are needed, showing that the main part of the calculation time is used for the derivative of the Cholesky factor. We want to stress again, that these results only refer to our implementation, however qualitatively the same results will be obtained regardless of the actual implementation. Since the accuracy of the approximation is influenced heavily by the choice of p, the results given in this paper lead either to more accurate calculations of the asymptotic variance with the same amount of computation time or to less computations holding the approximation accuracy fixed.

Second, the results show that the weighting of the past, $\hat{W}_p^-$ does not influence the asymptotic distribution of the estimates, if the order is specified correctly. Therefore, in order to obtain optimal estimates, one only has to choose an optimal future weighting $\hat{W}_f^+$. This might be of theoretical interest, since no analytical results on the optimal choice of this matrix exist, to the best of the authors knowledge. Theorem 3 restricts the possible design parameters to the choice of f and the weighting matrix $\hat{W}_f^+$.

However, the authors want to emphasize, that the optimization of the asymptotic accuracy of the transfer function estimates is not the only measure for the optimality of a procedure. Also the question of the approximation properties for misspecification e.g. of the order might be of interest. In this case one cannot simply neglect the influence of $\hat{W}_p^-$, see e.g. (Bauer *et al.*, 1998).

It should be remarked, that the above results may be generalized in many respects. If a different normalization for the factorization of

$$(\hat{W}_f^+)^{-T/2}\hat{U}_n\hat{\Sigma}_n\hat{V}_n'(\hat{W}_p^-)^{-1/2}$$

is chosen, then the asymptotic distribution of $(\tilde{A}, \tilde{K}, \tilde{C}, \tilde{\Omega})$ might depend on $\hat{W}_p^-$. However the asymptotic distribution of any system invariant (e.g. the poles of the system, or the transfer function) computed from these estimates, does not depend on the past weighting, since $(\tilde{A}, \tilde{K}, \tilde{C}, \tilde{\Omega})$ and $(\hat{A}, \hat{K}, \hat{C}, \hat{\Omega})$ describe the same system.

The fact, that the asymptotic distribution does not depend on the distribution of the noise sequence, adds value to the previously published plots, e.g. in (Bauer, 1998; Bauer *et al.*, 1998; Deistler *et al.*, 1995). It also is an important robustness property.

Finally we want to note that the case of additional observed inputs can be handled in a similar fashion. For the CCA/N4SID type of procedures the results are analogously, i.e. the weighting matrix $\hat{W}_p^-$ does not influence the asymptotic distribution and the asymptotic distribution is not changed, if weightings estimated from data are replaced by their expectations. For the MOESP type of algorithm, $\hat{\mathcal{O}}_f$ is used to estimate the matrices A and C. Thus by reversing the roles of $\hat{\mathcal{O}}_f$ and $\hat{\mathcal{K}}_p$ in Lemma 3, one can conclude that the asymptotic distribution of $\hat{\mathcal{O}}_f$ and thus of the estimate of A and C (if $\hat{\mathcal{O}}_f$ is normalized suitably) do not depend on the weighting $\hat{W}_f^+$. Carrying these results over to the system poles (i.e. of the eigenvalues of A) we obtain the result given in (Jansson, 1997). However in this case it is shown in (Bauer and Jansson, 2000), that the asymptotic distribution of the transfer function estimate might depend on both weightings.

5. CONCLUSIONS

In this paper the effect of the choice of the weighting matrices on the asymptotic distribution of the transfer function estimates obtained by subspace procedures has been investigated. The main result states, that the asymptotic distribution is not changed, when weightings, which are estimated from the data, are replaced by the corresponding expectations. As a byproduct we also obtain the result, that the asymptotic distribution of the estimated transfer function does not depend on the distribution of the noise. The implication of the main result is that the calculation of the asymptotic variance can be significantly speeded up. It is also shown, that one of the weighting matrices in the CCA/N4SID type of procedure does not influence the asymptotic distribution and thus can be set arbitrarily, while the other weighting matrix influences the asymptotic accuracy. The optimal choice of this matrix is still an open question. Finally we want to emphasize, that the whole discussion referred to the case, where the order is specified correctly. In the case, where a lower order is specified both weighting matrices influence the asymptotic bias and thus have to be considered, when searching for an optimal approximation.

6. REFERENCES

Bauer, D. (1998). Some Asymptotic Theory for the Estimation of Linear Systems Using Maximum Likelihood Methods or Subspace Algorithms. PhD thesis. TU Wien.

Bauer, D. and M. Jansson (2000). Analysis of the asymptotic properties of the moesp type of subspace algorithms. To appear in *Automatica*, April 2000.

Bauer, D., M. Deistler and W. Scherrer (1998). User choices in subspace algorithms. In: *Proceedings of the CDC'98 Conference*. Tampa, Florida. pp. 731–736.

Bauer, D., M. Deistler and W. Scherrer (1999). Consistency and asymptotic normality of some subspace algorithms for systems without observed inputs. *Automatica* **35**, 1243–1254.

Deistler, M., K. Peternell and W. Scherrer (1995). Consistency and relative efficiency of subspace methods. *Automatica* **31**, 1865–1875.

Jansson, M. (1997). On Subspace Methods in System Identification and Sensor Array Signal Processing. PhD thesis. KTH, Stockholm.

Jansson, M. and B. Wahlberg (1998). On consistency of subspace methods for system identification. *Automatica* **34**(12), 1507–1519.

Larimore, W. E. (1983). System identification, reduced order filters and modeling via canonical variate analysis. In: *Proc. 1983 Amer. Control Conference 2.* (H. S. Rao and P. Dorato, Eds.). Piscataway, NJ. pp. 445–451. IEEE Service Center.

Lewis, R. and G. Reinsel (1985). Prediction of multivariate time series by autoregressive model fitting. *Journal of Multivariate Analysis* **16**, 393–411.

Peternell, K., W. Scherrer and M. Deistler (1996). Statistical analysis of novel subspace identification methods. *Signal Processing* **52**, 161–177.

Van Overschee, P. and B. DeMoor (1994). N4sid: Subspace algorithms for the identification of combined deterministic-stochastic systems. *Automatica* **30**, 75–93.

Verhaegen, M. (1994). Identification of the deterministic part of mimo state space models given in innovations form from input-output data. *Automatica* **30**(1), 61–74.

Viberg, M., B. Ottersten, B. Wahlberg and L. Ljung (1993). Performance of subspace based state space system identification methods. In: *Proc. of the 12th IFAC World Congress*. Vol. 7. Sydney, Australia. pp. 369–372.

WEIGHTING IN SUBSPACE-BASED SYSTEM IDENTIFICATION

Tony Gustafsson *

* *University of California San Diego*
Department of Electrical and Computer Engineering
9500 Gilman Drive Mail Code 0407
La Jolla, CA 92093-0407, USA

Abstract: Subspace-based methods for system identification are often based on an estimate of the range space of the extended observability matrix. It is thus of great interest to investigate, and also optimize, the accuracy of the estimated subspace. Especially, the influence of certain weighting matrices is an unresolved issue. Here, an asymptotic analysis of the accuracy of the estimated subspace is presented. The main result of the analysis is that statistically sound weighting matrices are found.

Keywords: Subspace Methods, Identification, Linear Systems, State Space Models, Discrete-time Systems

1. INTRODUCTION

Methods for Subspace-based State Space System Identification (4SID) have lately been suggested as alternatives to traditional identification techniques, see e.g. (Larimore, 1983; VanOverschee, 1995; Verhaegen, 1994; Viberg, 1995). The 4SID methods are attractive since they estimate a state-space realization directly from input-output data, without any canonical parameterizations and non-linear searches required by the Prediction Error Method (PEM), cf. (Ljung, 1987; Söderström and Stoica, 1989). Although 4SID methods have been studied for a while and several successful applications have been reported, the theory is far from fully understood.

In the literature, several attempts to analyze the accuracy of 4SID methods have appeared, e.g. (Bauer, 1998; Viberg *et al.*, 1997). The available results are however complicated and little insight is gained, into how user-defined quantities affect the accuracy. The objective of the work in this paper is to present a statistical investigation of the accuracy of 4SID methods. More precisely, we will investigate the accuracy of the estimated observability matrix only. In the instrumental variable (IV) framework of (Viberg, 1995), certain weighting matrices appear. Applying the derived expressions for the asymptotic accuracy of the subspace estimate, it will be shown how these weighting matrices should be chosen.

[1] Support by the Swedish Foundation for International Cooperation in Research and Higher Education, and Telefonaktiebolaget LM Ericsson is gratefully acknowledged.

2. PROBLEM FORMULATION

Assume that the linear time-invariant dynamical system under study can be described by the following n^{th} order state-space model with p outputs and m inputs collected in $\mathbf{y}(t)$ and $\mathbf{u}(t)$ respectively:

$$\begin{aligned} \mathbf{x}(t+1) &= \mathbf{A}\mathbf{x}(t) + \mathbf{B}\mathbf{u}(t) + \mathbf{w}(t) \\ \mathbf{y}(t) &= \mathbf{C}\mathbf{x}(t) + \mathbf{D}\mathbf{u}(t) + \mathbf{v}(t) \end{aligned} \quad (1)$$

Here, $\mathbf{w}(t)$ denotes the n-dimensional process noise, $\mathbf{v}(t)$ denotes the p-dimensional measurement noise, and $\mathbf{x}(t)$ denotes the n-dimensional state vector. The following assumptions are considered to hold throughout the paper:

A1: The system described by (1) is asymptotically stable, i.e. all eigenvalues of $\mathbf{A} \in \mathbb{R}^{n \times n}$ lie strictly inside the unit-circle.

A2: The noise processes are stationary zero-mean white Gaussian noise processes and

$$E\left\{\begin{bmatrix}\mathbf{w}(t)\\ \mathbf{v}(t)\end{bmatrix}\begin{bmatrix}\mathbf{w}(t')\\ \mathbf{v}(t')\end{bmatrix}^T\right\} = \begin{bmatrix}\mathbf{R}_{ww} & \mathbf{R}_{wv}\\ \mathbf{R}_{wv}^T & \mathbf{R}_{vv}\end{bmatrix}\delta_{t,t'}, \tag{2}$$

where $\delta_{t,t'}$ denotes Kronecker's delta function. Furthermore, both $\mathbf{w}(t)$ and $\mathbf{v}(t)$ are uncorrelated with $\mathbf{u}(t')$, $\forall t, t'$. The assumption on gaussianity is introduced to simplify the derivation of the asymptotic distribution of the subspace estimate, and milder assumptions may be imposed without affecting the conclusions.

A3: The input $\mathbf{u}(t)$ is a quasi-stationary (Ljung, 1987) deterministic sequence, with correlation sequence $\mathbf{R}_{uu}(\tau) = \bar{E}\left\{\mathbf{u}(t+\tau)\mathbf{u}^T(t)\right\}$, where $\bar{E}$ is defined as

$$\bar{E}\{\cdot\} = \lim_{N\to\infty}\frac{1}{N}\sum_{t=1}^{N}E\{\cdot\}. \tag{3}$$

A4: The pair $\{\mathbf{A}, \mathbf{C}\}$ is observable, and the pair $\{\mathbf{A}, [\mathbf{B}\ \mathbf{R}_{ww}^{1/2}]\}$ is reachable. Here, $(\cdot)^{1/2}$ denotes a symmetric square root of $(\cdot)$.

Define the stacked "future" and "past" output vectors as

$$\mathbf{y}_f(t) = [\mathbf{y}^T(t), \ldots, \mathbf{y}^T(t+\alpha-1)]^T \tag{4}$$

$$\mathbf{y}_p(t) = [\mathbf{y}^T(t-1), \ldots, \mathbf{y}^T(t-\beta)]^T \tag{5}$$

where $\alpha > n$, $\beta > n/(m+p)$ are user-defined integers. The stacked vectors of the input and noise vectors are defined similarly. The following input-output relation is then easily obtained:

$$\mathbf{y}_f(t) = \mathbf{\Gamma}_\alpha\mathbf{x}(t) + \mathbf{\Phi}_\alpha\mathbf{u}_f(t) + \mathbf{n}_f(t), \tag{6}$$

where $\mathbf{n}_f(t)$ contains the contribution of the noise sources. The extended observability matrix $\mathbf{\Gamma}_\alpha$ is defined as $\mathbf{\Gamma}_\alpha = \left[\mathbf{C}^T,\ \mathbf{A}^T\mathbf{C}^T,\ \cdots(\mathbf{A}^{\alpha-1})^T\mathbf{C}^T\right]^T$, and $\mathbf{\Phi}_\alpha$ is a matrix of impulse response matrices from $\mathbf{u}(t)$ to $\mathbf{y}(t)$:

$$\mathbf{\Phi}_\alpha = \begin{bmatrix}\mathbf{D} & \mathbf{0} & \cdots & \mathbf{0}\\ \mathbf{CB} & \mathbf{D} & & \mathbf{0}\\ \vdots & \ddots & \ddots & \vdots\\ \mathbf{CA}^{\alpha-2}\mathbf{B} & \cdots & \mathbf{CB} & \mathbf{D}\end{bmatrix}. \tag{7}$$

Suppose that measurements for $t = 1, \ldots, N+\alpha+\beta-1$ are available. The key problem in subspace identification is that of estimating $\mathcal{R}\{\mathbf{\Gamma}_\alpha\}$, where $\mathcal{R}\{\mathbf{\Gamma}_\alpha\}$ denotes the (column) range space of $\mathbf{\Gamma}_\alpha$. A large class of 4SID methods attempts to solve this problem by computing an SVD:

$$\begin{aligned}\widehat{\mathbf{G}} &= \mathbf{W}_L\mathbf{Y}_f\mathbf{\Pi}_{U_f}^{\perp}\mathbf{P}^T\mathbf{W}_R = \widehat{\mathbf{Q}}\widehat{\mathbf{S}}\widehat{\mathbf{V}}^T\\ &= \begin{bmatrix}\widehat{\mathbf{Q}}_1 & \widehat{\mathbf{Q}}_2\end{bmatrix}\begin{bmatrix}\widehat{\mathbf{S}}_1 & \mathbf{0}\\ \mathbf{0} & \widehat{\mathbf{S}}_2\end{bmatrix}\begin{bmatrix}\widehat{\mathbf{V}}_1 & \widehat{\mathbf{V}}_2\end{bmatrix}^T,\end{aligned} \tag{8}$$

where $\mathbf{W}_{L,R}$ are weighting matrices. $\mathbf{W}_L$ is assumed to be non-singular, and $\mathbf{W}_R$ is assumed to be (symmetric and) positive definite. The partitioning of the SVD is such that $\widehat{\mathbf{Q}}_1 \in \mathbb{R}^{\alpha p\times n}$, and the matrix $\widehat{\mathbf{S}}$ is diagonal with non-increasing elements on its diagonal. The block Hankel matrix $\mathbf{Y}_f$ is defined as $\mathbf{Y}_f = [\mathbf{y}_f(1), \ldots, \mathbf{y}_f(N)]$. The corresponding definitions for the input and noise matrices are similar. Furthermore, the orthogonal projection onto the nullspace of $\mathbf{U}_f$ is defined as $\mathbf{\Pi}_{U_f}^{\perp} = \mathbf{I} - \mathbf{U}_f^T(\mathbf{U}_f\mathbf{U}_f^T)^{-1}\mathbf{U}_f$, provided $\mathbf{U}_f$ has full rank. The "IV-matrix" $\mathbf{P}$ should fulfill:

A5: The matrix[2]

$$\mathbf{\Delta} = \lim_{N\to\infty}\mathbf{X}_f\mathbf{\Pi}_{U_f}^{\perp}\mathbf{P}^T \tag{9}$$

is assumed to exist and to be full rank ($= n$). Here, $\mathbf{X}_f = [\mathbf{x}(\beta+1), \cdots, \mathbf{x}(\beta+N)]$.

A6: The "instruments" $\mathbf{p}(t)$ are chosen such they are uncorrelated with $\mathbf{n}_f(t)$.

Assumption **A6** is fulfilled for example when $\mathbf{P} = \mathbf{U}_p$ or $\mathbf{P} = [\mathbf{U}_p^T\ \mathbf{Y}_p^T]^T$, whereas **A5** is more difficult to guarantee (the input has to satisfy some kind of persistence of excitation requirement), cf. (Jansson and Wahlberg, 1998).

An estimate of $\mathcal{R}\{\mathbf{\Gamma}_\alpha\}$ can then be defined as $\mathbf{W}_L^{-1}\widehat{\mathbf{Q}}_1$. The system matrices $\mathbf{A}$ and $\mathbf{C}$ can be estimated from $\mathbf{W}_L^{-1}\widehat{\mathbf{Q}}_1$ using the inherent shift-invariant structure of $\mathbf{\Gamma}_\alpha$. In the literature several different methods can be found for estimating $\mathbf{B}$ and $\mathbf{D}$. To simplify the presentation, we will in the following assume that $\mathbf{W}_L = \mathbf{I}$. Later on, it is shown that $\mathbf{W}_L$ has no influence on the studied measure of quality. The key issues addressed in this paper are

(1) How should we measure the accuracy of $\widehat{\mathbf{Q}}_1$?
(2) Can we optimize the accuracy of $\widehat{\mathbf{Q}}_1$?

3. A STATISTICAL FRAMEWORK

3.1 *A Motivation of the Proposed Approach*

Desiring to investigate the accuracy of a system identification method, we first have to define a figure of merit. The most common approach in the literature is to study the asymptotic variance of estimated (canonical) parameters describing the system. Results in this direction for 4SID methods can be found in (Bauer, 1998). However, the available results are not very explicit and little insight is gained. In this section, we will revisit the problem of evaluating the asymptotic variance using a somewhat heuristic analysis.

---

[2] Throughout the paper it is assumed that the involved matrices are appropriately scaled (i.e. with $1/\sqrt{N}$), so that quantities like $\mathbf{Y}_f\mathbf{Y}_f^T \to \mathbf{R}_{y_fy_f}$, as $N \to \infty$.

Given the estimated observability matrix $\widehat{\Gamma}_\alpha$, the estimates of the system matrices $\mathbf{A}$ and $\mathbf{C}$ can be written as

$$\text{vec}\left\{\left[\widehat{\mathbf{A}}^c, \widehat{\mathbf{C}}^c\right]\right\} = \mathbf{f}(\text{vec}\{\widehat{\Gamma}_\alpha\}) \quad (10)$$

for some non-linear mapping $\mathbf{f}(\cdot)$ which is assumed to be differentiable. Here it is implicitly assumed that the system matrices are given using a canonical parameterization, denoted by superscript $(\cdot)^c$. Hence,

$$\text{vec}\{[\mathbf{A}^c, \mathbf{C}^c]\} = \mathbf{f}(\Gamma_\alpha) = \mathbf{f}(\Gamma_\alpha \mathbf{Z}) \quad (11)$$

for any non-singular matrix $\mathbf{Z}$. Depending on how $\mathbf{f}(\cdot)$ is defined, both subspace fitting and shift-invariance approaches are covered by the above formulation. In this contribution we consider only the accuracy of $\left[\widehat{\mathbf{A}}^c, \widehat{\mathbf{C}}^c\right]$.

The estimated system matrices can clearly be written as

$$\text{vec}\left\{\left[\widehat{\mathbf{A}}^c, \widehat{\mathbf{C}}^c\right]\right\} = \mathbf{f}(\text{vec}\{\Pi_{\Gamma_\alpha}\widehat{\Gamma}_\alpha + \Pi^{\perp}_{\Gamma_\alpha}\widehat{\Gamma}_\alpha\}), \quad (12)$$

where

$$\Pi_{\Gamma_\alpha} = \Gamma_\alpha\left(\Gamma_\alpha^T\Gamma_\alpha\right)^{-1}\Gamma_\alpha^T, \text{ and } \Pi^{\perp}_{\Gamma_\alpha} = \mathbf{I} - \Pi_{\Gamma_\alpha}. \quad (13)$$

Next, using the property of Eq. (11), and assuming N large, $\text{vec}\{[\mathbf{A}^c, \mathbf{C}^c]\} = \mathbf{f}(\text{vec}\{\Pi_{\Gamma_\alpha}\widehat{\Gamma}_\alpha\})$. Define also the matrix $\Gamma_\alpha^\perp$ (the orthogonal complement of Γ_α) through the relationships

$$\left(\Gamma_\alpha^\perp\right)^T\Gamma_\alpha^\perp = \mathbf{I},\ \Gamma_\alpha^T\Gamma_\alpha^\perp = \mathbf{0}. \quad (14)$$

Applying a Taylor series expansion, and since $\Pi^{\perp}_{\Gamma_\alpha} = \Gamma_\alpha^\perp(\Gamma_\alpha^\perp)^T$, the asymptotic estimation error can be written as

$$\begin{aligned}&\text{vec}\left\{\left[\widehat{\mathbf{A}}^c, \widehat{\mathbf{C}}^c\right]\right\} - \text{vec}\left\{[\mathbf{A}^c, \mathbf{C}^c]\right\}\\ &\simeq \mathbf{f}'(\Gamma_\alpha)\left(\mathbf{I}\otimes\Gamma_\alpha^\perp\right)\text{vec}\left\{(\Gamma_\alpha^\perp)^T\widehat{\Gamma}_\alpha\right\},\end{aligned} \quad (15)$$

where $\mathbf{f}'$ contains the partial derivatives of $\mathbf{f}(\cdot)$, evaluated at Γ_α. The asymptotic (normalized) estimation variance can hence be written as

$$\begin{aligned}&\mathbf{J}(\mathbf{W}_R)\\ &= \lim_{N\to\infty}\text{Var}(\sqrt{N}(\text{vec}\left\{\left[\widehat{\mathbf{A}}, \widehat{\mathbf{C}}\right]\right\} - \text{vec}\left\{[\mathbf{A}^c, \mathbf{C}^c]\right\}))\\ &= \mathbf{f}'(\Gamma_\alpha)\left(\mathbf{I}\otimes\Gamma_\alpha^\perp\right)\mathbf{R}_{\varepsilon\varepsilon}(\mathbf{W}_R)\left(\mathbf{I}\otimes\Gamma_\alpha^\perp\right)^T(\mathbf{f}'(\Gamma_\alpha))^T,\end{aligned} \quad (16)$$

where (assuming it exists)

$$\mathbf{R}_{\varepsilon\varepsilon}(\mathbf{W}_R) = \lim_{N\to\infty} E\left\{\text{vec}\left\{\varepsilon_N\right\}\text{vec}^T\left\{\varepsilon_N\right\}\right\}. \quad (17)$$

It can now be claimed that if a certain weighting matrix $\mathbf{W}_R^o$ is found, such that $\mathbf{R}_{\varepsilon\varepsilon}(\mathbf{W}_R^o) \leq \mathbf{R}_{\varepsilon\varepsilon}(\mathbf{W}_R)$ for any positive definite $\mathbf{W}_R$, it follows that $\mathbf{J}(\mathbf{W}_R^o) \leq \mathbf{J}(\mathbf{W}_R)$, irrespective of the chosen method for estimating $\mathbf{A}$ and $\mathbf{C}$. The goal should consequently be to minimize $\mathbf{R}_{\varepsilon\varepsilon}$ with respect to $\mathbf{W}_R$. Unfortunately we have not been able to achieve this. Instead, we have used a suboptimal procedure, as will be explained in the next section.

3.2 *Some Asymptotic Results*

Introduce the following general assumption, considered to hold throughout the paper.

A7: First, note that

$$\begin{aligned}\mathbf{G} &= \lim_{N\to\infty}\widehat{\mathbf{G}} = \lim_{N\to\infty}\mathbf{Y}_f\Pi^{\perp}_{U_f}\mathbf{P}^T\mathbf{W}_R\\ &= \Gamma_\alpha\underbrace{\left(\mathbf{R}_{xp} - \mathbf{R}_{xu_f}\mathbf{R}^{-1}_{u_fu_f}\mathbf{R}_{u_fp}\right)}_{\Delta}\mathbf{W}_R,\end{aligned} \quad (18)$$

exists with probability one (w.p.1). It is assumed that the state space basis implicitly used to define Γ_α (and Δ) is arbitrary, but *fixed.* For example, Γ_α could be defined using echolon coordinates. Denote the SVD of $\mathbf{G}$ as

$$\mathbf{G} = \mathbf{Q}_1\mathbf{S}_1\mathbf{V}_1^T, \quad (19)$$

where $\mathbf{Q}_1 \in \mathbb{R}^{\alpha p\times n}$ denotes the limiting left singular vectors. It is also assumed that all singular values in $\mathbf{S}_1$ are non-zero, cf. **A5**, and distinct. This assumption implies that the SVD in (19) is unique as long as the orientation of the singular vectors is fixed [3]. As a consequence of the above assumptions, $\lim_{N\to\infty}\widehat{\mathbf{Q}}_1 = \mathbf{Q}_1$ w.p.1.

Under the above assumption,

$$\Gamma_\alpha\mathbf{T}^{-1}_{W_R} = \mathbf{Q}_1,\ \mathbf{T}_{W_R}\Delta\mathbf{W}_R = \mathbf{S}_1\mathbf{V}_1^T, \quad (20)$$

for a (unique) non-singular similarity transformation $\mathbf{T}_{W_R}$ (determined among others by the SVD and $\mathbf{W}_R$). In the sequel, the following trick will be useful,

$$\begin{aligned}\mathbf{V}_1\mathbf{S}_1^{-1} &= (\mathbf{T}_{W_R}\Delta\mathbf{W}_R)^\dagger\\ &= \mathbf{W}_R\Delta^T\left(\Delta\mathbf{W}_R^2\Delta^T\right)^{-1}\mathbf{T}^{-1}_{W_R},\end{aligned} \quad (21)$$

where $(\cdot)^\dagger$ denotes the pseudo-inverse. This reformulation of $\mathbf{V}_1\mathbf{S}_1^{-1}$ relates the limiting singular vectors to correlation-like quantities, and will be helpful in the asymptotic analysis. Then the following important result can be shown (cf. (Gustafsson, 1999)):

Result 1. Define the residuals $\varepsilon_N = \sqrt{N}\,(\Gamma_\alpha^\perp)^T\widehat{\Gamma}_\alpha$. Then the elements of ε_N asymptotically have a

---

[3] Following the discussion in (Bauer, 1998), the involved SVD is unique for a generic subset of transfer functions. Thus, the following analysis is confined to such a generic subset.

zero mean Gaussian distribution, with an asymptotic covariance matrix

$$\mathbf{R}_{\varepsilon\varepsilon} = \lim_{N\to\infty} E\left\{\text{vec}\left\{\varepsilon_N\right\}\text{vec}^T\left\{\varepsilon_N\right\}\right\}$$
$$= \sum_{|\tau|<\alpha}\left(\mathbf{H}^T\mathbf{R}_{zz}(\tau)\mathbf{H}\right)\otimes\left(\left(\mathbf{\Gamma}_\alpha^\perp\right)^T\mathbf{R}_{n_f n_f}(-\tau)\mathbf{\Gamma}_\alpha^\perp\right), \tag{22}$$

where $\mathbf{z}(t) = \left[\mathbf{u}_p^T(t),\ \mathbf{y}_p^T(t),\ \mathbf{u}_f^T(t)\right]^T$, $\mathbf{H} = \mathbf{M}\mathbf{W}_R^2\mathbf{\Delta}^T\left(\mathbf{\Delta}\mathbf{W}_R^2\mathbf{\Delta}^T\right)^{-1}$, and $\mathbf{M} = \left[\mathbf{I},\ -\mathbf{R}_{u_f p}^T\mathbf{R}_{u_f u_f}^{-1}\right]^T$.

Remark 1. Previously we have stated that $\mathbf{W}_L$ has no effect on the accuracy of the estimated observability matrix, but it has not been proven. An outline of the proof could be as follows. When a weighting $\mathbf{W}_L$ is included, introduce the relationships

$$\mathbf{W}_L\mathbf{\Gamma}_\alpha\mathbf{T}_{W_{L,R}}^{-1} = \mathbf{Q}_1,\ \mathbf{T}_{W_{L,R}}\mathbf{\Delta}\mathbf{W}_R = \mathbf{S}_1\mathbf{V}_1^T \tag{23}$$

The estimated observability matrix must now be converted to a fixed state space basis. More precisely, unless the estimated observability matrix is converted to a fixed basis, $V(\cdot)$ is not a well-defined function of $\mathbf{W}_L$. The next step is thus to make a basis transformation to the arbitrary basis used to define $\mathbf{\Gamma}_\alpha$. Hence, a realization dependent transformation $\widehat{\mathbf{T}}_{W_{L,R}}$ is applied. If, for example, echolon coordinates are used, let $\widehat{\mathbf{T}}_{W_{L,R}} = \left[\widehat{\mathbf{Q}}_1\right]_{1:n,1:n}^{-1}$. In (Gustafsson, 1999), a more detailed discussion on these issues can be found. We thus define the estimated observability matrix as $\widehat{\mathbf{\Gamma}}_\alpha = \widehat{\mathbf{Q}}_1\widehat{\mathbf{T}}_{W_{L,R}}$. Then

$$\begin{aligned}\widehat{\mathbf{\Gamma}}_\alpha &= \mathbf{W}_L^{-1}\widehat{\mathbf{Q}}_1\widehat{\mathbf{T}}_{W_{L,R}}\\ &= \mathbf{\Gamma}_\alpha\mathbf{X}_f\mathbf{\Pi}_{U_f}^\perp\mathbf{P}^T\mathbf{W}_R\widehat{\mathbf{V}}_1\widehat{\mathbf{S}}_1^{-1}\widehat{\mathbf{T}}_{W_{L,R}}\\ &+ \mathbf{N}_f\mathbf{\Pi}_{U_f}^\perp\mathbf{P}^T\mathbf{W}_R\widehat{\mathbf{V}}_1\widehat{\mathbf{S}}_1^{-1}\widehat{\mathbf{T}}_{W_{L,R}},\end{aligned} \tag{24}$$

and it can be shown that $\widehat{\mathbf{V}}_1\widehat{\mathbf{S}}_1^{-1}\widehat{\mathbf{T}}_{W_{L,R}} \to \mathbf{W}_R\mathbf{\Delta}^T\left(\mathbf{\Delta}\mathbf{W}_R^2\mathbf{\Delta}^T\right)^{-1}$, as $N\to\infty$, and the assertion follows. Note however that, depending on the chosen algorithm, $\mathbf{W}_L$ can influence the accuracy of the estimates of $\mathbf{B}$ and $\mathbf{D}$, cf. the discussion in (Bauer and Jansson, 2000).

4. MAIN RESULTS

Let us first define the proposed figure of merit:

Definition 1. The scalar function

$$V(\mathbf{W}_R) = \text{Tr}\left\{\mathbf{R}_{\varepsilon\varepsilon}\right\} = \text{Tr}\left\{\mathbf{H}^T\mathbf{\Sigma}\mathbf{H}\right\} \tag{25}$$

is proposed to be used as the measure of accuracy of the estimated observability matrix. Here, $\text{Tr}\left\{\cdot\right\}$ denotes the trace operator, and

$$\mathbf{\Sigma} = \sum_{|\tau|<\alpha}\mathbf{R}_{zz}(\tau)\text{Tr}\left\{\mathbf{\Pi}_{\Gamma_\alpha}^\perp\mathbf{R}_{n_f n_f}(-\tau)\right\}. \tag{26}$$

The "optimal" weighting $\mathbf{W}_R^o$ is in this paper defined as $\mathbf{W}_R^o = \arg\min_{\mathbf{W}_R} V(\mathbf{W}_R)$.

Remark 2. Consider two arbitrary positive definite matrices $\mathbf{X}$ and $\mathbf{Y}$, Clearly, $\mathbf{X}\leq\mathbf{Y}$ implies $\text{Tr}\left\{\mathbf{X}\right\}\leq\text{Tr}\left\{\mathbf{Y}\right\}$. However, $\text{Tr}\left\{\mathbf{X}\right\}\leq\text{Tr}\left\{\mathbf{Y}\right\}$ does not imply that $\mathbf{X}\leq\mathbf{Y}$. Therefore we cannot claim that $\mathbf{J}(\mathbf{W}_R^o)\leq\mathbf{J}(\mathbf{W}_R)$ for all $\mathbf{W}_R$. However, since

$$\begin{aligned}\text{Tr}\left\{\mathbf{J}\right\} \leq \text{Tr}&\left\{\left(\mathbf{I}\otimes\mathbf{\Gamma}_\alpha^\perp\right)^\text{T}(\mathbf{f}'(\mathbf{\Gamma}_\alpha))^\text{T}\right.\\ &\left.\mathbf{f}'(\mathbf{\Gamma}_\alpha)\left(\mathbf{I}\otimes\mathbf{\Gamma}_\alpha^\perp\right)\right\}\text{Tr}\left\{\mathbf{R}_{\varepsilon\varepsilon}\right\}\end{aligned} \tag{27}$$

we can claim that the proposed weighting minimizes an upper bound of $\text{Tr}\left\{\mathbf{J}\right\}$. In (Gustafsson, 1999), the weighting matrix was found as the minimizing argument of

$$\widetilde{V}(\mathbf{W}_R) = \lim_{N\to\infty} N\,E\left\{\|\widehat{\mathbf{Q}}_1^T\mathbf{\Gamma}_\alpha^\perp\|_F^2\right\}, \tag{28}$$

which can be interpreted in terms of angles between subspaces. Interestingly enough, $\widetilde{V}(\mathbf{W}_R) = V(\mathbf{W}_R)$, cf. (Gustafsson, 1999).

4.1 *The Optimal Column Weighting*

Using the proposed measure of accuracy, i.e. $V(\mathbf{W}_R)$, the following result is straightforward to verify:

Result 2. Under the introduced assumptions, and if $\mathbf{\Sigma}$ is non-singular, the optimal weighting matrix $\mathbf{W}_R^o$ is

$$\left(\sum_{|\tau|<\alpha}\left(\mathbf{M}^T\mathbf{R}_{zz}(\tau)\mathbf{M}\right)\text{Tr}\left\{\mathbf{\Pi}_{\Gamma_\alpha}^\perp\mathbf{R}_{n_f n_f}(-\tau)\right\}\right)^{-1/2}$$

Here, "optimal" means that $V(\mathbf{W}_R^o)\leq V(\mathbf{W}_R)$ for all possible positive definite matrices $\mathbf{W}_R$.

The obvious drawback of our findings is that $\mathbf{W}_R^o$ depends on quantities that are unknown. To deal with this problem, a two-step algorithm is proposed:

(1) Use the PO-MOESP algorithm to obtain an estimate of $\mathcal{R}\{\mathbf{\Gamma}_\alpha\}$.
(2) Define the sequence

$$\mathbf{Y}_f\mathbf{\Pi}_{U_f}^\perp = \left[\mathbf{y}^\perp(\beta+1),\ \cdots,\mathbf{y}^\perp(\beta+N)\right],$$

and, for $\tau = 0, \cdots, \alpha - 1$, compute:

$$\widehat{\mathbf{R}}_{y^\perp y^\perp}(\tau) = \frac{1}{N} \sum_{t=\beta+1}^{N+\beta-\tau} \mathbf{y}^\perp(t+\tau)\mathbf{y}^\perp(t)^T, \tag{29}$$

and similarly for $\widehat{\mathbf{R}}_{zz}(\tau)$. The estimate of the observability matrix is then taken as the n principal left singular vectors of

$$\mathbf{Y}_f \mathbf{\Pi}_{U_f}^\perp \mathbf{P}^T \left(\sum_{|\tau|<\alpha} \left(\widehat{\mathbf{M}}^T \widehat{\mathbf{R}}_{zz}(\tau) \widehat{\mathbf{M}} \right) \mathrm{Tr}\left\{ \mathbf{\Pi}_{\widehat{\Gamma}_\alpha}^\perp \widehat{\mathbf{R}}_{y^\perp y^\perp}(-\tau) \right\} \right)^{-1/2} \tag{30}$$

In the above procedure quantities like $\widehat{\mathbf{R}}_{y^\perp y^\perp}$ appear instead of quantities like $\widehat{\mathbf{R}}_{n_f n_f}$. This is an approach to deal with the fact that it is difficult to directly estimate $\widehat{\mathbf{R}}_{n_f n_f}$. Replacing $\widehat{\mathbf{R}}_{n_f n_f}$ with $\widehat{\mathbf{R}}_{y^\perp y^\perp}$ has however asymptotically no effect on the accuracy, cf. (Gustafsson, 1999). For easy reference, the proposed algorithm for estimating $\mathcal{R}\{\mathbf{\Gamma}_\alpha\}$ will be referred to as the OW-4SID algorithm.

4.2 *An Approximation of the Optimal Weighting*

This section introduces approximations of $V(\mathbf{W}_R)$, which in turn will allow us to explain the sub-optimality of existing choices of weighting matrices $\mathbf{W}_R$. First, note that (cf. (Gustafsson, 1999)) $\mathbf{H}^T \mathbf{\Sigma} \mathbf{H} \leq \kappa \mathbf{H}^T \mathbf{R}_{zz}(0) \mathbf{H}$, where the scalar κ is independent of $\mathbf{W}_R$. For notational convenience, define $\widetilde{\mathbf{R}} = \kappa \mathbf{H}^T \mathbf{R}_{zz}(0) \mathbf{H}$. The idea now is to minimize $\widetilde{\mathbf{R}}$ with respect to the matrix $\mathbf{W}_R$. Another way to arrive at the same result would be to heuristically study the term corresponding to $\tau = 0$ of the summation in the expression for $\mathbf{\Sigma}$.

Result 3. Assume that $\mathbf{\Omega}$ is non-singular, and define

$$\mathbf{W}_R^{ao} = \mathbf{\Omega}^{-1/2} = \left(\mathbf{R}_{pp} - \mathbf{R}_{u_f p}^T \mathbf{R}_{u_f u_f}^{-1} \mathbf{R}_{u_f p} \right)^{-1/2} \tag{31}$$

Then $\widetilde{\mathbf{R}}(\mathbf{W}_R^{ao}) = \kappa \left(\mathbf{\Delta} \mathbf{\Omega}^{-1} \mathbf{\Delta}^T \right)^{-1} \leq \widetilde{\mathbf{R}}(\mathbf{W}_R)$, for all possible positive definite weighting matrices $\mathbf{W}_R$.

Note, $\mathbf{W}_R^{ao}$ is exactly the limiting expression of the weighting employed by CVA (Larimore, 1983) and PO-MOESP (Verhaegen, 1994). We have consequently provided a theoretically sound derivation of a weighting that in practice has been found to give accurate estimates.

5. NUMERICAL EXAMPLES

Consider identification of the following simple ARMAX SISO system:

$$(1 - q^{-1} + 0.5q^{-2})y(t) = (1 - 0.5q^{-2})u(t) + (1 + q^{-1} + 0.5q^{-2})e(t) \tag{32}$$

where $e(t)$ is white Gaussian noise. In the following, four different identification algorithms are investigated: 1) $\mathbf{W}_R = \mathbf{I}$, 2) The PO-MOESP algorithm (i.e. $\widehat{\mathbf{W}}_R = \widehat{\mathbf{W}}_R^{ao}$), 3) The OW-4SID algorithm, and 4) The ARMAX routine in Matlabs System Identification Toolbox. The 4SID algorithms all use the MOESP approach (Verhaegen, 1994) for finding the estimates of $\mathbf{B}$ and $\mathbf{D}$. The performance of the above algorithms is investigated using two kinds of measures. The first is the standard deviation of the pole estimation error, denoted as $P(N)$, where the 4SID methods find the poles as the eigenvalues of $\bar{\mathbf{\Gamma}}^\dagger \underline{\mathbf{\Gamma}}$. Here, $\bar{\mathbf{\Gamma}}$ denotes the matrix constructed from the first $\alpha - 1$ block-rows of $\widehat{\mathbf{Q}}_1$. Similarly, $\underline{\mathbf{\Gamma}}$ contains the last $\alpha - 1$ block-rows of $\widehat{\mathbf{Q}}_1$. The second performance index considered is

$$D(N) = \frac{1}{M} \sum_{k=1}^{M} \|\widehat{\boldsymbol{\theta}}_N^k - \boldsymbol{\theta}_0\|, \tag{33}$$

where $M = 200$ denotes the number of independent Monte Carlo simulations, and $\widehat{\boldsymbol{\theta}}_N^k$ contains all estimated parameters of the kth estimate of the transfer function from u to y, when N samples are used. Let u be a realization of zero mean white Gaussian noise with unit variance, and apply the same input in all realizations. Furthermore, let the standard deviation of e be $\sigma_e = 2$. The outcome of this experiment is shown in Fig. 1. Consider next Fig. 2. The simulation setup is identical to the previous one, with the distinction that the input signal is low-pass filtered with a Butterworth filter of order 3 and cut-off frequency 0.5π. Once again a notable difference between PO-MOESP and OW-4SID is observed, and once again the OW-4SID approach is close to the accuracy of PEM.

The OW-4SID algorithm has in several simulation examples other than those reported on here, achieved a similar accuracy improvement. One should however note that the accuracy in terms of the measure $D(N)$ quite often is dominated by errors in the estimates of $\mathbf{B}$ and $\mathbf{D}$. Hence, the improved accuracy offered by OW-4SID, may be small (in terms of $D(N)$) in comparison to the errors in the estimates of $\mathbf{B}$ and $\mathbf{D}$.

6. CONCLUSIONS

The main benefit of the paper is perhaps that a theoretical framework of *how* to analyze subspace

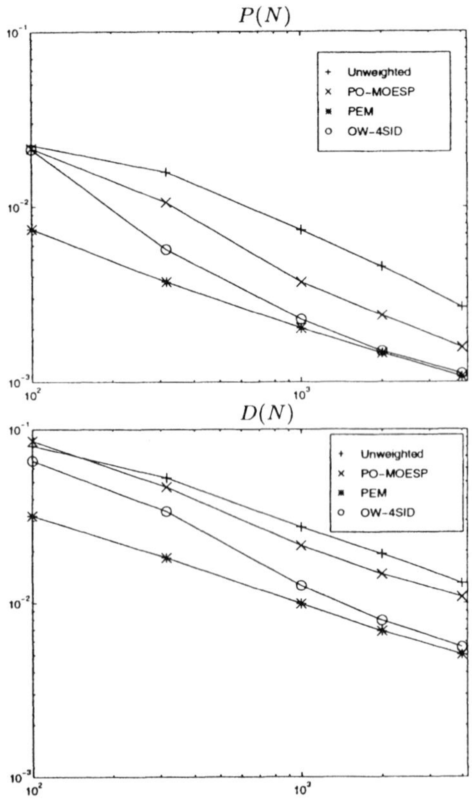

Fig. 1. Identification of the system (32). Top: $P(N)$. Bottom: $D(N)$. The input is white noise, $\sigma_e = 2$, $\alpha = \beta = 8$.

methods has been proposed. Once the (suboptimal) measure of quality was defined, several useful results on how to choose weighting matrices were found. A particularly important result is that a novel weighting matrix was found, which in our numerical examples showed superior accuracy compared to PO-MOESP. The weighting that minimized the upper-bounded criterion has previously appeared in the literature. It is however quite interesting to know what kind of approximations that are involved to arrive at this specific weighting.

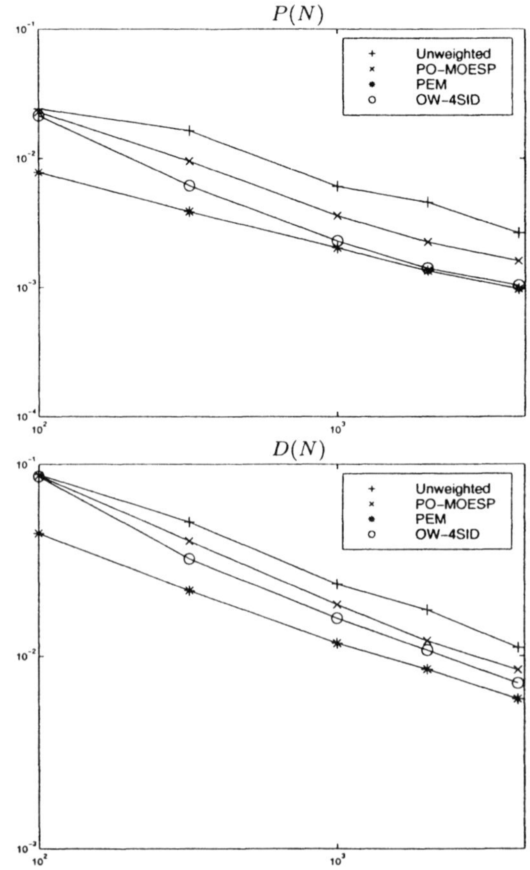

Fig. 2. Identification of the system (32). Top: $P(N)$. Bottom: $D(N)$. The input is low-pass filtered (Butterworth filter of order 3, with cut-off frequency 0.5π.) white noise, $\sigma_e = 1$, $\alpha = \beta = 8$.

7. REFERENCES

Bauer, D. (1998). Some Asymptotic Theory for the Estimation of Linear Systems Using Maximum Likelihood Methods or Subspace Algorithms. PhD thesis. TU Wien. Wien, Austria.

Bauer, D. and M. Jansson (2000). "Analysis of the Asymptotic Properties of the MOESP type of Subspace Algorithms". *Automatica* **36**(4), 497–509.

Gustafsson, T. (1999). Subspace Methods for System Identification and Signal Processing. PhD thesis. Chalmers University of Technology. Gothenburg, Sweden.

Jansson, M. and B. Wahlberg (1998). "On Consistency of Subspace Methods for System Identification". *Automatica* **34**(12), 1507–1519.

Larimore, W. E. (1983). "System Identification, Reduced-Order Filtering and Modeling via Canonical Variate Analysis". In: *Proc. ACC.* pp. 445–451.

Ljung, L. (1987). *System Identification: Theory for the User.* Prentice-Hall. Englewood Cliffs, NJ.

Söderström, T. and P. Stoica (1989). *System Identification.* Prentice-Hall. London, U.K.

VanOverschee, P. (1995). Subspace Identification – Theory, Implementation,Application. PhD thesis. Katholieke Universiteit Leuven. Leuven, Belgium.

Verhaegen, M. (1994). "Identification of the Deterministic Part of MIMO State Space Models given in Innovations Form from Input-Output Data". *Automatica* **30**(1), 61–74.

Viberg, M. (1995). "Subspace-based Methods for the Identification of Linear Time-Invariant Systems". *Automatica* **31**(12), 1835–1852.

Viberg, M., B. Wahlberg and B. Ottersten (1997). "Analysis of State Space System Identification Methods Based on Instrumental Variables and Subspace Fitting'. *Automatica* **33**(9), 1603–1616.

OPTIMAL WEIGHTING FOR MOESP TYPE OF PROCEDURES

D. Bauer * T. Gustafsson **

[1] *Institute f. Econometrics, Operations Research and System Theory, University of Technology Vienna, Argentinierstr. 8, A-1040 Wien , Austria, e-mail: Dietmar.Bauer@tuwien.ac.at*
[2] *University of California San Diego Depertment of Electrical and Computer Engineering 9500 Gilman Drive Mail Code 0407 La Jolla, CA 92093-0407, USA*

Abstract: In this paper we discuss the asymptotic properties of the MOESP type of subspace methods with respect to certain user-defined parameters. A new instrumental variables interpretation of these algorithms is given. This leads to a lower bound on the achievable estimation accuracy, which is different from the Cramér Rao lower bound. Some facts about the asymptotic covariance matrix are derived.

Keywords: Subspace Methods, Asymptotic Properties, Identification, Linear Systems, State Space Models, Discrete-time Systems

1. INTRODUCTION

Subspace algorithms for the estimation of linear systems are numerically attractive alternatives to the usual implementations of the more classical maximum likelihood or prediction error methods. A number of different algorithms have been developed in the past: One class of algorithms estimates the state in a first step and uses the estimated state sequence to obtain an estimate of the system. The first algorithm of this type has been proposed by (Larimore, 1983) and further refined by (Peternell *et al.*, 1996). Also one variant of N4SID (see (Van Overschee and DeMoor, 1994)) falls into this class. The statistical properties of this type of algorithms have been derived in (Peternell, 1995) and (Bauer *et al.*, 1999; Bauer, 1998), (Bauer *et al.*, 2000) derive some facts about the asymptotic covariance matrix of the estimated system. A second class of algorithms aims at estimating the (extended) observability matrix in the first step and calculating the estimates of the system matrices from the observability matrix. A popular algorithm in this area is MOESP (Verhaegen, 1994). The asymptotic distribution of this class of algorithms has been examined in (Jansson and Wahlberg, 1997; Bauer and Jansson, 2000). For this class of algorithms, an instrumental variables interpretation has been given by (Viberg, 1995) and further extended and utilized by (Gustafsson, 1999). The last reference also develops an 'optimal' weighting procedure, where optimal is not directly related to the accuracy of the estimated system matrices, but rather to the accuracy of the estimated observability matrix. The aim of this paper is to show the equivalence of the MOESP type of procedure to a new instrumental variables procedure for a particular choice of weighting. By deriving an optimal weighting, a lower bound on the achievable accuracy is obtained. It should be noted, that the

[1] Parts of the research has been done while this author was at the Department of Electrical and Computer Engineering, University of Newcastle, Australia.
[2] Support by the Swedish Foundation for International Cooperation in Research and Higher Education, and Telefonaktiebolaget LM Ericsson is gratefully acknowledged.

instrumental variables framework put forth in this paper differs from the one given in (Viberg, 1995), since the present one also includes the crucial SVD step.

The organization of the paper is as follows: In the next section we will state the model set and the assumptions and also briefly describe the studied algorithms. Section 3 then states the main results. Section 4 provides a discussion including a numerical example, and section 5 concludes.

2. MODEL SET AND ASSUMPTIONS

In this paper, linear, finite dimensional, discrete time, state space systems of the form

$$\begin{aligned} x_{t+1} &= Ax_t + Bu_t + K\varepsilon_t \\ y_t &= Cx_t + Du_t + E\varepsilon_t \end{aligned} \quad (1)$$

where y_t denotes the s-dimensional observed output, u_t denotes the m-dimensional observed input, and ε_t denotes the s-dimensional unobserved white noise. For simplicity, assume that ε_t is an i.i.d sequence with zero mean and covariance matrix unity and finite moments up to order four. For a discussion on minimal assumptions on the noise sequence for the asymptotic distribution of the estimated system matrices to be normal see (Bauer and Jansson, 2000). The matrices $A \in \mathbb{R}^{n\times n}, B \in \mathbb{R}^{n\times m}, K \in \mathbb{R}^{n\times s}, C \in \mathbb{R}^{s\times n}, D \in \mathbb{R}^{s\times m}$ and $E \in \mathbb{R}^{s\times s}$ are parameter matrices. The matrix E is assumed to be lower triangular with positive entries on the diagonal. This corresponds to the assumption of nonsingular innovations. The system is assumed to be stable (i.e. $|\lambda_{max}(A)| < 1$, where λ_{max} denotes an eigenvalue of maximum modulus). Furthermore $|\lambda_{max}(A - KE^{-1}C)| < 1$ is assumed, i.e. the transfer function relating the noise to the output is assumed to be strictly minimum-phase.

Concerning the observed input sequence u_t, it is assumed throughout that $u_t = v_t + \sum_{j=1}^{h} c_j e^{i\lambda_j t}$ for some vectors $c_j \in \mathbb{C}^m$, where

$$v_t = \sum_{j=0}^{\infty} K_u(j)\eta_{t-j}.$$

Here η_t is assumed to be i.i.d. and $\|K_u(j)\| \leq c\rho^j, |\rho| < 1$ is assumed to hold for some $0 < c < \infty$. The input process will furthermore be assumed to be independent of the noise process ε_t, and u_t is assumed to be persistently exciting of sufficient order (see Theorem 1 for details). Similar results can be obtained, when u_t is a pseudo-stationary process.

The MOESP type of subspace algorithms are based on the following property of the state space representation: Assume, that integers $\alpha > n$ and $\beta \geq n$ are provided by the user. Let $\mathbf{Y}_{t,\beta} = [y_{t-1}^T, y_{t-2}^T, \cdots, y_{t-\beta}^T]^T$ be the vector of the stacked (finite) past, and let $\mathbf{Y}_{t,\alpha} = [y_t^T, y_{t+1}^T, \cdots, y_{t+\alpha-1}^T]^T$ be the vector of the stacked (finite) future of the output process. Define $\mathbf{U}_{t,\beta}$ and $\mathbf{U}_{t,\alpha}$ analogously using u_t in place of y_t, and let $\mathbf{P}_{t,\beta} = [\mathbf{Y}_{t,\beta}^T, \mathbf{U}_{t,\beta}^T]^T$. Furthermore, let $\mathbf{\Gamma}_\alpha = [C^T, A^TC^T, \cdots, (A^T)^{\alpha-1}C^T]^T$ denote the extended observability matrix. Then the following equation can easily be shown to hold:

$$\mathbf{Y}_{t,\alpha} = \mathbf{\Gamma}_\alpha x_t + \mathbf{\Phi}_\alpha \mathbf{U}_{t,\alpha} + \mathbf{N}_{t,\alpha}. \quad (2)$$

Here $\mathbf{N}_{t,\alpha}$ is equal to the contribution due to the future of the noise and $\mathbf{\Phi}_\alpha$ is defined as

$$\mathbf{\Phi}_\alpha = \begin{bmatrix} D & 0 & \cdots & 0 \\ CB & D & \ddots & \vdots \\ \vdots & \ddots & \ddots & 0 \\ CA^{\alpha-2}B & \cdots & CB & D \end{bmatrix}.$$

Now the MOESP type of algorithms can be described as follows[3]: If the state x_t would be known, then equation (2) would be a simple linear regression problem, which can be solved using instrumental variables techniques. Introducing the notation $\langle a_t, b_t \rangle = \frac{1}{T}\sum_{j=1+\beta}^{T-\alpha} a_t b_t^T$, the estimator used in MOESP can be motivated by the following equation (using equation (2))

$$\begin{aligned} &\langle \mathbf{Y}_{t,\alpha}, \begin{bmatrix} \mathbf{P}_{t,\beta} \\ \mathbf{U}_{t,\alpha} \end{bmatrix} \rangle = \mathbf{\Gamma}_\alpha \langle x_t, \begin{bmatrix} \mathbf{P}_{t,\beta} \\ \mathbf{U}_{t,\alpha} \end{bmatrix} \rangle \\ &+ \mathbf{\Phi}_\alpha \langle \mathbf{U}_{t,\alpha}, \begin{bmatrix} \mathbf{P}_{t,\beta} \\ \mathbf{U}_{t,\alpha} \end{bmatrix} \rangle + \langle \mathbf{N}_{t,\alpha}, \begin{bmatrix} \mathbf{P}_{t,\beta} \\ \mathbf{U}_{t,\alpha} \end{bmatrix} \rangle. \end{aligned} \quad (3)$$

PO-MOESP chooses the IV-vector as $[\mathbf{P}_{t,\beta}^T \ \mathbf{U}_{t,\alpha}]^T$. Due to the independence of the noise and the observed inputs, the last term converges to zero almost surely (a.s.). Note that the column space of the first term is equal to the column space of the observability matrix. Note also that a change in the state space leads to a postmultiplication of the observability matrix with a nonsingular matrix. Therefore the range space of the observability matrix is a system invariant. Since the state vector can be written as $x_t = \mathbf{L}_1\mathbf{P}_{t,\beta} + \mathbf{L}_2 x_{t-p}$, equation (3) motivates to take

$$[\hat{\boldsymbol{\theta}}_{\alpha,\beta}, \hat{\mathbf{\Phi}}_\alpha] = \langle \mathbf{Y}_{t,\alpha}, \begin{bmatrix} \mathbf{P}_{t,\beta} \\ \mathbf{U}_{t,\alpha} \end{bmatrix} \rangle \langle \begin{bmatrix} \mathbf{P}_{t,\beta} \\ \mathbf{U}_{t,\alpha} \end{bmatrix}, \begin{bmatrix} \mathbf{P}_{t,\beta} \\ \mathbf{U}_{t,\alpha} \end{bmatrix} \rangle^{-1}$$

as an estimate of $(\mathbf{\Gamma}_\alpha \mathbf{L}_1, \mathbf{\Phi}_\alpha)$. Typically $\hat{\boldsymbol{\theta}}_{\alpha,\beta}$ will be full rank and a rank n approximation is sought, which is obtained from a singular value decomposition of $\hat{\mathbf{W}}_\alpha^+ \hat{\boldsymbol{\theta}}_{\alpha,\beta} \hat{\mathbf{W}}_\beta^- = \hat{\mathbf{U}}_n \hat{\mathbf{\Sigma}}_n \hat{\mathbf{V}}_n^T + \hat{\mathbf{R}}$. Here $\hat{\mathbf{W}}_\alpha^+$ and $\hat{\mathbf{W}}_\beta^-$ are weighting matrices. For common choices (see e.g. (Peternell, 1995)). The diagonal matrix $\hat{\mathbf{\Sigma}}_n$ contains the n principal singular values ordered decreasing in size. $\hat{\mathbf{U}}_n$ contains the corresponding left singular vectors. An estimate

[3] Only the method named PO-MOESP will be dealt with in this paper. An analogous analysis is possible for the method named PI-MOESP.

of the range space of $\mathbf{\Gamma}_\alpha$ is obtained as $\hat{\mathbf{\Gamma}}_\alpha = (\hat{\mathbf{W}}_\alpha^+)^{-1}\hat{\mathbf{U}}_n$. From $\hat{\mathbf{\Gamma}}_\alpha$, $\hat{C}$ is usually obtained as the first block row, and $\hat{A} = \hat{\mathbf{\Gamma}}_{\alpha-1}^\dagger \hat{\mathbf{\Gamma}}_\alpha^\dagger$, where $\hat{\mathbf{\Gamma}}_{\alpha-1}^\dagger$ denotes the Moore-Penrose pseudoinverse of the matrix $\hat{\mathbf{\Gamma}}_{\alpha-1}$, which is composed of the first $\alpha - 1$ block rows of $\hat{\mathbf{\Gamma}}_\alpha$. Further $\hat{\mathbf{\Gamma}}_\alpha^\dagger$ denotes the matrix, which is obtained from $\hat{\mathbf{\Gamma}}_\alpha$ by omitting the first block row. Finally $\hat{B}$ and $\hat{D}$ are obtained from $\hat{\mathbf{\Gamma}}_\alpha$ and $\hat{\mathbf{\Phi}}_\alpha$ (comp. (Verhaegen, 1994)) using some nonlinear mapping, which also depends on the weighting matrix $\hat{\mathbf{W}}_\alpha^+$. However, since it can be shown (see (Bauer *et al.*, 2000)) that the asymptotic distribution of the estimates is not altered, if we exchange the estimate $\hat{\mathbf{W}}_\alpha^+$ by the possibly system dependent (but deterministic) quantity $\mathbf{W}_\alpha^+$, we may assume that the deterministic weighting is used. The remaining matrices E and K corresponding to the transfer function acting on the white noise ε_t are usually not estimated in the MOESP framework, and therefore we refrain from giving any results on this aspect.

3. THE MAIN RESULT

Following the above description it is clear that the subspace methodology may be decomposed into two main steps: 1) estimation of $\hat{\mathbf{\Gamma}}_\alpha$ and $\hat{\mathbf{\Phi}}_\alpha$, and 2) extraction of the system matrices from these estimates using a nonlinear mapping. Therefore we might also decompose the analysis into two steps. In this paper, only the first step will be analyzed. The investigation of the second step still is an unsolved problem, to the best of the authors knowledge.

In order to compare the asymptotic variances of different methods, it is necessary to transform the estimated system to a specific canonical form. Also the estimate $\hat{\mathbf{\Gamma}}_\alpha$ has to be transformed to a unique representation, which is independent of the chosen state basis for analogous reasons. This will be done by restricting the first n rows to be equal to the identity matrix. For systems in the generic neighborhood of the echelon canonical form this is possible a.s. for large sample sizes. This will be assumed from now on. Thus, let $(\hat{A}_e, \hat{B}_e, \hat{C}_e, \hat{D}_e)$ denote the estimated system converted into echelon form and let $\hat{\mathbf{\Gamma}}_\alpha^e = \hat{\mathbf{\Gamma}}_\alpha[\hat{\mathbf{\Gamma}}_\alpha]_n^{-1}$ denote the transformed estimate of the observability matrix. Here $[.]_n$ denotes the matrix composed of the first n rows. Since $(\hat{A}_e, \hat{B}_e, \hat{C}_e, \hat{D}_e) = f(\hat{\mathbf{\Gamma}}_\alpha^e, \hat{\mathbf{\Phi}}_\alpha)$, it follows that any superior estimator of $[\hat{\mathbf{\Gamma}}_\alpha^e, \hat{\mathbf{\Phi}}_\alpha]$ relative another estimator also results in superior estimates of the system matrices, since

$$T\mathrm{Var}(\mathrm{vec}[\hat{A}_e, \hat{B}_e, \hat{C}_e, \hat{D}_e]) \doteq J_f \mathrm{Var}(\mathrm{vec}[\hat{\mathbf{\Gamma}}_\alpha^e, \hat{\mathbf{\Phi}}_\alpha]) J_f^T.$$

Here J_f denotes the matrix of partial derivatives of f with respect to the entries in $\mathbf{\Gamma}_\alpha, \mathbf{\Phi}_\alpha$ and Var(.) denotes the variance of the corresponding vector. Now consider the estimation error in $\hat{\mathbf{\Gamma}}_\alpha^e$. Using the linearisation of the SVD (see (Bauer *et al.*, 2000)), it can be shown that the estimation error

$$\hat{\mathbf{\Gamma}}_\alpha^e - \mathbf{\Gamma}_\alpha^e \doteq [\mathbf{I}_{s\alpha} - \mathbf{\Gamma}_\alpha^e[\mathbf{I}_n, 0]](\hat{\boldsymbol{\theta}}_{\alpha,\beta} - \boldsymbol{\theta}_{\alpha,\beta})\mathbf{L} \quad (4)$$

where $\mathbf{\Gamma}_\alpha^e$ denotes the limit of $\hat{\mathbf{\Gamma}}_\alpha^e$ and $\boldsymbol{\theta}_{\alpha,\beta}$ the limit of $\hat{\boldsymbol{\theta}}_{\alpha,\beta}$. Here $\mathbf{L} = \mathbf{W}_\beta^-(\mathbf{W}_\beta^-)^T(\mathbf{K}_\beta^e)^T(\mathbf{\Sigma}^e)^{-1}$. Further $\doteq$ means, that $\sqrt{T}$ times the difference converges to zero in probability. Finally $\boldsymbol{\theta}_{\alpha,\beta} = \mathbf{\Gamma}_\alpha^e \mathbf{K}_\beta^e$, where this equation defines $\mathbf{K}_\beta^e$. The $m \times m$ identity matrix is denoted by $\mathbf{I}_m$. Expression (4) builds the basis of the comparison to the IV estimator to be described below.

The main idea is to consider a general instrumental variables estimator for the vectorized version of equation (2). For notational convenience, introduce the following matrices:

$$\boldsymbol{\theta}^T = [\mathrm{vec}[\mathbf{\Gamma}_\alpha^e]^T, \mathrm{vec}[\mathbf{\Phi}_\alpha]^T]$$
$$\mathbf{X}_{t,\alpha} = [(x_t^e \otimes \mathbf{I}_{\alpha s}), (\mathbf{U}_{t,\alpha} \otimes \mathbf{I}_{\alpha s})].$$

Here $\otimes$ denotes the Kronecker product and x_t^e denotes the state vector in echelon coordinates. Using this notation, equation (2) can be rewritten as

$$\mathbf{Y}_{t,\alpha} = \mathbf{X}_{t,\alpha}^T \boldsymbol{\theta} + \mathbf{N}_{t,\alpha}.$$

If the echelon state x_t^e would be known, using a matrix $\mathbf{Z}_t$ as instruments, the following IV estimator would be obtained:

$$\tilde{\boldsymbol{\theta}} = \left(\langle \mathbf{X}_{t,\alpha}, \mathbf{Z}_t\rangle \hat{\mathbf{M}} \langle \mathbf{Z}_t, \mathbf{X}_{t,\alpha}\rangle\right)^{-1} \langle \mathbf{X}_{t,\alpha}, \mathbf{Z}_t\rangle \hat{\mathbf{M}} \langle \mathbf{Z}_t, \mathbf{Y}_{t,\alpha}^T\rangle$$

If the state would be known, different identification methods could of course be used. In our case the assumption is used only in order to compare MOESP with the above IV-type of estimator.

Here $\hat{\mathbf{M}}$ denotes a general weighting matrix. For the moment we only impose the restriction, that the limit $\mathbf{M}$ exists (a.s.) and is nonsingular (a.s.). Further assumptions are introduced, whenever needed. The assumptions on $\mathbf{Z}_t$ have to be such that $\tilde{\boldsymbol{\theta}}$ exists, is consistent and fulfills a central limit theorem (for a detailed discussion of sufficient conditions see (White, 1984)). Under appropriate conditions the estimation error $\tilde{\boldsymbol{\theta}} - \boldsymbol{\theta}$ is asymptotically equivalent to

$$(\langle \mathbf{X}_{t,\alpha}, \mathbf{Z}_t\rangle \mathbf{M} \langle \mathbf{Z}_t, \mathbf{X}_{t,\alpha}\rangle)^{-1} \langle \mathbf{X}_{t,\alpha}, \mathbf{Z}_t\rangle \mathbf{M} \langle \mathbf{Z}_t, \mathbf{N}_{t,\alpha}^T\rangle$$

where asymptotic equivalence means, that the error multiplied by $\sqrt{T}$ tends to zero in probability. Let the instruments be given by $\mathbf{Z}_t = [[(\mathbf{P}_{t,\beta})^T, (\mathbf{U}_{t,\alpha})^T]^T \otimes \mathbf{I}_{s\alpha}]$. Then, using the notation $\mathbf{Z}_t^1 = [(\mathbf{P}_{t,\beta})^T, (\mathbf{U}_{t,\alpha})^T]^T$, the variance of $\sqrt{T}\langle \mathbf{Z}_t, \mathbf{N}_{t,\alpha}^T\rangle$ is equal to

$$\mathcal{R}_n = \frac{1}{T}\sum_{t,s=1}^{T} [\mathbb{E}(\mathbf{Z}_t^1(\mathbf{Z}_s^1)^T) \otimes \mathbb{E}(\mathbf{N}_{t,\alpha}\mathbf{N}_{s,\alpha}^T)]$$
$$\doteq \sum_{|\tau|<\alpha} \mathbf{R}_{z^1z^1}(\tau) \otimes \mathbf{R}_{nn}(\tau)$$

Here we used the properties of the Kronecker product and the equality $\mathbf{N}_{t,\alpha} = (1 \otimes \mathbf{N}_{t,\alpha})$. Also $\mathbf{R}_a(\tau) = \mathbb{E}a_t a_{t+\tau}^T$. Note, that $\mathcal{R}_n$ can be estimated consistently from consistent estimates of the system $(\hat{A}_e, \hat{B}_e, \hat{C}_e, \hat{D}_e, \hat{E}_e, \hat{K}_e)$. The structure of $\mathcal{R}_n$ is discussed below.

In order to compare the estimates obtained by the studied IV framework given above with the MOESP type of procedures, let $\hat{\mathbf{M}} = (\hat{\mathbf{W}} \otimes \mathbf{I}_{s\alpha})$. Further let

$$\mathbf{L}_{\alpha,\beta} = \begin{bmatrix} \mathbf{I}_{\beta(m+s)} & -\langle \mathbf{P}_{t,\beta}, \mathbf{U}_{t,\alpha}\rangle\langle \mathbf{U}_{t,\alpha}, \mathbf{U}_{t,\alpha}\rangle^{-1} \\ 0 & \mathbf{I}_{\alpha m} \end{bmatrix}$$

and let $\mathbf{P}_{t,\beta}^{\Pi} = [\mathbf{I}_{\beta(m+s)}, 0]\mathbf{L}_{\alpha,\beta}\begin{bmatrix}\mathbf{P}_{t,\beta}\\ \mathbf{U}_{t,\alpha}\end{bmatrix}$. Then using $\mathbf{X}_{t,\alpha}^1 = [(x_t^e)^T, \mathbf{U}_{t,\alpha}^T]^T$ the IV estimate $\tilde{\boldsymbol{\theta}}$ is equal to

$$\{[(\langle \mathbf{X}_{t,\alpha}^1, \mathbf{Z}_t^1\rangle \hat{\mathbf{W}}\langle \mathbf{Z}_t^1, \mathbf{X}_{t,\alpha}^1\rangle)^{-1}\langle \mathbf{X}_{t,\alpha}^1, \mathbf{Z}_t^1\rangle\hat{\mathbf{W}}] \otimes \mathbf{I}_{s\alpha}\}$$
$$\text{vec}[\langle \mathbf{Y}_{t,\alpha}, \mathbf{Z}_t^1\rangle] = \text{vec}[\langle \mathbf{Y}_{t,\alpha}, \mathbf{Z}_t^1\rangle \hat{\mathbf{W}}\langle \mathbf{Z}_t^1, \mathbf{X}_{t,\alpha}^1\rangle$$
$$(\langle \mathbf{X}_{t,\alpha}^1, \mathbf{Z}_t^1\rangle \hat{\mathbf{W}}\langle \mathbf{Z}_t^1, \mathbf{X}_{t,\alpha}^1\rangle)^{-1}]$$

Consider the choice

$$\hat{\mathbf{W}} = \mathbf{L}_{\alpha,\beta}^T \text{diag}[\langle \mathbf{P}_{t,\beta}^{\Pi}, \mathbf{P}_{t,\beta}^{\Pi}\rangle^{-1}$$
$$\hat{\mathbf{W}}_\beta^-(\hat{\mathbf{W}}_\beta^-)^T\langle \mathbf{P}_{,t\beta}^{\Pi}, \mathbf{P}_{t,\beta}^{\Pi}\rangle^{-1}, \langle \mathbf{U}_{t,\alpha}, \mathbf{U}_{t,\alpha}\rangle^{-1}]\mathbf{L}_{\alpha,\beta} \tag{5}$$

and note, that $\mathbb{E}x_t^e(\mathbf{P}_{t,\beta}^{\Pi})^T(\mathbb{E}\mathbf{P}_{t,\beta}^{\Pi}(\mathbf{P}_{t,\beta}^{\Pi})^T)^{-1} = \mathbf{K}_\beta^e$. Straightforward calculations using the matrix inversion lemma show that $\left[\tilde{\boldsymbol{\theta}}\right]_{1:ns\alpha}$ is equivalent to

$$\text{vec}[\langle \mathbf{Y}_{t,\alpha}, \mathbf{P}_{t,\beta}^{\Pi}\rangle\langle \mathbf{P}_{t,\beta}^{\Pi}, \mathbf{P}_{t,\beta}^{\Pi}\rangle^{-1}$$
$$\mathbf{W}_\beta^-(\mathbf{W}_\beta^-)^T(\mathbf{K}_\beta^e)^T(\mathbf{\Sigma}_n^e)^{-1}] \tag{6}$$

where $\mathbf{\Sigma}_n^e = \mathbf{K}_\beta^e\mathbf{W}_\beta^-(\mathbf{W}_\beta^-)^T(\mathbf{K}_\beta^e)^T$ and $[X]_{1:r}$ denotes the vector of the first r coordinates of the matrix X. In order to compare the estimates to the ones obtained by the MOESP procedure, normalize also the estimate corresponding to the IV estimator to have the identity matrix in the first n rows of $\tilde{\mathbf{\Gamma}}_\alpha$, where $\text{vec}[\tilde{\mathbf{\Gamma}}_\alpha] = \left[\tilde{\boldsymbol{\theta}}\right]_{1:ns\alpha}$ and denote the resulting matrix with $\tilde{\mathbf{\Gamma}}_\alpha^e$. It can be shown (cf. (Bauer *et al.*, 2000)), that the asymptotic estimation error $\tilde{\mathbf{\Gamma}}_\alpha^e - \mathbf{\Gamma}_\alpha^e$ has the same distribution as

$$[\mathbf{I}_{s\alpha} - \mathbf{\Gamma}_\alpha^e[\mathbf{I}_n, 0]](\hat{\boldsymbol{\theta}}_{\alpha,\beta} - \boldsymbol{\theta}_{\alpha,\beta})$$
$$\mathbf{W}_\beta^-(\mathbf{W}_\beta^-)^T(\mathbf{K}_\beta^e)^T(\mathbf{\Sigma}^e)^{-1} \tag{7}$$

Comparing (4) and (7) then shows that the asymptotic distribution of the MOESP estimate is equivalent with the above IV method, with the previous special choice of the weighting matrix (i.e. equation (5)). Inserting the expression for the asymptotic variance of $\hat{\boldsymbol{\theta}}_{\alpha,\beta} - \boldsymbol{\theta}_{\alpha,\beta}$ into equation (7) leads to Result 2.1 in (Gustafsson, 1999).

For the estimation of B and D according to the MOESP procedure the estimates $\hat{\mathbf{\Phi}}_\alpha$ and

$$\langle \mathbf{Y}_{t,\alpha}, \mathbf{U}_{t,\alpha}\rangle\langle \mathbf{U}_{t,\alpha}, \mathbf{U}_{t,\alpha}\rangle^{-1}$$

respectively are used. For the IV method the use of any of the two matrices leads to identical estimates, since according to the algorithm this matrix is premultiplied by $\tilde{\mathbf{\Gamma}}_\alpha^\perp$. Then the equivalence follows from the matrix inversion formula.

Theorem 1. Let $y_t, t \in \mathbb{Z}$ be a process generated by a system of the form (1), which fulfills the assumptions of section 2 and let the input process u_t have the representation given in section 2. Furthermore u_t is assumed to be persistently exciting of order $f + p$. Assume that the true system is in the generic neighborhood of the echelon canonical form. Then the estimates of the normalized observability matrix $\hat{\mathbf{\Gamma}}_\alpha^e$ obtained from a MOESP type of procedure using the weighting matrix $\hat{\mathbf{W}}_\beta^-$ are asymptotically equivalent to the estimates of an instrumental variables method applied to equation (2) (assuming the state to be known), if the instruments are constructed from $\mathbf{P}_{t,\beta}$ and $\mathbf{U}_{t,\alpha}$ and are used together with the IV weighting chosen according to equation (5) (which is not the optimal choice). Furthermore the corresponding estimates of the system matrices are asymptotically equivalent.

Given the proposed IV interpretation, a natural question is how to optimally choose the weighting $\mathbf{M}$. To this end consider the matrix

$$\mathcal{R}_n = \sum_{\tau=-\alpha+1}^{\alpha-1} \mathbf{R}_{z^1z^1}(\tau) \otimes \mathbf{R}_{nn}(\tau) =$$
$$(\mathbf{I}_{s\alpha} \otimes \mathbf{\Phi}_\alpha^e)\left\{\sum_{-\alpha+1}^{\alpha-1} \mathbf{R}_{z^1z^1}(\tau) \otimes \mathbf{R}_{ee}(\tau)\right\}(\mathbf{I}_{s\alpha} \otimes \mathbf{\Phi}_\alpha^e)^T$$

Here $\mathbf{\Phi}_\alpha^e$ is defined analogously to $\mathbf{\Phi}_\alpha$ with E replacing D, and K replacing B. Further $\mathbf{R}_{ee}(\tau) = \mathbb{E}E_{t,\alpha}^+ E_{t+\tau,\alpha}^T$, where $E_{t,\alpha}^+$ is defined analogously to $Y_{t,\alpha}^+$ using the noise ε_t in place of the observations y_t. Thus $\mathbf{R}_{ee}(\tau)$ is a block matrix, where the $(i, i-\tau)$ blocks are equal to the identity matrix, the remaining ones being zero. The structure of the sum in the equation above can be shown to be of the form as presented in the Appendix. Here we only used the assumed stationarity of the input and output sequences. In the pseudo-stationary case the expectation has to be replaced by the limit of the sample averages.

It follows from the structure of the matrix given in the appendix that the rank of $\mathcal{R}_n$ is at most $(s+m)s(\alpha+\beta-1) + (2\alpha-1)ms$ and thus $\mathcal{R}_n$ is singular. The rank of $\mathcal{R}_n$ can also be derived from the fact, that asymptotically equivalently the regression could be based upon the sample covariances of $\mathbf{P}_{t,\beta}^-$ and $\mathbf{U}_{t,\alpha}^+$ with $\mathbf{Y}_{t,\alpha}^+$, which contain $\mathbb{E}y_t y_{t-\tau}^T, \mathbb{E}y_t u_{t-\tau}^T, \tau = 1, \cdots, \alpha+\beta-1$ and $\mathbb{E}u_t u_{t+\tau}^T, \tau = 0, \cdots, \alpha-1$ as linearly independent elements. The expression given in the appendix also shows, that for the actual calculation of the

$$
\left[\begin{array}{ccc|ccc|c|ccc}
g_{z,z}(0) & \cdots & g_{z,z}(\alpha-1) & g_{z,z}(1) & \cdots & g_{z,z}(\alpha) & \cdots & g_{z,u}(-\alpha) & \cdots & g_{z,u}(-1) \\
g_{z,z}(-1) & \cdots & g_{z,z}(\alpha-2) & g_{z,z}(0) & \cdots & g_{z,z}(\alpha-1) & & g_{z,u}(-\alpha-1) & \cdots & g_{z,u}(-2) \\
 & & & & & & & & & \\
g_{z,z}(-\alpha+1) & \cdots & g_{z,z}(0) & g_{z,z}(2-\alpha) & \cdots & g_{z,z}(1) & \cdots & g_{z,u}(-2\alpha+1) & \cdots & g_{z,u}(-\alpha) \\
\hline
g_{z,z}(-1) & \cdots & g_{z,z}(\alpha-2) & g_{z,z}(0) & \cdots & g_{z,z}(\alpha-1) & \cdots & g_{z,u}(-\alpha-1) & \cdots & g_{z,u}(-2) \\
 & & & & & & & & & \\
g_{z,z}(1-\alpha) & \cdots & g_{z,z}(0) & g_{z,z}(2-\alpha) & \cdots & g_{z,z}(1) & & g_{z,u}(-2\alpha+1) & \cdots & g_{z,u}(-\alpha) \\
g_{z,z}(-\alpha) & \cdots & g_{z,z}(-1) & g_{z,z}(1-\alpha) & \cdots & g_{z,z}(0) & \cdots & g_{z,u}(-2\alpha) & \cdots & g_{z,u}(-\alpha-1) \\
\hline
\vdots & & \vdots & \vdots & & \vdots & \vdots & \vdots & & \vdots \\
\hline
g_{u,z}(1) & \cdots & g_{u,z}(\alpha) & g_{u,z}(2) & \cdots & g_{u,z}(1+\alpha) & \cdots & g_{u,u}(1-\alpha) & \cdots & g_{u,u}(0) \\
g_{u,z}(0) & \cdots & g_{u,z}(\alpha-1) & g_{u,z}(1) & \cdots & g_{u,z}(\alpha) & & g_{u,u}(-\alpha) & \cdots & g_{u,u}(-1) \\
 & & & & & & & & & \\
g_{u,z}(2-\alpha) & \cdots & g_{u,z}(1) & g_{u,z}(3-\alpha) & \cdots & g_{u,z}(2) & \cdots & g_{u,u}(2-2\alpha) & \cdots & g_{u,u}(1-\alpha) \\
\hline
\vdots & & \vdots & \vdots & & \vdots & \vdots & \vdots & & \vdots \\
\hline
g_{u,z}(\alpha) & \cdots & g_{u,z}(2\alpha-1) & g_{u,z}(\alpha+1) & \cdots & g_{u,z}(2\alpha) & \cdots & g_{u,u}(0) & \cdots & g_{u,u}(\alpha-1) \\
g_{u,z}(\alpha-1) & \cdots & g_{u,z}(2\alpha-2) & g_{u,z}(\alpha) & \cdots & g_{u,z}(2\alpha-1) & & g_{u,u}(-1) & \cdots & g_{u,u}(\alpha-2) \\
 & & & & & & & & & \\
g_{u,z}(1) & \cdots & g_{u,z}(\alpha) & g_{u,z}(2) & \cdots & g_{u,z}(\alpha+1) & \cdots & g_{u,u}(-\alpha+1) & \cdots & g_{u,u}(0)
\end{array}\right] \otimes \mathbf{I}_s
$$

Table 1. Structure of the inner sum of the matrix $\mathcal{R}_n$, see Appendix.

covariance matrix only the second moments and the matrix Φ_α^e are needed. Therefore an implementation of the regressions involved in the first step of the MOESP type of procedures based on the sample covariance matrices can be used to reduce the cost of deriving the covariance matrix. This might be valuable reduction of computer time in the two step procedure proposed in (Gustafsson, 1999).

Since the asymptotical variance of the estimation error is singular, an optimal weight is equal to the limit of $[\mathcal{R}_n + \langle \mathbf{Z}_{t,\alpha}, \mathbf{X}_t\rangle\langle \mathbf{X}_t, \mathbf{Z}_{t,\alpha}\rangle]^\dagger$ where † denotes the Moore-Penrose pseudo-inverse. For a proof of this fact, see (Söderström and Stoica, 1989) complement C4.3.

Unfortunately, the optimal weight does not correspond to a special choice of the weighting matrix in the MOESP type of procedure. Therefore the significance of the result lies in the fact, that the optimal weighting sets a lower bound on the estimation accuracy in the first step of the procedure and thus also for the accuracy of the estimated system as explained above. Note that the IV interpretation merely removes the weighting matrix $\hat{\mathbf{W}}_\beta^-$ as a potential factor for not achieving optimal accuracy in the following sense: If for given integers α, β and for a given weighting $\hat{\mathbf{W}}_\alpha^+$ the lower bound obtained using the optimal IV weight is higher than the Cramér Rao bound, then no choice of the weighting $\hat{\mathbf{W}}_\beta^-$ (including the 'optimal weighting' of Gustafsson) can achieve the Cramér Rao bound. In the next section such an example is given.

4. NUMERICAL EXAMPLE

In this section a numerical example is presented to demonstrate the previously derived lower bound. The second aim of this section is to give an example that the result obtained in (Jansson, 1997), that the row weighting $\hat{\mathbf{W}}_\alpha^+$ does not influence the asymptotic variance of the estimated poles does not carry over to the estimates of the system matrices, as has been pointed out already in (Bauer and Jansson, 2000). The choice of the row weighting also influences the lower bound corresponding to the IV weighting, as it affects the accuracy obtained in the second step of the MOESP type of procedures, i.e. the extraction of the system matrix estimates.

Consider then the following system:

$$
y_t = \frac{1 - 0.5q^{-2}}{1 - q^{-1} + 0.5q^{-2}} u_t + \frac{1 + q^{-1} + 0.5q^{-2}}{1 - q^{-1} + 0.5q^{-2}} \varepsilon_t
$$

The inputs are taken as an i.i.d. Gaussian sequence filtered by $1/(1 - 0.5q^{-1})$. Figure 1 shows the asymptotic variance of the estimates of the transfer function relating the observed inputs to the output, at 50 equally spaced angular frequencies in the range $[0, \pi)$ for $\alpha = 8, \beta = 3$. The five curves shown correspond to the Cramér Rao lower bound, the original MOESP procedure, the MOESP type of procedure using the weighting derived by (Gustafsson, 1999), the lower bound derived in this paper and the variance corresponding to the MOESP type of procedure using the CCA type of row weighting $\hat{\mathbf{W}}_\alpha^+$. It can be observed, that in this case the lower bound provided by the IV interpretation is in fact higher than the Cramér Rao bound and thus adds something to the knowledge. This is however not the case in all situations. It also shows, that the asymptotic variance of the transfer function estimate depends on the choice of the weighting $\hat{\mathbf{W}}_\alpha^+$, although the accuracy of the estimated poles does not.

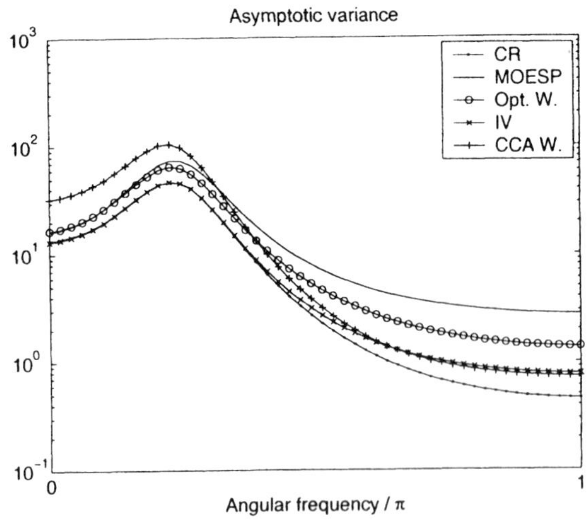

Fig. 1. Asymptotic variance evaluated at 50 equally spaced points in the angular frequency range $[0, \pi)$. The various lines correspond to the Cramér Rao bound (.), the MOESP weighting (-), the optimal weighting scheme (o), the lower bound based on the IV interpretation (x) and the CCA weighting (+). $\alpha = 8, \beta = 3$.

5. CONCLUSIONS

This paper deals with the accuracy of the MOESP type of subspace methods. It is shown, that the asymptotic accuracy can be investigated by decomposing the estimation procedure into two steps, where the first step essentially produces the same results (in asymptotic terms) as an instrumental variables approach, where the true state is assumed to be known. This leads to an optimal IV weight. However the optimally weighted IV estimate cannot be realized since it relies on knowledge of the state and thus the interpretation only leads to an additional lower bound on the achievable accuracy, when using a specified procedure. The information of this lower bound is of course only of use, if it is higher than the Cramér Rao lower bound. The main contribution of the paper however lies in the new IV interpretation of the MOESP type of procedures, which is related to the accuracy of the estimated system matrices directly, rather than only on the first step of the estimation as has been done in (Viberg, 1995). This might lead to some insights in the effects of the user choices.

6. REFERENCES

Bauer, D. (1998). Some Asymptotic Theory for the Estimation of Linear Systems Using Maximum Likelihood Methods or Subspace Algorithms. PhD thesis. TU Wien.

Bauer, D. and M. Jansson (2000). Analysis of the asymptotic properties of the moesp type of subspace algorithms. To appear in *Automatica*, April 2000.

Bauer, D., M. Deistler and W. Scherrer (1999). Consistency and asymptotic normality of some subspace algorithms for systems without observed inputs. *Automatica* **35**, 1243–1254.

Bauer, D., M. Deistler and W. Scherrer (2000). On the impact of weighting matrices in subspace algorithms. Proceedings of the SYSID 2000 conference, Sta. Barbara, USA.

Gustafsson, T. (1999). Subspace Methods for System Identification and Signal Processing. PhD thesis. Chalmers University, Gothenburg, Sweden.

Jansson, M. (1997). On Subspace Methods in System Identification and Sensor Array Signal Processing. PhD thesis. KTH, Stockholm.

Jansson, M. and Bo Wahlberg (1997). Counterexample to general consistency of subspace system identification methods. In: *Proceedings of SYSID'97*. Fukuoka, Japan. pp. 1677 – 1682.

Larimore, W. E. (1983). System identification, reduced order filters and modeling via canonical variate analysis. In: *Proc. 1983 Amer. Control Conference 2*. (H. S. Rao and P. Dorato, Eds.). Piscataway, NJ. pp. 445–451. IEEE Service Center.

Peternell, K. (1995). Identification of Linear Dynamic Systems by Subspace and Realization-Based Algorithms. PhD thesis. TU Wien.

Peternell, K., W. Scherrer and M. Deistler (1996). Statistical analysis of novel subspace identification methods. *Signal Processing* **52**, 161–177.

Söderström, T. and P. Stoica (1989). *System Identification*. Prentice Hall.

Van Overschee, P. and B. DeMoor (1994). N4sid: Subspace algorithms for the identification of combined deterministic-stochastic systems. *Automatica* **30**, 75–93.

Verhaegen, M. (1994). Identification of the deterministic part of mimo state space models given in innovations form from input-output data. *Automatica* **30**(1), 61–74.

Viberg, M. (1995). Subspace-based methods for the identification of linear time-invariant systems. *Automatica* **31**(12), 1835–1851.

White, H. (1984). *Asymptotic Theory for Econometricians*. Academic Press.

APPENDIX

This appendix demonstrates the structure of the inner sum in the equation defining $\mathcal{R}_n$. Using the notation $g_{z,u}(j) = \mathbb{E} z_t u_{t-j}^T, g_{u,u}(j) = \mathbb{E} u_t u_{t-j}^T, g_{z,z}(j) = \mathbb{E} z_t z_{t-j}^T$ we obtain that

$$\sum_{-\alpha+1}^{\alpha-1} \mathbf{R}_{z^1 z^1}(\tau) \otimes \mathbf{R}_{ee}(\tau)$$

is equal to the expression given in Table 1.

SUBSPACE ALGORITHMS FOR DATA FROM VARYING SENSOR LOCATIONS [1]

Albert Benveniste [*,2] **Michèle Basseville** [*,3]
Maurice Goursat [**] **Laurent Mevel** [*,2]

[*] IRISA, *Campus de Beaulieu, 35042 Rennes Cedex, F.-*
benveniste@irisa.fr - http//www.irisa.fr/sigma2/
[**] INRIA, *BP 105, Rocquencourt, 78153 Le Chesnay Cedex, F.*

Abstract: Subspace identification algorithms currently emerge as an efficient tool. In this paper, we investigate their use for the output-only identification of the eigenstructure of a linear MIMO system. We focus on the following situation: several successive data sets are recorded, with sensors at different locations in the structure; for doing this, some of the sensors, called the *reference sensors,* are kept fixed, while the other ones are moved for the different records. The interest of this setup is to emulate a situation in which hundreds of sensors are available, while in fact only, say, ten are actually at hand. This situation is typical in structural analysis in vibration mechanics, a case which motivated our study. One additional difficulty here is that the input, besides being not observed, is turbulent in nature and nonstationary. The purpose of this paper is to show how subspace methods can be adapted to such a situation.

Keywords: Subspace algorithms.

1. INTRODUCTION, PRACTICAL MOTIVATIONS

Subspace identification algorithms currently emerge as an efficient tool for many applications (Ottersten, Viberg and Kailath, 1992; Van Overschee and De Moor, 1996; Viberg and Ottersten, 1991). Our main motivation in this paper is output-only (Hermans and Van der Auweraer, 1999; Abdelghani *et al.*, 1999) structural identification in vibration mechanics. This problem consists in identifying the modes and mode shapes of a structure subject to ambient, non measured, vibrations, by using accelerometer measurements or strain gauges. This is output-only identification, as the excitation input is unknown and not measured [4]. As shown in section 2, subspace methods allow to identify the eigenstructure of a linear MIMO system, which turns out to be equivalent to modes and mode shapes.

We wish to analyze how subspace methods behave when the following, frequently used, setup is used:

> Several successive data sets are recorded, with sensors at different locations in the structure; for doing this, some of the sensors, called the *reference sensors,* are kept fixed, while the other ones are moved.

[1] This work has been carried out within the framework of Eureka project 1562 SINOPSYS (model based **S**tructural monitoring using **IN-OP**eration **SYS**tem identification), coordinated by LMS, Leuven, Belgium. During this work, L.M. was with LMS, Leuven.
[2] Also with INRIA.
[3] Also with CNRS.

[4] Examples are: offshore structures subject to swell, bridges subject to wind and traffic, etc.

This setup is frequently referred to as *polyreference setup* in vibration mechanics (Deblauwe, Allemang and Brown, 1987), and we keep this denomination. Of course, the interest of polyreference setup is to mimic a situation in which hundreds of sensors are available, while in fact only, say, ten are actually at hand.

However, there is one unpleasant feature of structural identification of structures subject to ambient excitation, namely

> The excitation is typically turbulent in nature, and is nonstationary [5].

The relevance of merging successive records, and its implementation in the case of nonstationary excitation, are the subject of this paper. In section 2 we briefly review covariance based subspace methods. The polyreference setup is stated in section 3. Then we discuss how data should be processed in case of polyreference setup. In section 4 we study the simpler case in which we assume the excitation noise to be stationary within each individual record (but not globally stationary). The corresponding implementation is discussed in section 5. Finally, the truly nonstationary excitation case is discussed in section 6. Conclusions and discussions follow.

2. SUBSPACE METHODS: QUICK VISIT

As vibration analysis is our motivation, we use structural identification in vibration analysis as our context for discussing modelling issues. We assume a continuous time linear system of the form:

$$\begin{cases} M\ddot{\mathcal{Z}}(s) + C\dot{\mathcal{Z}}(s) + K\mathcal{Z}(s) = \nu(s) \\ Y(s) = L\mathcal{Z}(s) \end{cases}$$
$$\mathbf{E}\nu(s)\nu^T(s') = Q^{\nu}\delta(s'-s) \tag{1}$$

where (M, C, K) are the mass, damping, and stiffness matrices, respectively, $\mathcal{Z}$ is the state vector (positions or accelerations), Y is the measurement vector (matrix L indicates which components of the state are actually measured), and ν is the excitation [6]. As reflected by (1), we assume that the ambient excitation is white noise, see (Prevosto *et al.*, 1991) for a justification of (or an excuse for) this. The case of nonstationary input excitation will be discussed subsequently. Structural identification (e.g., eigenstructure identification) consists in identifying the pairs (μ, ψ_μ), solutions of

$$(M\mu^2 + C\mu + K)\Psi_\mu = 0\,, \psi_\mu = L\Psi_\mu \tag{2}$$

Now, sampling (1,2) at period δ yields in the usual way the following equivalent discrete time state space form:

$$\begin{cases} X_{k+1} = FX_k + V_k \\ Y_k = HX_k \end{cases}$$
$$e^{\delta\mu} = \lambda\ ,\ \psi_\mu = \varphi_\lambda \stackrel{\Delta}{=} H\Phi_\lambda \tag{3}$$

where the pairs (λ, Φ_λ) are eigenvalues and eigenvectors of state transition matrix F. Note that the collection of pairs $(\lambda, \varphi_\lambda)$ form a canonical parametrization of the pole part of system (3).

Recall subspace identification of eigenstructure $(\lambda, \varphi_\lambda)$ consists of the following steps [7]:

$$R_i \stackrel{\Delta}{=} \mathbf{E}\ (Y_k\, Y_{k-i}^T)$$
$$\mathcal{H} \stackrel{\Delta}{=} \begin{pmatrix} R_0 & R_1 & R_2 & \dots & R_q \\ R_1 & R_2 & R_3 & \dots & \vdots \\ R_2 & R_3 & R_4 & \dots & \vdots \\ \vdots & \vdots & \ddots & \dots & \vdots \\ R_p & \vdots & \ddots & \dots & \vdots \end{pmatrix}$$

$$\text{factorize}\quad \mathcal{H} = \mathcal{O}\,\mathcal{C}\quad \text{where}$$

$$\mathcal{O} \stackrel{\Delta}{=} \begin{pmatrix} H \\ HF \\ HF^2 \\ \vdots \\ HF^p \end{pmatrix}$$
$$\mathcal{C} \stackrel{\Delta}{=} (G\ FG\ F^2G\ \dots\ F^qG)$$
$$G \stackrel{\Delta}{=} \mathbf{E}\,(X_k Y_k^T)$$

$$\text{finally:}\quad \mathcal{O} \longmapsto (H, F) \longmapsto (\lambda, \varphi_\lambda) \tag{4}$$

The actual implementation is sketched now. Substitute

$$\widehat{R}_i = \frac{1}{N}\sum_{k=1}^{N} Y_k Y_{k-i}^T \tag{5}$$

in (4), this yields the corresponding empirical Hankel matrix we denote by $\widehat{\mathcal{H}}$. Then compute

[5] Fluid/structure interaction in offshore structures results in shock effects causing nonstationary excitation, and the same holds for wind and traffic on bridges.

[6] To simplify the notations, we do not consider measurement noise here. Minor modifications are needed for subspace algorithms to take measurement noise into account, this is well known and we refer the reader to (Van Overschee and De Moor, 1996) or (Basseville, Abdelghani and Benveniste, 2000).

[7] We give here a form of subspace algorithm in which we start from output covariance matrices, instead of matrices of data; the difference with the usual, data based, form is minor, at least for eigenstructure identification, see (Van Overschee and De Moor, 1996).

SVD($\widehat{\mathcal{H}}$) and truncate the SVD to the desired model order, this yields an estimate $\widehat{\mathcal{O}}$ for the observability matrix $\mathcal{O}$. From $\widehat{\mathcal{O}}$, get $(\widehat{H}, \widehat{F})$, and then $(\widehat{\lambda}, \widehat{\varphi}_\lambda)$.

3. THE POLYREFERENCE SETUP: MATHEMATICAL MODELLING

Excitation with constant covariance: a simple solution. We consider again system (3). We partition observation matrix H and measurement vector Y as

$$H = \begin{pmatrix} H^0 \\ H^1 \\ \vdots \\ H^J \end{pmatrix}, \quad Y_k = \begin{pmatrix} Y_k^{(0)} \\ Y_k^{(1)} \\ \vdots \\ Y_k^{(J)} \end{pmatrix} \tag{6}$$

We assume the following polyreference setup: instead of a single record for (Y_k) we collect, successively, J different records:

$$\underbrace{\begin{pmatrix} Y_k^{(0)} \\ Y_k^{(1)} \end{pmatrix}}_{\text{record 1}} \underbrace{\begin{pmatrix} Y_k^{(0)} \\ Y_k^{(2)} \end{pmatrix}}_{\text{record 2}} \cdots \underbrace{\begin{pmatrix} Y_k^{(0)} \\ Y_k^{(J)} \end{pmatrix}}_{\text{record } J} \tag{7}$$

More precisely, the situation is the following. For $j = 1, \ldots, J$ different records, we have:

$$\begin{cases} X_{k+1}^j &= F X_k^j + V_k^j, \ \text{cov}(V_k^j) = Q \\ Y_k^{0,j} &= H^0 X_k^j \qquad \text{(the reference)} \\ Y_k^j &= H^j X_k^j \qquad \text{(sensor pool n}^o j) \end{cases} \tag{8}$$

meaning that, for record j, sensor pools 0 and j are used. Thus, for $j = 1, \ldots, J$, we have access to empirical estimates of the following covariances, for $i \geq 0$:

$$\begin{array}{lll} R_i^{0,j} &= \mathbf{E} Y_k^{0,j} (Y_{k-i}^{0,j})^T &= H^0 F^i G \\ R_i^j &= \mathbf{E} Y_k^j (Y_{k-i}^{0,j})^T &= H^j F^i G \end{array} \tag{9}$$

where

$$G \stackrel{\Delta}{=} \mathbf{E} X_k^j (Y_k^{0,j})^T \tag{10}$$

is *independent from* j since we assumed the input excitation noise V to be stationary. Thus stacking the R_i^j as follows

$$R_i = \begin{pmatrix} R_i^0 \\ R_i^1 \\ \vdots \\ R_i^J \end{pmatrix}, \quad R_i^0 \stackrel{\Delta}{=} R_i^{0,j} \ \forall j \tag{11}$$

yields the basic factorization

$$R_i = H F^i G, \tag{12}$$

so we can proceed again as if we had a full set of sensors, instead of sensor pools with varying locations (compare (6) and (7)). Thus direct application of the subspace algorithm on stacked R_i^j as in (11) works in the case of stationary input excitation noise.

Excitation with time-varying covariance. Unfortunately, a problem arises when, in (8), $\text{cov}(V_k^j) = Q_j$ varies with j, since

$$G_j \stackrel{\Delta}{=} \mathbf{E} X_k^j (Y_k^{0,j})^T \tag{13}$$

also varies with j in this case. Hence, factorizations (9) still hold, but with a varying G_j, hence we loose factorization (12) as the latter requires G to be constant. As a consequence, bruteforce application of subspace algorithm by stacking the different R_i^j will not work in the case of input excitation noise with time-varying covariance. In the next section, we explain how to circumvent this difficulty.

The idea is to *renormalize* the covariance matrices of the different records to make them looking as if they were obtained with the same excitation, i.e., the same $G_j \equiv G_1, \forall j$. This is explained next.

4. RENORMALIZING THE DIFFERENT RECORDS

In the sequel, we use the generic notations (4). As $R_i^{0,j} = H^0 F^i G_j$ we have

$$\mathcal{H}^{0,j} \stackrel{\Delta}{=} \begin{pmatrix} R_0^{0,j} & R_1^{0,j} & R_2^{0,j} & \ldots \\ R_1^{0,j} & R_2^{0,j} & R_3^{0,j} & \ldots \\ R_2^{0,j} & R_3^{0,j} & R_4^{0,j} & \ldots \\ \vdots & \vdots & \ddots & \vdots \end{pmatrix} = \mathcal{O}(H^0, F)\ \mathcal{C}(F, G_j)$$

because of the definition (13) of G_j. Setting

$$\begin{array}{l} R_i^0 = (\, R_i^{0,1} \ R_i^{0,2} \ \ldots \ R_i^{0,J} \,) \\ G = (\, G_1 \ G_2 \ \ldots \ G_J \,) \end{array} \tag{14}$$

we also have

$$\mathcal{H}^0 \stackrel{\Delta}{=} \begin{pmatrix} R_0^0 & R_1^0 & R_2^0 & \ldots \\ R_1^0 & R_2^0 & R_3^0 & \ldots \\ R_2^0 & R_3^0 & R_4^0 & \ldots \\ \vdots & \vdots & \ddots & \vdots \end{pmatrix} = \mathcal{O}(H^0, F)\ \mathcal{C}(F, G) \tag{15}$$

Assume that pair (H^0, F) is observable. Then, using (14), from factorization (15) we obtain

$\mathcal{C}(F, G_j)$ for $j = 1, \ldots, J$. On the other hand, we have

$$\mathcal{H}^j \triangleq \begin{pmatrix} R_0^j & R_1^j & R_2^j & \ldots \\ R_1^j & R_2^j & R_3^j & \ldots \\ R_2^j & R_3^j & R_4^j & \ldots \\ \vdots & \vdots & \ddots & \vdots \end{pmatrix} = \mathcal{O}(H^j, F)\ \mathcal{C}(F, G_j) \qquad (16)$$

Assume that pair (F, G_j) is controllable for all j [8]. Consider

$$\overline{\mathcal{H}}^j = \mathcal{H}^j\ \left(\mathcal{C}^T(F, G_j)\ [\mathcal{C}(F, G_j)\mathcal{C}^T(F, G_j)]^{-1}\ \mathcal{C}(F, G_1)\right). \qquad (17)$$

Note that $\overline{\mathcal{H}}^j$ is just the Hankel matrix corresponding to

$$\overline{R}_i^j \triangleq H^j F^i G_1 \qquad (18)$$

i.e., $\overline{R}_i^j$ is obtained by substituting G_1 for G_j in $R_i^j = H^j F^i G_j$. This is the desired renormalization for the covariance matrices of Y, for the J different records. Thus we can proceed as in section 3: build, for $i \geq 0$:

$$\overline{R}_i = \begin{pmatrix} \overline{R}_i^1 \\ \overline{R}_i^2 \\ \vdots \\ \overline{R}_i^J \end{pmatrix}$$

Then, using (18), we get

$$\overline{\mathcal{H}} \triangleq \begin{pmatrix} \overline{R}_0 & \overline{R}_1 & \overline{R}_2 & \ldots \\ \overline{R}_1 & \overline{R}_2 & \overline{R}_3 & \ldots \\ \overline{R}_2 & \overline{R}_3 & \overline{R}_4 & \ldots \\ \vdots & \vdots & \ddots & \vdots \end{pmatrix} = \mathcal{O}(H, F)\ \mathcal{C}(F, G_1) \qquad (19)$$

and we continue as for the usual case.

Note that we do not need to compute explicitly the G_j matrices, as only the right factors are needed from factorizations (16) in order to compute $\overline{\mathcal{H}}^j$ in (17). Then $\overline{\mathcal{H}}$ is directly obtained by stacking the $\overline{\mathcal{H}}^j$ on a row-per-row manner.

5. IMPLEMENTATION ISSUES

Throughout this section, $\widehat{M}$ denotes the empirical estimate for quantity M. Based on (17), we need to compute

[8] This is a reasonable assumption, it just tells that, while the excitation may change, we need it to excite all the modes of the system, for each data set.

$$\widehat{\overline{\mathcal{H}}}^j = \widehat{\mathcal{H}}^j\ \left(\mathcal{C}^T(F, G_j)\ [\mathcal{C}(F, G_j)\mathcal{C}^T(F, G_j)]^{-1}\ \mathcal{C}(F, G_1)\right) \qquad (20)$$

from which $\widehat{\overline{\mathcal{H}}}$ follows by proper stacking, as in (19). For performing (20) we propose the following algorithm:

STEP 1 : Build the empirical version of $\mathcal{H}^0$ (see (15)), we denote it by $\widehat{\mathcal{H}}^0$.

STEP 2 : SVD-decompose $\widehat{\mathcal{H}}^0$, truncate at convenience by setting to 0 the smallest singular values. Keep only the right factor $\mathcal{C}(F, G)$, see (15). Since at this stage we do not try to recover the pair (F, G), we call this right factor just $\mathcal{C}$.

STEP 3 : Following the corresponding decomposition of G in (14), decompose $\mathcal{C}$ as

$$\mathcal{C} = (\ \mathcal{C}_1\ \mathcal{C}_2\ \ldots\ \mathcal{C}_J\ )$$

STEP 4 : Compute the renormalizing matrix factors

$$(\mathcal{C}_j^T\ [\mathcal{C}_j \mathcal{C}_j^T]^{-1}\ \mathcal{C}_1)$$

STEP 5 : Renormalize

$$\widehat{\overline{\mathcal{H}}}^j = \widehat{\mathcal{H}}^j\ \ (\mathcal{C}_j^T\ [\mathcal{C}_j \mathcal{C}_j^T]^{-1}\ \mathcal{C}_1)$$

STEP ∞ : We are back to the standard problem.

6. THE TRULY NONSTATIONARY EXCITATION CASE

In section 4 we have assumed that, within each given record, the excitation is stationary, but the excitation noise covariance matrix is different for different records. This is a reasonable approximation when the environmental conditions vary slowly with time. But we can do much better, indeed, by building upon the work of (Benveniste and Fuchs, 1985).

We again consider the polyreference setup (8). But now we assume instead that $Q^\nu = Q^\nu(s)$ is time varying in model (1). Assume a sampling period δ in building the discrete time model (3) from continuous time model (1). We have

$$\mathbf{E} V_k V_{k'}^T = Q_k\ \delta(k - k')\ , \qquad (21)$$

where

$$Q_k = \int_{k\delta}^{(k+1)\delta} e^{As} \widetilde{Q}^\nu(s) e^{A^T s} ds\ ,$$

$$A = \begin{pmatrix} 0 & I \\ -M^{-1}K & -M^{-1}C \end{pmatrix},$$

$$\text{and}\ \ \widetilde{Q}^\nu(s) = \begin{pmatrix} 0 & 0 \\ 0 & M^{-1}Q^\nu(s)M^{-T} \end{pmatrix}. \qquad (22)$$

We thus have again (7), but, now,

$$H = \begin{pmatrix} H^0 \\ H^1 \\ \vdots \\ H^J \end{pmatrix} \quad (23)$$

$$\begin{cases} X^j_{k+1} &= FX^j_k + V^j_k, \ \mathrm{cov}(V^j_k) = Q_k \\ Y^{0,j}_k &= H^0 X^j_k \qquad \text{(the reference)} \\ Y^j_k &= H^j X^j_k \qquad \text{(sensor pool n}^o j) \end{cases}$$

where the important point is that Q_k *is time-varying within each record* (see (22)), hence Y_k cannot be assumed stationary within each individual record. Consequently we cannot apply the argument in section 4.

Nevertheless, this technique can still be applied, with little modification, as we show now. For $j = 1, \ldots, J$, set

$$A^j_{K_j} = \frac{1}{K_j} \sum_{k \in \mathcal{I}_j} \|Y^{0,j}_k\|^2 + \|Y^j_k\|^2 \quad , \ K_j = |\mathcal{I}_j|$$

and $|\mathcal{I}|$ is the cardinal of interval $\mathcal{I}$. Consider also

$$\begin{aligned} G^j_{K_j} &= \frac{1}{A^j_{K_j}} \sum_{k \in \mathcal{I}_j} X^j_k \, (Y^{0,j}_k)^T \\ R^{0,j}_i(K_j) &= \frac{1}{A^j_{K_j}} \sum_{k \in \mathcal{I}_j} Y^{0,j}_{k+i} \, (Y^{0,j}_k)^T \\ R^j_i(K_j) &= \frac{1}{A^j_{K_j}} \sum_{k \in \mathcal{I}_j} Y^j_{k+i} \, (Y^{0,j}_k)^T \end{aligned} \quad (24)$$

Then, under conditions (C-1,2,3) of section II.A of (Benveniste and Fuchs, 1985), it is proved in formula (III.9) of the same paper, that

$$\begin{aligned} R^{0,j}_i(K_j) &= H^0 F^i G^j_{K_j} + o(K_j) \\ R^j_i(K_j) &= H^j F^i G^j_{K_j} + o(K_j) \end{aligned} \quad (25)$$

where matrix term $o(K)$ tends to zero when K tends to infinity. Formula (25) shows that the method of section 4 applies, provided that we use $R^j_i(K_j)$ defined in (24) as an empirical estimate for the covariance of measurement vectors in non-stationary polyreference setup.

7. DISCUSSION

We have presented the polyreference setup for eigenstructure identification. This type of setup is popular for instance in vibration mechanics.

Processing polyreference data for eigenstructure identification is already performed today in vibration mechanics by 1/ performing eigenstructure identification for each record, separately, and then 2/ manually gluing the results for the different records. More precisely, denote by $(\lambda^j, \varphi^j_{\lambda^j})$ the generic pair {eigenvalue, observed part of eigenvector} for the different records $j = 1, \ldots, J$. Note first that the poles λ^j should be pairwise identical for the different records, at least theoretically — denote by λ such a generic pole. Then, to relate the different eigenstructures for the different records, one uses the fact that there are reference sensors common to all records, namely those corresponding to observation matrix H^0. Fix an eigenvalue λ and denote by φ^j_λ the corresponding (observed part of) eigenvector for record j. Partition φ^j_λ into

$$\varphi^j_\lambda = \begin{pmatrix} \varphi^{0,j}_\lambda \\ \varphi^{j,j}_\lambda \end{pmatrix}$$

according to the block rows H^0 and H^j of matrix H. For $j' \neq j$ we must have $\varphi^{0,j'}_\lambda = \alpha_{j,j'} \ \varphi^{0,j}_\lambda$ for some complex scalar factor $\alpha_{j,j'}$, since such eigenvectors are defined up to a multiplicative complex scalar factor. Since one knows $\varphi^{0,j'}_\lambda$ and $\varphi^{0,j}_\lambda$, one can compute $\alpha_{j,j'}$. Thus one can glue φ^j_λ and $\varphi^{j'}_\lambda$ by constructing

$$\begin{pmatrix} \varphi^{0,j}_\lambda \\ \varphi^{j,j}_\lambda \\ \alpha^{-1}_{j,j'} \varphi^{j',j'}_\lambda \end{pmatrix}$$

and, clearly, performing in this way one can reconstruct a global eigenvector, as if one had all sensors corresponding to observation matrix H simultaneously working.

However all this is not so easy in practice, as appairment of poles for different records turns out to be sometimes not easy, and thus the result of eigenvector gluing may not be truly consistent.

The method we have proposed is an interesting alternative, as it consists in merging the input data of the subspace algorithm instead of gluing the results obtained for each individual record. The difficulty consisted in finding the appropriate renormalization of the data, prior to merge. But we have shown that a simple and robust solution exists.

REFERENCES

A. Benveniste and J-J. Fuchs (1985). Single sample modal identification of a nonstationary stochastic process. *IEEE Tr. Automat. Contr.*, vol.AC-30, pp.66-74.

B. Ottersten, M. Viberg and T. Kailath (1992). Analysis of subspace fitting and ML techniques for parameter estimation. *IEEE Tr. Sign. Proc.*, vol.SP-40, pp. 590-599.

P. Van Overschee and B. De Moor (1996). *Subspace Identification for Linear Systems: Theory, Implementation, Applications.* Kluwer.

M. Viberg and B. Ottersten (1991). Sensor array processing based on subspace fitting. *IEEE Tr. Sign. Proc.*, vol.SP-39, no 5, pp.1110-1121.

L. Hermans and H. Van der Auweraer (1999). Modal testing and analysis of structures under operational conditions: industrial application'. *Mechanical Systems and Signal Processing*, vol.13, no 2, pp.193-216.

M. Abdelghani, M. Goursat, T. Biolchini, L. Hermans and H. Van der Auweraer (1999). Performance of output-only identification algorithms for modal analysis of aircraft structures. *IMAC-XVII, International Modal Analysis Conference*, Kissimmee, FL.

M. Prevosto, M. Olagnon, A. Benveniste, M. Basseville and G. Le Vey (1991). State-space formulation, a solution to modal parameter estimation. *Jal Sound and Vibration*, vol. 148, no 2, pp.329-342.

F. Deblauwe, R.J. Allemang and D.L. Brown (1987). The polyreference time domain technique, *IMAC-V, International Modal Analysis Conference,* pp. 832-845.

M. Basseville, M. Abdelghani and A. Benveniste (2000). Subspace-based fault detection algorithms for vibration monitoring. *Automatica,* vol.36, no 1, pp.101-109.

STATE AND PARAMETER ESTIMATION IN BIOTECHNICAL BATCH REACTORS

Karel J. Keesman

Systems and Control Group, Wageningen University, Bomenweg 4, 6703 HD Wageningen, The Netherlands

Abstract. In this paper the problem of state and parameter estimation in biotechnical batch reactors is considered. Models describing the biotechnical process behaviour are usually non-linear with time-varying parameters. Hence, the resulting large dimensions of the augmented state vector, roughly > 7, in recursive estimation is a crucial problem. However, by decomposition techniques on the basis of singular perturbation analysis or batch phase analysis one is able to reduce the dimension of the augmented state vector. Furthermore, prior knowledge of parameters and initial states is essential. It is therefore shown how these initial values can be effectively obtained from the data. The approach will be demonstrated by two examples, a wastewater sludge treatment and a beer fermentation process, using real data. *Copyright © 2000 IFAC*

Keywords. state estimation, parameter estimation, singular perturbation method, observers, biotechnology, batch reactors

1. INTRODUCTION

In the last decades much research has been focused on parameter and state estimation in dynamical systems, in particular for linear systems. For non-linear systems the estimation problem becomes more complicated, but still good procedures, usually based either on linearization or on (local) optimisation, became available. However, it is widely recognised that when using optimisation-based indirect estimation schemes the number of decision variables must be limited to avoid bad convergence, local minima or high computational demands. Clearly, this limitation also holds for the Extended Kalman filter or, in general, for non-linear observer schemes. In practice, the number of decision variables should preferably be smaller that 5-7, say.
Bioprocesses, in particular due to the living organisms present in the system, are often described by non-linear dynamical models with time-varying parameters (see Bastin and Dochain, 1990, for a general introduction into on-line estimation in bioreactors). Hence, there is a need to simultaneously estimate state variables and model parameters, leading to augmented state vectors of large dimension. In the following, it is demonstrated how among other techniques singular perturbation analysis may help to reduce the dimension of the estimation problem. In this paper the focus is on processes in *batch bioreactors*. Therefore, one has the opportunity to distinguish between different batch phases, like phases of unlimited growth or phases with only decay. Consequently, the estimation of states and parameters should preferably be limited to these specific phases. Furthermore, since prior knowledge about initial state and parameter values is essential in solving non-linear estimation problems, to avoid biased estimates to a large extent, methods to extract this information from

prior knowledge and available experimental data will also be proposed.

The objective of this paper is to demonstrate (i) the application of decomposition methods and (ii) how initial estimates can effectively be obtained for recursive non-linear state and parameter estimation in batch bioreactors. The outline of the paper is as follows. In section 2 first some theoretical background material is presented. Then, in sections 3 and 4 the theory is further illuminated and illustrated to two real-world examples: a wastewater sludge treatment and a beer fermentation process, which are typical examples of biodegradation and production processes. Finally, section 5 contains some concluding remarks.

2. THEORETICAL BACKGROUND

2.1 *Bioprocess modelling*

The general dynamical model of bioprocesses in stirred tank batch reactors (see Bastin and Dochain (1990) for the general case) is given by

$$\frac{d\xi}{dt} = Kr(\xi) \tag{1}$$

where ξ is the n-dimensional state vector, K the biochemical conversion matrix and $r(\xi) := [r_1(\xi), r_2(\xi),\ldots, r_l(\xi)]^T$ is the vector of the reaction kinetics.

For the specific applications dealt with in the next sections, one can write (1) in a more detailed form, that is

$$\frac{dS}{dt} = -\frac{r_1}{Y_{XS}}(S,X) + r_2(R,X) \tag{2}$$

$$\frac{dX}{dt} = r_3(S,X) - r_4(X) \tag{3}$$

$$\frac{dP}{dt} = r_5(S,X) \tag{4}$$

$$\frac{dR}{dt} = r_6(X) - r_2(R,X) \tag{5}$$

where S is the substrate, X the biomass, R the intermediate substrate and P the product concentration. Hence,

$$K = \begin{bmatrix} -1/Y_{XS} & 1 & 0 & 0 & 0 & 0 \\ 0 & 0 & 1 & -1 & 0 & 0 \\ 0 & 0 & 0 & 0 & 1 & 0 \\ 0 & -1 & 0 & 0 & 0 & 1 \end{bmatrix} \tag{6}$$

Furthermore, r_1 and r_3 are related to growth, r_2 to conversion of the intermediate substrate, r_4 to decay of the biomass and r_5, r_6 to production of the intermediate and final product. Bastin *et al.* (1997) proposed a method to decompose the estimation problem by separately estimating the unknowns in K and in $r_i(S,R,X)$. However, as can be seen from (6), in this case K contains only one unknown, which could also have been determined from a simple batch respiration test. Therefore, other decomposition techniques are needed to effectively deal with these kind of estimation problems.

2.2 *Model decomposition techniques*

As will be illustrated by the example of endogenous respiration in a batch reactor, the state vector can be decomposed into a state vector with slow components (x), usually related to the biomass in the bioreactor, and a state vector with fast components (z), as *e.g.* substrate concentrations. In other words, the dynamics in the bioreactor show a two time-scale behaviour. In particular, $x = [X\ P]^T$ and $z = [S\ R]^T$.

Analysis of this type of behaviour can be performed along the line of singular perturbation theory (see, e.g. Kokotovic *et al.*, 1986; Khalil, 1992). Let us therefore rewrite the model (Eqn. 1) in the form of the standard singular perturbation model with $\xi = [x\ z]^T$

$$\dot{x} = f(t,x,z,\varepsilon) \qquad x(t_0) = x_0(\varepsilon) \tag{7}$$

$$\varepsilon\dot{z} = g(t,x,z,\varepsilon) \qquad z(t_0) = z_0(\varepsilon) \tag{8}$$

for $t \in [t_0,t_1]$, $\varepsilon \in [0,\varepsilon_1]$ with ε_1 a small number, $x \in \mathbb{R}^n$ and $z \in \mathbb{R}^m$. It is assumed that the functions f and g are continuously differentiable. Let $x(t,\varepsilon)$ denote the solution to the equations with slow dynamics (7) and $z(t,\varepsilon)$ the 'fast' solution to (8).

Let us first assume that $\varepsilon = 0$, so that (8) degenerates to an algebraic equation with solution $\bar{z}(t) = h(t,\bar{x}(t))$ where $\bar{x}(t)$ denotes the solution to (7) for $\varepsilon = 0$. Hence, for the endogenous process model this means that some of the derivatives are set to zero. Consequently, the unique roots, representing the 'quasi-steady state' values of S and R, can be found from this algebraic equation. Omitting the proof of satisfying certain stability conditions over the time interval $[t_b, t_1]$, the 'quasi-steady-state' or 'slow' model becomes

$$\frac{dx}{dt} = f(t,x,h(t,x),0) \qquad x(t_0) = x_0(0) \tag{9}$$

an autonomous system description with solution $\bar{x}(t)$, which approximately holds on $[t_b, t_1]$ with t_b properly chosen.

For the study of the fast dynamics the following relationship has to be analysed,

$$\frac{dz}{d\tau} = g(t,x,z,\varepsilon) \qquad z(0) = z_0(\varepsilon) \tag{10}$$

where $\tau = (t-t_0)/\varepsilon$ so that $d\tau/dt = 1/\varepsilon$. Note then that $t = t_0 + \varepsilon\tau$ and $x = x(t_0 + \varepsilon\tau, \varepsilon)$ become slowly varying variables. Setting again $\varepsilon = 0$, so that $t = t_0$ and $x = x_0(\varepsilon)$, reduces (10) to

$$\frac{dz}{d\tau} = g(t_0, x_0(\varepsilon), z, 0) \qquad z(0) = z_0(0) \tag{11}$$

with frozen parameters (t_0, $x_0(\varepsilon)$). For our example with $z = [S \;\; R]^T$, (11) is the so-called 'boundary-layer' or 'fast' model, which approximately describes the dynamics of S and R over $[t_0, t_b]$.

In conclusion, during the initial batch phase or after extinction of the fast responses, the responses of the corresponding 'fast' and 'slow' system can usually be described by simpler functions with less parameters. Consequently, the identification problem can be formulated as two subproblems with reduced models.

For the second example of beer fermentation in a batch reactor, which does not show this two time-scales behaviour, another type of decomposition based on distinction of different batch reaction phases can be used. For instance, one can distinguish between the non-limiting growth phase in the beginning of the reaction and the death phase when all substrate has been used. Formally, these phases can also be investigated via parameter sensitivity analysis (see *e.g.* Holmberg, 1982). Clearly, for each of these phases different models can be used, which in general results in a much simpler estimation problem formulation.

Hence, in both applications the key issue is *model decomposition in time*. Once the decomposition has been realised and prior knowledge has been used to obtain proper initial vales, the unknown state and parameters can be estimated. In the following, a general framework for recursive estimation of states and parameters on the basis of Extended Kalman filter (EKF) techniques will be presented.

2.3 *EKF-based observer theory*

Let us start by writing the estimation problem in state-space notation with stochastic components. That is, define the (augmented) state vector

$$\mathbf{x} := [\xi_1 \;\; \ldots \;\; \xi_n \;\; \theta_1 \;\; \ldots \;\; \theta_p]^T \tag{12}$$

where $\xi_1, \ldots, \xi_n$ denote the n states of (1) and $\theta_1, \ldots, \theta_p$ the p unknown, possibly time-varying parameters. Notice then that in practice, as a result of model decomposition, ξ may be limited to just x or z with the corresponding parameters $\theta_1, \ldots, \theta_q$, where $q \le p$. Furthermore, the uncertain relationship between states and observed outputs ($\mathbf{y} \in \mathbb{R}^s$), which usually are discrete-time variables, is given by

$$\dot{\mathbf{x}}(t) = \mathbf{f}(\mathbf{x}) + \mathbf{w}(t) \tag{13a}$$

$$\mathbf{y}(t_k) = \mathbf{h}(\mathbf{x}) + \mathbf{v}(t_k) \quad \text{for } k = 0, 1, \ldots, N \tag{13b}$$

where $\mathbf{w}$ is the system noise, representing modelling error and unknown inputs, and $\mathbf{v}$ is the measurement noise vector. Define the Jacobi matrix $\mathbf{F}$ with elements $f_{ij} = \left[\frac{\partial f_i(\hat{\mathbf{x}})}{\partial \hat{x}_j}\right]_{\hat{\mathbf{x}}=\hat{\mathbf{x}}_{k-1}}$ and from this the transition matrix $\mathbf{A}_k = \mathbf{A}(\hat{\mathbf{x}}_{k-1}) := e^{\mathbf{F}\Delta t}$. Furthermore, define the matrix $\mathbf{H}$ with elements $h_{ij} = \left[\frac{\partial h_i(\hat{\mathbf{x}})}{\partial \hat{x}_j}\right]_{\hat{\mathbf{x}}=\hat{\mathbf{x}}_{k-1}}$.

Then, the Extended Kalman filter equations (see e.g. Jazwinski, 1970; Gelb, 1974) for this continuous-discrete-time dynamical system are given by

prediction:

$$\hat{\mathbf{x}}_{k/k-1} = \hat{\mathbf{x}}_{k-1} + \int_{t_{k-1}}^{t_k} \mathbf{f}(\hat{\mathbf{x}})d\tau \tag{14}$$

$$\mathbf{P}_{k/k-1} = \mathbf{A}_k \mathbf{P}_{k-1} \mathbf{A}_k^T + \mathbf{Q}_k \tag{15}$$

correction:

$$\hat{\mathbf{x}}_k = \hat{\mathbf{x}}_{k/k-1} + \mathbf{P}_{k/k-1}\mathbf{H}_k^T [\mathbf{R}_k + \mathbf{H}_k \mathbf{P}_{k/k-1} \mathbf{H}_k^T]^{-1} \{\mathbf{y}_k - \mathbf{h}(\hat{\mathbf{x}}_{k/k-1})\} \tag{16}$$

$$\mathbf{P}_k = \mathbf{P}_{k/k-1} - \mathbf{P}_{k/k-1}\mathbf{H}_k^T [\mathbf{R}_k + \mathbf{H}_k \mathbf{P}_{k/k-1} \mathbf{H}_k^T]^{-1} \mathbf{H}_k \mathbf{P}_{k/k-1} \tag{17}$$

where $\mathbf{P}_k$ is the covariance matrix of the estimates at time instant t_k, $\mathbf{Q}_k$ is the covariance matrix associated with the system noise ($\mathbf{w}_k$) and $\mathbf{R}_k$ the covariance matrix of the measurement noise ($\mathbf{v}_k$) at t_k. The index "$k/k-1$" denotes prediction from time instant t_{k-1} to t_k. Notice then from the definition of $\mathbf{F}$ that $\mathbf{A}_k$ is evaluated at each new sampling instant. However, application of this algorithm (14-17) will not generally guarantee stability. Therefore, a somewhat modified EKF, based on regularisation theory (see Boutayeb *et al.*, 1998 for details), will be applied, where the noise terms are omitted in Eqns. (13a-b). Hence, EKF is used here as an observer with $\mathbf{Q}_k = 0$. Furthermore,

$$\mathbf{R}_k = \lambda \mathbf{H}_k \mathbf{P}_{k/k-1} \mathbf{H}_k^T + \zeta \mathbf{I}_p \tag{18}$$

where $\lambda > 0$ and $\zeta > 0$ are fixed by the user and $\mathbf{I}_p$ denotes the $p \times p$ unity matrix. In the following biotechnical examples, where λ is set to 3 and 6, respectively, and ζ is set to 1, the full identification approach will be demonstrated and illuminated.

3. WASTEWATER SLUDGE TREATMENT EXAMPLE

Let us start with presenting the mathematical model describing the endogenous process. The following set of differential equations describe the activated sludge process under endogenous and non-limiting dissolved oxygen conditions in a batch reactor

$$\frac{dS(t)}{dt} = -\frac{\mu}{Y}\frac{S(t)}{K_S + S(t)}X(t) + k_H R(t) \qquad (19)$$

$$\frac{dR(t)}{dt} = (1 - f_p)bX(t) - k_H R(t) \qquad (20)$$

$$\frac{dX(t)}{dt} = \mu\frac{S(t)}{K_S + S(t)}X(t) - bX(t) \qquad (21)$$

$$\frac{dP(t)}{dt} = f_p bX(t) \qquad (22)$$

The first equation describes the readily biodegradable substrate (S) dynamics in terms of a loss due to growth of biomass and production by hydrolysis of slowly biodegradable substrate (R). In the second equation the change in R is governed by a fraction of the decay of heterotrophic biomass (X) and hydrolysis. The change in biomass due to growth and decay is described in the third equation. Finally, the fourth equation represents the accumulation of inert material (P) as a result of decay of the biomass.

The relation between the measurable outputs, (endogenous) *respiration rate* (r in mg/l.h), and *mixed liquor volatile suspended solids* (Ξ in g COD/m^3), and states of this autonomous system are given by

$$r(t) = \frac{1-Y}{Y}\mu\frac{S(t)}{K_S + S(t)}X(t) \qquad (23)$$

$$\Xi(t) = R(t) + X(t) + P(t) \qquad (24)$$

As was shown by Keesman *et al.* (1998) this model shows a two time-scales behaviour. The 'slow' or 'quasi-steady state' model, describing the fast processes by the algebraic equations: $\bar{S}(t) = S^*$ (a constant depending on parameter values only) and $\bar{R}(t) = \frac{(1-f_p)b}{k_H}\bar{X}(t)$, and the slow processes by

$$\frac{d\bar{X}(t)}{dt} = [(1 - f_p)bY - b]\bar{X}(t) \qquad (25)$$

$$\frac{d\bar{P}}{dt} = f_p b\bar{X}(t) \qquad (26)$$

with solutions denoted by an overbar, approximately holds for $t > 0.5$ d. After some manipulation the following approximation is obtained

$$r(t) \approx (1 - Y)(1 - f_p)b\bar{X}(t) \qquad (27)$$

Taking the derivative on both sides, substituting (25) and solving the differential equation leads to

$$r(t) \approx r_0\, e^{\beta t} \qquad (28)$$

with $\beta := [(1 - f_p)Y - 1]\ b$, which for given $f_p = 0.08$ (fixed value in literature) and $Y = 0.70$ (from batch respiration test) may be used to obtain an estimate of b from the tail of the available respirometric data. Notice from (23) that from 'fast' response data, available in the first few hours of the experiment, one can try to get estimates of $(1-Y)/Y{*}\mu X(0)$, K_S and $S(0)$. It appears (Keesman *et al.*, 1998), however, that for given Y the direction in the parameter space given by $0.51\mu X(0) + 0.86S_S(0)$ is badly identifiable and that only K_S can be uniquely identified from this data.

Let us now focus on the 'slow' response data, which for the estimation of the endogenous behaviour is of most interest. In a first step, the parameter values of b, μ and k_H, for given values of f_p and Y, as well as the state trajectories are reconstructed from prior parameter knowledge and from the experimental data (presented in Fig. 1) by direct model inversion (see Keesman and Spanjers, 2000 for further details) to obtain reliable initial estimates for the next formal optimisation step.

Then, in the second step the recursive scheme (Eqns. 14-17) is applied to estimate the state trajectories. From the estimation results (not shown here, see Keesman and Spanjers, 2000) it appeared that $R(t) >> 0.01\ X(t)$, so that hydrolysis should be modelled as $k_h\,X(t)$ with k_h roughly 0.01 h^{-1} instead of the commonly used expression $k_H\ R(t)$, which is only valid if $R(t) << 0.01\ X(t)$. However, a good fit for Ξ is obtained, but not for r which strongly depends on the decay coefficient b (see Eqn. 27). Augmenting the state vector with the "unobserved state" b leads to more acceptable results (see Fig. 1 and 2), although further investigations are needed to improve the estimation results.

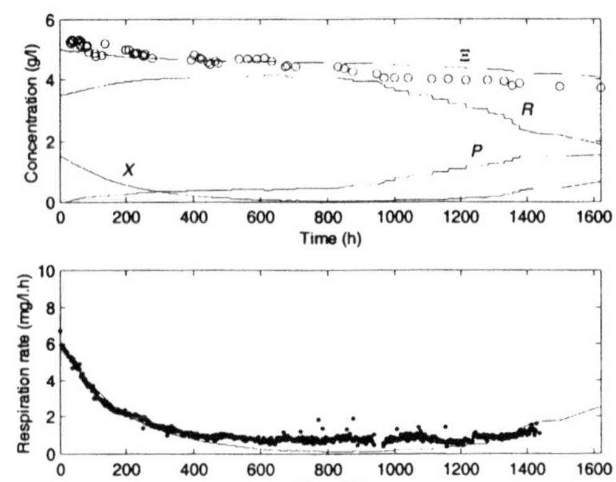

Fig. 1. Recursively estimated states/outputs for time-varying decay coefficient.

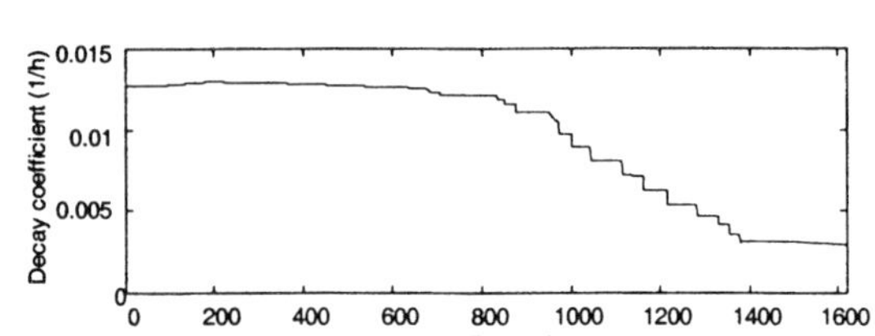

Fig. 2. Recursively estimated decay coefficient.

4. BEER FERMENTATION EXAMPLE

Beer fermentation, that is in terms of the reaction mechanism,

biomass (yeast) + sugar + $H_2O \rightarrow$
alcohol + CO_2 + H_2O

can, under batch conditions, be described by the following differential equations,

$$\frac{dS(t)}{dt} = -\frac{\mu_S S(t)}{K_S + S(t)} X(t) \tag{29}$$

$$\frac{dX(t)}{dt} = \frac{\mu_X S(t)}{K_X + S(t)} X(t) - bX(t) \tag{30}$$

$$\frac{dP(t)}{dt} = \frac{\mu_P S(t)}{K_P + S(t)} X(t) \tag{31}$$

(see *e.g.* Torija-Isasa *et al.*, 1997). The first equation describes the substrate (glucose) dynamics in terms of a loss due to growth of the yeast. In the second equation the change in active biomass (X) is governed by growth and death of the yeast. The production of ethanol is described in the third equation.
The measurable output of this autonomous system is the yeast concentration (Ξ in mg/l), obtained from a dielectric measuring procedure (Meijer, 1998), *i.e.*

$$\Xi(t) = X(t) \tag{32}$$

In a first step we try to determine some of the parameter values from prior knowledge and the experimental data. Let us therefore assume that the chemical formula for yeast is given by: $CH_{1.8}O_{0.5}$. Furthermore, glucose is $C_6H_{12}O_6$ and ethanol is C_2H_5OH. On the basis of these formulae one can write the overall algebraic reaction equation

$$r_X\, CH_{1.8}O_{0.5} + r_S\, C_6H_{12}O_6 + r_P\, C_2H_5OH + r_C\, CO_2 + r_W\, H_2O = 0$$

Hence, for the specific elements one can derive relationships between the reaction rates in moles/h (see *e.g.* Van 't Riet and Tramper, 1991). That is,

$$\begin{aligned} &\text{C-balance:} && r_X + r_S + r_P + r_C = 0 \\ &\text{H-balance:} && 1.8r_X + 2r_S + 3r_P + 2r_W = 0 \\ &\text{O-balance:} && 0.5r_X + r_S + 0.5r_P + 2r_C + r_W = 0 \end{aligned} \tag{33}$$

which gives us three equations with five unknowns. Consequently, the following three relationships in terms of r_X and r_S can be obtained:

$$\begin{aligned} r_C &= -0.2r_X - 0.33r_S \\ r_P &= -0.8r_X - 0.67r_S \\ r_W &= 0.3r_X \end{aligned} \tag{34}$$

Assume that the relationship between r_S and r_X is given by

$$-r_S = \frac{r_X}{Y} + m_S X_{tot} \tag{35}$$

with Y a yield factor, m_S a maintenance coefficient (moles/moles.h) and X_{tot} the total amount of biomass (moles). Initially, there is no growth limitation and no decay, so that

$$r_X = \frac{dX_{tot}}{dt} = \mu_X X_{tot} \tag{36}$$

where μ_X can be estimated from the measured biomass data in the initial stage of the fermentation. Given this estimate $\hat{\mu}_X$ together with values for Y and m_S from literature, that is Y = 0.14 moles/moles and m_S = 0.15 moles S/moles X.h (Van 't Riet and Tramper, 1991), and the mole masses M_X = 21.8, M_S = 30, M_P = 23 g, the growth coefficients μ_S and μ_P (1/h) can be found from

$$\hat{\mu}_S = \left(\frac{\hat{\mu}_X}{Y} + m_S\right)\frac{M_S}{M_X} \tag{37}$$

$$\begin{aligned} \hat{\mu}_P &= \frac{r_P}{X_{tot}} \cdot \frac{M_P}{M_X} \\ &= \left(-0.8\hat{\mu}_X + 0.67\frac{\hat{\mu}_X}{Y} + 0.67 m_S\right)\cdot\frac{M_P}{M_X} \end{aligned} \tag{38}$$

It appears that for a specific experiment $\hat{\mu}_X = 0.058$ (1/h), so that the other estimates become $\hat{\mu}_S = 0.78$ and $\hat{\mu}_P = 0.35$ (1/h). Parameter sensitivity analysis using sensitivity functions reveals that especially μ_X is a dominant parameter, which also determines the other conversion parameters as shown before. Furthermore, the death parameter and the Michealis-Menten coefficients are of minor importance. Therefore, the Michaelis-Menten coefficients have been taken from literature, that is K_X = 0.7464, K_S = 0.0252, and K_P = 3.2155. And, finally, the death parameter has been estimated from the last part of the data when all substrate has been used.
A typical result of an application of the observer (Eqn. 14-17) for an extended beer fermentation experiment, when *only* dielectric measuring data were directly available, is presented in Fig. 3.

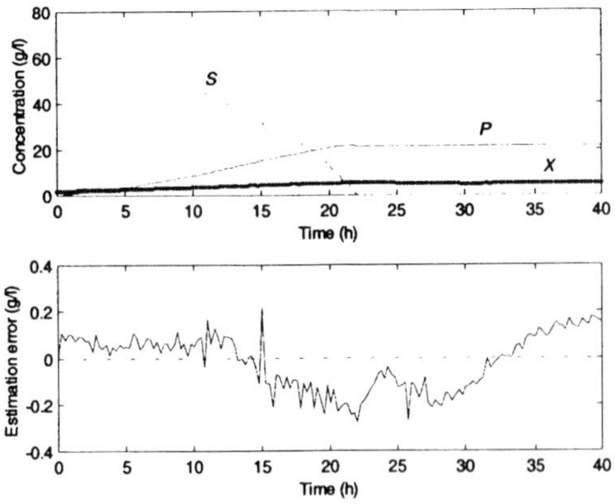

Fig. 3. Estimation results of extended beer experiment no. 1.

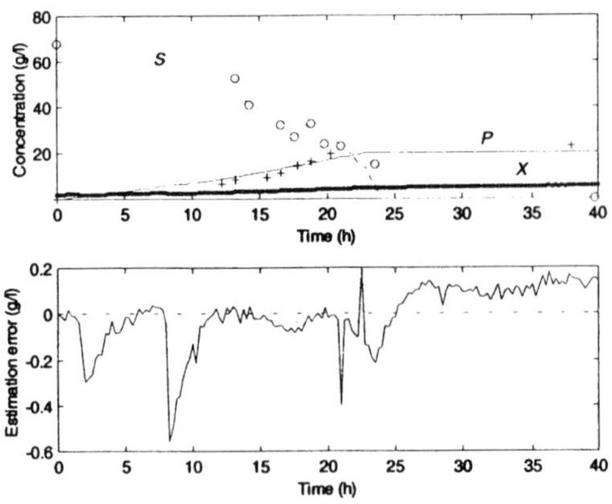

Fig. 4. Validation of estimation results of extended beer experiment no. 2.

However, in an additional experiment laboratory analyses of both sugar (o) and ethanol (+) concentrations have been used to *validate* the estimation procedure. These results are presented in Fig. 4, where only the estimation error of X is shown.

5. CONCLUDING REMARKS

It has been shown how to reduce a non-linear estimation problem for biotechnological processes on the basis of model decomposition using singular perturbation theory and batch phase analysis. Furthermore, since reliable initial values are of key importance in non-linear estimation, methods using direct model inversion and chemical structure relationships have been presented. The approach has been successfully applied to two real-world examples of biological wastewater sludge treatment and beer fermentation.

ACKNOWLEDGEMENT

We like to thank Marianne Busch for collecting the endogenous respiration data and Gerben Brans for performing the beer experiments at IMAG-DLO (Wageningen).

REFERENCES

Bastin, G., O. Bernard, L. Chen and V. Chotteau (1997). Identification of mathematical models for bioprocess control. In: *Proc. 11th Forum of Applied Biotechnology*, Ghent, Belgium, 25-26 Sept. 1997, pp. 1559-1571.

Bastin, G. and D. Dochain (1990). *On-line Estimation and Adaptive Control of Bioreactors*. Elsevier Sci. Publ., Amsterdam.

Boutayeb, M., H. Rafaralahy, M. Darouach (1998). Convergence analysis of the Extended Kalman filter used as an observer for nonlinear deterministic discrete-time systems. *IEEE Trans. Autom. Control*, **42(4)**: 581-586.

Gelb, A. (Ed.) (1974). *Applied Optimal Estimation*. The MIT Press, Cambridge, Mass.

Holmberg, A. (1982). On the practical identifiability of microbial growth models incorporating Michaelis-Menten nonlinearities. *Math. Biosc.* **62**:23-43.

Jazwinski, A.H. (1970). *Stochastic Processes and Filtering Theory*, Academic Press, New York.

Keesman K.J., H. Spanjers and G. van Straten (1998). Analysis of endogenous process behavior in activated sludge. *Biotechnol. & Bioeng.* **57(2):** 155-163

Keesman, K.J. and H. Spanjers (2000). Endogenous model state and parameter estimation from an extensive batch experiment, To appear in *Biotechnol. & Bioeng.*

Khalil, H.K. (1992). *Non-linear Systems*. Macmillan Publishing Company, New York.

Kokotovic, P.V., Khalil, H.K., O'Reilly, J. (1986). *Singular Perturbations Methods in Control: Analysis and Design*. Academic Press, New York.

Meijer, S. (1998). *On-line biomass concentration determination with a dielectric measuring method.* MSc Thesis Systems and Control Group, Wageningen University.

Riet, K. van 't and J. Tramper (1991). *Basic bioreactor design*. Marcel Dekker, New York.

Torija-Isasa, M.E., C. Diez-Marques, M.C. Camara-Hurtado, B. Andres-y-Toro, J.M. Giron-Sierra and C. Fernandez-Conde (1997). Changes in soluble sugars during experimental fermentation for production of beer. *Alimentacion Equipos y Tecnologia*, **16**, 59-71.

PARAMETER ESTIMATION THAT IMPROVES VALIDITY OF MODEL FOR A CERTAIN MODEL USE

Mario Zec, Nadja Hvala[1] and Stanko Strmčnik

J. Stefan Institute, Jamova 39, SI-1000 Ljubljana, Slovenia

Abstract: The paper considers the problem of parameter estimation in the case that model parameters are not uniquely identifiable from measurements. A simulation analysis indicates that from measurements it is possible to extract data that is important for the estimation of model parameters relative to a certain model use. If this data is improved by a proper measurement campaign (proper choice of measured variables, better accuracy, higher measurement frequency) it is to expect that a valid model for a certain model use is obtained. Simulation study is performed for a simple activated sludge model from wastewater treatment, while the estimation of model parameters is done by Monte-Carlo simulation. *Copyright © 2000 IFAC*

Keywords: parameter estimation, validation, sensitivity analysis, Monte-Carlo simulation, water pollution.

1. INTRODUCTION

Simulation study presented in the paper was inspired by specific characteristics of wastewater treatment plant models. Despite of a widespread model use that is usually based on one of the standard models in this area, like ASM1 (Henze, *et al.*, 1987), the design of valid and reliable models of wastewater treatment processes is still relatively difficult. One of the obstacles for the model use in practice is the estimation of model parameters. They must be fit to the individual plant and the conditions in the plant, because some of the model parameters show strong dependency upon plant operation and wastewater composition.

Despite of research efforts spent by several research groups during the past years, still no established procedure is agreed upon for model calibration (Weijers, *et al.*, 1997). The reason is that mathematical models of wastewater treatment processes are known as *poorly defined systems* (Sperling, 1993), where a large number of parameters and state variables is usually included in the model structure, the model equations express high non-linearities and they lack *identifiability* in some equations. Due to identifiability problems unique and reliable estimates of model parameters cannot be obtained. It is known that if the model is found unidentifiable, certain parameters can only be estimated in combination with other parameters. In that case different parameter sets can yield almost the same level of performance. Although a good match in terms of the model responses and measured variables is obtained, it may mask significant errors in the individual parameters (Murray-Smith, 1998).

Identifiability of wastewater treatment models has been studied by several authors (Weijers, *et al.*, 1997; Godfrey and DiStefano, 1985; Ayesa, *et al.*, 1993). For example, Jeppsson (1996) has shown that even very simple activated sludge model that is

[1]Corresponding author. Phone: +386 61 1773 606, Fax: +386 61 219 385, e-mail: nadja.hvala@ijs.si

theoretically identifiable is practically unidentifiable if a relatively small noise is added to the measured variables. Despite of this fact, the methods developed for parameter estimation of well defined systems are still widely applied in this area. Most often, parameter estimation results in a unique set of parameter values, and gives a single result when the model is used for a certain purpose. However, it would be more appropriate if the parameter estimation procedure took into account the model uncertainty and identifiability problems, and gave the confidence in parameter estimates. In this way, also the application of the model for a certain purpose would enable to ascertain the confidence in the obtained results. In case of large confidence limits that indicate unidentifiability of a certain parameter, also large confidence limits of the obtained result are obtained, if that parameter affects the given model use significantly.

The above reasoning represents the starting point of the research work presented in this paper. The aim of the study was to find out whether it is possible to reduce the parameter estimates confidence limits, so that small range of results is obtained when the model is used for a certain purpose. The testing hypothesis is that this can be done by taking into account the model use already in the parameter estimation procedure. In the paper, the relations between the model use and parameter estimates are studied by observing characteristic shapes of those model responses that produce narrower range of model results. The study was performed by the aid of Monte-Carlo simulation, which was used to represent parameters values in the model parameter space.

The paper is organised as follows. Section 2 presents the simulation model. In section 3 different sets of model parameters and the range of model results are defined. The obtained simulation results are shown in section 4. The paper ends with the conclusion.

2. SIMULATION MODEL

The study deals with a simple activated sludge model from wastewater treatment which describes the bacterial growth and decay in a batch reactor (Jeppsson, 1996; Godfrey and DiStefano, 1985):

$$\frac{dX}{dt} = \mu_{\max} \frac{S}{K_S + S} X - bX \quad (1)$$

$$\frac{dS}{dt} = -\frac{1}{Y} \mu_{\max} \frac{S}{K_S + S} X \quad (2)$$

where X represents the biomass concentration [g/m^3], S represents the concentration of substrate [g/m^3], μ_{max} is maximum specific growth rate [h^{-1}], K_S is the substrate half-saturation coefficient [g /m^3], Y is yield factor [(g cell COD formed)/(g COD oxidised)] and b is biomass decay rate [h^{-1}].

In the following it will be assumed that the model structure has been already chosen and is not a subject of identification, it is only necessary to estimate the model parameters. The chosen model (1)-(2) contains 4 parameters and 2 initial conditions. It has been assumed that the values of two parameters (Y and b) are already known. It is necessary to estimate parameters μ_{max} and K_S and two initial conditions (X_0 and S_0), further addressed as parameters. The expected ranges for μ_{max} and K_S were adopted from ASM1, while the ranges for X_0 and S_0 were taken from (Jeppsson, 1996) and are shown in Table 1.

To calibrate the model parameters it is necessary to dispose of plant measurements. For the simulation study the measurements were obtained artificially by simulating the model at parameters values as shown in the last column of Table 1. It has been assumed that measurements of S are available for model calibration.

In the study, the model will be used for two different purposes:

- for setting the optimal batch duration, *i.e.* the time when substrate concentration S during the batch drops bellow a certain value (*i.e.* $S < 5\text{g/m}^3$), and
- for determining the maximum biomass concentration X obtained during the batch.

In the first case the model use is interesting in an on-line control of a batch reactor for determining the most suitable time for completion of a certain batch phase. The second model use gives the information about the maximum biomass produced during the batch.

Table 1. The values of model parameters and initial conditions used in the simulation study

Model parameters and initial conditions		Expected range	Values used for simulation
Known parameters	b [h^{-1}]	-	0.02
	Y [g/g]	-	0.66
Estimated parameters	μ_{max} [h^{-1}]	0.15-0.35	0.25
	K_S [g/m^3]	5-55	10
Initial conditions	X_o [g/m^3]	2-6	4
	S_o [g/m^3]	20-60	40

3. PARAMETER ESTIMATION

3.1 Objective function

The parameter estimation procedure is based on the agreement between the computed and measured process variable, and is evaluated by an objective function condidering the following assumptions:

- Around the measured values of process variable it is possible to assign a certain region. Due to measurement error and model uncertainty it is to expect that the true values of process variable lie within this region.
- The probability density within the region is supposed to be uniform, *i.e.* the probability that a certain value within the region represents the true value of process variable is equal for all the values in the region.
- The model is considered of a good quality if the model response lies within the specified region.

To take into account the above assumptions the objective function based on maximum absolute deviation can be used:

$$e_{abs\max} = e(\boldsymbol{y_m}) = \left|\max_k(c_k)\right| \quad (3)$$

with

$$\boldsymbol{c} = [c_1,...,c_k,...,c_n]^T = \boldsymbol{U}(\boldsymbol{y}_p - \boldsymbol{y}_m) \quad (4)$$

where $\boldsymbol{c}$ denotes a vector of weighted deviations between the measured and computed process output, $\boldsymbol{y}_p$ is a vector of process measurements, $\boldsymbol{y}_m$ is a vector of computed values of model output at corresponding sampling points, n denotes the number of measurements, $\boldsymbol{U}$ is an nxn diagonal matrix with u_{ii} representing the weight of the i-th measured value. Throughout this paper the weighting will be performed in such a way that a certain measurement is included or excluded in the objective function, *i.e.* $u_{ii} \in \{0,1\}$.

In contrast to more widely used least squares methods, parameter estimation based on $e_{abs\max}$ enables to set bounds on maximum deviation between the computed and measured value at each sampling point. The chosen parameters values are considered as appropriate if the objective functions satisfies the following criterion:

$$e_{abs\max} \le K \quad (5)$$

where constant K denotes the maximum allowable deviation between computed and measured value, and defines a symmetrical region around the measured values inside which the response of models with acceptable parameters values is to be expected.

3.2 Definition of different parameter sets and the region of model use

For the estimation of model parameters different parameters sets will be used. Let us first define the vector of model parameters $\boldsymbol{\theta}$, which is defined as $\boldsymbol{\theta} = [\theta_1,...,\theta_j,...,\theta_k]^T$, where θ_j, j=1,...,k, represent individual parameters that need to be estimated, k is the number of parameters. In our case k=4, and

$$\boldsymbol{\theta} = [X_0, S_0, \mu_{\max}, K_S]^T \quad (6)$$

A choice of parameters values for X_0, S_0, μ_{max} and K_s defines a parameter vector $\boldsymbol{\theta}_i$ in the parameter space. For the parameter estimation, it is reasonable to search for appropriate parameters values only on a subspace Θ that includes reasonable parameters values. In our case the process model (1)-(2) is based on biological reactions where model parameters have certain physical meanings. Therefore, their reasonable values are already known and are expected to lie within the ranges given in Table 1.

Parameter vectors $\boldsymbol{\theta}_i$ can be classified into different parameter sets. Let us define a set $\mathcal{A}$ as a set of vectors in the parameter subspace Θ. From $\mathcal{A}$ it is possible to define a subset $\mathcal{B}$, $\mathcal{B} \subseteq \mathcal{A}$, which includes only those vectors in $\mathcal{A}$ which have acceptable parameters values. Parameters values are considered as acceptable if, when applied in the model, the model response lies in the allowable region around the measured values, as specified in (5).

These relations can be mathematically expressed in the following way. If we define transformation T that maps a vector of model parameters $\boldsymbol{\theta}$ in the model response $y(t)$:

$$T : \boldsymbol{\theta} \rightarrow y(t) \quad (7)$$

then according to (5) it holds:

$$\begin{aligned} &T_{\mathcal{A}} : \boldsymbol{\theta}_i \in \mathcal{A} \rightarrow y_i(t) \quad (8)\\ &\boldsymbol{y}_{m,i} = [y_i(t_1),...,y_i(t_j),...,y_i(t_n)]^T \\ &\text{If } e(\boldsymbol{y}_{m,i}) < K \text{ then } \boldsymbol{\theta}_i \in \mathcal{B} \text{ for } \forall \boldsymbol{\theta}_i \in \mathcal{A} \end{aligned}$$

where $\boldsymbol{y}_{m,i}$ is an n dimensional vector representing the values of the model response $y_i(t)$ at sampling points t_j, j=1,...,n, and $e(\boldsymbol{y}_{m,i})$ is defined according to (3)-(4). The above defined relations between the parameter vectors and model responses are schematically shown in Fig. 1.

Let us define another transformation U that maps a vector of model parameters $\boldsymbol{\theta}$ in the model result z, when the model is applied for a certain use:

$$U : \boldsymbol{\theta} \rightarrow z \quad (9)$$

z is a real value, $z \in \Re$, and is computed from the model output obtained when transformation U is applied on a parameter vector $\boldsymbol{\theta}$.

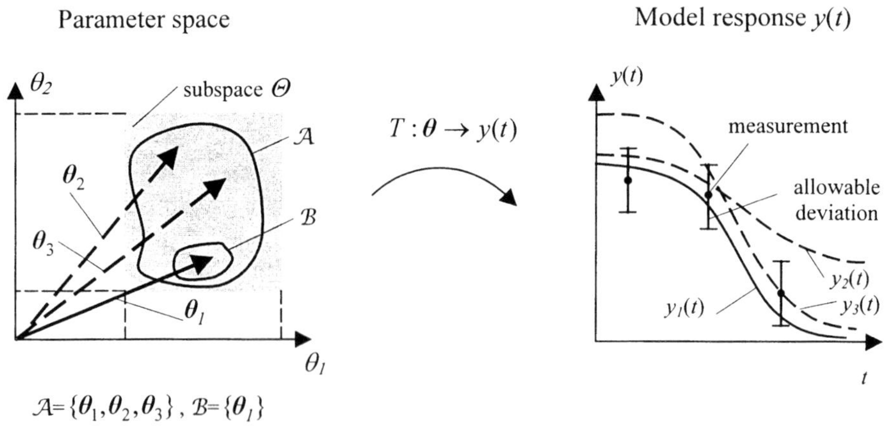

Fig. 1. Transformation of model parameters vectors from 2-dimensional parameter space into the model response, and their classification into sets $\mathcal{A}$ and $\mathcal{B}$.

If U is performed on a set of vectors θ_i, then different values z_i are obtained that spread over a certain region R. This region represents an interval inside which the values of the model use can be expected. R can be specified by its lower (r_{min}) and upper (r_{max}) bounds, *i.e.*

$$R = [r_{\min}, r_{\max}] = [\min_i(z_i), \max_i(z_i)] \tag{10}$$

If U is performed on different parameter sets, also different regions of model use are obtained. In the parameter estimation procedure it is especially interesting to find out whether transformation $U_{\mathcal{B}}$ and the thus obtained $R_{\mathcal{B}}$ give a satisfactory (small enough) region of the model use. This question is considered in the validation process. If the region of model results is not satisfactory, it is necessary to improve the parameter estimation procedure in such a way that a set of vectors $\mathcal{V}$ with valid parameters values is obtained. In that case $U_{\mathcal{V}}$ produces a valid region of model use $R_{\mathcal{V}}$.

A possible way of reducing the range of model results is by lowering the value of K in (5). Such an approach can be justified only if the model uncertainty and measurement error is improved, so that the uncertainty region around the measured values is reduced. In this case a subset C can be defined, $C \subseteq \mathcal{B}$, which is defined as a set of parameters vectors corresponding to a reduced value of K_r ($K_r < K$):

$$\begin{aligned} &T_{\mathcal{B}} : \theta_i \in \mathcal{B} \to y_i(t) \\ &y_{m,i} = [y_i(t_1),\ldots,y_i(t_j),\ldots,y_i(t_n)]^T \\ &\text{If } e(y_{m,i}) < K_r \text{ then } \theta_i \in C \text{ for } \forall \theta_i \in \mathcal{B} \end{aligned} \tag{11}$$

For C it is to expect that U_C produces a narrower range of model results than $U_{\mathcal{B}}$, *i.e.* $r_{C\min} \geq r_{\mathcal{B}\min}$, $r_{C\max} \leq r_{\mathcal{B}\max}$, and $r_{\mathcal{B}\max} - r_{\mathcal{B}\min} \geq r_{C\max} - r_{C\min}$. By a proper choice of K_r inequality holds, and for some parameter vectors in C', $C' = \mathcal{B} \backslash C$, transformation U produces results inside $R_{\mathcal{B}}$ but outside R_C. Parameter vectors with the above properties can be denoted as a set $\mathcal{M}$, which is a subset of C', $\mathcal{M} \subseteq C'$. In the parameter estimation procedure, parameter vectors in $\mathcal{M}$ should be excluded, so that narrower and thus valid range of model use is obtained. The above defined relations between the parameter sets and the range of model results are shown in Fig. 2.

Considering the so defined $\mathcal{M}$, it is interesting to know what are the characteristics of the model responses of the parameter vectors in $\mathcal{M}$ obtained by $T_{\mathcal{M}}$. The hypothesis of the simulation analysis described in the next section is that they include

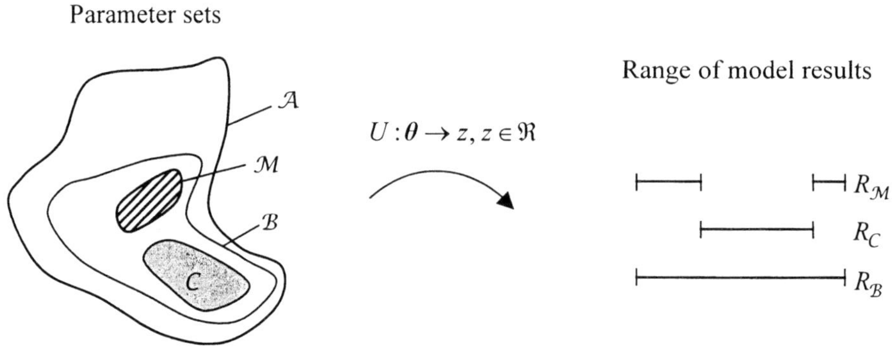

Fig. 2. Relations between different parameter sets and the range of model results.

some important features that can be usefully applied in the parameter estimation. With this information, measurement of process variables can be designed in such way that estimates of vectors with valid parameters values $\mathcal{V}$ are obtained, so that they are appropriate for a certain model use.

4. SIMULATION ANALYSIS

Simulations were performed in Matlab-Simulink. In the subspace Θ, parameter vectors that determine the set $\mathcal{A}$ were chosen by Monte-Carlo simulation. To represent the parameter space $8x10^6$ simulations were performed. The constant K for selecting parameter vectors that constitute the subset $\mathcal{B}$ was set to 2.76, while its reduced value K_r to determine the subset C was set to 2.208 (80% of K).

The parameter estimation procedure starts with the determination of parameter vectors in the subspace Θ (set $\mathcal{A}$), and then according to the course of model responses potentially classifies them into $\mathcal{B}$ (8). Let us first examine model responses of the parameter vectors in $\mathcal{B}$ (Fig. 3). As expected, they lie inside the area that is defined by the measurements of S and the maximum allowable deviation around each measurement. Model responses in Fig. 3 are represented by grey area which indicates high number of parameter vectors in $\mathcal{B}$.

For the estimation of parameter values that give small range of model results, it is important to consider the characteristics of the model responses for the parameter vectors in $\mathcal{M}$. From Fig. 4 it can be seen that these model responses show some characteristic features depending on the model use. For model use related to batch duration, the model responses are in the first part distributed over the whole range between the boundary lines, while at approx. 10h they split into two groups that produce results solely close to the boundary lines. Similar conclusions can be derived also for the model use related to the biomass concentration. In this case, model responses form two groups of responses in the first part of the model response, while at approx. 8h they spread over the whole range between the boundary lines. From observations above the following can be concluded: model responses for parameter vectors in $\mathcal{B}$ and $\mathcal{M}$ have a significantly different distribution in one part of the model response. Therefore it is to expect that the range of model results is affected by the influence of model parameters on the shape of model response in that part of the response.

The above reasoning was examined in simulation experiments where parameter estimates were obtained based on only one part of measurements of S. The obtained results are shown in Fig. 5. The left

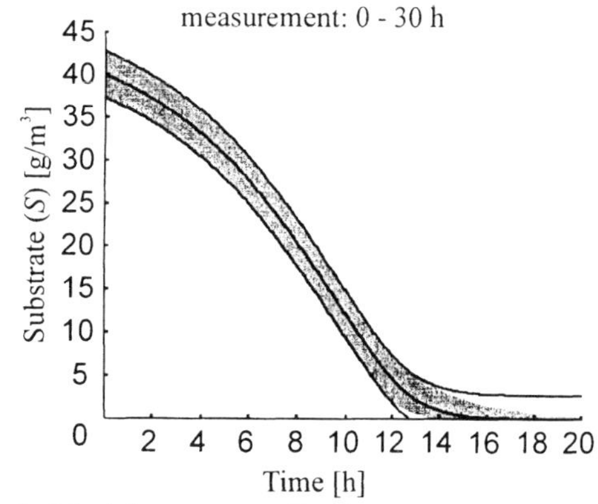

Fig. 3. Model responses for the parameter vectors in $\mathcal{B}$ (grey area), measurements of S (middle black line) and the boundary lines of the maximum allowable deviation from measurements (black outer lines).

diagram shows an example where the measured part of the model response was adjusted to using the model for determining the batch duration. For that purpose parameter vectors were classified into set $\mathcal{B}$ depending on their response during 9-20h. The right diagram shows the case where only the model response during 0-9h was taken into account and was adjusted to use the model for determining the biomass concentration. The two cases were obtained by setting diagonal elements of $\boldsymbol{U}$ to 0 in the appropriate part of the model response (4).

From Fig. 5 it can be seen that with so obtained parameter vectors in $\mathcal{B}$, model responses lie between the boundary lines in that part of the response where measurements were taken into account, but spread over much wider area in the other part. However, as in the most significant part of the model response (relative to a certain model use) the responses lie within the boundary limits, it is to expect that the range of model results is comparable to those obtained when the whole response is taken into account. This can be confirmed by the results shown in Table 2. It can be seen that if the measured part of the model response is chosen in accordance with a certain model use, almost equal range of model results is obtained despite of taking into account only the measurements of one part of the model response (comparing simulations 1 and 2 for batch duration, and simulations 1 and 3 for biomass concentration). Also, it can be seen that the results for the other model use deteriorate (much wider range).

Table 2. The range of model results when estimating parameters from different sets of measurements

Case		1	2	3
Measured part		0h-30h	9h-20h	0h-9h
Model use:	r_{min}	11.10	11.10	10.40
batch duration	r_{max}	13.05	13.05	16.25
Model use:	r_{min}	22.49	15.39	22.29
biomass conc.	r_{max}	29.25	39.17	29.08

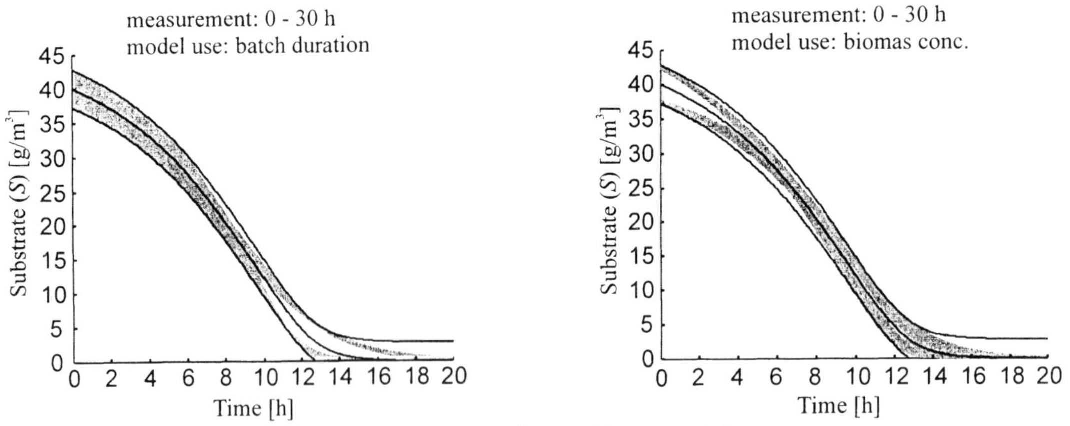

Fig. 4. Model responses for the parameter vectors in $\mathcal{M}$ for different model use.

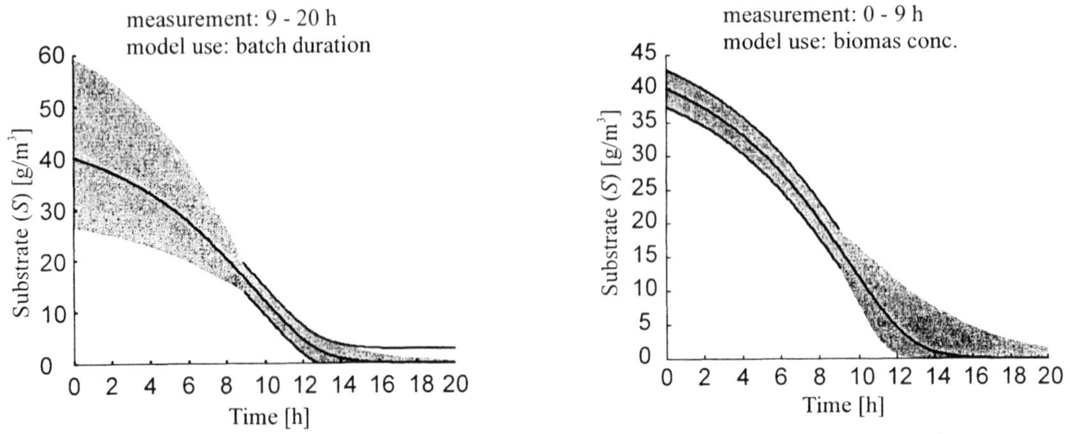

Fig. 5. Model responses for the parameter vectors in $\mathcal{B}$ (grey area) when only one part of measurements is considered in the parameter estimation. Middle black line represents all measurements of *S*, dark outer lines represent boundary lines for model response in that part, where measurements of *S* are taken into account.

5. CONCLUSIONS

The results of the simulation study indicate that better estimates of parameters values relative to a certain model use can be obtained if the purpose of the model is taken into account. The approach can be used to prepare for measurement of process variables that give data for model calibration. If the purpose of the model is known in advance and interrelations between the parameter values and measurements of process variables can be established, it is possible to adjust the measurement campaign to a certain model use, so that better parameter estimates are obtained. The approach can be specially important in cases where measurements of process variables are expensive, long-lasting and/or difficult to perform. Our future work will consider examples with more complex models, with additional measurement noise and lower measurement frequency. Also, interrelation between different measured variables and a certain model use will be considered.

REFERENCES

Ayesa, E., J. Florez, L. Larrea and J.L. Garcia-Heras (1993). Evaluation of sensitivity and observability of the state vector for system identification and experimental design. *Water Sci. Tech.*, **28**/11-12, 209-218.

Godfrey, K.R. and J.J. DiStefano (1985). Identifiability of model parameters. *Proc. 7th IFAC/IFORS Symposium Identification and System Parameter Estimation*, York, 3-7 July 1985, Pergamon Press, Oxford, 89-114.

Henze, M., C.P.L. Grady Jr, W. Gujer, G.v.R. Marais and T. Matsuo (1987). *Activated sludge model No. 1*, IAWPRC, London, England.

Jeppsson, U. (1996). *Modelling aspects of wastewater treatment processes*. PhD Thesis. Lund University, Lund.

Murray-Smith, D.J. (1998). Methods for the external validation of continuous system simulation models: a review. *Mathematical and Computer Modelling of Dynamical Systems*, **4**/1, 5-31.

Sperling, M.v. (1993). Parameter estimation and sensitivity analysis of an activated sludge model using Monte Carlo simulation and the analyst's involvement. *Wat. Sci. Tech.*, **28**/11-12, 219-229.

Weijers, S.R., H.A. Preisig, A. Buunen and T.W.M. Wouda (1997). Parameter estimation of activated sludge model No. 1 from full scale plant input/output data. *European Control Conference ECC 97*, 1-4 July 1997, Brussels, Belgium, 657-663.

RECURSIVE IDENTIFICATION OF ACTIVATED SLUDGE WASTEWATER TREATMENT PLANTS

Oscar A. Zanabria Sotomayor *[,1] **Song W. Park** *
Claudio Garcia **

* *LSCP - Department of Chemical Engineering*
** *LAC - Department of Electronic Engineering*
Polytechnic School of the University of São Paulo, Brazil
E-mail: oscar@lscp.pqi.ep.usp.br

Abstract: The activated sludge wastewater treatment plants are complex systems, incorporating a large number of biological, physical-chemical and biochemical processes that are difficult to supervise and control. Hence, there is a great interest in obtain adequate models for purpose of control. In this paper is performed on-line identification of two important variables of the activated sludge process: the nitrate and the dissolved oxygen concentrations. The identification is made employing the continuous-discrete Extended Kalman Filter. The obtained model parameters are applied to compare the process response and shows a good fitting to real data. The models may be used to implement advanced control strategies.

Keywords: Waste treatment, Water pollution, Identification, Extended Kalman filters, Nonlinear systems, Reduced-order models

1. INTRODUCTION

The activated sludge process is the biological process more used to treat the wastewater, either domestic or industrial. Its objective is to achieve a sufficiently low biodegradable matter concentration and low nutrients concentration in the outflow together with minimal sludge production, minimizing the use of energy and operational cost.

Theses process are large and very complex systems subject to strong variations in the inflow and composition of the incoming wastewater that are not well known. The process is multivariable, with many cross couplings and nonlinear. The dynamics depends on the variability of the influent, the complexity of the biological, physical-chemical and bio-chemical -phenomena. Other problem is that the process has a large range of time constants (from a few minutes to several days).

Two important variables of the process to be controlled are (Cost-624, 1998): the dissolved oxygen concentration (DOC) and the nitrate concentration. The DOC in the aerobic zone of an activated sludge process must be sufficient to supply enough oxygen to the microorganisms, in such a way that the organic matter be degraded and the ammonium be converted in nitrate. the nitrate concentration in an anoxic zone must be controlled to keep a sufficiently low effluent nitrate concentration.

This paper demostrates the application of the Extended Kalman Filter in the iterative adjusts of unknown parameters of reduced order models corresponding to the nitrate concentration and DOC in activated sludge processes. The objective is obtaining adequate models that may be used to implement advanced control strategies.

[1] Finantial support of FAPESP (98/12375-7)

2. THE ACTIVATED SLUDGE PROCESS

The activated sludge process (ASP) is a biological process in which microorganisms oxidize and mineralize the organic matter. Those microorganisms are kept by the air that is injected in the tank and by the use of agitators. The oxygen is used by the microorganisms the oxidize the organic material. Aiming to keep the microbial population, the sludge of the secondary settler is recirculated to the aeration tank. The biomass concentration is controlled by the flow rate of sludge in excess. Besides, the sludge in excess. Besides, the sludge in excess is removed to avoid sludge in the flowing water and to keep constant the concentration of solids in suspension.

An ASP, in a configuration with pre-denitrification for removal of organic matter and nitrogen from domestic wastewater, is shown in figure (1). The configuration is formed by a bioreactor, composed of an anoxic zone and two aerobic zones, and a secondary settler. The compartments of the bioreactor are considered to have constant volume (15 m^3 each) and to be ideally mixed whereas the secondary settler (20 m^3) is modeled employing a series of 10 layers (one-dimensional model). The influent flow Q_{int} is 4.17 m^3/h, with a proportion of biodegradable organic matter of 224 $mgCOD/l$ and a hydraulic retention time of 15.6 h. The internal recycle flow rate $Q_{int} = 2Q_{in}$, the external recycle flow rate $Q_{sl} = 0.8Q_{in}$ and the wastage flow rate is Q_w=0.0258 m^3/h. The air flow rate in aerobic zones are: Q_{air_1}=10.15 m^3/h and Q_{air_2}=6.7 m^3/h, respectively.

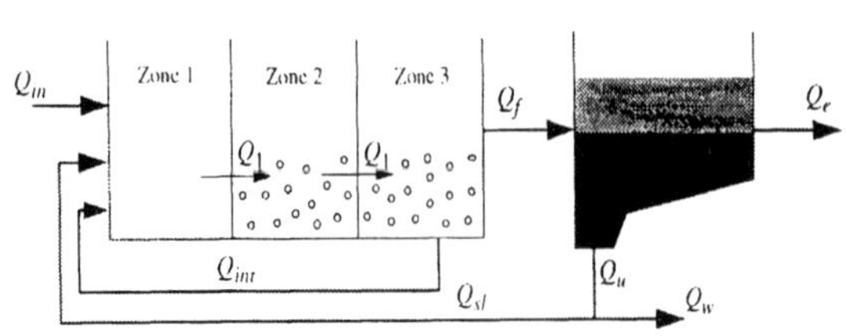

Fig. 1. ASP configuration

The model for each bioreactor zone is based on the IAWQ model No. 1 (Henze *et al.*, 1987) and for the settler on the double exponential settling velocity model (Takács *et al.*, 1991). The values of the process parameters are here omitted but can be found on Zanabria *et al.* (1999).

3. THE EXTENDED KALMAN FILTER

The Kalman filter (Kalman, 1960) is a set of mathematical equations that provides an efficient computational (recursive) solution of least-squares method.

Extended Kalman Filters (EKF) are a logical generalization of linear Kalman filters for the case where the system dynamics vary with operating and control point in non-linear systems.

Next are presented the set of equations that describe the EKF. For more details, see (Ljung, 1979) and (Ljung and Soderstrom, 1983).

Prediction error:

$$e(t) = y(t) - \hat{y}(t)$$

Kalman gain:

$$K(t) = \frac{P(t-1)\phi(t)}{r_2 + \phi^T(t)P(t-1)\phi(t)}$$

Prediction error covariance matrix:

$$P(t) = [I - K(t)\phi^T(t)]P(t-1) + R_1$$

Estimated parameter vector:

$$\hat{\theta}(t) = \hat{\theta}(t-1) + K(t)e(t)$$

4. IDENTIFICATION OF THE NITRATE CONCENTRATION

The IAWQ model No. 1 is very complex, containing 13 nonlinear coupled differential equations and 10 parameters to be used in on-line identification and control.

The first principles equation, corresponding to IAWQ model No. 1, that describe the concentration of nitrate in the anoxic zone is presented. Assuming zero dissolved oxygen concentration, which is reasonable for an anoxic zone, the following equation is obtained:

$$\frac{dS_{NO}}{dt} = -\left(\frac{1-Y_H}{2.86Y_H}\right)\left(\frac{S_S}{K_S+S_S}\right)\left(\frac{S_{NO}}{K_{NO}+S_{NO}}\right) \mu_H \eta_g X_{BH} + \frac{Q_{in}}{V}(S_{NO_{in}} - S_{NO}) + \frac{Q_{int}}{V}(S_{NO_{int}} - S_{NO}) + \frac{Q_{sl}}{V}(S_{NO_{sl}} - S_{NO}) \qquad (1)$$

As can be seen in equation (1) the gain of the process varies with the S_S, S_{NO} and X_{BH} concentrations. It also depends on time varying parameters as K_S and K_{NO}, and flow rates. Since there are so many uncertainties in the process, we prefer to estimate the process using a reduced order model.

Jeppsson (1996) proposed a reduced order model of the IAWQ model, with 5 state veriables and 6 parameteres that have to be estimated.

The correspondiong reduced order model equation that describes the nitrate concentration dynamics in the anoxic zone is given by the following expression:

$$\frac{dS_{NO}}{dt} = -\frac{1-Y_H}{2.86Y_H} r_H X_{COD} X_{BH} + \frac{Q_{in}}{V}(S_{NO_{in}} - S_{NO}) + \frac{Q_{int}}{V}(S_{NO_{int}} - S_{NO}) + \frac{Q_{sl}}{V}(S_{NO_{sl}} - S_{NO}) \quad (2)$$

This equation is used for identification of the nitrate concentration considering the following assumptions:

- the X_{COD} variable is measurable and equal to S_S variable. In practice COD measurements are often considered to be equal to the amount of S_S;
- the X_{BH} variable is considered constant and equal to their respective mean value. This is assumed because of the non-existence of heterotrophic biomass sensors. Besides, the relative variation of this variable is minimum (with exception of a strong change in the concentration of S_S, pH or toxic elements in the influent, in which case X_{BH} considerably varies);
- the flows and the nitrate concentration measurements in Q_{in}, Q_{int} and Q_{sl} are available. The parameters Y_H and r_H need are estimated. All sensor dynamics are neglected.

The identification procedure is processed on-line, in open loop, through a continuous-discrete EKF with a variable forgetting factor (Fortescue *et al.*, 1981). The parameter vector gradient of the prediction error are determined by:

$$\theta^{\mathbf{T}}(\mathbf{t}) = [Y_H, r_H]$$

and

$$\phi(\mathbf{t}, \theta) = \frac{\Delta t X_{BH} X_{COD}}{2.86} \begin{bmatrix} r_H \frac{1}{Y_H} \\ \frac{Y_H - 1}{Y_H} \end{bmatrix}$$

respectively, and the input signal corresponds to random variations in the internal recycle flow (rich in nitrate), having as minimum and maximum values 0.1 and 1.9 times the nominal value of Q_{int} respectively. The initial parameter vector is $\widetilde{\theta}(0) = [1, 1]^T$ and the initial covariance matrix is $P(0) = diag(10^6, 10^6)$. The observed noise covariance matrix is $R_1 = 0$ and the minimal forgetting factor is 0.95. The selected sampling time is $\Delta t = 3.12$ min., that is approximately $1/10$ of the open loop response time constant to a 50% step variation in the internal recycle flow, in nominal operational conditions. The concentrations of the influent and the flows Q_{in} and Q_{sl} are kept constant.

The disturbing flow used in the identification, the output (continuous line), the identified output (dotted line) and the estimated parameters are shown in figure (2).

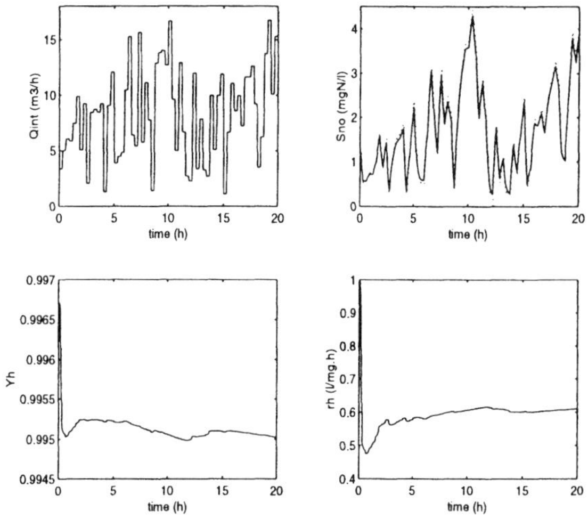

Fig. 2. Identification of the reduced order model for nitrate concentration

Aiming to compare the obtained model with the process, in order to be used in control, the process and the model were submitted to a disturbing sequence in Q_{int}, considering $Y_H = 0.995$ and $r_H = 0.6\ l/(mg.h)$. the reduced order model (dotted line) adequately adjusts itself to the process response (continuous line) as shown in figure (3).

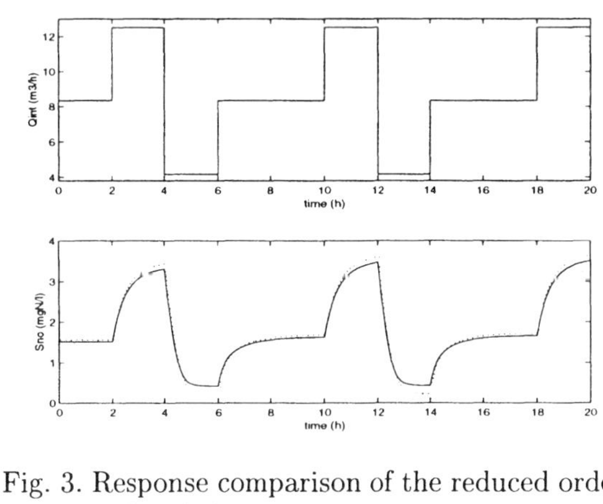

Fig. 3. Response comparison of the reduced order model and the process

This model has been successfully used to control the nitrate recycle flow rate in a pre-denitrifying activated sludge process (Zanabria *et al.*, 2000).

5. IDENTIFICATION OF THE DISSOLVED OXYGEN CONCENTRATION

The rigorous IAWQ model that describe the dissolved oxygen concentration, in aerobic environment, is given by the following equation:

$$\frac{dS_O}{dt} = -(\frac{1-Y_H}{Y_H})(\frac{S_S}{K_S+S_S})(\frac{S_O}{K_{OH}+S_O})$$
$$\mu_H X_{BH} - (\frac{4.57-Y_A}{Y_A})(\frac{S_{NH}}{K_{NH}+S_{NH}})$$
$$(\frac{S_O}{K_{OH}+S_O})\mu_A X_{BA} + K_l a(S_{O_{sat}} - S_O)$$
$$+\frac{Q_1}{V}(S_{O_1} - S_O) \qquad (3)$$

As can be observed, this equation depends on several time varying variables, parameters and flow rates, and other uncertainties in the system.

Due to the complexity of equation (3), it was chosen the following expression proposed by Bastin and Dochain (1990):

$$\frac{dS_O}{dt} = -OUR + K_l a(S_{O,at} - S_O) + \frac{Q_1}{V}(S_{O_1} - S_O) \qquad (4)$$

The equation (4) is used for identification of the DOC in the last aerobic zone , taking the following assumptions:

- the OUR (oxygen utilization rate) or bacterial respiration rate is available;
- the oxygen transfer function $K_l a$ is assumed to depend nonlinearly on the airflow rate and has the following form (Lindberg, 1997):

$$K_l a = k_1\left(1 - exp\left(\frac{-k_2 Q_{air_2}}{1000}\right)\right)$$

- DOC measurement in the aerobic zones are available. the sensor dynamics are neglected.

The procedure of identification applies a continuous-discrete EKF algorithm with variable forgetting factor, where the parameter vector and the gradient of the prediction error are:

$$\theta^{\mathbf{T}}(\mathbf{t}) = [k_1, k_2]$$

and

$$\phi(\mathbf{t}, \theta) = \Delta t(S_{O_{sat}} - S_O(t-1)) \cdot \begin{bmatrix} 1 - exp(-\alpha(t)) \\ \beta(t)exp(-\alpha) \end{bmatrix}$$

respectively, with:

$$\alpha(t) = \frac{k_2 Q_{air_2}(t-1)}{1000}$$

$$\beta(t) = \frac{k_1 Q_{air_2}(t-1)}{1000}$$

The input signal correspond to random perturbation in the air injection flow, between 0.5 and 1.5 times the nominal value of Q_{air_2}. The initial parameter vector is $\widehat{\theta}(0) = [10, 10]^T$, the initial covariance matrix is $P(0) = diag(10^6, 10^6)$, the observed noise covariance matrix is $R_1 = 0$ and the minimal forgetting factor is 0.95. The selected sampling time is $\Delta t = 16.56$ sec., that is approximately 1/10 of the open loop response time constant to a 50% step variations in the air injection flow rate. The concentrations of the influent and the flows $Q_1 (= Q_{in} + Q_{int} + Q_{sl})$ and Q_{air_1} are kept constant.

The input data, the output (continuous line), the identified output (dotted line) and the estimated parameters are shown in figure (4).

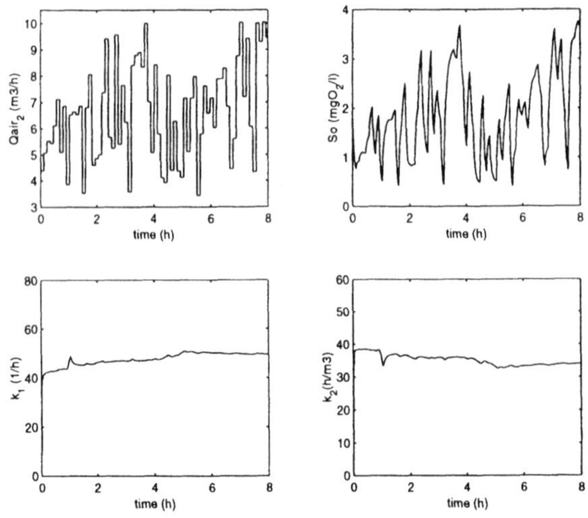

Fig. 4. Identification of DOC dynamics

To compare the response of the identified model with the response of the process, each model and process were submitted to a disturbance sequence in Q_{air_2}, considering the parameter values $k_1 = 49.45\ h^{-1}$ and $k_2 = 34.12\ h/m^3$. the obtained model (dotted line) adjusts perfectly to the process response (continuous line) as shown in figure (5). Figure 6 show that the estimated oxygen transfer function (dotted line) is very close to the value $K_l a$ of the process (continuous line).

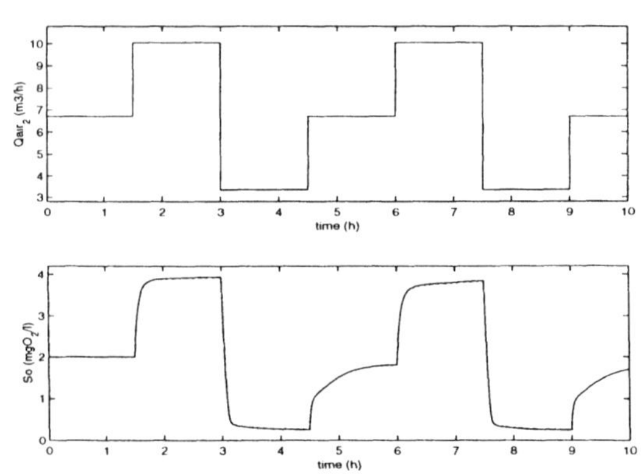

Fig. 5. Real and estimated values of the DOC

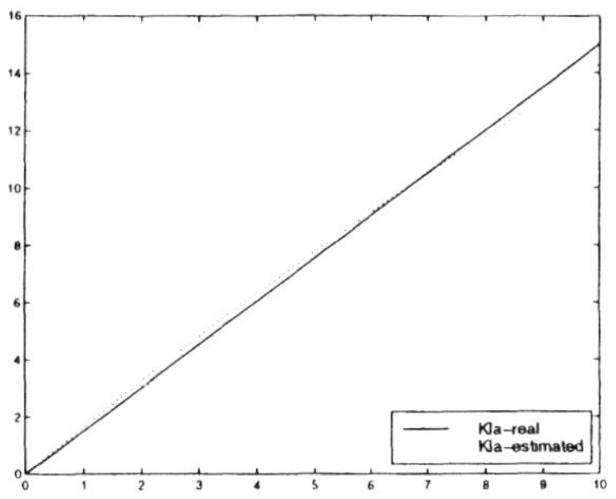

Fig. 6. Real and estimated values of the $K_l a$ function

6. CONCLUSIONS

According to the results obtained, the identification technique employed was successful and it is a promising tool for the implementation of advanced control strategies when regulating the nitrate and the dissolved oxygen concentrations in the wastewater treatment plants.

Nomenclature:

X_{BH} Heterotrophic biomass $[mg/l]$

X_{BA} Autotrophic biomass $[mg/l]$

X_{COD} Biodegradable organic matter $[mg/l]$

S_{NH} Soluble ammonium nitrogen $[mg/l]$

S_S Soluble biodegradable substrate $[mg/l]$

S_{NO} Soluble nitrate nitrogen $[mg/l]$

S_O Dissolved oxygen concentration (DOC) $[mg/l]$

S_{O_1} DOC in the first aerobic zone $[mg/l]$

$S_{O_{sat}}$ Saturated oxygen concentration $[mg/l]$

OUR Respiration rate $[mg/l.h]$

K_S Half saturation coefficient (hsc) for heterotrophs $[mg/l]$

K_{NH} Ammonium hsc for autotrophs $[mg/l]$

K_{NO} Nitrate hsc for denitrifying heterotrophs $[mg/l]$

K_{OH} Oxygen hsc for heterotrophs $[mg/l]$

K_{OA} Oxygen hsc for autotrophs $[mg/l]$

Y_H Yield factor for heterotrophic biomass

Y_A Yield factor for autotrophic biomass

r_H Reaction rate for heterotrophs $[l/mg.h]$

μ_H Maximum growth rate for heterotrophs $[h^{-1}]$

μ_A Maximum growth rate for autotrophs $[h^{-1}]$

η_g Correction factor for μ_H under anoxic conditions

Q Flow rate $[m^3/h]$

Q_{air_1} Air flow rate in the first aerobic zone $[m^3/h]$

Q_{air_2} Air flow rate in the second aerobic zone $[m^3/h]$

V Volume of the zone $[m^3]$

$K_l a$ Oxygen transfer function $[h^{-1}]$

k_1 Oxygen transfer function parameter $[h^{-1}]$

k_2 Oxygen transfer function parameter $[h/m^3]$

subscript:

in Wastewater inflow

int Internal recycle

sl Sludge recycle

7. REFERENCES

Bastin, G. and D. Dochain (1990). *On-line Estimation and Adaptive Control of Bioreactors.* Elsevier, Amsterdan.

Cost-624 (1998). Optimal management of wastewater treatment system. *The European Cooperation in the Field of Scientific and Technical Research, Website: (www.ensic.u-nancy.fr/COSTWWTP/).*

Fortescue, T., L. Kershenbaum and B. Ydstie (1981). Implementation of self-tuning regulators with variable forgetting factor. *Automatica* **17**(6), 831–835.

Henze, M., W. Gujer, G. Marais and Matsuo M. (1987). Activated sludge model n.1. Technical Report No.1. IAWQ. London.

Jeppsson, U. (1996). Modelling aspects of wastewater treatment processes. PhD thesis. Lund Institute of Technology. Lund, Sweden.

Kalman, R.E. (1960). A new approach to linear filtering and prediction theory. *Trans. ASME, J. Basic Engineering* **82**, 35–45.

Lindberg, C.-F. (1997). Control and estimation strategies applied to the activated sludge processes. PhD thesis. Uppsala University. Uppsala, Sweden.

Ljung, L. (1979). Asymptotic behavior of the extended kalman filter as a parameter estimator for linear systems. *IEEE Trans. on Automatic Control* **AC24**(1), 36–50.

Ljung, L. and T. Soderstrom (1983). *Theory and practice of recursive identification.* MIT Press. Cambridge, Mass.

Takács, I., G. Patry and D. Nolasco (1991). A dynamic model of clarification-thickening process. *Water Research* **25**(10), 1263–1271.

Zanabria, O., S. Park and C. Garcia (1999). A model reference for evaluating control strategies in activated sludge wastewater treatment plants. *Proceedings of the 3rd Int. Res. Conf.*

on Water Reuse and Water Treatment Plants Operation, Toulouse-France. pp. 295–300.

Zanabria, O., S. Park and C. Garcia (2000). Nitrate concentration-base control of a pre-denitrifying activated sludge system. *IFAC-ADCHEM'2000, Pisa-Italy (Accepted).*

PRACTICAL IMPLEMENTATION OF THE OPTIMAL EXPERIMENT DESIGN METHODOLOGY FOR ESTIMATION OF MICROBIAL HEAT RESISTANCE PARAMETERS

Karina J.E. Versyck, Kristel Bernaerts, and Jan F.M. Van Impe [1]

BioTeC – *Bioprocess Technology and Control*
Department of Food and Microbial Technology
Katholieke Universiteit Leuven
Kardinaal Mercierlaan 92, B-3001 Leuven (Belgium)
Tel. +32-16-32.19.47, Fax +32-16-32.19.60
jan.vanimpe@agr.kuleuven.ac.be

Abstract: In this paper, the design and the implementation of an experiment aimed at optimal estimation of the microbial kinetic parameters in a thermal inactivation model is discussed. Traditionally, the parameters are obtained by applying (linear or nonlinear) regression procedures on data which are collected in a time-consuming and labor-intensive series of inactivation experiments at constant temperatures (so-called *static* experiments). Opposed to the static approach, in this contribution, the temperature is considered as a time-varying (*dynamic*) control input for identification experiments. The temperature-time profile is optimized with respect to a Fisher information matrix based objective function. During this design on simulation level, parameter values estimated from preliminar experiments are considered as the nominal values. The optimal dynamic temperature profile obtained as such has been implemented in practice. The quality of the resulting parameter estimates is discussed.

Keywords: Optimal experiment design, Parameter estimation, Biotechnology (Food Processing), Implementation

1. INTRODUCTION

The concept of *predictive microbiology* is that a detailed knowledge of the responses of microorganisms to environmental conditions enables the objective evaluation of the microbial safety and quality of food products (McMeekin, *et al.*, 1993). This paper deals with the practical implementation of the *Optimal Experiment Design* (OED) methodology for parameter estimation which has recently been introduced in predictive microbiology (Versyck, *et al.*, 1999). In the latter paper, it is shown –by checking the structural identifiability of the inactivation model and by numerical simulations– that *dynamic temperature profiles* are needed in inactivation experiments aimed at simultaneous estimation of the kinetic parameters. Traditionally, the parameters are estimated from population density data collected in a series of time-consuming and labor-intensive *static* experiments in which the temperature is kept constant, e.g., (Juneja, *et al.*, 1997; Line, *et al.*, 1991). Opposed to this static approach, during OED for parameter estimation the temperature is defined as the *time-varying input* to be optimized. The objective functions during optimization are derived from the *Fisher information*

[1] Corresponding author

matrix (Mehra, 1974). Such *dynamic experiments* are favorable from different points of view. First, the microorganisms are *excited* and, as such, are forced to reveal the influence of the temperature history on the microbial rate. Second, the model can be tested on its *validity* for time-varying temperatures. Note that parameters which have been determined under static conditions mostly yield no satisfactory predictions under dynamic conditions (Van Impe, *et al.*, 1992). Third, the fact that the organisms are exposed to several temperatures within one single experiment implies a considerable *saving of time and experimental effort*.

The paper is organized as follows. In Section 2 an appropriate model structure for description of the inactivation dynamics is presented. Subsequently, the parameter estimation problem is stated in terms of the OED methodology. In Section 3 it is explained how numerical values for the unknowns in the inactivation model are obtained from some preliminar static experiments. These values are needed for the optimal design of a dynamic temperature profile on simulation level. The application of the OED methodology and the implementation of the resulting optimal temperature-time profile in practice are discussed in Section 4. Conclusions and suggestions for further research are stated in Section 5.

2. PROBLEM STATEMENT

2.1 *Mathematical model*

With N [cfu/mL] (colony forming units per mL) the microbial population density, Q_p a measure for the physiological state of the population, N_{res} [cfu/mL] the residual population density, and n defined as $\ln N$, q_p as $\ln Q_p$, and n_{res} as $\ln N_{\text{res}}$, a dynamic model has been constructed for microbial inactivation by heating (Geeraerd, *et al.*, 1999):

$$\frac{dn}{dt} = -k_{\max}(T) \cdot \frac{1 - \exp(n_{\text{res}} - n)}{1 + \exp q_p} \tag{1}$$

$$\frac{dq_p}{dt} = -k_{\max}(T) \tag{2}$$

The model (1)-(2) is a so-called *primary* model which describes the population dynamics as function of time (Figure 2). The second factor in (1) describes a shoulder and/or a tail if present. Further, an Arrhenius-type (*secondary*) model is considered for the maximum inactivation rate $k_{\max}$ [1/min] as function of the temperature T:

$$k_{\max}(T) = \frac{\ln 10}{D_{\text{ref}}} \exp[\frac{\ln 10}{z}(T - T_{\text{ref}})] \tag{3}$$

with D_{ref} [min] the decimal reduction time corresponding to the reference temperature T_{ref} [°C], and z [°C] the number of degrees change of temperature required to effect a tenfold change in the D-value.

2.2 *Optimal experiment design problem*

When the model structure (1)–(2)–(3) is assumed to be correct, the remaining identification problem is the parameter estimation based on experimental data. In this contribution, the goal of OED is to find the best possible admissible temperature profile $T^*(t)$ for estimation of the parameters D_{ref} and z from survivor count data. Hence, the output vector and the vector of model parameters to be estimated are defined as $\mathbf{y} \equiv y \triangleq n$ and $\mathbf{p}^{\mathbf{T}} \triangleq [D_{\text{ref}} \;\; z]$ respectively.

The cost functions used during the OED are related to the shape of the error functional $\mathcal{J}_I$ (4) near the true process parameters:

$$\mathcal{J}_I(\mathbf{p}) \triangleq \int_0^{t_f} \mathbf{e}^{\mathbf{T}} \, \mathbf{Q} \, \mathbf{e} \, dt \tag{4}$$

with the error vector $\mathbf{e}$ defined as $(\mathbf{y}(\mathbf{p}, t) - \mathbf{y}_m(t))$; $\mathbf{y}_m$ is the vector of measured outputs, $\mathbf{y}$ is the vector of model predictions, and the weighting matrix $\mathbf{Q}$ is the inverse of the measurement error covariance matrix. A zero-mean gaussian white measurement noise is assumed which is reasonable when the model output is defined as the logarithmic transform of the cell density (see (1)) (Jarvis, 1989). For the single output system under study, the $\mathbf{Q}$ matrix reduces to a constant scalar $\sigma_n^{-2} = 0.33$ which represents the inverse of the (experimentally determined) measurement error variance. The linearization of the outputs $\mathbf{y}$ near the true process parameters leads (via the error functional $\mathcal{J}_I$ (4)) to the Fisher information matrix $\boldsymbol{\mathcal{F}}$ (5) as a tool to approximate the parameter estimation quality that can be attained in a certain experiment, e.g., (Munack, 1989):

$$\boldsymbol{\mathcal{F}} \triangleq \int_0^{t_f} \left(\frac{\partial \mathbf{y}}{\partial \mathbf{p}}(t) \bigg|_{\mathbf{p}_o} \right)^{\mathbf{T}} \mathbf{Q} \left(\frac{\partial \mathbf{y}}{\partial \mathbf{p}}(t) \bigg|_{\mathbf{p}_o} \right) dt \tag{5}$$

As the true parameters are not known, during the OED a nominal set $\mathbf{p}_o$ is to be used. This set is assumed to be sufficiently close to the true process parameters such that the methodological errors –as consequence of the underlying linear approximations– can be considered negligible. Scalar functions of the matrix $\boldsymbol{\mathcal{F}}$ can be used as performance indices for OED (Mehra, 1974; Goodwin and Payne, 1977; Walter and Pronzato, 1997). In the present study, the condition number of the matrix $\boldsymbol{\mathcal{F}}$ (5), i.e., the modified E-optimal design cost is used. Mathematically, this optimal design problem reads as follows:

$$\min_{T(t)\in\mathcal{T}} J[T(t)] \text{ with } J[T(t)] \triangleq \frac{\lambda_{\max}(\mathcal{F})}{\lambda_{\min}(\mathcal{F})} \quad (6)$$

with λ an eigenvalue of the matrix $\mathcal{F}$ (5) and $\mathcal{T}$ the set of admissible temperature-time profiles. The cost (6) should be as near as possible to unity, which corresponds to circular lines of constant functional values, and a cone-like shape of the functional $\mathcal{J}_I$ (4) in the parameter space.

3. PRELIMINAR EXPERIMENTS

The OED methodology requires numerical values for the initial states $n(t_0)$ and $q_p(t_0)$, and both the primary and secondary model parameters, i.e., n_{res} in (1), and (D_{ref}, z) in (3) respectively. In the present study, these values are obtained from static experiments at four temperature levels (each experiment was performed in triplicate). For the experimental protocol reference is made to the Appendix.

3.1 *Values for* $D_{\text{ref,o}}$ *and* z_o

The initial guesses for the model parameters D_{ref} and z are obtained in a traditional way. A *global nonlinear regression* is applied on the *log-linear part* of all the experimentally determined survivor curves at constant temperatures. In the log-linear part of the survivor curves the second factor at the right hand side in the model (1) becomes equal to 1. This follows directly from the fact that, in that particular phase, the shoulder and tail related variables –q_p and n_{res} respectively– do not interfere in the population dynamics. Consequently, the model equation (1) reduces to the following first-order kinetics:

$$\frac{dn}{dt} = -k_{\max}(T) \quad (7)$$

where the constant $k_{\max}$ is given by (3). Using Equation (3) the following static version of the first-order model (7) can be written (remind that $n = \ln N$):

$$\ln\frac{N}{N_{\text{OL}}} = -\ln 10 \cdot \frac{t - t_{\text{OL}}}{D_{\text{ref}}} \cdot 10^{\left(\frac{T - T_{\text{ref}}}{z}\right)} \quad (8)$$

This nonlinear regression function is traditionally used to estimate the parameters D_{ref} and z from the log-linear parts of static inactivation data at several temperatures. To account for the differences in (*i*) cell densities N, and in (*ii*) time instants at the start of the log-linear parts of the survivor curves, relative cell densities are considered during the global fitting. The cell densities are rescaled by the densities of survivors at the start of the log-linear part of the curve (denoted as N_{OL}). The initial time t_{OL} is also chosen at this point. The experimental data of the static inactivation experiments are plotted in Figure 1. The symbols t_{OL} and t_{fL} on the time axes indicate the start and the end time of the log-linear parts in the curves respectively.

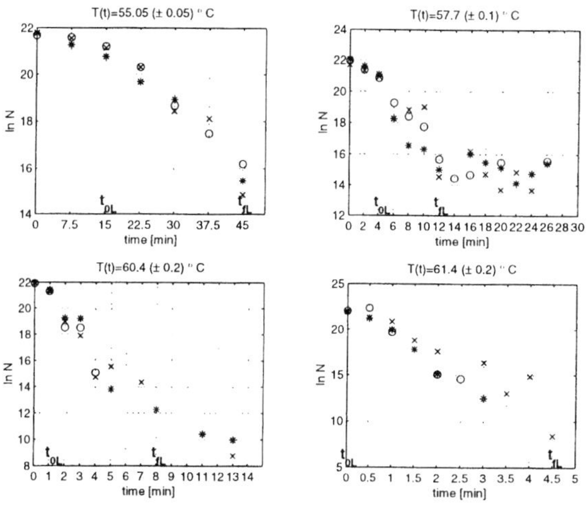

Fig. 1. Static experiments.

3.2 *Values for initial states and* n_{res}

The following static version of model (1) is used as a regression function on each of the four static inactivation data sets depicted in Figure 1. During these regressions also the *shoulders* and *tails* are considered to find realistic values for the initial states and the n_{res}-value (to be used further during OED):

$$\begin{aligned} n(t) &= n(t_0) - k_{\max}A(t) \\ &\quad + \ln\left[1 + \frac{e^{k_{\max}\cdot A(t)} - 1}{e^{n(t_0)-n_{\text{res}}}}\right] \\ A(t) &= \frac{1}{k_{\max}} \cdot \ln\left[\frac{e^{k_{\max}\cdot t} + e^{q_p(t_0)}}{1 + e^{q_p(t_0)}}\right] \end{aligned} \quad (9)$$

The parameters in the secondary model for the maximum inactivation rate $k_{\max}$ (3) are fixed to the values obtained by the above-discussed fitting procedure (via (8)): $D_{\text{ref}} = 2.49$ min and $z = 5.67$°C.

From the plots of the static inactivation data (Figure 1) it can be seen that the initial cell densities are approximately identical at all temperatures as the prehistory of the cell cultures (as described in the Appendix) is well reproducible. Consequently, the initial cell density $n(t_0)$ for OED is chosen equal to the mean of the best-fit values (obtained when using (9)), namely, $n(t_0) = 21.88$. Moreover, this value corresponds to the mean of the experimentally observed $n(t_0)$-values.

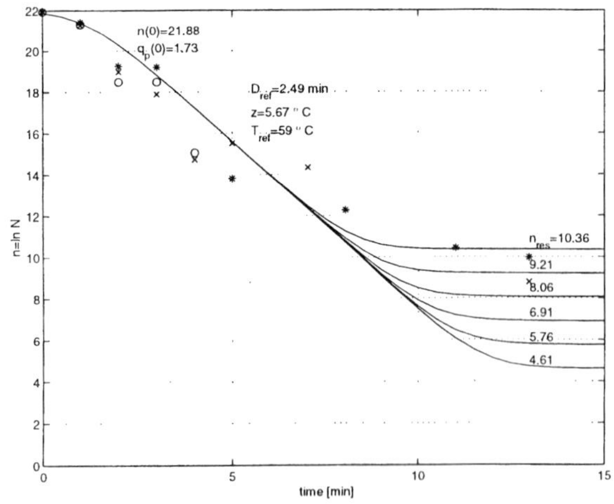

Fig. 2. Experimental data (symbols) and simulation of static inactivation at $T = 60.4°C$ with indicated values in model (1)–(2)–(3).

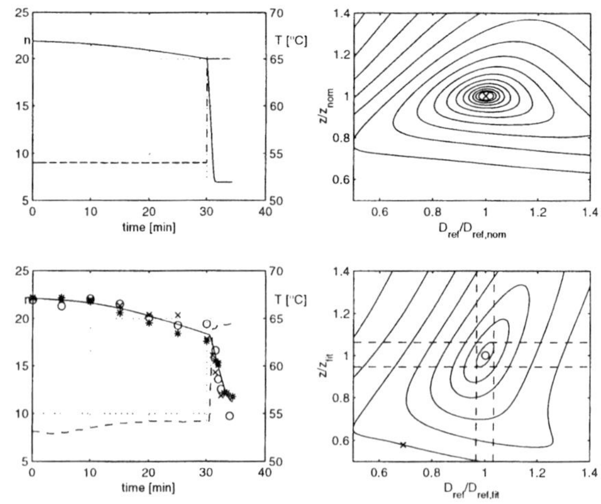

Fig. 3. Optimal experiment. Upper plots: simulation. Lower plots: experiment and best-fit.

The reproducibility of the inoculum growth conditions also implies that the microorganisms can be assumed to be in the same physiological state at the start of each inactivation experiment. This is reflected by the values for the initial second state variable $q_p(t_0)$: the best-fit estimates obtained with (9) are in the same order of magnitude for the different temperatures. During OED the value corresponding to the inactivation curve at 57.7°C is used, namely, $q_p(t_0) = 1.73$. From Figure 2 it is obvious that also the data collected at another temperature (here, $T = 60.4°C$) can be described fairly well when using this value for $q_p(t_0)$.

Figure 2 also illustrates the little effect of the n_{res}-value on the log-linear part of the static inactivation curve where the inactivation rate reaches its maximum value of k_{max} (which is determined by the kinetic parameters D_{ref} and z via (3)). During OED a value near the detection limit of the experimental set-up is used, namely, $n_{res} = 6.91$.

4. OPTIMAL DYNAMIC EXPERIMENT

For ease of practical implementation a *step profile* is proposed as the structure of the time-varying temperature profile $T(t)$. The initial temperature T_1, the switch time t_1, and the temperature T_2 during the second phase are the three Degrees Of Freedom (DOF) to be optimized with respect to the design criterion (6).

4.1 *OED and implementation*

Table 1 summarizes (i) the numerical values for the initial states and parameters used in model (1)–(2)–(3) during OED (upper part), (ii) the constraints imposed during the optimization and the optimal values of the DOF (center part), and (iii) the corresponding modified E-cost of 2.39 which is close to the theoretical minimum of 1 for this cost (see (6)), and the length of the optimal experiment (lower part).

The optimal input (T^*) and output (n^*) are simulated in Figure 3 (upper left plot). The contours of the corresponding cone-like error functional $\mathcal{J}_I$ (4) are also plotted in Figure 3 (upper right plot). The experimentally applied temperature profile is given in Figure 3 (lower left plot) together with the experimental survivor counts. Obviously, the applied temperature profile is in fact sub-optimal (due to restrictions on the set-up). Though, for the nominal parameters and initial conditions of the OED still a satisfactory modified E-cost (6) of 3.04 is obtained when this *truly experimental input* is applied to the model (1)–(2)–(3).

Table 1. Numerical values for OED.

Values for OED				
$n(t_0)$	21.88	T_{ref}	[°C]	59
$q_p(t_0)$	1.73	$D_{ref,o}$	[min]	2.49
n_{res}	6.91	z_o	[°C]	5.67
Optimal DOF for $T^*(t)$				
T_1	[°C]	$\in$	[50, 60]	54
t_1	[min]	$\in$	[5, 30]	30
T_2	[°C]	$\in$	[50, 65]	65
			Criterion (6)	2.39
			t_f [min] (stop if $n = n_{res}$)	34.5

4.2 *Results and discussion*

During dynamic fitting (by model (1)–(2)–(3)) of the data collected during the dynamic experiment, the initial state $n(t_0)$ is fixed to the mean of the measured initial cell densities: $n(t_0) = 22.07$.

The full line in the lower left plot of Figure 3 is the model output of (1)–(2)–(3) for the best-fit values which are summarized in Table 2.

In the contour plot (lower right plot of Figure 3) of the corresponding sum of squared errors between modeled and measured outputs (SSE), the inner ellipse represents the boundary of the 90% joint confidence region which contains the SSEs obeying the following inequality condition (Beale, 1960):

$$SSE \leq LSE \cdot \{1 + \frac{n_p}{m - n_p} F(n_p, m - n_p, 1 - \alpha)\} \quad (10)$$

with LSE the least-squares error, n_p the number of parameters estimated simultaneously (here, 4, since apart from the parameters D_{ref} and z, the initial state $q_p(t_0)$ and the n_{res}-value are considered as DOF for minimization of the SSE), m the number of observations (here, 34), and F the value of the Fisher-Snedecor F-distribution with n_p and $m - n_p$ DOF and a confidence level α (here, 0.01). The F-values are tabulated in statistical text-books, and can also be provided by statistical software (here, F equals 2.14). As n_p equals 4, a grid search in a four-dimensional space is necessary to determine the SSE-values lower than the threshold value in (10). This results in a four-dimensional hypervolume as joint confidence region, which can however not be visualized as such. Therefore, during construction of the two-dimensional contour plot (lower right plot in Figure 3), $q_p(t_0)$ and n_{res} have been fixed to their optimal values. It should be noted that this method delivers an underestimation of the actual joint confidence region. The computationally intensive alternative of constructing nonlinear confidence regions via Monte-Carlo simulations has not been considered thus far for this case study.

Table 2. Fitting results for OE.

Least-squares error (LSE) estimates				
$q_p(t_0)$	1.74	D_{ref}	[min]	3.55
n_{res}	11.04	z	[°C]	9.73
		LSE	14.99	
	Criterion (6)		27.80	

For the given experimental input and the best-fit values (Table 2), the modified E-cost (6) – calculated with respect to estimation of the parameters D_{ref} and z– equals 27.8, which is sub-optimal, but, still of an acceptable order of magnitude. From the contour plot it can be seen that there is indeed some correlation between the estimates for D_{ref} and z.

The variances on the parameter estimates can be obtained by computationally intensive jackknife and bootstrap methods. Alternatively, they can be approximated by considering the extremes of the 90% joint confidence region which are indicated by the horizontal and vertical lines in Figure 3 (lower right plot). These boundaries correspond approximately to the ends of the individual 95% confidence intervals of the parameters. However, due to the non-linearity of the model output in the parameters, the intervals indicated on the normalized parameter axes are slightly asymmetric around the best-fit values. Here, the half of the approximate 95% confidence interval is considered as an approximation of $1.96 \cdot \sigma$ (with σ the standard error). This results in the following approximate standard errors: $\sigma(D_{\text{ref}}) = 0.06$ min and $\sigma(z) = 0.28$ °C. Note that these values are too optimistic as the two-dimensional joint confidence region depicted in Figure 3 is already an under-estimation of the true joint confidence region as mentioned above.

When the standard errors on the parameter estimates are approximated as follows, e.g., (Vanrolleghem and Dochain, 1998):

$$\sigma(p_i) = s\sqrt{\mathcal{F}_{ii}^{-1}} \text{ with } s^2 = \frac{LSE}{m - n_p} \quad (11)$$

(with the matrix $\mathcal{F}$ as defined in (5) calculated for the experimental input), the following values are obtained: $\sigma(D_{\text{ref}}) = 0.11$ min $\sigma(z) = 0.32$ °C. These standard errors are also statistically optimistic as the equations in (11) imply a linear approximation.

Both ways of calculating approximate standard errors (via the joint confidence region or via the Fisher information matrix) yield values that are in the same order of magnitude. Furthermore, these errors are comparable to the errors obtained when performing a whole set of static experiments and applying the classic (linear or nonlinear) regression procedures (results not shown). Though, here, the same information is obtained by much less experimental effort. Nevertheless, there is a gap between the nominal value (marked with ×) and the best-fit value (marked with o) for the parameter D_{ref} and particularly, for the z-value (see right-hand plots of Figure 3). This discrepancy suggests that the initial guesses which were used for the design of the experiment are not yet close enough to the true biological parameters. Hence, the performed so-called *optimal* experiment is in fact *sub-optimal* with respect to the real-life system. Therefore, further iteration through the cycle of OED and subsequent parameter estimation is needed to check the convergence of this approach towards an experiment which is truly optimal with respect to the biological system.

5. CONCLUSIONS

Up till recently, the system-theory based OED methodology was an unexplored area in predictive microbiology. To illustrate the potential power of this methodology, the practical implementation of an optimal temperature-time profile as exciting control input for parameter estimation is presented for a model describing the microbial decay during heat treatments. The numerical values for the design of the optimal temperature profile on simulation level are extracted from preliminar experiments at constant temperatures. The features of the best-fit parameter estimates suggest that further iterative OED is needed in order to converge to the true parameters of the biological system under study. This is a topic of on-going research.

REFERENCES

Beale, E.M.L. (1960). Confidence regions in non-linear estimation. *J. Roy. Stat. Soc.*, **B22**, 41-88.

Geeraerd, A., C. Herremans and J.F. Van Impe (1999). Structural model requirements to describe microbial inactivation. In: *Predictive Microbiology applied to Chilled Food Preservation* (C.M. Bourgeois and T.A. Roberts. (Ed)), 280-287. Office for Official Publications of the European Communities, Luxembourg (ISBN 92-828-5750-6).

Goodwin, G.C. and R.L. Payne (1977). *Dynamic System Identification. Experiment Design and Data Analysis.* Academic Press, New York.

Jarvis, B. (1989). *Statistical Aspects of the Microbiological Analysis of Foods.* Elsevier, Amsterdam.

Juneja, V.K., O.P. Snyder Jr. and B.S. Marmer (1997). Thermal destruction of *Escherichia coli* O157:H7 in beef and chicken: determination of *D*- and *z*-values. *Int. J. Food Microbiol.*, **35**, 231-237.

Line, J.E., A.R. Fain Jr., A.B. Moran, L.M. Martin, R.V. Lechowich, J.M. Carosella and W.L. Brown (1991). Lethality of heat of *Escherichia coli* O157:H7: *D*-value and *z*-value determinations in ground beef. *J. Food Prot.*, **54(10)**, 762-766.

McMeekin, T.A., J.N. Olley, T. Ross and D.A. Ratkowsky (1993). *Predictive Microbiology: Theory and Application.* Research Studies Press Ltd., Taunton.

Mehra, R.K. (1974). Optimal input signals for parameter estimation in dynamic systems: survey and new results. *IEEE Transactions on Automatic Control*, **19(6)**, 753-768.

Munack, A. (1989). Optimal feeding strategy for identification of Monod-type models by fed-batch experiments. In: *Computer Applications in Fermentation Technology Modelling and Control of Biotechnological Processes*, 195-204. Elsevier, Amsterdam.

Van Impe, J.F., B. Nicolaï, T. Martens, J. De Baerdemaeker and J. Vandewalle (1992). Dynamic mathematical model to predict microbial growth and inactivation during food processing. *Appl. Environ. Microbiol.*, **58(9)**, 2901-2909.

Vanrolleghem, P.A. and D. Dochain (1998). Bioprocess model identification. In: *Advanced Instrumentation, Data Interpretation and Control of Biotechnological Processes* (J.F. Van Impe, P.A. Vanrolleghem and D.M. Iserentant. (Ed)), 251-318. Kluwer Academic Publishers, Dordrecht.

Versyck, K.J., K. Bernaerts, A.H. Geeraerd and J.F. Van Impe (1999). Introducing optimal experimental design in predictive microbiology: a motivating example. *Int. J. of Food Microbiol.*, **51(1)**, 39-51.

Walter, E. and L. Pronzato (1997). *Identification of Parametric Models from Experimental Data.* Springer, London.

ACKNOWLEDGMENTS

Author Karina Versyck is a research assistant with the Fund for Scientific Research-Flanders (FWO). Author Kristel Bernaerts is a research assistant with the Institute for the Promotion of Innovation by Science and Technology in Flanders (IWT). Work supported in part by FWO Project G.0267.99, Projects OT/99/24 and COF/98/008 of the Research Council of the Katholieke Universiteit Leuven, EU-Project CT97-3129, and the Belgian Program on Interuniversity Poles of Attraction, initiated by the Belgian State, Prime Minister's Office for Science, Technology and Culture. The scientific responsibility is assumed by its authors.

APPENDIX: MATERIAL AND METHODS

Organism
The heat resistance parameters are estimated for an *Escherichia coli* K12 MG1655 strain which is chosen as a representative for the foodborne pathogen *E. coli* O157:H7. Starter cultures were prepared by transfering a loopful of the stock culture (kept at -80°C) to 20 mL Brain Heart Infusion (BHI) broth. After aerobic incubation at 37°C during 9.5 h, 20 μL of the cell suspension was transferred to 20 mL fresh BHI broth. The broth was subsequently incubated at 37°C for 15 h. The cultures were harvested by centrifugation of three 1 mL portions. The cells were washed with BHI broth and resuspended in equal volumes of BHI broth. From each 1 mL cell suspension a series of capillaries was prepared by pipetting volumes of 45 or 50 μL into sterile glass capillar tubes which were sealed by a gas flame.

Thermal inactivation procedure
The capillaries were immersed in a constant temperature water bath in which the temperature was monitored by a data-logging thermometer (portable microprocessor HI92840 C, HANNA Instruments). For the implementation of the temperature step profile two water baths at the respective temperatures were used. It was assumed that there is no come-up time for the temperature in the capillaries. At regular times triplicate capillaries were removed from the water bath and plunged into an ice-water bath until analyzed.

Enumerating procedure
After aseptically opening of the capillaries, decimal serial dilutions were prepared in a saline peptone solution. Appropiate dilutions were surface-plated on Plate Count Agar (PCA) using a Spiral Plater (Spiral Systems Inc., Eddy Jet IUL Instruments). Plates were incubated for 24 h at 37°C and enumerated.

Software
The regressions and statistical analyses were performed with the statistical software package `SAS` 6.12 (Copyright (c) 1989-1996 by SAS Institute Inc.). The programs for simulation, optimization, and dynamic fitting were written in `FORTRAN77` (Version 5.1). Several routines from the `NAG Mark 17` Library (NAG Ltd., Oxford, UK) were used.

SYSTEM IDENTIFICATION OF ELECTRONIC NOSE DATA FOR MONITORING CYANO-BACTERIA IN POTABLE WATER: BLACK-BOX MODELLING

G.E. Searle, J.W. Gardner, M.J. Chappell, K.R. Godfrey, M.J. Chapman*

School of Engineering, University of Warwick, Coventry CV4 7AL, UK.
**School of MIS-Mathematics, University of Coventry, Coventry CV1 5FB, UK.*

Abstract: Linear black-box modelling techniques are applied to data collected from an electronic nose experiment. FIR, ARX, MAX and ARMAX inverse models of the nose system are used to discriminate between two different strains of cyanobacteria. Models for both pre-processed and raw data are proposed, and successful classification rates of 100% and 99.3% are obtained, respectively. *Copyright © 2000 IFAC*

Keywords: chemical sensors, data processing, linear multivariable systems, moving average models, pre-processing, signal processing.

1. INTRODUCTION

Electronic noses are now commercially available and are being used to analyse a variety of different odours from many different sources such as foodstuffs and beverages (e.g. classification of 'quality' of tomatoes (Di Natale *et al.*, 1998), and discrimination between blends of coffee (Gardner *et al.*, 1992)). Current research includes the application of electronic nose techniques to the analysis of biological samples, with the aim of using them for bioprocess monitoring and medical diagnosis. The data analysed in this paper come from a biological experiment involving cyanobacteria (blue-green algae, or more precisely microsystis aeruginosa).

Data from electronic nose experiments have traditionally been analysed using the steady-state responses of the sensors. Principal components analysis, cluster analysis, discriminant function analysis and artificial neural networks have all been applied to the steady-state sensor responses with a significant degree of success (Gardner and Bartlett, 1998). In this paper the data are analysed using linear black-box models (Ljung, 1987). These techniques have been shown to work well when applied to the static (steady-state) responses of the sensors (Marco *et al.*, 1996). Here we form a variety of black-box models to analyse both static and dynamic sensor responses. The degrees of success obtained with the different model forms are reported and compared with those previously obtained using artificial neural networks.

2. EXPERIMENTAL DETAILS

The data analysed in this paper were collected from an electronic nose experiment designed to investigate the possibility of monitoring cyanobacteria (blue-green algae) in water. The headspaces of two different strains of cyanobacteria, one toxic, the other non-toxic, grown in a nutrient medium (BG11) were repeatedly sampled by an electronic nose system for a period of 40 days. The nose consisted of 6 commercial metal oxide resistive odour sensors (Shin *et al.*). The repeated exposure cycle was as follows:

- 23 mins. 20 secs. - medium (no bacteria)
- 2 mins. - toxic microcystis aeruginosa PCC 7806 strain
- 23 mins. 20 secs. - medium (no bacteria)
- 2 mins. - non-toxic microcystis aeruginosa PCC 7941 strain.

The outputs from the sensors were recorded every ten seconds, producing 2300 static-sample vectors for the models for pre-processed data, and 350358 data vectors for the dynamic models. The aim of the experiment was to investigate how well the electronic nose system could discriminate between the two species of cyanobacteria. The data have already been analysed using artificial neural networks with considerable success (Shin *et al.*). Here we use system identification techniques to analyse the same data and compare the efficacy of linear time-invariant black-box models with the non-linear neural network-based methods. A plot of a short section of the data recorded is given in Figure 1.

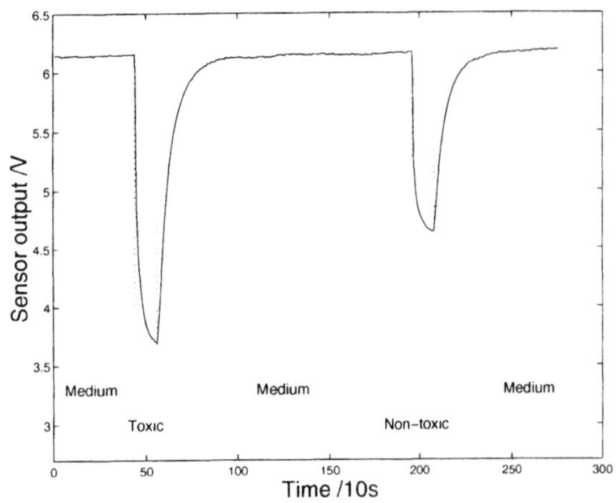

Figure 1. A section of the raw data, showing the response of a single sensor when exposed to toxic and non-toxic bacteria.

3. BLACK-BOX MODELLING

The *System Identification Toolbox* within MATLAB provides a multitude of functions designed to enable the user to create and evaluate black-box models. The general form for the discrete time multi-input, single-output models considered is given below (Ljung, 1987):

$$A(q^{-1})y(t) = \frac{B_1(q^{-1})}{F_1(q^{-1})}u_1(t-k_1)+\ldots + \frac{B_n(q^{-1})}{F_n(q^{-1})}u_n(t-k_n)+\frac{C(q^{-1})}{D(q^{-1})}e(t), \quad (1)$$

where y is the model output, u_i is the i-th input (for $i = 1, \ldots, n$), e is white noise and k_i is the delay from input u_i to the system. The functions $A(q^{-1})$, $B_i(q^{-1})$, $C(q^{-1})$, $D(q^{-1})$ and $F_i(q^{-1})$ are polynomials in the backwards shift operator q^{-1}, which is defined by:

$$q^{-1}x(t) = x(t-1). \quad (2)$$

Table 1 The model structures considered.

Model structure:	Polynomials used:
FIR	B
ARX	A, B
MAX	B, C
ARMAX	A, B, C

Not all of the polynomials A, B, C, D and F are used in a particular model. The models considered in this paper are listed in Table 1, together with details of the polynomials used. The model structures considered were selected due to their simplicity and previously reported success (Marco *et al.*, 1996).

The models considered in this paper have 6, 7 or 8 inputs and a single output. The inputs correspond to the outputs of the six gas sensors in the test chamber (and possibly also the outputs of temperature and humidity sensors present in the chamber), and the output corresponds to a classification of the substance presented.

Hence we are considering 'inverse models' for the nose system. See Figure 2 for a pictorial representation of this.

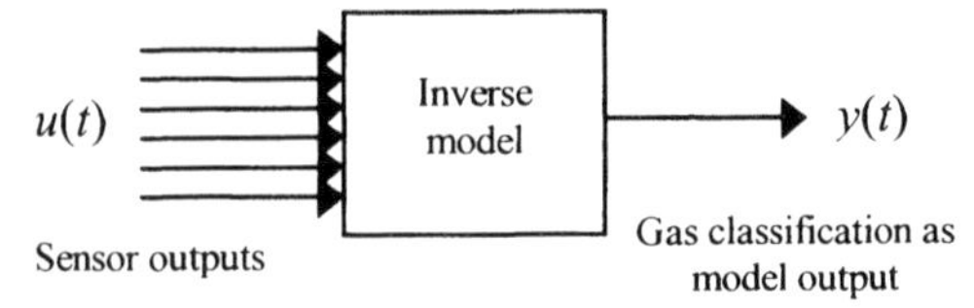

Figure 2. Diagrammatic representation of the electronic nose modelling problem.

For the experiment considered here, the classification system must simply place the odour into one of two categories: 'toxic' or 'non-toxic'. This classification was encoded into a numerical output of +1 for 'toxic' and −1 for 'non-toxic'.

Two distinct classes of models were analysed: models for steady-state data in Section 4 (extracted static information) and models for the full data-set in Section 5 (dynamic information intact).

In both sections the effects of normalising the gas sensor data are also considered, i.e. at each data point (data sample at a particular time, or pre-processed response vector), the 6-dimensional (corresponding to the six gas sensors used) data vector is normalised so as to have unit length. This is done in order to attempt to remove the effects of variation in concentration on the discrimination performance. Hence reducing the possibility that the model will learn to discriminate between the two bacteria based on differences between the magnitudes of the sensor responses rather than the differing patterns of responses.

4. STATIC MODELS

Previous work has concentrated on using pre-processing algorithms to extract information from data, which is then passed on to pattern recognition algorithms or artificial neural networks (Gardner and Bartlett, 1998; Gardner and Bartlett, 1994). The main reasons for applying the pre-processing algorithms are:

- Information compression - all of the pre-processing algorithms tested here actually reduce the response of a sensor over a single exposure cycle to a single value.
- Baseline drift removal - the sensors used suffer to varying degrees from baseline drift - the sensors do not return exactly to their initial outputs when the odour is removed. Over long periods this drift can seriously affect discrimination ability. The pre-processed values (e.g. difference) can be less susceptible to these long-term changes.
- Reduction of temperature dependence - some models for the temperature dependence of the sensor outputs predict that the effects of variations in temperature should be removed by some of the pre-processing algorithms.

In this section black-box models for these pre-processed (or 'static') data are analysed. The four pre-processing algorithms tested are:

- Static response: V_b
- Difference : $V_b - V_m$
- Relative difference: $\dfrac{V_b}{V_m}$
- Fractional difference: $\dfrac{V_b - V_m}{V_m}$

where V_b is the sensor voltage when the sensor is exposed to bacteria (final value at end of exposure period) and V_m is the voltage in the medium (immediately before the sensor is exposed to bacteria). The pre-processed sensor outputs for each of the six gas sensors are considered to be the inputs to MISO black-box models.

The evaluation of each model structure and pre-processing algorithm was carried out using the following procedure:

- The pre-processing algorithm was applied to the data to produce a sequence of pre-processed data vectors (6-dimensional), one for each bacteria exposure cycle.
- The resulting data-set was split into halves. The first half was used to train a model of the specified structure (i.e. estimate the polynomial coefficients).
- The model was used to produce a simulated output from the second half of the input data.
- The simulated output for each response cycle was converted to a classification by simply taking its sign.
- The sequence of these classifications was compared with the output data to evaluate the performance of the model.

4.1 FIR models

The simplest of the model structures considered is the FIR (finite impulse response) structure, where only the B polynomials are used. A six-input single-output FIR model has six B polynomials, so the structure of such a model is described in terms of twelve quantities: the orders of the six polynomials and the six delays from the inputs to the system. For simplicity and to greatly reduce computation time, our survey of the model structures was restricted to those having identical orders for all of the B polynomials and identical delays for all of the inputs. Thus our FIR model structures are effectively described by just two quantities: the order and the delay.

For this pre-processed data, the most appropriate delay for the models was found to be zero. This is to be expected since a non-zero delay would force the model to classify an odour using only the information gained from past odour inputs, and not the current odour. The optimal orders (chosen as a compromise between maximising success rate and minimising model complexity) for the B polynomials were either 2 or 3 for each of the different pre-processing algorithms. The results obtained using the optimal model orders are detailed in Table 2.

Table 2 Percentage success rates obtained using FIR models for pre-processed data.

	Percentage success rate using	
Pre-processing algorithm used:	Pre-processed data:	Normalised pre-processed data:
Static response	98.8	97.6
Static difference	99.7	97.4
Fractional difference	99.5	97.0
Relative difference	99.5	98.4

It should be noted that the normalisation process seemed to reduce the effectiveness of the models. This could be due to the normalisation process removing some of the concentration information from the data and thus forcing the models to learn to distinguish between the different 'types' of 'smell' rather than the different 'strengths' of 'smell'.

4.2 More complex model forms

The addition of a low order A polynomial to the model structures to form an ARX (auto-regressive with extra input) model was found to generally produce an improvement in the maximum success rates. By selecting the appropriate orders for each pre-processing algorithm, perfect classification of the test data was obtained with all of the pre-processing algorithms. The orders required were mostly 1 for the B polynomial and between 1 and 3 for the A polynomial.

The addition of a C polynomial to FIR models to form MAX (moving average with extra input) models increased the success rates to between 99.9% and 100% for all of the pre-processing algorithms; normalised and not. The orders necessary for these MAX models were approximately 2 for the *B* polynomials and between 0 and 3 for the *C* polynomial. Thus the MAX models generally involved more parameters than the ARX models, suggesting that the ARX models were preferable. No significant increase in model performance was gained by the inclusion of both *A* and *C* polynomials to form ARMAX (auto-regressive moving average with extra input) models.

5. DYNAMIC MODELS

The pre-processing techniques detailed in the previous section reduce the dynamically sampled signal to a single value for each exposure cycle (for each sensor). This seems to work well, but it has been suggested (Hines *et al.*) that the dynamic response might contain information which could be used to improve the classification accuracy. In this section we apply black-box modelling techniques to the full dynamic data. As with the static data, we also apply the techniques to a normalised version of the same data-set.

For this analysis the data-set was split into two equally-sized sets. The data from the first 20 days of the experiment were used to train the models and the remaining data were used to evaluate the performance of the models.

Once a model was constructed, it was evaluated using the following procedure:

- The model was used to produce simulated output from the second half of the input data.
- The simulated output was sampled at the data point corresponding to the end of each exposure cycle, producing a single value for each cycle.
- The numerical value was converted to a classification by simply taking its sign.
- The sequence of these classifications was compared with the actual known values to evaluate the performance of the model.

The output of the model was encoded as for the static models. Thus if the model was to work perfectly, the resulting classification vector would be an alternating sequence of +1's and −1's, matching the known sequence. The percentage of correct classifications formed the basis for comparison between different model structures.

5.1 FIR models

The first model structures to be considered were FIR models. Orders from 1 to 25 and delays from 0 to 5 time units (corresponding to 0 to 50 seconds) were tested. It was found that the optimum delay was zero for both the raw and normalised data-sets, and the optimum orders were 12 for the raw data and 11 for the normalised data. The appropriate delay for these dynamic models reflects the physical characteristics of the system. A non-zero delay might be expected due to the time taken between the switching of a valve to the arrival of the new substance in the sensor chamber, and for mixing or saturation to occur in the chamber. Hence the lengths of tubing and flow-rates used could be expected to have a significant effect on the delay in the system. The lack of an observable delay in these data can be attributed to the relatively slow sampling rate (0.1 Hz) and relatively fast kinetics present in the system (i.e. relatively short tubes, fast flow-rate and small sensor chamber).

The normalised data produced significantly higher success rates (99.3% vs. 91.3%) than the raw data. In fact, models with order as low as 8 produced a very high success rate (97.4%) for the normalised data.

5.2 ARX models

With the addition of an auto-regressive polynomial *A* to form an ARX model, the output of the model becomes a function of past values of the model output as well as past (and present) values of the inputs. The delays were again fixed at zero, since the inputs were the same as for the FIR models. The orders of the *B* polynomials were again all matched, so that the model structures were described by just two quantities: the order of the *A* polynomial and the order of the *B* polynomials.

The order of the *A* polynomial was varied for each model between 0 (to include the FIR case) and 15 and the orders of the *B* polynomials were varied between 1 and 15 for both the raw and normalised data. The percentage success rates obtained for the raw data are plotted in Figure 3.

From Figure 3 (for the raw data) we see that for low order (orders 2 to 7) *B* polynomials, the addition of a first order *A* polynomial (allowing the model to take into account the model output from the previous sample, 10 seconds prior to the current time) improves the classification success. However, none of the true ARX structures produced a classification rate to match the 91.3% obtained with a 12th order FIR structure.

Similar results were obtained using the normalised data-set. The addition of the *A* polynomial generally decreased the success rate, and all of the true ARX structures performed worse than the optimum (11th order - giving a 99.3% success rate) FIR model.

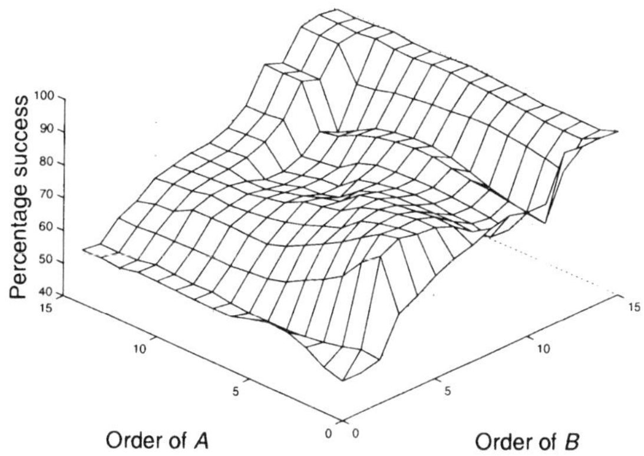

Figure 3. Plot of percentage successful classification versus orders of *A* and *B* polynomials in ARX models with zero delays. The models were trained on the first half of the raw data-set and tested on the second half.

Figure 4. Plot of percentage success versus orders of *B* and *C* polynomials for MAX models with zero delays. The models were trained on the first half of the raw data set and tested on the second half.

5.3 MAX and ARMAX models

The addition of a moving average term to the FIR and ARX models to produce MAX and ARMAX models, respectively, corresponds to the inclusion of a *C* polynomial in the model structure (see Section 3). For MAX models, the order of *A* is zero, so the quantities determining the structure are the order of *B* and the order of *C* (the delay is again set to zero). Orders from 0 to 15 for the *C* polynomial and from 1 to 15 for the *B* polynomial were tested. The percentage success rates obtained using the raw data are plotted in Figure 4.

Once again it appears that an FIR model of order 12 is a good choice for the models for the raw data. A slightly better classification rate can be obtained with order 14 for the *B* polynomials and order 9 for the *C* polynomials (93.8% vs. 91.3%) but this is not a sufficiently significant improvement to warrant the adoption of the more complex model.

Similarly, the success rates obtained using models for the normalised data were generally decreased with the addition of the *C* polynomial.

As with the move from FIR to ARX models, the addition of an autoregressive term to the MAX models to form ARMAX models did not improve the classification performance.

Table 3 Results for FIR models with extra inputs.

Percentage successful classification with extra inputs:	Data used:	
	Raw	Normalised
None	91.3	99.3
Temperature and humidity	92.6	99.4
Temperature	91.1	99.3
Humidity	86.3	99.4

5.4 Dynamic models with temperature and humidity as extra inputs

All of the models considered thus far have had 6 inputs: the outputs from the 6 gas sensors. In the odour sensor chamber there was also a temperature sensor and a humidity sensor. The outputs from these can be included as extra inputs to the models, producing models that may effectively compensate for variations in temperature and humidity.

Table 3 shows the percentage successful classification results obtained using FIR models having temperature and / or humidity as extra inputs. All models for raw data used *B* polynomials of order 12, and those for normalised data used order 11 (note: the extra inputs were not included in the normalisation process).

It can be seen that the addition of extra input(s) to the models for the normalised data made little difference to the success rate. However, the results for the raw data are somewhat inconclusive. The addition of a single extra input of either temperature or humidity decreased the success rate (substantially in the case of humidity), but the addition of both increased the success rate slightly.

6. DISCUSSION AND CONCLUSIONS

The static models dealing with pre-processed data were able to produce success rates of 99.7% (un-normalised) and 98.4% (normalised) using low order (order 2 or 3) FIR models. These success rates were increased to 100% when the models were extended to form ARX, MAX or ARMAX structures. The dynamic FIR models produced slightly lower success rates using un-normalised data (91.3%) but comparable success with normalised data (99.3%). The extension of the dynamic models to form ARX,

MAX and ARMAX models was investigated, but was found to produce no significant improvement. The addition of chamber temperature and humidity information as extra inputs to the models was also found to produce no significant improvement.

The use of black-box models for pre-processed (static) data, whilst successful here, could be considered less feasible in real-world applications than the use of dynamic models. This is because a static model of order, say, 3 requires inputs from 2 previous exposure cycles as well as the current cycle. In some applications it may not be desirable to wait for 3 whole exposure cycles before obtaining meaningful results. However in a dynamic model of order 12 say, the 11 past inputs correspond to the outputs of the system for the previous 110 seconds. Thus there should be little or no delay before the results of the analysis are forthcoming.

The normalisation of the data seemed to have different effects on the static and dynamic models. The success rates using normalised static data were noticeably less than those using data which had not been normalised (e.g. FIR model for fractional difference algorithm: 99.7% vs. 97.0%). This could be attributed to the models for un-normalised data effectively 'learning' to discriminate between the two bacteria based partly on the strength of the signals rather than the pattern of the responses across the sensor array. Once this 'strength' information was removed, the models were less effective. However the normalisation process significantly improved the performances of the dynamic models. This opposite effect could be explained in a similar manner. The models for un-normalised dynamic data could be partly learning to classify the bacteria based on the strengths of the signals. But this could be 'confusing' the models when they were applied to data gathered much later than the training data. The sensor signals have a tendency to drift somewhat. This could be perceived as a change in strength of the signal by the model, resulting in misclassifications. The normalisation of the data forces the model to classify the bacteria according to the pattern of sensor outputs rather than the strengths and thus improves the long-term ability of the model to classify the bacteria correctly.

The data analysed in this paper have also been analysed elsewhere using artificial neural networks (Shin *et al.*) and a 100% successful classification rate was achieved using a Fuzzy ARTMAP applied to pre-processed data. Here we have shown that comparable success rates can be achieved using simple models that could be easily and cheaply incorporated into handheld digital electronic nose devices.

Further work is in progress to use linear black-box models to identify the current growth phase of the bacterial sample. The bacteria go through four distinct growth phases during their lifetime, and it has been shown that neural networks are able to achieve a high success rate when classifying the growth phase of the bacteria. It is hoped that the black-box models will be able to produce a performance to rival, or even surpass that of the neural networks.

ACKNOWLEDGEMENTS

The authors would like to thank both Mr. H.W. Shin and Dr. C.S. Dow for their work carrying out the biological experiments.

REFERENCES

Di Natale, C., Macagnano, A., Paolesse, R., Mantini, A., Tarizzo, E., D'Amico, A., Sinesio, F., Bucarelli, F.M., Moneta, E., and Quaglia, G.B. (1998). Electronic nose and sensorial analysis: comparison of performances in selected cases. *Sensors and Actuators B*, **50**, 246-252.

Hines, E.L., Llobet, E., and Gardner, J.W. Electronic noses: a review of signal processing techniques. *Proc. IEE: Circuits, Devices and Systems*, in press.

Gardner, J.W., and Bartlett, P.N. (1994). A brief history of electronic noses. *Sensors and Actuators B*, **18-19**, 211-220.

Gardner, J.W., and Bartlett, P.N., (1998). *Electronic Noses*, Oxford University Press, Oxford.

Gardner, J.W., Shurmer, H.V. and Tan, T.T. (1992). Application of an electronic nose to the discrimination of coffees. *Sensors and Actuators B*, **6**, 71-75.

Ljung, L. (1987). *System Identification - Theory for the User*. Prentice-Hall, New Jersey.

Marco, S., Pardo A., Davide, F.A.M., Di Natale, C., D'Amico, A., Hierlemann, A., Mitrovics J., Schweizer M., Weimar, U. and Gopel, W. (1996). Different strategies for the identification of gas sensing systems. *Sensors and Actuators B*, **34**, 213-223.

Shin, H.W., Llobet, E., Gardner, J.W., Hines, E.L. and Dow, C.S. The classification of the strain and growth phase of cyanobacteria in potable water using an electronic nose system. Submitted to *Proc. IEE - Sci. Meas. Technol.*

METABOLISM AND CELL CYCLE MODELLING BY MEANS OF NEURAL NETWORK BASED STRUCTURES

Andreia Hanomolo, Philippe Bogaerts, Raymond Hanus

**Université Libre de Bruxelles, Faculté des Sciences Appliquées,
Service d'Automatique et d'Analyse des Systèmes, Av. F.D. Roosevelt 50,
CP 165/55, 1050 Bruxelles, Belgique
e-mail: ahanomol@labauto.ulb.ac.be**

Abstract
Due to their importance to the pharmaceutical industry and implicitly to the health, the animal cell cultures and the research carried out over their behaviour has gained recently an increasing attention. The paper presents a case study concerning the application of the neural networks in the modelling of the main parameters of an animal cell culture: the metabolic components and the cell cycle phases. For the metabolism a hybrid (neural-classical) structure is described for the building of a continuous simulator capable to reconstruct the trajectory of the main components from the initial conditions. A special attention is paid to the choice of the cost function and to the "amount" of the classical contribution to the hybrid model. For the cell cycle the outputs of the above continuous simulator are used and a neural network black box approach is employed in order to obtain the evolution of the three phases. The structures were tested on two batch animal cell cultures for which few and asynchronous measurements are available, namely one culture growing on microcarriers, the other in suspension. *Copyright © 2000 IFAC*

Keywords: continuous simulator, neural-classical structure, bioprocesses modelling, likelihood cost function, decoupling strategy

1. INTRODUCTION

Animal cell cultures are necessary for the production of many biological-active proteins like therapeutic and diagnostic antibodies, vaccines and hormones. For optimising process control the use of a mathematical model describing the relations between the relevant parameters of an animal cell culture is an essential tool (Martens, *et al.*, 1992).

In what concerns the analytical approaches in the literature there is a number of structured and more or less unstructured models. The unstructured models make no assumptions with respect to cellular processes and in most cases a Monod type growth equation is used. Structured models have been reported based on cell cycle and assumptions concerning metabolic processes. There are also models that describe only the metabolic pathway.

There are essentially two problems encountered in the bioprocesses:

- their kinetics is poorly understood making it difficult to develop an analytical model;
- there are no cheap and reliable sensors capable to provide direct real-time measurements; typically only some of the variables are determined by slow infrequent off-line laboratory analysis (Bastin *et al.*, 1990).

There are few classical approaches that led to good metabolic models like observer based methods (Bogaerts, 1999b, Bogaerts *et al*, 1999). Due to their ability to capture non-linear characteristics, neural networks (NN) have been used also for the bioprocess monitoring. In the same time they served as software sensors for the estimation of biological variables difficult to measure on-line.

For the metabolism simulation neural networks (NN) have been employed in both black box and hybrid approaches. The black box models use no knowledge about the process assumed (di Massimo *et al*, 1992) (Willis *et al.*, 1992) while the hybrid ones (neural+classical) combine NN with first principles or process constraints related to the bioprocess (Dors *et al.*, 1995, Psichogios *et al.*, 1992).

The paper describes the results of a research based on a hybrid approach in which the NN is of the radial basis function (RBF) type. The classical part consists in a first step in the use of the mass balances and in the second step it is improved by adding the decoupling of the pseudo-stoichiometry estimation from the kinetics.

For the modelling of the cell cycle a static neural network (NN) is employed whose inputs are the estimated values of the metabolic components obtained with the hybrid model described above.

The paper is structured in two main parts: the metabolism modelling and the cell cycle modelling. In the first part the two types of hybrid models are described and the results of their application to real bioprocesses are presented.

In the second part the static neural model used for the cell cycle estimation is described and the results of its application to one animal cell culture are presented.

2. METABOLISM MODELLING

2.1 *General dynamical model of a bioprocess metabolism*

The general dynamical model on which is based the hybrid approach is described by (Bastin *et al.*, 1990):

$$\frac{d\xi}{dt} = K\varphi(\xi,t) - D\xi - Q(\xi) + F \qquad (1)$$

with:

1) $K\varphi(\xi,t)$ - the pseudo-stoichiometry and the nonlinear kinetics of biochemical reactions where:

   ξ - vector of component concentrations (biomass, nutrients, metabolites);
   $\varphi(\xi,t)$ - reaction rates;
   K – yield coefficient matrix

2) $-D\xi - Q(\xi) + F$ - the transport dynamics of the components through the bioreactor where:

   D - dilution rate;
   Q – vector of mass outflow rates in gaseous phase;
   F – vector of feed mass flow rate.

In the case of batch cultures the model (1) becomes:

$$\frac{d\xi}{dt} = K\varphi(\xi,t) \quad \text{with} \quad \xi(0) = \xi_0 \qquad (2)$$

The use of NN is justified precisely by the properties of the term $K\varphi(\xi,t)$: the reaction rates are complex non-linear functions of the operating conditions and of the state of the process which makes their analytical modelling extremely cumbersome.

2.2 *Description of the applications*

The first batch culture is a CHO-K1 line growing on microcarriers for which 7 experiments with 4 measured metabolic components were available: biomass (*X*), glucose (*G*), lactate (*L*), glutamine (*Gln*). The biomass is the mass of living organisms, *G* and *Gln* are nutrients and *L* is a metabolite. The experiments were characterised by asynchronous measurements and different initial conditions for *G* and *Gln* distributed like in fig. 1.

The second culture is CHO 320 growing in suspension, which produces the interferon γ and for which 5 experiments with asynchronous measurements for metabolism (*X*, *G*, *L*, *Gln*) as well as for the cell cycle are available. Contrary to the first application, all experiments have the same initial conditions. They differ only by their concentration of fœtal calf serum (FCS) introduced in the medium which is respectively: 1%, 2.5%, 5%, 7.5% and 10%.

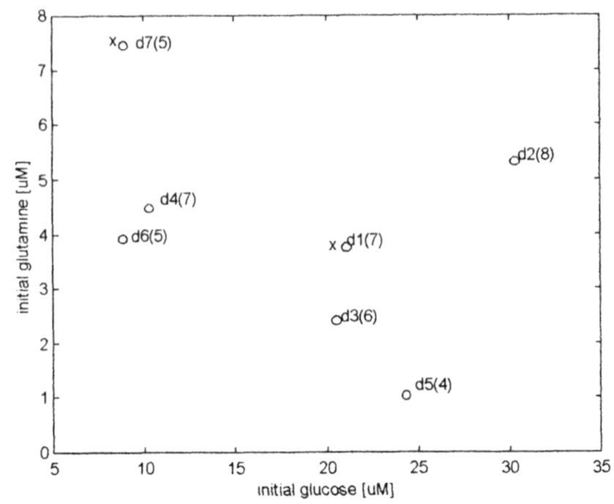

Fig 1 Initial conditions for the first batch culture

2.3 *Training algorithm*

In both cases an RBF network is used which has the particularity of a linear output layer whose output value is given by the weighted sum of the hidden neurons outputs. The activation functions of the hidden neurons are gaussians. As training algorithm a hybrid training is performed: unsupervised for the centres and supervised for the weights determination.

The centres of the gaussians are placed in the data space so that to represent possible groups by using the *k*-means algorithm (Duda, *et al.*, 1973). The width of a hidden neuron was chosen as the mean value of the distances between the centre of the respective neuron and the other gaussian centres.

Usually, in the literature, for the weights estimation a cost function based on the euclidian distance between real and estimated values is considered (Haykin, 1994, Dors *et al.*, 1995, Psichogios *et al.*, 1992). This approach is incorrect as long as it is well known that measurements in bioprocesses are provided with important errors. Thus in the present approach the confidence intervals are computed for each measurement point and a maximum likelihood is employed which takes into account the measurement errors at each point of each metabolic component (Hanomolo *et al.*, 2000):

$$J_1 = \sum_{i=1}^{s} \sum_{k_i \in K_i} \sum_{\alpha \in M} \left(\frac{\xi_\alpha(k_i) - \xi_{m\alpha}(k_i)}{\sigma_\alpha(k_i)} \right)^2 \quad (3)$$

with :
s – number of training sets
k_i - the measurement moments for the i-th training series.
K_i - the set of k_i for the i-th training series.
$\xi_{m\alpha}(k_i)$ - the measured value of concentration α at moment k_i;
$\xi_\alpha(k_i)$ - the simulated value of concentration α at moment k_i;
$\sigma_\alpha(k_i)$ - the standard deviation associated to the concentration α belonging to the set of concentrations M, at moment k_i.

The errors are assumed to be white noise normally distributed with zero mean (Bogaerts, 1999a). The weights, W are obtained by the minimisation of J_1:

$$\Theta = W = \arg\min(J_1) \quad (4)$$

On the other hand, a rigorous and realistic approach should also take into account the errors associated to initial conditions and estimate also the most likely initial conditions (Bogaerts, 1999a, Hanomolo *et al.*, 2000). This is accomplished by the minimisation of the same cost function with the difference that the parameters to be estimated are now both the weights and the initial conditions:

$$\Theta = \{W, \xi_0\} = \arg\min(J_1) \quad (5)$$

The cost function is nonquadratic in the parameters, which implies the use of a non-linear algorithm for the parameter optimisation accomplished with the Levenberg - Marquardt algorithm.

In order to determine the NN parameters and to validate the proposed models, for the first culture a typical identification procedure was carried out. The whole data set was divided in two, one for training (simple validation), the other for the test (cross validation).

For the second culture because of the only varying parameter (the FCS concentration) a leave one out strategy was employed. The latter consisted in leaving one of the data sets out and using the others for the learning. The procedure was applied five times, corresponding to the five values for FCS.

2.4 First hybrid model

In a first step the neural network was employed for the modelling of the pseudo-stoichiometry and the kinetics. The discretized model in terms of NN is:

$$\xi(k+1) = NN(\xi(k)) + \xi(k) \quad \text{with } \xi(0) = \xi_0 \quad (6)$$

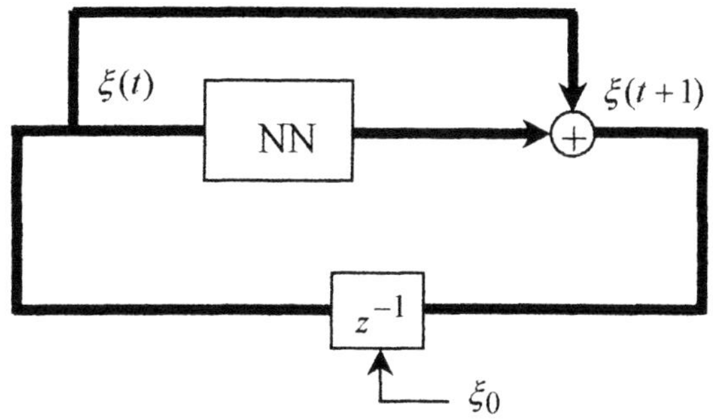

Fig. 2 Structure of the first simulator

and the corresponding architecture is represented in fig. 2.

As learning set the data from 5 experiments : d2, d3, d4, d5, d6 were chosen, the rest being kept for the test phase (marked with a cross in fig. 1). Results were very good in both phases, estimated data being inside the confidence intervals. This was the case even for d7 (fig. 3) whose initial conditions were far from the training surface. The model proved thus, good extrapolation capabilities (Hanomolo *et al.*, 2000).

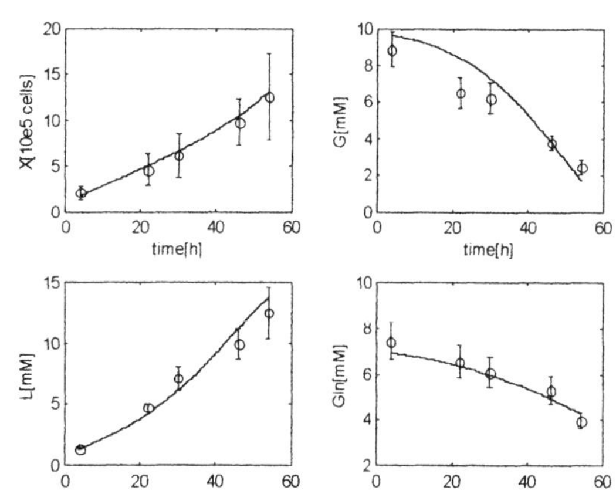

Fig. 3 Simulation results in cross validation for d7

2.5 Second hybrid model

When tested on the second cell culture even in simple validation phase one of the metabolic components was badly estimated. The estimated lactate had an opposite evolution to the real one (fig. 4).

The model was not appropriate and more prior knowledge has been added. This consisted in the decoupling of the pseudo-stoichiometry from the kinetics estimation (fig. 5).

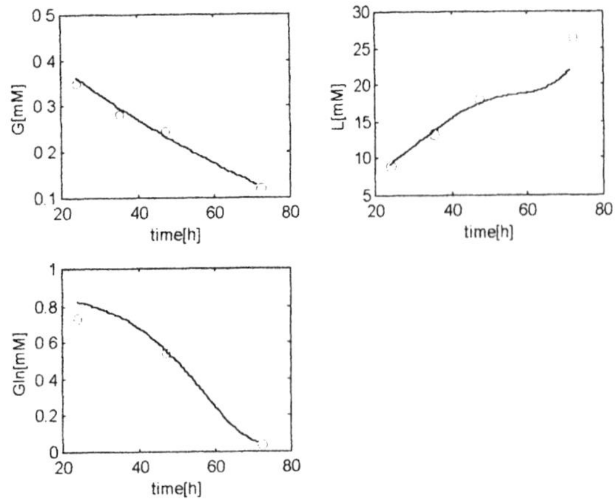

Fig. 4 Simple validation for the second culture, first model

The NN is used this time only for the kinetics estimation and the model in terms of NN becomes:

$$\xi(t+1) = \xi(t) + K * NN(\xi(t), FCS) \tag{7}$$

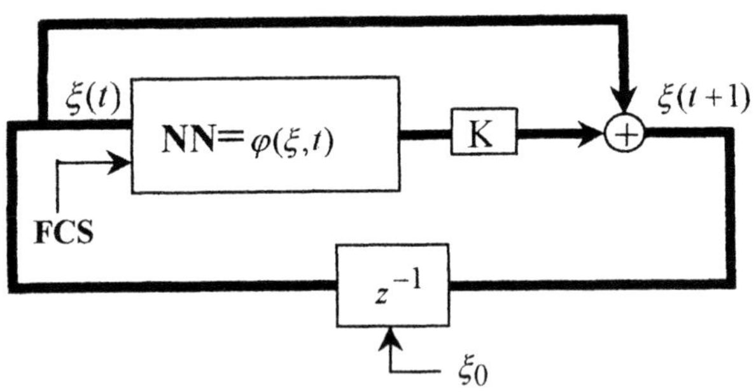

Fig. 5 Structure of the second simulator

The yield coefficient matrix is determined by classical approaches. The implementation is based on an idea proposed in (Chen *et al.*, 1996). It consists in identifying the yield coefficient matrix K by performing the partitioning of the state vector ξ. A new state variable is defined so that the evolution of the dynamics of this last one doesn't depend on the kinetics. Together with the state, a partition of the matrix K - the yield coefficient matrix - is also performed: $\xi = [\xi_a \ \ \xi_b]$ and $K = [K_a \, K_b]$ so that $\text{rank}(K_a) = $ full row rank $= p$. A unique matrix C can be obtained from the following relation:

$$CK_a + K_b = 0 \tag{8}$$

An auxiliary state is defined as: $z \overset{\Delta}{=} C\xi_a + \xi_b$ which leads for a batch bioprocesses, to:

$$\frac{dz}{dt} = 0 \tag{9}$$

In the case of a batch bioprocess the full row rank condition is easily fulfilled. In our particular case the yield matrix is:

$$K = \begin{bmatrix} 1 & 0 \\ 0 & -1 \\ -gln_g & -gln_m \\ l_g & l_m \end{bmatrix}$$ and the state partition is:

$\xi_a = [\text{X G}]^T$ and $\xi_b = [\text{Gln L}]^T$. The yield coefficients are determined by the minimisation of least squares cost functions. Once the pseudo-stoichiometric coefficients determined the reaction rates are estimated by means of the NN and of the algorithm described in paragraph 2.3.

It could be noticed that the values for the pseudo-stoichiometric coefficients are similar from one leave-one-out experiment to the other which confirms the correctness of our model for the culture considered. For each leave one out procedure several configuration with hidden neurons were tried. A good hybrid model should lead (apart similar yield coefficient matrices) to similar configurations. Indeed the obtained structures had 2 or 3 hidden neurons.

In fig. 6 are presented the results for the case when the experiment with 1 % FCS was left-out. It is interesting to notice that in this case, too, the model behaved satisfactorily also in extrapolation (1 % FCS is the extremity of the FCS concentration interval).

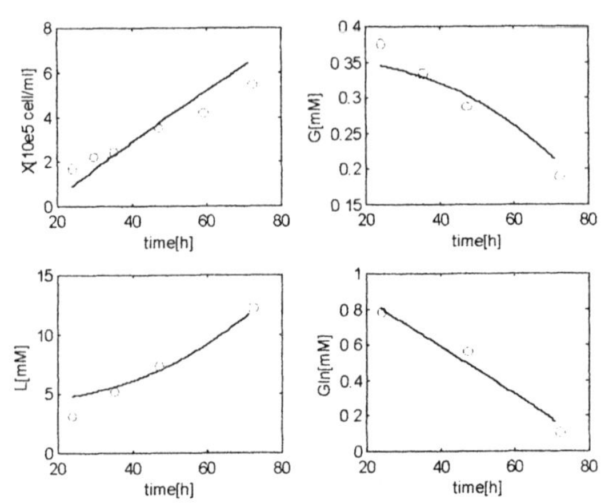

Fig. 6 Cross validation results for leave 1 % FCS out

3. MODELLING THE CELL CYCLE

The cell cycle is the time interval between the formation of a cell by the division of the mother cell and the subsequent division of the daughter cell. The three phases of the cycle are :

- *G1* – division (30-40 % of the cycle)
- *S* – DNA synthesis (30-50 % du cycle)
- *G2-M* (*G2*) – division of the cell in 2 (15-20 % of the cycle).

The concentration of biomass is the sum of the biomass concentrations of each of the cell cycle phases: $X = X_{G1} + X_S + X_{G2}$.

Usually modelling the cell cycle is performed by analytical methods and a lot of assumptions are made and particular cases are considered (Martens *et al.*, 1995). The present paper makes use of the results provided by the metabolism simulator and makes no supplementary assumptions.

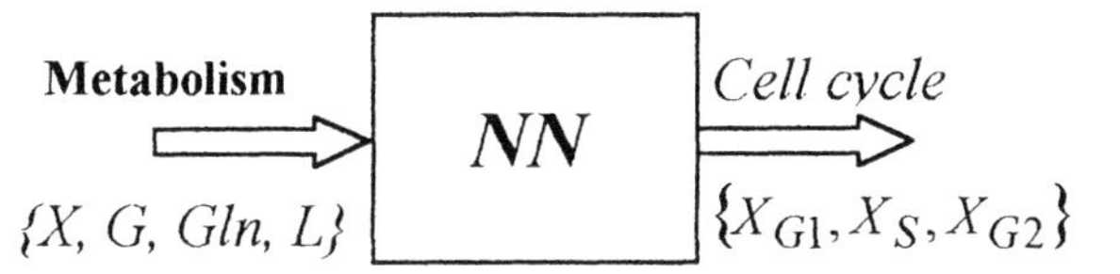

Fig.7 Structure for the cell cycle estimation from the metabolism

Using the second hybrid simulator as the real process the estimated values of the metabolism components are taken. A static NN is used to reconstruct from these values the cell cycle, precisely, the three phases (fig. 7).

The leave one out strategy was applied and the results in cross validation from the hidden neurons point of view are similar: between 10 and 12. The discrepancies that can be noticed in fig. 8 are quite normal as the weak point of the NN is the extrapolation capabilities (1 % FCS is at the extremity of the interval of FCS values). And this is the case also for leave 10 % FCS out (the other extremity).

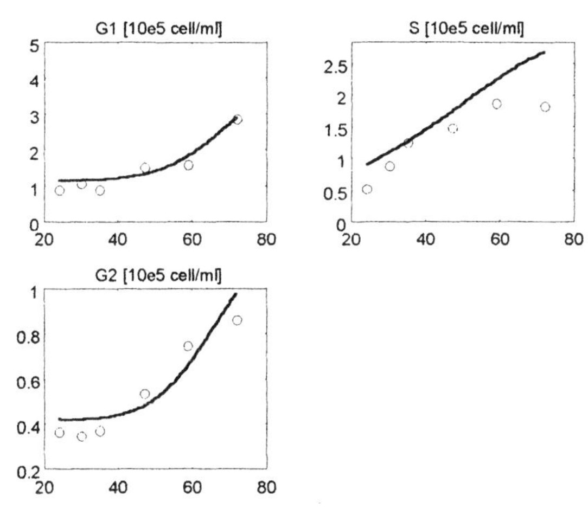

Fig.8 Cross validation for leave 1 % FCS out

On the other hand, for the other experiments, results are quite satisfactory even with the few data that were available.

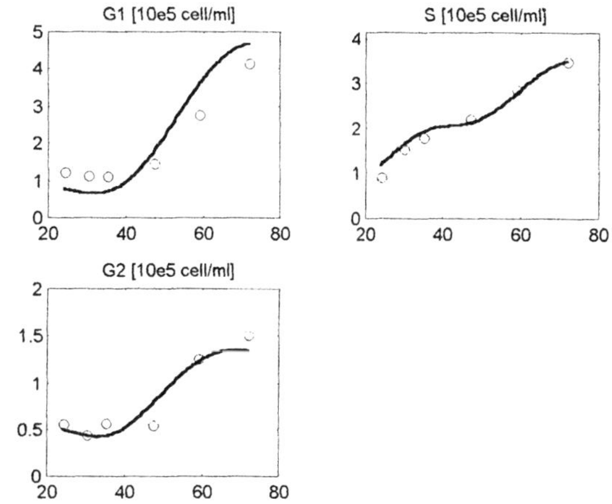

Fig. 9 Cross validation for leave 7.5 % FCS out

The model needs improvement and more tests should be carried out. Nevertheless we can conclude that this simple model could be a possible solution to the cell cycle modelling provided a good model is available for the metabolic components. This means that no more difficult measurements will be necessary for the 3 phases, except of course of those needed for the training.

4. CONCLUSIONS

Two hybrid structures have been compared for the simulation of the metabolism evolution for batch bioprocesses. When the NN models both the pseudo-stoichiometry and the kinetics results are good for one culture and not for the other. This structure is thus not applicable to any culture, is restricted to some.

When decoupling the pseudo-stoichiometry from the kinetics, better results are obtained, the hybrid structure being capable to reconstruct the evolution of metabolic components from only the initial conditions. In the second approach the difficulty encountered is related to the yield coefficients matrix (K) identification. It must also be mentioned that both methods were tested only on batch cultures. Nevertheless, results are encouraging and the second structure should be tested on more complex cultures like semi-continuous or fed-batch.

The preliminary results in the cell cycle modelling are quite promising as in case these performances confirm for other cultures, too, it means that a good model for the metabolism simulation is enough for the estimation of all parameters of the animal cell culture. This would simplify a lot the measurement procedure, as no more time expensive measurements should be necessary ; the model could "replace" the measurement procedure.

Acknowledgements - This work was sponsored by the European project: "Efficient Cell". Special thanks go to Dr. John Wérenne, Dr. Mohamed Al Rubeai and to their teams of biologists who helped at the fulfilment of the present research.

REFERENCES

Bastin, G., Dochain, D. (1990). *On-line Estimation and Adaptive Control of Bioreactors*, Elsevier,

Bogaerts, Ph. (1999a), *Thése de doctorat*, Université Libre de Bruxelles

Bogaerts, Ph. (1999b). A hybrid asymptotic-Kalman observer for bioprocesses. In: *Bioprocesses Engineering*, **vol. 20,** n° 3, pp. 249-255

Bogaerts, Ph., Castillo J., and Hanus R. (1999). A general mathematical modelling technique for bioprocesses in engineering applications. In: *System Analysis Modeling Simulation*, **vol. 35**, pp.87-113

Chen L., Bastin, G. (1996). Structural identifiability of the yield coefficients in bioprocess models when the reaction rates are unknown. In: *Mathematical Biosciences,* **vol 132**, pp. 35-67

Di Massimo, C., Montague, M. J., Tham, M.T., Morris A.J. (1992). Towards improved Penicilin Fermentation via Artificial Neural Networks. In: *Comp. Chem. Eng.*, **vol. 16**, no. 4, pg. 283-291

Dors, M., Simutis, R., Lubert, A.. (1995). Advanced Supervision of mammalian Cell Cultures Using Hybrid Process Models. In: *Preprints of the 6th Int. Conf. on Comp. Appl. in Biotechnology* , pp. 72-77, Germany

Duda, R.O, Hart, P.E. (1973). *Pattern Classification and Scene Analysis*, John Wiley & Sons

Hanomolo, A., Bogaerts, Ph., Graefe, J., Cherlet, M., Wérenne, J., Hanus, R. (2000). Maximum likelihood parameter estimation of a hybrid neural-classical structure for the simulation of bioprocesses, *Mathematics and Computers in Simulation*, **vol 51**, pg. 375-385

Haykin, S. (1994). *Artificial Neural Networks. A Comprehensive Foundation*, MacMillan College Publishing Company

Martens, D.E., Sipkema, E. M., de Gooijer, C. D., Beuvery, E. C., Tramper, J. (1995). A combined Cell-Cycle and metabolic model for the growth of hybridoma cells in steady-state continuous culture. In: *Biotechnology and Bioengineering*, **vol. 48,** pg. 49-65

Psichogios, D.C., Ungar, L. H. (1992). A Hybrid Neural Network-First Principles Approach to Process Modeling. In: *AIChE Journal*, **vol. 38**, pg. 1499-1511

Willis, M.J., Montague, G.A., di Massimo, C., Tham, M.T., Morris, A.J. (1992). Artificial Neural Networks in Process Estimation and Control. In: *Automatica*, **vol.28**, no. 6, pp. 1181-1187

A NONPARAMETRIC APPROACH TO MODEL ERROR MODELING

Anders Stenman *,[1] **Fredrik Tjärnström** *

* *Department of Electrical Engineering, Linköpings universitet, SE-581 83 Linköping, SWEDEN. Fax: +46 13 28 26 22*

Abstract: To validate an estimated model and evaluate its reliability is an important part of the system identification process. Recent work on model validation has shown that the use of explicit model error models provide a better way of visualizing the possible deficiencies of the nominal model. Previous contributions have mainly focused on parametric black-box models for estimating the error model. However, this requires that a correct model order for the error model has to be selected. Here we suggest an adaptive and nonparametric frequency-domain method that estimates the frequency response of the model error by an automatic procedure. A benefit with this approach is that the tuning can be done locally, i.e., that different resolutions can be used in different frequency bands. The ideas are based on local polynomial regression and utilize a statistical criterion for selecting the optimal resolution. *Copyright ©2000 IFAC*

Keywords: model error modeling, non-parametric regression, frequency-response methods

1. INTRODUCTION

The concept of model validation has received an increased attention in recent years. Approaches with unknown but bounded disturbances (UBB), stochastic embedding, control oriented model validation, $\mathcal{H}_\infty$-identification etc, typically have the ambition of providing more reliable models for control design. Another approach in this direction is the concept of *model error modeling* (Ljung, 1999*a*), where parametric black-box models have been utilized with the intention to reveal the model error's properties more in detail. However, this requires that the parameterization and the order of the error model are correctly selected.

In this contribution we suggest a nonparametric frequency-domain approach that estimates the model error's frequency response by an automatic and adaptive procedure. The method is based on local polynomial modeling ideas and is an application of the procedure proposed in Stenman *et al.* (1999). It has the advantage that the estimation can be tuned locally for each frequency, i.e., that different resolutions can be used in different frequency bands. Moreover, it also provides an easy way of getting pointwise confidence bands for the estimate. This can be a very useful tool when visualizing the properties of the model error.

The organization of the paper is as follows. Section 2 gives the general setup. Section 3 shows how the concept of local polynomial models can be used for providing more detailed information about the model error. Section 4 illustrates the concept with a numerical example, and in Section 5, finally, some concluding remarks are drawn.

2. PRELIMINARIES

The setup for *model validation* (or model invalidation) can be described as follows: We assume that we have at hand a nominal input-output model $\hat{G}(q)$ along with a validation data set,

[1] Corresponding author. Email: stenman@isy.liu.se

$$Z^N \triangleq \{(u(1), y(1)), \ldots, (u(N), y(N))\}, \quad (1)$$

where $y(t)$ and $u(t)$ represent the output and the input of the system. We would now like to construct a testing procedure by which we may *falsify* or *invalidate* the given model. That is, devise a procedure that makes us able to say that it is unlikely (or even impossible) that the model has generated the validation data. If the model is not falsified by this test, however, it seems reasonable to say that the model has passed the validation.

A common theme among a majority of the traditional approaches for model validation, is the idea of evaluating the model's quality by testing its performance on fresh data. This has typically involved a whiteness test on the residuals,

$$\varepsilon(t) \triangleq y(t) - \hat{G}(q)u(t), \quad t = 1, \ldots, N, \quad (2)$$

(or an examination of the correlation between the residuals and past inputs) in combination with a qualitative comparison of the true and predicted outputs. In the next section, though, we shall see how more accurate information can be obtained by proper modeling of the model error.

3. DIRECT MODEL ERROR MODELING

An alternative to the above statistical approaches for validation, is to try to estimate the parts in $\varepsilon(t)$ that depend on unmodeled dynamics with black-box models. This was referred to as *model error modeling* by Ljung (1999*a*). The idea was however first introduced in Ljung *et al.* (1991).

The most general error model is perhaps the nonlinear model

$$\varepsilon(t) = m_\Delta(\varepsilon^{t-1}, u^t) + w(t), \quad (3)$$

where $m_\Delta(\cdot)$ is an arbitrary nonlinear mapping that takes care of the unmodeled dynamics, and $w(t)$ is a disturbance. However, to keep things simple, it appears to be reasonable to specialize into linear models of the form

$$\varepsilon(t) = \Delta(q)u(t) + w(t). \quad (4)$$

3.1 *Modeling Approaches*

An obvious way of modeling $\Delta(q)$ in (4) is of course to utilize high order FIR or ARX models. This was the approach adopted by Ljung (1999*a*). The problem is then of course how to choose the right model structure and model order. If these were known this approach would be impossible to beat since its variance is then known to meet the Cramèr-Rao bound.

Another possibility is to use the standard non-parametric frequency domain methods, based on the Blackman-Tukey method (Blackman and Tukey, 1958). The idea is to compute an estimate of the model error's frequency response $\Delta(e^{i\omega})$ by smoothing the ratio

$$\hat{\hat{\Delta}}(e^{i\omega_k}) \triangleq \frac{\Xi(\omega_k)}{U(\omega_k)} = \frac{\sum_{t=1}^{N} \varepsilon(t)e^{-i\omega_k t}}{\sum_{t=1}^{N} u(t)e^{-i\omega_k t}}. \quad (5)$$

The quantity (5) is normally referred to as the empirical transfer function estimate, ETFE, of the error model. This is a very "raw" and noisy estimate of the frequency response, and typically needs to be further refined. The classical solution is to average over neighboring ETFE values using a fixed window $W_\gamma(\cdot)$ (Ljung, 1999*b*);

$$\hat{\Delta}(e^{i\omega_k}) = \frac{\int_{-\pi}^{\pi} W_\gamma(\omega - \xi)|U_N(\xi)|^2 \hat{\hat{\Delta}}(e^{i\xi})d\xi}{\int_{-\pi}^{\pi} W_\gamma(\omega - \xi)|U_N(\xi)|^2 d\xi}. \quad (6)$$

One can show that the variance of $\hat{\Delta}(e^{i\omega_k})$ then approaches

$$\operatorname{Var} \hat{\Delta}(e^{i\omega_k}) \approx \frac{0.75\gamma}{N} \cdot \frac{\Phi_v(\omega)}{\Phi_u(\omega)},$$

provided the standard Hamming window is used. The main drawback here is that one usually needs a high γ to get a good frequency resolution. This will typically give a very noisy estimate (Stenman *et al.*, 1999). A remedy is to use the nonparametric frequency-domain approach that was introduced in Stenman *et al.* (1999). A short description of this procedure will be given next.

3.2 *A Local Polynomial Approach*

The basic idea behind the local polynomial approach is to replace the averaging (6) with a more general smoothing procedure. The frequency function of the model error, $\Delta(e^{i\omega})$, is assumed to be smooth and is estimated by fitting a pth degree polynomial to the data located within a sliding window. That is, for each frequency ω, an estimate is obtained via the weighted regression problem; minimize

$$V(\omega, \beta) = \sum_{k=1}^{N} \left| \hat{\hat{\Delta}}(e^{i\omega_k}) - \sum_{j=0}^{p} \beta_j(\omega_k - \omega)^j \right|^2 \times w_k(\omega) \cdot \frac{1}{\sigma_k^2}, \quad (7)$$

subject to $\beta_j \in \mathbb{C}$, where the $w_k(\omega)$'s are the window weights. If $\hat{\beta}_j$, $j = 0, \ldots, p$, denote the minimizers of (7), an estimate of the model error frequency response is given by

$$\hat{\Delta}(e^{i\omega}) = \hat{\beta}_0. \quad (8)$$

The window weights are usually chosen according to

$$w_k(\omega) = K\left(\frac{\omega_k - \omega}{h}\right), \quad (9)$$

where h is a smoothness parameter (usually referred to as the *bandwidth*) controlling the spread of the window, and $K(\cdot)$ is a *kernel* function with support on $[-1, 1]$ assigning weights to each remote frequency-domain measurement according to its distance from ω. However, it has been shown that it is also possible to select the weights via an explicit optimization stage that aims at minimizing the mean square error of (8) subject to the weights. Consult Stenman (1999) for details around this.

We have in (7) also included the inverse variance of the measurements as an extra weighting. It can be estimated either directly from the frequency domain data (5) using standard variance estimation techniques for nonparametric models (Loader, 1999), or by utilizing theoretical formulas (Guillaume *et al.*, 1996).

3.2.1. *Local Model Tuning* It is well known that the resolution parameter h has a critical impact on the resulting estimate since it governs a trade-off between bias and variance. Methods that use the available data to produce a bandwidth are usually referred to as *bandwidth selectors*, and have been extensively studied within the statistical literature, see, for instance, Fan and Gijbels (1996).

As in previous applications we here consider an adaptive, localized and generalized C_p approach with variance penalty α (Cleveland and Loader, 1994)

$$C(\omega, h) \triangleq \sum_{k=1}^{N} \frac{w_k(\omega)}{\sigma_k^2 \operatorname{tr}(\mathbf{W})} \left|\hat{\hat{\Delta}}(e^{i\omega_k}) - \hat{\Delta}(e^{i\omega_k})\right|^2 - 1 + \alpha \frac{\operatorname{tr}((\boldsymbol{\Omega}^T\mathbf{W}\mathbf{V}\boldsymbol{\Omega})^{-1}(\boldsymbol{\Omega}^T\mathbf{W}^2\mathbf{V}\boldsymbol{\Omega}))}{\operatorname{tr}(\mathbf{W})}. \quad (10)$$

Here $\boldsymbol{\Omega}$ is a Vandermonde-type matrix containing the basis functions for the polynomial model, and $\mathbf{W}$ and $\mathbf{V}$ are diagonal matrices containing the weights $w_k(\omega)$ and the inverse variances $1/\hat{\sigma}_k^2$, respectively. The parameter α is an extra tuning knob that provides the possibility of changing the penalty on the variance term in (10). It defaults to 2, but can be increased if the variability of estimates gets too high (see Stenman (1999)). Thus, for each frequency ω, a local choice of resolution $h = h(\omega)$ can be obtained by minimizing $C(\omega, h)$ subject to h.

3.2.2. *Confidence Bands for the Estimate* The local polynomial smoothing method provides point estimates of the frequency function at specified frequencies. However, in the context of model error modeling, it is in general of more interest to assess the *uncertainty* associated with the error model rather than the model itself. (The estimated frequency response of the model error can in fact be quite bad as a consequence of the low signal-to-noise ratio of the residuals).

Nevertheless, it turns out that computing pointwise confidence bands is a rather straightforward thing to do. Note that the frequency response via (7) effectively is estimated by means of a linear smoothing operation,

$$\hat{\Delta}(e^{i\omega}) = \sum_{k=1}^{N} W_k(\omega)\hat{\hat{\Delta}}(e^{i\omega_k}), \quad (11)$$

where the linear weights $\{W_k(\omega)\}$ are determined either explicitly by the optimization approach, or implicitly by a kernel and a polynomial fitting procedure. Hence it follows that the variance of (11) is approximately

$$\operatorname{Var} \hat{\Delta}(e^{i\omega}) = \hat{\sigma}^2(\omega) \cdot \|\mathbf{w}(\omega)\|^2, \quad (12)$$

where $\hat{\sigma}^2(\omega)$ denotes an estimate of the complex noise variance at the frequency ω, and

$$\mathbf{w}(\omega) \triangleq \begin{pmatrix} W_1(\omega) & \ldots & W_N(\omega) \end{pmatrix}^T \quad (13)$$

is the vector of smoothing weights. It is thus straightforward to construct approximate confidence bands for $\Delta(e^{i\omega})$ (or more precisely, $\operatorname{E} \hat{\Delta}(e^{i\omega})$).

It can be shown that the real and imaginary parts of $\hat{\Delta}(e^{i\omega})$ are asymptotically uncorrelated and jointly normally distributed with variances equal to half the value of (12). See, e.g., Ljung (1999*b*). Using Gauss' approximation formula we find that the absolute value of the smoothed estimate is asymptotically normally distributed with variance

$$\operatorname{Var} \left|\hat{\Delta}(e^{i\omega})\right| \approx \frac{\hat{\sigma}^2(\omega) \cdot \|\mathbf{w}(\omega)\|^2}{2}$$

(Ljung, 1999*b*, Exercise 9G.1). From this, confidence bands along the amplitude curve can be constructed. A confidence band of degree 95% at each frequency will be formed as

$$\left|\hat{\Delta}(e^{i\omega})\right| \pm \frac{1.96}{\sqrt{2}} \hat{\sigma}(\omega) \cdot \|\mathbf{w}(\omega)\|.$$

Remember, however, that these values are only approximations. To obtain more accurate confidence bands, a more detailed analysis of the distributions of the amplitude and phase must be performed and some bias adjustment might be required (see, e.g., Fan and Gijbels (1996)). Different possibilities also exist depending on which quantities we want to display (the amplitude, the logarithm of the amplitude and so on). It might also be possible to utilize numerical methods such as point-mass approaches or bootstrapping (Efron and Tibshirani, 1993).

4. AN EXAMPLE

Let us illustrate the above model error modeling approach with an example (adapted from Ljung

(1999*a*)). A data set of length 2000 was generated using the "system"

$$y(t) = \frac{B(q)}{F(q)}u(t) + v(t) \tag{14a}$$

with

$$B(q) = q^{-1} + 0.5q^{-2}, \tag{14b}$$
$$F(q) = 1 - 2.2q^{-1} + 2.42q^{-2} - 1.87q^{-3} + 0.7225q^{-4}, \tag{14c}$$

where the input $u(t)$ was chosen as a zero mean Gaussian sequence with unit variance, and the disturbance $v(t)$ was an additive Gaussian noise sequence with zero mean and variance 0.5. The data set was divided into two equally sized parts, one for estimation and one for validation. A second order output-error model $\hat{G}(q)$ was identified using the estimation dataset and the system identification toolbox in MATLAB (Ljung, 1997). This model will be referred to as the nominal model in the remainder of the section.

Figure 1 shows the frequency response of the nominal model (dashed curve) along with its 95% confidence bands (shaded). The frequency response of the true system (14) has also been included (solid line). It is clear that the confidence bands do not contain the true system. This is consistent

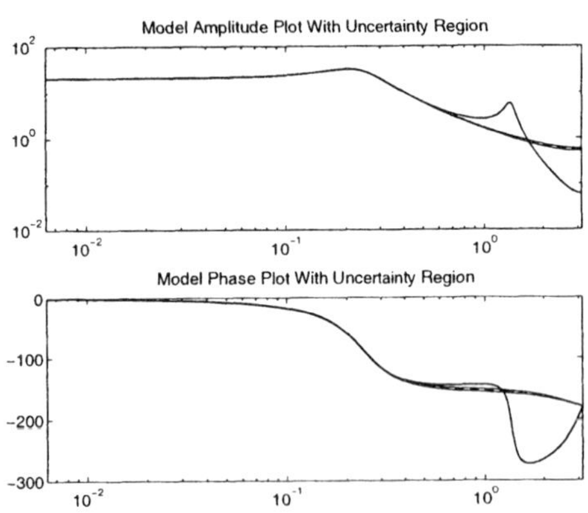

Fig. 1. The frequency response of the nominal model (dashed), the true system (solid) and its 95% confidence bands (shaded).

with Figure 2, where the correlation function of the residuals and the cross-correlation between residuals and past inputs are displayed. We see that the values of the cross-correlation function do not seem to pass a traditional-type validation test. The conclusion is therefore that there are some significant errors present in the nominal model. It is however not obvious what types of errors they represent.

Let us see what conclusions we can draw by using the procedure outlined in Section 3.2. Figure 3 shows the amplitude curve of the model error model (solid line) along with its uncertainty regions (shaded). For comparison, the amplitude

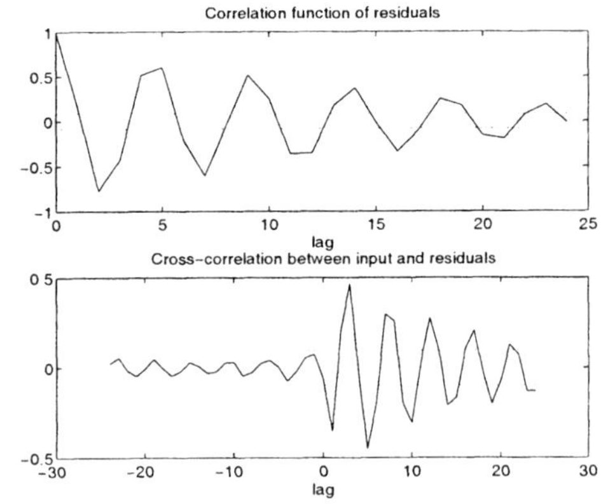

Fig. 2. Correlation and cross-correlation functions with 95% confidence bands.

curve of the "true" model error has also been included (solid line). The uncertainty regions tell us that there are significant model errors in the frequency range above 0.2 rad/s. The frequency

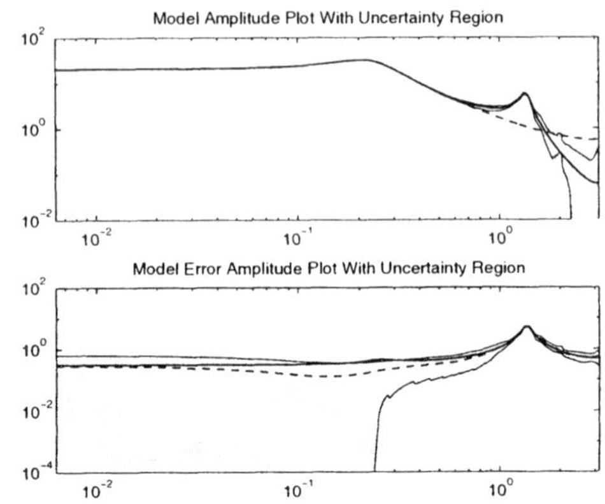

Fig. 3. Lower plot: The amplitude curve of the estimated model error (dashed) using local smoothing methods along with its uncertainty bands (shaded). The solid curve is the true model error. Upper plot: The amplitude curve of the nominal model (dashed) and of the true system (solid) together with the uncertainty region obtained as the sum of the nominal model and the model error model uncertainty (shaded).

response has been estimated and tuned locally according to (7) and (10), using a local quadratic model structure ($p = 2$), the tricube kernel

$$K(u) = \begin{cases} (1 - |u|^3)^3, & |u| \leq 1, \\ 0, & |u| > 1, \end{cases}$$

and a variance estimate $\hat{\sigma}_k^2$ obtained as

$$\hat{\sigma}_k = \frac{\hat{\Phi}_v(\omega_k)}{|U_N(\omega_k)|^2},$$

where $U_N(\omega_k)$ is the discrete Fourier transform of $u(t)$ and $\hat{\Phi}_v(\omega_k)$ is an initial estimate of the noise spectrum. For more details see Stenman *et al.* (1999). The selected bandwidths are displayed in Figure 4. It is clearly seen that the bandwidth

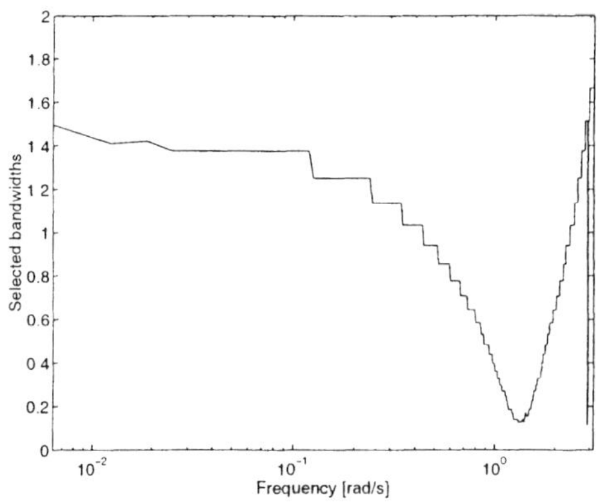

Fig. 4. The selected bandwidths according to the C_p criterion, here $\alpha = 5$ was used.

selector (10) adapts the resolution to the properties of the data.

We compare this result with what is obtained using parametric and standard frequency domain methods.

In the nonparametric case we use $\gamma = 100$ to make sure that no resonance peaks are missed. The result can be found in Figure 5. We see that we find the resonance peak quite accurately, but the noisiness makes it difficult to accurately interpret the type of model errors present in the nominal model. The "size" of the confidence regions using these nonparametric methods are more or less the same. This is to be expected since the use the same a priori information.

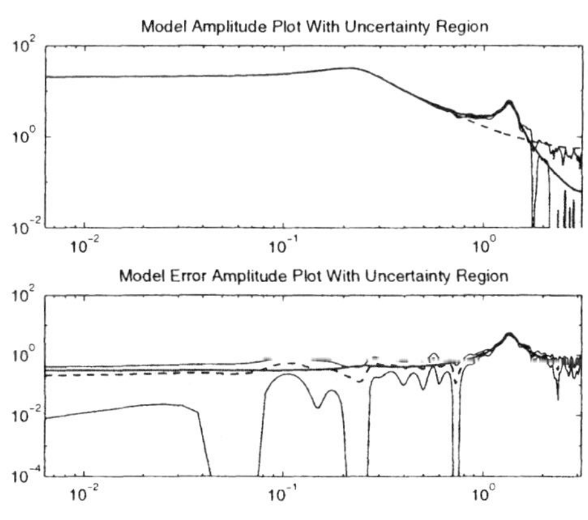

Fig. 5. Lower plot: The amplitude curve of the estimated model error (dashed) using traditional nonparametric frequency domain methods with $\gamma = 100$ along with its uncertainty bands (shaded). The solid curve is the true model error. Upper plot: The amplitude curve of the nominal model (dashed) and of the true system (solid) together with the uncertainty region obtained as the sum of the nominal model and the model error model uncertainty (shaded).

When using parametric models, the problem is always to decide the structure and order of the model. Here we have used an ARX model of order 15, see Figure 6. We used an ARX model to eliminate the risk of ending up in local minima. The order was chosen more at random, but high enough to pass a standard validation test. The regions are a little bit smaller than those obtained with the other methods. Note that the ARX structure is not correct in this example, the correct structure is actually an OE model of order 6. This implies that there is some bias in the estimate, which in turn implies that the variance estimate of the parametric model might not be correct. Therefore one can argue that the local smoothing methods presented here is a better alternative since its parameter free and produce very good estimates, i.e., has high resolution and high smoothing at the same time.

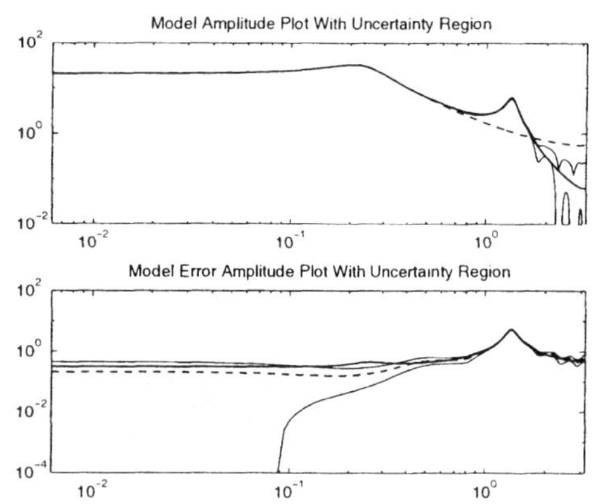

Fig. 6. Lower plot: The amplitude curve of the estimated model error (solid) using an ARX model of order 15 along with its uncertainty bands (shaded). The solid curve is the true model error. Upper plot: The amplitude curve of the nominal model (dashed) and of the true system (solid) together with the uncertainty region obtained as the sum of the nominal model and the model error model uncertainty (shaded).

5. CONCLUSIONS

To validate an estimated model and evaluate its reliability is an important part of the system identification process. We have demonstrated that the use of explicit *model error models* is a convenient way of visualizing the possible deficiencies of the nominal model, and provides a useful tool for the validation procedure.

Based on local polynomial modeling and a statistical bandwidth selection scheme, we have proposed a method that estimates the model error's frequency response using an adaptive and automatic procedure. This allows for a frequency-dependent choice of frequency resolution and gives several advantages over traditional parametric and non-parametric approaches. The frequency response of the error model typically exhibit fine details to different degrees in different frequency bands. The results are more easily interpreted than standard frequency domain methods and have variance properties similar to parametric models. Our adaptive approach thus gives a useful alternative compared to other multi-resolution techniques like, e.g., wavelets.

6. REFERENCES

Blackman, R.B. and J.W. Tukey (1958). *The Measurement of Power Spectra.* Dover, New York.

Cleveland, W.S. and C. Loader (1994). Smoothing by local regression: Principles and methods. Technical report. AT&T Bell Laboratories. 600 Mountain Avenue, Murray Hill, NJ 07974, USA.

Efron, B. and R. J. Tibshirani (1993). *An Introduction to the Bootstrap.* Chapman & Hall.

Fan, J. and I. Gijbels (1996). *Local Polynomial Modelling and Its Applications.* Chapman & Hall.

Guillaume, P., I. Kollár and R. Pintelon (1996). Statistical analysis of nonparametric transfer function estimates. *IEEE Transactions on Instrumentation and Measurement* **45**(2), 594–600.

Ljung, L. (1997). *System Identification Toolbox – User's Guide.* The Mathworks Inc. 24 Prime Park Way, Natick, MA.

Ljung, L. (1999*a*). Model validation and model error modeling. In: *The Åström Symposium on Control* (B. Wittenmark and A. Rantzer, Eds.). Studentlitteratur, Lund, Sweden.

Ljung, L. (1999*b*). *System Identification: Theory for the User.* 2nd ed.. Prentice-Hall.

Ljung, L., B. Wahlberg and H. Hjalmarsson (1991). Model quality: The roles of prior knowledge and data information. In: *Proceedings of the 30th IEEE Conference on Decision and Control.* Brighton, U.K.. pp. 273–278.

Loader, C. (1999). *Local Regression and Likelihood.* Springer, New York.

Stenman, A. (1999). Model on Demand: Algorithms, Analysis and Applications. PhD thesis. Dept of EE, Linköping University. SE-581 83, Linköping, Sweden.

Stenman, A., F. Gustafsson, D.E. Rivera, L. Ljung and T. McKelvey (1999). On adaptive smoothing of empirical transfer function estimates. In: *Preprints of the 14th World Congress of IFAC, Beijing, P.R. China* (H.F. Chen and B. Wahlberg, Eds.). Vol. H. Elsevier Science. pp. 415–420.

MODEL ERROR MODEL FROM IDENTIFICATION IN CLOSED-LOOP

Alina Besançon-Voda

Laboratoire d'Automatique de Grenoble
Bp 46, 38402-St.Martin d'Heres, France
Alina.Besancon@inpg.fr

Abstract: Prediction errors (or residuals) obtained with a model identified in closed-loop contain information on the distance between the model and the "true" system. In this paper it is shown how this distance can be estimated, using either spectral analysis or parametric identification, for plant models obtained with several identification in closed-loop methods. Asymptotic variance expressions are also given. *Copyright ©2000 IFAC*

Keywords: Identification in closed-loop, residual analysis, model error model, sensitivity functions.

1. IDENTIFICATION IN CLOSED-LOOP OF PLANT MODEL

Given a model G of a plant, an important question is how far this model might be from a true description. The analysis of the residuals (error between the measured output and model's simulated response to the measured input signal) can be used to obtain a size of the unmodeled dynamics (Ljung and Guo 1997, Ljung 1987), and it is pointed out that there are two separate sources for the model residual : *model error* and *disturbances*.
In this paper, it is considered that the model G has been identified in closed-loop and the prediction errors are available. For some identification in closed-loop methods, it is shown that from the expression of the prediction error, the difference $G_0 - G$ between the true description of the system and the model (called *model error model* in (Ljung and Guo 1997, Gevers 1998)) can be estimated and the asymptotic variance of this model error model is calculated for models identified in closed-loop with several methods . Moreover, the uncertainty region of the model can be expressed.

It has to be pointed out that very similar results have recently been published, dealing with asymptotic variance calculation of the model G (Forssel and Ljung 1999). The difference is that the present paper treats asymptotic variance calculation of the model of $G_0 - G$. For the case of identification in open-loop of the model G, results of identification of the error model $G_0 - G$ and if its variance have been presented in (Zhu 1989).

Consider the closed-loop configuration of a LTI system, presented usually as in figure 1, where C is the controller stabilizing the plant G_0. The external excitation signal r_t is independent of the output noise v_t and is placed either on the reference (r_2) or at the plant input (r_1):

$$r = r_1 + C.r_2 \tag{1}$$

The output noise is considered to be a filtered white noise $e(t)$ with variance λ_0 and filter H_0.

For notation simplicity, the argument q of the transfer function operator or $e^{i\omega}$ (in frequency domain) are dropped out, as well as parameter vector θ. In the remainder of the paper, G represents the plant model, H the noise model, $S = \dfrac{1}{1+C.G}$ is the output sensitivity of the identified model

G and controller C, $S_0 = \dfrac{1}{1+C.G_0}$ is the real output sensitivity function. The complementary sensitivity function CGS is denoted by T and the real one CG_0S_0 is denoted by T_0. The plant input can be expressed in closed-loop as :

$$u_t = S_0.r_t - C.S_0.v_t \tag{2}$$

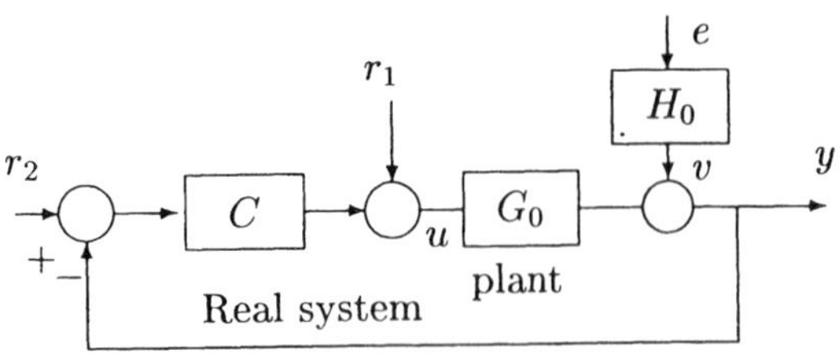

Fig. 1. Closed loop configuration

1.0.0.1. *Plant model G obtained with the direct method* .
The expression of the prediction error ε_t^f obtained from an identification in closed loop of the model G, H and with data filter L, is

$$\varepsilon_t^f = L.H^{-1}\{(G_0 - G)S_0r_t + S^{-1}S_0v_t\} \tag{3}$$

1.0.0.2. *Plant model G obtained with the CLOE method* (Landau and Karimi 1997).
The expression of prediction error is (like shown in (Karimi and Landau 1998)):

$$\varepsilon_t = (G_0 - G)S_0Sr_t + S_0v_t \tag{4}$$

1.0.0.3. *Plant model G obtained with the two-stage method* (Van den Hof and Schrama 1993).
The expression of the prediction error is (Karimi and Landau 1998):

$$\varepsilon_t = (G_0.S_0 - G.\hat{S}_0)r_t + S_0v_t \tag{5}$$

where $\hat{S}_0$ represents the estimate of the real sensitivity function realized in the first stage and (for simplicity), it is considered that in the second stage OE method has been used ($H = 1$).

2. MODEL ERROR MODEL

2.1 *Spectral estimation of filtered model error model*

2.1.1. *Plant model G obtained with the direct method* The following computations are first necessary: simulated output $\hat{y}_t = G.S.r_t = \frac{G}{1+CG}.r_t$, cross spectra $\Phi_{\varepsilon^f,r}$ and $\Phi_{\hat{y},r}$. Then the normalized model error model $\Delta = \dfrac{G_0 - G}{G}$, filtered by the real output sensitivity S_0 can be estimated from :

$$\frac{\Phi_{\varepsilon^f,r}}{\Phi_{\hat{y},r}} = \frac{L}{H}.\frac{G_0 - G}{G}.\frac{S_0}{S} = \frac{L}{H}\frac{1}{S}\Delta.S_0 \tag{6}$$

It can also be seen that expression (6) becomes simpler if OE method ($H = 1$) is used, with no data filter ($L = 1$):

$$\frac{\Phi_{\varepsilon^f,r}}{\Phi_{\hat{y},r}} = \frac{G_0 - G}{G}.\frac{S_0}{S} = \frac{1}{S}\Delta.S_0 \tag{7}$$

This idea has been sketched in (Besançon-Voda 1998, Besançon-Voda and Rey 1999) and is inspired from (Kosut 1996) (validation and spectral estimation).

2.1.2. *Plant model G obtained with the CLOE method* Consider the prediction error expression (4), and apply the same computations ($\hat{y}_t = G.S.r_t$ and spectra) as previously, then (7) can be derived.

2.1.3. *Plant model G obtained with the two-stage method* If in the first stage, a consistent estimate of S_0 has been identified, expression (7) still holds.

Details on computations necessary to obtain (6) and (7) have been omitted.

2.2 *Parametric identification of model error model*

2.2.1. *Identification of filtered model error model* For a plant model obtained with the direct method, and using equation (3), the model error model filtered by the real sensitivity function $(G_0 - G).S_0$ can be estimated using a parametric identification scheme, with r_t as input signal and ε_t^f as output, as r_t is independent of v_t. The same idea can be applied to the model obtained with CLOE method using the prediction error (4) as output and $S.r_t$ as input. As for the two-stage method, if in the first stage a consistent estimate of S_0 has been obtained, then it can be considered that prediction error (5) can be used to identify $(G_0 - G).S_0$, with r_t as input.

2.2.2. *Identification of model error model* Model error $G_0 - G$ can be identified using the two-stage method (Van den Hof and Schrama 1993) modified to cope with our problem. Consider a plant model identified with the direct method. Then

- In the first stage, identify a model $\hat{S}_0$ of the real sensibility function S_0, using (2) and measured signals u_t and r_t. An instrumental variable is then constructed as $(L.H^{-1}).\hat{S}_0 r_t$, which is independent of the noise in (3).

• In the second stage, identify a model $\tilde{G}$ of $G_0 - G$, using the instrumental variable as input and ε_t^f as output, and an OE or BJ model (Box-Jenkins), in order to parameterize independentely the noise model.

The same idea can be used in case of a plant model identified in closed-loop with the two-stage identification method, using prediction errors (5) as output in the second stage of identification of model error model.

In case of model identified with CLOE method, one can use either the instrumental variable $\hat{S}_0.S.r_t$ as input and (4) as output in the second stage, or the instrumental variable $\hat{S}_0 r_t$ as input and $S^{-1}\varepsilon_t$ as ouput.

For any of these methods, the quality of the estimate $\tilde{G}$ in the second stage depends on the quality of the estimate $\hat{S}_0$ in the first stage.

2.3 *Asymptotic variance expression of the model error model estimate*

The asymptotic variance of the model error model identified with the two-stage method (described in paragraph 2.2.2) can be calculated following similar lines as in (Gevers *et al.* 1996). The following expressions depend on the hypothesis that in the first stage, a consistent estimate of S_0 has been obtained.

In case of a plant model obtained with the direct method (and considering $L.H^{-1} = 1$ for simplicity), the asymptotic variance expression results as (after some computations):

$$cov(\tilde{G}) \approx \frac{n_1}{N} \frac{1}{\mid S_0 \mid^2} \frac{\Phi_v}{\Phi_r} \tag{8}$$

where n_1 is the number of parameters of $\tilde{G}$to be identified.

The same expression is obtained in the case of a model identified by CLOE method, if a parametric identification of $\tilde{G}$ is realized, from (4), with $S^{-1}\varepsilon_t$ as output and $\hat{S}_0 r_t$ as input.

In case of the 2-stage method, the asymptotic variance for $\tilde{G}$ becomes (after some computations, see the sketch of proof in the appendix):

$$cov(\tilde{G}) \approx \frac{n_1}{N} \mid S_0^{-1} - S^{-1} + 1 \mid^2 \frac{\Phi_v}{\Phi_r} \tag{9}$$

In (Gevers 1998), a different method to determine $\tilde{G}$ has been presented, using an identification of the closed loop transfer function model error, from the difference between the measured and simulated output:

$$\varepsilon_t = y_t - \hat{y}_t = (G_0 S_0 - GS)r_t + S_0 v_t$$

in the case when $r_t = Cr_2$. Then the model error $\tilde{T}$ of the identified complementary sensitivity function has the property:

$$cov(\tilde{T}) \approx \frac{n_2}{N} \frac{\mid S_0 \mid^2 \Phi_v}{\Phi_{r_2}}$$

where n_2 is the number of parameters of $\tilde{T}$to be identified. It follows that (see the sketch of proof in the appendix):

$$\begin{aligned} cov(\tilde{G}) &\approx \frac{n_2}{N} \frac{\mid S_0 \mid^2}{\mid S \mid^4} \frac{\Phi_v}{\Phi_r} \\ &\approx \frac{n_2}{N} \frac{\mid S_0 \mid^2}{\mid S \mid^4} \frac{1}{\mid C \mid^2} \frac{\Phi_v}{\Phi_{r_2}} \end{aligned} \tag{10}$$

3. AN EXAMPLE

An example of identification of the model error model from a system simulated as a third order OE structure and modeled by a second order (identified in closed-loop with the direct method) is shown. The true plant is:

$$G_0 = \frac{0.1138z^2 - 0.1691z^2 + 0.0621}{z^3 - 2.68z^2 + 2.43z - 0.74}$$

and the identified model is

$$G = \frac{0.09z - 0.0284}{z^2 - 1.74z + 0.77}$$

and it is shown in figure 2. The true model error

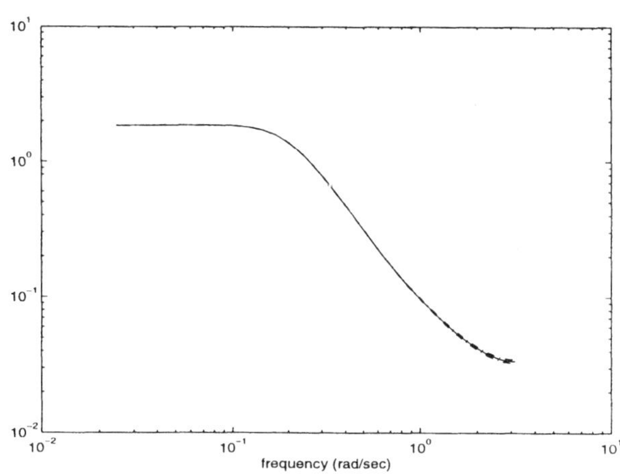

Fig. 2. Model G identified with a direct method in closed-loop.

is then $G_0 - G$

$$\frac{0.02z^5 - 0.019z^4 - 0.03z^3 + 0.017z^2 + 0.02z - 0.01}{z^7 - 0.96z^6 - 0.6z^5 + 0.35z^4 + 0.23z^3 + 0.5z^2 - 0.5z}$$

The identification of the model error model is realized as a 30th order OE model as described in paragraph 2.2.2. Concerning the asymptotic variance presented in section 2.3, the results obtained are shown in figures 3 to 5 as follows: in figure 3 the identified closed loop transfer function T is presented, while in figure 4 the model error of the closed loop transfer function $\tilde{T}$ is shown.

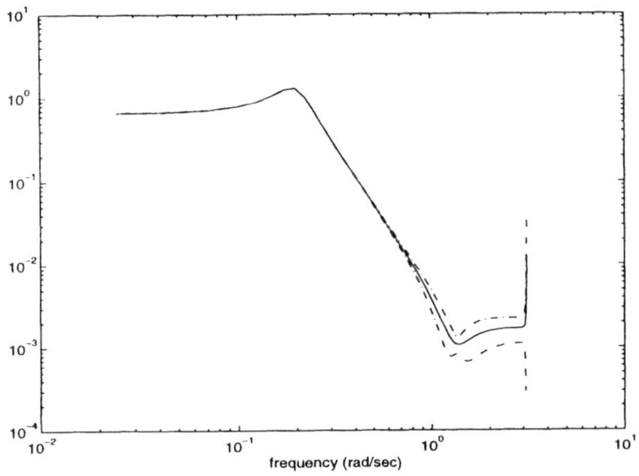

Fig. 3. Identified closed-loop transfer function, T.

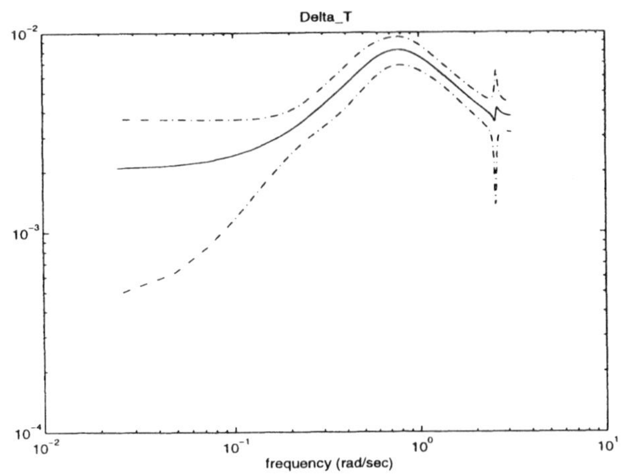

Fig. 4. Model error of the identified closed-loop transfer function, $\tilde{T}$.

The comparison of the covariances of the model error model with expressions (8) and (10) are plotted together in figure 5.

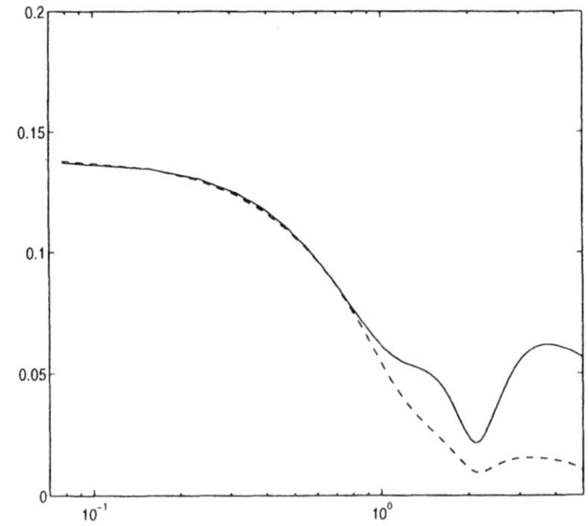

Fig. 5. Covariance of model error model $cov(\tilde{G})$: identified (solid) and derived from closed loop model error model (dashed).

4. CONCLUSION

The present paper presents some issues in identification of the model error for the case when the model has been identified in closed-loop, using one of the three well-known methods: direct, 2-stage or CLOE method. The identification of the model G is considered to be achieved and it is not treated in this paper. However, it is considered that the prediction errors obtained during the identification of the model G are available. Based on this knowledge and depending on the identification method used for obtaining the model G, it is shown that it is possible to identify also the additive model error $G_0 - G$, using a two-stage method. The expression of the variance of this model is presented for several identification in closed-loop methods. The estimated model error model is further useful for model validation purposes or for robust control design, as shown in (Gevers 1998).

APPENDIX

Derivation of (8)

If we apply the two-stage method to identify $\tilde{G}$, and using $L.H^{-1} = 1$ for simplicity, we have the following equations for both stages:

$$\begin{aligned} \varepsilon_t &= (G_0 - G)S_0 r_t + S^{-1} S_0 v_t \\ u_t &= S_0 r_t - C S_0 v_t \end{aligned} \qquad (11)$$

Then a coprime factor structure is applied on the system above

$$\begin{aligned} \varepsilon_t &= N_0 r_t + S^{-1} S_0 v_t \\ u_t &= D_0 r_t - C S_0 v_t \end{aligned} \qquad (12)$$

with $N_0 = (G_0 - G)S_0$ and $D_0 = S_0$, that are a coprime factor representation of $G_0 - G$ as $G_0 - G = N_0 D_0^{-1}$. Furthermore, applying the varinace expressions to the multivariable situation of (12) and following similar lines as in (Gevers *et al.* 1996) where $cov(G)$ is calculated, we obtain

$$cov \begin{pmatrix} N \\ D \end{pmatrix} = \frac{n}{N} \frac{\lambda_0}{\Phi_r} \begin{pmatrix} S^{-1} S_0 H_0 & 0 \\ 0 & -C S_0 H_0 \end{pmatrix} \cdot \cdot \begin{pmatrix} 1 & 1 \\ 1 & 1 \end{pmatrix} \begin{pmatrix} (S^{-1} S_0 H_0))^{\star} & 0 \\ 0 & (-C S_0 H_0)^{\star} \end{pmatrix}$$

Then

$$cov\begin{pmatrix} N \\ D \end{pmatrix} \approx \frac{n}{N}\frac{\lambda_0}{\Phi_r} \mid S_0 H_0 \mid^2 \cdot \cdot \begin{pmatrix} \mid S^{-1} \mid^2 & -C^\star S^{-1} \\ -C(S^{-1})^\star & \mid C \mid^2 \end{pmatrix} \quad (13)$$

For the model error model we obtain:

$$cov(\tilde{G}) \approx \frac{1}{\mid D_0 \mid^2} \frac{n}{N} \frac{\mid S_0 \mid^2 \Phi_v}{\Phi_r} \{\mid S^{-1} \mid^2 + \\ + \mid (G_0 - G)C \mid^2 \\ -2Re((G_0 - G)C(S^{-1})^\star)\} \\ \approx \frac{n}{N}\frac{\Phi_v}{\Phi_r}\{(S^{-1} + (G_0 - G)C)^2\}$$

which finally gives

$$cov(\tilde{G}) \approx \frac{n}{N}\frac{\mid S_0 \mid^2 \Phi_v}{\Phi_r} \quad (14)$$

Derivation of (9)

In this case the following equations are considered

$$\varepsilon_t = (G_0 - G)S_0 r_t + S_0 v_t \\ u_t = S_0 r_t - CS_0 v_t \quad (15)$$

which gives

$$\varepsilon_t = N_0 r_t + S_0 v_t \\ u_t = D_0 r_t - CS_0 v_t \quad (16)$$

with $N_0 = (G_0 - G)S_0$ and $D_0 = S_0$. Then

$$cov\begin{pmatrix} N \\ D \end{pmatrix} = \frac{n}{N}\frac{\lambda_0}{\Phi_r}\begin{pmatrix} S_0 H_0 & 0 \\ 0 & -CS_0 H_0 \end{pmatrix} \cdot \cdot \begin{pmatrix} 1 & 1 \\ 1 & 1 \end{pmatrix}\begin{pmatrix} (S_0 H_0)^\star & 0 \\ 0 & (-CS_0 H_0)^\star \end{pmatrix}$$

or, briefly:

$$cov\begin{pmatrix} N \\ D \end{pmatrix} = \frac{n}{N}\frac{\lambda_0}{\Phi_r} \mid S_0 H_0 \mid^2 \begin{pmatrix} 1 & -C^\star \\ -C & \mid C \mid^2 \end{pmatrix} (17)$$

For the model error model we obtain:

$$cov(\tilde{G}) = \frac{1}{\mid D_0 \mid^2} \frac{n}{N} \frac{\mid S_0 \mid^2 \Phi_v}{\Phi_r}\{1 + \\ + \mid (G_0 - G)C \mid^2 + 2Re((G_0 - G)C)\} \\ = \frac{n}{N}\frac{\Phi_v}{\Phi_r} \mid 1 + (G_0 - G)C \mid^2$$

which finally gives

$$cov(\tilde{G}) = \frac{n}{N}\frac{\Phi_v}{\Phi_r} \mid {S_0}^{-1} - S^{-1} + 1 \mid^2 \quad (18)$$

Derivation of (10)

In order to derive $cov(\tilde{G})$, first see that in that case, $cov(G) = \mathcal{E}((G - \mathcal{E}G).(G - \mathcal{E}G)^T)$, with the zero mean of the estimate $\mathcal{E}G = 0$, it follows that $cov(G) = \mathcal{E}((G(T)).(G(T))^T)$ and approximating $G(T)$ by its Taylor expansion of first order,

$$G(T) = G(0) + \frac{\partial G}{\partial T}(T - 0) = \frac{\partial G}{\partial T}T$$

because $G(0) = 0$ then

$$cov(G) = \mathcal{E}(\frac{\partial G}{\partial T}T).(\frac{\partial G}{\partial T})T)^{\star,T} \\ = (\frac{\partial G}{\partial T})\mathcal{E}(T.T^{\star,T})(\frac{\partial G^\star}{\partial T}) \quad (19)$$

If we take into account that $G = \frac{T}{C(1-T)}$ and that

$$\frac{\partial G}{\partial T} = \frac{1}{C(1-T)^2} = \frac{1}{CS^2}$$

then

$$cov(\tilde{G}) \approx \frac{n_2}{N}\frac{\mid S_0 \mid^2 \Phi_v}{\mid S \mid^4 \Phi_r} \\ \approx \frac{n_2}{N}\frac{\mid S_0 \mid^2}{\mid S \mid^4}\frac{1}{\mid C \mid^2}\frac{\Phi_v}{\Phi_{r_2}} \quad (20)$$

5. REFERENCES

Besançon-Voda, A. (1998). Identification and validation in closed loop. Applications (from LAG): flexible arm and pneumatic actuator. In: *Intern. Summer School on Interaction between Identification and Control.* Grenoble, France.

Besançon-Voda, A. and D. Rey (1999). Issues in identification, validation in closed-loop and control design. In: *5th European Control Conference (ECC 99).* Duisburg, Germany.

Forssel, U. and L. Ljung (1999). Closed-loop identification revisited. *Automatica* **35**(7), 1215–1241.

Gevers, M. (1998). On the role of experimental conditions in identification for control. In: *Intern. Summer School on Interaction between Identification and Control.* Grenoble, France.

Gevers, M., L. Ljung and P. Van den Hof (1996). Asymptotic variance expressions for closed-loop identification and their relevance in identification for control. *Selected Topics in Identification, Modelling and Control* **9**, 9–15.

Karimi, A. and I. D. Landau (1998). Comparison of closed-loop identification methods in terms of bias distribution. *Systems and Control Letters* **34**, 159–168.

Kosut, R. (1996). Adaptation, learning and validation. In: *1996 IFAC Workshop: Identification and Robust Control Interplay.* San Francisco, CA.

Landau, I. D. and A. Karimi (1997). An output error recursive algorithm for unbiased identification in closed-loop. *Automatica* **33**, 933–938.

Ljung, L. (1987). *System Identification - Theory for the user.* Prentice-Hall, Englewood Cliffs, New Jersey.

Ljung, L. and L. Guo (1997). Role of model validation for assessing the size of the unmodeled dynamics. *IEEE Transactions on Automatic Control* **42**(9), 1230–1239.

Van den Hof, P. and R. Schrama (1993). An indirect method for transfer function estimation from closed loop data. *Automatica* **29**(6), 1523–1527.

Zhu, Y.C. (1989). Estimation of transfer functions: asymptotic theory and a bound of model uncertainty. *Int. Journal of Control* **49**(6), 2241–2258.

ON MODEL ERROR MODELING IN SET MEMBERSHIP IDENTIFICATION

A. Garulli * and W. Reinelt **,[1]

* *Dipartimento di Ingegneria dell'Informazione*
Università di Siena, Via Roma 56, 53100 Siena, Italy
E-mail: garulli@ing.unisi.it

** *Department of Electrical Engineering*
Linköping University, S-581 83 Linköping, Sweden
E-mail: wolle@isy.liu.se

Abstract: A recent perspective on model error modeling is applied to set membership identification techniques in order to highlight the separation between unmodeled dynamics and noise. Model validation issues are also easily addressed in the proposed framework. The computation of the minimum noise bound for which a nominal model is not falsified by i/o data, can be used as a rationale for selecting an appropriate model class. Uncertainty is evaluated in terms of the frequency response, so that it can be handled by $\mathcal{H}_\infty$ control techniques. *Copyright ©2000 IFAC*

Keywords: Identification for robust control, set membership, model error modeling, unmodeled dynamics, model validation.

1. INTRODUCTION

One of the main objectives of control-oriented identification is to estimate models that are suitable for robust control design techniques. To this purpose, the identification procedure must deliver not only a nominal model, but also a reliable estimate of the uncertainty associated to the model. Different paradigms for the description of uncertainty have been addressed in the literature (see e.g. the special issues (Kosut *et al.*, 1992*a*; Söderström and Åström, 1995) and the recent book (Garulli *et al.*, 1999)). One of the most popular relies on the hypothesis that the identification error is unknown-but-bounded (UBB) and it is known as *bounded error* or *set membership* identification (see (Milanese and Vicino, 1991; Milanese *et al.*, 1996) and the references therein).

In standard identification problems, the error originates from two different sources: a "variance" term, due to noise affecting the data, and a "bias" term, due to system dynamics which is not captured by the estimated nominal model (often addressed also as the *model error*). Clearly, the nature of these two error terms is quite different: the former is generally uncorrelated with the input signal (when open-loop experiments are considered, while the latter strongly depends on the model class and on the input used in the identification experiment. The "size" of the model error is not negligible in most practical situations, especially those in which the order of the nominal model must be small (a typical requirement of robust design techniques). Moreover, while a priori information on measurement noise is often available, similar hypotheses on the unmodeled dynamics seem to be less realistic.

The first works on set membership identification (Fogel and Huang, 1982; Norton, 1987; Walter and Piet-Lahanier, 1990) provided efficient algorithms for estimating the set of feasible models, compatible with the UBB error assumption, without separating the contribution from unmodeled dynamics and noise (or equivalently, assuming that the plant generating the data belong to the considered model family). Model errors were explicitly accounted for in (Wahlberg and Ljung, 1992; Kosut *et al.*, 1992*b*). Estimation of the size of the model error from residual data has been adopted in the set membership context in (Giarrè *et al.*, 1997), for nominal models obtained via least-squares identification. Separation of unmodeled dynamics

[1] Supported by the European Union within the European Research Network in System Identification (ERNSI), which is gratefully acknowledged.

and noise in a setting including both stochastic and deterministic uncertainties has been studied in (Venkatesh and Dahleh, 1997).

In this paper, it is shown how model error modeling concepts adopted by classical identification techniques (Ljung, 1999*a*; Ljung, 1999*b*) can be easily extended to set membership identification, providing a general strategy for separating error contributions. A model error model identification procedure based on set membership techniques is formulated. In this context, error bounds can be used to extract information about the size of the unmodeled dynamics and the quality of the nominal model. Frequency domain uncertainty regions are considered in order to provide adequate models for $\mathcal{H}_\infty$ control design.

The paper is organized as follows. In Section 2, model error modeling ideas are briefly summarized. Section 3 introduces the main concepts of set membership identification and describes a procedure for identification of both nominal model and model error. The features of the proposed approach are discussed in Section 4, while Section 5 reports an example which clarifies the effectiveness of set membership model error modeling. Finally, Section 6 gives some concluding remarks.

2. MODEL ERROR MODELING

In this section, the basic ideas of model error modeling are briefly resumed (see (Ljung, 1999*b*) for a thorough treatment). Let (u_m, y_m) be a collection of i/o measurements and assume that a nominal model G_n of the system that generated the data has been estimated, according to some identification procedure. Then, the model error modeling strategy can be summarized as follows.

(1) Compute the residual $\epsilon = y_m - G_n u_m$.
(2) Consider the "error" system, with input u_m and output ϵ, and identify a model G_e for this system. This is an estimate of the error due to undermodeling, the so-called *model error model*. Standard statistical properties of the estimated model lead to the definition of the *uncertainty region of the model error* (e.g., given by the 99% confidence region of the estimated model error).
(3) The *nominal model uncertainty region* is obtained by adding up the nominal model and the uncertainty region of the model error.
(4) Model validation: the nominal model is not falsified if and only if it lies inside its own uncertainty region or, equivalently, if and only if zero is an element of the model error uncertainty region. This can be easily checked by looking at the corresponding uncertainty bands in the Bode or Nyquist plots (model validation in the frequency domain).

Identification of the model error from residual data provides a separation between noise and unmodeled dynamics. In fact, G_e can be seen as an estimate of the dynamic system $\Delta(\cdot)$ (possibly nonlinear, time-varying, etc.), such that

$$\epsilon = \Delta(u) + v \tag{1}$$

where v is the noise term, which is assumed to be uncorrelated with the input u.

A few observations can be made.

- The model error G_e must be selected among a pre-specified class of models. Unfortunately, there is not a standard procedure for selecting the structure of the model error. This is rather based on a priori knowledge on the system generating the data, and on the purposes for which the model uncertainty is estimated. For example, a typical requirement is that unmodeled dynamics is captured at those frequencies that are significant for control design purposes.
- The size of the uncertainty region clearly depends on the selected nominal model. When a severe undermodeling occurs (due for example to the fact that we are required to deliver a low order nominal model to the robust control designer), one cannot expect the model error uncertainty region to be small at all frequencies, i.e. the nominal model will be falsified, whatever the structure of the model error.
- A falsified nominal model may still be accepted, provided that the related uncertainty band is small at frequencies of interest. However, it would be useful to evaluate the "minimum" model error uncertainty such that the nominal model is not falsified. As we will see, this can be easily done in the framework of set membership identification.

3. SET MEMBERSHIP IDENTIFICATION

The model error modeling procedure described in the previous section can be applied to any nominal model G_n. In this paper, nominal models provided by set membership identification techniques will be considered. In the following, we summarize the main features of these techniques.

Assume that the measured i/o data are generated by a system S_0, according to

$$y_m = S_0(u_m) + v$$

where S_0 belongs to a set $\mathcal{K}$ (the a priori information on the system) and the noise v is bounded in some norm $\mathcal{Y}$, i.e. $\|v\|_\mathcal{Y} \leq \delta$, for given $\delta > 0$. Then it is possible to define the *feasible system set*

$$FSS = \{S \in \mathcal{K} : \ \|y_m - S(u_m)\|_\mathcal{Y} \leq \delta\} \tag{2}$$

which is the set of systems that are compatible with the measured data and the a priori assumptions. The set FSS can be very complicated, depending on the structure of $\mathcal{K}$ and the norm $\|\cdot\|_{\mathcal{Y}}$. For example, if $\mathcal{K}$ is the class of LTI systems whose impulse response has an assigned exponential decay and noise is ℓ_∞-bounded, then FSS is an infinite dimensional polytope. Obviously, for real-world systems, including nonlinearities or time-varying drifts, FSS may be much more complex.

Since the set FSS contains all the information provided by data and a priori assumptions, it is natural to evaluate the quality of a nominal model G_n according to its worst-case error with respect to elements of FSS. In other words, the *identification error* associated to G_n is given by

$$E(G_n) = \sup_{S \in FSS} \|S - G_n\|_S \qquad (3)$$

where $\|\cdot\|_S$ is a suitable norm in the system space.

In order to identify a nominal model, a model class must be chosen. A common requirement is that the nominal model must be simple (low dimensional, linearly parameterized, etc.). Hence, a typical structure for G_n is

$$G_n(q;\theta) = \sum_{i=1}^{n} \theta_i B_i(q) \qquad (4)$$

where the $B_i(q)$ are user defined basis functions, such as FIR filters, Laguerre or Kautz filters (Wahlberg, 1991), orthonormal basis functions (Van den Hof *et al.*, 1995), etc. If we denote by $\mathcal{M}$ the set of nominal models parameterized as in (4), the problem of selecting a model in $\mathcal{M}$ according to the criterion (3) is usually addressed as *conditional* set membership identification (Garulli *et al.*, 1997). The optimal nominal model is given by the conditional central estimate

$$G_n^* = \arg \inf_{G \in \mathcal{M}} \sup_{S \in FSS} \|S - G\|_S. \qquad (5)$$

In most practical situation, finding an exact solution of the min-max optimization problem (5) is a prohibitive task and computationally tractable alternatives must be looked for. These may include recursive approximations of the set FSS (employing different approximating regions like ellipsoids, orthotopes, parallelotopes, see e.g. (Fogel and Huang, 1982; Walter and Piet-Lahanier, 1990; Chisci *et al.*, 1998)), and/or suboptimal pointwise estimators like projection algorithms, interpolatory algorithms, etc. The evaluation of bounds on the identification error of these algorithms has been tackled in different contexts and several results are now available (see e.g. (Garulli *et al.*, 1998; Milanese and Taragna, 1999)). On the other hand, these bounds may be conservative in many situations of interest for several reasons. First, the identification error (3) depends on FSS, whose computation is often untractable. Moreover, the tightness of these bounds is related to that of the noise bound δ: if the latter is overestimated, the former will also be excessively large.

Whatever nominal model has been identified using the above procedures, the actual size of the unmodeled dynamics can be evaluated from available data via the error modeling technique described in Section 2. The identification of the error system (1) can be performed using set membership identification algorithms, exploiting the noise bound on v. If the structure of the model error model is chosen as

$$G_e(q;\bar{\theta}) = \sum_{i=1}^{\bar{n}} \bar{\theta}_i \bar{B}_i(q) \qquad (6)$$

the feasible set for the model error parameters is given by

$$FES = \{\bar{\theta} \in \mathbb{R}^{\bar{n}} : \|\epsilon - \sum_{i=1}^{\bar{n}} \bar{\theta}_i \bar{B}_i(q) u_m\|_{\mathcal{Y}} \leq \delta\}. \qquad (7)$$

Then, one may select the worst-case optimal error model by computing the Chebishev center of FES

$$\bar{\theta}^* = \arg \inf_{\tilde{\theta} \in \mathbb{R}^{\bar{n}}} \sup_{\bar{\theta} \in FES} \|\tilde{\theta} - \bar{\theta}\| \qquad (8)$$

where $\|\cdot\|$ denotes the Euclidean norm.

As for (5), problem (8) may be computationally unfeasible, and suboptimal solutions are sought. For example, let us consider the case of ℓ_∞-bounded noise. Since FES is a polytope in $\mathbb{R}^{\bar{n}}$, it can be recursively approximated by ellipsoids or parallelotopes, and the center of these approximating sets may be chosen as an estimate of $\bar{\theta}^*$.

If robust identification is oriented to $\mathcal{H}_\infty$ control, an uncertainty band associated to the nominal model frequency response must be delivered to the control designer. In this framework, the frequency plot of the error model uncertainty region can be obtained by mapping the set FES onto the complex plane for each frequency of interest. This leads to the frequency domain uncertainty set

$$\mathcal{V}(FES) = \bigcup_{\omega} \mathcal{V}_\omega(FES)$$

where

$$\mathcal{V}_\omega(FES) = \{z \in \mathbb{C} : \; z = A_\omega \bar{\theta}, \; \bar{\theta} \in FES\} \qquad (9)$$

and $A_\omega = [\bar{B}_1(e^{j\omega}) \ldots \bar{B}_{\bar{n}}(e^{j\omega})] \in \mathbb{C}^{1\times\bar{n}}$. Once again, the computation of $\mathcal{V}(FES)$ may be a formidable task if the exact FES defined by (7) is considered. Hence, set approximations are useful also in this respect.

The overall set membership identification strategy can be summarized as follows.

(1) Identify the nominal model $G_n(q;\theta)$, using conditional estimators based on the feasible system set (2).
(2) Compute the residual $\epsilon = y_m - G_n(q;\theta)u_m$.
(3) Select a model error structure and compute (or approximate) the feasible model error set (7).
(4) Identify a nominal model error $G_e(q;\bar{\theta})$, using optimal or suboptimal estimators based on FES.
(5) Map the nominal model plus the model error and its uncertainty region onto the frequency domain.

Remark. The use of residual data in set membership identification for evaluating the worst-case $\mathcal{H}_\infty$ norm of the unmodeled dynamics has been introduced in (Giarrè *et al.*, 1997), for standard least-squares nominal models. The above set membership model error modeling strategy can be seen as a general framework in which the structures of the nominal and error models, and the corresponding identification algorithms must be chosen by the user according to the specific problem (a priori knowledge, noise bound, error norm, etc.) In the next section, it will be shown how it is possible to exploit this framework to obtain a useful measure of the size of the unmodeled dynamics. This will also provide useful information on the selection of the nominal model structure.

4. MODEL VALIDATION AND MODEL CLASS SELECTION

When identification of the model error model is performed via standard prediction error methods, one can choose the 99% confidence region of the model error as the uncertainty set to be added to the nominal model (Ljung, 1999*b*). However, it is very likely that the nominal model does not lie inside the resulting uncertainty region, whenever a remarkable undermodeling occurs.

For this reason, it is useful to evaluate the "distance from validation" of the nominal model. In the following, this is done in the set membership setting of Section 3. Let us consider the set $\mathcal{V}_\omega(FES)$ in (9) and define

$$d(\omega) = \min_{z \in \mathcal{V}_\omega(FES)} |z|. \tag{10}$$

If we set

$$d = \sup_\omega d(\omega) \tag{11}$$

then it is clear that the nominal model can be deemed to be unfalsified if $d = 0$. If ellipsoidal approximations of FES are considered, the sets $\mathcal{V}_\omega(FES)$ are ellipses in the complex plane and the optimization problem (10) can be easily solved. Then, d in (11) can be approximated by taking the maximum over a finite number of frequencies in the range of interest.

It is worth observing that d is a function of δ, since FES (and its frequency image $\mathcal{V}_\omega(FES)$) depend on the noise bound. Then, one can look for the minimum value of δ for which the nominal model is still not falsified by data, i.e.

$$\delta^* = \min_{\{\delta:\ d=0\}} \delta. \tag{12}$$

Since d is an increasing function of δ, the computation of δ^* can be easily performed within the desired precision, using a standard bisection on δ.

It is believed that the quantity δ^* is important in several respects.

- It allows one to deliver the smallest frequency domain uncertainty region associated to the model error model, which does not falsify the nominal model. This meets a typical requirement of the control designer: the "tightest" uncertainty band around the nominal model, without contradicting the information given by data.
- It can be used to correct the a priori noise bound. In fact, if δ^* turns out to be smaller than the a priori noise bound δ, one may conclude that this bound was overestimated.
- It gives a rationale for selecting the structure of the nominal model. Let us denote by p a vector of parameters that define a model class for G_n (for example, the number and/or the pole locations of the family of basis functions $B_i(q)$). For each p, one can perform set membership identification of nominal model and model error model as suggested in Section 3, and compute $\delta^*(p)$. Then, at least in principle, one can choose the optimal model structure by computing

$$p^* = \arg\inf_p \delta^*(p). \tag{13}$$

In practice, the optimization in (13) can be performed over a finite set $\mathcal{P} = \{p_1 \dots p_m\}$ (see the example in the next section, where p is the pole of a Laguerre expansion of fixed order).

5. EXAMPLE

Let us consider the same plant as in (Goodwin *et al.*, 1999), fed with an harmonic input signal. The output is corrupted by additive white noise, normally distributed with variance 0.0625. Notice that this is an unfavorable situation in the set membership framework, since the ℓ_∞ norm of the noise v is about 1, but only 10 samples out of 4000 used in the identification experiment are larger than 0.7. This means that, except for a few "outliers" the noise bound $\delta = 1$ is overestimated, while it is well known that set membership identification algorithms enjoy nice properties when the noise is boundary visiting.

Laguerre expansions of order $n = 4$ and $\bar{n} = 10$ are selected as nominal model class $\mathcal{M}$ and model error model class, respectively. The pole of both Laguerre bases is chosen as $p = 0.96$ (this choice will be discussed later). As a nominal model, we take $G_n(q, \theta^r)$, where

$$\theta^r = \arg \inf_{\theta \in \mathbb{R}^n} \|y_m - \sum_{i=1}^{n} \theta_i B_i(q) u_m\|_\infty.$$

This is known as restricted projection estimate and can be easily computed by linear programming (its properties in the worst-case setting are discussed in (Garulli *et al.*, 1998)). The model error model $G_e(q; \bar{\theta})$ is estimated by recursively approximating the set FES in (7) by ellipsoids, and then taking $\bar{\theta}$ as the center of the ellipsoid (in (7), we set $\|\cdot\|_{\mathcal{Y}} = \ell_\infty$ and $\delta = 1$).

Figure 1 shows the resulting frequency uncertainty region for both the nominal model and the model error, as suggested by the model error modeling approach outlined in Section 2. The true plant is also reported (dashed line in the upper plot) in order to show that a real undermodeling occurs at high frequencies. Nevertheless, the uncertainty is pretty narrow at frequencies up to $\omega = 5\ rad/s$, and from the lower plot it is evident that the nominal model is not falsified by the data.

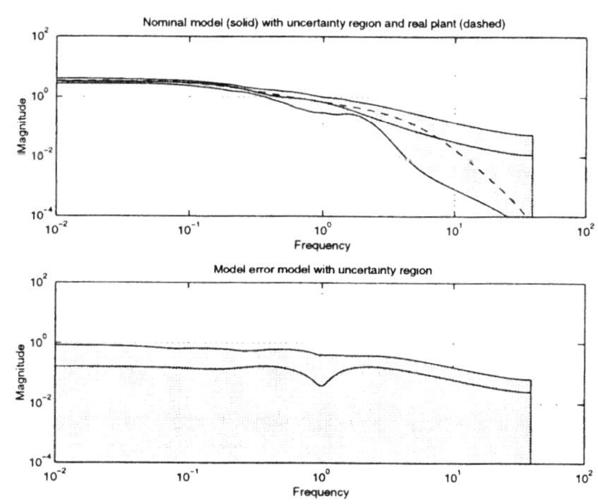

Fig. 1. Nominal model and model error model for a Laguerre expansion with pole $p = 0.96$ ($\delta = 1$). The true plant is shown in the upper plot (dashed line).

Let us consider δ^* in (12). From numerical optimization over δ, one gets $\delta^* \approx 0.82$. Loosely

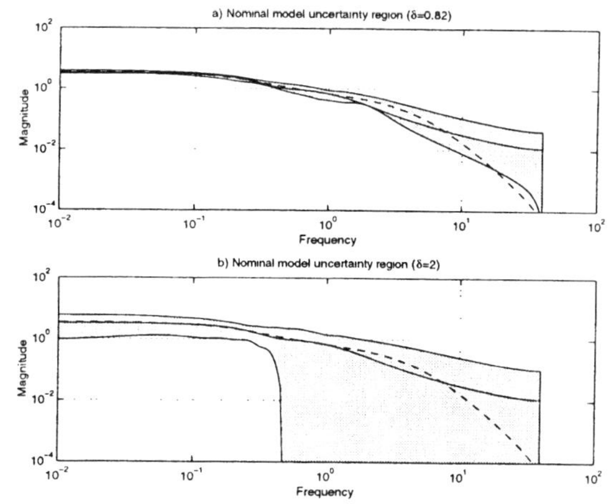

Fig. 2. Nominal model uncertainty bands for a Laguerre expansion with pole $p = 0.96$. (a) $\delta = \delta^*$, (b) $\delta = 2$.

speaking, this "confirms" the a priori assumption on the noise bound. The nominal model and uncertainty band for $\delta = \delta^*$ are shown in Figure 2a, while the corresponding plot for an overestimated noise bound ($\delta = 2$) is reported in Figure 2b. In the latter case, the actual value of δ^* allows one to deem the obtained uncertainty region unreasonably large.

An obvious criticism is that a rigorous way for selecting the model classes for both the nominal and the model error model is lacking. As said above, the structure of the model error model should rely on the purpose for which the model is used. For example, if one wants to estimate more carefully the unmodeled dynamics at high frequencies, a possible choice is a Kautz model with two real poles in 0.96 and 0.7. The result is shown in Figure 3 and it presents a wider uncertainty band for $\omega > 5\ rad/s$, as one might expect.

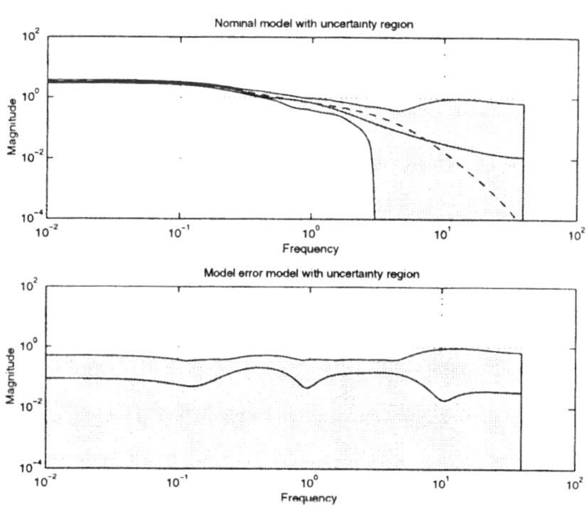

Fig. 3. Nominal model: Laguerre expansion with pole $p = 0.96$; model error model: Kautz expansion with poles $p = 0.96, 0.7$ ($\delta = 1$).

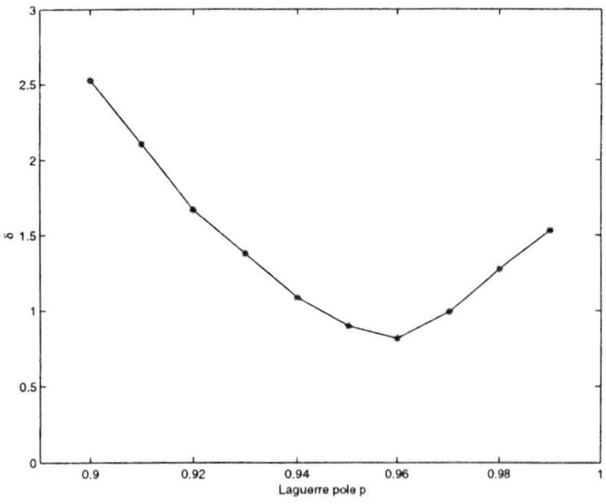

Fig. 4. Different values of δ^* as a function of the Laguerre pole p.

As long as the nominal model structure is concerned, the choice can be based on the value of δ^*, as explained in Section 4. In Figure 4, δ^* is plotted for different values of the Laguerre pole p, in the case when a Laguerre expansion is used for both nominal and model error model ($n = 4$, $\bar{n} = 10$). It can be seen that $p = 0.96$ guarantees model validation with the lowest noise bound ($\delta \approx 0.82$). This corresponds to the tightest uncertainty band around the nominal model, among the considered nominal model structures (see Figs. 1 and 2).

6. CONCLUSIONS AND FUTURE WORK

General model error modeling ideas and their application in the context of set membership identification have been analyzed. In particular, it has been shown that the separation of noise and unmodeled dynamics is quite natural in this framework, and that the minimum noise bound for which a nominal model is not falsified by the data can be easily computed.

A lot of work is still to be done, in several directions. For example, more complicated model error model structures are needed, to cope with nonlinearities or time-varying drifts. Obviously, this requires more sophisticated identification algorithms and smarter approximations in the computation of the uncertainty regions. Moreover, accurate criteria for the a priori selection of the structure of the model error model, depending on the specific control-oriented identification problem, must still be investigated.

7. REFERENCES

Chisci, L., A. Garulli, A. Vicino and G. Zappa (1998). Block recursive parallelotopic bounding in set membership identification. *Automatica* **34**(1), 15–22.

Van den Hof, P. M. J. Van, P. S. C. Heuberger and J. Bokor (1995). System identification with generalized orthonormal basis functions. *Automatica* **31**(12), 1821–1834.

Fogel, E. and F. Huang (1982). On the value of information in system identification - bounded noise case. *Automatica* **18**, 229–238.

Garulli, A., A. Tesi and A. Vicino (Eds.) (1999). *Robustness in Identification and Control.* Lecture Notes in Control and Information Sciences. Springer. London.

Garulli, A., A. Vicino and G. Zappa (1997). Conditional central algorithms for worst-case estimation and filtering. In: *Proc. of 36th IEEE CDC.* San Diego. To appear in *IEEE Trans. on Aut. Control.*

Garulli, A., B. Z. Kacewicz, A. Vicino and G. Zappa (1998). Properties of conditional algorithms in restricted complexity set-membership identification. In: *Proc. of 37th IEEE CDC.* Tampa. To appear in *IEEE Trans. on Aut. Control.*

Giarrè, L., M. Milanese and M. Taragna (1997). $\mathcal{H}_\infty$ identification and model quality evaluation. *IEEE Transactions on Automatic Control* **42**(2), 188–199.

Goodwin, G. C., J. H. Braslavsky and M. M. Seron. (1999). Non-stationary stochastic embedding for transfer function estimation. In: *Proc. of the 14th IFAC World Congress.* Bejing, China.

Kosut, R. L., G. C. Goodwin and M. P. Polis (Eds.) (1992*a*). Special issue on system identification for robust control design. *IEEE Transactions on Automatic Control* **37**, 899–1008.

Kosut, R. L., M. K. Lau and S. P. Boyd (1992*b*). Set-membership identification of systems with parametric and nonparametric uncertainty. *IEEE Transactions on Automatic Control* **37**(7), 929–941.

Ljung, L. (1999*a*). Comments on model validation as set membership identification. In: *Robustness in Identification and Control* (A. Garulli, A. Tesi and A. Vicino, Eds.). Springer. pp. 7–16.

Ljung, L. (1999*b*). Model validation and model error modeling. In: *Proc. of the Åström Symposium on Control.* Lund, Sweden. pp. 15–42.

Milanese, M. and A. Vicino (1991). Optimal estimation theory for dynamic systems with set membership uncertainty: an overview. *Automatica* **27**(6), 997–1009.

Milanese, M. and M. Taragna (1999). SM identification of model sets for robust control design from data. In: *Robustness in Identification and Control* (A. Garulli, A. Tesi and A. Vicino, Eds.). Springer. pp. 17–34.

Milanese, M., J. P. Norton, H. Piet-Lahanier and E. Walter (eds.) (1996). *Bounding Approaches to System Identification.* Plenum Press. New York.

Norton, J. P. (1987). Identification and application of bounded-parameter models. *Automatica* **23**, 497–507.

Söderström, T. and K. J. Åström (Eds.) (1995). Special issue on trends in system identification. *Automatica.*

Venkatesh, S. R. and M. A. Dahleh (1997). Identification in the presence of classes of unmodeled dynamics and noise. *IEEE Transactions on Automatic Control* **42**(12), 1620–1635.

Wahlberg, B. (1991). System identification using Laguerre models. *IEEE Transactions on Automatic Control* **39**, 1276–1282.

Wahlberg, B. and L. Ljung (1992). Hard frequency-domain model error bounds from least-squares like identification techniques. *IEEE Transactions on Automatic Control* **37**(7), 900–912.

Walter, E. and H. Piet-Lahanier (1990). Estimation of parameter bounds from bounded-error data: a survey. *Mathematics in Computer and Simulation* **32**, 449–468.

PRELIMINARY TEST FOR NONLINEAR INPUT OUTPUT RELATIONS IN SISO SYSTEMS

Torben Knudsen

Aalborg University, Institute of Electronic Systems, Department of Control Engineering, Fredrik Bajersvej 7, DK-9220 Aalborg Ø, Denmark, E-mail: tk@control.auc.dk

Abstract: This paper discusses and develops preliminary statistical tests for detecting nonlinearities in the deterministic part of SISO systems with noise. The most referenced method is unreliable for common noise processes as e.g. colored. Therefore two new methods based on superposition and sinus input are developed. They are much more robust, especially the sinus method which is reliable also for colored, heavy tailed or skew distributed noise. *Copyright © 2000 IFAC*

Keywords: nonlinearity test; identification; Subba Rao test; robust methods;

1. INTRODUCTION

Deriving nonlinear models from experimental data are much more complicated than deriving linear models. Consequently, if there is doubt about the existence or significance of nonlinear effects a good preliminary test for linearity would be very useful (Haber and Unbehauen, 1990, p. 652). Existing tests of this kind (Billings and Voon, 1983) suffers from lack of robustness and do not have well defined statistical properties.

Therefore the objectives in this paper are: To develop simple robust tests based on basic properties for linear systems. And to compare the statistical properties for these test with the existing ones.

Below the problem is first discussed in more details. Then existing methods are explained. This is followed by the development of new methods. The statistical properties are then analyzed and finally a conclusion is drawn.

2. CHOICE OF PROBLEM

Some notation are first introduced. Assume N samples of input u and output y are measured from a SISO system (3a) and collected in vectors U and Y (1)-(2).

$$U \triangleq \left(u(1)\ \dots\ u(N)\right)^T \tag{1}$$

$$Y \triangleq \left(y(1)\ \dots\ y(N)\right)^T \tag{2}$$

To get a solvable problem the basic assumptions (3) seems necessary i.e. noise additive, stationary and independent of input and initial conditions known in some way (3c).

$$y(t) = f_t(u_1^t) + v(t)\ ,\ t \in \mathbb{N} \tag{3a}$$

$$u_1^t \triangleq \left(u(1)\ \dots\ u(t)\right)^T \tag{3b}$$

$$u(t) = u(1) \quad \forall\, t < 1 \tag{3c}$$

$$u, v \text{ are independent} \tag{3d}$$

$$v \text{ stationary} \tag{3e}$$

$$\mathrm{E}(v(t)) = 0\ ,\ \mathrm{Var}(v(t)) = \sigma_v^2 \tag{3f}$$

The choices specifying the problem are the following:

- *Preliminary* analysis methods i.e. method based on data before a candidate model is found or *validation* methods based on a candidate model.
- Assumptions about u.
- Assumptions about v as e.g. white noise and normal distribution.

- Proper *statistical* test or tests based on *indicators* with no precise statistical properties.

In this paper the choices are a proper preliminary statistical test method whit as few assumptions on u and v as possible.

3. EXISTING METHODS

For time series this problem is well covered and there has been developed a number of methods starting with the Subba Rao (SR) method (Tong, 1990, sec. 5.3). This method is based on the fact that third order moments of linear combinations of white noise e with symmetrical distribution f_e are zero.

$$v(t) = \sum_{i=0}^{\infty} a_i e(t-i) \ , \ f_e(x) = f_e(-x) \Rightarrow \quad (4)$$
$$\mathrm{E}(v(t_1)v(t_2)v(t_3)) = 0 \quad \forall t_1, t_2, t_3 \in \mathbb{Z}$$

There also exist a well known statistical method of validation type where specific nonlinear terms can be tested (Ljung, 1999; Söderström and Stoica, 1989; Billings and Voon, 1986). A very different validation test of indicator type is found in Knudsen (1996).

The literature covering preliminary methods for systems with measurable input is more limited. The reference text books on system identification (Ljung, 1999; Söderström and Stoica, 1989) only gives a brief discussion on the problem and some references. For purely deterministic systems 10 indicator methods are presented in Haber (1985). The statistical test methods for noisy systems are based on the ideas from the Subba Rao test. In Billings and Voon (1983) it is shown that the third order correlation (6) are zero for linear systems under the assumptions (5).

$$y(t) = f_t(u_1^t) + v(t) \quad (5a)$$
$$u(t) = u_c(t) + b \ , \ b \neq 0 \quad (5b)$$
$$u_c, v \text{ independent} \quad (5c)$$
$$\mathrm{E}(v) = \mathrm{E}(u_c) = 0 \quad (5d)$$
$$\mathrm{E}(v^{2n+1}) = \mathrm{E}(u_c^{2n+1}) = 0 \quad \forall n \in \mathbb{N} \quad (5e)$$
$$v(t) \text{ is a linear stochastic process (4)} \quad (5f)$$

$$f_t \text{ is linear in } u_1^t \Leftrightarrow$$
$$\mathrm{E}(y_c(t+\tau)y_c(t)^2) = 0 \quad \forall \tau \in \mathbb{Z} \ , \quad (6)$$
$$y_c(t) = y(t) - \mathrm{E}(y(t))$$

The test suggested in Billings and Voon (1983) is then to reject linearity if the estimated correlation $|\hat{\phi}_{y_c y_c^2}(\tau)|$ corresponding to (6) is to large (7). The correlation estimate is defined in the usual way (8).

$$\text{reject linearity if } |\hat{\phi}_{y_c y_c^2}(\tau)| > \frac{1.96}{\sqrt{N}} \quad (7)$$

$$\hat{\phi}_{xy}(\tau) \triangleq \frac{\widehat{\mathrm{Cov}}_{xy}(\tau)}{\sqrt{\widehat{\mathrm{Cov}}_{xx}(0)\widehat{\mathrm{Cov}}_{yy}(0)}} \ , \quad (8)$$

$$\widehat{\mathrm{Cov}}_{xy}(\tau) \triangleq$$
$$\begin{cases} \frac{1}{N}\sum_{t=1}^{N-\tau}(x(t)-\bar{x})(y(t+\tau)-\bar{y}) & ,\tau \geq 0 \\ \widehat{\mathrm{Cov}}_{yx}(-\tau) & ,\tau < 0 \end{cases} ,$$
$$\bar{x} \triangleq \frac{1}{N}\sum_{t=1}^{N} x(t)$$

Notice that the test (7) is only approximate because it requires the additional assumption that y_c and y_c^2 must be white noise (Ljung, 1999, sec. 16.6) which normally is not the case.

It is not clear in Billings and Voon (1983) which τ to use and what to do if only one or few correlations falls outside the limits. In this paper $|\tau| \leq 12$ is used because this is the number used in the examples in Billings and Voon (1983). To obtain a p-value it is chosen to use the χ^2 test (9)–(10) where $F^{-1}_{\chi^2(2\tau_m+1)}$ is the inverse distribution function for a χ^2 distribution with $2\tau_m+1$ degrees of freedom. This test is consistent with (7) because it requires the same assumptions.

$$T = N \sum_{|\tau| \leq \tau_m} \hat{\phi}_{y_c y_c^2}(\tau)^2 \ , \ \tau_m = 12 \quad (9)$$
$$p = 1 - F^{-1}_{\chi^2(2\tau_m+1)}(T) \quad (10)$$

The main problem with the method above is that only the limiting conditions (6) is proven. Moreover there is no results concerning the statistical properties of the suggested test.

4. NEW METHODS

As seen in (7) the input u does not explicitly enter the test at all, this is because it is based on a time series approach. The methods developed here should make more use of the input and the basic properties for linear systems because this will probably be more power full.

4.1 *Non-Linearity test based on superposition*

The simplest solution is first to choose an appropriate input sequence U. Run this two times through the system and run $2U$ through the system. This produces the following signals where V_i is the noise sequence corresponding to U_i.

$$U_2 \triangleq U_1 \triangleq U \ , \ U_3 \triangleq 2U$$
$$Y_i = F(U_i) + V_i \ , \ i = 1,2,3$$

Introduce also the differences (11) and (12) then the following basic theorem is easily obtained.

$$D_1 \triangleq Y_2 - Y_1 = V_2 - V_1 \quad (11)$$

$$D_2 \triangleq Y_3 - Y_2 - Y_1 = F(2U) - 2F(U) + V_3 - V_2 - V_1 \quad (12)$$

Theorem 1. (Sup. method with 3 seq. (SUP3)). Assume the system is given by (3) and the sequences V_1, V_2, V_3 are independent, then:

$$\begin{aligned} &D_1, D_2 \text{ are uncorrelated} \\ &D_1, D_2 \text{ are independent if } V_i \text{ are normal} \end{aligned} \quad (13)$$

$$\frac{\mathrm{E}(D_1^T D_1)}{2N} = \sigma_v^2 \quad (14)$$

$$\frac{\mathrm{E}(D_2^T D_2)}{3N} \geq \sigma_v^2 \quad (15)$$

$$F(2U) = 2F(U) \Leftrightarrow \frac{\mathrm{E}(D_2^T D_2)}{3N} = \sigma_v^2 \quad (16)$$

Proof. (13) is proven by using independence for V_1, V_2, V_3, U and $\mathrm{E}(V_2 V_2^T) = \mathrm{E}(V_1 V_1^T)$ which gives

$$\mathrm{E}(D_1 D_2^T) = \mathrm{E}((V_2 - V_1)(F(2U) - 2F(U) + V_3 - V_2 - V_1)^T) = 0$$

(15)–(16) are proved by

$$\mathrm{E}(D_2^T D_2) = \mathrm{E}((F(2U) - 2F(U))^T (F(2U) - 2F(U))) + N3\sigma_v^2$$

Finally (14) follows directly from (11). □

The test for linearity is now performed by testing superposition (17) which reduces to comparing variances for the stochastic processes $d_1(t)$ and $\sqrt{2/3}d_2(t)$ (18) where the means are assumed zero and $d_i(t)$ are the elements in D_i. This variance test is discussed after the next section.

$$\begin{aligned} H_0 &: F(2U) = 2F(U) \\ H_1 &: F(2U) \neq 2F(U) \end{aligned} \quad (17)$$

$$\begin{aligned} H_0 &: \mathrm{Var}(\sqrt{2/3}\, d_2) = \mathrm{Var}(d_1) \\ H_1 &: \mathrm{Var}(\sqrt{2/3}\, d_2) > \mathrm{Var}(d_1) \end{aligned} \quad (18)$$

There is however at least one type of nonlinearities (38) and input which the above test is insensitive to. In this case the similar method below with 5 sequences can be used.

$$U_2 \triangleq U_1 \triangleq U \ , \ U_3 \ , \ U_4 \ , \ U_5 \triangleq U_3 + U_4$$
$$Y_i = F(U_i) + V_i \ , \ i = 1,2,3,4,5$$

$$D_1 \triangleq Y_2 - Y_1 = V_2 - V_1 \quad (19)$$

$$\begin{aligned} D_2 &\triangleq Y_5 - Y_4 - Y_3 \\ &= F(U_5) - F(U_4) - F(U_3) \quad (20) \\ &\quad + V_5 - V_4 - V_3 \end{aligned}$$

Theorem 2. (Sup. method with 5 seq. (SUP5)). Assume the system is given by (3) and the sequences V_1, V_2, V_3, V_4, V_5 are independent, then theorem 1 also applies to D_1, D_2 defined for five sequences (19)–(20) except that D_1, D_2 is always independent and the linearity condition in (16) is now $F(U_5) = F(U_4) + F(U_3)$.

Proof. Similar to the proof for theorem 1 □

4.2 *Non-Linearity test based on sinus input*

The sinus response method is very similar to the superposition method above. Here the difference between the output and a estimated sinus is compared to the noise variance.

Assume that the input is now a sinus (21) with appropriate choices of a and ω. Below some of the previous notation are reused.

$$u(t) = a\sin(\omega t) \ , \ t \in \mathbb{Z} \quad (21)$$
$$U_2 \triangleq U_1 \triangleq U$$
$$Y_i = F(U_i) + V_i \ , \ i = 1,2$$

Introduce also the differences (22) and (24) below where $\widehat{Y}_i$ is a simple LS estimate. This then gives the next theorem.

$$D_1 \triangleq Y_2 - Y_1 = V_2 - V_1 \quad (22)$$

$$X \triangleq \begin{pmatrix} \sin(\omega) & \cos(\omega) \\ \sin(2\omega) & \cos(2\omega) \\ \vdots & \vdots \\ \sin(N\omega) & \cos(N\omega) \end{pmatrix}$$

$$\hat{\theta}_i \triangleq (X^T X)^{-1} X^T Y_i \ , \ i = 1,2 \quad (23)$$
$$\widehat{Y}_i \triangleq X\hat{\theta}_i \ , \ i = 1,2$$
$$D_2 \triangleq (Y_2 - \widehat{Y}_2) + (Y_1 - \widehat{Y}_1) \quad (24)$$

Theorem 3. (Sinus input method (SIN)). Assume the system is given by (3) except for (3c) whish is replaced by (21) and the sequences V_1, V_2 are independent, then:

$$\frac{\mathrm{E}(D_1^T D_1)}{2N} = \sigma_v^2 \quad (25)$$

assymtotically for $N \to \infty$

$$\frac{\mathrm{E}(D_2^T D_2)}{2N} \geq \sigma_v^2 \quad (26)$$

$$\text{linear system} \Rightarrow \frac{\mathrm{E}(D_2^T D_2)}{2N} = \sigma_v^2 \quad (27)$$

$$D_1, D_2 \text{ are uncorrelated} \quad (28)$$

Remark 3.1. In the statistical tests it is convenient to have D_1 and D_2 uncorrelated which is the reason for the specific choice (24) of D_2.

Proof. If the system is linear then $f_t(u_1^t) = \alpha\sin(\omega t) + \beta\cos(\omega t)$ for some α, β and, because regressors are uncorrelated with noise, the LS estimate is consistent i.e. the following mean square (ms) convergence applies:

$$\hat{\theta}_i \to (\alpha \ \beta)^T \ , \ \hat{y}_i(t) \to f_t(u_1^t) \ ,$$
$$y(t) - \hat{y}_i(t) \to v_i(t) \ , \ d_2(t) \to v_2(t) + v_1(t)$$
$$\text{(ms) for } N \to \infty$$

which proves (27) and (28). (25) and (26) are now straight forward. □

The test (31) following from theorem 3 is similar to (18) except no scaling is required.

4.3 *Statistical test comparing two variances*

The above developed linearity tests are both turned into test of equal variances for two stochastic processes which are zero mean under H_0. Adopting the notation from the sinus input method the problem is the following: Given data (29) and basic assumptions (30) find tests for (31).

$$d_i(t) \ , \ t = 1, \ldots N \ , \ i = 1, 2 \tag{29}$$

$$\begin{aligned} &d_i \text{ stationary } , \ \mathrm{E}(d_i(t)) = 0 \ , \\ &d_1(t_1), d_2(t_2) \text{ uncorrelated } \quad \forall t_1, t_2 \end{aligned} \tag{30}$$

$$\begin{aligned} H_0 &: \mathrm{Var}(d_2) = \mathrm{Var}(d_1) \\ H_1 &: \mathrm{Var}(d_2) > \mathrm{Var}(d_1) \end{aligned} \tag{31}$$

In all cases d_i are linear combinations of independent system noise processes v_i. Thus under H_0 d_i's are all zero mean and with the same autocorrelation function as $v(t)$. Therefore the assumption about second order properties for the noise v_i also applies to d_i under H_0 which is very convenient.

To develop a proper statistical test a test statistic and its distribution under H_0 are required. The basic assumptions (30) are really minimal for this purpose. Below a number of test are presented with decreasing restrictive assumptions i.e. increasing robustness.

Normal white (NW): The noise samples $v_i(t)$ are normal and mutually independent with variance σ_v^2 ($\mathrm{NID}(0, \sigma_v^2)$) then the standard F test (32) and p-value (33) can be applied.

$$H_0 \Rightarrow d_i(t) \in \mathrm{NID}(0, \sigma_d^2) \wedge d_1, d_2 \text{ independent} \Rightarrow$$

$$T = \frac{\sum_{t=1}^{N} d_2(t)^2}{\sum_{t=1}^{N} d_1(t)^2} \in F(N, N) \Rightarrow \tag{32}$$

$$p = 1 - F^{-1}_{F(N,N)}(T) \tag{33}$$

Normal colored (N): The noise v_i is a linear normal process (4) with a unknown correlation structure. In this case the results above does not hold. The alternative approach used here is to find a correlation length n_r where correlation between $d_i(t)$ and $d_i(t+\tau)$ is small for $\tau \geq n_r$ and then resample every n_r sample and finally use the NW method on the decimated sequences. The correlation length n_r is estimated by (34) where the lag one autocorrelation $\hat{\rho}_1$ is the standard correlation coefficient estimate for d_1. This method is based on an AR(1) approximation. However it has proven robust in simulation experiments at the cost of reduced number of samples and thereby power. The value $\rho_l = 0.31$ is of cause subject for adjustment.

$$|\hat{\rho}_1|^{n_r} = \rho_l \Leftrightarrow n_r = \frac{\log(\rho_l)}{\log(|\hat{\rho}_1|)} \ , \ \rho_l = 0.31 \tag{34}$$

White (W): The noise samples $v_i(t)$ is mutually independent with variance σ_v^2 and equal but unknown distribution ($\mathrm{ID}(0, \sigma_v^2)$). Then the non-parametric squared ranks test (35) is used with the normal approximation for the test statistic. d_1 and d_2 are ranked together and $R(d_2(t))$ are the ranks for d_2, see Conover (1999, sec. 5.3) for details.

$$T = \frac{S - \mu_S}{\sigma_S} \ , \ S = \sum_{t=1}^{N} R(d_2(t))^2 \ , \tag{35a}$$

$$\mu_S = \frac{N(2N+1)(4N+1)}{6} \ , \tag{35b}$$

$$\sigma_S = \sqrt{\frac{N^2(2N+1)(4N+1)(16N+11)}{180}} \tag{35c}$$

$$\begin{aligned} &F_{d_1}(x) = F_{d_2}(x) \ \forall x \ \wedge \\ &d_1, d_2 \text{ independent} \end{aligned} \quad \Rightarrow \tag{35d}$$

$$T \underset{N \to \infty}{\in} \mathrm{N}(0,1) \Rightarrow \tag{35e}$$

$$p = 1 - F^{-1}_{\mathrm{N}(0,1)}(T) \tag{35f}$$

Basic assumptions (B): The noise v_i has unknown correlation structure and distribution i.e. only the basic assumptions (30) applies. Then decimation followed by the non-parametric method (W) can be used.

5. NON-LINEARITY TEST PROCEDURES

Candidate nonlinearity test procedures are combinations of the three new methods and the four variance tests. The SUP3 method will however be excluded from further investigation because it is insensitive to at least the nonlinearity in (38). Further the distribution free rank test in (W) and (B) are extremely sensitive to deviations between the distributions of d_1 and d_2 (35d). Consequently they can not be combined with the SUP methods as they produce different distributions for d_1 and d_2 when the noise v is non normal. If v has a symmetrical distribution the SIN method produces equal distributions of d_1 and d_2 but they will be dependent all through uncorrelated. Even through independence is also a condition in the rank test (35d) this specific dependence fortunately improves rather than spoils the performance of the SIN method. This leaves the following 7 combined methods for simulation analysis: SR, SUP5-NW, SUP5-N, SIN-NW, SIN-N, SIN-W and SIN-B.

6. STATISTICAL PROPERTIES

Statistical tests can be compared by their power function i.e. the probability of rejecting H_0 as a function of the degree of violation of H_0 towards H_1. When H_0 is true the power is also known as false alarm probability and must be less or equal to α the level of significance. When H_1 is true the power, or detection probability, should be as high as possible. The only safe conclusion from the theory is that the power given H_0 and NW noise assumptions will equal α_0 for the SUP5-NW and SIN-NW methods. Below the power is therefore estimated by simulation of different cases of H_0 and H_1 combined with all the different assumptions.

6.1 *Examples and experimental conditions*

Three examples will be considered for the deterministic part. A basic linear first order system (36) with gain one. Next only the input part is made nonlinear to give a Hammerstein type of nonlinearity (37) and finally the linear system is perturbed with a changing time constants depending on whether the output increases or decreases (38) as e.g. with direction dependent friction.

$$y_d(t) = \frac{b}{1+aq^{-1}}u(t-1)\ , \quad a=-0.9\ ,\ b=1+a \qquad (36)$$

$$y_d(t) = \frac{b}{1+aq^{-1}}\tan^{-1}(ku(t-1))\ ,\ k=2 \qquad (37)$$

$$y_d(t) = -a(t)y_d(t-1) + b(t)u(t-1)\ ,\quad a(t) = \begin{cases} a_f = -(-a)^{k_\tau} & \text{if } y_d(t) \geq y_d(t-1) \\ a_s = -(-a)^{\frac{1}{k_\tau}} & \text{otherwise} \end{cases}\ ,\ b(t) = 1 + a(t)\ ,\ k_\tau = 1.25 \qquad (38)$$

The measured output follows (39) where there are six examples of noise from normal white (40) to non normal and correlated (45). The MA(10) process (41) is chosen in favor of AR(1) processes to test the robustness of the decimation method. The non normal cases (42)–(45) are with heavy tails as with "outliers" and skewed in the two last cases. The scaling results in equal standard deviation σ_v at 0.1. For the heavy tailed noise it was necessary to use a robustified LS (Staudte and Sheather, 1990, sec. 7.6) for estimating the sinus (23).

$$y(t) = y_d(t) + v(t) \qquad (39)$$

$$v(t) = \sigma e(t)\ ,\ e(t) \in \text{NID}(0,1)\ ,\ \sigma = 0.1 \qquad (40)$$

$$v(t) = \sigma w(t)\ ,\ w(t) = M(q)e(t)\ ,\quad M(q) = \frac{1+q^{-1}+\cdots+q^{-9}}{\sqrt{10}} \qquad (41)$$

$$v(t) = \sigma\,\text{sign}(e(t))e(t)^4\kappa_1\ ,\quad \kappa_1 = \frac{1}{\sqrt{7\times 5\times 3}} \qquad (42)$$

$$v(t) = \sigma\,\text{sign}(w(t))w(t)^4\kappa_1 \qquad (43)$$

$$v(t) = \sigma(e(t)^2-1)\kappa_2\ ,\ \kappa_2 = \frac{1}{\sqrt{2}} \qquad (44)$$

$$v(t) = \sigma(w(t)^2-1)\kappa_2 \qquad (45)$$

As suggested in Billings and Voon (1983) the input for the SR method is a sinus with a level different from zero. The amplitude and level are 0.5 to make it vary between 0 and 1 and thus give approximately 10% noise. The frequency equals the bandwidth of the deterministic system (36). Input to the SUP5 method is a modified PRBS signal (Söderström and Stoica, 1989, sec. 5.3) with the same level and RMS as the sinus and with a bandwidth as the system. The input to SIN and SIN has zero level and an amplitude given the same RMS as above. The total number of samples is set to 500 i.e. $1\times 500, 5\times 100$ and 2×250 for the SR, SUP5 and SIN methods respectively. The level of the test α_0 is set to the widely used 5% and the observed power can then be estimated as (46). Further it is reasonable to have $2\sigma_{\hat{\alpha}} = 0.01$ for the real power at 0.05 which is obtained with the number of simulations N_s at 1900.

$$\hat{\alpha} = \frac{\#\{p_i < 0.05\}}{N_s} \qquad (46)$$

6.2 *Results*

For every combination of deterministic systems and noise 1900 simulations of 500 samples are performed. For every simulation all 7 p-values corresponding to the chosen methods are calculated. The results consisting of $3\times 6\times 7 = 126$ estimated powers are presented in separate bar plots for each deterministic system. In the plots bars are grouped by method and noise process.

A carefull examination off all figures shows that the decimation based methods SUP5-N, SIN-N and SIN-B are in general superior to the corresponding methods whitout decimation. Thus we can consentrate on comparing these with the SR method.

Figure 1 shows that the existing SR method only works for normal white noise and the false alarm probabilities are unacceptable high in all other cases especially the colored ones. The SUP5-N method is a little better as it works also for normal colored noise. The SIN-B method also works for symmetrical distributed noise as expected. The

SIN-N method is perfectly reliable in all cases even the non normal symmetrical and skewed ones. The reason for this is the special dependence between d_1 and d_2 arrising from heavy tailed noise, the full explanation is however to long and complicated to be included here.

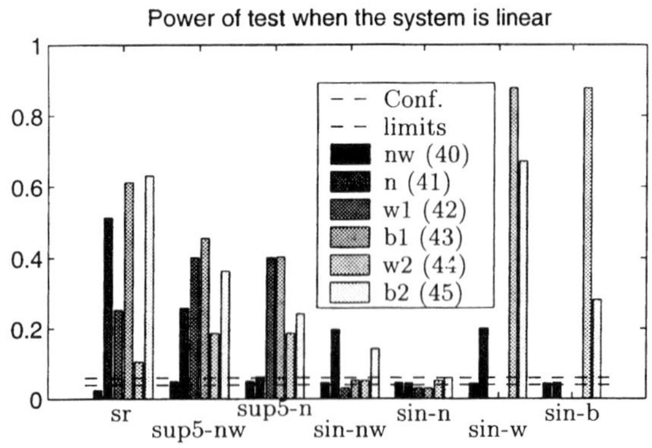

Fig. 1. False alarm probabilities when the system is linear.

Figure 2–3 shows detection probabilities for methods having false alarms less than 0.1. For the Hammerstein sytem the SR and SUP5-N methods are more powerfull than the SIN-N and SIN-B methods. The picture in figure 3 where the dynamical part is nonlinear is reversed in the sense that the SIN methods are very good in contrast to the poor performence of the SR and SUP5-N methods.

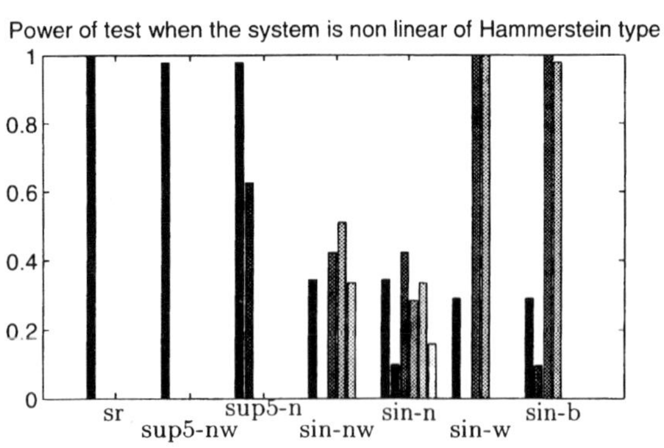

Fig. 2. Detection probabilities when the system is non linear of the Hammerstein type (37). Bars corresponding to $\hat{\alpha} > 0.1$ when the system is linear (figure 1) are omitted.

7. CONCLUSION

This paper discusses and developes preliminary statistical tests for detecting nonlinearities in the deterministic part of SISO systems with noise. The existing method is based on third order correlations. It is shown to work for normal white noise but for other common noise processes e.g. colored, it is absolutely unreliable. Therefore two new methods are developed. One is based on superposition and the other on sinus input. They are superior to the existing method. Especially the sinus method whis is reliable also for colored, heavy tailed or skew distributed noise is a significant improvement to the existing method.

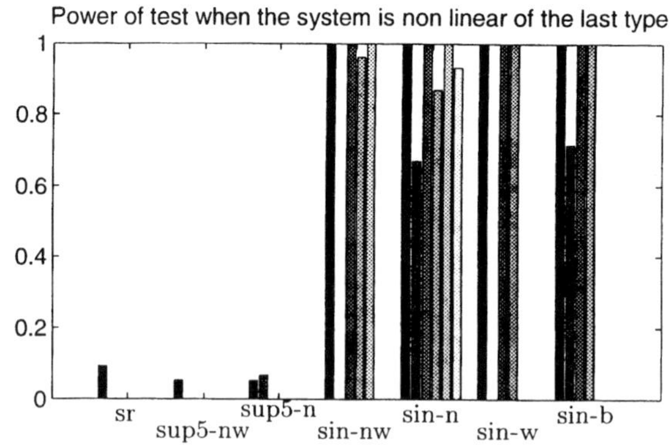

Fig. 3. Detection probabilities when the system is non linear of the last type (38). Bars are omitted as in figure 2.

8. REFERENCES

Billings, S.A. and W.S.F. Voon (1983). Structure detection and model validity tests in the identification of nonlinear systems. *IEE Proceedings D* **130**(4), 193–199.

Billings, S.A. and W.S.F. Voon (1986). Correlation based model validity tests for nonlinear models. *International Journal of Control* **44**(1), 235–244.

Conover, W. J. (1999). *Practical Nonparametric Statistics*. John Wiley & Sons.

Haber, R. (1985). Nonlinearity tests for dynamic processes. In: *7th IFAC Symposium on System Identification*. pp. 409–413.

Haber, R. and H. Unbehauen (1990). Structure identification of nonlinear dynamic systems - a survey on input/output approaches. *Automatica* **26**(4), 651–677.

Knudsen, Morten (1996). Determination of parameter estimation errors due to noise and undermodelling. In: *IEEE Instrumentation and Measurement Technology Conference*.

Ljung, Lennart (1999). *System Identification, Theory for the User*. Prentice-Hall information and system sciences series. second ed.. Prentice-Hall.

Söderström, Torsten and Petre Stoica (1989). *System Identification*. Prentice Hall International Series in System and Control Engineering. Prentice Hall. New York.

Staudte, Robert G. and Simon J. Sheather (1990). *Robust Estimation and Testing*. John Wiley & Sons, Inc.

Tong, Howell (1990). *Non-linear Time Series - A Dynamical System Approach*. Oxford Statistical Science Series. Clarendon Press Oxford.

EXACT BOUNDS FOR THE FREQUENCY RESPONSE OF AN UNCERTAIN PLANT WITH ELLIPSOIDAL PERTURBATIONS

G. Chesi * A. Garulli * A. Tesi ** A. Vicino *

* *Dipartimento di Ingegneria dell'Informazione*
Università di Siena, Via Roma 56, 53100 Siena, Italy
E-mail: {chesi,garulli,vicino}@ing.unisi.it

** *Dipartimento di Sistemi e Informatica*
Università di Firenze, Via di S. Marta 3, 50139 Firenze, Italy
E-mail: atesi@dsi.unifi.it

Abstract: This paper deals with the frequency domain properties of an ellipsoidal family of rational functions, i.e. a family of rational functions whose coefficients depend affinely on an ellipsoidal parameter set. The problems considered are relevant to several recently developed techniques in the identification for control research area. The frequency plots of such a family are characterized and an efficient algorithm for computing the envelope of the Bode plots is devised. In particular, it is shown that the extremal values of the magnitude and phase of the value set at each frequency, which are in general non-convex optimization problems, can be computed via the solution of a sequence of Linear Matrix Inequalities (LMIs). *Copyright ©2000 IFAC*

Keywords: Identification for control, parametric uncertainty, frequency response, linear matrix inequalities.

1. INTRODUCTION

Frequency domain properties of families of rational functions are fundamental issues in the area of robust control against parametric perturbations since long time. Beyond the importance of frequency domain approaches for controller design in classical (e.g, Quantitative Feedback Theory (QFT), (Horowitz, 1963)) as well as modern robust control approaches (e.g., μ theory (Doyle *et al.*, 1982; Doyle, 1982)), further motivations may be found in the recently developed area of identification for robust control (Kosut *et al.*, 1992*a*; Garulli *et al.*, 1999). Typical identification algorithms, in both probabilistic and deterministic framework, provide a plant model described by an ellipsoidal family of rational functions, i.e., a set of rational functions whose coefficients depend affinely on a number of parameters belonging to an ellipsoidal set in the parameter space (Ljung, 1999; Fogel and Huang, 1982). In order to exploit robust control techniques, the envelope of the Nyquist and/or Bode plots of the rational functions of the family is needed (Wahlberg and Ljung, 1992; Kosut *et al.*, 1992*b*; Reinelt *et al.*, 1999).

Despite the large amount of work in the parametric robustness area, a complete characterization of the Nyquist and Bode plots of an ellipsoidal family of rational functions has not been given yet. Indeed, while several results are available for the class of interval as well as polytopic plants (see (Tesi and Vicino, 1991; Bartlett, 1990), and (Bhattacharyya *et al.*, 1995; Dahleh *et al.*, 1993) for more extensive overviews), no efficient methods have been proposed for the ellipsoidal case.

The present paper is an attempt to fill this gap providing a description of the frequency response of an ellipsoidal family of rational functions. Particular attention is devoted to the role of the value set of uncertain rational functions, a well-known concept in the parametric robustness area. Moreover, an efficient algorithm for computing the envelope of the Bode plots of the uncertain rational functions is devised. In particular, it is shown that the computation of the extremal values of the magnitude and phase of the ellipsoidal plant family at each frequency, which in general amounts to solving non-convex optimization problems, can be performed through the solution of a one-parameter family of Linear Matrix Inequalities (LMIs).

The paper is organized as follows. In Section 2 the problem formulation is introduced. Section 3 is devoted to the characterization of the value set, while Section 4 deals with some relevant optimization problems. Some examples are illustrated in Section 5 to show the application of the proposed method. Section 6 reports some concluding remarks. Proofs of the results in the paper are not reported due to lack of space; the interested reader can find them in (Chesi *et al.*, 1999*a*).

Notation. In the following, some notation adopted in the paper is reported. $\Re(x)$, $\Im(x)$: real and imaginary part of $x \in \mathbb{C}$; $|x|$, $\arg(x)$: magnitude and argument of $x \in \mathbb{C}$; $\mathrm{Im}_{p\in\mathcal{A}} f(p)$: image of the set $\mathcal{A}$ through the function $f(\cdot)$; $\mathrm{adj}(M)$ and $\ker(M)$: adjoint and kernel of matrix M; $\lambda_{max}(M)$: maximum real eigenvalue of matrix M; $f(x;y)|_{x=\bar{x}}|_{y=\bar{y}} = \lim_{y\to\bar{y}} [\lim_{x\to\bar{x}} f(x;y)]$.

2. PROBLEM FORMULATION

Let us consider the family of plants

$$G(s;p) = \frac{N(s;p)}{D(s;p)} = \frac{N_0(s) + \sum_{i=1}^{k} p_i N_i(s)}{D_0(s) + \sum_{i=1}^{k} p_i D_i(s)} \qquad (1)$$

where $N_i(s)$ and $D_i(s)$, $i = 0, \cdots, k$, are given polynomials, and the parameter vector $p = (p_1 \cdots p_k)'$ belongs to the ellipsoid $\mathcal{E} \subset \mathbb{R}^k$, defined as

$$\mathcal{E} = \{p \in \mathbb{R}^k : (p - p_c)'E(p - p_c) \le 1\} \qquad (2)$$

being $E \in \mathbb{R}^{k\times k}$ a positive definite matrix and $p_c \in \mathbb{R}^k$.

The aim of this paper is to provide a characterization of the envelope of the frequency plots for the family of plants (1)-(2). The set of frequency plots of the ellipsoidal family is defined as follows

$$\mathcal{V} = \bigcup_{\bar{\omega}\ge 0} \mathcal{V}_{\bar{\omega}} \qquad (3)$$

where

$$\mathcal{V}_{\bar{\omega}} = \bigcup_{\bar{p}\in\mathcal{E}} \left(G(j\omega;p)|_{p=\bar{p}}|_{\omega=\bar{\omega}}\right) \qquad (4)$$

is the image of the plant family at the frequency $\bar{\omega}$.
We observe that without loss of generality it can be assumed $p_c = 0_k$ and $E = I_k$. Indeed, it is always possible to define new polynomials $\tilde{N}_i(s), \tilde{D}_i(s)$ such that the family of plants is defined by $G(s;p) = \tilde{N}(s;p)/\tilde{D}(s;p)$, with p belonging to the set

$$\mathcal{E} = \{p \in \mathbb{R}^k : p'p \le 1\}. \qquad (5)$$

We will also assume that $N(s;p)$ and $D(s;p)$ have no common roots on the imaginary axis for all values of $p \in \mathbb{R}^k$, i.e.

$$\forall\omega \ge 0\ \exists p \in \mathbb{R}^k : N(j\omega;p) \ne 0 \text{ or } D(j\omega;p) \ne 0. \qquad (6)$$

Before we proceed with the characterization of the set $\mathcal{V}_{\bar{\omega}}$ in (4), it is worth pointing out that in general $G(j\omega;p)|_{p=\bar{p}}|_{\omega=\bar{\omega}} \ne G(j\omega;p)|_{\omega=\bar{\omega}}|_{p=\bar{p}}$.

Now, let us define the function

$$G_{\bar{\omega}}(p) = G(j\omega;p)|_{\omega=\bar{\omega}}. \qquad (7)$$

Extending the concept of *value set* of a polynomial (Bhattacharyya *et al.*, 1995) to transfer functions, the value set of the plant family at a fixed frequency $\bar{\omega}$ is defined as

$$\hat{\mathcal{V}}_{\bar{\omega}} = \mathop{\mathrm{Im}}_{p\in\mathcal{E}} G_{\bar{\omega}}(p), \qquad (8)$$

i.e., it is the image of the set $\mathcal{E}$ under the transformation $G_{\bar{\omega}}(\cdot)$. By definitions (4) and (8) of the sets $\hat{\mathcal{V}}_{\bar{\omega}}$ and $\mathcal{V}_{\bar{\omega}}$, it easily follows that they coincide if $N(j\bar{\omega};p)$ and $D(j\bar{\omega};p)$ do not admit any common factor. This property is stated in the next lemma.

Lemma 1. Suppose that

$$\nexists \bar{p} \in \mathcal{E} : N(j\bar{\omega};\bar{p}) = 0 \text{ and } D(j\bar{\omega};\bar{p}) = 0. \qquad (9)$$

Then, $\mathcal{V}_{\bar{\omega}} = \hat{\mathcal{V}}_{\bar{\omega}}$.

Throughout the rest of the paper, it will be assumed that condition (9) holds, so that $\hat{\mathcal{V}}_{\bar{\omega}} = \mathcal{V}_{\bar{\omega}}$. When (9) is not verified, the characterization of the set (3) is slightly more involved. Details can be found in (Chesi *et al.*, 1999*a*).

3. CHARACTERIZATION OF THE VALUE SET

In this section, the value set $\hat{\mathcal{V}}_{\bar{\omega}}$ in (8) will be characterized. To proceed, consider the matrices $A(u,v) \in \mathbb{R}^{2\times k}$ and $B(u,v) \in \mathbb{R}^2$ given by

$$A(u,v) = \begin{pmatrix} a_1(u,v) \\ a_2(u,v) \end{pmatrix} = \begin{pmatrix} \phi_R(1) & \dots & \phi_R(k) \\ \phi_I(1) & \dots & \phi_I(k) \end{pmatrix}$$
$$B(u,v) = \begin{pmatrix} b_1(u,v) \\ b_2(u,v) \end{pmatrix} = \begin{pmatrix} \phi_R(0) \\ \phi_I(0) \end{pmatrix}, \qquad (10)$$

where

$$\phi_R(i) = \Re(D_i(j\bar{\omega}))u - \Im(D_i(j\bar{\omega}))v - \Re(N_i(j\bar{\omega})),$$
$$\phi_I(i) = \Im(D_i(j\bar{\omega}))u + \Re(D_i(j\bar{\omega}))v - \Im(N_i(j\bar{\omega})), \qquad (11)$$

From (1) and (7), it follows that

$$G_{\bar{\omega}}(p) = u + jv \in \hat{\mathcal{V}}_{\bar{\omega}} \text{ if and only if } A(u,v)p + B(u,v) = 0. \quad (12)$$

Now, let us define the sets

$$\mathcal{Q}_i = \{u + jv \in \mathbb{C} : \text{rank}(A(u,v)) = i\} \quad (13)$$

for $i = 0, 1, 2$. For each $\mathcal{Q}_i$, we define the corresponding subset $\mathcal{Q}_i^*$ of the value set $\hat{\mathcal{V}}_{\bar{\omega}}$

$$\mathcal{Q}_i^* = \{z \in \mathcal{Q}_i : G_{\bar{\omega}}(p) = z \text{ for some } p \in \mathcal{E}\}. \quad (14)$$

Hence, we can write

$$\hat{\mathcal{V}}_{\bar{\omega}} = \bigcup_{i=0}^{2} \mathcal{Q}_i^*. \quad (15)$$

The computation of the sets $\mathcal{Q}_i^*$ will be discussed in the next subsections. First, let us introduce the following standard result (see e.g. (Wahlberg and Ljung, 1992)).

Lemma 2. Let $\mathcal{E}$ be defined as in (5), $A \in \mathbb{R}^{q \times k}$, and $B \in \mathbb{R}^q$. Let $\text{rank}(A) = q$ with $q \leq k$. Then,

$$\underset{p \in \mathcal{E}}{\text{Im}}\, Ap + B = \left\{w \in \mathbb{R}^q : (w - B)' (AA')^{-1} (w - B) \leq 1\right\}.$$

3.1 *Computation of $\mathcal{Q}_2^*$*

For $u + jv \in \mathcal{Q}_2$ we have that $A(u,v)$ has full rank and hence, from (12) and Lemma 2,

$$\mathcal{Q}_2^* = \{z = u + jv \in \mathcal{Q}_2 : f(u,v) \leq 0\},$$

where the function

$$f(u,v) = B'(u,v) \left(A(u,v)A'(u,v)\right)^{-1} B(u,v) - 1$$

can be characterized according to the following lemma.

Lemma 3. Let $f(u,v) = h(u,v)/g(u,v)$ with

$$h(u,v) = B'(u,v)\text{adj}\left(A(u,v)A'(u,v)\right) B(u,v) - g(u,v), \quad (16)$$

$$g(u,v) = \det\left(A(u,v)A'(u,v)\right), \quad (17)$$

and $A(u,v)$, $B(u,v)$ defined by (10). Then, $h(u,v)$ and $g(u,v)$ are polynomials in u, v of degree four, i.e.

$$h(u,v) = \sum_{i=0}^{4} h_i(u,v),$$

$$g(u,v) = \sum_{i=0}^{4} g_i(u,v),$$

where $h_i(u,v)$ and $g_i(u,v)$ are homogeneous forms of degree i. Furthermore,

$$h_4(u,v) = a(u^2 + v^2)^2, \quad (18)$$

$$g_4(u,v) = b(u^2 + v^2)^2, \quad (19)$$

for appropriate $a, b \in \mathbb{R}$.

It is obvious that $g(u,v) > 0$ for all $u, v \in \mathcal{Q}_2$. Thus, we have the following characterization of $\mathcal{Q}_2^*$.

Proposition 1. The set $\mathcal{Q}_2^*$ is given by

$$\mathcal{Q}_2^* = \{z = u + jv \in \mathcal{Q}_2 : h(u,v) \leq 0\}. \quad (20)$$

3.2 *Computation of $\mathcal{Q}_1^*$ and $\mathcal{Q}_0^*$*

First, let us consider $\mathcal{Q}_1^*$. From (12) and (13)-(14), it is clear that $\mathcal{Q}_1^*$ is not empty if and only if the matrix

$$M(u,v) = [B(u,v)\ A(u,v)] = \begin{pmatrix} \phi_R(0) & \dots & \phi_R(k) \\ \phi_I(0) & \dots & \phi_I(k) \end{pmatrix}$$

satisfies $\text{rank}(M(u,v)) = 1$ for some u, v. Then, we can write

$$\mathcal{Q}_1^* = \mathcal{M}_1 \bigcap \mathcal{Z}_1$$

where

$$\mathcal{M}_1 = \{u + jv : \ \text{rank}(M(u,v)) = 1\}$$

and

$$\mathcal{Z}_1 = \{u + jv : b_i^2(u,v) - a_i(u,v)a_i'(u,v) \leq 0 \quad i = 1, 2\}.$$

The inequalities in $\mathcal{Z}_1$ follow from the application of Lemma 2 to each row of (12).
Let us first analyze the set $\mathcal{M}_1$. Denote the minors of $M(u,v)$ by

$$m_{il}(u,v) = \det \begin{pmatrix} \phi_R(i) & \phi_R(l) \\ \phi_I(i) & \phi_I(l) \end{pmatrix}$$

for $i, l = 0, \dots, k$, $i < l$. For each minor, the following topological characterization holds.

Lemma 4. The set $\mathcal{N}_{il} = \{u + jv : \ m_{il}(u,v) = 0\}$ is either the complex plane, a circle, a line, or the empty set.

The above lemma allows one to give a complete characterization of $\mathcal{Q}_1^*$.

Proposition 2. The set $\mathcal{Q}_1^*$ is either the empty set or one of the following sets:
- one or two arcs of circle (possibly degenerating to one circle);
- one or two line segments (possibly degenerating to one line);
- one or two points.

Consider now $u + jv \in \mathcal{Q}_0$. Since in this case $A(u,v)$ is identically zero, we have

$$\mathcal{Q}_0^* = \mathcal{Q}_0 \bigcap \mathcal{Z}_0$$

where

$$\mathcal{Z}_0 = \{u + jv : \; b_1(u,v) = b_2(u,v) = 0\}.$$

From (10)-(11), it turns out that $\mathcal{Q}_0$ is given by the intersection of sets that are either the complex plane, or a point, or the empty set, while the constraints in $\mathcal{Z}_0$ define either the complex plane or a point. Notice however that $\mathcal{Q}_0$ and $\mathcal{Z}_0$ cannot be both the whole complex plane, because of (6). Therefore, $\mathcal{Q}_0^*$ has the following topological structure.

Proposition 3. The set $\mathcal{Q}_0^*$ is either the empty set or it is a singleton.

As a final remark, we observe that the computation of the value set (8) requires essentially the solution of the inequality $h(u,v) \leq 0$, in the subset of $\mathbb{C}$ where the full rank condition for the matrix $A(u,v)$ is satisfied. The contributions of $\mathcal{Q}_1^*$ and $\mathcal{Q}_0^*$ are easy to compute and can be treated separately. The above characterization will be exploited in the following to compute useful extreme values of the frequency envelope of the plant family.

4. COMPUTATION OF THE EXTREMES OF THE VALUE SET

In this section we describe an algorithm for the computation of some extreme points of the value set $\hat{\mathcal{V}}_{\bar{\omega}}$. Let us start with the problem of finding the maximum and minimum absolute value of an element in this set, i.e.

$$\underline{m} = \inf_{z \in \hat{\mathcal{V}}_{\bar{\omega}}} |z| \quad , \quad \overline{m} = \sup_{z \in \hat{\mathcal{V}}_{\bar{\omega}}} |z|.$$

From (15), we get

$$\underline{m} = \min_{i=0,1,2} \underline{q}_i \quad , \quad \overline{m} = \max_{i=0,1,2} \overline{q}_i$$

where

$$\underline{q}_i = \inf_{z \in \mathcal{Q}_i^*} |z| \quad , \quad \overline{q}_i = \sup_{z \in \mathcal{Q}_i^*} |z|.$$

Let us concentrate on the computation of $\underline{q}_2$ and $\overline{q}_2$. Indeed, from Propositions 2 and 3 we know that $\mathcal{Q}_1^*$ and $\mathcal{Q}_0^*$ are very elementary sets and hence the computation of $\underline{q}_1, \overline{q}_1, \underline{q}_0, \overline{q}_0$ is straightforward.

Consider the set $\mathcal{Q}_2^*$. From Proposition 1 we have that

$$\mathcal{Q}_2^* = \mathcal{Q}_2 \bigcap \{u + jv : h(u,v) \leq 0\}$$

where $\mathcal{Q}_2$ represents the full rank condition on $A(u,v)$. In order to get rid of this condition in the optimization problems to be solved for, we seek an alternate way for computing $\underline{q}_2$ and $\overline{q}_2$. First, note that except for the cases $\underline{q}_2 = 0$ and $\overline{q}_2 = +\infty$, these extremal values are obtained on the boundary of $\mathcal{Q}_2^*$. Since these special cases can be easily checked a priori, we look for a set having the same boundary of $\mathcal{Q}_2^*$, while possessing a more simple expression. Since from (16)-(17) it is easy to verify that

$$h(u,v) \geq 0 \qquad \forall u + jv \in \mathbb{C} \setminus \mathcal{Q}_2,$$

let us introduce the set

$$\mathcal{Q}_2^{\#} = \{u + jv : h(u,v) < 0\}.$$

Optimization over $\mathcal{Q}_2^{\#}$ is equivalent to optimization over $\mathcal{Q}_2^*$, provided that $\mathcal{Q}_2^*$ and $\mathcal{Q}_2^{\#}$ have the same boundary. This is guaranteed by the following lemma.

Lemma 5. The set $\mathcal{Q}_2^{\#}$ is such that

$$\partial \mathcal{Q}_2^{\#} = \partial \mathcal{Q}_2^*.$$

From the above lemma and the previous discussion, it can be concluded that

$$\underline{q}_2 = \inf_{(u,v):\; h(u,v)<0} \sqrt{u^2 + v^2}, \tag{21}$$

$$\overline{q}_2 = \sup_{(u,v):\; h(u,v)<0} \sqrt{u^2 + v^2}. \tag{22}$$

Equations (21)-(22) define two non-convex optimization problems. However, exploiting some recent results on convex optimization (Chesi *et al.*, 1999*b*), it can be shown that they can be solved via a sequence of LMIs. To proceed, let us introduce the polynomial

$$\tilde{h}(u,v) = -h(u,v)h(-u,-v).$$

Exploiting Lemma 3, it can be easily verified that

$$\tilde{h}(u,v) = \sum_{i=0}^{3} \tilde{h}_{2i}(u,v) - h_4^2(u,v) \tag{23}$$

where $\tilde{h}_{2i}(u,v)$ are homogeneous forms of degree $2i$. The following result is a consequence of the symmetry of the functions to be optimized in (21)-(22) and the definition of $\tilde{h}(u,v)$.

Lemma 6. Consider the optimization problems

$$\underline{\rho} = \inf_{(u,v):\; \tilde{h}(u,v)>0} u^2 + v^2 , \tag{24}$$

$$\overline{\rho} = \sup_{(u,v):\; \tilde{h}(u,v)>0} u^2 + v^2 . \tag{25}$$

Then,

$$\underline{q}_2 = \sqrt{\underline{\rho}} \quad , \quad \overline{q}_2 = \sqrt{\overline{\rho}}.$$

Now, let us introduce the following functions

$$\begin{aligned} \underline{h}(u,v;c) &= \sum_{i=0}^{3} \tilde{h}_{2i}(u,v) \frac{(u^2+v^2)^{3-i}}{c^{3-i}} \\ &\quad - h_4^2(u,v) \frac{c}{u^2+v^2} \\ \overline{h}(u,v;c) &= \sum_{i=0}^{3} \tilde{h}_{2i}(u,v) c^{3-i} (u^2+v^2)^{3-i} \\ &\quad - h_4^2(u,v) \frac{1}{c(u^2+v^2)}. \end{aligned}$$

From (18) and (23), it is easily verified that $\underline{h}(u,v;c)$ and $\overline{h}(u,v;c)$ are homogeneous forms of degree 6 having the following property.

Lemma 7. Let c be a positive scalar and consider the circles

$$\mathcal{Z}_l(c) = \{(u,v):\ u^2+v^2=c\},$$
$$\mathcal{Z}_h(c) = \left\{(u,v):\ u^2+v^2=\frac{1}{c}\right\}.$$

Then, the following statements hold:

i) $\tilde{h}(u,v) > 0 \quad \forall (u,v) \in \mathcal{Z}_l(c)$ if and only if $\underline{h}(u,v;c) > 0 \quad \forall (u,v) \in \mathbb{R}^2 \setminus \emptyset$;

ii) $\tilde{h}(u,v) > 0 \quad \forall (u,v) \in \mathcal{Z}_h(c)$ if and only if $\overline{h}(u,v;c) > 0 \quad \forall (u,v) \in \mathbb{R}^2 \setminus \emptyset$.

Lemma 6 and Lemma 7 suggest that we can compute $\underline{\rho}$ (resp., $\overline{\rho}$) as the minimum c such that $\underline{h}(u,v;c)$ (resp., $\overline{h}(u,v;c)$) loses its negativity, i.e.,

$$\underline{\rho} = \sup\{\ \underline{c} \geq 0:\ \underline{h}(u,v;c) < 0 \quad \forall (u,v) \in \mathbb{R}^2 \setminus \emptyset,\ \ \forall c \in [0,\underline{c}]\}$$
$$\overline{\rho} = \sup\{\ \overline{c} \geq 0:\ \overline{h}(u,v;c) < 0 \quad \forall (u,v) \in \mathbb{R}^2 \setminus \emptyset,\ \ \forall c \in [0,\overline{c}]\}.$$

Therefore, the original problems (24)-(25) amount to solving a one-parameter family of positivity tests on homogeneous forms.

It is well known that the study of positivity of a homogeneous form is far from being a trivial task. However, exploiting the special structure of $\underline{h}(u,v;c)$ and $\overline{h}(u,v;c)$, it can be shown that each positivity test can be performed via a standard LMI.

Let us introduce the vector $x = (u^3, u^2v, uv^2, v^3)' \in \mathbb{R}^4$. It is easy to verify that there exist symmetric matrices $M_l(c), M_h(c) \in \mathbb{R}^{4\times4}$ such that for all $x \in \mathbb{R}^4$ we have

$$\underline{h}(u,v;c) = x'M_l(c)x\ , \qquad \overline{h}(u,v;c) = x'M_h(c)x.$$

Moreover, let $\alpha = (\alpha_1,\alpha_2,\alpha_3)' \in \mathbb{R}^3$ and consider the following symmetric matrix

$$L(\alpha) = \begin{pmatrix} 0 & 0 & \alpha_1 & \alpha_2 \\ 0 & -2\alpha_1 & -\alpha_2 & \alpha_3 \\ \alpha_1 & -\alpha_2 & -2\alpha_3 & 0 \\ \alpha_2 & \alpha_3 & 0 & 0 \end{pmatrix}.$$

Notice that $L(\alpha)$ is the most general symmetric matrix satisfying $x'L(\alpha)x = 0$, $\forall x$ and $\forall \alpha$. We are ready to state the main result of this section.

Theorem 1. Let

$$\underline{c}^* = \inf\ c \quad \text{s.t.} \quad \min_{\alpha} \lambda_{max}\left[M_l(\alpha;c)\right] > 0$$
$$\overline{c}^* = \inf\ c \quad \text{s.t.} \quad \min_{\alpha} \lambda_{max}\left[M_h(\alpha;c)\right] > 0$$

where

$$M_l(\alpha;c) = L(\alpha) + M_l(c),$$
$$M_h(\alpha;c) = L(\alpha) + M_h(c).$$

Then,

$$\underline{q}_2 = \sqrt{\underline{c}^*}\ , \quad \overline{q}_2 = \frac{1}{\sqrt{\overline{c}^*}}.$$

The above theorem states that $\underline{q}_2$ and $\overline{q}_2$ can be computed via the solution of a one-parameter family of convex optimization problems. Specifically, for a fixed value of the parameter c, each problem requires the minimization of the maximum eigenvalue of a symmetric matrix with respect to the vector α, a problem that can be solved via a standard LMI (Nesterov and Nemirovsky, 1993). Moreover, a bisection algorithm on the scalar c can be employed to speed up the convergence.

The maximum and minimum phase of elements in the value set $\hat{\mathcal{V}}_{\bar{\omega}}$ can be computed by solving the optimization problems

$$\underline{\varphi} = \inf_{z\in\hat{\mathcal{V}}_{\bar{\omega}}} \arg z\ , \quad \overline{\varphi} = \sup_{z\in\hat{\mathcal{V}}_{\bar{\omega}}} \arg z\ .$$

Consider the computation of $\underline{\varphi}_2$ and $\overline{\varphi}_2$ where

$$\underline{\varphi}_2 = \inf_{z\in\mathcal{Q}_2^*} \arg z\ , \quad \overline{\varphi}_2 = \sup_{z\in\mathcal{Q}_2^*} \arg z$$

(once again, the corresponding optimizations over $\mathcal{Q}_1^*$ and $\mathcal{Q}_0^*$ can be easily solved separately). From Lemma 5 we have that

$$\underline{\varphi}_2 = \inf_{(u,v):\ h(u,v)<0} \arg(u+jv),$$
$$\overline{\varphi}_2 = \sup_{(u,v):\ h(u,v)<0} \arg(u+jv).$$

Now, let us introduce the following function

$$\hat{h}(\gamma;\theta) = h(u,v)|_{u=\gamma\cos\theta, v=\gamma\sin\theta}\ .$$

Then, we have that

$$\underline{\varphi}_2 = \sup\{\ \hat{\theta} \in (-2\pi,0]:\ \hat{h}(\gamma;\theta) > 0\ \ \forall\gamma>0,\ \ \forall\theta \in (-2\pi,\hat{\theta}]\}$$
$$\overline{\varphi}_2 = \inf\{\ \hat{\theta} \in (-2\pi,0]:\ \hat{h}(\gamma;\theta) > 0\ \ \forall\gamma>0,\ \ \forall\theta \in [\hat{\theta},0]\}.$$

and hence

$$\underline{\varphi}_2 = \inf_{\theta\in\Theta} \theta\ , \qquad \overline{\varphi}_2 = \sup_{\theta\in\Theta} \theta$$

where $\Theta = \{\theta:\ \exists\gamma > 0,\ \hat{h}(\gamma;\theta) < 0\}$. So, we can compute $\underline{\varphi}_2$ and $\overline{\varphi}_2$ via the solution of a one-parameter family of positivity tests on the polynomial $\hat{h}(\gamma;\theta)$ for any fixed θ. As for the magnitude, a bisection algorithm can be employed on the scalar θ in order to find the minimum and maximum phase.

5. EXAMPLES

Example 1 Consider the function

$$G(s;p) = \frac{s+1+0.5p_1(s^2+1)+0.5p_2(s+10)}{s(s^2+0.4s+1)+0.5p_3 s^2}$$

where $p'p \leq 1$. Figure 1 shows the Bode plots of this uncertain plant in the frequency range $[0.1, 10]$ rad/s (for ease of illustration, the phase plot is reported in degrees, in the interval $[-360, 0]$). Condition (9) is satisfied for all frequencies.

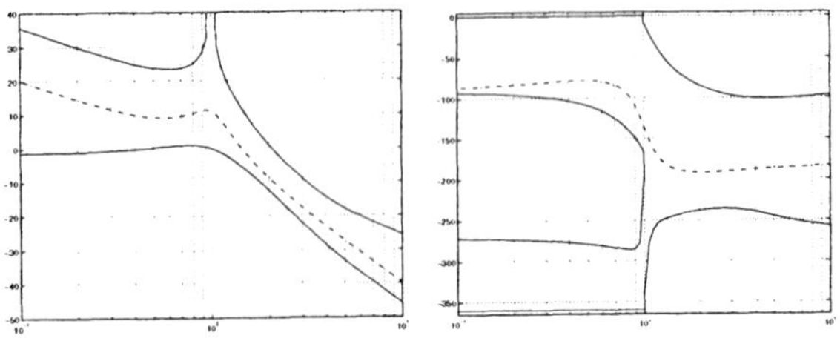

(a) magnitude (b) phase

Fig. 1. Example 1. Central plant (dashdot) and upper and lower bound (solid).

Example 2 Consider the function

$$G(s;p) = \frac{s^2+0.3s+1.2+0.5p_1+0.5p_2 s}{s(s^2-0.2s+1)+0.5p_2(s+1)+0.5p_3 s^2+0.5p_4 s}$$

and its behaviour shown in Figure 2. In this case condition (9) is not satisfied for $\omega = [0.933, 1.023]$ rad/s. From the characterization of the general case, described in (Chesi *et al.*, 1999*a*), it turns out that for all these frequencies the value set is equal to $\mathbb{C}$, as it can be seen in Figure 2.

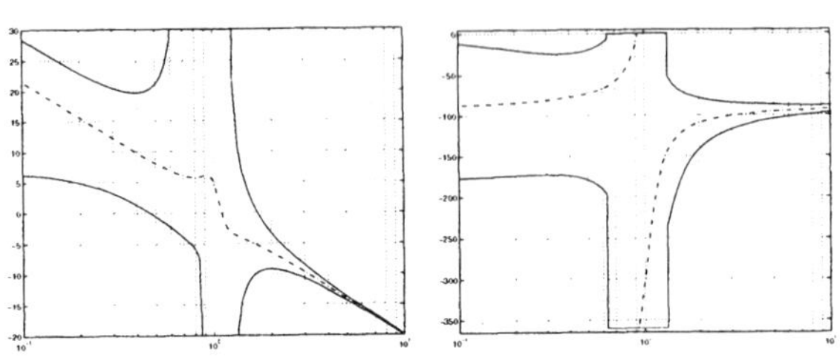

(a) magnitude (b) phase

Fig. 2. Example 2. Central plant (dashdot) and upper and lower bound (solid).

6. CONCLUSIONS

In this paper the problem of characterizing the frequency plots of an ellipsoidal family of rational functions has been considered. Special attention is devoted to the computation of the envelope of the Bode plots of such a family, a key problem in the area of identification for robust control. In particular, it is shown that the extremal values of the magnitude and phase at each frequency can be obtained via the solution of a one-parameter family of LMIs.

7. REFERENCES

Bartlett, A. C. (1990). Nyquist, bode, and nichols plots of uncertain systems. In: *Proc. 1990 American Control Conference*. San Diego, CA.

Bhattacharyya, S. P., H. Chapellat and L. H. Keel (1995). *Robust Control: The Parametric Approach*. Prentice Hall. NJ.

Chesi, G., A. Garulli, A. Tesi and A. Vicino (1999*a*). Exact bounds for the frequency response of an uncertain plant with ellipsoidal perturbations. Technical report. Dipartimento di Ingegneria dell'Informazione. Università di Siena.

Chesi, G., A. Tesi, A. Vicino and R. Genesio (1999*b*). A convex approach to a class of minimum norm problems. In: *Robustness in Identification and Control* (A. Garulli, A. Tesi and A. Vicino, Eds.). Springer-Verlag. London. pp. 359-372.

Dahleh, M., A. Tesi and A. Vicino (1993). An overview on extremal properties for robust control of interval plants. *Automatica* **29**, 707–721.

Doyle, J. C. (1982). Analysis of feedback systems with structured uncertainties. *IEE Proceedings-D Control Theory and Applications* **129**, 242–250.

Doyle, J. C., J. E. Wall and G. Stein (1982). Performance and robustness analysis for structured uncertainty. In: *Proc. 21st IEEE Conf. on Decision and Control*. Orlando, Florida. pp. 629–636.

Fogel, E. and F. Huang (1982). On the value of information in system identification - bounded noise case. *Automatica* **18**, 229–238.

Garulli, A., A. Tesi and A. Vicino (Eds.) (1999). *Robustness in Identification and Control*. Lecture Notes in Control and Information Sciences. Springer. London.

Horowitz, I. (1963). *Synthesis of Feedback Control Systems*. Academic Press. New York.

Kosut, R. L., G. C. Goodwin and M. P. Polis (Eds.) (1992*a*). Special issue on system identification for robust control design. *IEEE Transactions on Automatic Control* **37**, 899–1008.

Kosut, R. L., M. K. Lau and S. P. Boyd (1992*b*). Set-membership identification of systems with parametric and nonparametric uncertainty. *IEEE Transactions on Automatic Control* **37**(7), 929–941.

Ljung, L. (1999). *System Identification: Theory for the User*. Prentice Hall. Englewood Cliffs, NJ. 2nd edition.

Nesterov, Y. and A. Nemirovsky (1993). *Interior Point Polynomial Methods in Convex Programming: Theory and Applications*. Siam. Philadelphia.

Reinelt, W., A. Garulli, L. Ljung, J. H. Braslavsky and A. Vicino (1999). Model error concepts in identification for control. In: *Proc. of the 38th IEEE Conference on Decision and Control*. Phoenix, AZ. pp. 1240–1245.

Tesi, A. and A. Vicino (1991). Kharitonov segments suffice for frequency response analysis of interval plant-controller families. In: *Control of Uncertain Dynamic Systems* (S. P. Bhattacharyya and L. H. Keel, Eds.). CRC Press. pp. 403-415.

Wahlberg, B. and L. Ljung (1992). Hard frequency-domain model error bounds from least-squares like identification techniques. *IEEE Transactions on Automatic Control* **37**(7), 900–912.

A NEW ORTHOGONAL BASIS FUNCTIONS FOR ERROR QUANTIFICATION IN SYSTEM IDENTIFICATION

R. Ouvrard, P. Coirault, T. Poinot, J.C. Trigeassou

L.A.I.I.-E.S.I.P., 40 Av. du Recteur Pineau 86022 Poitiers cedex. France
tel: 33 (0)5 49 45 35 10 Fax: 33 (0)5 49 45 40 34
e-mail: ouvrard@esip.univ-poitiers.fr

Abstract: This paper describes a new method for quantifying errors in continuous time system identification. The key idea is to estimate a global model constituted of a nominal model and a model of bias error. The latter is composed of particular orthogonal basis functions and gives an information about the true bias error. An application to a chemical reaction column is used to illustrate this new approach. *Copyright © 2000 IFAC*

Keywords: Continuous time systems, Error estimation, Orthogonal basis functions.

1. INTRODUCTION

The main difficulty in linear system identification is that usual models give partial representation of physical complexity of the true system.

Generally, low order model has the same behavior as the true system in low frequencies. Difference between system and its model has two components.

The first component, often called bias error, comes from relationships describing physical phenomena that are an approximation of reality. This undermodelling is located mainly in transient dynamics, i.e. in high frequencies.

The second component, often called variance error, is caused by the noise disturbing data.

In many cases, the physical complexity of the system is such that it is not reasonable to consider a model having a structure close to that the system, particularly for the purposes of control design. In such cases, the model order is necessarily inferior to the system one, i.e. an undermodelling is realized. The direct consequence of this approach is the presence, in addition to the variance error, of bias error.

For the purpose of robust control design, it is necessary to quantify these two errors. There exists different approaches (Mäkilä *et al.*, 1995; Ninness and Goodwin, 1994, 1995) to solve this problem such as bounded error estimation, estimation in H_∞ or stochastic embedding approach.

The key idea (Ouvrard, 1997; Ouvrard *et al.*, 1997, 1999) developed in this paper is to estimate a global model constituted of a nominal model and a model of bias error.

In linear system identification, it is more common to use discrete time models. However, in order to enhance physical insight, the study in continuous time case has been chosen.

2. PROBLEM DESCRIPTION

The aim is to identify a linear system belonging to the Hardy space $H_2(C_+)$ and to quantify simultaneously the bias error and the variance error.

Let $G_\Delta(s)$ that characterizes the bias error. There exists in the literature (Kosut *et al.*, 1992) different forms of $G_\Delta(s)$, such that additive, multiplicative, inverse multiplicative or a combination of these, to characterize the undermodelling. In this paper, the additive form is considered, i.e. the true system can be decomposed as follows:

$$G(s) = G_N(s,\underline{\theta}_o) + G_\Delta(s). \qquad (1)$$

$G_N(s,\underline{\theta})$ is a stable and strictly proper transfer function of restricted complexity. $\underline{\theta} \in \Re^p$ is the p dimension parameter vector.

This transfer function $G_N(s,\underline{\theta})$ represents the slow modes and the steady state of the system. In other words, in the frequency domain, it represents the system behavior in low frequencies.

The part $G_\Delta(s)$ represents the rapid modes unmodelled by $G_N(s,\underline{\theta})$. In steady state, $G_\Delta(s)$ has a null value, i.e. a behavior essentially situated in high frequencies.

The aim is then to realize the parametric estimation of $G_N(s,\underline{\theta})$ and $G_\Delta(s)$. The estimated models $G_N(s,\hat{\underline{\theta}}_N)$ and $G_\Delta(s,\hat{\underline{\theta}}_\Delta)$ are respectively called nominal model and bias error model.

The global model is defined as follows:

$$G(s,\hat{\underline{\theta}}) = G_N(s,\hat{\underline{\theta}}_N) + G_\Delta(s,\hat{\underline{\theta}}_\Delta). \qquad (2)$$

The nominal model $G_N(s,\underline{\theta}_N)$ has to allow the control design. Its complexity must be reasonable and is generally reduced in comparison with the system one. A criterion of choice of the structure to $G_N(s,\underline{\theta}_N)$ can be for example the bias/variance tradeoff or the a priori knowledge of a 'gray box' model.

Consider the following nominal model:

$$G_N(s,\underline{\theta}_N) = \frac{b_0 + b_1 s + \cdots + b_m s^m}{a_0 + a_1 s + \cdots + a_{n-1} s^{n-1} + s^n} \qquad (3)$$

with $n > m$.

Let $\underline{\theta}_N$ be the parameter vector to estimate:

$$\underline{\theta}_N^T = [a_0 \; \ldots \; a_{n-1} \; b_0 \; \ldots \; b_m]. \qquad (4)$$

To avoid conflicts between the nominal model $G_N(s,\underline{\theta}_N)$ and the bias error model $G_\Delta(s,\underline{\theta}_\Delta)$, it is essential to allocate at each of them a distinct frequency domain.

The nominal model, having for role to represent dominant modes and the static gain of the system, presents therefore a low-pass frequency behavior. Consequently, the bias error model must have a band-pass or high-pass frequency behavior with a null static gain.

The key idea of this approach is to estimate the impulse response of the bias error $G_\Delta(s)$ by using a finite linear combination of orthogonal basis functions. The model $G_\Delta(s,\underline{\theta}_\Delta)$ is defined by:

$$G_\Delta(s,\underline{\theta}_\Delta) = \sum_{i=0}^{r-1} \theta_i B_i(s), \; \underline{\theta}_\Delta^T = [\theta_0 \; \ldots \; \theta_{r-1}], \qquad (5)$$

where $\{B_i(s)\}$ is a particular basis with band-pass or high-pass orthogonal functions.

Thus, the model $G_\Delta(s,\underline{\theta}_\Delta)$ gives an information on the bias error and the variance error can be quantified by classical statistical tools.

3. PARTICULAR ORTHOGONAL BASIS CONSTRUCTION

The method used in this section is similar to that introduced by Ninness and Gustafsson (1994) for the unifying construction of orthonormal bases in the discrete time case, that has been extended to the continuous time case (Ouvrard, 1997).

Consider the continuous Parseval-Plancherel's relation linking time and frequency domains, as follows:

$$\int_{-\infty}^{+\infty} x(t)\, y^*(t)\, dt = \frac{1}{2\pi} \int_{-\infty}^{+\infty} X(j\omega)\, Y^*(j\omega)\, d\omega \qquad (6)$$

where $(.)^*$ defines the complex conjugate transpose. It is well known that (6) defines an inner product on H_2 which allows to examine the orthogonality properties of two basis functions $B_n(s)$ and $B_m(s)$:

$$< B_n, B_m > = \frac{1}{2\pi} \int_{-\infty}^{+\infty} B_n(j\omega)\, B_m^*(j\omega)\, d\omega. \qquad (7)$$

The purpose is to construct a basis $\{B_i\}$ of band-pass orthogonal functions centered on pulsations $\{p_i\}$.

Firstly, consider the following function:

$$B_0(s) = \frac{A s}{(s + p_0)^2}. \qquad (8)$$

It remains to choose the constant A to achieve the normalization $\|B_0\|^2 = < B_0, B_0 > = 1$.

By using (7) and Cauchy's residue theorem, we obtain:

$$\|B_0\|^2 = \frac{A^2}{4p_0} . \quad (9)$$

So, $A = 2\sqrt{p_0}$

and $B_0(s) = \dfrac{2\sqrt{p_0}\, s}{(s+p_0)^2}$.

Now, we wish to construct a band-pass function centered on p_1. Then, the second function $B_1(s)$ is given by:

$$B_1(s) = \frac{A' s}{(s+p_1)^2}\left(\frac{p_0 - s}{s+p_0}\right)^2 . \quad (10)$$

Note that the structure of $B_1(s)$ allows to ensure the orthogonality property:

$$< B_0, B_1 >= 0 . \quad (11)$$

Again the choice of A' achieves the normalization $\|B_1\|^2 =< B_1, B_1 >= 1$:

$$\|B_1\|^2 = \frac{A'^2}{4p_1} . \quad (12)$$

Then, $B_1(s)$ becomes:

$$B_1(s) = \frac{2\sqrt{p_1}\, s}{(s+p_1)^2}\left(\frac{p_0 - s}{s+p_0}\right)^2 . \quad (13)$$

Continuing in this fashion for pulsations $\{p_0, p_1, \cdots, p_{n-1}\}$ then provides the nth band-pass orthogonal basis function:

$$\boxed{B_n(s) - \frac{2\sqrt{p_n}\, s}{(s+p_n)^2}\prod_{k=0}^{n-1}\left(\frac{p_k - s}{s+p_k}\right)^2} . \quad (14)$$

Remark: In the same manner, it can be constructed band-pass orthogonal functions with complex modes or with a more important selectivity.

The same method is used to construct high-pass orthogonal functions, but this time, high-pass functions are introduced, i. e.:

$$B_0(s) = \frac{A s}{s+p_0}$$
$$B_1(s) = \frac{A' s}{s+p_1}\frac{p_0 - s}{s+p_0} . \quad (15)$$

Then the nth high-pass orthogonal basis function is obtained:

$$\boxed{B_n(s) = \frac{\sqrt{2/p_n}\, s}{s+p_n}\prod_{k=0}^{n-1}\frac{p_k - s}{s+p_k}} . \quad (16)$$

These two new orthogonal bases present the particularity to have a null static gain. For this reason, they are not complete bases. But this particularity is used in the quantification of the bias error.

4. SYSTEM IDENTIFICATION

Let $\{y(t), u(t)\}$, $t=0 \,..\, N\text{-}1$, be observations generated by a stable time invariant system with additive stochastic disturbance:

$$y(t) = G(q)u(t) + e(t) \quad (17)$$

where $G(q)$ is the true system, q is shift operator and $e(t)$ is a stochastic disturbance which is uncorrelated with $u(t)$.

Consider the global model $G(s,\underline{\theta})$ constituted of a nominal model $G_N(s,\underline{\theta}_N)$ defined by (3) and a bias error model $G_\Delta(s,\underline{\theta}_\Delta)$ defined by (5) with band-pass or high-pass orthogonal functions respectively given by (14) and (16).

The parametric estimate is performed by minimization of the following quadratic criterion:

$$J(\underline{\theta}) = \frac{1}{N}\sum_{t=0}^{N-1}(y(t) - \hat{y}(\underline{\hat{\theta}}, t))^2 , \quad (18)$$

where $\hat{y}(\underline{\hat{\theta}}, t)$ is the response of the global model $G(s,\underline{\hat{\theta}})$ to $u(t)$ and $\underline{\theta}^T = \lfloor \underline{\theta}_N^T \quad \underline{\theta}_\Delta^T \rfloor$ is the parameter vector.

The solution is obtained by a nonlinear optimization algorithm. Consider the Levenberg-Marquardt algorithm (Marquardt, 1963) given by:

$$\underline{\theta}_{i+1} = \underline{\theta}_i + \Delta\underline{\theta} \quad (19)$$

where $\Delta\underline{\theta}$ is the solution of:

$$\left[\mu_i + J''_{\theta\theta}(\underline{\theta}_i)\right]\Delta\underline{\theta} = -\underline{J}'_\theta(\underline{\theta}_i) . \quad (20)$$

Define parametric sensitivity functions $\underline{\sigma}_N(t)$ of the nominal model $G_N(s,\underline{\hat{\theta}}_N)$:

$$\underline{\sigma}_N(t) = \frac{\partial\, \hat{y}_N(\underline{\hat{\theta}}_N, t)}{\partial \underline{\theta}_N} . \quad (21)$$

Let $\underline{\Psi}_\Delta(t)$, be the regressor of the linear regression $\hat{y}_\Delta(\underline{\hat{\theta}}_\Delta, t) = \underline{\Psi}_\Delta^T(t)\underline{\hat{\theta}}_\Delta$ given by:

$$\underline{\Psi}_\Delta^T(t) = [x_0(t) \quad x_1(t) \quad \ldots \quad x_{r-1}(t)] \quad (22)$$

where $x_i(t)$ is response of $B_i(s)$ to $u(t)$.

Define the vector $\underline{\sigma}(t)$ by:

$$\underline{\sigma}(t)=\begin{bmatrix}\underline{\sigma}_N(t)\\ \underline{\Psi}_\Delta(t)\end{bmatrix}. \qquad (23)$$

Thus, the gradient and the Hessian of the criterion $J(\underline{\theta})$ are respectively given by:

$$J'_\theta(\underline{\theta})=\frac{\partial J(\underline{\theta})}{\partial \underline{\theta}}=-\frac{2}{N}\sum_{k=0}^{N-1}\underline{\sigma}(t)(y(t)-\hat{y}(\hat{\underline{\theta}},t)) \quad (24)$$

$$J''_{\theta\theta}(\underline{\theta})=\frac{\partial^2 J(\underline{\theta})}{\partial \underline{\theta}\,\partial \underline{\theta}^T}\sim\frac{2}{N}\sum_{k=0}^{N-1}\underline{\sigma}(t)\underline{\sigma}^T(t)\ . \quad (25)$$

Remark: This estimator can be initialized by applying successively the State Variable Filter method for the nominal model, and the least squares estimator for the bias error model.

Suppose that the order r of the model $G_\Delta(s,\hat{\underline{\theta}}_\Delta)$ is sufficiently large to consider that the bias error between the global model $G(s,\hat{\underline{\theta}})$ and the true system $G(s)$ is negligible.

Then the prediction error given by:

$$\varepsilon(\hat{\underline{\theta}}_{optimum},t)=y(t)-\hat{y}(\hat{\underline{\theta}}_{optimum},t) \qquad (26)$$

tends to the perturbation $e(t)$ when N tends to infinity. If the perturbation $e(t)$ is a zero mean white noise, then the estimator $\hat{\underline{\theta}}_{optimum}$ has a minimal variance.

Consider now the colored noise case, i.e. $e(t)=H(q)v(t)$ in (17) where $v(t)$ is a zero mean stationary white noise process and $H(q)$ is an ARX structure. Thereby, the estimator $\hat{\underline{\theta}}_{optimum}$ is unbiased for a sufficiently large r, but has not a minimal variance. In order to minimize this variance, it is necessary to whiten the residues $\varepsilon(\hat{\underline{\theta}}_{optimum},t)$

Using least squares technique on the residue sequence, an estimate of $H(q)$ can be obtained. Then, an inverse filter is used to compute new filtered input-output sequences $\{y_f(t),u_f(t)\}$, as follows:

$$y_f(t)=(\hat{H}(q))^{-1}y(t);\ u_f(t)=(\hat{H}(q))^{-1}u(t). \quad (27)$$

From these sequences $\{y_f(t),u_f(t)\}$, a new parameter vector $\hat{\underline{\theta}}_{optimum}$ can be obtained via (19). This procedure must be repeated while the residue sequence $\varepsilon(\hat{\underline{\theta}}_{optimum},t)$ is not white.

So, it is possible to estimate the covariance matrix of the estimator $\hat{\underline{\theta}}_{optimum}$ by:

$$P_{\underline{\theta}}\sim\frac{2\hat{\sigma}_f^2}{N}\left[J''_{\theta\theta}(\hat{\underline{\theta}}_{optimum})\right]^{-1}. \qquad (28)$$

$\hat{\sigma}_f^2$ is the unbiased variance estimate of the white sequence $v(t)$ given by (Ljung, 1987):

$$\hat{\sigma}_f^2=\frac{J(\hat{\underline{\theta}}_{optimum})}{N-dim(\underline{\theta})} \qquad (29)$$

where

$$J(\hat{\underline{\theta}}_{optimum})=\frac{1}{N}\sum_{t=0}^{N-1}(y_f(t)-\hat{y}_f(\hat{\underline{\theta}}_{optimum},t))^2\ .$$

5. ERROR QUANTIFICATION

5.1. Bias error

The undermodelling of the true system $G(s)$ by the nominal system $G_N(s,\underline{\theta})$ is estimated by the bias error model $G_\Delta(s,\hat{\underline{\theta}}_\Delta)$.

The estimated frequency response $G_\Delta(j\omega,\hat{\underline{\theta}}_\Delta)$ gives an information on the distribution of the bias error in frequency domain. For example, a H_∞ norm bound of the bias error can be deduced.

5.2. Variance error

The estimator (19) provides not only an estimate $\hat{\underline{\theta}}$ of the global model parameter vector, but also an estimation of the covariance matrix $P_{\underline{\theta}}$ given by (28), provided that residues are white and that the stochastic perturbation is centered and stationary.
Thus, as it has been shown in (Ljung, 1987; Goodwin et al., 1992; Ninness and Goodwin, 1994), it is possible to quantify the variance error by confidence intervals.

In the parametric domain, a confidence ellipsoid is given by:

$$(\underline{\theta}_*-\hat{\underline{\theta}})^T P_{\underline{\theta}}^{-1}(\underline{\theta}_*-\hat{\underline{\theta}})\sim\chi^2_{dim(\hat{\underline{\theta}})} \qquad (30)$$

where $\underline{\theta}_*=\arg\min_{\underline{\theta}}\left\{\lim_{N\to\infty}\frac{1}{N}E\{J(\underline{\theta})\}\right\}$ is the best estimation of $\underline{\theta}$ (Ljung, 1987).

In the frequency domain, notice that the global model response $G(j\omega,\hat{\underline{\theta}})$ is constituted of:

- a part $G_N(j\omega,\hat{\underline{\theta}}_N)$ nonlinear in parameters,
- a part $G_\Delta(j\omega,\hat{\underline{\theta}}_\Delta)$ linear in parameters.

Therefore, it is necessary for the nonlinear part to realize an approximation by using Taylor's Theorem. Thus, the confidence ellipses of the global model frequency response $G(j\omega,\hat{\underline{\theta}})$ are given by:

$$g^T(f)\,P_g^{-1}\,g(f) \sim \chi_2^2 \qquad (31)$$

with $g(f)=\begin{bmatrix} Re\{G(j\omega,\underline{\theta}_*)-G(j\omega,\hat{\underline{\theta}})\} \\ Im\{G(j\omega,\underline{\theta}_*)-G(j\omega,\hat{\underline{\theta}})\} \end{bmatrix}$,

$$P_g=\Omega(f)\,P_{\underline{\theta}}\,\Omega^T(f)$$

$$\Omega^T(f)=\begin{bmatrix} Re\left\{\left.\dfrac{\partial G_N}{\partial \underline{\theta}_N^T}\right|_{\hat{\underline{\theta}}_N}\right\} & Re\{[B_0\ \dots\ B_{r-1}]\} \\ Im\left\{\left.\dfrac{\partial G_N}{\partial \underline{\theta}_N^T}\right|_{\hat{\underline{\theta}}_N}\right\} & Im\{[B_0\ \dots\ B_{r-1}]\} \end{bmatrix}$$

6. APPLICATION

An application illustrates which is presented in above sections. The process studied in this section is the chemical reaction column shown in figure 1 and used in a vegetable oil esterification process.

The conversion rate of the reaction depends on the temperature of the product. Therefore, we are concentrated on the thermic model of the reactor, particularly the non-heated zone between *T3* and *T2*.

Oil, alcohol and catalyst are introduced at the top of the column and the product of the reaction is extracted at the bottom. Between these two points, the product has crossed three heating zones and two non-heated intervals.

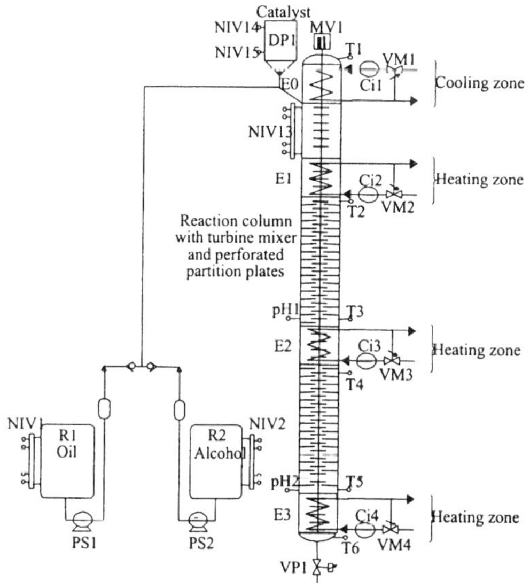

Fig. 1. Chemical reactor of a vegetable oil esterification process.

An acquisition of a temperature measure set with sampling interval $T = 30\,s$ and length $N=1000$ is performed.

The aim is to estimate the thermic transfer $T3(s)/T2(s)$. A preliminary study using the model estimation method (Ouvrard, 1997; Ouvrard *et al.*, 1999) has been led. But, there is not model structure allowing to represent all the underlying dynamics of the system. Then, we use the error quantification method described in this paper.

The global model is constituted by a 3rd order nominal model and a bias error model given by a sum of four high-pass orthogonal functions with $p_i = 8\,10^{-3}\ rad/s$, $i = 0,\dots,3$.

The estimated nominal model is given by:

$$G_N(s,\hat{\underline{\theta}}_N)=\frac{5.3\,10^{-6}\,s+5.6\,10^{-9}}{s^3+7.4\,10^{-3}\,s^2+1.7\,10^{-5}\,s+6.9\,10^{-9}} \qquad (32)$$

The frequency responses of the global model and the nominal model are shown in figure 2 and figure 3. In figure 4, an indication of the bias error in the frequency domain is given. A nominal model simulation in the measures is given in figure 5.

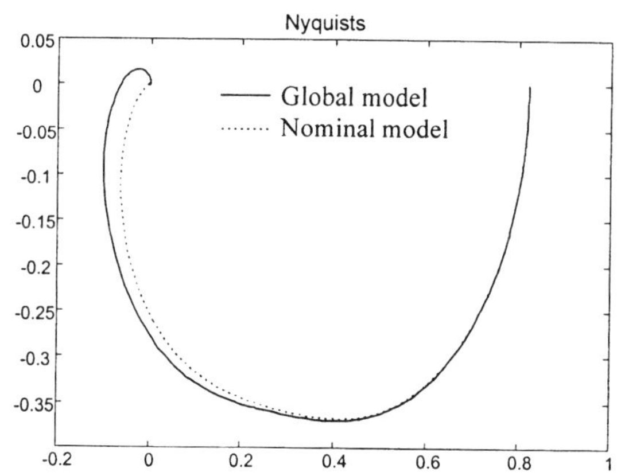

Fig. 2. Frequency responses of global and nominal models.

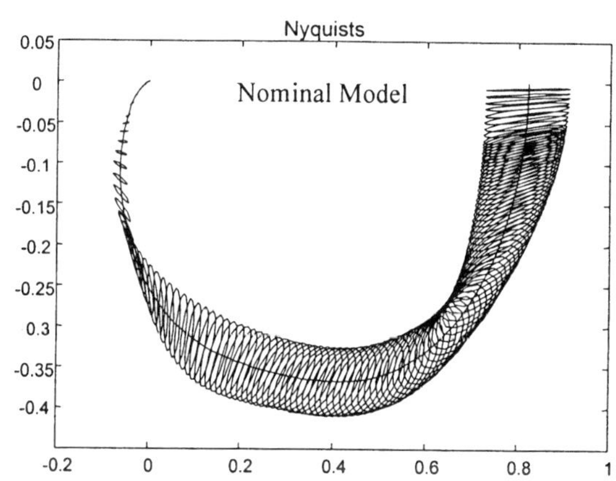

Fig. 3. Frequency response nominal model with confidence ellipses.

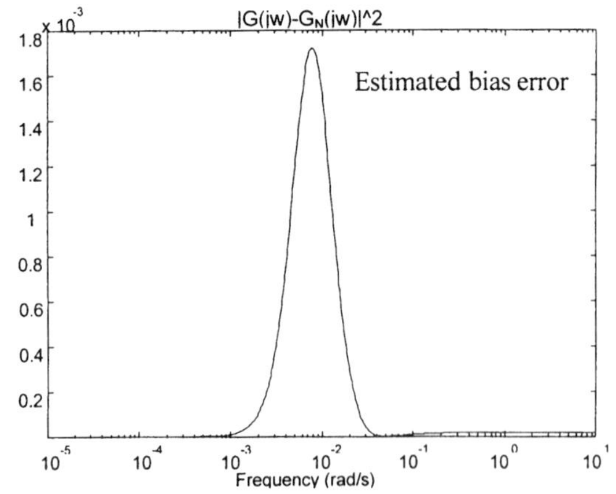

Fig. 4. Estimated bias error in the frequency domain.

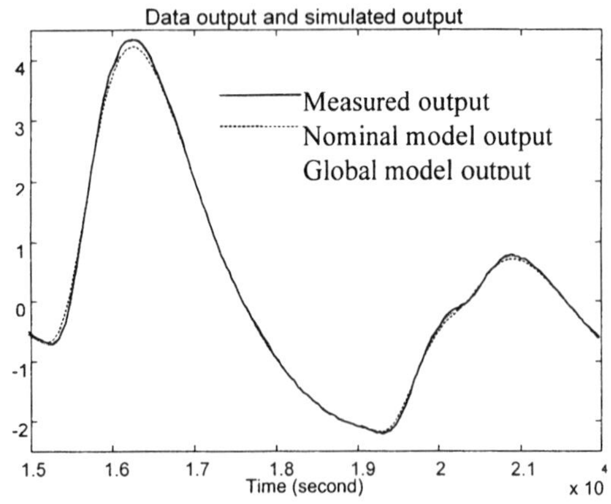

Fig. 5. Model simulations in the measures.

7. CONCLUSION

A continuous time system identification with quantification of both bias and variance errors has been presented.

This technique is based on the estimation of a global model constituted of a nominal model that represents the system behavior in low frequencies, and a bias error model that represents the rapid modes unmodelled by the nominal model.

The bias error model gives an information on the true bias error. The variance error is quantified by classical statistical tools.

Two new particular band-pass and high-pass orthogonal functions has been introduced.

REFERENCES

Goodwin G., M. Gevers and B.M. Ninness (1992). Quantifying the error in estimated transfer functions with application to model order selection. *IEEE Trans. Automat. Contr.*, Vol. **AC-37**, pp. 913-928.

Kosut R., M. Lau and S. Boyd (1992). Set-membership identification of systems with parametric and nonparametric uncertainty. *IEEE Trans. Automat. Contr.*, Vol. **AC-37**, pp. 929-941.

Ljung L. (1987). System identification: theory for the users. (Prentice Hall, Ed.).

Mäkilä P., J.R. Partington and T.K. Gustafsson (1995). Worst-case control-relevant identification. *Automatica*, Vol. **31**, pp. 1799-1819.

Marquardt D.W. (1963). An algorithm for Least-Squares estimation of non-linear parameters. *J. Soc. Indust. Appl. Math.*, Vol. **11**, N° 2.

Ninness B. and G. Goodwin (1994). Estimation of model quality. *10th IFAC Symposium SYSID'94, Copenhagen*, Vol. **1**, pp. 25-44.

Ninness B. and G. Goodwin (1995). Estimation of model quality. *Automatica*, Vol. **31**, No. 12, pp. 1771-1797.

Ninness B. and F. Gustafsson (1994). A unifying construction of orthonormal bases for system identification. *Technical Report EE9432*, Department of Electrical and Computer Engineering, University of Newcastle, Australia.

Ouvrard R. (1997). Contribution à l'identification robuste des systèmes à temps continu - Application au Génie des Procédés-. *Thèse de l'Université de Poitiers*, France.

Ouvrard R., P. Coirault and J.C. Trigeassou (1997). Continuous-time system identification by overparametrized model and orthonormal functions. *11th IFAC Symposium SYSID'97, Fukuoka*, Vol. **3**, pp. 1355-1360.

Ouvrard R., P. Coirault, T. Poinot and J.C. Trigeassou (1999). Continuous time system identification with orthonormal functions -A new approach-. *European Control Conference ECC'99, Karlsruhe.*

BOX-JENKINS CONTINUOUS-TIME MODELING

R. PINTELON, J. SCHOUKENS and Y. ROLAIN

Vrije Universiteit Brussel, Department ELEC, Pleinlaan 2, 1050 Brussel, Belgium

Abstract. This paper treats the identification of continuous-time models using arbitrary band-limited excitation signals. A modeling approach is presented that has the two following advantages: 1) asymptotically (the amount of data tends to infinity) there is no approximation error over the complete frequency band from DC to Nyquist, 2) it allows to identify general parametric noise models. The key idea is to combine a continuous-time plant model with a discrete-time noise model (= hybrid Box-Jenkins model structure). *Copyright © 2000 IFAC*

Key Words. system identification, continuous-time systems and models, frequency domain

1. INTRODUCTION

This paper treats the identification of linear, time invariant, continuous-time systems (see Fig. 1)

$$\sum_{n=0}^{n_a} a_n y^{(n)}(t) = \sum_{m=0}^{n_b} b_m u^{(m)}(t) \quad (1)$$

from N samples $u(rT_s)$ and $y(rT_s)$, $r = 0, 1, ..., N-1$, of the input $u(t)$ and output $y(t)$ signals (T_s stands for the sampling period). The basic problem of parametric continuous-time modeling is the correct reconstruction of the time-derivatives in differential equation (1) from the observed samples $u(rT_s)$ and $y(rT_s)$, $r = 0, 1, ..., N-1$. In practise the output observations are always disturbed by noise and, besides a plant model, one should also identify a noise model for $v(t)$ (see Fig. 2), which complicates the problem. The solutions proposed in the literature can be classified according to the input signal used.

If the *input u(t) is periodic,* then differential equation (1) can be transformed without systematic errors to a set of linear algebraic equations, and the identification can be done in the frequency domain (see Schoukens and Pintelon, 1991; Pintelon *et al.*, 1994, and the references therein). A non-parametric noise model is obtained independently of the plant model in a pre-processing step (Schoukens *et al.*, 1997).

If the *input u(t) is non-periodic and piecewise constant,* then differential equation (1) can be transformed without systematic errors to a difference equation which is parametrized in the original continuous-time model parameters, and the identification can be done using standard time domain methods (see Ljung, 1999 and the references therein). A parametric noise model is identified simultaneously with the plant model (Ljung, 1999).

If the *input u(t) is non-periodic and band-limited,* then differential equation (1) is transformed to a discrete-time equation by explicit or implicit approximation of the time-derivatives using digital filters. According to the digital filters used these methods are known as the Poisson moment functional approach, the integrated sampling approach, the instantaneous sampling approach, the Laguerre approach ... (see Chou *et al.*, 1999; Sinha and Rao, 1991, and the references therein). These modeling approaches work reasonably well (small systematic errors) for frequencies below $f_s/4$, but in the neighbourhood of the Nyquist frequency ($f > f_s/4$) they either introduce large errors or require (very) high order digital filters (Van hamme *et al.*, 1991). Weighted (non-)linear least squares (Sinha and Rao, 1991), maximum likelihood (Van hamme *et al.*, 1991), instrumental variable (Sinha and Rao, 1991) and subspace identification (Johansson *et al.*, 1999) methods have been developed. All these methods, except the ARMAX-modeling in Johansson (1994), either do not identify a noise model or assume that it is known. Also the systematic approximation errors are mostly not taken into account in the consistency proofs.

We conclude from this brief overview that the existing continuous-time modeling approaches for non-periodic band-limited signals firstly do not allow accurate modeling near to the Nyquist frequency and hence always suffer from a (small) bias, and secondly are unable to identify a general parametric noise model. The contribution of this paper is to fill these two gaps:

This work is supported by the Belgium National Fund for Scientific Research; the Belgian Government as a part of the Belgian Programme on Interuniversity Poles of Attraction (IUAP4/2) initiated by the Belgian State, Prime Minister's Office, Science Policy Programming; and the Flemmish Community (GOA-IMMI).

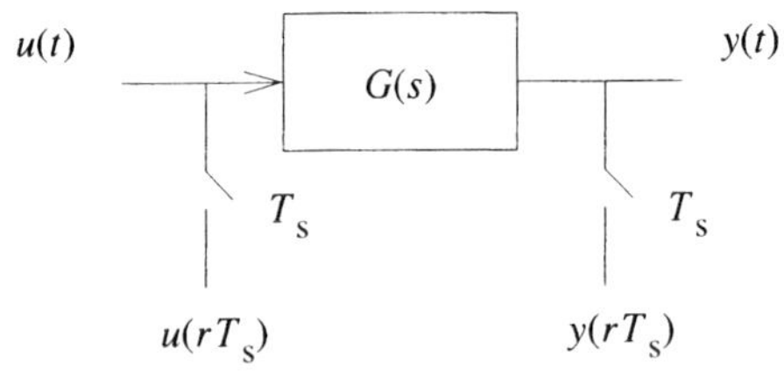

Figure 1: Identification of a continuous-time system: problem statement.

- It is shown that the approximation (alias) errors of the direct approach (Pintelon and Schoukens, 1997) in the complete frequency band from DC ($f = 0$) to Nyquist ($f = f_s/2$) converge to zero at the rate $O(N^{-1/2})$ as $N \to \infty$.

- A consistent estimator for a hybrid Box-Jenkins model structure is developed: it combines a parametric continuous-time plant model with a parametric discrete-time noise model.

Note that the same hybrid modeling idea has been used in Johansson (1994) for continuous-time ARMAX models.

Because of lack of space the proofs of the theorems have been omitted. The reader is referred to Pintelon and Schoukens (2000) and Pintelon *et al.* (2000) for more details.

2. CONTINUOUS-TIME MODEL FOR ARBITRARY SIGNALS

2.1. Introduction

Consider the set up shown in Fig. 1. The relationship between the scaled discrete Fourier transforms (DFT) $U(k)$, $Y(k)$ of the input and output samples $u(rT_s)$, $y(rT_s)$

$$X(k) = N^{-1/2}\sum_{r=0}^{N-1} x(rT_s)z_k^{-r} \tag{2}$$

($z_k = \exp(j2\pi k/N)$, $X = U, Y$ and $x = u, y$), and the plant transfer function

$$G(s, \theta) = \frac{B(s, \theta)}{A(s, \theta)} = \frac{\sum_{m=0}^{n_b} b_m s^m}{\sum_{n=0}^{n_a} a_n s^n} \tag{3}$$

is given by (Pintelon and Schoukens, 1997)

$$Y(k) = G(s_k, \theta)U(k) + T(s_k, \theta) + \delta(s_k) \tag{4}$$

where $\delta(s_k)$ is the alias error, $s_k = j2\pi k/(NT_s)$, and $T(s, \theta)$ the plant transient term

$$T(s, \theta) = \frac{I(s, \theta)}{A(s, \theta)} = \frac{\sum_{m=0}^{n_i} i_m s^m}{\sum_{n=0}^{n_a} a_n s^n} \tag{5}$$

with $n_i = \max(n_a, n_b) - 1$. The parameter vector θ contains the plant model parameters and the equivalent initial conditions of the plant model: $\theta^T = [a^T b^T i^T]$ with $a^T = [a_0 a_1 \ldots a_{n_a}]$, $b^T = [b_0 b_1 \ldots b_{n_b}]$ and $i^T = [i_0 i_1 \ldots i_{n_i}]$. Transfer function model (4) is made identifiable by constraining θ, for example, $a_{n_a} = 1$ or $\|a\|_2^2 + \|b\|_2^2 + \|i\|_2^2 = 1$. Formula (4) is exactly true without any restriction on the bandwidth of the excitation signal $u(t)$. In this section the asymptotic behaviour ($N \to \infty$) of the alias term $\delta(s_k)$ is studied for band-limited excitations signals with finite power.

Definition 1: A deterministic (random) signal $x(t)$ is band-limited if its amplitude (power) spectrum $|X(j\omega)|$ ($S_{xx}(j\omega)$) satisfies: $|X(j\omega)| = 0$ ($S_{xx}(j\omega) = 0$) for any $|\omega| > \omega_B$ with $\omega_B < \omega_s/2$.

Definition 2: A band-limited signal $x(t)$ has finite non-zero power if,

$$\frac{1}{NT_s}\int_{-NT_s/2}^{+NT_s/2} E\{x^2(t)\}dt = O(N^0) > 0 \tag{6}$$

for any N, ∞ included ($E\{\}$ is the expected value).

Definition 3: A normalised periodic signal has the form

$$x(t) = \sum_{k=1}^{F} \frac{A_k}{\sqrt{N}}\sin(\omega_k t + \phi_k) \tag{7}$$

where $A_k > 0$ and where F increases with N. The peak value of $x(t)$ is uniformly bounded ($\max_t |x(t)| \le C < \infty$ for any N, ∞ included).

For stationary random signals (6) implies that $x(t)$ should have a finite non-zero variance.

2.2. Results

The convergence rate to zero of the plant transient term $T(s_k, \theta)$ and the alias term $\delta(s_k)$ is established in the following two lemmas.

Lemma 1 - convergence rate $T(s_k, \theta)$: For bounded excitations $u(t)$ with finite left $(n_b - 1)$th order derivative applied to stable plants $G(s, \theta)$ or unstable plants captured in a stabilizing feedback loop, the plant transient term $T(s_k, \theta)$ tends to zero as $O(N^{-1/2})$.
Proof: See Pintelon *et al.* (2000).

Lemma 2 - convergence rate $\delta(s_k)$: Consider band-limited signals with finite non-zero power. The residual alias error $\delta(s_k)$ tends to zero as $O(N^{-1/2})$ in probability for random excitations with differentiable power spectrum $S_{uu}(j\omega)$, $dS_{uu}(j\omega)/d\omega < \infty$ for $|\omega| \le \omega_B$, and as $O(N^{-1/2})$ for (normalised) periodic excitations.
Proof: See Pintelon and Schoukens (2000) and Pintelon *et al.* (2000).

Lemmas 1 and 2 allow to calculate how fast the extended transfer function model (4) converges to

$$Y(k) = G(s_k, \theta)U(k) \tag{8}$$

as $N \to \infty$. This is summarized in the following theorem.

Theorem 1 - convergence rate extended transfer function model: Under the assumptions of Lemmas 1 and 2 the relative convergence rate of (4) to (8) is $O(N^{-1/2})$ in probability for random excitations with differentiable power spectrum, $O(N^{-1/2})$ for normalised periodic signals (see Definition 3), and $O(N^{-1})$ at the excited DFT frequencies for periodic signals with a fixed number of frequencies ((7) with $F = O(N^0)$ and $A_k/\sqrt{N} = O(N^0)$).

Proof: Follows immediately from Lemmas 1 and 2 and the fact that the DFT spectrum of band-limited signals with finite non-zero power is an $O(N^0)$ for stationary random signals, $O(N^0)$ for normalised periodic signals, and $O(N^{1/2})$ at the excited DFT frequencies for periodic signals with a fixed number of frequencies.

2.3. Discussion

Theorem 1 shows that the classical transfer function model (8) contains asymptotically $(N \to \infty)$ no approximation errors in the complete frequency band from DC to Nyquist for band-limited input signals with finite non-zero power.

The transient term $T(s, \theta)$ is zero if the initial and final conditions of the experiment are equal. This is the case for the steady state response of periodic signals observed during an integer number of periods and for time-limited signals. For the band-limited versions of these signals the alias term $\delta(s_k)$ is also zero.

From Lemmas 1 and 2 it follows that the residual alias term $\delta(s_k)$ and the plant the transient term $T(s_k, \theta)$ converge to zero at the same rate. Hence, $\delta(s_k)$ cannot be neglected w.r.t. $T(s_k, \theta)$, even for "large" N. However, practise has shown that the alias error $\delta(s_k)$ can be approximated well by a polynomial (Pintelon and Schoukens, 1997). Therefore to reduce $\delta(s_k)$ in (4), the order of the polynomial $I(s, \theta)$ is increased: $n_i \geq \max(n_a, n_b) - 1$.

3. HYBRID BOX-JENKINS MODEL STRUCTURE

3.1. The model

The key idea to solve the noisy continuous-time modeling problem (see Fig. 2) is to assume that the disturbing noise $v(t)$ can be written at the sampling instances as filtered white discrete time noise. It avoids the mathematical difficult problem of treating continuous-time random processes and is equivalent to saying that $v(t)$ is a filtered piecewise constant continuous-time noise source (see Fig. 3). Note that exactly the same noise assumption is made in the classical discrete-time Box-Jenkins model (Ljung, 1999). The continuous-time and discrete-time noise transfer functions are related by $H(z^{-1}) = (1 - z^{-1})Z\{s^{-1}H_c(s)\}$ with $Z\{\ \}$ the Z-transform (Middleton and Goodwin, 1990). The relationship between the DFT spectrum $V(k)$ of the samples $v(rT_s)$, $r = 0, 1, \ldots, N-1$, and the discrete-time noise transfer function

$$H(z^{-1}, \theta) = \frac{C(z^{-1}, \theta)}{D(z^{-1}, \theta)} = \frac{1 + \sum_{m=1}^{n_c} c_m z^{-m}}{1 + \sum_{n=1}^{n_d} d_n z^{-n}} \qquad (9)$$

is given by (Pintelon *et al.*, 1997),

$$V(k) = H(z_k^{-1}, \theta)E(k) + T_H(z_k^{-1}, \theta) \qquad (10)$$

where $E(k)$ is the DFT spectrum of the white noise sequence $e(rT_s)$, $r = 0, 1, \ldots, N-1$, and $T_H(z^{-1}, \theta)$

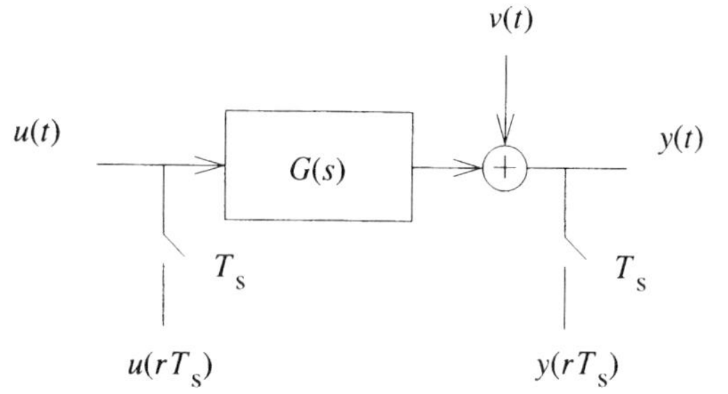

Figure 2: Identification of a continuous-time system from noisy output observations.

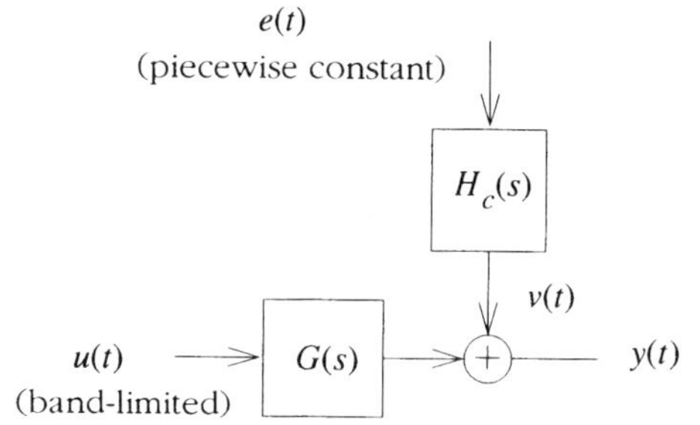

Figure 3: Hybrid Box-Jenkins model structure: band-limited excitation signal $u(t)$ with piecewise constant disturbing noise source $e(t)$.

the noise transient term

$$T_H(z^{-1}, \theta) = \frac{J(z^{-1}, \theta)}{D(z^{-1}, \theta)} = \frac{\sum_{m=0}^{n_j} j_m z^{-m}}{1 + \sum_{n=1}^{n_d} d_n z^{-n}} \qquad (11)$$

with $n_j = \max(n_c, n_d) - 1$. The parameter vector θ contains the noise model parameters and the equivalent initial conditions of the noise model: $\theta^T = [c^T d^T j^T]$ with $c^T = [c_1 \ldots c_{n_c}]$, $d^T = [d_1 \ldots d_{n_d}]$ and $j^T = [j_0 j_1 \ldots j_{n_j}]$. Note that transfer function model (9) is identifiable. The convergence rate to zero of the noise transient term $T_H(z_k^{-1}, \theta)$ is established in the following lemma.

Lemma 3 - convergence rate $T_H(z_k^{-1}, \theta)$: For independently distributed noise $e(rT_s)$ with continuous probability density function (pdf) the noise transient term $T_H(z_k^{-1}, \theta)$ tends to zero as $O(N^{-1/2})$ with probability one.

Proof: See Pintelon *et al.* (2000).

Combining the continuous-time plant model (4) with the discrete-time noise model (10) gives the hybrid Box-Jenkins model structure of Fig. 3

$$\begin{aligned} Y(k) &= G(s_k, \theta)U(k) + H(z_k^{-1}, \theta)E(k) \\ &\quad + T(s_k, \theta) + T_H(z_k^{-1}, \theta) + \delta(s_k) \end{aligned} \qquad (12)$$

with $\theta^T = [a^T b^T c^T d^T i^T j^T]$ and, for example, $a_{n_a} = 1$ or $\|a\|_2^2 + \|b\|_2^2 + \|i\|_2^2 = 1$. Theorem 1 and Lemma 3 allow to calculate the convergence rate of (12) to

$$Y(k) = G(s_k, \theta)U(k) + H(z_k^{-1}, \theta)E(k) \qquad (13)$$

as $N \to \infty$. This is summarized in the following

theorem.

Theorem 2 - convergence rate hybrid Box-Jenkins model: Under the assumptions of Lemmas 1 to 3 the relative convergence rate of (12) to (13) is $O(N^{-1/2})$ in probability for random excitations with differentiable power spectrum, $O(N^{-1/2})$ for normalised periodic signals, and $O(N^{-1})$ at the excited DFT frequencies for periodic signals with a fixed number of frequencies.
Proof: Follows immediately from Theorem 1 and Lemma 3.

3.2. The estimator

The frequency domain estimate $\hat{\theta}_{BJ}$ of the hybrid Box-Jenkins model parameters minimizes

$$V(\theta, z) = \frac{1}{N}\sum_{k=0}^{N-1}\left|\frac{Y(k) - G(s_k, \theta)U(k) - T(s_k, \theta) - T_H(z_k^{-1}, \theta)}{H(z_k^{-1}, \theta)}\right|^2 \qquad (14)$$

w.r.t. θ, where z stands for the vector of the observed input and output DFT spectra. Some additional assumptions are required to prove the consistency of $\hat{\theta}_{BJ}$.

Assumption 1 - disturbing noise: The noise sequence $e(rT_s)$ is independently distributed with zero mean, variance σ^2 and bounded moments of order 4. The noise $e(t)$ is independent of the excitation $u(t)$.

Assumption 2 - existence of a true model: The data is generated according to $y(t) = G(s, \theta_0)u(t) + H_c(s)e(t)$ with $u(t)$ band-limited, $e(t)$ piecewise constant, $G(s, \theta_0)$ stable, and $H(z^{-1}, \theta_0) = (1 - z^{-1})Z\{s^{-1}H_c(s)\}$ a stable and inversely stable monic filter. $G(s, \theta_0)$ and $H(z^{-1}, \theta_0)$ have no cancelling poles and zeros.

Assumption 3 - identifiability condition: There exists an N_0 such that for any $N \geq N_0$, ∞ included, the expected value of the cost function (14), $E\{V(\theta, z)\}$, has a unique global minimum θ_*.

Assumption 4 - regularity condition cost function: There exists an N_0 such that for any $N \geq N_0$, ∞ included, the cost function (14), $V(\theta, z)$, is a continuous function of θ in a closed and bounded neighbourhood of θ_*.

Theorem 3 - consistency plant and noise model parameters: Under the conditions of Lemmas 1 to 3 and Assumptions 1 to 4 the estimates of the plant and noise model parameters are consistent:

$$\underset{N \to \infty}{\text{plim}} \begin{bmatrix} \hat{a}_{BJ} \\ \hat{b}_{BJ} \end{bmatrix} = \begin{bmatrix} a_0 \\ b_0 \end{bmatrix} \text{ and } \underset{N \to \infty}{\text{plim}} \begin{bmatrix} \hat{c}_{BJ} \\ \hat{d}_{BJ} \end{bmatrix} = \begin{bmatrix} c_0 \\ d_0 \end{bmatrix} \qquad (15)$$

Proof: See Pintelon *et al.* (2000).

3.3. Discussion

Although the input signal $u(t)$ is band-limited, the summation in the cost function (14) is taken over all the DFT frequencies. Otherwise the estimate of the noise model parameters would not be consistent. The estimate of the plant model parameters is consistent even if the summation is only taken over a subset of the DFT frequencies, for example, the bandwidth of the excitation (see the proof of Theorem 3). The

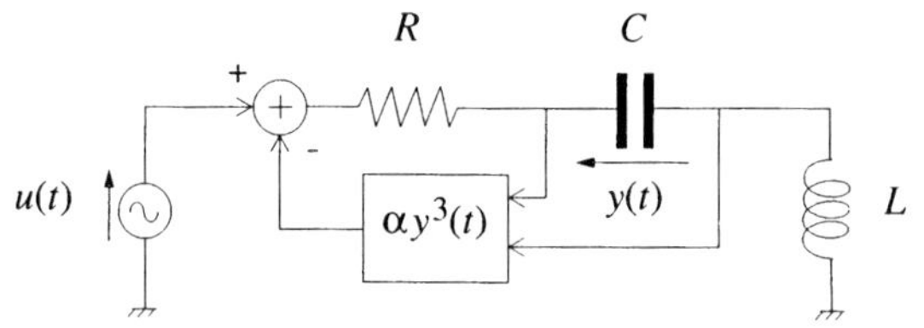

Figure 4: Block diagram of the electrical circuit with $R = 16\ \Omega$, $C = 9.4\ \mu F$, $L = 1$ H and $\alpha \sim 1000\ V^{-2}$. The inductance is constructed using a capacitive loaded gyrator (Temes and LaPatra, 1977). The circuit basically acts as a second order plant with on top some disturbances due to the non-linear feedback. For broad band random excitations and breadboard periodic excitations with random phases $u(t)$ these disturbances act as noise (Schoukens *et al.*, 1998). The rms and peak value of the input signal used equal 16 mV and 60 mV respectively.

consistency of the noise model parameters is maintained if the summation in (14) is replaced by the union of disjoint sets of indices k such that the corresponding z_k values of each set cover uniformly the unit circle (see Pintelon and Schoukens, 1999). This allows filtering of the input and output spectra (for example, removal of every second DFT line) without affecting the consistency of the noise model parameters.

With some additional effort the presented method allows also hybrid ARMAX modeling. The denominators of the plant and the noise model are then parametrized in their poles, and the poles of $H(z^{-1}, \theta)$ are related to those of $G(s, \theta)$ by the impulse invariant transformation $z = \exp(sT_s)$. Starting values are obtained via the hybrid Box-Jenkins model.

The approach is not consistent in feedback. The basic reason for this is that the observed input signal, which is a mixture of a band-limited signal (originating from the reference signal) and a filtered zero-order-hold signal (originating from the process noise), is multiplied in the hybrid Box-Jenkins model structure by a continuous-time plant model.

In practise the continuous-time noise $e(t)$ will mostly not be piecewise constant (Lowen and Teich, 1990; Pyati, 1992) and the discrete-time noise model will be an approximation. This only influences the uncertainty, but not the consistency, of the estimated plant model parameters.

Although the equivalent initial conditions of the plant and noise model are not consistently estimated, they are added in the cost function (14) to improve the finite sample behaviour of the estimates of the plant and noise model parameters.

Although the influence of the alias term $\delta(s_k)$ in the cost function (14) is hidden in the observed output (see (12) with $\theta = \theta_0$), it is present and causes a (small) bias in the estimated model parameters. This bias term tends to zero as N tends to infinity (see Theorem 3). To suppress its influence for finite values of N, the order of the numerator polynomial $I(s, \theta)$ in the plant transient term $T(s, \theta)$ is increased $n_i \geq \max(n_a, n_b) - 1$.

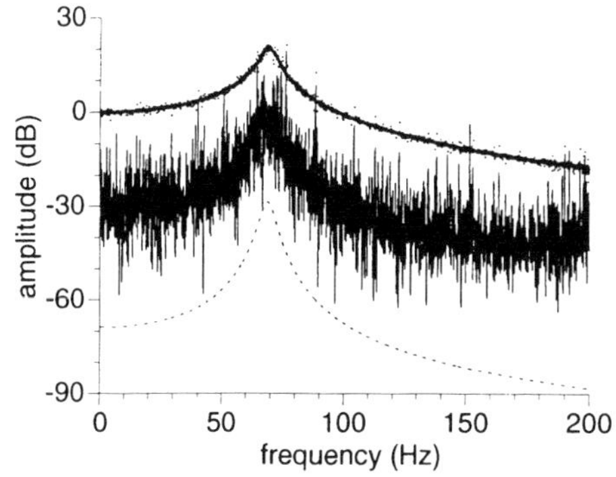

Figure 5: Validation of identified plant model with random excitation measurements. Dots: frequency response function (FRF) $\hat{G}_r(s_k)$ obtained from the random excitation experiment, solid line: difference between FRF and plant model $\hat{G}_r(s_k) - G(s_k, \hat{\theta}_{BJ})$, dashes: variance of the plant model $\mathrm{var}(G(s_k, \hat{\theta}_{BJ}))$.

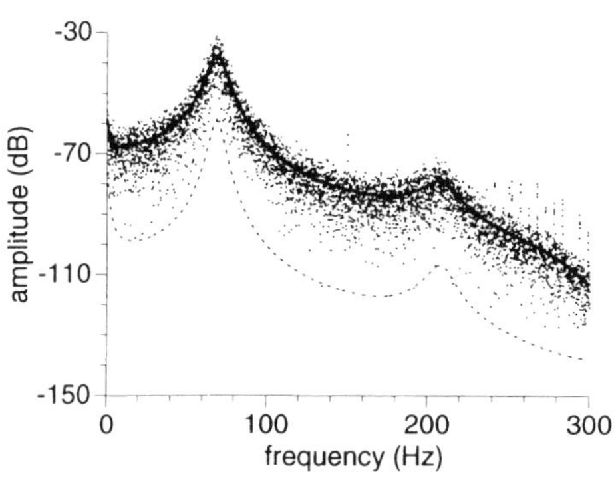

Figure 6: Validation of the identified noise model with the random excitation measurements. Dots: noise residuals $\hat{V}(k)$, solid line: noise model $\hat{\sigma}H(z_k^{-1}, \hat{\theta}_{BJ})$, and dashes: variance of the noise model $\mathrm{var}(\hat{\sigma}H(z_k^{-1}, \hat{\theta}_{BJ}))$.

4. EXPERIMENTAL RESULTS

The hybrid modeling approach is illustrated on real measurements of an electrical circuit which acts as a second order plant (see Fig. 4). Two experiments have been performed, the first with a band limited random signal and the second with a band-limited periodic signal with random phases. Both signals have the same power and the same bandwidth (200 Hz). For both experiments the sampling frequency f_s is about 600 Hz ($20\,\mathrm{MHz}/2^{15}$). The hybrid Box-Jenkins model is estimated using the random excitation experiment and is cross-validated using the periodic excitation experiment.

$N = 10400$ input/output samples of the random excitation experiments are used to identify the Box-Jenkins model (12). It turns out that a model structure with second order plant model ($n_a = 2$, $n_b = 0$, $n_i = 2$) and an eighth order noise model ($n_c = 8$, $n_d = 8$, $n_j = 7$) explains very well the data (see Figures 5 and 6). Figure 5 compares the estimated plant model $G(s_k, \hat{\theta}_{BJ})$ to the frequency response function $\hat{G}_r(s_k)$ obtained from the random excitation experiment

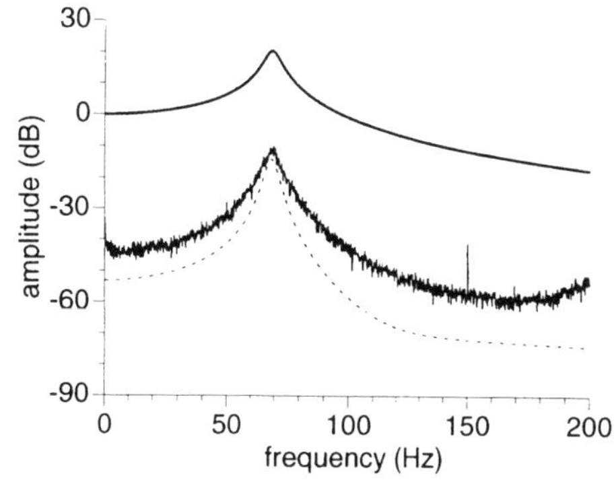

Figure 7: Cross-validation of the identified plant model with the periodic excitation measurements. Bold line: frequency response function (FRF) $\hat{G}_p(s_k)$ obtained from the periodic excitation experiment, solid line: variance measured FRF $\mathrm{var}(\hat{G}_p(s_k))$, and dashes: difference between the plant model $G(s_k, \hat{\theta}_{OE})$ identified from the periodic excitation measurements and the plant model $G(s_k, \hat{\theta}_{BJ})$ identified from the random excitation measurements.

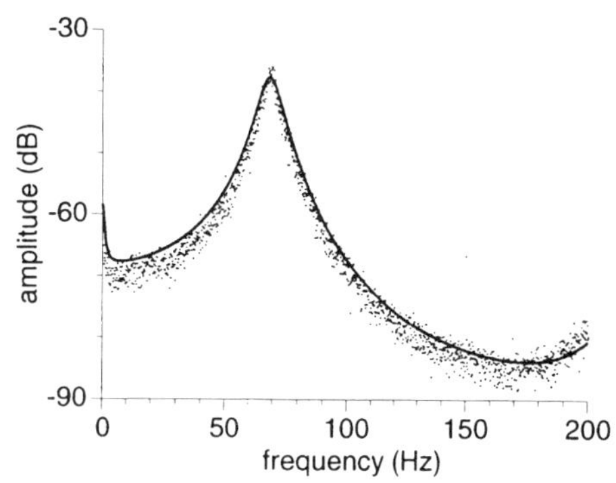

Figure 8: Cross-validation of the identified noise model with the periodic excitation measurements. Dots: non-parametric noise model $\mathrm{var}(\hat{G}_p(s_k))|U(k)|^2$ obtained from the periodic excitation experiment, and solid line: parametric noise model $\hat{\sigma}H(z_k^{-1}, \hat{\theta}_{BJ})$ obtained from the random excitation experiment.

$$\hat{G}_r(s_k) = \frac{Y(k) - T(s_k, \hat{\theta}_{BJ}) - T_H(z_k^{-1}, \hat{\theta}_{BJ})}{U(k)} \tag{16}$$

while Figure 6 compares the estimated noise model $H(z_k^{-1}, \hat{\theta}_{BJ})$ to the noise residuals

$$\hat{V}(k) = Y(k) - G(s_k, \hat{\theta}_{BJ})U(k) - T(s_k, \hat{\theta}_{BJ}) - T_H(z_k^{-1}, \hat{\theta}_{BJ}) \tag{17}$$

Note that the plant model is only shown within the bandwidth of the excitation signal while the noise model is shown from DC ($f = 0$) to Nyquist ($f = f_s/2$).

The estimated Box-Jenkins model is cross-validated with the periodic measurements in Figures 7 and 8. Figure 7 compares the estimated plant model $G(s_k, \hat{\theta}_{BJ})$ to the frequency response function $\hat{G}_p(s_k)$ obtained from the periodic excitation experiment, while Figure 8 compares the estimated noise model $H(z_k^{-1}, \hat{\theta}_{BJ})$ to the non-parametric noise model $\mathrm{var}(\hat{G}_p(s_k))|U(k)|^2$ obtained from the periodic measurements. It follows that the estimated Box-Jenkins model explains very well the periodic excitation experiment. Note that the cross-validation of the noise model is only possible within the bandwidth of the excitation signal.

5. CONCLUSION

The simplicity of the presented continuous-time modeling is very appealing: the method merely boils down to a classical processing of the input and output DFT spectra and does not require the design of any digital filter. The proposed hybrid Box-Jenkins model structure contains asymptotically no plant approximation errors and allows general parametric noise models. A disadvantage of the method is that it is not consistent in feedback.

6. REFERENCES

Brillinger, D. R., *Time Series: Data Analysis and Theory*. New York: McGraw-Hill, 1981.

Chou, C. T., M. Verhaegen and R. Johansson (1999). Continuous-Time Identification of SISO Systems Using Laguerre Functions. *IEEE Trans. Signal Processing,* Vol. 47, no 2, pp. 349-362.

Johansson, R. (1994). Identification of Continuous-Time Models. *IEEE Trans. Signal Processing,* Vol. 42, no 4, pp. 887-897.

Johansson, R., M. Verhaegen and C. T. Chou (1999). Stochastic Theory of Continuous-Time State-Space Identification. *IEEE Trans. Signal Processing,* Vol. 47, no 1, pp. 41-51.

Kailath, T. (1980). *Linear Systems*. Prentice-Hall, Englewood Cliffs, NJ (USA).

Ljung, L. (1999). *System Identification, Theory for the User*. Prentice-Hall, Upper Saddle River, NJ (USA).

Lowen, S. B. and M. C. Teich (1990). Power-Law Shot Noise, *IEEE Trans. Inform. Theory,* Vol. IT-36, n° 6, pp. 1302-1318.

Lukacs, E. (1975). *Stochastic Convergence*. Academic Press, New York (USA).

Middleton, R. H. and G. C. Goodwin (1990). *Digital Control and Estimation*. Prentice-Hall, London.

Pintelon, R., P. Guillaume, Y. Rolain, J. Schoukens and H. Van hamme (1994). Parametric Identification of Transfer Functions in the Frequency Domain - A Survey. *IEEE Trans. Automat. Contr.,* Vol. 39, no 11, pp. 2245-2260.

Pintelon, R. and J. Schoukens (1997). Identification of Continuous-Time Systems Using Arbitrary Signals. *Automatica,* Vol. 33, no 5, pp. 991-994.

Pintelon, R. and J. Schoukens (1999). Time Series Analysis in the Frequency Domain, *IEEE Trans. Sign. Proc.,* Vol. 47, no 1, pp. 206-210.

Pintelon, R. and J. Schoukens (2000). Box-Jenkins Continuous-time modeling, Internal Note 2000/01.

Pintelon, R., J. Schoukens and Y. Rolain (2000). Box-Jenkins Continuous-time modeling, *Automatica,* Vol. 36, no 7.

Pintelon, R., J. Schoukens and G. Vandersteen (1997). Frequency Domain System Identification Using Arbitrary Signals, *IEEE Trans. Autom. Contr.,* Vol. AC-42, no 12, pp. 1717-1720.

Pyati V. P. (1992). An Exact Expression for the Noise Voltage Across a Resistor Shunted by a Capacitor, *IEEE Trans. Circuits and Systems-I,* Vol. CAS-39, n° 12, pp. 1027-1029.

Schoukens, J. and R. Pintelon (1991). *Identification of Linear Systems: A Practical Guideline to Accurate Modeling*. Pergamon, London (GB).

Schoukens, J., G. Vandersteen, R. Pintelon and P. Guillaume (1997). Frequency Domain System Identification Using Non-parametric Noise Models Estimated from a Small Number of Data Sets, *Automatica,* Vol. 33, no 6, pp. 1073-1086.

Schoukens J., T. Dobrowiecki and R. Pintelon (1998). Parametric Identification of Linear Systems in the Presence of Nonlinear Distortions. A Frequency Domain Approach, *IEEE Trans. Autom. Contr.,* Vol. AC-43, no 2, pp. 176-190.

Sinha, N. K. and G. P. Rao, eds. (1991). *Identification of Continuous-time Systems*. Kluwer Academic Publishers, Dordrecht (The Netherlands).

Söderström, T. (1974). Convergence properties of the generalized least squares identification method, *Automatica,* Vol. 10, pp. 617-626.

Temes G. C. and J. W. LaPatra (1977). *Circuit Synthesis and Design*. McGraw-Hill, New York (USA).

Van hamme, H., R. Pintelon and J. Schoukens (1991). Discrete-time modeling and identification of continuous-time systems: a general framework. *Chapter 2 of 'Identification of Continuous-time Systems', edited by Sinha and Rao,* Kluwer Academic Publishers, Dordrecht (The Netherlands), pp. 17-77.

IDENTIFICATION METHODS OF DYNAMIC SYSTEMS IN PRESENCE OF INPUT NOISE

Umberto Soverini * Torsten Söderström **

* *Department of Electronics, Information Science and Systems*
University of Bologna
Viale del Risorgimento 2, I-40136 Bologna, Italy
phone: +390512093026; e-mail: usoverini@deis.unibo.it
** *Department of Systems and Control, Information Technology*
Uppsala University
SE-75103 Uppsala, Sweden
phone: +46184713075; e-mail: ts@SysCon.uu.se

Abstract: In many practical situations the observed input–output data of an identified process are corrupted by noise both on the input and output. It is well-known that in this case many identification methods give biased results. In this work three different approaches characterized by consistent estimates of the system parameters are briefly described and their performances are compared on the basis of numerical simulations.

Keywords: System identification, parameter estimation, stochastic systems, linear systems, errors–in–variables models, correlation methods, prediction error methods

1. INTRODUCTION

The parameter estimation of linear dynamic systems from noise–corrupted measurements is a fundamental research problem which has been deeply investigated in the past decades and for which many different solutions have been presented (Ljung, 1987; Söderström and Stoica, 1989). Representations where errors or measurement noises are present on both the inputs and outputs are usually called "errors–in–variables" models and play an important role when the identification purpose is the determination of the inner laws that describe the process, rather than the prediction of its future behaviour.

One of the first solutions for identification of dynamic errors–in–variables models was presented in the works of Koopmans (1937) and Levin (1964). In more recent years the Koopman–Levin method has been reproposed by several authors with slightly different variants (Fernando and Nicholson, 1985; Van Huffel, 1997).

In Beghelli *et al.* (1990) a classical algebraic estimation problem, the so called "Frisch scheme", was extended to the identification of linear dynamic errors–in–variables systems in presence of white and uncorrelated additive noises with unknown (and possibly different) variances.

From this point of view, the compensated–bias least–squares techniques (Stoica and Söderström, 1982) are even more flexible since they can work under more general noise conditions (Zheng and Feng, 1989; Zheng, 1999).

In Söderström (1981) the performance of different identification methods was compared in an errors–in–variables context. One of these approaches is the classical instrumental variable method, with suitably delayed inputs as instruments. One of the principal advantages of the method is its applicability under fairly general noise conditions; moreover it is rather inexpensive from a computational point of view. However, with this procedure a poor accuracy of the parameter estimates is often obtained.

A particular and interesting method which was proposed in Söderström (1981) is the so called "Joint-Output" approach. In this case both the inputs and outputs of the errors-in-variables model are regarded as outputs of a multivariable system whose parameters are then estimated by using a prediction error method. This technique needs to extend the system representation by including also models for the noises and for the undisturbed input signal. The coefficients of these parts of the process are then to be estimated together with the system parameters.

In this paper the performances of the Instrumental Variable method, the Frisch Scheme approach and the Joint-Output method are analyzed on the basis of numerical simulations. The content of this work is organized as follows. Section 2 gives the definition of the identification problem for errors-in-variables models. Then, the main characteristics of the identification approaches are briefly recalled and described in Section 3. A simulated system already proposed in the literature is used to compare the different identification methods. The discussion of the obtained results is reported in Section 4. Finally, some concluding remarks are given in Section 5.

2. STATEMENT OF THE PROBLEM

Let us consider the single-input, single-output, linear, discrete, time-invariant system described in Figure 1.

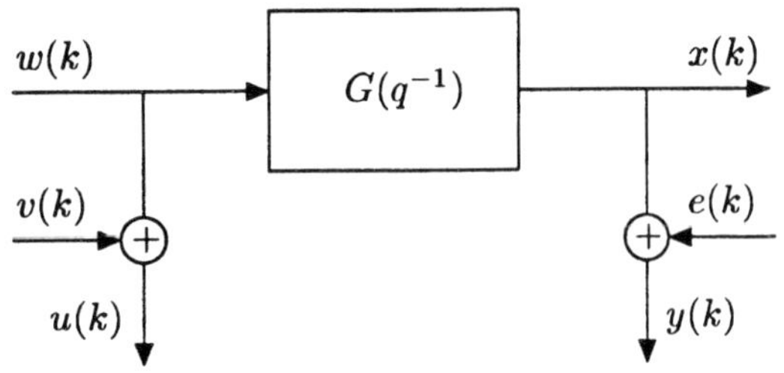

Fig. 1. Errors-in-variables description

The noise-free input is denoted by $w(k)$ and the undisturbed output by $x(k)$. They are linked through the linear difference equation

$$A(q^{-1})\, x(k) = B(q^{-1})\, w(k) \ , \qquad (1)$$

where $A(q^{-1})$ and $B(q^{-1})$ are polynomials of the type

$$\begin{aligned} A(q^{-1}) &= 1 + a_1\, q^{-1} + \ldots + a_n\, q^{-n} \qquad (2) \\ B(q^{-1}) &= b_1\, q^{-1} + \ldots + b_n\, q^{-n} \end{aligned}$$

and q^{-1} is the backward shift operator. It is not restrictive to assume that the polynomials $A(q^{-1})$, $B(q^{-1})$ have equal degree n, which represents the order of the system.

In an errors-in-variables environment the observations are corrupted by additive measurement noises, i.e.

$$\begin{aligned} u(k) &= w(k) + v(k) \qquad (3) \\ y(k) &= x(k) + e(k) \ . \end{aligned}$$

The following assumptions are introduced.

A1. The dynamic system (1) is asymptotically stable, i.e. $A(z)$ has all zeros outside the unite circle; all the system modes are observable and controllable, i.e. $A(z)$ and $B(z)$ have no common factors. Moreover, the order n of the system is *a priori* known.

A2. The true input $w(k)$ is a zero-mean stationary ergodic random signal with rational spectrum.

A3. The sequences $v(k)$ and $e(k)$ are zero-mean white noises with unknown variances λ_v^* and λ_e^*, respectively. These processes are mutually uncorrelated and uncorrelated with the noise-free signals $w(k)$ and $x(k)$.

The problem under investigation can be stated as follows.

Problem 1. Let the available measurements $u(k)$ and $y(k)$ be generated in accordance with the previous assumptions. Determine the system characteristics, i.e. the transfer function

$$G(q^{-1}) = \frac{B(q^{-1})}{A(q^{-1})} \qquad (4)$$

and the variances λ_v^* and λ_e^* of the noise sequences.

3. ESTIMATION METHODS

In this section we describe the principal features of three different identification methods: the Instrumental Variable method, the Joint-Output method and the Frisch scheme approach. The latter method is rather novel and a complete consistency analysis for it is not yet available.

3.1 *Instrumental Variable method*

For the dynamic system (1) the following *Yule-Walker* equations hold for all k

$$A(q^{-1})\, r_{xw}(k) = B(q^{-1})\, r_w(k) \ , \qquad (5)$$

where $r_w(i) = \mathrm{E}[w(k+i)\, w(k)]$ is the auto-correlation function of the signal $w(k)$ and $r_{xw}(i) = \mathrm{E}[x(k+i)\, w(k)]$ is the cross-correlation function between $x(k)$ and $w(k)$.

Since the measurement noises are white and uncorrelated, for the noisy signals $u(k)$ and $y(k)$ we have $r_{yu}(k) = r_{xw}(k)$ for all k, and $r_u(k) = r_w(k)$ for $k \neq 0$.

It is then possible to determine the coefficients of polynomials $A(q^{-1})$ and $B(q^{-1})$ by using the following matrices obtained with the samples $r_u(k)$, $r_{yu}(k)$ $(1 \leq k \leq 3n+1)$

$$\Sigma^s(uu) = \begin{bmatrix} r_u(n) & \dots & r_u(1) \\ r_u(n+1) & \dots & r_u(2) \\ \vdots & \vdots & \vdots \\ r_u(3n) & \dots & r_u(2n+1) \end{bmatrix}, \quad (6)$$

$$\Sigma^s(yu) = \begin{bmatrix} -r_{yu}(n+1) & \dots & -r_{yu}(1) \\ -r_{yu}(n+2) & \dots & -r_{yu}(2) \\ \vdots & \vdots & \vdots \\ -r_{yu}(3n+1) & \dots & -r_{yu}(2n+1) \end{bmatrix}. \quad (7)$$

Thus, from the Yule–Walker equations (5) it follows that

$$\Sigma^s p_n = 0 , \quad (8)$$

where Σ^s is the following $(2n+1) \times (2n+1)$ matrix

$$\Sigma^s = \begin{bmatrix} \Sigma^s(yu) & | & \Sigma^s(uu) \end{bmatrix} \quad (9)$$

and

$$p_n = [1, \theta^T]^T , \quad \theta = [a_1, \dots, a_n, b_1, \dots, b_n]^T \quad (10)$$

is the vector containing the coefficients of $G(q^{-1})$.

This approach can be interpreted as an Instrumental Variable (IV) method using delayed inputs as instruments; it is not difficult to see, in fact, that previous relations can be settled in standard IV decsription (Söderström and Stoica, 1989).

Remark 1. This method leads to a consistent estimate of the parameters. However, it is characterized by a bad numerical accuracy (Söderström, 1981; Söderström and Stoica, 1981). In spite of this disadvantage, the required computational load is quite small since this technique involves simple calculations only. Moreover, it can be noted that, with a suitable choice of the instrumental vector $z(k)$, the IV method can be easily extended for more general noise conditions.

3.2 *Frisch Scheme approach*

With reference to the noisy–free processes $w(k)$ and $x(k)$, let us consider the $\ell \times \ell$ symmetric Toeplitz matrix $\hat{\Sigma}_\ell(ww)$, built with the samples $r_w(k)$ $(k = 0, \dots, \ell-1)$

$$\hat{\Sigma}_\ell(ww) = \begin{bmatrix} r_w(0) & \dots & r_w(\ell-1) \\ r_w(1) & \dots & r_w(\ell-2) \\ \vdots & \vdots & \vdots \\ r_w(\ell-1) & \dots & r_w(0) \end{bmatrix}, \quad (11)$$

and similarly

$$\hat{\Sigma}_\ell(xx) = \begin{bmatrix} r_x(0) & \dots & r_x(\ell) \\ r_x(1) & \dots & r_x(\ell-1) \\ \vdots & \vdots & \vdots \\ r_x(\ell) & \dots & r_x(0) \end{bmatrix}. \quad (12)$$

In the same way, with the samples $r_{xw}(k)$ $(k = -\ell+1, \dots, \ell)$ the following $\ell \times (\ell+1)$ matrix is introduced

$$\hat{\Sigma}_\ell(xw) = \begin{bmatrix} -r_{xw}(1) & \dots & -r_{xw}(-\ell+1) \\ -r_{xw}(2) & \dots & -r_{xw}(-\ell+2) \\ \vdots & \vdots & \vdots \\ -r_{xw}(\ell) & \dots & -r_{xw}(0) \end{bmatrix}. \quad (13)$$

We can now define the $(2\ell+1) \times (2\ell+1)$ symmetric matrix $\hat{\Sigma}_\ell$ partitioned in the following way

$$\hat{\Sigma}_\ell = \begin{bmatrix} \hat{\Sigma}_\ell(xx) & \hat{\Sigma}_\ell(wx) \\ \hat{\Sigma}_\ell(xw) & \hat{\Sigma}_\ell(ww) \end{bmatrix}, \quad (14)$$

where $\hat{\Sigma}_\ell(wx) = \hat{\Sigma}_\ell(xw)^T$ and $r_{wx}(k) = r_{xw}(-k)$.

It can be easily seen that the matrix $\hat{\Sigma}_\ell$ becomes singular when $\ell \geq n$. In particular, when $\ell = n$ we have $\ker \hat{\Sigma}_n = p_n$, where p_n is the vector defined in (10). With reference to the case $\ell = n+1$ we have

$$\ker \hat{\Sigma}_{n+1} = \text{im}\,[p'_{n+1}\ p''_{n+1}] , \quad (15)$$

where

$$\begin{aligned} p'_{n+1} &= [0, 1, a_1, \dots, a_n, 0, b_1, \dots, b_n]^T \quad (16) \\ p''_{n+1} &= [1, a_1, \dots, a_n, 0, b_1, \dots, b_n, 0]^T . \end{aligned}$$

In presence of additive and uncorrelated white noises, we can introduce the covariance matrix of $u(k)$ and $y(k)$, satisfying the relation

$$\Sigma_\ell = \hat{\Sigma}_\ell + \tilde{\Sigma}^*_\ell , \quad (17)$$

where

$$\tilde{\Sigma}^*_\ell = \text{diag}\,[\lambda^*_e I_{\ell+1},\ \lambda^*_v I_\ell] \quad (18)$$

is the covariance matrix of the noises. In Beghelli *et al.* (1990) it was shown that, for every ℓ, the matrices $\tilde{\Sigma}_\ell = \text{diag}\,[\lambda_e I_{\ell+1},\ \lambda_v I_\ell]$ such that

$$\hat{\Sigma}_\ell = \Sigma_\ell - \tilde{\Sigma}_\ell \geq 0 \quad (19)$$

are defined by the points of a convex curve belonging to the first quadrant of the noise plane (λ_e, λ_v). Every point $P = (\lambda_e, \lambda_v)$ of the curve of order ℓ is related to a possible model $p_\ell(P)$, given by

$$\hat{\Sigma}_\ell(P) p_\ell(P) = 0 \ . \qquad (20)$$

In particular the point $P^* = (\lambda_e^*, \lambda_v^*)$, corresponding to the actual variances of the noises, belongs to all curves related to orders $\ell \geq n$. Moreover, the corresponding model, $p_n(P^*)$, is characterized by the actual coefficients of the system. The search for the solution of Problem 1 may thus start from the determination, in the plane (λ_e, λ_v), of the common point P^*.

For this purpose, we can search among the solutions $p_n(P)$ satisfying condition (19) the one which minimizes the cost function

$$\begin{aligned} J(P) = \text{trace}\, & \left([p'_{n+1}(P)\ p''_{n+1}(P)]^T \right. \qquad (21) \\ & \left. \hat{\Sigma}_{n+1}(\bar{P})\, [p'_{n+1}(P)\ p''_{n+1}(P)] \right) , \end{aligned}$$

where $\bar{P} = (\bar{\lambda}_e, \bar{\lambda}_v)$ and its coordinates are given by

$$\bar{\lambda}_v = \frac{\lambda_v}{\sigma_M} \qquad \bar{\lambda}_e = \frac{\lambda_e}{\sigma_M} \ , \qquad (22)$$

where (Guidorzi and Pierantoni, 1995)

$$\sigma_M = \max \text{eig} \left(\Sigma_{n+1}^{-1} \, \text{diag}[\lambda_e I_{n+1}, \lambda_v I_n] \right) . \quad (23)$$

It can be easily seen that $J(P^*) = 0$ when all the assumptions are satisfied.

Remark 2. Even if a theoretical consistency analysis of this procedure is not available at the moment, the simulation results have shown that the method has good statistical properties and is robust with respect to violation of the assumptions.

3.3 *Joint–Output approach*

In this method the errors–in–variable model is regarded as a multivariable system with both $u(k)$ and $y(k)$ as outputs. The signal $w(k)$ is decribed as an ARMA process of the type

$$w(k) = F(q^{-1})\, \rho(k) = \frac{C(q^{-1})}{D(q^{-1})}\, \rho(k) \ , \quad (24)$$

where $\rho(k)$ is a white noise with variance λ_ρ^* and the polynomials $C(q^{-1})$, $D(q^{-1})$ are of the form

$$\begin{aligned} C(q^{-1}) &= 1 + c_1 q^{-1} + \ldots + c_p q^{-p} \qquad (25) \\ D(q^{-1}) &= 1 + d_1 q^{-1} + \ldots + d_p q^{-p} \ , \end{aligned}$$

with known degree p; moreover, $C(z)$ and $D(z)$ are relatively prime and asymptotically stable polynomials. In this way, the whole errors–in–variables model can be considered as a system with a two–dimensional output vector $Y(k) = [u(k)\, y(k)]^T$ and three mutually uncorrelated white noise sources $e(k)$, $v(k)$ and $\rho(k)$:

$$\begin{bmatrix} u(k) \\ y(k) \end{bmatrix} = \begin{bmatrix} F(q^{-1}) & 0 & 1 \\ G(q^{-1})\, F(q^{-1}) & 1 & 0 \end{bmatrix} \begin{bmatrix} \rho(k) \\ e(k) \\ v(k) \end{bmatrix} . \, (26)$$

Note that the model $F(q^{-1})$ of the undisturbed input is a part of the errors–in–variables representation and its coefficients must be estimated together with the parmeters of $G(q^{-1})$.

Using well–known techniques the model (26) can be converted into the innovations form

$$Y(k) = S(q^{-1})\, \varepsilon(k) \ , \qquad (27)$$

where $S(z)$ is the square (2×2) canonical (stable and minimum–phase) spectral factor of $Y(k)$ and $\varepsilon(k)$ is the two–dimensional innovations vector with identity covariance matrix. One way to compute (27) from (26) is to convert the model (26) into a state space form. Of course, this step can be performed in many different way. Then, the innovations form (27) can be computed from the solution of a Riccati equation (Söderström, 1981).

Since the innovations $\varepsilon(k)$ of model (27) can also be interpreted as the one step ahead prediction errors, a prediction error method can be applied for the identification of the system parameters. The parameter vector θ includes the coefficients of polynomials $A(q^{-1})$, $B(q^{-1})$, $C(q^{-1})$, $D(q^{-1})$ and the variances λ_ρ, λ_v, λ_e . The solution of Problem 1 is then determined by minimizing the loss function

$$J(\theta) = \frac{1}{N} \sum_{k=1}^{N} \varepsilon(k)^T\, \varepsilon(k) \ , \qquad (28)$$

where N is the number of input–output samples.

Remark 3. In Söderström (1981) it was proved that this method yields consistent parameter estimates under quite mild conditions; its main disadvantage is that the numerical optimization procedure is, in general, quite involved since, at every step, a Riccati equation must be solved in order to find the innovations $\varepsilon(k)$ used in (28). Moreover, the procedure may fail to give good results when only poor initial parameter estimates are available.

4. SIMULATION RESULTS

In this section a standard example described in Söderström (1981) has been used in the simulations. Let us consider the following system

$$A(q^{-1})\,x(k) = B(q^{-1})\,w(k)\ , \qquad (29)$$

where

$$A(q^{-1}) = 1 - 1.5\,q^{-1} + 0.7\,q^{-2} \qquad (30)$$
$$B(q^{-1}) = 1.0\,q^{-1} + 0.5\,q^{-2}\ .$$

The undisturbed input $w(k)$ is modelled as

$$w(k) = \frac{C(q^{-1})}{D(q^{-1})}\,\rho(k)\ , \qquad (31)$$

where $\rho(k)$ is a white noise with $\lambda_\rho^* = 1$ and

$$C(q^{-1}) = 1 \qquad (32)$$
$$D(q^{-1}) = 1 - 0.9\,q^{-1}\ .$$

The variances of the noise–free signals are

$$E\,[w(k)^2] = 5.26\ , \quad E\,[x(k)^2] = 320.2\ . \quad (33)$$

Example 1. As a first comparative study of the described identification methods, 10 different realizations of $N = 200$ data points have been generated; the measurement noise variances are $\lambda_v^* = \lambda_e^* = 1$.

Table 1 reports the mean values and the standard deviations of the estimated system parameters with the Instrumental Variable method (IV), the Frisch Scheme approach (FR) and the Joint–Output approach (JO), respectively. Three runs with the IV method gave unstable models; the failing runs have not been included in the results. It can be seen from Table 1 that the parameters estimate obtained with the IV method is not very accurate. The FR and JO methods are characterized, approximately, by the same level of accuracy even if the JO method is to be preferred. Moreover, it can be noted that both the methods yield a better estimate for the coefficients of $A(z)$ rather than of $B(z)$. The JO method also gives a good estimate of the noise–free input parameters.

As a measure of the quality of the estimated models, the variance of the equation error $\zeta(k) = y(k) - \varphi(k)^T\,\theta$ has been considered, where

$$\varphi(k)^T = [-y(k-1),\ldots,-y(k-n), \qquad (34)$$
$$u(k-1),\ldots,u(k-n)]\ .$$

It can be easily verified that

$$V = \mathrm{E}\,[\zeta(k)\,\zeta(k)^T] \qquad (35)$$
$$= \lambda_e^* + \theta^T\,\mathrm{diag}\,[\lambda_e^* I_n,\ \lambda_v^* I_n]\,\theta.$$

In Table 2 the estimates of the noise variances and the values of

$$V(\hat\theta_h) = \frac{1}{N}\sum_{k=1}^{N}\left[y(k) - \varphi(k)^T\,\hat\theta_h(N)\right]^2 \qquad (36)$$

have been reported, where $\hat\theta_h(N)$ denotes the estimate of the parameter vector obtained with the different identification methods (h = IV, FR, JO) and $N = 200$.

Method	λ_e	λ_v	$V(\hat\theta_h)$
IV	—	—	27.898 ± 37.557
FR	1.024 ± 0.111	0.936 ± 0.144	5.435 ± 0.749
JO	0.988 ± 0.223	0.975 ± 0.157	5.182 ± 0.613
true value	1.0	1.0	4.99

Table 2. Means and standard deviations for 10 runs (N=200)

The bad performance of the IV method is pointed out in Table 2 by $V(\hat\theta_{IV})$ which is very far from the true value. Good estimates are obtained with the FR and JO methods, with quite similar results.

A comparison between the accuracy of the parameter estimates and the computational load required by the different methods has also been performed. The simulations have shown that the IV and FR methods are characterized by a similar computational load while the JO method is, approximately, 20 times slower. On the other hand, the JO method yields very good estimates of the parameters, while the results obtained with the IV method are very inaccurate.

Example 2. The robustness of the described approaches with respect to different amounts of input and output noise has been tested on 15 noise conditions, associated to all possible pairs of noise variances when $\lambda_v^* = \{1,2,3\}$ and $\lambda_e^* = \{1,5,10,20,50\}$. For a direct comparison of the Frisch Scheme approach and the Joint–Output method, the results of a Monte Carlo simulation based on 20 independent realizations with sample size $N = 200$ are summarized in Table 3, where the estimates of the noise variances have been reported. As expected, the results get worst when the amounts of noise on the data increase. However, the JO approach is quite robust also in presence of low signal–to–noise ratios. From Table 3, it can be noted that the FR method gives poor estimates of λ_v, expecially in presence of unbalanced amounts of noise on the data.

5. CONCLUSIONS

In this paper the identification problem of linear dynamic systems from noise–corrupted output measurements has been addressed by using three different methods: the Instrumental Variable method, the Joint–Output approach and the Frisch Scheme approach. The performances of these methods have been compared on the basis of a simulated system already proposed in the literature.

Method	a_1	a_2	b_1	b_2	d_1	λ_ρ
IV	-1.356 ± 0.325	0.595 ± 0.223	2.015 ± 2.601	0.045 ± 2.196	—	—
FR	-1.539 ± 0.052	0.720 ± 0.034	1.087 ± 0.308	0.260 ± 0.430	—	—
JO	-1.514 ± 0.052	0.704 ± 0.040	1.014 ± 0.239	0.393 ± 0.262	-0.844 ± 0.074	1.225 ± 0.450
true value	-1.5	0.7	1.0	0.5	-0.9	1.0

Table 1. Means and standard deviations for 10 runs (N=200)

	$\lambda_e^* = 1$	$\lambda_e^* = 5$	$\lambda_e^* = 10$	$\lambda_e^* = 20$	$\lambda_e^* = 50$	Method
$\lambda_v^* = 1$	1.004 ± 0.133	5.031 ± 0.666	10.002 ± 1.300	20.005 ± 2.704	50.040 ± 6.863	FR
	0.941 ± 0.163	0.720 ± 0.437	0.653 ± 0.459	0.581 ± 0.468	0.618 ± 0.535	
	1.023 ± 0.254	5.218 ± 1.322	10.714 ± 2.843	22.796 ± 7.622	84.973 ± 7.341	JO
	0.990 ± 0.131	0.984 ± 0.143	0.971 ± 0.150	0.952 ± 0.172	0.917 ± 0.246	
$\lambda_v^* = 2$	1.020 ± 0.135	5.055 ± 0.663	10.011 ± 1.306	19.965 ± 2.960	50.086 ± 7.549	FR
	1.851 ± 0.238	1.372 ± 0.723	1.261 ± 0.867	1.238 ± 0.852	1.227 ± 0.854	
	1.028 ± 0.267	5.208 ± 1.294	10.494 ± 2.521	21.484 ± 5.446	61.513 ± 26.427	JO
	2.005 ± 0.258	1.995 ± 0.278	1.977 ± 0.278	1.957 ± 0.286	1.889 ± 0.398	
$\lambda_v^* = 3$	1.033 ± 0.142	5.072 ± 0.656	10.016 ± 1.411	19.934 ± 2.733	49.844 ± 8.217	FR
	2.746 ± 0.322	2.10 ± 0.971	1.931 ± 1.215	1.867 ± 1.223	1.965 ± 1.136	
	1.025 ± 0.282	5.180 ± 1.392	10.434 ± 2.611	21.266 ± 5.298	56.011 ± 17.076	JO
	3.041 ± 0.419	3.038 ± 0.453	3.004 ± 0.442	2.963 ± 0.437	2.924 ± 0.525	

Table 3. Estimates of (λ_e, λ_v) in different noise conditions for 20 runs (N=200)

The comparative analysis has confirmed a well-known fact that the IV method yields, in general, very inaccurate parameter estimates in spite of the low computational load required. On the contrary, the JO approach gives a very good accuracy of the parameter estimates, even in presence of high amounts of noise on the data. It must be noted, however, that this approach relies on computationally demanding iterative minimization procedures which may fail to give good results when only poor initial parameter estimates are available. In this respect, the FR approach might be the method to be preferred since its behaviour is characterized by a good compromise between the required computational load and the accuracy in the estimates.

The simulation results show that the FR method has good statistical properties and is robust with respect to violation of the assumptions. At the moment these results are not supported by a theoretical consistency analysis of the procedure. This topic will be the object of future investigations.

References

S. Beghelli, R.P. Guidorzi and U. Soverini, The Frisch scheme in dynamic system identification, *Automatica*, Vol. 26, pp. 171–176, 1990.

K.V. Fernando and H. Nicholson, Identification of linear systems with input and output noise: the Koopmans–Levin method, *IEE Proc. Part D*, Vol. 132, No. 1, pp. 30–36, Jan. 1985.

R. Frisch, *Statistical confluence analysis by means of complete regression systems*, Pub. No. 5, University of Oslo, Economics Institute, 1934.

R.P. Guidorzi and M. Pierantoni, A new parametrization of Frisch scheme solutions, *Proc. 12–th Int. Conf. on Systems Science*, Wroclaw, Poland, pp. 114–120, Sept. 1995.

T.J. Koopmans, *Linear Regression Analysis of Economic Time Series*, The Netherlands: N. V. Haarlem, 1937.

M.J. Levin, Estimation of a system pulse transfer function in the presence of noise, *IEEE Trans. Automat. Contr.*, Vol. AC–9, pp. 229–235, July 1964.

L. Ljung, *System Identification: Theory for the User*, Englewood Cliffs, NJ: Prentice–Hall, 1987.

T. Söderström, Identification of stochastic linear systems in presence of input noise, *Automatica*, Vol. 17, No. 5, pp 713–725, 1981.

T. Söderström and P. Stoica, Comparison of some instrumental variable methods – Consistency and accuracy aspects, *Automatica*, Vol. 17, No. 1, pp. 101–115, 1981.

T. Söderström and P. Stoica *System Identification*, Hemel Hempstead, U.K.: Prentice–Hall Int., 1989.

P. Stoica and T. Söderström, Bias correction in least–squares identification, *Int. J. Contr.*, Vol. 35, No. 3, pp. 449–457, 1982.

S. Van Huffel (Ed), *Recent advances in total least squares techniques and errors–in–variables modelling*, Philadelphia, PA: SIAM, 1997.

W.X. Zheng and C.B. Feng, Unbiased parameter estimation of linear systems in presence of input and output noise, *Int. J. Adaptive Contr. Signal Processing*, Vol. 3, pp. 231–251, 1989.

W.X. Zheng, On least–squares identification of stochastic linear systems with noisy input–output data, *Int. J. Adaptive Contr. Signal Processing*, Vol. 13, pp. 131–143, 1999.

FREQUENTIAL MOMENTS : APPLICATION TO LINEAR SYSTEM IDENTIFICATION

E. Etien, J.D Gabano, T Poinot, J.C Trigeassou

Laboratoire d'Automatique et d'Informatique Industrielle.
Ecole Supérieure d'Ingénieurs de Poitiers
Université de Poitiers (France)
Tel : 33 (0)5.49.45.35, FAX : 33(0)5.49.45.40.34
email : Jean-Claude.Trigeassou@esip.univ-poitiers.fr

Abstract: A new class of linear system invariant is presented. An estimation of frequential moments is performed by a least squares algorithm. The linears relationships between moments and transfer function parameters are also used to estimate local frequency models. *Copyright © 2000 IFAC*

Keywords: Frequency estimation, least squares method, Parameter estimation, local frequency models

1. INTRODUCTION

The objective of linear continuous time systems identification can be double : to give a mathematical model available for control or to determine relationships between structural parameters and physical ones for a better knowledge of the process. However, in control topics, the complete knowledge of the system is not necessary in all the spectrum domain. It can be limited in the closed loop bandwith or in the vicinity of the critical point in the Nyquist plot. We must also provide a reduced order model. Mc Laurin series approximations are very used in identification (Subrahmanyam, et al, 1995; Jonckheere, et al, 1989) and reduction order methods (Bultheel, et al, 1986; Sreeram, et al, 1993). The invariants most employed are time moments and Markov parameters which represent the comportment of the system respectively in low and high frequencies.

In this paper, we present a new class of invariants : frequentials moments. These coefficients are derived from a Taylor series expansion at $s = j\omega_o$ and constitute a generalization of Markov parameters and time moments. These new invariants permit to obtain reduced order models localized in a desired frequency domain.

2. FREQUENTIAL MOMENTS

Let us consider a SISO linear system characterized by its continuous transfer function :

$$H(s)=\frac{Y(s)}{U(s)}=\frac{B(s)}{A(s)}=\frac{b_0+b_1\cdots+b_m s^m}{a_0+a_1 s+\cdots+a_{n-1}s^{n-1}+s^n}$$

$$m \leq n \qquad (1)$$

H(s) can be expanded as a Taylor series expansion at $s = j\omega_o$:

$$\hat{H}(s)=\sum_{n=0}^{\infty}(-1)^n (s-j\omega_o)^n . A_{n,\omega_o} \qquad (2)$$

in which $\{A_{n,\omega_o}, n=0,\cdots\infty\}$ are the Frequential Moments (Etien, et al, 2000).

We make the following comment about theses moments :
$\hat{H}(s)$ admits the Mac Laurin series expansion

$$\hat{H}(s)=\sum_{n=0}^{\infty}(-1)^n A_n s^n \qquad (3)$$

and the asymptotic series expansion

$$\hat{H}(s)=\sum_{n=1}^{\infty}A_{-n}s^{-n} \qquad (4)$$

In these two cases ($\omega_o=0$ and $\omega_o=\infty$), frequential moments can be intrepreted respectively as the time moments and Markov parameters of *H(s)*.
We restrict the Taylor series to a finite dimension model of order *N* as

$$\hat{H}_N(s)=\sum_{n=0}^{N-1}(-1)^n A_{n,\omega_o}(s-j\omega_o)^n \qquad (5)$$

Let s_{dom} be the *H(s)* closest pole to $j\omega_o$. It is possible to define a domain *D* where $\hat{H}_N(s)$ can be considered as a local frequential model of *H(s)*. *D* is a disk centered on $j\omega_o$ and its radius is $R=|s_{dom}-j\omega_o|$. In the frequency domain $(s=j\omega)$, the limits of this domain are defined by

$$D_\omega=[\omega_m;\omega_M]=[\omega_o+R;\omega_o-R] \qquad (6)$$

Figures 1 and 2 shows the convergence domain and the Nyquist plot of a 2th order system and its reduced model.

$\hat{H}_N(s)$ appears as a local model localized in the domain $D\omega$, in the vinicity of the pulsation ω_o. In order to use this model, the input signals must have a spectrum limited in this area. This condition will ensure the convergence of the model $\hat{H}_N(s)$.

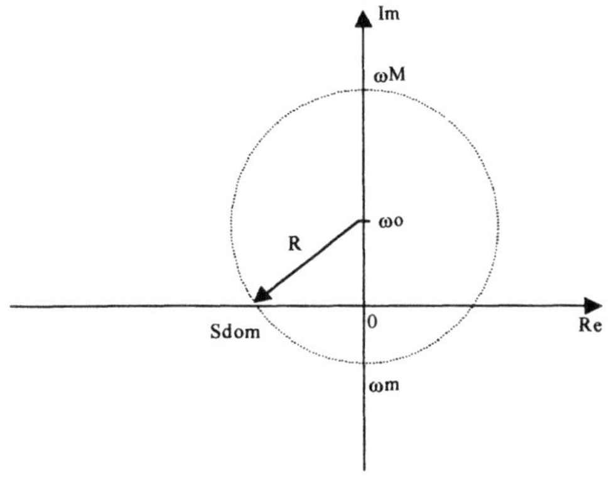

Fig. 1. Convergence domain

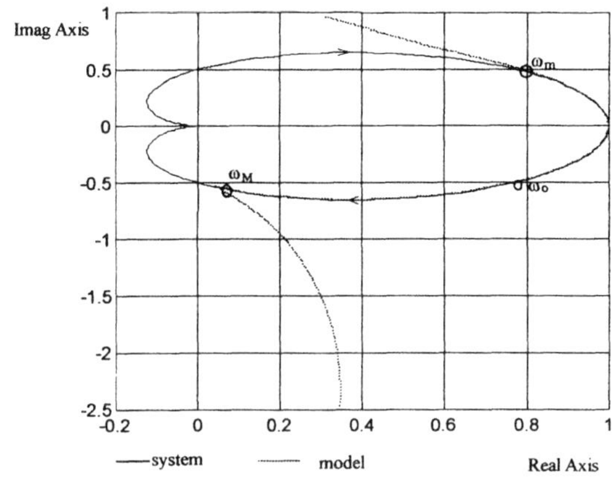

Fig.2. Nyquist plot

3. FREQUENTIAL MOMENTS ESTIMATION

In this section, we define several operations wich allows us to use $\hat{H}_N(s)$ as a model of the system. (Etien, 1999). An translation operation and a low order filtering permit to condition the signals used in the procedure of identification.

We define the translated input $u_m(t)$ as

$$u_m(t)=u(t)\exp^{-j\omega_o t} \qquad (7)$$

Similary, the translated output of the model is

$$\hat{x}_m(t)=x(t)\exp^{-j\omega_o t} \qquad (8)$$

Considering these signals, we obtain a translated model define by :

$$\hat{H}_{N,m}(s+j\omega_o)=\frac{\hat{X}_m(s)}{U_m(s)}=\sum_{n=0}^{N-1}(-1)^n A_{n,\omega_o}s^n \qquad (9)$$

Where $U_m(s)$ and $\hat{X}_m(s)$ are the Laplace transform of $u_m(t)$ and $\hat{x}_m(t)$.
We can also write :

$$\hat{x}_m(t)=\sum_{n=0}^{N-1}(-1)^n u_m^{(n)}(t).A_{n,\omega_o} \qquad (10)$$

where $u_m^{(n)}(t)$ is the nth derivative of $u_m(t)$. To compute these derivatives, we use a state variable filter technique.
We define the filtered signals

$$U_{mf}(s)=F(s).U_m(s)$$
$$\hat{X}_{mf}(s)=F(s).\hat{X}_m(s) \qquad (11)$$

F(s) is a NF^{th} order filter :

$$F(s)=\frac{\omega_f^{N_F}}{(s+\omega_f)^{N_F}} \qquad (12)$$

F(s) permits to limit the spectum of input/output signals according to the convergence domain defined by (6). Moreover, the simulation of *F(s)* allows to compute the successive derivatives of $u_{mf}(t)$

The estimated output is given by

$$\hat{x}_{mf}(t) = \sum_{n=0}^{N-1} (-1)^n u_{mf}^{(n)}(t).A_{n,\omega_o} = \underline{\varphi}_{\omega_o}^T \underline{A}_{\omega_o} \qquad (13)$$

Which is linear in the parameters. Using the regressor matrix

$$\phi_{\omega_o}^T = \left[\underline{\varphi}_{\omega_o,1}^T \cdots \underline{\varphi}_{\omega_o,k}^T \cdots \varphi_{\omega_o,K}^T \right] \qquad (14)$$

The least squares estimation of frequential moments vector can be computed :

$$\hat{\underline{A}}_{\omega_o} = \left[\phi_{\omega_0}{}^H \phi_{\omega_0}\right]^{-1} . \phi_{\omega_0}{}^H \underline{Y}_{mf} \qquad (15)$$

Figure3 shows the principle scheme of the estimation

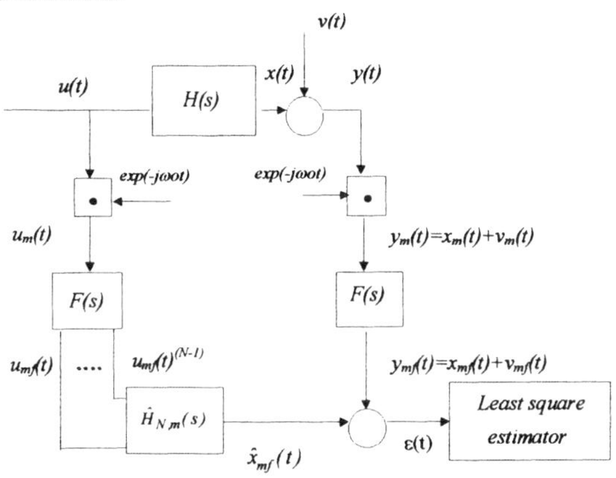

Fig.3. Principle of frequential moments estimation

In the case of an independant and zero mean output disturbance *v(t)*, $\hat{\underline{A}}_{\omega_o}$ is an unbiased estimation of *H(s)* frequential moments (Ljung, 1987).

4. PARAMETRIC ESTIMATION

The following procedure has been used in fault detection topics with time moments (Etien, et al, 1999). We generalize this principle to frequential moments.

Considering equations (1) and (2), a relationship beetween moments and parameters is obtained by

$$\underline{A}_{\omega o}(\theta) = \left[a'_{\omega_0}\right]^{-1} \underline{b}'_{\omega_0} \qquad (16)$$

Where $\left[a'_{\omega_0}\right]$and $\underline{b}'_{\omega_0}$ are matrix and vector containing the parameters of the translated transfer function

$$H(s+j\omega_o) = \frac{B(s+j\omega_o)}{A(s+j\omega_o)} = \frac{b'_0 + b'_1 \cdots + b'_m s^m}{a'_0 + a'_1 s + \cdots + a'_{n-1} s^{n-1} + s^n}$$

(17)

The continuous parameter vector

$$\hat{\underline{\theta}}^T = \left[\hat{a}_0, \hat{a}_1, \cdots, \hat{a}_{n-1}, \hat{b}_0, \hat{b}_1, \cdots, \hat{b}_m\right] \qquad (18)$$

is estimated by minimizing the following quadratic weighted criterion

$$J(\underline{\theta}) = \sum_{q=0}^{N_\omega} \mu_q \underline{\varepsilon}_{A_{\omega_q}}{}^H R_{\omega_q} \underline{\varepsilon}_{A_{\omega_q}} \qquad (19)$$

according to Marquardt's algorithm (Marquardt, 1963).

$\underline{\varepsilon}_{A_{\omega q}}$ is a residual constitute by the difference between the frequencial moments Least Squares estimate $\hat{\underline{A}}_{\omega_q}$ and the moments determined from equation (15)

$$\underline{\varepsilon}_{A_{\omega_q}} = \hat{\underline{A}}_{\omega_q} - \underline{A}_{\omega_q}(\underline{\theta}) \qquad (20)$$

Remark 1

Criterium (19) is made of N_ω residuals corresponding to $(N_\omega + 1)$ pulsations ω_q. For every pulsation, we consider *N* frequential moments. The choice of these pulsations permit to only conserv the spectral areas of the system which are interesting for us. In comparaison, time moments methods [4] use one pulsation $(N_\omega = 0)$: $\omega_o = 0$. Consequently, they are not adapted to correctly study transient comportments and are used for aperiodics systems.

Remark 2

$J(\underline{\theta})$ is a weighted criterium. In the ideal case $J(\underline{\theta})$ would be weighted by the covariance matrix

$$R_{\text{cov}} = \text{cov}(\hat{\underline{A}}_{\omega_q}) \qquad (21)$$

We should obtain a minimal variance Markov estimator (Poinot, et al, 1997) Because of the difficulties to calculate R_{cov}, we choose the simple weighting matrix

$$R_{\omega_q} = \left[\phi_{\omega q}^H \phi_{\omega q}\right] \qquad (22)$$

R_{ω_q} is computed from the frequential Moments least squares estimate and takes into account moments precision. The coefficients μ_q permit to wheight $J(\underline{\theta})$ with regard to the $N\omega$ pulsations ω_q. We choose μ_q as the trace of the matrix R_{ω_q}.

Considering the input and output datas obtained from the system, the following operations can be performed to find the desired model

1) We define a spectral domain where we want to perform a reduced model . We choose $(N_\omega + 1)$

pulsations evenly distributed in this domain : $\Omega = \lfloor \omega_1, \omega_2, ..., \omega_{N_\omega} \rfloor$

2) For each pulsation, we form the tranlated input signal defined in (7) and translated output signal

$$y_m(t) = y(t)\exp^{-j\omega_o t} \quad (23)$$

filtred signals are

$$U_{mf}(s) = F(s).U_m(s)$$

$$Y_{mf}(s) = F(s).Y_m(s) \quad (24)$$

3) We perform, for each pulsation, a vector of frequential moments ($dim: N \times 1$) by the relation (15) and we calulate the weighting matrix defined in (22)
4) With these $(N_\omega + 1)$ vectors, the quadratic weighted criterium (19) can be formed
5) The optimal value of parameters $\hat{\underline{\theta}}$ is calculated by the Marquardt algorithm :

$$\hat{\underline{\theta}}_{k+1} = \hat{\underline{\theta}}_k + \Delta\underline{\theta} \quad (25)$$

With

$$\Delta\underline{\theta} = -\left(J''_{\underline{\theta\theta}}\left(\hat{\underline{\theta}}_k\right) + \mu_k I \right)^{-1} J'_{\underline{\theta}}\left(\hat{\underline{\theta}}_k\right) \quad (26)$$

And

$$J'_{\underline{\theta}} = -2\Re\left\{ \sum_{q=1}^{N_\omega} \mu_q \left(\Sigma_{\underline{\theta}}^{\omega_q} \right)^H R_{\omega_q} \underline{\varepsilon}_{A_{\omega_q}} \right\} \quad (27)$$

$$\left[J''_{\underline{\theta\theta}} \right] = 2\Re\left\{ \sum_{q=1}^{N_\omega} \mu_q \left(\Sigma_{\underline{\theta}}^{\omega_q} \right)^H R_{\omega_q} \Sigma_{\underline{\theta}}^{\omega_q} \right\} \quad (28)$$

$$\Sigma_{\underline{\theta}}^{\omega_q} = \frac{\partial \underline{A}_{\omega_q}(\underline{\theta})}{\partial \underline{\theta}} \quad (29)$$

5. NUMERICAL EXAMPLE

We propose, in this example, to identify two reduced models of a fourth order simulated system corresponding to different spectrum domains.
Consider a SISO system,

$$H_n(s) = \frac{\omega_1^2}{s^2 + 2m_1\omega_1 s + \omega_1^2} + \frac{\omega_2^2}{s^2 + 2m_2\omega_2 s + \omega_2^2}$$

with $m_1 = 0.1, m_2 = 0.08, \omega_1 = 1rd/s, \omega_2 = 3rd/s$

This system is simulated with a sample period $T_e = 0.05s$. The measured output $y(k)$ is perturbated by a white noise with a signal to noise ratio $S/N = 10$.

The parameters of the filter $F(s)$ are $\omega_f = m_1\omega_1$ and we consider a Taylor series composed of eight moments ($N_F = 8$). The two following cases are considered :

(i) In the vicinity of the first resonance, $\omega \in \lfloor 1e^{-2};1 \rfloor$ with ω the pulsation in *rd/s*. We define, in this case, a low frequencie reduced order model $H_{r,l}(s)$ as :

$$H_{r,l}(s) = \frac{b_{0,l}}{a_{0,l} + a_{1,l}s + s^2}$$

The number of moments kept for the parametric estimation is $N_\omega = 3$.

(ii) In the vicinity of the second resonance, $\omega \in [2;10]$. A high frequencies reduced order model $H_{r,h}(s)$ is given by

$$H_{r,h}(s) = \frac{b_{0,h}}{a_{0,h} + a_{1,h}s + a_{2,h}s^2 + s^3}$$

In this case $N_\omega = 4$.

Tables 1 and 2 exhibit the estimates for the two models $H_{r,l}(s)$ and $H_{r,h}(s)$.

Table1. Parameters estimates – low frequencies reduced model

	$b_{o,l}$	$a_{o,l}$	$a_{1,l}$
$H_{r,l}(s)$	2,2615	3,2805e^{-1}	1,1937

Table 2. Parameters estimates – high frequencies reduced model

	$b_{o,h}$	$a_{o,h}$	$a_{1,h}$	$a_{2,h}$
$H_{r,h}(s)$	2,244e^2	2,354e^2	2,128e^1	26,33

Figures 4 and 5 show the bode plots of the system and of the two reduced model.

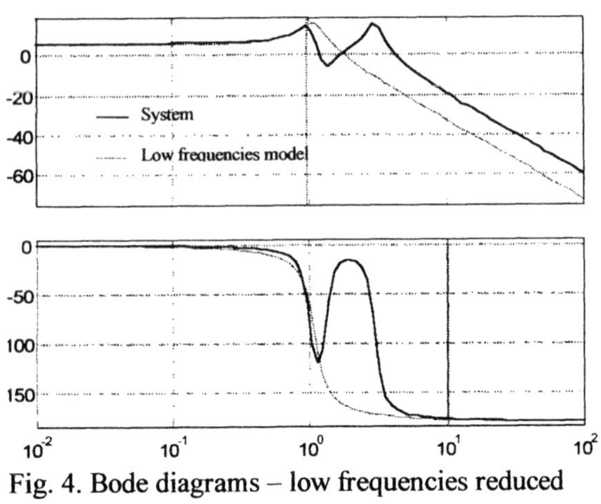

Fig. 4. Bode diagrams – low frequencies reduced model

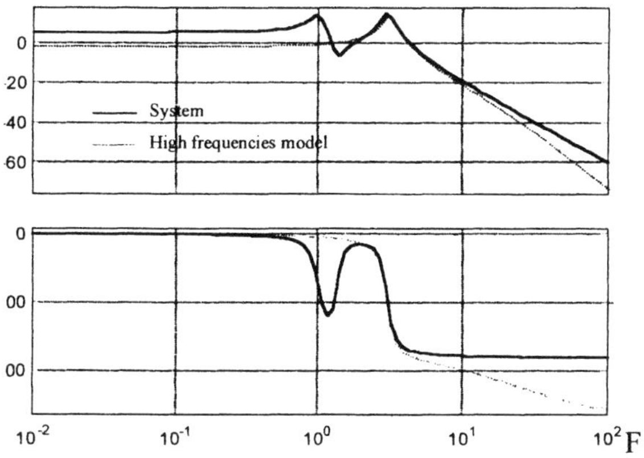

Fig.5. Bode diagrams – high frequencies reduced model

figures 6 and 7 show the relatives frequency errors defined by

$$\varepsilon_i(\omega) = \left|H_n(j\omega) - H_{r,i}(j\omega)\right| / \left|H_n(j\omega)\right|$$

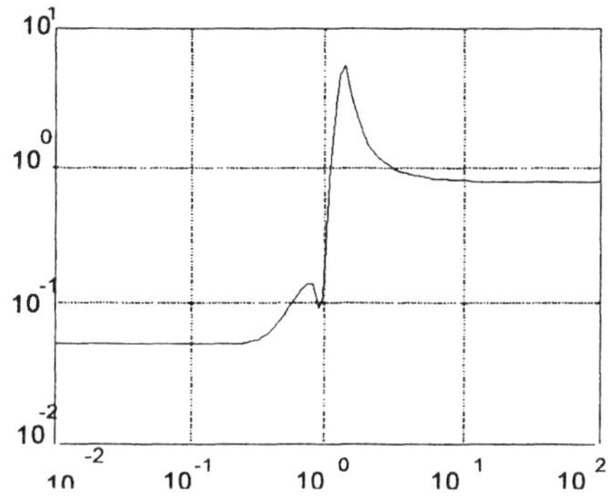

Fig. 6. Frequency relative error – low frequencies model

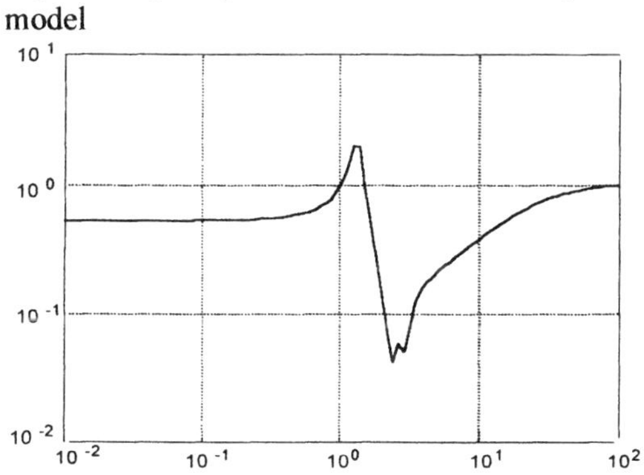

Fig.7. Frequency relative error – high frequencies

We can conclude that the two reduced models provide good approximations of $H_n(s)$ in the two specified spectrum domains.

CONCLUSION

A new idenfication method is proposed in this paper. Based on a Taylor series expansion of the system in the vicinity of $s = j\omega_o$,it is able to provide good approximations of the considered process in the specified spectrum domain. Moments estimation permits to build a weighted criterium taking into account moments precision and spectral distribution.

REFERENCES

A.V.B. Subrahmanyam, D.C Saha and G.P Rao 'Identification of continuous time MIMO systems via Time Moments', *Control – Theory and advanced technology,* **vol.10**, No 4, Part 3, pp 1359-1378, September, 1995

[2] E. Jonckheere and C.Ma, 'Combined sequence of Markov parameters and moments in linear systems', *IEEE Trans. Automat., Contro.,* **vol. AC-34**, pp. 379-382, Mar. 1989.

Bultheel, A. and M. Van Barel , 'Pade techniques for model reduction in linear system theory – A survey', *Jou. Of Computational and Applied Mathematics,* **Vol. 14** pp. 401-438, 1986

V. Sreeram and P. Agathokly, ' On the computation of the Gram matrix in time domain and its applications', *IEEE Trans. Automat. Control.* **Vol.38** pp 1516-1520, Oct.1993

E. Etien, J.D Gabano, L. Rambault, J.C Trigeassou, "frequential moments : application to controller reduction", *to appear in ACC'2000 Conference.*

E. Etien, *"Caractérisation des systèmes à représentation continue par les moments fréquentiels : application à l'identification et à la réduction de modèle "*, Thèse de l'Université de Poitiers, 20 Juillet 1999.

Ljung L., "System identification – *Theory for the users"* Prentice Hall, 1987

E. Etien, J.D. Gabano, " DC motor monitoring using time moments modelling ", *IEEE International Symposium on Diagnostic for Electrical Machines, Power Electronics and Drives.IEEE-SDEMPED'99,* Gijon, Spain, September 1-3, 1999

Marquardt W. D., "An algorithm for least squares estimation of non-linear parameters" *Journ. Soc. Indust . Appl. Math.,* **Vol.11** / USA, N°2, June 1963, pages 431-441.

T. Poinot, J.C. Trigeassou, P. Coirault, " Overparametrization : a tool for system identification". *System Identification SYSID'97,* Japon 8-11 July 1997.

ON THE STATISTICAL ACCURACY OF THE EMPIRICAL TRANSFER FUNCTION ESTIMATOR

P.M.T. Broersen

Department of Applied Physics Delft University of Technology
P.O.Box 5046, 2600 GA Delft The Netherlands
phone +31 15 2786419 email broersen@tn.tudelft.nl

Abstract: At its introduction, the accuracy of the Empirical Transfer Function Estimator ETFE has been described for periodical input and for a realization of a stationary stochastic process as input. Later, an infinite value for the variance of the ETFE has been derived by treating the input as a stationary stochastic process. The results for the three different types of input are compared. An example clarifies the experimental conditions for each type of input. It shows under which conditions the theory of a single realization of a stochastic process applies and when a full stochastic treatment is necessary. This has consequences for the method of averaging that has to be applied to improve the accuracy. *Copyright © 2000 IFAC*

Keywords: ETFE, frequency response, random input, system identification, non-parametric model, spectral analysis, stochastic process

1. INTRODUCTION

Transfer functions can be estimated non-parametric or with parametric models. Both methods have been studied extensively for different experimental conditions. Non-parametric methods estimate the impulse response, the step response or the transfer function, which is the Fourier transform of the impulse response. For deterministic signals, the transfer function can be defined easily in the frequency domain as the quotient of the Fourier transforms of output and input signal. Ljung (1985) has introduced the Empirical Transfer Function Estimate ETFE as an estimator for deterministic input-output relations, which can also be applied to stochastic signals. He studied the accuracy of the ETFE for periodic input and for a realization of a stochastic process as input, with the restriction that the input signal is regarded as a given sequence. For periodic input signals, the transfer function is only considered at the limited number of frequencies that is present in the input. The variance of the ETFE diminishes if more periods are available, because the number of observations increases while the number of frequencies for evaluation remains the same. The situation is different for stationary stochastic processes with a continuous spectral density. For a single realization of a stochastic process, the number of independent frequencies at witch the ETFE can be estimated increases proportional to the number of observations. However, the accuracy at each frequency is finite and does not improve. Ljung (1999) regards the input as a given sequence and probabilistic quantities of the ETFE refer exclusively to the properties of the additive noise in the measurements. A stochastic input is considered as a *given* realization of a stochastic process. In this way, deterministic approaches can be applied to stochastic processes.

The interpretation of Fourier transforms of stochastic processes requires care. Theoretically, the Fourier transform of a stationary stochastic process does not exist as an ordinary function. Priestley (1981) showed that a stationary stochastic process must have infinite length in the time domain and hence the Fourier transform would not converge. The spectral representation of a stationary stochastic process is defined as a Fourier-Stieltjes transform that uses a special class of functions. Those functions are not differentiable and properties can only be contributed to their expectations. That might be one of the reasons (Broersen, 2000) that time series analysis gives a better performance in spectral estimation than methods based on the Fourier transform of the data or their estimated covariance. Broersen (1994) investigated whether this problem has practical consequences for the estimation of the transfer function. He treated the input signal as a stochastic signal, so the variability of the input signal will contribute to the statistical accuracy of the transfer function. The expressions for bias are not severely influenced. But the expressions for the variance differ considerably from the deterministic approach (Broersen 1995). By considering the input signal as stochastic, the variance of the elementary ETFE becomes *infinite*. The reason is that the denominator can become zero. Moreover, averaging raw ETFE's does not improve the result sufficiently, because the variance remains infinite. Only estimating average spectra in the numerator and in the denominator before taking their quotient is an acceptable method of averaging. However, one degree of freedom is lost in the accuracy of this averaging (Broersen 1995). That absorbs the single degree of freedom that is present in the quotient that defines the ETFE.

Guillaume, et al., (1996) considered a different type of noisy input signals. Measurements of both the input and the output signal are considered to be disturbed by additive noise. In this way, a non-zero input signal at a certain frequency can be canceled by the noise. They made the ETFE more robust by excluding the outcome at the frequencies where the denominator turns out to be too small. The bias and the variance of their estimator become finite. The statistical properties of their estimator depend on a subjective choice: the value of denominator that is excluded as being too small. Using a similar approach to stochastic inputs, a practical estimator can be constructed that never becomes infinite. This solution, just neglecting undesirable large estimates, has been used in practice. The variance of such a robust estimator, however, is no longer the same as the variance of the ETFE. The tails of the probability density function of the ETFE are cut of.

The approaches of Ljung (1999) and Broersen (1995) give different answers for the variance of the ETFE. Nevertheless, both approaches are correct theoretically, but both have their own area of applicability. The variance of the estimator ETFE describes the accuracy if the experiment is repeated and the difference is a consequence of exactly what will be repeated. Such interpretation problems are known in the literature. Matheron (1989) provided a guide for the determination of the suitability and the range of applicability of probabilistic modeling. It is perfectly legal to model a unique but unknown quantity, about which partial information is available, as if it were drawn from an ensemble of similar quantities and to apply probabilistic modeling to this unique quantity. Matheron (1989) often uses practical examples to illustrate the almost philosophical problems, as mathematical definitions are difficult to grasp the details or the interpretation. This paper will describe the experimental conditions for the different variance expressions of the ETFE in a railway example.

The reasoning in transfer function analysis relies on asymptotic properties. However, this may not always be sound. As an example, take estimation of the coherency, which should be a measure for the linear relationship between input and output signal. Priestley (1981) shows that estimating N frequencies of a coherency function from N data points of input and output yields the result 1 for all coherency points. This does not change if N increases and it is independent of the characteristics of the input signal. It means that the coherency will not reveal any relevant information about the process in this case, even not asymptotically. Only one degree of freedom for each frequency in the estimated coherency is not sufficient to get useful information. The coherency estimator uses cross and auto spectral estimates. It has some resemblance with the ETFE for one single batch of data that also has one degree of freedom. This is an indication that the treatment of the statistical properties of estimators like the ETFE requires care.

2. DEFINITIONS AND NOTATION

Theoretical results for the expectation and the variance of quotients of, possibly complex, stochastic variables have been given by Broersen (1994, 1995). The causal processes that are considered are given by

$$y(t) = \sum_{k=0}^{\infty} g(k)u(t-k) + v(t), \tag{1}$$

where y(t) is the observed output signal, g(k) the impulse response with the exponentially stable transfer function

$$G(e^{j\omega}) = \sum_{k=0}^{\infty} g(k)e^{-j\omega k}. \tag{2}$$

The additive output noise v(t) is assumed to be independent of u(t), the observed input process.

The Fourier transform of M observations of the output signal y(t) is defined as:

$$Y_M(\omega) = \frac{1}{\sqrt{M}} \sum_{t=0}^{M-1} y(t)e^{-j\omega t}. \tag{3}$$

The values for $\omega=2\pi k/M$, k=0, 1, ..., M-1 form the familiar discrete Fourier Transform (DFT). Likewise, u(t) with $U_M(\omega)$ and v(t) with $V_M(\omega)$ are DFT transform pairs. The relation between the transforms is given by (Ljung, 1999)

$$\mathbf{Y}_M(\omega) = \mathbf{G}(e^{j\omega})\mathbf{U}_M(\omega) + \mathbf{V}_M(\omega) + \mathbf{R}_M(\omega). \quad (4)$$

The term $R_M(\omega)$ represents **two** end effects:

- the transform of the response to the initial conditions which is the input signal prior to t=0
- the continuation of the response for t > M-1 on the input signal for t<M.

Together, they are given by (Priestley 1981, Broersen 1995):

$$\mathbf{R}_M(\omega) = \frac{1}{\sqrt{M}} \sum_{k=0}^{\infty} \mathbf{g}(k) e^{-j\omega k} \sum_{t=-k}^{-1} \{\mathbf{u}(t) - \mathbf{u}(t+M)\} e^{-j\omega t}. \quad (5)$$

Ljung (1985) defined the Empirical Transfer Function Estimate ETFE as:

$$\mathbf{ETFE} = \hat{\mathbf{G}}(e^{j\omega}) = \frac{\mathbf{Y}_M(\omega)}{\mathbf{U}_M(\omega)}. \quad (6)$$

For averaging of ETFE's, two different approaches will be studied here, that are both described by Ljung (1999). Suppose that N=KM observations of input and output are available and that the data are split in K batches of M observations with transforms $Y_M^{(k)}(\omega)$ and $U_M^{(k)}(\omega)$ for the k'th batch. The first estimator is defined as the average of the ETFE:

$$\mathbf{avETFE} = \frac{1}{K} \sum_{k=1}^{K} \frac{\mathbf{Y}_M^{(k)}(\omega)}{\mathbf{U}_M^{(k)}(\omega)}. \quad (7)$$

Firstly, quotients are taken, which are averaged afterwards. The second estimator starts with Welch's method to determine the average of the cross spectral estimate of input and output signal and the auto spectrum of the input signal. Only afterwards, the averaged spectral estimates are divided:

$$\mathbf{avSPEC} = \frac{\frac{1}{K} \sum_{k=1}^{K} \mathbf{Y}_M^{(k)}(\omega) \mathbf{U}_M^{(k)}(\omega)^*}{\frac{1}{K} \sum_{k=1}^{K} \mathbf{U}_M^{(k)}(\omega) \mathbf{U}_M^{(k)}(\omega)^*}. \quad (8)$$

The * denotes the complex conjugate. It is possible to introduce windows in the averaging analysis like Ljung (1999) did, but for clarity of presentation that is not considered here. Broersen (1995) described the influence of a Daniell window for averaging the numerator and denominator of the ETFE. The prime intention of this paper is to describe as clearly as possible the influence of the experimental conditions on the result of averaging with (7) or with (8).

3. VARIANCE OF STOCHASTIC QUOTIENTS

The quotient of two independent *real* normally distributed zero mean stochastic variables has a Cauchy distribution. This has no finite moments, so the mean and variance do not exist (Kendall and Stuart, 1963).

The expectation of the quotient of two zero mean, *complex*, normally distributed variables a and b is given by Miller (1974) as:

$$\mathbf{E}\left[\frac{\mathbf{a}}{\mathbf{b}}\right] = \mathbf{E}\left[\frac{\mathbf{ab}^*}{\mathbf{bb}^*}\right] = \frac{\mathbf{E}[\mathbf{ab}^*]}{\mathbf{E}[\mathbf{bb}^*]}. \quad (9)$$

So the expectation of the quotient a/b exists and is finite, even with E[b]=0. The variance of a/b can be described with Fisher's variance-ratio distribution, known as the F-distribution. The sum of the squares of the real and imaginary parts gives a total of 2 contributions in the numerator and in the denominator of a/b. However, the expectation of an F distribution requires at least 3 degrees of freedom or three independent contributions in the denominator to give a finite answer (Kendall and Stuart, 1963). So the theoretical variance of a/b is ∞. These results can be applied to the EFTE by substituting $Y_M(\omega)$ for a and $U_M(\omega)$ for b.

Broersen (1995) has derived with some calculus an exact variance expression for *real* independent zero mean normally distributed stochastic variables x_i, i=1,...,K:

$$\mathbf{var} \frac{\mathbf{x}_i}{\sum_{j=1}^{K} \mathbf{x}_j^2} = \frac{1}{\mathbf{K(K-2)}}, \quad \mathbf{K} \geq 3. \quad (10)$$

From that result an exact theoretical formula can be derived for *complex*, normally distributed independent zero mean stochastic variables a_i and b_i, with variances σ_a^2 and σ_b^2 respectively, each with equal variance contributions of the real and imaginary parts of half their total variance:

$$\mathbf{var} \frac{\sum_{j=1}^{K} \mathbf{a}_j \mathbf{b}_j^*}{\sum_{j=1}^{K} \mathbf{b}_j \mathbf{b}_j^*} = \frac{1}{\mathbf{K}-1} \frac{\sigma_a^2}{\sigma_b^2}, \quad \mathbf{K} \geq 2. \quad (11)$$

For K=1, this variance becomes infinite.

The Fourier transform of a finite number of observations of a stochastic process is for every frequency a complex stochastic variable with expectation zero and with equal variances for the real and the imaginary parts; (Priestley, 1981). Hence, the variance of (11) can be applied to the different estimators for average transfer functions, replacing a_j by $V_M^{(j)}(\omega)+R_M^{(j)}(\omega)$ and b_j by $U_M^{(j)}(\omega)$. The first term of (4) yields the constant $G(e^{j\omega})$ in the transfer function estimates and does not contribute to the variance.

4. RAILWAY EXAMPLE

The dynamic behavior of a railway vehicle on a straight track is dominated by parasitic motions in the lateral plane, commonly referred to as hunting, a coupled translation and rotation. Broersen (1973) showed that this can be seen as a response to the stochastic input of lateral irregularities in the center-line of the rails. So a transfer function can be defined between the lateral rail irregularities as input and the lateral motions of a railway vehicle as output. By considering experiments with a constant speed of the vehicle, a simple relation exists between the lateral stochastic input as a function of the *distance* and the lateral input as a function of the *time*. Due to the fixed position in space, this signal can be used as input in repeated experiments. The irregularities are mainly influenced by the level of maintenance of the railway track. They can be considered as almost identical for consecutive passages of the same part of the track by the same vehicle at the same speed. This gives the opportunity to repeat an experiment with the same stochastic input. The experimental conditions for the three different cases for the EFTE can be easily defined for the railway dynamics at a constant forward speed.

Case 1.
A periodic excitation is created by constructing a loop as a test track where the railway vehicle can circle around. The input signal can be stochastic due to inevitable small irregularities in the position of the rails. It is also possible to add a desired input signal, by giving certain parts of the test track a deliberate lateral displacement. In this way, it is possible to create a pseudo-random excitation. This situation is also called Case 1 in Ljung (1999).

Case 2.
A deterministic input or a single realization of a stochastic process can be obtained by using a predefined or an arbitrary part of the track. However, the bias and variance of the estimator in case 2 describe the statistical accuracy of repeating exactly this experiment: using precisely the same (originally arbitrary) part of the track for repeated experiments. Care should be taken to use exactly the same length of the track, to assure that $R_M(\omega)$ in (4) remains the same. So the initial conditions before the test part of the single realization should be identical. Taking only a few observations more in both input and output signal may heavily disturb the accuracy of the bias and variance expressions. Case 2 has the same name in Ljung (1999). The experimental conditions allowing to consider a stochastic input signal as a given sequence have been described here. Note that in the absence of noise, the output would be identical for repeated experiments and the variance of the ETFE would be zero. Averaging over repeated experiments means taking the average transfer function over exactly the same input signal, with only a new realization of the measurement noise.

Case 3.
A stochastic process as input is obtained by using arbitrary parts of the track. The bias and variance of the estimator in case 3 describe the statistical accuracy of repeating this experiment by using different parts of the track for repeated experiments. Averaging over repeated experiments means taking the average transfer function for different parts of the track. These experimental conditions have been used in the derivation of Broersen (1994, 1995).

Matheron (1989) has discussed the applicability of the concepts of probability to practical situations. He discussed the use of probabilistic models and techniques in apparently unique situations or structures. Randomness is in no way uniquely defined, but it is only a property of the models we choose. Given a unique phenomenon which is completely determined but largely unknown to the experimenter, is a deterministic approach to be preferred or is it possible and meaningful to conceptualize it as a probabilistic model? That means representing the phenomenon as drawn at random from a family of virtual phenomena. Those problems are difficult in many specific application areas. Matheron (1989) gives an example of a mineral deposit, for which fragmentary information is available but an estimate about the total amount is necessary to make a decision about commercial exploitation. Although there is a single deterministic but unknown amount in this specific deposit, it is possible that an accurate answer can be found by using a probabilistic concept.

Matheron (1989) mentions the application to time series. The interpretation problem is representative for time series analysis, since there the data consists in just the unique realization of a random process and repeated trials are usually unavailable. As a matter of fact, no single answer can be given to the practical problem of the variance of the ETFE. Different mathematically correct results are obtained for the three cases described above. The experimental conditions for the three cases make clear which result is applicable for a given situation. Case 1 is applicable for periodic input, case 2 for repeatedly using exactly the same input signal without making it periodic and case 3 for averaging over different realizations of a stochastic process.

5. BIAS of ETFE

The bias will be treated shortly here. For periodic excitation it is zero because u(t) and u(t+M) in (5) cancel if the period is M. The single realization case 2 gives a 'bias' that better can be considered as a constant systematic error in the ETFE, because only one realization of $R_M(\omega)$ is involved, with the specific values for u(t) substituted. Finally, the bias in case 3 is described by Broersen (1994) and can be written as:

$$\mathbf{bias[ETFE]} = \mathbf{E[Y_M(\omega)\,U_M(\omega)^*]/h_u(\omega)\text{-}G(e^{j\omega})}\,.$$

With Eq. 4and the assumption that additive noise and input signal are independent, the bias becomes:

$$\mathbf{bias[ETFE]} = \mathbf{E[R_M(\omega)\,U_M(\omega)^*]/\sigma_u^2}\,.$$

$U_M(\omega)$ contains only input data from t= 0 to M-1. Looking in $R_M(\omega)$ for terms with the same index, Broersen (1994) derived after some exercise:

$$\textbf{bias[ETFE]} = -\frac{1}{M}\left(\sum_{k=0}^{\infty} kg(k)e^{-j\omega k} - \sum_{k=M+1}^{\infty}(k-M)g(k)e^{-j\omega k}\right)$$

The derivative $G'(e^{j\omega})$ of the transfer function (2) with respect to its argument is given by:

$$G'(e^{j\omega}) = -\sum_{k=0}^{\infty} kg(k)e^{-j\omega(k+1)},$$

and the first term in the bias summation can be written as:

$$\textbf{bias[ETFE]} \approx \frac{1}{M}e^{j\omega}G'(e^{j\omega}). \tag{12}$$

It can be a good approximation to the bias if M is at least as long as the significant non-zero part of the impulse response. That is anyhow a prerequisite for accurate non-parametric estimation of a transfer function.

6. VARIANCE EXPRESSIONS for ETFE

Expressions for the variance of the ETFE have been derived by Ljung (1999) and by Broersen (1995).

Case 1: Periodic, Ljung (1999).

$$\textbf{var[ETFE]} = \frac{h_v(\omega)}{|U_M(\omega)|^2}$$
$$\textbf{var[avETFE]} = \textbf{var[avSPEC]} = \frac{1}{K}\frac{h_v(\omega)}{|U_M(\omega)|^2} \tag{13}$$

The term $|U_M(\omega)|^2$ is a known fixed number for every frequency and is just a scale factor in this variance expression, $h_v(\omega)$ is the power spectral density of the noise.

Case 2: Deterministic or given stochastic sequence, Ljung (1999).

$$\textbf{var[ETFE]} = \frac{h_v(\omega)}{|U_M(\omega)|^2}$$
$$\textbf{var[avETFE]} = \textbf{var[avSPEC]} = \frac{1}{K}\frac{h_v(\omega)}{|U_M(\omega)|^2} \tag{14}$$

The term with $R_M(\omega)$ in (4) is each time the same if the experiment is repeated K times with exactly the same input signal of length M. Therefore; this term does not contribute to the variance, only to the bias. The variance situation is similar to Case1. The variance for K repetitions is improved with a factor K if the number of evaluation points for ω is kept M. Another possibility for increasing the sample size would be to consider the K times M observations as a new, longer, input signal of length N=KM that is evaluated at KM frequencies ω. In that case, the first formula applies, with a new longer but fixed input: $\text{var[ETFE]}=h_v(\omega)/|U_{KM}(\omega)|^2$.

Case 3: Stochastic input, Broersen (1995).

$$\textbf{var[ETFE]} = \infty$$
$$\textbf{var[avETFE]} = \infty \tag{15}$$
$$\textbf{var[avSPEC]} = \frac{1}{K-1}\frac{h_v(\omega)+|R_M(\omega)|^2}{h_u(\omega)}$$

An expression for the contribution of $R_M(\omega)$ is given in Broersen (1995); it is proportional to 1/M and hence its influence diminishes for increasing M.

Broersen (1995) also described an improved estimator, based on averaging the spectra of numerator and denominator with a Daniel window has a smaller variance because it uses all KM data in one transform and has only one transient with initial conditions. His simulations confirmed for small K that the divisor in var[avSPEC] is 1/(K-1) in (15); also the influence of $R_M(\omega)$ was verified in those simulations.

7. DISCUSSION

This paper concentrates on the question how to average the ETFE to obtain a good estimator. In Case1 and Case 2, the variance of the ETFE would be zero if no noise would be present. In Case 3, the contribution of $R_M(\omega)$ to the variance of the ETFE is ∞, even without any additive noise. Furthermore, one degree of freedom is lost in averaging, according to (11). Averaging the ETFE directly in avETFE, so taking quotients of Fourier transforms and averaging the quotients is taking the average of contributions which each have infinite variance. The variance of the average will remain infinite. In all three cases, avSPEC with first averaging of numerator and denominator gives a good result.

The variance of the ETFE has

two different values,

a finite and an infinite one. The ETFE describes the estimation of M frequencies of the transfer function from M values of input and output signal. Which expression for the variance is correct depends on the experimental conditions: what exactly is repeated. However, those *experimental* conditions are not necessarily specified in a *written* text. So, without further specifications, both expressions can be applicable. If one would consider the variance in repeating the same experiment with the same input signal, the finite value is valid. If one would consider the variance in repeating the experiment with a new input sequence, the infinite value applies.

This rather arbitrary situation becomes unambiguous if exactly is described which averages are taken: averaging quotients or averaging numerator and denominator separately before taking the quotient. The averages avETFE and avSPEC are defined in (7) and (8), respectively. The sequence of averaging and division is important, and the experimental conditions become explicit by the averaging procedure that is used. This uniquely determines which of the three cases applies.

- By using a periodic repetition of the M data points of the input signal, the experimental conditions determine Case 1.
- By using exactly the same input signal and exactly the same initial conditions for every repetition of the experiment, the input can be considered as a given realization of a stationary stochastic process. This is Case 2. The initial conditions may be zero. Otherwise, care should be taken to have the same initial conditions in each repetition. The main difference with Case 1 is that $R_M(\omega)$ plays a role in the 'bias'. As $R_M(\omega)$ is identical for each repetition, a better name for the bias would be 'fixed systematic error'. As also the denominator $U_M(\omega)$ is identical for each repetition, there is no difference between avETFE and avSPEC.
- By averaging over different realizations of a stochastic input signal, the experimental conditions prescribe that Case 3 is applicable. Averaging ETFE estimates gives still infinite variance. Some type of spectral averaging of numerator and denominator is necessary prior to division to obtain an accurate estimator for the transfer function.

At first sight, it might be curious that the variance in Case 3 is infinite, in contrast with the other cases. However, in using the deterministic given value $|U_M(\omega)|^2$ in the denominator, this can be very small for certain values of ω. If a single realization of a stochastic process is used as input, the standard deviation of $|U_M(\omega)|^2$ is equal to its expectation. Hence, the appearance of $|U_M(\omega)|^2$ will be very erratic, also if the spectrum of u(t), $h_u(\omega)$ is smooth or even a constant, like in white noise excitation. Therefore, the variance in cases 1 and 2 can for some points, with small $|U_M(\omega)|^2$, become extremely large. This is for Guillaume, et al., (1996) a reason to leave such points out of the estimation, with the argument that the excitation at that frequency was too weak. Those accidental values which are found in a single realization of the input signal play no role in Case 3; the variance is either infinite for K=1 or it is determined by the *true* expectation $h_u(\omega)$ of the input spectrum.

CONCLUSIONS

The literature gives two different expressions for the variance of the Empirical Transfer Function Estimate ETFE. One finite outcome applies to periodic input, deterministic input or a realization of a stochastic input considered as a given sequence. The other infinite variance is applicable if the input is considered as stochastic. For estimation from a single batch of data, without further specifications, both expressions are correct and only a better description of the experimental conditions would decide which expression is applicable.

The situation becomes completely different for the variance of the average transfer function from different batches of input and output data. If the input signal is the same for every batch, averaging ETFE's or averaging spectra and afterwards estimating the transfer function are identical. Averaging over K batches gives a variance reduction with a factor K.

A different stochastic input signal in every batch of data gives an infinite variance of the ETFE, before and also after averaging. The estimator with average cross spectrum in numerator and the average input spectrum as denominator gives a finite variance then. In this expression for the variance, one degree of freedom is lost in comparison with the situation that the input is the same for every batch.

REFERENCES

Broersen, P.M.T. (1973). Estimation of multivariable railway dynamics from normal operating records. *Preprints of the 3rd IFAC Symposium on Identification and System Parameter Estimation*, the Hague/Delft, 425-433.

Broersen, P.M.T. (1994). On bias in transfer functions estimated with stochastic excitation. *PROC ICASSP Conf., Adelaide*, **vol. IV**, 49-52

Broersen, P.M.T. (1995). A comparison of transfer function estimators. *IEEE Trans. on Instrumentation and Measurement*, **vol. 44**, 657-661.

Broersen, P.M.T. (2000). Facts and fiction in spectral analysis. *IEEE Trans. on Instrum. and Meas.*, **vol. 49,** no. 4, August 2000.

Guillaume, P., I. Kollar and R. Pintelon (1996). Statistical analysis of nonparametric transfer function estimates. *IEEE Trans. on Instrumentation and Measurement*, **vol. 45**, 594-600.

Kendall, M.G. and A. Stuart (1963). *The advanced theory of statistics*, London, Charles Griffin, second edition.

Ljung, L. (1985). On the estimation of transfer functions. *Automatica*, **vol. 21**, 677-696.

Ljung, L. (1999). *System Identification, Theory for the user.* Second Edition, Prentice-Hall, Upper saddle River. (First edition 1987).

Matheron, G. (1989), *Estimating and Choosing, An essay on probability on practice.* Springer-Verlag, Berlin. (translated from French by A.M. Hasofer).

Miller, K.S. (1974). *Complex stochastic processes, an Introduction to theory and application.* Reading, MA: Addison-Wesley.

Priestley, M.B. (1981). *Spectral Analysis and Time Series*, London, Ac. Press.

APPROACHES FOR IDENTIFYING CONTINUOUS-TIME AR PROCESSES FROM UNEVENLY SAMPLED DATA

Erik Larsson and Torsten Söderström

Department of Systems and Control,
Information Technology, Uppsala University
P O Box 27, SE-751 03 Uppsala, Sweden.
Email: ekl@syscon.uu.se.

Abstract: When identifying a continuous-time AR process from discrete-time data, an obvious approach is to replace the derivative operator in the continuous-time model by an approximation. In some cases, a linear regression model can then be formulated. The well-known least squares method would be very desirable to apply, since it enjoy good numerical properties and low computational complexity, in particular for fast or nonuniform sampling. The focus of this paper is the latter, *i.e.* nonuniform sampling. Two consistent least squares schemes for the case of unevenly sampled data are presented. The precise choice of derivative approximation turns out to be crucial.

Keywords: system identification, autoregressive process, continuous-time, uneven sampling.

1. INTRODUCTION

Parameter estimation of continuous time systems is an important subject which has numerous applications ranging from control, signal processing, to astrophysics and economics, Parzen (1984); Phadke and Wu (1974); Unbehauen and Rao (1990). Research in control and identification of continuous-time systems has been focussed on the corresponding discretized models which describe the relations between samples of the inputs and outputs. One particularly interesting and practical scenario is identification of continuous-time systems using discrete-time data.

The objective of this paper is identification of continuous-time AR systems, where the data are observed unevenly in time. We will use a "direct approach", where the derivatives are substituted by discrete-time differences, for example by replacing differentiation by a delta operator. In this fashion the model is casted into a (discrete-time) linear regression. Some possible advantages of such a direct approach include:

- It has good numerical properties, especially for fast sampling, compared with methods solely based on discrete-time models, Fan and Liu (1994).
- It is computationally efficient.

In order to get consistent estimates some special care has to be taken when choosing the derivative approximation. In this paper two least squares methods that will give consistent estimates for our posed problem are introduced.

This paper present an extension of previous research done at the Department of Systems and Control at Uppsala University regarding ARX models, in the case of evenly sampled data. See *e.g.* Bigi *et al.* (1994); Söderström *et al.* (1995, 1997*b*).

2. BACKGROUND - CONSTANT SAMPLING RATE

Consider a continuous-time AR process

$$\begin{aligned}(p^n + a_1 p^{n-1} + \ldots + a_n)y(t) &= e(t) \\ Ee(t)e(s) &= \sigma^2\delta(t-s)\end{aligned} \tag{1}$$

where p denotes the differentiation operator, $y(t)$ is the output and $e(t)$ a (continuous-time) white noise source. The time series is observed at $t = h, 2h, 3h, \ldots, Nh$. The model order n is supposed to be known. A way to estimate n is described in Söderström *et al.* (1997*a*). It is of interest to estimate the parameter vector

$$\theta = (a_1 \ldots a_n)^T \tag{2}$$

from the available data.

To rewrite the model (1) we substitute the differentiation operator p^j by the following discrete approximation

$$p^j \approx D^j = \frac{1}{h^j}\sum_{\mu}\bar{\beta}_{j,\mu}q^{\mu} \tag{3}$$

where q is the shift operator. Needless to say we have to impose some conditions on the $\bar{\beta}$-weights in order to make the approximation (3) meaningful. We introduce what we will refer to as the *natural conditions*

$$D^j y(t) = p^j y(t) + O(h) \tag{4a}$$

which, by using (3), can be expressed as

$$\sum_{\mu}\bar{\beta}_{j,\mu}\mu^{\nu} = \begin{cases} 0 & \nu = 0, \ldots, j-1 \\ j! & \nu = j \end{cases} \tag{4b}$$

If (4b) is fulfilled, h is sufficiently small and $y(t)$ is smooth enough the approximation "error", introduced by using the operator D^j instead of p^j, will be small.

Remark 2.1. Notice that the well known delta forward operator

$$\delta = \frac{q-1}{h} \tag{5}$$

satisfies the condition (4a).

After substituting the derivatives in (1) by approximations (3), the following linear regression model can be formed

$$\begin{aligned} w(t) &= \varphi^T(t)\theta + \varepsilon(t) \\ w(t) &= D^n y(t) \\ \varphi^T(t) &= [-D^{n-1}y(t) \ldots - D^0 y(t)] \\ \theta &= (a_1 \cdots a_n)^T \end{aligned} \tag{6}$$

It turns out that a standard least squares estimate of θ

$$\hat{\theta}_N = [\Sigma_{t=1}^N \varphi(t)\varphi^T(t)]^{-1}[\Sigma_{t=1}^N \varphi(t)w(t)] \tag{7}$$

or an instrumental variable (IV) estimate of θ

$$\hat{\theta}_N = [\Sigma_{t=1}^N z(t)\varphi^T(t)]^{-1}[\Sigma_{t=1}^N z(t)w(t)] \tag{8}$$

where $z(t)$ is the vector of instruments, will in general be heavily biased, also for N large and h small. There are several simple means to modify (7) and (8) to avoid the bias, see Bigi *et al.* (1994); Söderström *et al.* (1997*a*); Söderström *et al.* (1997*b*):

- An LS scheme, with a shifted structure for estimating the nth order derivative of $y(t)$.
- An LS scheme, with a bias compensation feature.
- An instrumental variable scheme, with delayed values of the output signal as instruments.

These three approaches all lead to an estimate $\hat{\theta}$ with a small bias of order $O(h)$.

3. TIME-VARYING SAMPLING RATE

For a uniform sampling interval it is fairly easy to find derivative approximations that fulfill the *natural conditions* (4a), *e.g.* the delta forward operator. For a time varying sampling rate this question becomes more intricate. In this section a general recipe on how to choose the approximations in order to limit the inevitable error will be introduced.

First a simple notation and a basic assumption are introduced:

N1: The distance between to sampling instances is denoted by h_k, *i.e.*

$$t_{k+1} - t_k = h_k \tag{9}$$

A1: It is assumed that

$$\underline{h} \leq h_k \leq \bar{h} \qquad \forall k \tag{10}$$

where $\underline{h} > 0$ and $\bar{h}$ is sufficiently bounded.

Now, consider the following linear approximation of the differentiation operator p

$$p^j y(t_k) \approx D^j y(t_k) \triangleq \sum_{\mu=0}^{j}\beta_k(j,\mu)y(t_{k+\mu}) \tag{11}$$

Remark 3.1. We have chosen to include $j+1$ terms in the weighted sum, *i.e.* $\mu = 0, \ldots, j$. For the special case of constant h_k, this will lead to the well known delta forward approximation.

In order to make the approximation (11) meaningful some conditions on the β-weights have to be imposed. A natural request would be

$$p^j y(t_k) = D^j y(t_k) + O(\bar{h}) \qquad j = 0, \ldots, n \tag{12a}$$

We will, in analogue with (4a), refer to (12a) as the *natural conditions*. For our derivative approximation scheme (11) these conditions are in fact only fulfilled for a unique set of β-weights, as is shown later. By using (11) the *natural conditions*(12a) can be reformulated as

$$\sum_{\mu=0}^{j} \beta_k(j,\mu)\lambda_k^{\nu}(\mu) = \begin{cases} 0 & \nu = 0,\ldots,j-1 \\ j! & \nu = j \end{cases} \tag{12b}$$

where $\lambda_k(\mu)$ is given as

$$\lambda_k(\mu) = (t_{k+\mu} - t_k) = \begin{cases} 0 & \mu = 0 \\ \sum_{s=0}^{\mu-1} h_{k+s} & \text{otherwise} \end{cases} \tag{13}$$

The result (12b) can easily be derived by a Taylor expansion of (11) around $y(t_k)$, see Larsson (1999); Larsson and Söderström (1999). Equation (12b) can be thought of as a linear system of equations with $j+1$ equations and $j+1$ unknowns. In order to get the derivatives in question, this system of equations has to be solved for every $j = 0,\ldots,n$, where n is the system order, as well as for every $k = 1,\ldots,N$, where N is the number of sampling points. Using a standard tool for solving linear systems of equations would require quite a lot of computation power. Fortunately, there exists a simple analytical expression for the solution to (12b), which is stated in the following lemma:

Lemma 3.1. Assuming a positive sampling interval ($h_k > 0 \quad \forall k$), the solution to (12b) exists and is unique. Further the solution is given by

$$\beta_k(j,\mu) = \frac{j!}{\prod_{\substack{s=0 \\ s\neq\mu}}^{j} \left(\lambda_k(\mu) - \lambda_k(s)\right)} \qquad \mu = 0,\ldots,j \tag{14}$$

Proof. See Larsson (1999); Larsson and Söderström (1999).

The analytical expression for $\beta_k(j,\mu)$ implies that the system of linear equations (12b) do not have to be solved numerically, which saves a lot of time and effort. For further properties of the β-weights, *e.g.* recursive formulas, see Larsson (1999); Larsson and Söderström (1999).

3.1 *Computing the derivatives*

In order to compute the sought derivatives we will imitate the recursion formula for delta forward, which in the case of an equidistant sampling satisfies:

$$D^j y(t) = \frac{1}{h}\left[D^{j-1}y(t+h) - D^{j-1}y(t)\right] \tag{15}$$

The extension of equation (15) for the unevenly sampled case does exist and is stated in the following lemma.

Lemma 3.2. The derivatives that fulfill the *natural conditions* (12a) can be generated by the following recursion formula

$$D^j y(t_k) = \frac{1}{\tilde{h}_k(j)}\left[D^{j-1}y(t_{k+1}) - D^{j-1}y(t_k)\right] \tag{16}$$

where $\tilde{h}_k(j) = \frac{1}{j}\lambda_k(j)$.

Proof. See Larsson (1999); Larsson and Söderström (1999).

Note that in the special case h_k constant, equation (16) and equation (15) coincides. The nice structure of (16) will make an implementation straightforward and computationally very efficient as compared to a direct solution of (12b) for each k.

4. ANALYSIS

We are now ready to define our identification problem for the time-varying case. Reconsider the continuous-time AR process defined in (1). The time series is observed at discrete time unevenly spaced, *i.e.* we have two known sets:

$$\{t_k\}_{k=1}^{N} \text{ – moments of observation} \tag{17}$$

$$\{y(t_k)\}_{k=1}^{N} \text{ – the observed data} \tag{18}$$

From these data the goal is to estimate the autoregressive parameters

$$\theta = (a_1 \ldots a_n)^T \tag{19}$$

Assume that A1 defined in section 3 is fulfilled. Further, let us introduce the following additional assumptions:

A2: The distance between two sampling instances is stochastic, *i.e.*

$$h_k \sim p_h(x) \tag{20}$$

independent and identically distributed (i.i.d.).

A3: Finally, assume that the system noise e_k and the sampling interval h_k are independent.

By substituting the differentiation operator p^j by approximations (11), where the β-weights fulfill the *natural conditions* (12b), the original system (1) can be reformulated as the following discrete linear regression

$$\begin{aligned} w(t_k) &= \varphi^T(t_k)\theta + \varepsilon(t_k) \\ w(t_k) &= D^n y(t_k) \\ \varphi^T(t_k) &= [-D^{n-1}y(t_k) \dots - D^0 y(t_k)] \\ \theta &= (a_1 \cdots a_n)^T \end{aligned} \tag{21}$$

The parameter vector θ is estimated by a standard least squares method as (7), which in the asymptotic case ($N \to \infty$) can be written as, see Larsson (1999)

$$[E\varphi(t_k)\varphi^T(t_k)]\hat{\theta} = [E\varphi(t_k)w(t_k)] \tag{22}$$

Remark 4.1. There are two stochastic processes to deal with, the actual system and the sampling process. This implies that some special care has to be taken when dealing with expectations of system parameters, as the expectation might be with respect to either, or both of the two stochastic processes. The following convention is used: Writing E will mean $E_{y,h}$ (expectation with respect to both processes) if nothing else is said. Further, we stress that our discrete-time system (21) is not stationary. This is a consequence of the fact that the time series is observed irregularly in time.

The least squares approach (22) will in most cases give rise to estimates with a severe bias. For the case with evenly sampled data a number of methods to overcome this bias problem has been proposed, see *e.g.* the references in Section 2. Here two of these methods, the *Shifted Least Squares* and the *Bias Compensated Least Squares* are extended.

We will consider the asymptotic behavior of the algorithms, that is when the number of data goes to infinity. To prove that the methods are consistent it is sufficient to show that equation (22) is unbiased when $\bar{h} \to 0$. First rewrite equation (22) as

$$\sum_{i=0}^{n} \hat{a}_{n-i} E\{D^i y(t_k) D^j y(t_k)\} = 0 \qquad j = 0, \dots, n-1 \tag{23}$$

where $\hat{a}_0 = a_0 = 1$. To proceed we make use of the following lemma:

Lemma 4.1. It holds that

I)

$$\begin{aligned} E\{D^i y(t_k) D^j y(t_k)\} = \\ (-1)^j p^{i+j} r(0^+) + O(\bar{h}) \\ {\scriptstyle 1 \le i,j \le n, \quad i+j < 2n-1} \end{aligned} \tag{24}$$

II)

$$\begin{aligned} E\{D^n y(t_k) D^{n-1} y(t_k)\} = \\ \xi_n (-1)^{n-1} p^{2n-1} r(0^+) + O(\bar{h}) \end{aligned} \tag{25}$$

where

$$\xi_n = \frac{(-1)^{n-1}}{(2n-1)!} \sum_{l=0}^{n} \sum_{m=0}^{n-1} E_h \Big\{ \beta_k(n,l)\beta_k(n-1,m)|\lambda_k(l) - \lambda_k(m)|^{2n-1} \Big\} \tag{26}$$

Proof. See Larsson (1999); Larsson and Söderström (1999).

Remark 4.2. One would expect that ξ_n should depend on k, as the system is time varying. This is not the case, because of the expectation with respect to h that appears in the expression for ξ_n. The troublesome element turns out to be $E\{D^n y(t_k) D^{n-1} y(t_k)\}$, which is natural as the derivative $p^n y(t)$ does not have a finite variance.

As a result of the lemma we can rewrite equation (23) as

$$\begin{aligned} &(-1)^{n-1}\hat{a}_0 (\xi_n p^{2n-1} r(0^+) + O(\bar{h})) \\ &+(-1)^j \sum_{i=0}^{n-1} \hat{a}_{n-i}(p^{i+j} r(0^+) + O(\bar{h})) = 0 \\ &\text{when} \quad j = n-1 \end{aligned} \tag{27}$$

$$(-1)^j \sum_{i=0}^{n} \hat{a}_{n-i}(p^{i+j} r(0^+) + O(\bar{h})) = 0 \qquad \text{when} \quad j = 0, \dots, n-2 \tag{28}$$

Assuming that $\xi_n = 1$, equation (27) and equation (28) can be jointly written as

$$\sum_{i=0}^{n} \hat{a}_{n-i} p^{i+j} r(0^+) = O(\bar{h}) \qquad j = 0, \dots, n-1 \tag{29}$$

By using $a_0 = 1$ and comparing the well known Yule-Walker relation, see Söderström *et al.* (1997*a*),

$$\sum_{i=0}^{n} a_{n-i} p^{i+j} r(t) = 0 \qquad t \ge 0^+ \quad j \ge 0 \tag{30}$$

to equation (29) it follows that

$$\hat{a}_i = a_i + O(\bar{h}) \tag{31}$$

which implies that in this case the estimates are consistent. The conclusion is that in order to get consistent estimates we must require that $\xi_n = 1$. Next, two algorithms that gives estimates that fulfill

$$\hat{\theta} = \theta + O(\bar{h}) \tag{32}$$

are introduced.

4.1 *Shifted Least Squares*

In this method we force $\xi_n = 1$ by restricting the derivatives a bit further.

Lemma 4.2. If the β-weights satisfy the *natural conditions* (12a) and $w(t_k)$ in the regression (21) is chosen as:

$$w(t_k) = D^n y(t_{k+s}) \qquad s \geq n-1 \tag{33}$$

then a least squares approach will give estimates that fulfill (32).

Proof. See Larsson (1999); Larsson and Söderström (1999).

Nominally we choose $s = n - 1$, that is, $w(t_k) = D^n y(t_{k+n-1})$. This is, in some sense, the same as shifting the nth order derivative $n - 1$ time-steps forward. Thus the name shifted least squares.

4.2 *Bias Compensated Least Squares*

Another way of getting $\xi_n = 1$ is to divide the element $D^{n-1}y(t_k)D^n y(t_k)$ in the normal equations (22) by ξ_n.

Lemma 4.3. If the β-weights satisfy the *natural conditions* (12a) and the normal equations are modified in the following way

$$\Big[\frac{1}{N}\sum_{t=1}^{N}\varphi(t)\varphi^T(t)\Big]\hat{\theta} = D\Big[\frac{1}{N}\sum_{t=1}^{N}\varphi(t)w(t)\Big] \tag{34}$$

where $D = diag(\frac{1}{\xi_n}, 1, \ldots, 1)$, then the parameter estimate will fulfill (32).

Proof. See Larsson (1999); Larsson and Söderström (1999).

The result seems straightforward to implement. Weight all the terms $D^{n-1}y(t_k)D^n y(t_k)$ by $1/\xi_n$, which is possible as ξ_n can be precalculated. It is, though, worth noticing that the factor ξ_n contains an expectation with respect to h, that might not be trivial to compute. The term ξ_n will also vary between different sampling processes.

5. NUMERICAL EXAMPLES

In this section the two estimation methods introduced in Section 4.1 and Section 4.2 are compared in a simulation study. Consider the second order system

$$(p^2 + a_1 p + a_2)y(t_k) = e(t_k) \tag{35}$$

For our simulations we have used $a_1 = 2$ and $a_2 = 2$. The system is observed $N = 10000$ times. We have used Monte Carlo simulations with 100 realizations, *i.e.*, the presented result is the mean value of 100 independent realizations. When a standard deviation is presented it is calculated from these realizations. In all simulations we have fixed $\underline{h} = 0.01$ and varied $\bar{h}$. The theoretical estimates are found by solving the normal equations (22) for fixed $\underline{h}$ and $\bar{h}$. Two different probability density functions to describe the variation of the sampling interval is used, *i.e.* $h_k \sim p_h(x)$, where $p_h(x)$ is either:

Uniform The sampling interval is uniformly distributed over an interval, *i.e.*

$$p_h(x) \sim U(\underline{h}, \bar{h}) \tag{36}$$

Two Point The sampling interval is two point distributed, *i.e.*

$$\begin{aligned} p_h(x) &= \theta\,\delta(x-\underline{h}) + (1-\theta)\,\delta(x-\bar{h}) \\ \theta &= \bar{h}/(\underline{h}+\bar{h}) \end{aligned} \tag{37}$$

The estimate of a_1 is presented in Figure 1, while the estimate of a_2 is presented in Figure 2. The results indicate that the two methods give estimates of comparable quality. It is also clear from the figures that there is a good match between the theoretical and the experimental values. Further we see that the underlying sampling distribution can have a significant influence on the estimates. Another observation is that for a two point sampling distribution the estimation error variance is not a monotonically decreasing function with respect to $\bar{h}$. It should be emphasized that the "direct"-methods introduced in this paper are computationally much more efficient than "indirect"-methods, *e.g.* PEM, while the properties of the estimates, bias and variance, are comparable. This statement is verified by several numerical examples, see Larsson (1999); Larsson and Söderström (1999).

6. CONCLUSIONS

The problem of estimating continuous-time autoregressive models from discrete-time data, where the data are observed irregularly in time, has been the focus of this paper. By replacing the continuous differentiation operator by some approximations a linear regression model were formed and solved by some least squares schemes. To get parameter estimates with negligible bias, some care had to be taken when making the approximation of the derivative operator, as well as when solving the least squares problem. In this paper a general recipe on how to choose the approximations in order to limit the inevitable "error" is given. The proposed algorithm is straightforward to implement and computationally efficient. Two least squares schemes have been introduced and analyzed, namely, the *shifted least squares* and the *bias compensated least squares*. Both methods lead to an estimate $\hat{\theta}$ with a small bias of order $O(\bar{h})$. The shifted least squares algorithm is very efficient and straightforward to implement,

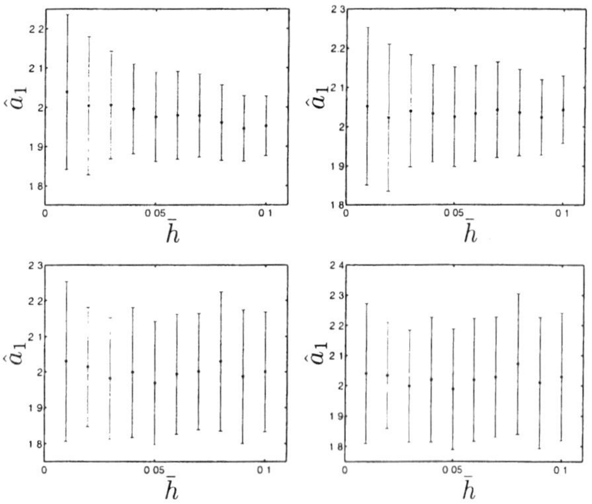

Fig. 1. Estimation results for a_1. True $a_1 = 2$. $(\cdots)$ - theoretical estimates. $(*)$ - Experimental estimates. The standard deviation of the estimates are shown by the vertical lines. (Upper left) - the shifted least squares, see Lemma 4.2, h_k is uniformly distributed. (Upper right) - the bias compensated least squares, see Lemma 4.3, h_k is uniformly distributed. (Lower left) - the shifted least squares, h_k is two point distributed. (Lower right) - the bias compensated least squares, h_k is two point distributed.

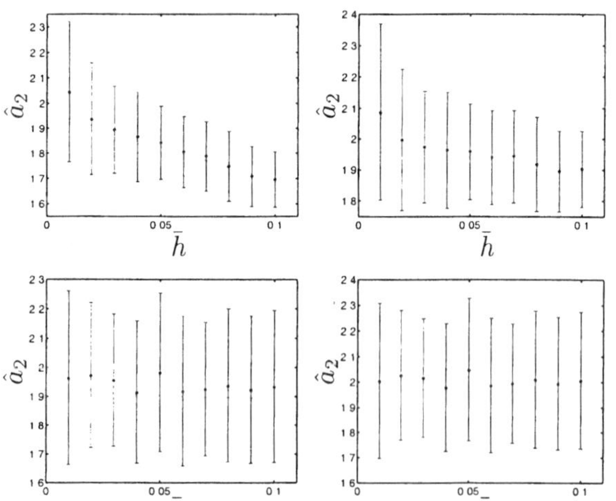

Fig. 2. Estimation results for a_2. True $a_2 = 2$. $(\cdots)$ - theoretical estimates. $(*)$ - Experimental estimates. The standard deviation of the estimates are shown by the vertical lines. (Upper left) - the shifted least squares, see Lemma 4.2, h_k is uniformly distributed. (Upper right) - the bias compensated least squares, see Lemma 4.3, h_k is uniformly distributed. (Lower left) - the shifted least squares, h_k is two point distributed. (Lower right) - the bias compensated least squares, h_k is two point distributed.

while the bias compensated least squares algorithm might be impractical due to the need of an evaluated expectation. By numerical examples the methods are demonstrated to work well. They give estimates of comparable quality. It was also shown that the underlying sampling distribution can have a great influence of the properties of the estimates. Finally, it should be emphasized that "direct"-methods are computationally very efficient compared with "indirect"-methods, *e.g.* PEM, especially when we have a time-varying sampling rate. Although the least squares methods introduced here are suboptimal, they seem to produce estimates with almost the same quality, *e.g* bias and variance, as estimates obtained from a prediction error method, especially for small values of $\bar{h}$.

References

Bigi, S., T. Söderström and B. Carlsson (1994). An IV-scheme for estimating continuous-time stochastic models from discrete-time data. In: *Proc. 10th IFAC Symposium on System Identification*. Copenhagen, Denmark.

Fan, H. and X. Liu (1994). Delta levinson and schur-type rls algorithms for adaptive signal processing. *IEEE Trans. on Signal Processing* **Vol. SP-42, No. 7**, 1629–1639.

Larsson, E. (1999). Identification of continuous-time AR systems in the case of unevenly sampled data. Master's thesis. Uppsala University, School of Engineering. Uppsala, Sweden.

Larsson, E. and T. Söderström (1999). Identification of continuous-time AR processes from unevenly sampled data. *Submitted for publication.*

Parzen, E. (1984). *Time Series Analysis of Irregularly Observed Data.* Springer-Verlag. New York, NY.

Phadke, M. S. and S. M. Wu (1974). Modeling of continuous stochastic processes from discrete observations with application to sunspots data. *J. American Stat. Assoc.*

Söderström, T., H. Fan, B. Carlsson and S. Bigi (1997*a*). Least squares parameter estimation of continuous-time ARX models from discrete-time data. *IEEE Trans. on Automatic Control* **AC-42**, 659–673.

Söderström, T., H. Fan, M. Mossberg and B. Carlsson (1997*b*). Bias-compensating schemes for estimating continuous-time AR process parameters. In: *Proc 11th IFAC Symposium on System Identification.* Kitakyushu, Japan.

Söderström, T., H. Fan, S. Bigi and B. Carlsson (1995). Can a least-squares fit be feasible for modelling continuous-time autoregressive processes from discrete-time data?. In: *Proc. IEEE Conference on Decision and Control.* New Orleans.

Unbehauen, H. and G. P. Rao (1990). Continuous-time approaches to system identification - a survey. *Automatica* **26**, 23–35.

TRANSFER FUNCTION AND FREQUENCY RESPONSE ESTIMATION USING RESONANT FILTERS

Peter J Gawthrop * Liuping Wang **

* *Centre for Systems and Control and Department of Mechanical Engineering, University of Glasgow, Glasgow. G12 8QQ Scotland*
** *Centre for Integrated Dynamics and Control, Dept. of Electical and Computer Engineering, The University of Newcastle, University Drive, Callaghan, 2308, Australia*

Abstract: A resonant filter approach is proposed for direct identification of continuous-time transfer functions from input-output data when the input contains periodic components. The asymptotic properties of the method are analysed; in particular the noise reduction properties are emphasised. Some illustrative simulations are provided.

Keywords: Parameter estimation; frequency response; continuous-time identification

1. INTRODUCTION

Periodic excitation has been used for estimation of the plant frequency response information since the early days of system identification (Eykhoff, 1974; Godfrey, 1993; Goodwin and Payne, 1977). The main advantages of using periodic input excitation are: (i) it offers improved signal to noise ratio at the selected excitation frequency regions, hence it reduces experimental time; (ii) it leads to data reduction when we take average of the data over its period. There are two basic means to generate periodic excitation signals for system identification purposes. One way is to use a combination of a set of sinusoidal signals where each sinusoid carries one required periodic information and the addition of this set of signals leads to the multi-frequency periodic signal. A multi-frequency periodic signal can be designed to achieve a "near" binary form by carefully choosing the initial phases of the sinusoids to minimize its peak-to-peak values (van den Bos and Krol, 1979; der Ouderaa *et al.*, 1988). An alternative way that has been found since the 80's is to use the relay feedback mechanism to generate periodic excitation to a plant (Åström and Hagglund, 1984; Åström and Hagglund, 1988). The main advantage of using relay feedback control lies in that this procedure automatically finds the important frequency regions of the plant when the plant dynamics are unknown. This nonlinear mechanism also prevents the output measurement noise from being fed back into the control signal, i.e. breaks the correlation relationship that exists in a standard feedback control environment. It is possible to obtain a multi-frequency periodic signal by conducting several experiments where each experiment is performed with relay in series with other dynamic elements (Wang *et al.*, 1999).

Up to now, most of the approaches for periodic input signals and relay feedback control signals are proposed for estimating plant frequency response information, a second step is then used to obtain a parametric model for the plant (Lundh and Åström, 1994; Wang *et al.*, 1997; Schoukens *et al.*, 1997; Goodwin and Welsh, 1999). A time domain approach is proposed by (Schei, 1994) for discrete time transfer function models based on relay feedback data. A more general approach for using periodic input signal has been discussed

[1] This work was accomplished whilst the first author was a visitor at CIDAC

very recently by Forssell *et al.* (1999) for identification of dynamic errors-in-variable systems where a discrete time model was identified from periodic signals using a so called compensated least squares method (Zheng and Feng, 1992).

In contrast, this paper brings together two strands of system identification research:

- direct estimation of process frequency response using frequency-sampling filters (Wang and Cluett, 1997; Wang *et al.*, 1999)
- continuous-time identification of plants under transient operating conditions (Gawthrop, 1984*a*; Gawthrop, 1984*b*)

The approach we propose is based on continuous time transfer function model using resonant filters. The proposed estimator has the following features:

- The algorithm estimates a (continuous-time) rational transfer function.
- The input signal need not be periodic, but must have significant periodic components.
- The estimator aims for a good frequency response fit at a finite number of frequencies corresponding to the discrete frequency components of the input signal; signals at other frequencies, or with a continuous spectrum, are asymptotically rejected.
- System initial conditions are explicitly considered so that data with transient components can be used; initial transient data does not have to be discarded
- Non-contiguous data sets from different experiments can be used to provide a combined transfer function estimate

The outline of the paper is as follows. Section 2 shows how an ordinary differential equation model can be transformed into a model appropriate for direct frequency response estimation. Section 3 shows how these models can be used. Section 4 gives the relation of the approach with frequency-domain methods and section 5 evaluates the method using the data of Wang *et al.* (1999). Section 6 concludes the paper.

2. SYSTEM MODELS

Following Gawthrop (1984*a*) and (1984*b*), the class of systems considered is those SISO systems with input $u(t)$ and output $y_0(t)$ described by an ordinary differential equation with real constant coefficients:

$$\sum_{i=0}^{n} a_i \frac{d^{n-i}y_0}{dt^{n-i}} = \sum_{i=0}^{n} b_i \frac{d^{n-i}u}{dt^{n-i}} \tag{1}$$

No assumptions are made about the values of the coefficients, and so the order of the system is not necessarily n – thus n represents an upper bound on the system order. The *measured* system output $y(t)$ is corrupted by additive white noise ψ of variance σ^2:

$$y(t) = y_0(t) + \psi(t) \tag{2}$$

Equation 1 can be Laplace transformed to give:

$$a(s)Y_0(s) = b(s)U(s) + d(s) \tag{3}$$

where s is the Laplace transform variable

$$a(s) = \sum_{i=0}^{n} a_i s^{n-i} \tag{4}$$

$$b(s) = \sum_{i=1}^{n} b_i s^{n-i} \tag{5}$$

where $d(s)$ is given by

$$d(s) = \sum_{i=1}^{n} d_i s^{n-i} \tag{6}$$

and represents the effect of initial conditions.

Following Gawthrop (1984*a*) and (1984*b*), introducing a polynomial $c(s)$ where:

$$c(s) = \sum_{i=0}^{n} c_i s^{n-i} \tag{7}$$

Equation 3 can be rewritten as:

$$\frac{a(s)}{c(s)}Y_0(s) = \frac{b(s)}{c(s)}U(s) + \frac{d(s)}{c(s)} \tag{8}$$

or in implicit form:

$$\frac{a(s)}{c(s)}Y_0(s) - \frac{b(s)}{c(s)}U(s) - \frac{d(s)}{c(s)} = 0 \tag{9}$$

Rewriting equation 9 in terms of the system output Y gives:

$$\frac{a(s)}{c(s)}Y(s) - \frac{b(s)}{c(s)}U(s) - \frac{d(s)}{c(s)} = E(s) \tag{10}$$

where

$$E(s) = \frac{a(s)}{c(s)}\Psi(s) \tag{11}$$

At this point, it is assumed (without loss of generality) that

(1) n is even
(2) $c(s)$ is monic ($c_n = 1$)

and so $c(s)$ can be chosen as the product of second-order factors as:

$$c(s) = \prod_{i=1}^{\frac{n}{2}} s^2 + 2\zeta\omega_i s + \omega_i^2 \tag{12}$$

where the $\frac{n}{2}$ *different* values of ω_i, together with ζ are chosen by the user as discussed in Section 3. In particular, the values of ω_i can be chosen to correspond to the frequencies found in the input signal.

Equation 10 can also be written in second-order partial fraction form as:

$$\sum_{i=1}^{\frac{n}{2}} \frac{\alpha_i^a \omega_i s + \beta_i^a}{s^2 + 2\zeta_i\omega_i s + \omega_i^2} Y(s) - \sum_{i=1}^{\frac{n}{2}} \frac{\alpha_i^b \omega_i s + \beta_i^b}{s^2 + 2\zeta_i\omega_i s + \omega_i^2} U(s)$$
$$- \sum_{i=1}^{\frac{n}{2}} \frac{\alpha_i^d \omega_i s + \beta_i^d}{s^2 + 2\zeta_i\omega_i s + \omega_i^2} = E(s)$$

Note that Equation 2 can be viewed as the continuous-time equivalent of the *frequency-sampled filter* used by Wang and Cluett (1997) and Wang *et al.* (1999). This form, and its relation to frequency-domain methods, will be considered further in Section 4.

Equation 10 can be transformed back into the time domain and rewritten in linear-in-the parameters form as:

$$X^T(t)\Theta = e(t) \tag{13}$$

where:

$X(t)$ is composed of the states of three SVFs:

$$X(t) = \begin{pmatrix} X^y(t) \\ -X^u(t) \\ -X^z(t) \end{pmatrix} \tag{14}$$

where:

$$\dot{X}^y = AX^y + By \tag{15}$$
$$X^y(0) = 0_n \tag{16}$$
$$\dot{X}^u = AX^u + Bu \tag{17}$$
$$X^u(0) = 0_n \tag{18}$$
$$\dot{X}^z = AX^z \tag{19}$$
$$X^z(0) = B \tag{20}$$

where 0_n is a zero column vector of length n.

Two state-space realisations of the transfer function representation of Equation 10 in the form of Equations 13–15 will be considered. Each SVF state is of dimension n.

Firstly, the *transfer-function* form (indicated by subscript TF) corresponds to the form of Equation 10 and is defined by:

$$A = A_{TF} = \begin{pmatrix} -c_1 & -c_2 & \dots & c_n \\ 1 & 0 & \dots & 0 \\ 0 & 1 & \dots & 0 \\ \dots & \dots & \dots & 0 \\ 0 & 0 & 1 & 0 \end{pmatrix}; \; B = B_{TF} = \begin{pmatrix} 1 \\ 0 \\ 0 \\ \dots\dots \\ 0 \end{pmatrix} \tag{21}$$

$$\Theta = \Theta_{TF} = \begin{pmatrix} a_1 \\ a_2 \\ \dots \\ a_n \\ b_1 \\ b_2 \\ \dots \\ b_n \\ d_1 \\ d_2 \\ \dots \\ d_n \end{pmatrix} \tag{22}$$

Secondly, the *frequency-response* form (indicated by subscript FR) which corresponds to the form of Equation 2

$$A = A_{FR} = \begin{pmatrix} A_1 & 0_{2\times2} & \dots & 0_{2\times2} \\ 0_{2\times2} & A_2 & \dots & 0_{2\times2} \\ \dots & \dots & \dots & \dots \\ 0_{2\times2} & 0_{2\times2} & \dots & A_{\frac{n}{2}} \end{pmatrix}; \; B = B_{FR} = \begin{pmatrix} \omega_1 \\ 0 \\ \omega_2 \\ 0 \\ \dots\dots \\ \omega_{\frac{n}{2}} \\ 0 \end{pmatrix} \tag{23}$$

$$\Theta = \Theta_{FR} = \begin{pmatrix} \alpha_1^a \\ \beta_1^a \\ \alpha_2^a \\ \beta_2^a \\ \dots \\ \alpha_{\frac{n}{2}}^a \\ \beta_{\frac{n}{2}}^a \\ \dots \\ \alpha_1^b \\ \beta_1^b \\ \alpha_2^b \\ \beta_2^b \\ \dots \\ \alpha_{\frac{n}{2}}^b \\ \beta_{\frac{n}{2}}^b \\ \alpha_1^d \\ \beta_1^d \\ \alpha_2^d \\ \beta_2^d \\ \dots \\ \alpha_{\frac{n}{2}}^d \\ \beta_{\frac{n}{2}}^d \end{pmatrix} \tag{24}$$

where

$$A_i = \omega_i \begin{pmatrix} -\zeta_i & -1 \\ 1 & 0 \end{pmatrix} \tag{25}$$

The corresponding SVF states are, in each case, linearly related by the *transformation matrix* $\mathfrak{T}$:

$$x_{FR} = \mathfrak{T} x_{TF} \tag{26}$$

Following Kailath (1980), the *transformation matrix* $\mathfrak{T}$ can be expressed in terms of the controllability matrices $\mathfrak{C}_{TF}$ $\mathfrak{C}_{FR}$ as:

$$\mathfrak{T} = \mathfrak{C}_{FR}\mathfrak{C}_{TF}^{-1} \quad (27)$$

In certain cases, elements of the parameter vector Θ_{TF} are known to be zero. There are three such cases relevant here:

(1) The degree n_a of $a(s)$ is less than $n = n_c$ the degree of $c(s)$. In this case (with reference to Equation 22) $a_i = 0$ for $1 \leq i \leq n_c - n_a$.
(2) The degree n_b of $b(s)$ is less than $n = n_c$ the degree of $c(s)$. In this case (with reference to Equation 22) $b_i = 0$ for $1 \leq i \leq n_c - n_b$.
(3) Following Gawthrop (1984*a*) and (1984*b*), a constant output offset is modelled within the context of Equation 1 by constraining $a(0) = b(0) = 0$. In this case, $a_n = b_n = 0$.

In all these three cases, it is convenient to define *constrained* state and parameter vectors X_C and Θ_C as the result of deleting the zero elements of Θ_{TF} and the corresponding elements of X_{TF}. Equation 28 then becomes

$$X_C^T(t)\Theta_C = e(t) \quad (28)$$

3. IDENTIFICATION

This section focuses on system identification when the input signal is composed of a number of periodic signals. This allows the LIP equation 13 to be *tuned* so that the resonances due to the $c(s)$ match the frequencies of the periodic components of the input signal. Moreover, the fact that Θ_{TF} is independent of the choice of $c(s)$ is used to allow multiple passes though the same data (with different $c(s)$ and thus tuned frequencies) to make full use of the frequency content of the data. Finally, the method is extended to allow multiple data sets (generated from the same system) to be combined using ideas from Gawthrop (1984*a*). Throughout this section, it is assumed that the transfer function form the SVF (Equations 21) is used.

In particular, the following assumptions are made:

Assumption 1 The system input contains n_ω sinusoidal signals at the frequencies $\omega_1 \dots \omega_{n_\omega}$:

$$u(t) = \sum_{i=1}^{n_\omega} A_i^u \sin \omega_i t + \phi_i^u \quad (29)$$

where the n_ω frequencies are *known*.

Assumption 2 The degrees n_a and n_b of $a(s)$ and $b(s)$ are known.

For a given set of n_ω frequencies represented as elements of the vector ω:

$$\omega = \begin{pmatrix} \omega_1 & \omega_2 & \dots & \omega_{n_\omega} \end{pmatrix} \quad (30)$$

the estimation algorithms are parameterised by the $n_\Omega \times \frac{n}{2}$ matrix Ω with the following constraints:

(1) $n_\Omega m_\Omega \geq n_\omega$
(2) Each element is an element of ω
(3) All elements in each row are different

Thus, for example, if $n_\omega = 6$ and $n = 4$, Ω can be chosen as:

$$\begin{pmatrix} \omega_1 & \omega_2 \\ \omega_3 & \omega_4 \\ \omega_5 & \omega_6 \end{pmatrix} \text{ or } \begin{pmatrix} \omega_1 & \omega_2 & \omega_3 \\ \omega_4 & \omega_5 & \omega_6 \end{pmatrix} \text{ or } \omega = \begin{pmatrix} \omega_1 & \omega_2 & \omega_3 & \omega_4 & \omega_5 & \omega_6 \end{pmatrix} \quad (31)$$

The following algorithms are proposed for a single dataset:

Algorithm 1

(1) Design an input signal comprising periodic components at n_ω different frequencies of interest – possibly using a relay experiment to generate the data (Wang and Cluett, 1997; Wang *et al.*, 1999).
(2) Choose the system order n, and hence the degree n_c of $c(s)$ so that $n_c > n$ and n_c is even.
(3) Choose the *constrained* parameter vector X_C (see Equation 28 and the preceding text).
(4) Create n_Ω polynomials $c_i(s)$ of the form of Equation 12.
(5) Run the SVF filter on the *same* data with each of the n_Ω filters and generate the corresponding average *information matrices*

$$S_i = \frac{1}{T}\int_{t=0}^{T} X(t)^T X(t) dt \quad (32)$$

(6) Create the composite information matrix

$$S = \frac{1}{n_\Omega}\sum_{i=0}^{n_\Omega} S_i \quad (33)$$

(7) Perform a singular-value decomposition of S and pick the vector corresponding to the *minimal* singular value; this is the estimate $\hat{\Theta}_C$ of Θ_C.

For multiple data sets, the *splicing* approach (Gawthrop, 1984*a*) is proposed. Briefly, for K datasets indexed by $1 \leq k \leq K$this involves:

Algorithm 2

(1) Perform steps 1–4 of Algorithm 1 for each data set.
(2) Replace equation 14 by

$$X(t) = \begin{pmatrix} X^y(t) \\ -X^u(t) \\ -X_1^z(t) \\ \dots \\ -X_K^z(t) \end{pmatrix} \quad (34)$$

where

$$\dot{X}_i^z = AX_i^z \tag{35}$$

and for dataset k

$$X_i^z(0) = \begin{cases} X_i^z 1_n \text{ if } i = k \\ 0_n \text{ if } i \neq k \end{cases} \tag{36}$$

(3) Using this extended data vector X, perform steps 5–6 of algorithm 1.
(4) Sum the information matrices from each dataset

$$S = \sum_{k=0}^{K} S_k \tag{37}$$

(5) Perform step 7 of algorithm 1.

4. RELATION WITH FREQUENCY DOMAIN METHODS

This section looks at the relation between the frequency response and transfer function representation in the unconstrained, single-pass case.

(1) In the *unconstrained* case the two representations are related by the (nonsingular) transformation of Equation 26.
(2) In the case of a single data pass (Algorithm 1 and $\Omega = \omega$) the parameter vectors are related by

$$\Theta_{FR} = \Theta_{TF} \mathfrak{T}^{-1} \tag{38}$$

where the transformation matrix $\mathfrak{T}$ is given by Equation 27.
(3) Thus the two representations are essentially equivalent although formally different.
(4) The FR formulation uses a bank of tuned filters. When $\zeta = 0$ this is equivalent to convolution with a sine and a cosine function.
(5) The FR formulation gives the real part β_i and imaginary part α_i of the numerator and denominator of the transfer function at each frequency ω_i.
(6) Both formulations allow for transient terms in the output and non-harmonically related frequencies (ie non-periodic input) – if these were dropped, must be essentially the same as using DFT techniques.

In the *constrained* case, the relation of Equation 26 can not be used. In this situation, the TF method can still be viewed as estimating a frequency response but *with the constraint that the frequency response corresponds to a transfer function of the chosen order.*

5. EXAMPLE

The data for this example is identical to that of Wang *et al.* (1999). It corresponds to two separate relay experiments conducted on the system:

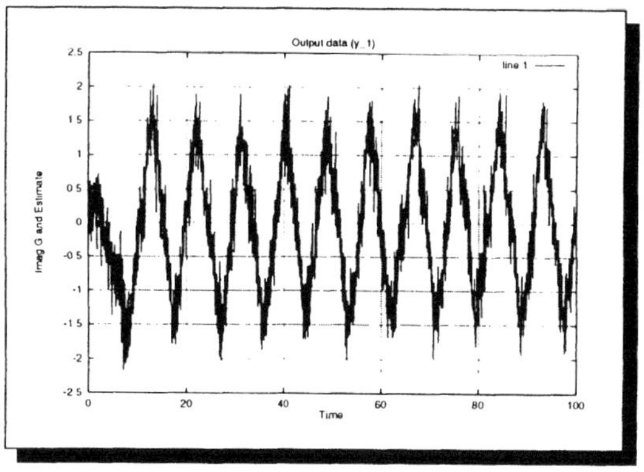

Fig. 1. Dataset 1: Wang/Cluett example

$$\frac{Y_0(s)}{U(s)} = G(s) = \frac{-2s+1}{(s+1)^3} \tag{39}$$

The first experiment (Fig 1) corresponds to the usual relay feedback and the frequency thus approximately corresponds to the critical frequency ω_1 (phase $= -\pi$); the second experiment includes an integrator in the feedback loop to give oscillation at a frequency ω_2 corresponding to a phase of $-\frac{\pi}{2}$.

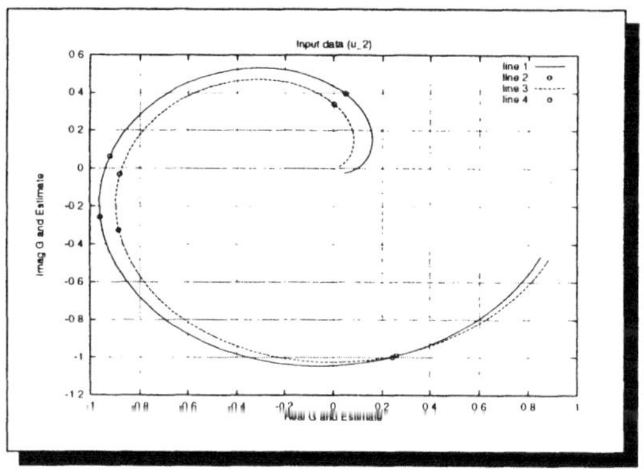

Fig. 2. Nyquist: Wang/Cluett example

The two data sets were spliced and analysed using filters at the fundamental ω_i and second harmonic $3\omega_i$ appropriate to each dataset. The results are shown in Nyquist form in Figure 2 where:

(1) The firm line is the estimated frequency response (with the 4 frequency points marked and
(2) The dashed line is the true frequency response (with the 4 frequency points marked.

6. CONCLUSION

A parameter estimation algorithm which uses resonant filters to enhance the periodic components of the input and output signals has been presented in a number of forms. One such form is related by a linear invertible transformation to a frequency-response estimator; but, in general, the method constrains the frequency response to be that of a rational transfer function with given numerator and denominator degrees. This is shown to lead to better results when the degree information is known. Moreover, this provides the basis for tests for determining the degree and relative degree of a system; this will be reported on in a future paper.

The infinite data length analysis confirms the intuitive idea that the resonant filters enhance signal to noise ratio when the the resonant frequencies are chosen to correspond to periodic components of the system input data.

The properties of the method have been verified by simulated examples and the usefulness of the method confirmed by application to real data.

7. ACKNOWLEDGEMENTS

This work was accomplished whilst the author was a visitor at the Centre for Integrated Dynamics and Control, University of Newcastle, New South Wales. He would like to thank Prof. Graham Goodwin for providing an excellent work environment.

8. REFERENCES

Åström, K. J. and T. Hagglund (1984). Automatic tuning of simple regulators with specifications on phase and amplitude margins. *Automatica* **Vol.20**, 645–651.

Åström, K. J. and T. Hagglund (1988). *Automatic Tuning of PID Controllers.* Instrument Society of America, Research Triangle Park, NC.

der Ouderaa, E. Van, J. Schoukens and J. Renneborg (1988). Peak factor minimization of input and output signals of linear systems. *IEE Trans. Instrumentation and Measurement* **Vol.37**, 207–212.

Eykhoff, P. (1974). *System identification.* Wiley, New York.

Forssell, U., F. Gustafsson and T. McKelvey (1999). Time-domain identification of dynamic errors-in-variables systems using periodic excitation signals. *Proceedings of the 14th IFAC World Congress, Beijing, China.*

Gawthrop, P. J. (1984*a*). Parameter identification from non-contiguous data. *Proceedings IEE* **131 Pt.D**(6), 261–265.

Gawthrop, P. J. (1984*b*). Parametric identification of transient signals. *IMA journal of Mathematical Control and Information* **1**, 117–128.

Godfrey, K. (1993). *Perturbation Signals for System Identification.* Prentice-Hall, Englewood Cliffs, NJ.

Goodwin, G. C. and R. L. Payne (1977). *Dynamic System Identification.* Academic Press, New York.

Goodwin, Graham C and James Welsh (1999). Analysis of a novel method of autotuning a multivariable plant based on quantization. In: *American Control Conference.*

Kailath, T. (1980). *Linear Systems.* Prentice-Hall.

Lundh, M. and K. J. Åström (1994). Automatic initialization of a robust self-tuning controller. *Automatica* **Vol.30**, 1649–1662.

Schei, T.S. (1994). Automatic tuning of PID controllers based on transfer function estimation. *Automatica* **Vol.30**, 1983–1989.

Schoukens, J., R. Pintelon, G. Vandersteen and P. Guillaume (1997). Frequency domain system identification using non-parametric noise models estimated from a small number of data sets. *Automatica* **Vol. 33**, 1073–1086.

van den Bos, A. and R. G. Krol (1979). Synthesis of discrete-interval binary signals with specified fourier amplitude spectra. *Int. J. Control* **Vol.30**, 871–884.

Wang, L. and W. R. Cluett (1997). Frequency-sampling filters: an improved model structure for step-response identification. *Automatica* **Vol. 33**, 939–944.

Wang, L., M. Desarmo and W. R. Cluett (1999). Recursive estimation of process frequency response and step response from relay feedback experiments. *Automatica* **Vol. 35**, no. 8.

Wang, Q.-G., C.-C. Hang and B. Zou (1997). Low-order modeling from relay feedback. *Ind. Eng. Chem. Res.* **Vol. 36**, 375–381.

Zheng, W. X. and C. B. Feng (1992). Identification of a class of dynamic errors-in-variable models. *International Journal of Adaptive and Signal Processing* **Vol.6**, 431–440.

SUBSPACE IDENTIFICATION METHOD FOR COMBINED DETERMINISTIC-STOCHASTIC BILINEAR SYSTEMS

Huixin Chen and Jan Maciejowski [1]

Department of Engineering
University of Cambridge
Cambridge CB2 1PZ, U.K.

Abstract: In this paper, a 'four-block' subspace system identification method for combined deterministic-stochastic bilinear systems is developed. Estimation of state sequences, followed by estimation of system matrices, is the central component of subspace identification methods. The prominent difference of our new approach is a 'four-block' arrangement of data matrices which leads to a linearization of the system state equation, when written in block form. A major advantage of this approach, over a previous bilinear subspace algorithm, is that the measured input is not restricted to be white. We show that, providing a certain data-dependent eigenvalue condition is met, our algorithm provides asymptotically unbiased estimates, and we indicate the rate at which the bias decreases. Simulation results show that this algorithm requires a smaller sample size than earlier algorithms (for comparable performance) and that the computational complexity is significantly lower. *Copyright*© 2000 *IFAC*

Keywords: System Identification, Bilinear Systems, Subspace Methods, State Sequnce Estimation, State-space Models.

1. INTRODUCTION

There has been much study of the identification problem of bilinear systems. A three-block subspace identification method for deterministic bilinear system is presented in (Chen and Maciejowski, 1999). In this paper we present a very closely related method for identifying a combined deterministic-stochastic bilinear state-space model. A method for subspace identification of combined deterministic-stochastic bilinear systems subject to white noise inputs was derived in (Favoreel *et al.*, 1996), (Favoreel *et al.*, 1997), (Favoreel *et al.*, 1999), which was the first effort to use subspace methodology to identify the state-space model of a bilinear system in the stochastic case. A subspace identification algorithm which can be used with a non-white input was proposed in (Favoreel and De Moor, 1998), but it results in biased estimates of the system. A 'three-block' subspace identification algorithm is given in (Chui and Maciejowski, 1998), which can remove the effects of unmeasured noise sources and obtain accurate estimates. In the linear system case, the three-block algorithm is linked with the system's Markov parameters. Unfortunately, the realization

[1] Corresponding author.

theory for bilinear systems is more complicated than for linear systems. In our 'four-block' subspace method for bilinear systems, the four-block data matrix arrangement and the observation matrix equation are used to linearise the system equation in the block matrix form using linear and bilinear algebra. This allows bilinear models to be obtained from row and column spaces of certain matrices, calculated from the input-output data by means of some bilinear-algebraic operations. All the proofs are omitted here, but can be found in (Chen and Maciejowski, 2000).

2. NOTATIONS

The use of much specialised notation seems to be unavoidable in the current context. Mostly we follow the notation used in (Favoreel *et al.*, 1999), (Chen and Maciejowski, 1999), but we introduce all the notation here for completeness.

We use $\otimes$ to denote the Kronecker product and $\odot$ the Khatri-Rao product of two matrices with $F \in \mathbf{R}^{t\times p}$ and $G \in \mathbf{R}^{u\times p}$ defined in (Khatri and Rao, 1968), (Suda *et al.*, 1973):

$$F \odot G \triangleq [f_1 \otimes g_1, f_2 \otimes g_2, \cdots, f_p \otimes g_p]$$

$+, \oplus$ and $\cap$ denote the sum, the direct sum and the intersection of two vector spaces, $\cdot^{\perp}$ denotes the orthogonal complement of a subspace with respect to the predefined ambient space, the Moore-Penrose inverse is written as $\cdot^{\dagger}$, and the Hermitian as $\cdot^{*}$.

In this paper we consider combined deterministic-stochastic time-invariant bilinear system of the form:

$$\begin{aligned} x_{t+1} &= Ax_t + Nu_t \otimes x_t + Bu_t + w_t \\ y_t &= Cx_t + Du_t + v_t \end{aligned} \qquad (1)$$

where $x_t \in \mathbf{R}^n, y_t \in \mathbf{R}^l, u_t \in \mathbf{R}^m$, and $N = [N_1\ N_2\ ...N_m] \in \mathbf{R}^{n\times nm}$, $N_i \in \mathbf{R}^{n\times n}$ $(i = 1, ..., m)$.

The input u_t is assumed to be independent of the measurement noise v_t and the process noise w_t. The covariance matrix of w_t and v_t is:

$$\mathbf{E}\left[\begin{pmatrix} w_p \\ v_p \end{pmatrix}\begin{pmatrix} w_q \\ v_q \end{pmatrix}^T\right] = \begin{bmatrix} Q & S \\ S^T & R \end{bmatrix}\delta_{pq} \geq 0$$

We assume that the sample size is $\tilde{N}$, namely that input-output data $\{u(t), y(t) : t = 0, 1, \ldots, \tilde{N}\}$ are available. For arbitrary t we define

$$X_t \triangleq [x_t\ x_{t+1}\ \ldots\ x_{t+j-1}] \in \mathbf{R}^{n\times j}$$

but for the special cases $t = 0$, $t = k$, $t = 2k$ and $t = 3k$ we define, with some abuse of notation,

$$\begin{aligned} X_p &\triangleq [x_0\ x_1\ \ldots\ x_{j-1}] \in \mathbf{R}^{n\times j} \\ X_c &\triangleq [x_k\ x_{k+1}\ \ldots\ x_{k+j-1}] \in \mathbf{R}^{n\times j} \\ X_f &\triangleq [x_{2k}\ x_{2k+1}\ \ldots\ x_{2k+j-1}] \in \mathbf{R}^{n\times j} \\ X_r &\triangleq [x_{3k}\ x_{3k+1}\ \ldots\ x_{3k+j-1}] \in \mathbf{R}^{n\times j} \end{aligned}$$

where k is the *row block size.* The suffices p, c, f and r are supposed to be mnemonic, representing 'past', 'current', 'future' and 'remote future' respectively. This division of the state history into four (overlapping) segments is the reason for referring to our algorithm as a 'four-block' method. In the deterministic case the 'remote future' segment is not necessary, and a simpler 'three-block' method results (Chen and Maciejowski, 1999).

We define U_t, U_p, U_f, U_r, Y_t, Y_p, Y_f, Y_r, W_t, U_p, W_f, W_r, V_t, V_p, V_f, V_r, similarly. These matrices will later be used to construct larger matrices with a 'generalised block-Hankel' structure. In order to use all the available data in these, the number of columns j is such that $\tilde{N} = 3k + j - 1$ and let $d_i = \Sigma_{p=1}^{i}(m+1)^{p-1}l$, $e_i = \Sigma_{p=1}^{i}(m+1)^{p-1}m$ and $f_k = e_k + \frac{m}{2}(m+1)^k + l[(m+1)^k - 1] + e_k^2$.

For arbitrary q and $i \geq q+2$, we define

$$X_{q|q} \triangleq \begin{pmatrix} X_q \\ U_q \odot X_q \end{pmatrix} \in \mathbf{R}^{(m+1)n\times j}$$

$$X_{i-1|q} \triangleq \begin{pmatrix} X_{i-2|q} \\ U_{i-1} \odot X_{i-2|q} \end{pmatrix} \in \mathbf{R}^{(m+1)^{i-q}n\times j}$$

$$Y_{q|q} \triangleq Y_q$$

$$Y_{i-1|q} \triangleq \begin{pmatrix} Y_{i-1} \\ Y_{i-2|q} \\ U_{i-1} \odot Y_{i-2|q} \end{pmatrix} \in \mathbf{R}^{d_{i-q}\times j}$$

$$U^{+}_{q|q} \triangleq U_q$$

$$U^{+}_{i-1|q} \triangleq \begin{pmatrix} U^{+}_{i-2} \\ U_{i-1} \\ U_{i-1} \odot U^{+}_{i-2|q} \end{pmatrix} \in \mathbf{R}^{((m+1)^{i-q}-1)\times j}$$

$$U^{++}_{q|q} \triangleq \begin{pmatrix} U_{q,1} \odot U_q \\ U_{q,2} \odot U_q(2:m,:) \\ U_{q,3} \odot U_q(3:m,:) \\ \vdots \\ U_{q,m} \odot U_{q,m} \end{pmatrix} \in \mathbf{R}^{\frac{m(m+1)}{2}\times j}$$

$$U^{++}_{i-1|q} \triangleq \begin{pmatrix} U^{++}_{i-2|q} \\ U_{i-1} \odot U^{++}_{i-2|q} \end{pmatrix} \in \mathbf{R}^{\frac{m}{2}(m+1)^{i-q}\times j}$$

$$U^y_{i-1|q} \triangleq U^+_{i-1|q} \odot Y_q$$

$$U^{u,y}_{i+k-1|k+q} \triangleq \begin{pmatrix} U_{i+k-1|k+q} \\ U^{++}_{i+k-1|k+q} \\ U^{y}_{i+k-1|k+q} \\ U^{+}_{i+k-1|k+q} \odot U_{i-1|q} \end{pmatrix}$$

$U_{i-1|q}, W_{i-1|q}$ and $V_{i-1|q}$ and can be defined similarly.

Remark 1. The meaning of $U^+_{i-1|q}$ is different from that in (Favoreel *et al.*, 1996). $U^{++}_{i-1|q}$ is newly introduced here.

$$X^c \triangleq X_{2k-1|k},\ X^f \triangleq X_{3k-1|2k},\ X^r \triangleq X_{4k-1|3k}$$
$$U^p \triangleq U_{k-1|0},\ U^c \triangleq U_{2k-1|k},\ U^f \triangleq U_{3k-1|2k}$$
$$U^{p,y} \triangleq U^{+p} \odot Y_p,\ U^{c,y} \triangleq U^{+c} \odot Y_c$$
$$U^{f,y} \triangleq U^{+f} \odot Y_f,\ U^{r,y} \triangleq U^{+r} \odot Y_r,$$
$$U^{c,u,y} \triangleq \begin{pmatrix} U^c \\ U^{++c} \\ U^{c,y} \\ U^{+c} \odot U^p \end{pmatrix},\ U^{f,u,y} \triangleq \begin{pmatrix} U^f \\ U^{++f} \\ U^{f,y} \\ U^{+f} \odot U^c \end{pmatrix}$$

$U^r, Y^p, Y^c, Y^f, Y^r, W^c, W^f, W^r, V^c, V^f, V^r, U^{+c}, U^{+f}, U^{+r}, U^{++c}, U^{++f}, U^{++r}$ and $U^{r,u,y}$ can be defined similarly. Finally, we denote by $\mathcal{U}_p$ the space spanned by all the rows of the matrix U_p. That is,

$$\mathcal{U}_p := \text{span}\{\alpha^* U_p, \quad \alpha \in \mathbf{R}^{km}\}$$

$\mathcal{U}_c, \mathcal{U}_f, \mathcal{U}_r, \mathcal{Y}_p, \mathcal{Y}_c, \mathcal{Y}_f, \mathcal{Y}_r, \mathcal{U}^p, \mathcal{Y}^p, \mathcal{U}^f, \mathcal{Y}^f, \mathcal{U}^{p,u,y}, \mathcal{U}^{f,u,y}$ *etc* are defined similarly.

3. ANALYSIS

Lemma 1. The system (1) can be rewritten in the following matrix equation form:

$$\begin{aligned} X_{t+1} &= AX_t + NU_t \odot X_t + BU_t + W_t \\ Y_t &= CX_t + DU_t + V_t \end{aligned} \qquad (2)$$

Lemma 2. For $j \geq 0$, and the block size k, we have

$$X_{k-1+j|j} = \begin{pmatrix} X_j \\ U^+_{k-1+j|j} \odot X_j \end{pmatrix}$$

Lemma 3. For F, G, H, J of compatible dimensions, $F \in \mathbf{R}^{k\times l}$, $G \in \mathbf{R}^{l\times m}$, $H \in \mathbf{R}^{p\times l}$, $J \in \mathbf{R}^{l\times m}$:

$$(FG \otimes HJ) = (F \otimes H)(G \otimes J)$$
$$(FG \odot HJ) = (F \otimes H)(G \odot J)$$

Lemma 4. (Input-Output Equation). For the combined deterministic-stochastic system (1) and $j \geq 0$, we have the following Input-Output Equation

$$\begin{aligned} X_{k+j} &= \Delta^X_k X_{k-1+j|j} + \Delta^U_k U_{k-1+j|j} + \Delta^W_k W_{k-1+j|j} \\ Y_{k-1+j|j} &= \mathcal{L}^X_k X_{k-1+j|j} + \mathcal{L}^U_k U_{k-1+j|j} + \mathcal{L}^W_k W_{k-1+j|j} \\ &\quad + \mathcal{L}^V_k V_{k-1+j|j} \end{aligned}$$

where

$$\Delta^X_n \triangleq [A\Delta^X_{n-1}, N_1\Delta^X_{n-1}, \dots, N_m\Delta^X_{n-1}]$$
$$\Delta^X_1 \triangleq [A, N_1, \dots, N_m]$$
$$\Delta^U_n \triangleq [B, A\Delta^U_{n-1}, N_1\Delta^U_{n-1}, \dots, N_m\Delta^U_{n-1}],\ \Delta^U_1 \triangleq B$$
$$\Delta^W_n \triangleq [I_{n\times n}, A\Delta^W_{n-1}, N_1\Delta^W_{n-1}, \dots, N_m\Delta^W_{n-1}]$$
$$\Delta^W_1 \triangleq I_{n\times n}$$
$$\mathcal{L}^X_k \triangleq \begin{bmatrix} C\Delta^X_{k-1} & 0 \\ \mathcal{L}^X_{k-1} & 0 \\ 0 & \mathcal{L}^X_{k-1} \end{bmatrix},\ \mathcal{L}^U_k \triangleq \begin{bmatrix} D & C\Delta^U_{k-1} & 0 \\ 0 & \mathcal{L}^U_{k-1} & 0 \\ 0 & 0 & \mathcal{L}^U_{k-1} \end{bmatrix}$$
$$\mathcal{L}^W_k \triangleq \begin{bmatrix} 0 & C\Delta^W_{k-1} & 0 \\ 0 & \mathcal{L}^W_{k-1} & 0 \\ 0 & 0 & \mathcal{L}^W_{k-1} \end{bmatrix},\ \mathcal{L}^V_k \triangleq \begin{bmatrix} I_{l\times l} & 0 & 0 \\ 0 & \mathcal{L}^V_{k-1} & 0 \\ 0 & 0 & \mathcal{L}^V_{k-1} \end{bmatrix}$$

with

$$\mathcal{L}^X_1 \triangleq [C, 0_{l\times m}],\ \mathcal{L}^U_1 \triangleq D,\ \mathcal{L}^W_1 \triangleq 0_{l\times n},\ \mathcal{L}^V_1 \triangleq I_{l\times l}$$

Lemma 5. For system (1), if

$$\lambda = \max_{j=0,\dots,\tilde{N}} |\text{eig}(A + \sum_{i=1}^{n} u_{j,i}N_i)| < 1, \qquad (3)$$

then

$$X_c = E(Y_c - DU_c - V_c) + (I - EC)\Delta^U_n U^p + o(\lambda^k)$$
$$X_f = E(Y_f - DU_f - V_f) + (I - EC)\Delta^U_n U^c + o(\lambda^k)$$
$$X_r = E(Y_r - DU_r - V_r) + (I - EC)\Delta^U_n U^f + o(\lambda^k)$$

where $o(\lambda^k)$ is used to denote a matrix M, such that $||M||_1 = o(\lambda^k)$.

Remark 2 This holds for any matrix E of compatible dimensions. In particular, it holds for $E = C^\dagger$, where $CC^\dagger = I$, and if $l \geq n$, then $I - C^\dagger C = 0$ and the expression become exact. In the sequel, we will assume the Moore-Penrose pseudo-inverse is used.

Theorem 1. The system (1) can be written in the following form if the condition (3) holds:

$$
\begin{aligned}
Y^c &= \mathcal{O}_k X_c + \mathcal{T}_k^u U^{c,u,y} + \mathcal{T}_k^v U^{+c} \odot V_c \\
&\quad + \mathcal{L}_k^W W^c + \mathcal{L}_k^V V^c + o(\lambda^k) \quad (4)\\
Y^f &= \mathcal{O}_k X_f + \mathcal{T}_k^u U^{f,u,y} + \mathcal{T}_k^v U^{+f} \odot V_f + \\
&\quad \mathcal{L}_k^W W^f + \mathcal{L}_k^V V^f + o(\lambda^k) \quad (5)\\
Y^r &= \mathcal{O}_k X_r + \mathcal{T}_k^u U^{r,u,y} + \mathcal{T}_k^v U^{+r} \odot V_r \\
&\quad + \mathcal{L}_k^W W^r + \mathcal{L}_k^V V^r + o(\lambda^k) \quad (6)\\
X_f &= \mathcal{F}_k X_c + \mathcal{G}_k^u U^{c,u,y} + \mathcal{G}_k^v U^{+c} \odot V_c \\
&\quad + \Delta_k^W W^c + o(\lambda^k) \\
X_r &= \mathcal{F}_k X_f + \mathcal{G}_k^u U^{f,u,y} + \mathcal{G}_k^v U^{+f} \odot V_f \\
&\quad + \Delta_k^W W^f + o(\lambda^k)
\end{aligned}
$$

where $\mathcal{O}_k, \mathcal{T}_k^u, \mathcal{T}_k^v, \mathcal{F}_k, \mathcal{G}_k^u$ and $\mathcal{G}_k^v$ are system-dependent constant matrices.

We make the assumption that due to ergodicity and stationarity of the variables, all the covariances used in this paper can be estimated by replacing ensemble means by time means. Therefore, we define the expectation operator $\mathbf{E}$ is equivalent to the operator $\mathbf{E}_j$ which is defined as follows:

$$\mathbf{E}[\bullet] = \lim_{j\to\infty} \mathbf{E}_j[\bullet] = \lim_{j\to\infty} \frac{1}{j}\sum_{k=0}^{j}[\bullet]$$

The orthogonal projection operator $\Pi_B A$ is defined as in (Favoreel *et al.*, 1996): $\Pi_B A \triangleq AB^T(BB^T)^\dagger B$

Theorem 2. If the linear part of the system (1) is observable and

$$\begin{pmatrix} Y^c \\ U^{c,u,y} \\ U^{f,u,y} \\ U^{r,u,y} \end{pmatrix} \quad (7)$$

is a full row rank matrix, denoting $\mathcal{S} := \mathcal{Y}^c + \mathcal{U}^{c,u,y} + \mathcal{U}^{f,u,y} + \mathcal{U}^{r,u,y}$ and $\mathcal{R} = \Pi_\mathcal{S}\mathcal{Y}_f + \mathcal{U}^{f,u,y}$, then

$$\Pi_{\mathcal{R}^\perp}\Pi_\mathcal{S}\mathcal{Y}^r = \mathcal{T}_k^u \Pi_{\mathcal{R}^\perp} U^{r,u,y} + o(\lambda^k) \quad (8)$$

4. ALGORITHM

1. Decompose Y^r into $\mathcal{O}_k X_r$ and $\mathcal{T}_k^u U^{r,y,u}$ using orthogonal projection: from (8) of Theorem 2, estimated $\mathcal{T}_k^u$

$$\hat{\mathcal{T}}_k^u = (\Pi_{\mathcal{R}^\perp}\Pi_\mathcal{S}\mathcal{Y}^r)(\Pi_{\mathcal{R}^\perp}U^{r,u,y})^\dagger \quad (9)$$

2. Obtain the SVD decomposition and partition as

$$
\begin{aligned}
&[\Pi_\mathcal{S} Y_{2k-1|k}\ \Pi_\mathcal{S} Y_{2k|k+1}] - \hat{\mathcal{T}}_k^u \left[U_{2k-1|k}^{u,y}\ U_{2k|k+1}^{u,y}\right] \\
&=: \Gamma\Sigma\Omega^* = \begin{bmatrix}\Gamma_1 & \Gamma_2\end{bmatrix}\begin{bmatrix}\Sigma_1 & 0\\ 0 & \Sigma_2\end{bmatrix}\begin{bmatrix}\Omega_1^*\\ \Omega_2^*\end{bmatrix}
\end{aligned}
$$

Since we expect

$$\Gamma\Sigma\Omega^* = \Gamma_1\Sigma_1\Omega_1^* = \mathcal{O}_k\,[X_{k-1}\quad X_k]$$

from (4-6) ($\text{rank}(\Sigma_1) = n$ and $\text{rank}(\Sigma_2) = 0$), form the estimates $\hat{\mathcal{O}}_k = \Gamma_1\Sigma_1^{1/2}$ and $[\hat{X}_{k-1}\quad \hat{X}_k] = \Sigma_1^{1/2}\Omega_1^*$, retaining only $\hat{n}$ significant singular values in Σ_1. ($\hat{\mathcal{O}}_k$ is not needed later.)

3. Estimate the parameters A, B, C, D, N on the basis of equation (2), as

$$\begin{bmatrix}\hat{X}_k\\ Y_{k-1}\end{bmatrix} = \begin{bmatrix}A & N & B\\ C & 0 & D\end{bmatrix}\begin{bmatrix}\hat{X}_{k-1}\\ U_{k-1}\odot\hat{X}_{k-1}\\ U_{k-1}\end{bmatrix} \quad (10)$$

in a least-squares sense.

Remark 3 Other estimates could be obtained by using other right-inverses in steps 1 and 3, and another factorisation in step 2. In (Favoreel *et al.*, 1999) it is suggested that constrained least-squares could be used in step 3, because of the known structure of the solution. Our initial simulation experience is that this does not have much effect on the estimated eigenvalues of matrix A and N.

Estimate the covariance matrix by calculating

$$\begin{bmatrix}\epsilon_w\\ \epsilon_v\end{bmatrix} = \begin{bmatrix}\hat{X}_{k+1}\\ Y_k\end{bmatrix} - \begin{bmatrix}\hat{A} & \hat{N} & \hat{B}\\ \hat{C} & 0 & \hat{D}\end{bmatrix}\begin{bmatrix}\hat{X}_k\\ U_k\odot\hat{X}_k\\ U_k\end{bmatrix}$$

$$\begin{bmatrix}\hat{Q} & \hat{S}\\ \hat{S}^T & \hat{R}\end{bmatrix} = \mathbf{E}_j\left[\begin{pmatrix}\epsilon_w\\ \epsilon_v\end{pmatrix}\begin{pmatrix}\epsilon_w\\ \epsilon_v\end{pmatrix}^*\right]$$

5. EXAMPLES

In this section, two simple second order bilinear systems introduced in (Favoreel *et al.*, 1996), (Favoreel *et al.*, 1999) are given to see how the new algorithm works and how it compares with the algorithm in (Favoreel *et al.*, 1996), (Favoreel *et al.*, 1999).

Example 1. The system matrices are

$$A = \begin{pmatrix} 0 & 0.5 \\ -0.5 & 0 \end{pmatrix}, \; B = \begin{pmatrix} 1 \\ 1 \end{pmatrix}, \; C = \begin{pmatrix} 1 & 1 \end{pmatrix},$$

$$D = 2, \quad N_1 = \begin{pmatrix} 0.4 \\ 0 \end{pmatrix}, \quad N_2 = \begin{pmatrix} 0 \\ 0.3 \end{pmatrix}$$

and the noise covariance matrices

$$Q = \begin{pmatrix} 0.16 & 0 \\ 0 & 0.04 \end{pmatrix}, \; R = 0.09, \; S = \begin{pmatrix} 0 \\ 0 \end{pmatrix}$$

In (Favoreel *et al.*, 1996), the input is a white noise series and $k = 3, j = 8191$. In both Table 1 and Table 2, the system input of case I has a uniform distribution with mean value zero and variance 1 and $\lambda = 0.7809$. In case II, we adjust the system noise as follows:

$$Q = \begin{pmatrix} 0.0016 & 0 \\ 0 & 0.0004 \end{pmatrix}, \; R = 0.0009, \; S = \begin{pmatrix} 0 \\ 0 \end{pmatrix}$$

to increase the noise ratio between the input signal and system noise. The estimation results are also given in Table 1. It is shown that the greater the ratio, the better the convergence. For cases III and IV, we introduce a colored noise input signal z, with mean 0, standard deviation 1.1664, $\lambda = 0.7906$ and $r_q = Ez_k z_{k+q} = 0.5^q, \quad q = 0, 1, 2, ...$ and the system noise is taken the same as in cases I and II respectively. The simulation results are also shown in Table 1. The values $k = 2, j = 595$ were used in all our simulations. These compare favourably with the values $k = 3, j = 8191$ used by (Favoreel *et al.*, 1996) for the bilinear N4SID algorithm.

Table 1. Example 1: Results with different inputs, noise ratios and algorithms

	eig(A)	eig(N)
True	$\pm 0.5i$	0.4, 0.3
N4SID	$-0.0027 \pm 0.4975i$	0.4011, 0.3055
Case I	$-0.0078 \pm 0.4864i$	0.4128, 0.3035
Case II	$0.0000 \pm 0.5000i$	0.4005, 0.3030
Case III	$0.0089 \pm 0.4945i$	0.3906, 0.3149
Case IV	$0.0005 \pm 0.4980i$	0.4006, 0.2976

The simulation results with different sample sizes for cases I and II are given in Table 2, to show the relationship between the accuracy and the sample size.

Example 2. The system matrices and the noise covariance matrices are

Table 2. Example 1: Effect of sample size

	eig(A)	eig(N)
True	$\pm 0.5i$	0.4, 0.3
j=95 (I)	$-0.0027 \pm 0.5018i$	0.3729, 0.3473
j=95 (II)	$0.0059 \pm 0.4982i$	0.4026, 0.3095
j=295 (I)	$0.0074 \pm 0.4966i$	0.4298, 0.3022
j=295 (II)	$0.0031 \pm 0.5008i$	0.4012, 0.2988
j=595 (I)	$0.0078 \pm 0.4864i$	0.4128, 0.3035
j=595 (II)	$0.0000 \pm 0.5000i$	0.4005, 0.3030

$$A = \begin{pmatrix} 0.5 & 0 \\ 0 & 0.3 \end{pmatrix}, B = \begin{pmatrix} 0 & 1 \\ -1 & 0 \end{pmatrix}, C = \begin{pmatrix} 1 & 0 \\ 0 & 2 \end{pmatrix}$$

$$D = \begin{pmatrix} 1 & 0 \\ 0 & 1 \end{pmatrix}, N_1 = \begin{pmatrix} 0.6 & 0 \\ 0 & 0.4 \end{pmatrix}, N_2 = \begin{pmatrix} 0.2 & 0 \\ 0 & 0.5 \end{pmatrix}$$

$$Q = \begin{pmatrix} 0.01 & 0 \\ 0 & 0.01 \end{pmatrix}, R = \begin{pmatrix} 0.01 & 0 \\ 0 & 0.01 \end{pmatrix}, S = \begin{pmatrix} 0 & 0 \\ 0 & 0 \end{pmatrix}$$

The input we selected here has a two-dimensional uniform distribution. Table 3 shows the true and estimated eigenvalues of A and N, as found by (Favoreel *et al.*, 1999) using the bilinear N4SID algorithms with $k = 2, j = 4095$, and using our new algorithm with $k = 2, j = 597, \lambda = 0.7803$, and solving equation (10) using both ordinary least-squares (Model I) and constrained least-squares (Model II).

Table 3. Example 2: Comparisons with different algorithms.

	eig(A)	eig(N_1)	eig(N_2)
True	0.5, 0.3	0.6, 0.4	0.2, 0.5
N4SID	0.5001	0.5994	0.5016
(j=4095)	0.2979	0.4020	0.1914
Model I	0.4998	0.5998	0.5000
(j=597)	0.3002	0.4000	0.2001
Model II	0.5004	0.5998	0.4999
(j=597)	0.2990	03997	0.1997

It is seen that the eigenvalues of the system matrices of the model and the original system are very close to each other. From Table 1 and Table 3, it is seen that the algorithm presented here has a quicker convergence rate and smaller sample size requirement compared to the bilinear N4SID algorithm. As stated in (Favoreel *et al.*, 1996), (Favoreel *et al.*, 1999), the noise covariance estimates are not very accurate since the k-parameter is too small ($k = 2$).

Remark 4 Our new algorithm has considerably lower computational complexity than the one given in (Favoreel and De Moor, 1998). The major computational load is involved in finding the right-inverse in (9). The matrix involved here has dimensions $f_k \times j$, where $f_k = e_k + (m/2)(m+1)^k + l[(m+1)^k - 1] + e_k^2$. For Example 2, with $k = 2$, we have $f_k = 136$. In (Favoreel and De Moor, 1998) , equation (9), the matrix whose right-inverse has to be found has dimensions $(d_k + 2e_k + e_k d_k + e_k^2) \times j$. The row dimension for both algorithms increases exponentially with k. For Example 2 with $k = 2$ this row dimension is 152. Furthermore, since our algorithm seems to require much smaller values of j for comparable performance, the column dimension is also much smaller for our algorithm in practice.

6. CONCLUSION

A new subspace identification algorithm has been presented for the identification of a general combined deterministic-stochastic bilinear system. The main result in this paper is that the system input does not have to be white noise, which is the main restriction and assumption in the current bilinear subspace system identification algorithms. Simulation results and comparisons with subspace identification algorithm have been given.

7. ACKNOWLEDGEMENT

The work reported in this paper was supported by the UK Engineering and Physical Science Research Council under the GrantGR/MO8332, and by the European Research Network on System Identification (ERNSI) under TMR contract ERB FMRX CT98 0206.

8. REFERENCES

Chen, H. and J.M. Maciejowski (1999). Subspace identification of deterministic bilinear systems. Accepted for ACC 2000, Chicago, Illinois, USA.

Chen, H. and J.M. Maciejowski (2000). New subspace identification method for bilinear systems. Technical Report CUED/F-INFENG/TR.357. Depatment of Engineering, University of Cambridge.

Chui, N.L.C. and J.M. Maciejowski (1998). Subspace identification – a Markov parameter approach. Technical Report CUED/F-INFENG/TR. 337. Department of Engineering, University of Cambridge.

Favoreel, W. and B. De Moor (1998). Subspace identification of bilinear systems. In: *Proc. MTNS*. pp. 787–790.

Favoreel, W., B. De Moor and P. Van Overschee (1996). Subspace identification of bilinear systems subject to white inputs. Technical report. Katholieke Universiteit Leuven.

Favoreel, W., B. De Moor and P. Van Overschee (1997). Subspace identification of balanced deterministic bilinear systems subject to white inputs. In: *ECC*.

Favoreel, W., B. De Moor and P. Van Overschee (1999). Subspace identification of bilinear systems subject to white inputs. *IEEE Trans on AC* **44**(6), 1157–1165.

Khatri, C.G. and C.R. Rao (1968). Solutions to some functional equations and their applications to characterization of probability distributions. *The Indian J. Stat. series A.* **30**, 167–180.

Suda, N., S. Kodama and M. Ikeda (1973). *Matrix Theory in Automatical Control.* Japanese Automatical Control Association.

IDENTIFICATION OF COLINEAR AND COINTEGRATED MULTIVARIABLE SYSTEMS USING CANONICAL VARIATE ANALYSIS

Wallace E. Larimore

Adaptics, Inc, 1717 Briar Ridge Road, McLean, VA 22101
Phone: 703 532-0062, Fax: 703 536-3319, Email: larimore@adaptics.com

Abstract. This paper addresses the problem of identification of a multivariable system where the output time series may be colinear and perhaps integrating processes. Such processes can occur when numerous measurement series are available for observing basically the same process with slight changes. Examples include financial series from the same economic sector, high dimensional series in the measurement of sheet forming processes such as sheet aluminum or paper machines, and highly instrumented vibrating structures. Basically, some of the measurements are providing information already contained in other measurements. By explicitly dealing with the possible rank deficiency of the measurements, a reduced rank process can be determined that can have far fewer parameters to estimate. Since modeling accuracy as measured by variance is proportional to the number of estimated parameters, the resulting models can be considerably more accurate. Maximum likelihood tests of hypotheses are constructed to determine the best reduced rank structure by optimal statistical tests. *Copyright © 2000 IFAC*

Key Words: Canonical variate analysis, System identification, Colinear and cointegrated process, Subspace identification, Tests on rank.

1. INTRODUCTION

The canonical variate analysis (CVA) of time series data collected from dynamical processes was originally proposed by Akaike (1974, 1975, 1976) for determining the approximate structure of multiple time series. It has recently been developed to the point of providing the automatic and accurate identification of complex multivariable dynamical systems (Larimore, 1983b, 1990, 1999). The procedure has been shown in a number of cases to be very close to maximum likelihood (ML) even in the case of small samples, feedback, and multivariable systems of high state order (Larimore, 1996, 1997, Deistler et al, 1995). Although autoregressive moving average (ARMAX) models are in principle equivalent, it is difficult to determine the appropriate model structure. Additionally, the parameter estimation requires nonlinear iterative parameter optimization that often encounters convergence problems especially for multivariable systems. The CVA provides an alternative method for finding state space or equivalent ARMAX models using a computationally efficient and stable algorithm.

The topic of system identification of colinear and cointegrated processes has been much discussed in the literature. For a general survey of the identification of such processes see Lütkepohl (1993) and Engle and Granger (1987). Here the topic is restricted to ML identification because the resulting models can be used for optimal maximum likelihood tests concerning alternative hypotheses about the system rank and structure. This fits nicely into the CVA procedure that produces models that are very close to ML.

2. MODELS OF COLINEAR PROCESSES

Consider the identification problem where the observed system inputs u_t and observed outputs y_t are given and it is desired to identify the input-output system and noise model for the system disturbances. The CVA procedure developed in Larimore (1983b, 1990) identifies the optimal state order k and a linear Markov process in the following general state space form

$$x_{t+1} = \Phi x_t + G u_t + w_t \qquad (1)$$

$$y_t = H x_t + A u_t + B w_t + v_t \qquad (2)$$

Here x_t is a k-order Markov state and w_t and v_t are white noise processes that are independent with covariance matrices Q and R respectively. These state equations are more general than typically used since the noise $Bw_t + v_t$ in the output equation may be correlated with the noise w_t in the state equation. This is a consequence of requiring that the state equation (1) be of minimal dimension k. Requiring the noises in (1) and (2) to be uncorrelated may result in a state space model were the state is higher dimensional than the Markov order k so that it is not a minimal order realization (Lindquist and Pavon, 1981). The CVA procedure provides the optimal estimates of the matrices Φ, G, H, A, B, Q, and R.

In a rank deficient or cointegrated measurement process, the measurement matrix H and/or the measurement covariance matrix R may be rank deficient. Specifically, in a cointegrating process, a linear combination of integrating output processes will be only white noise with no autocorrelation or dynamic structure. The process state has no information for prediction of such a linear combination of the outputs. The process can be viewed as occurring on a subspace of the measurement space. If such a subspace can be determined, then the process can be described using fewer parameters on such a subspace with a resulting increase in model accuracy. While this is conceptually simple, the technical problem is to describe processes on such subspaces, determine this subspace in the presense of process and measurement noise, and obtain a process description on the subspace. That is, the problem is to obtain a good statistical procedure for the determination of the subspace dimension, good estimators of the subspace and of parameters describing the process on the subspace.

In particular, we consider the case that on some subspace $s = L_r y$ of y, the processes s is full rank with $rank(L_r[H\ A]) = r$, where the columns of L_r span the subspace of y on which the process is of full rank r. This definition will include both the hypothesis of $A = 0$ as well as the case of A unrestricted. Now in addition, the white noise process in the measurement equation can have a singular covariance matrix. Let $n_t = Bw_t + v_t$ be the noise in the measurement equation. Then we assume that the covariance matrix of the noise n_t on the subspace s is nonsingular so that $L_r \Sigma_{nn} L_r^T$ is nonsingular. This is equivalent to the assumption that on the subspace s, all linear combinations of the states and inputs are observed with noise. Otherwise, we assume that the covariance Σ_{nn} can have arbitrary rank. Using an innovations representation of the process, it can be shown that B can always be expressed in the form $B = L_r B_r$ so that B is null except on s. Then all noise not on the subspace s is described by the covariance matrix R.

The question of rank deficiency of R is purely a numerical issue that is dealt with directly in the CVA method below. Since CVA is scale invariant (Larimore, 1997), each observation is first scaled by the root sum square (RMS) of its total variation. This is analogous to computing a correlation matrix where the diagonal elements are 1. If there are linear combinations of the observations that have variance less than a threshold near machine percision, then such a subspace is discarded in the computations. This procedure is done automatically in the reduced rank calculation procedure.

The basic problem is to devise a method for inferring the rank r and a basis L_r of the subspace s, inferring the covariance matrix R and its rank, and estimating the parameters for the various state space matrices. A rank deficient H maps the state vector x onto a subspace of the measurement vector y. Before developing the identification procedure, first a brief summary of the canonical variate analysis procedure is provided (see also the indepth tutorial paper Larimore (1999), and then the extension to rank deficient processes is discussed.

3. CANONICAL VARIATE SYSTEM IDENTIFICATION

A fundamental concept in the CVA approach is the *past* and *future* of a process. Suppose that data are given consisting of observed outputs y_t and observed inputs u_t at time points labeled $t = 1, \ldots, N$ that are equal spaced in time. Associated with each time t is a past vector p_t consisting of the past outputs and inputs occurring prior to time t as well as a future vector f_t consisting of outputs at time t or later, specifically

$$p_t = (y_{t-1}^T, u_{t-1}^T, y_{t-2}^T, u_{t-2}^T, \ldots)^T, \quad f_t = (y_t^T, y_{t+1}^T, \ldots)^T \tag{3}$$

For simplicity, first consider a purely stochastic process with no observed deterministic input to the system. A fundamental property of a linear, time invariant, gaussian Markov process of finite state order is the existence of a finite dimensional state x_t which is a linear function of the past p_t

$$x_t = Cp_t \tag{4}$$

The state x_t has the property that the conditional probability density of the future f_t conditioned on the past p_t is identical to that of the future f_t conditioned on the finite dimensional state x_t so

$$p(f_t|p_t) = p(f_t|x_t) \tag{5}$$

Thus, only a finite number of linear combinations of the past are relevant to the future evolution of the process.

To extend this concept to processes involving deterministic controls or inputs, the effects of future inputs must first be removed from the future outputs. Let q_t denote the future inputs $q_t = (u_t^T, u_{t+1}^T, \ldots,)^T$. Now consider the conditional random vector

$$f_t|q_t = f_t - Mq_t \tag{6}$$

where the regression coefficient for predicting f_t from q_t, is $M = S_{fq}S_{qq}^{-1}$.

Then the process is a *controlled Markov processes* of order k if there exists a k-order state such that the conditional probability density of $f_t|q_t$ given the past p_t is identical to the probability of $f_t|q_t$ given the state x_t so

$$p((f_t|q_t)|p_t) = p((f_t|q_t)|x_t) \tag{7}$$

This is equivalent of the statement that

$$p(f_t|(q_t, p_t)) = p(f_t|(q_t, x_t)) \tag{8}$$

In terms of the discussion of CVA in Appendix A, the random vectors x_t and y_t are replaced respectively by the random vectors p_t and $f_t|q_t$ of this section.

In the computational problem given finite data, the past and future of the process are taken to be finite of length ℓ lags so

$$p_t = (y_{t-1}^T, u_{t-1}^T, \cdots, y_{t-\ell}^T, u_{t-\ell}^T)^T, \quad f_t = (y_t^T, \cdots, y_{t+\ell-1}^T)^T \tag{9}$$

and similarly for the future inputs q_t. Akaike (1976) proposed choosing the number ℓ of lags by least squares autoregressive modeling using recursive least squares algorithms and choosing the number of lags as that minimizing the AIC criterion discussed below. This insures that a sufficient number of lags are used to capture all of the statistically significant behavior in the data. This procedure is easily generalized to include the case with inputs u_t by using ARX models.

The above discussion of Markov processes gives a heuristic argument for using a CVA between past and future to determine the state. A rigerous derivation of the following expression for the log likelihood function is developed in Larimore (1997)

$$\log p(Y|U, \theta, x_{\ell+1}) = \frac{1}{\ell} \sum_{t=\ell+1}^{N} ((f_t|u_t)|p_t, \theta) \tag{10}$$

where $x_{\ell+1}$ is the initial state at time $t = \ell + 1$. This expression is precise except for 'end effects' that disappear with large sample. This provides the basis for maximizing the likelihood function using CVA and thus avoiding nonlinear optimization methods previously employed in ML identification.

In this and the next paragraph, the CVA between the past p_t and future $f_t|q_t$ with the effect of future inputs q_t is discussed. Note that the notation in these two paragraphs, particularly J and J_k, are not related to other CVA and similar notation developed elsewhere in this paper. From Appendix A, the CVA from (23) and (24) on the past and future gives the transformation matrices J and L respectively specifying the canonical variables c and d associated with the past p_t and future f_t. For each choice k of state order (the rank r used in Appendix A), the "memory" of the process (the intermediate variables z_t of Appendix A) is defined in terms of the past as

$$m_t = J_k p_t = [I_k \; 0] J p_t \tag{11}$$

where the vector m_t consists of the first k canonical variables. The vector m_t is intentionally called 'memory' rather than 'state'. A given selection of memory m_t may not correspond to the state of any well defined k-order Markov process since truncating states of a Markov process will not generally result in a Markov process for the remaining state variables. In particular, the memory m_t does not usually contain all of the information in the past for prediction of the future values of m_t, i.e. $m_{t+1}, m_{t+2}, \ldots$ For the system identification problem, this is not a problem since many orders k will be considered and the one giving the best prediction will be chosen as the optimal order. This optimal order memory will correspond to the state of a Markov process within the sampling variability of the problem.

In particular, consider the state equations (1) and (2) with the state x_t replaced with the memory m_t determined from CVA. The multivariate regression equations are expressed in terms of covariances, denoted by Σ, among various vectors as (Anderson, (1984), pp. 291)

$$\begin{pmatrix} \Phi & G \\ H & A \end{pmatrix} = \Sigma \begin{pmatrix} m_{t+1} \\ y_t \end{pmatrix}, \begin{pmatrix} m_t \\ u_t \end{pmatrix} \Sigma^{-1} \begin{pmatrix} m_t \\ u_t \end{pmatrix}, \begin{pmatrix} m_t \\ u_t \end{pmatrix} \tag{12}$$

where computation is obtained by the substitution of $m_t = J_k p_t$. The covariance matrix of the prediction error as well as matrices Q, R, and B have similar simple expressions (Larimore, 1983b).

The above description of CVA system identification applies without modification to the point including determining the state of the process. However modifications in the above multivariate regression step (12) to compute the coefficients of the state equations are required to deal efficiently with potential colinearity of the process. To do this, we need the solution to the reduced rank regression problem summarized in Appendix A and derived in detail in Larimore (1997).

Now to apply the solution of Appendix A for the reduced rank regression problem to the colinear system identification problem of this paper, consider the reduced rank regression from the state x to the outputs y. The solution is given by the CVA between x and y to find J, L, and P. Then for any choice of rank r, the optimal reduced rank regression is given by J_r, L_r, and P_r. These matrices then define the optimal regression coefficients for H of rank r in the case that we assume $A = 0$. In the case that A is not null, the estimate of $(H \; A)$ is obtained by a CVA between $(x; u)$ and y. In either case, the subspace on which the process is full rank is given by the first r columns of $L^\dagger$. The estimate for the one step prediction error covariance matrix R is the estimate of the covariance of the error of regression, S_{ee}.

4. TESTING HYPOTHESES ON REDUCED RANK

A major advantage of the CVA method is that it is close to maximum likelihood in terms of parameter estimation error and tests of hypotheses (Larimore, 1996). This permits the use of maximum likelihood tests among diverse hypotheses including tests on rank. Maximum likelihood tests have optimal properties for asymptotically large sample sizes.

To decide on the model state order or model structure, recent developments based upon an objective information measure is used. Such a method was originally developed by Akaike (1973) and involves the use of the Akaike Information Criterion (AIC) for deciding the appropriate order of a statistical model. Considerations of the fundamental statistical principles of sufficiency and repeated sampling provide a sound justification for the use of an information criterion as an objective measure of model fit (Larimore 1983a, Larimore and Mehra, 1985). In particular, suppose that the true probability density is p_* and an approximating density is p_1, then the measure of approximation of the model p_1 to the truth p_* is given by the Kullback discrimination information (Kullback, 1957)

$$I_Y(p_*, p_1) = \int p_*(Y) \log \frac{p_*(Y)}{p_1(Y)} dY \quad (13)$$

It can be shown that for large sample the AIC is an optimal estimator of the Kullback information and achieves optimal decisions on model order (Shibata, 1981).

The AIC for each order k is defined by

$$AIC(k) = -2 \log p(Y^N, U^N; \hat{\theta}_k) + 2f M_k \quad (14)$$

where p is the likelihood function based on the observations (Y^N, U^N) at N time points, where $\hat{\theta}_k$ is the maximum likelihood parameter estimate using a k-order model with M_k parameters. The small sample correction factor f is equal to 1 for Akaike's original AIC, and is discussed below for the small sample case. The model order k is chosen corresponding to the minimum value of AIC(k). For the model state order k taken greater than or equal to the true system order, the CVA procedure provides an approximate maximum likelihood solution. For k less than the true order, the CVA estimates of the system are suboptimal so the likelihood function may not be maximized. However this will only accentuate the difference between the calculated AIC of the lower order models and the model of true order so that reliable determination of the optimal state order for approximation is maintained.

The number of parameters in the state space model (1) and (2) is

$$M_k = k(2n + m) + mn + n(n + 1)/2 \quad (15)$$

where k is the number of states, n is the number of outputs, and m is the number of inputs to the system. Candy et al (1979) develop this result by considering the size of the equivalence class of state space models having the same input/output and noise characteristics. Thus the number of functionally independent parameters in a state space model is far less than the number of elements in the various state space matrices. The AIC provides an optimal procedure for model order selection in large sample sizes.

A small sample correction to the AIC has been recently developed for model order selection (Hurvich et al, 1989). The small sample correction factor f is

$$f = \frac{N}{N - (\frac{M_k}{n} + \frac{n+1}{2})} \quad (16)$$

The effective sample size N is the number of time points at which one-step predictions are made using the identified model. For a large sample N, the small sample factor f approaches 1, the value of f originally used by Akaike in defining AIC. The small sample correction has been shown to produce model order selection that is close to the optimal as prescribed by the Kullback information measure of model approximation error.

The AIC is very easy to apply to the testing of hypotheses if the maximum likelihood parameter estimates are available and the number of estimated parameters can be determined. To determine the number of estimated parameters for testing hypotheses on rank requires some detailed analysis. In computation of the CVA as in Appendix A between two sets x $(k \times 1)$ and y $(n \times 1)$ of regression variables where a regression of rank r is used, the number of estimated parameters is as follows. Let J_r $(r \times k)$ and L_r $(r \times n)$ be the first r rows respectively of the matrices J and L of the CVA. Now under the hypothesis of rank r in the CVA, the rk elements of J_r are subject to the $r(r+1)/2$ constraints imposed by the equality $J_r S_{xx} J_r^T = I_r$, taking into account the symmetry of the equations. Thus there are $rk - r(r+1)/2$ parameters that are independently estimated. Similarly, there are $rn - r(r+1)/2$ parameters estimated in the matrix L_r. Finally there are r canonical correlations p_i estimated so that the total is $r(k+n-r)$ estimated parameters. Note that in the full rank case where $r = min(k, n)$ the number of estimated parameters reduces to kn as it should.

Now consider the case of an arbitrary rank on both the $[H\ A]$ and R matrices except that we will require that the range space of R contains that of $[H\ A]$. The reason for this is that we assume that all linear combinations of the output process $Hx_t + Au_t$ are observed with nonzero noise. We can also assume without loss of generality that R is full rank, since if it is not, we can always transform y to a set of variables on the full rank part of the range of R. Finally, we will consider both of the hypotheses $A = 0$ and A unrestricted.

The easiest way to count parameters is to consider Case 1 of restricting attention to the range space of $[H\ A]$ and computing the number of parameters for the full rank case on that subspace. Case 2 will be the general unrestricted case. Then we can determine the difference in the counting of parameters estimated in L_r between this case and the general case. Finally, we then consider the number of additional parameters needed to estimate R or equivalently S_{ee}.

Theorem 1. Let k be the state order, n the rank of the covariance matrix S_{ee} of the prediction error of the output measurements. Then for a given choice r for a possibly reduced rank r of $[H\ A]$, the number of identifiable parameters in the process (1) and (2) is

$$k(2r + m) + r(m + n - r) + n(n + 1)/2 \quad (17)$$

In the case that the constraints $A = 0$ are assumed, the number of estimated parameters are reduced by rm.

Proof. We consider the difference between Case 1 of the process on the image of $[H\ A]$, and Case 2 of the process on the full rank part of S_{ee}. These two cases differ only in the number of parameters estimated in L_r and S_{ee}. In Case 1, the process is full rank and the total number of estimated parameters from Candy et al (1979) is

$$k(2r + m) + rm + r(r + 1)/2 \quad (18)$$

where the term rm is deleted if $A = 0$ is imposed on the model. In Case 1, the number of estimated parameters in L_r is $rm + r^2 - r(r+1)/2$ if A is estimated and $r(r-1)/2$ if it is not. Now in Case 2, the number of estimated parameters for L_r is $r(m+n) - r(r+1)/2$

Model Rank	Number of Parameters	AIC	Δ Number of Parameters	ΔAIC	Significance Level
1	43	-379.23			
2	58	**-2154.28**			
3	71	-2135.61	13	18.67	0.20
4	82	-2113.91	11	21.70	0.05

Table 1: Full and reduced rank models with optimal (minimum AIC model) shown in **bold)**

if A is estimated and $rn - r(r+1)/2$ if it is not. So the number of additional estimated parameters in L_r for Case 2 verses Case 1 is $r(n-r)$. Thus in Case 2 for either hypothesis, we need to add an additional $r(n-r)$ parameters for L_r to Equation (18) giving the number parameters for Case 1. In Case 1 there are $r(r+1)/2$ parameters estimated in S_{ee} verses Case 2 with $n(n+1)/2$ estimated parameters on the full rank part of S_{ee}. Thus adding $r(n-r)$ and $n(n+1)/2 - r(r+1)/2$ to Equation (18) give the total number of parameters in Equations (17) for Case 2.

5. SIMULATION EXAMPLE

Simulated data for a colinear example was used to demonstrate the procedure. The observed output data y_t is of the form

$$y_t = T_n \overline{y}_t + e_t \tag{19}$$

where T_n is a 4×2 transformation, $\overline{y}_t$ is the 2×1 vector of outputs from the simulation example in Larimore (1996), and e_t is a 4×1 vector of independent normal random vectors with variance 0.01. Thus the dynamics and autocorrelation structure of y_t is nontrivial only on a 2-dimensional subspace of y_t.

In the CVA model fitting, the state order of 6 was chosen using AIC. The reduced rank procedure for computing $(H\ A)$ was implemented for ranks 1 through 4. For each rank, the resulting state space model was used to compute the Kalman filter innovations that were then used in evaluation of the log likelihood function (Larimore, 1996). The AIC was then computed as shown in Table 1.

The incremental change in number of additional parameters and the corresponding difference in the AIC is also shown between adjacent models with ranks higher than the true rank of 2. The ΔAIC's are expected to be independent Chi-squared random variables with degrees of freedom equal to the number of additional parameters shown in column 4 of the table. The observed significance levels are within the expected statistical variability.

APPENDIX A: OPTIMAL REDUCED RANK REGRESSION

In the discussion below we use the notation y for the observations and x for the state, or in the case that A in equation (2) is nonzero, for the state also augmented with u, i.e. $(x; u)$. Also, in the rest of this Appendix, the notation A and B will refer to matrices associated with the reduced rank discussion and will *not* refer to the state space matrices in the rest of the paper. See also the tutorial discussion in Larimore(1999).

Consider the following reduced rank regression problem with two vectors of random variables, $x(k\times1)$ and $y(n\times1)$. We consider the vector x as the set of predictor variables and the vector y as the set of variables to be predicted. Assume that x and y are jointly distributed as normal random variables with mean zero and covariance matrices Σ_{xx}, Σ_{yy}, Σ_{xy}, and that the relationship giving the optimal prediction of y from x is linear. The extension to the case of a nonzero mean is trivial, and assuming a zero mean will simplify the derivation. Suppose that N observations (x_t, y_t) for $1 \leq t \leq N$ are given and write $X = (x_1, \ldots, x_N)$ and $Y = (y_1, \ldots, y_N)$. Then define the sample covariance matrix $S_{yx} = YX^T/N$ and similarly for S_{xx} and S_{yy}.

To consider the reduced rank regression problem, we consider an intermediate set of variables z of dimension r that may be fewer in number than dim(x) such that z contains all of the information in x relevant to predicting y. This problem includes the determination of the dimension or rank r of z, as well as estimation of the parameters defining the linear relationships between x and z and between z and y.

To be precise, consider the model described by the equations

$$y_t = Bz_t + e_t \tag{20}$$

$$z_t = Ax_t \tag{21}$$

that holds for the t-th observations (x_t, y_t) for $t = 1, \ldots, N$. The vector e_t is the error in the linear prediction of y_t from x_t given matrices $A(r\times k)$ and $B(n\times r)$, and $\Sigma_{ee}(n\times n)$ is the covariance matrix of the prediction error e_t. It can be assumed for the present derivation that for different observation indexes s and t, the errors e_s and e_t are uncorrelated. In Larimore (1997), this is extended to the general case of a Markov process. These equations are of course equivalent to predicting y from x as

$$y_t = BAx_t + e_t = Cx_t + e_t \tag{22}$$

where the matrix $C = BA$ has the rank constraint rank(C) $\leq r$.

In Larimore (1997), the maximum likelihood solution to the above reduced rank multivariate regression problem is given in terms of the following canonical variate analysis.

CVA Theorem: Let $S_{xx}(k\times k)$ and $S_{yy}(n\times n)$ be non-negative definite (satisfied by sample covariance matrices). Then there exist matrices $J(\overline{k} \times k)$ and $L(\overline{n} \times n)$ such that

$$JS_{xx}J^T = I_{\overline{k}} \; ; \quad LS_{yy}L^T = I_{\overline{n}} \qquad (23)$$

$$JS_{xy}L^T = P = diag(p_1, ..., p_r, 0, ..., 0), \qquad (24)$$

where $\overline{k} = \text{rank}(S_{xx})$ and $\overline{n} = \text{rank}(S_{yy})$.

The equations (23) and (24) are a generalized singular value decomposition (GSVD) involving the sample covariance matrices S_{xx} and S_{yy} as weightings. An optimal choice for A, B, and Σ_{ee} for a given choice of r are derived in Larimore (1997). The solution is expressed directly in terms of J, L, and P of the GSVD. The optimal choice for A is

$$\hat{A} = [I_r \; 0]J \; ; \quad \text{thus} \;\; z = [I_r \; 0]Jx = [I_r \; 0]c \qquad (25)$$

So $\hat{A}$ is the first r rows of J. Equivalently z, the optimal rank r predictors of y, consists of the first r canonical variables $c_1, \ldots, c_r$. The predictable (with a reduction in error) linear combinations of y are given by the random variables $d = [I_r \; 0]Ly$, the first r canonical variables $d_1, \ldots, d_r$. Denoting P_r as P with p_k set to zero for $k \geq r$, the optimal prediction of d from c is simply $\hat{d} = P_r c$. The covariance matrix of the error in predicting d from c is $I - P_r^2$. The optimal estimates of B and Σ_{ee} are obtained by applying the inverse (or, in the case that S_{yy} is singular, the pseudoinverse denoted by $\dagger$) of L to go from d back to y and are

$$\hat{B} = L^{\dagger}P[I_r \; 0]^T \; ; \quad \hat{\Sigma}_{ee} = L^{\dagger}(I - P_r^2)L^{\dagger T} \qquad (26)$$

Note that the relationship between X and Y is completely symmetric so that if X and Y are interchanged at the beginning, the roles of A and J are interchanged with B and L in the solution.

The log likelihood maximized over A, B, and Σ_{ee}, or equivalently over $C = BA$ with constrained rank r, is given by the first r canonical correlations p_i between c_i and d_i.

$$\max \log p(Y|X; C, \Sigma_{ee}) = \frac{N}{2}\sum_{i=1}^{r} \log |S_{yy}|^{-1}(1 - p_i^2) \qquad (27)$$

Optimal statistical tests on rank involve likelihood ratios. Thus the optimal rank or order selection depends only on the canonical correlations p_i. A comparison of potential choices of rank can thus be determined from a single GSVD computation on the covariance structure.

REFERENCES

Akaike, H. (1973), "Information Theory and an Extension of the Maximum Likelihood Principle," *2nd International Symposium on Information Theory*, Eds. B.N. Petrov and F. Csaki, pp. 267-281. Budapest: Akademiai Kiado.

Akaike, H. (1974), "Stochastic Theory of Minimal Realization," *IEEE Trans. Automatic Control*, Vol. 19, pp. 667-674.

Akaike, H. (1975), "Markovian Representation of Stochastic Processes by Canonical Variables," *SIAM J. Control.*, Vol. 13, pp. 162-173

Akaike, H. (1976). "Canonical Correlation Analysis of Time Series and the Use of an Information Criterion," *System Identification: Advances and Case Studies*, R.K. Mehra and D.G. Lainiotis, eds., New York: Academic Press, pp. 27-96.

Candy, J.V., Bullock, T.E., and Warren, M.E. (1979), "Invariant Description of the Stochastic Realization," *Automatica,* Vol. 15, pp. 493-5.

Deistler, M., K. Peternell and W. Scherrer (1995), "Consistency ahd Relative Efficiency of Subspace Methods," *Automatica*, Vol. 31, pp. 1865-1875.

Engle, R.F., and C.W.J. Granger (1987), "Co-Integration and Error Correction: Representation Estimation and Testing," *Econometrica* Vol. 55, pp. 251-276.

Hurvich, C.M. and C.L. Tsai (1989), "Regression and Time Series Model Selection in Small Samples," *Biometrika* Vol. 76, pp. 297-307.

Larimore, W.E. (1983a). "Predictive Inference, Sufficiency, Entropy, and an Asymptotic Likelihood Principle", *Biometrika*, Vol. 70, pp. 175-81.

Larimore, W.E. (1983b). "System Identification, Reduced-Order Filtering and Modeling Via Canonical Variate Analysis", *Proc. 1983 American Control Conference*, H.S. Rao and T. Dorato, Eds., pp. 445-51. New York: IEEE.

Larimore, W.E. (1990), "Canonical Variate Analysis for System Identification, Filtering, and Adaptive Control, " *Proc. 29th IEEE Conference on Decision and Control*, Honolulu, Hawaii, December, Vol. 1, pp. 635-9.

Larimore, W.E. (1996), "Statistical Optimality and Canonical Variate Analysis System Identification," *Signal Processing*, Vol. 52, pp. 131-144.

Larimore, W.E. (1997), "Optimal Reduced Rank Modeling, Prediction, Monitoring, and Control Using Canonical Variate Analysis," *IFAC Internat. Symp. on Advanced Control of Chemical Processes*, Banff, Canada, June 9-11, 1997.

Larimore, W.E. (2000), *ADAPTx Automated System Identification Software Users Manual*, Adaptics, Inc, 1717 Briar Ridge Road, McLean, VA 22101

Larimore, W.E. (1999), "Tutorial Paper: Automated Multivariable System Identification and Industrial Applications," *Proc. 1999 American Control Conference*, June 24, 1999, San Diego, CA, pp. 1148-1162.

Larimore, W.E. and R.K. Mehra (1985), "The Problem of Overfitting Data," *Byte*, Vol. 10, pp. 167-80.

Lindquist, A., and M. Pavon (1981), "Markovian Representation of Discrete-Time Stationary Stochastic Vector Processes," *20th IEEE Conf. on Decision and Control*, Vol. 3, pp. 1345-56.

Lütkepohl, H. (1993), *Introduction to Multiple Time Series Analysis*, New York: Springer-Verlag.

Shibata, R. (1981), "An Optimal Autoregressive Spectral Estimate," *Ann. Statistics*, Vol. 9, pp. 300-6.

ON STATE ESTIMATION OF UNKNOWN SYSTEMS BASED ON SUBSPACE STATE-SPACE SYSTEM IDENTIFICATION

Hiroshi Oku *,[1] **Hidenori Kimura** **

* *Department of Mathematical Engineering and Information Physics, Graduate School of Engineering, University of Tokyo, 7-3-1, Hongo, Bunkyo-ku, Tokyo 113-8656, Japan*
** *Graduate School of Frontier Sciences, University of Tokyo, 7-3-1, Hongo, Bunkyo-ku, Tokyo 113-8656, Japan*

Abstract: In this paper, we consider a state estimation problem of a MIMO model with which a MIMO system is identified by using Subspace State-Space System IDentification (4SID). This problem is very important because the knowledge of the present state of a model can be a clue to design an adaptive control law for a system. First, we review the recursive 4SID algorithm based on the matrix inversion lemma shown in our previous paper. Then, we propose a scheme for estimating the state of models obtained by the recursive 4SID algorithm. A major advantage of 4SID schemes is that we can select the order of an identified model as the number of dominant singular values of a certain matrix which consists of I/O data. However, it becomes a bottleneck on considering the recursive state estimation because the dimension of state space changes from one time to another during the identification proceeds. To circumvent this problem, we introduce a new kind of *state* (named "*quasi-state*") which is still meaningful over different update steps. Finally, we illustrate the usefulness of our idea by considering a tracking control problem. *Copyright© 2000IFAC*

Keywords: identification, state estimation, recursive algorithms, subspace methods, state-space models, adaptive control

1. INTRODUCTION

A great deal of attention has been paid to subspace state-space system identification (4SID) methods in the last decade, and a number of papers have been published, e.g., Verhaegen and Dewilde(1992*a*; 1992*b*), Verhaegen(1993; 1994) and Van Overschee and De Moor(1994; 1996). These methods are specially suitable for identification of MIMO systems in the sense that the identified models are represented in the state space. Also, several results with respect to recursive algorithms of 4SID methods, especially MOESP (MIMO Output-Error State-sPace system identification) methods, have been reported by Verhaegen and Deprettere(1991), Gustafsson et al.(1998) and Oku and Kimura(1999).

In this paper, we consider recursive state estimation of a MIMO model obtained by a recursive 4SID algorithm, i.e., combined identification/estimation of unknown MIMO systems. This problem is very important because the knowledge of the present state of a model can be a clue to design an adaptive control law for unknown systems.

However, there are two major difficulties in considering this problem. First, a principal advantage of 4SID methods seems to be its flexibility in the selection of the order of a model. It can be chosen as the number of dominant singular values of a certain matrix which consists of I/O data. This means that, when we carry out recursive 4SID for a system, we may take different choices of the order of models at different update steps. Second,

[1] Partially supported by Grant-in-Aid for JSPS Fellows from the Ministry of Education, Science, Sports and Culture

suppose we can select the correct order, that is, the same order as a system to be identified. The system and an identified model are related by a similarity transformation. On considering recursive identification, a similarity transformation by which an identified model is related to the system at an update step may *not* be identical to one at another update step.

Unfortunately, conventional state estimation methods, e.g., Kalman filter or so-called *observer* in control theory, are not available directly for the situation which we mentioned above because they need a fixed estimate of the order of a system and a fixed similar transformation commonly over all update steps.

To overcome these difficulties, we introduce a new kind of "*state*" which maintains its meaning as a state over different update steps. This "*state*" must reflect some intrinsic properties of the system which are invariant with respect to choices of the order of models and the degree of freedom of similarity transformations. We identify such a "*state*" which possesses such properties.

This paper is organized as follows. In Section 2, some assumptions required in the paper are introduced. Section 3 gives a brief review of the basic 4SID method. In Section 4, we introduce a recursive algorithm of the basic 4SID which is based on the matrix inversion lemma (Oku and Kimura, 1999). In Section 5, we consider a state estimation problem of a MIMO model derived by the 4SID method. A vector made by premultiplying the state of a system to be identified by the extended observability matrix of it is employed as the "*state*" mentioned above. We also present estimates of such a new "*state*" at the past, the present and the one-step ahead steps, respectively, in this section. Finally, an illustrative example is shown in Section 6.

2. PRELIMINARIES

In this section, some assumptions required in this paper are introduced.

A discrete-time LTI system to be identified is described as follows:

$$x_{k+1} = Ax_k + Bu_k, \qquad (1a)$$

$$y_k = Cx_k + Du_k, \qquad (1b)$$

where $x_k \in \mathbb{R}^n$, $u_k \in \mathbb{R}^r$ and $y_k \in \mathbb{R}^m$. Let the matrix A be stable and the pair (A, C) be observable. The initial state of the system x_0 is assumed to be 0. A sequence of input data u_i is assumed to be a realization of a bounded stochastic process which is independent of the initial state x_0. We assume that there exists an integer $N_0 > 0$ such that for a given integer ν satisfying $\nu \geq n+1$, a given input sequence $\{u_i\}$ and $\forall N \geq N_0$

$$\sum_{j=1}^{N} u_{\nu(j)} u_{\nu(j)}^T \qquad (2)$$

is nonsingular, where the notation $u_{\nu(\cdot)}$ is defined as follows:

$$u_\nu(i) := \begin{bmatrix} u_{i-\nu+1} \\ \vdots \\ u_i \end{bmatrix} \Bigg\} \nu \text{ vectors} \qquad (3)$$

If the output measurement of the system (1) is contaminated by the noise term v_k, then the contaminated measurement $\tilde{y}_k$ is described as

$$\tilde{y}_k = y_k + v_k. \qquad (4)$$

Let the noise v_k be a zero-mean Gaussian white noise with covariance

$$\mathbf{E} v_i v_i^T = Q_v.$$

It is assumed that v_k is independent of u_i for any i and the initial state x_0.

Remark 1. If u is a quasi-stationary signal(Ljung, 1999), replace the expectation $\mathbf{E}$ by $\overline{\mathbf{E}}$ where

$$\overline{\mathbf{E}}\overline{u}(k) := \lim_{M\to\infty} \frac{1}{M} \sum_{k=1}^{M} \mathbf{E}\overline{u}(k).$$

3. BRIEF REVIEW OF BASIC 4SID

In this section, we give a brief review of basic 4SID for noise-free systems. For details, see Verhaegen and Dewilde(1992*a*).

For a finite sequence of input-output (for short, "I/O") data $\{(u_j,\ y_j)\}$, $j = 1, \cdots, N$, $u_\nu(i) \in \mathbb{R}^{r\nu}$, $y_\nu(i) \in \mathbb{R}^{m\nu}$ are defined as follows:

$$u_\nu(i) := \begin{bmatrix} u_{i-\nu+1} \\ \vdots \\ u_i \end{bmatrix}, \quad y_\nu(i) := \begin{bmatrix} y_{i-\nu+1} \\ \vdots \\ y_i \end{bmatrix}. \qquad (5)$$

Then, from (1), it follows that

$$y_\nu(i) = \Gamma_\nu x_{i-\nu+1} + H_\nu u_\nu(i), \qquad (6)$$

where

$$\Gamma_\nu := \left[C^T \ (CA)^T \ \cdots \ (CA^{\nu-1})^T \right]^T, \qquad (7)$$

$$H_\nu := \begin{bmatrix} D & & & 0 \\ CB & D & & \\ \vdots & \vdots & \ddots & \\ CA^{\nu-2}B & CA^{\nu-3}B & \cdots & D \end{bmatrix}. \qquad (8)$$

The equation (6) yields the following matrix equation with respect to the input-state-output relations:

$$Y_N = \Gamma_\nu X_{N-\nu+1} + H_\nu U_N, \qquad (9)$$

where

$$U_N := \left[u_{\nu(\nu)} \ u_{\nu(\nu+1)} \ \cdots \ u_{\nu(N)} \right] \qquad (10)$$

$$Y_N := \left[y_{\nu(\nu)}\ y_{\nu(\nu+1)}\ \cdots\ y_{\nu(N)} \right] \tag{11}$$

$$X_{N-\nu+1} := \left[x_1\ x_2\ \cdots\ x_{N-\nu+1} \right],$$

Note that U_N and Y_N are $r\nu \times (N-\nu+1)$ and $m\nu \times (N-\nu+1)$ matrices which consist of N I/O data, respectively. $\Gamma_\nu \in \mathcal{R}^{m\nu \times n}$ and $H_\nu \in \mathcal{R}^{m\nu \times r\nu}$ are said to be an *extended observability matrix* and a *lower triangular block Toeplitz matrix containing the Markov parameters*, respectively.

Notice that the matrix $U_N U_N^T$ is invertible because (3) is invertible. We define the following two projection matrices

$$\Pi_{U_N} := U_N^T \left(U_N U_N^T \right)^{-1} U_N, \tag{12}$$

$$\Pi_{U_N}^{\perp} := I - \Pi_{U_N}. \tag{13}$$

Then, postmultiplication of the equation (9) by $\Pi_{U_N}^{\perp}$ yields the relation

$$Y_N \Pi_{U_N}^{\perp} = \Gamma_\nu X_N \Pi_{U_N}^{\perp}. \tag{14}$$

Note that $Y_N \Pi_{U_N}^{\perp}$ in the left hand side of (14) can be made of the input and output Hankel matrices, (10) and (11). If the singular value decomposition of the left side of (14) is obtained as

$$Y_N \Pi_{U_N}^{\perp} = \left[E_N\ E_N^{\perp} \right] \begin{bmatrix} S & \\ & 0 \end{bmatrix} \begin{bmatrix} F_N^T \\ (F_N^{\perp})^T \end{bmatrix}, \tag{15}$$

it is known that for a sufficient large N there exists a nonsingular matrix T such that

$$\Gamma_\nu = E_N T. \tag{16}$$

From E_N and $E_N^{\perp}$ a state-space representation of the resulting model (A_T, B_T, C_T, D_T) can be calculated such that

$$\begin{aligned} A_T &= TAT^{-1}, & B_T &= TB, \\ C_T &= CT^{-1}, & D_T &= D. \end{aligned} \tag{17}$$

A detailed procedure of the derivation of such matrices has been shown by Verhaegen and Dewilde(1992*a*).

The recursive scheme of identification is reduced to obtaining the recursive computation of $Y_N \Pi_{U_N}^{\perp}$.

4. A RECURSIVE ALGORITHM OF THE BASIC 4SID

A recursive 4SID algorithm based on the RQ factorization has already been proposed by Verhaegen and Deprettere(1991). In this paper, we introduce a recursive algorithm which is based on the matrix inversion lemma (Oku and Kimura, 1999). It is well known that algorithms based on the matrix inversion lemma sometimes exhibit numerically poor performance with respect to the round-off error. Algorithms based on matrix factorization (e.g. QR or UD factorization methods) were devised as a remedy for this problem initially in state estimation (Bierman, 1977). Therefore, from the numerical point of view, the algorithm proposed in this paper is inferior to the one due to Verhaegen and Deprettere.

On the other hand, from a theoretical point of view, the algorithm can clarify the fundamental structure of the recursive 4SID algorithm, specially the relations among input, output and state vectors of a system to be identified. This advantage led us to recusive PO-MOESP (or PI-MOESP) algorithms (Oku and Kimura, 1999). It is also useful to analyze some properties of the 4SID schemes which will be stated in the following sections. For the details of MOESP(MIMO Output-Error State sPace model identification) schemes, see Verhaegen and Dewilde(1992*a*; 1992*b*) and Verhaegen(1993; 1994).

From (15), postmultiplication of $Y_N \Pi_{U_N}^{\perp}$ by its transpose yields

$$\begin{aligned} \Xi_N &:= Y_N \Pi_{U_N}^{\perp} \left(Y_N \Pi_{U_N}^{\perp} \right)^T = Y_N \Pi_{U_N}^{\perp} Y_N^T \\ &= \left[E_N\ E_N^{\perp} \right] \begin{bmatrix} S^2 & \\ & 0 \end{bmatrix} \begin{bmatrix} E_N^T \\ (E_N^{\perp})^T \end{bmatrix} \end{aligned} \tag{18}$$

The relation (18) means that both E_N and $E_N^{\perp}$ can be derived from computation of eigenvalues of the real symmetric matrix $\Xi_N := Y_N \Pi_{U_N}^{\perp} Y_N^T$, which is called *compressed I/O data matrix* in this paper. A recursive algorithm of updating this compressed I/O data matrix is derived in the sequel. Let

$$P_N := \left(U_N U_N^T \right)^{-1}. \tag{19}$$

Algorithm 1. (Oku and Kimura, 1999) Suppose Ξ_{N-1}, P_{N-1} and $Y_{N-1} U_{N-1}^T$ have already been obtained. When a pair of the N-th I/O data (u_N, y_N) is obtained, we can construct $u_{\nu(N)}$ and $y_{\nu(N)}$ from (5). Then, the N-th compressed I/O data matrix Ξ_N can be updated recursively by the following procedure:

$$\Xi_N = \Xi_{N-1} + \alpha_N e_N e_N^T, \tag{20}$$

where

$$P_N = P_{N-1} - \alpha_N P_{N-1} u_{\nu(N)} u_{\nu(N)}^T P_{N-1}, \tag{21}$$

$$\alpha_N := \left(1 + u_{\nu(N)}^T P_{N-1} u_{\nu(N)} \right)^{-1}, \tag{22}$$

$$e_N := y_{\nu(N)} - Y_{N-1} U_{N-1}^T P_{N-1} u_{\nu(N)}, \tag{23}$$

$$Y_N U_N^T = Y_{N-1} U_{N-1}^T + y_{\nu(N)} u_{\nu(N)}^T. \tag{24}$$

Proof . See Oku and Kimura(1999). □

Remark 2. This algorithm works well also for the case where the contaminated measurement $\tilde{y}_k$ is observed because

$$\lim_{N \to \infty} \frac{1}{N - \nu + 1} V_N U_N^T = 0, \tag{25}$$

where V_N is a matrix composed of the noise v_k in the same manner as (10). In this case, this algorithm is modified by replacing $y_{\nu(N)}$, Y_i, Ξ_i

($i=N-1$ or N) and e_N by $\tilde{y}_{\nu(N)}$, $\tilde{Y}_i$, $\tilde{\Xi}_i := \tilde{Y}_i \Pi_{U_i}^{\perp} \tilde{Y}_i^T$ and

$$\tilde{e}_N := \tilde{y}_{\nu(N)} - \tilde{Y}_{N-1} U_{N-1}^T P_{N-1} u_{\nu(N)}. \qquad (26)$$

5. ON STATE ESTIMATION OF MIMO MODELS

A principal advantage of 4SID schemes is that the order of a model, denoted by $\hat{n}$, can be chosen to be the number of dominant singular values in (15). This means that, when we apply recursive 4SID for the system, we may take different choices of the order of models at different update steps.

If we can select the correct order, that is, the same order as the system, from (17) the system and an identified model are related by a similarity transformation matrix T. On considering recursive identification, a similarity transformation matrix T_1 by which an identified model are related to the system at an update step N_1 may *not* be identical to one T_2 at another update step N_2, i.e., $T_1 \neq T_2$ for $N_1 \neq N_2$.

In this section we consider a state estimation problem of unknown multivariable systems utilizing the recursive 4SID. The problem is very important because the knowledge of the present state of the identified models can be a clue to design an adaptive control law for the system. Unfortunately, conventional state estimation methods, e.g. Kalman filter or *observer*, are not available directly for our case because they assume that the order of the system is fixed and similarity transformations are common over all update steps.

To overcome this difficulty, we introduce a new kind of *state* which is still meaningful over different update steps. Let

$$z_i := \Gamma_\nu x_i, \quad z_i \in \mathbb{R}^{m\nu}. \qquad (27)$$

We take z_i as a *state* at step i. The term z_i is named "the *quasi-state* at step i" in this paper. Note that the dimension of z_i is independent of the order of the system n and of the order of a model $\hat{n}$. From the assumption that Γ_ν is of full column rank, the state of the system is denoted by

$$x_i = \Gamma_\nu^\dagger z_i, \qquad (28)$$

where $\Gamma_\nu^\dagger$ is the Moore-Penrose inverse of Γ_ν.

If it were not for the measurement noise and if the lower triangular block Toeplitz matrix containing the Markov parameters H_ν in (8) were known, we could easily obtain the quasi-state z_i from (6), i.e.,

$$z_i = y_{\nu(i+\nu-1)} - H_\nu u_{\nu(i+\nu-1)}.$$

Since the matrix H_ν is unknown but can be estimated by the recursive identification, the quasi-state z_i can also be estimated recursively.

When a pair of data (u_N, y_N) is obtained at step N, $u_{\nu(N)}$ and $\tilde{y}_{\nu(N)}$ are composed of (5). Let estimates of Γ_ν, H_ν and (A, B, C, D) at the step N be denoted as

$$E_N, \quad \hat{H}_N, \quad (\hat{A}_N, \ \hat{B}_N, \ \hat{C}_N, \ \hat{D}_N),$$

respectively. These matrices are obtained from the Algorithm 1, the eigenvalue decomposition of (20).

Now, based on (6), we consider that an estimate $\hat{z}_{N-\nu+1|N}$ of the quasi-state $z_{N-\nu+1}$ is given by the following equation:

$$\hat{z}_{N-\nu+1|N} := \tilde{y}_{\nu(N)} - \hat{H}_N u_{\nu(N)} - E_N^{\perp} (E_N^{\perp})^T \alpha_N \tilde{e}_N. \qquad (29)$$

The third term of the right hand side of (29) plays a role of cancelling the noise effect. We shall explain the reason. From (1) the estimates $\hat{z}_{N|N}$ and $\hat{z}_{N+1|N}$ of the quasi-states z_N and z_{N+1}, respectively, are calculated respectively as

$$\hat{z}_{N|N} := E_N (\hat{A}_N)^{\nu-1} E_N^\dagger \hat{z}_{N-\nu+1|N} + E_N \sum_{i=1}^{\nu-1} (\hat{A}_N)^{i-1} \hat{B}_N u_{N-i}, \qquad (30)$$

$$\hat{z}_{N+1|N} := E_N \hat{A}_N E_N^\dagger \hat{z}_{N|N} + E_N \hat{B}_N u_N. \qquad (31)$$

From (28), the states of the estimated model ($\hat{A}_N$, $\hat{B}_N$, $\hat{C}_N$, $\hat{D}_N$) are calculated as

$$\hat{x}_{N-\nu+1|N} = E_N^\dagger \hat{z}_{N-\nu+1|N}, \qquad (32)$$

$$\hat{x}_{N|N} = E_N^\dagger \hat{z}_{N|N}, \qquad (33)$$

$$\hat{x}_{N+1|N} = E_N^\dagger \hat{z}_{N+1|N}. \qquad (34)$$

The rest of this section is devoted to the explanation of why the third term of the right hand side of (29) eliminates the noise effect. From (4) and (9), we obtain

$$\tilde{Y}_N = \Gamma_\nu X_{N-\nu+1} + H_\nu U_N + V_N. \qquad (35)$$

It is assumed that there exists a nonsingular matrix T such that $\Gamma = E_N T$. The premultiplication of (35) by $(E_N^{\perp})^T$ and the postmultiplication by U_N^T yield

$$(E_N^{\perp})^T \tilde{Y}_N U_N^T = (E_N^{\perp})^T H_\nu U_N U_N^T + (E_N^{\perp})^T V_N U_N^T. \qquad (36)$$

Because of (25), we can neglect the second term of the right hand side of (36) for sufficient large N, i.e.,

$$(E_N^{\perp})^T \tilde{Y}_N U_N^T \approx (E_N^{\perp})^T H_\nu U_N U_N^T.$$

From (19),

$$(E_N^{\perp})^T \left(\tilde{y}_{\nu(N)} - H_\nu u_{\nu(N)}\right) \approx (E_N^{\perp})^T \left(\tilde{y}_{\nu(N)} - \tilde{Y}_N U_N^T P_N u_{\nu(N)}\right).$$

Note that, from (22),

$$
\begin{aligned}
&\tilde{y}_{\nu(N)} - \tilde{Y}_N U_N^T P_N u_{\nu(N)} \\
&= \tilde{y}_{\nu(N)} - \left[\tilde{Y}_{N-1}\ \tilde{y}_{\nu(N)}\right] \begin{bmatrix} U_{N-1}^T \\ u_{\nu(N)}^T \end{bmatrix} \\
&\quad \cdot \left(P_{N-1} - \alpha_N P_{N-1} u_{\nu(N)} u_{\nu(N)}^T P_{N-1}\right) u_{\nu(N)} \\
&= \tilde{y}_{\nu(N)} - \tilde{Y}_{N-1} U_{N-1}^T P_{N-1} u_{\nu(N)} \\
&\quad - \tilde{y}_{\nu(N)} u_{\nu(N)}^T P_{N-1} u_{\nu(N)} \\
&\quad + \alpha_N \tilde{Y}_{N-1} U_{N-1}^T P_{N-1} u_{\nu(N)} u_{\nu(N)}^T P_{N-1} u_{\nu(N)} \\
&\quad + \alpha_N \tilde{y}_{\nu(N)} u_{\nu(N)}^T P_{N-1} u_{\nu(N)} u_{\nu(N)}^T P_{N-1} u_{\nu(N)} \\
&= \tilde{y}_{\nu(N)} \\
&\quad - \tilde{y}_{\nu(N)} u_{\nu(N)}^T P_{N-1} u_{\nu(N)} (1 - \alpha_N u_{\nu(N)}^T P_{N-1} u_{\nu(N)}) \\
&\quad - \tilde{Y}_{N-1} U_{N-1}^T P_{N-1} u_{\nu(N)} (1 - \alpha_N u_{\nu(N)}^T P_{N-1} u_{\nu(N)}) \\
&= \tilde{y}_{\nu(N)} - \alpha_N \tilde{y}_{\nu(N)} u_{\nu(N)}^T P_{N-1} u_{\nu(N)} \\
&\quad - \alpha_N \tilde{Y}_{N-1} U_{N-1}^T P_{N-1} u_{\nu(N)} \\
&= \tilde{y}_{\nu(N)} (1 - \alpha_N u_{\nu(N)}^T P_{N-1} u_{\nu(N)}) \\
&\quad - \alpha_N \tilde{Y}_{N-1} U_{N-1}^T P_{N-1} u_{\nu(N)} \\
&= \alpha_N \left(\tilde{y}_{\nu(N)} - \tilde{Y}_{N-1} U_{N-1}^T P_{N-1} u_{\nu(N)}\right) \\
&= \alpha_N \tilde{e}_N. \qquad (37)
\end{aligned}
$$

Therefore, we have

$$
(E_N^{\perp})^T (\tilde{y}_{\nu(N)} - H_\nu u_{\nu(N)}) = (E_N^{\perp})^T \alpha_N \tilde{e}_N. \quad (38)
$$

On the other hand, without noise we could derive

$$
\begin{aligned}
&(E_N^{\perp})^T (y_{\nu(N)} - H_\nu u_{\nu(N)}) \\
&\quad = (E_N^{\perp})^T \left(y_{\nu(N)} - Y_N U_N^T P_N u_{\nu(N)}\right) \\
&\quad = (E_N^{\perp})^T \alpha_N \left(y_{\nu(N)} - Y_{N-1} U_{N-1}^T P_{N-1} u_{\nu(N)}\right) \\
&\quad = (E_N^{\perp})^T \alpha_N e_N.
\end{aligned}
$$

Since $\Gamma_\nu = E_n T$, we obtain

$$
\begin{aligned}
(E_N^{\perp})^T \alpha_N e_N &= (E_N^{\perp})^T \alpha_N (\Gamma_\nu x_{N-\nu+1} \\
&\quad - \Gamma_\nu X_{N-\nu} U_{N-1}^T P_{N-1} u_{\nu(N)}) \\
&= 0. \qquad (39)
\end{aligned}
$$

Hence, in comparison with (38) to (39) we can regard the term $E_n^{\perp}(E_n^{\perp})^T \alpha_N \tilde{e}_N$ as estimation error of $z_{N-\nu+1}$ due to the mesurement noise.

Now, we summarize the estimation scheme of the state.

- A model of the system at step N results from recursive 4SID.

$$
E_N,\ \hat{H}_N,\ (\hat{A}_N, \hat{B}_N, \hat{C}_N, \hat{D}_N)
$$

- Past states of the model $\hat{z}_{N-\nu+1|N}$ and $\hat{x}_{N-\nu+1|N}$ are obtained from the following equations:

$$
\begin{aligned}
\hat{z}_{N-\nu+1|N} &= \tilde{y}_{\nu(N)} - \hat{H}_N u_{\nu(N)} \\
&\quad - E_N^{\perp}(E_N^{\perp})^T \alpha_N \tilde{e}_N, \\
\hat{x}_{N-\nu+1|N} &= E_N^{\dagger} \hat{z}_{N-\nu+1|N}.
\end{aligned}
$$

- The present states of the model $\hat{z}_{N|N}$ and $\hat{x}_{N|N}$ are obtained from the following equations:

$$
\begin{aligned}
\hat{z}_{N|N} &= E_N \hat{A}_N^{\nu-1} E_N^{\dagger} \hat{z}_{N-\nu+1|N} \\
&\quad + E_N \sum_{i=1}^{\nu-1} \hat{A}_N^{i-1} \hat{B}_N u_{N-i}, \\
\hat{x}_{N|N} &= E_N^{\dagger} \hat{z}_{N|N}.
\end{aligned}
$$

- The one-step-ahead predictors of the model $\hat{z}_{N+1|N}$ and $\hat{x}_{N+1|N}$ are obtained from the following equations:

$$
\begin{aligned}
\hat{z}_{N+1|N} &= E_N \hat{A}_N E_N^{\dagger} \hat{z}_{N|N} + E_N \hat{B}_N u_N, \\
\hat{x}_{N+1|N} &= E_N^{\dagger} \hat{z}_{N+1|N}.
\end{aligned}
$$

Remark 3. We have yet neither analyzed the convergence property of the estimated quasi-state, nor evaluated the covariance of the estimation error. These are our future works.

6. A NUMERICAL EXAMPLE

In order to illustrate the usefulness of our idea presented in the previous section, we consider a tracking control problem.

Consider an unknown system described as (1) with

$$
A = \begin{bmatrix} 0.8 & 0.1 & 0.3 \\ 0.1 & 0.55 & 0.2 \\ -0.1 & 0.3 & 0.7 \end{bmatrix},\ B = \begin{bmatrix} -0.45 & 0.3 \\ 0.15 & 0.6 \\ 0.3 & 0.2 \end{bmatrix},
$$
$$
C = \begin{bmatrix} -0.18 & -0.135 & 0.18 \\ -0.18 & 0.3 & -0.3 \end{bmatrix}, \qquad (40)
$$

and $D = 0$. Our control objective is to drive the tracking output y to a desired constant reference signal r. Here, $r = [8,\ -5]^T$. Our control strategy is shown in Figure 1.

The measurement $\tilde{y}$ is contaminated by noise v which is a zero mean white noise with covariance $0.25 * \mathtt{eye}(2)$, where $\mathtt{eye}(j)$ is a MATLAB command which makes the $j \times j$ identity matrix.

Suppose we have N pairs of I/O data $(u_i, y_i), i = 1, \cdots, N$. To solve this problem, one-step-ahead control input u_{N+1} is decided according to the following procedure.

Define

$$
\xi_{k+1} - \xi_k := r - \tilde{y}_k\ (-: \varepsilon_k).
$$

Since the parameters of the system (40) A, B and C are assumed to be unknown, we use parameters of an estimated model $\hat{A}_N$, $\hat{B}_N$ and $\hat{C}_N$ for estimating the state. The model is obtained by the recursive 4SID identification with taking $\nu = 5$. Let the order of the model be described as $\hat{n}_N$. Then, we obtain the following extended state-space representation:

$$
\begin{bmatrix} \hat{x}_{k+1} \\ \xi_{k+1} \end{bmatrix} = \begin{bmatrix} \hat{A}_N & 0 \\ -\hat{C}_N & I \end{bmatrix} \begin{bmatrix} \hat{x}_k \\ \xi_k \end{bmatrix} + \begin{bmatrix} \hat{B}_N \\ 0 \end{bmatrix} u_k + \begin{bmatrix} 0 \\ I \end{bmatrix} r. \qquad (41)
$$

We apply the linear-quadratic regulator design for the extended model with weighting matrices $Q = \mathtt{eye}(\hat{n}_N)$ and $R = \mathtt{eye}(2)$. Let the resulting controller be described as

$$
u_k = -K_N x_k. \qquad (42)
$$

We use one-step-ahead predictor $\hat{x}_{N+1|N}$ instead of x_k in (42). The one-step-ahead control input u_{N+1} is given by

$$u_{N+1} = -K_N \hat{x}_{N+1|N}.$$

We use MATLAB ver.5.2 for Windows and Control Systems Toolbox 4.0 for numerical computation. First 100 I/O data are used to calculate initial values for recursive identification. Between time step 1000 and 2000 we examine whether the controller works well provided that we make wrong choices of the order of the models. On $1000 \leq N < 1500$ we estimate it as $\hat{n}_N = 2$, while on $1500 \leq N < 2000$ $\hat{n}_N = 4$. On others we select the correct order $\hat{n}_N = 3$. The simulation results of the tracking output y are plotted in Figure 6. It is known that the controller does not perform well if we take a smaller estimate than the true order.

At $N = 3000$, we compare the eigenvalues of $\hat{A}_N$ with the ones of A. The former are 0.8956, 0.7582 and 0.3918, while the latter are 0.8579, 0.7776 and 0.4145. Therefore, it turns out that recursive identification also works well.

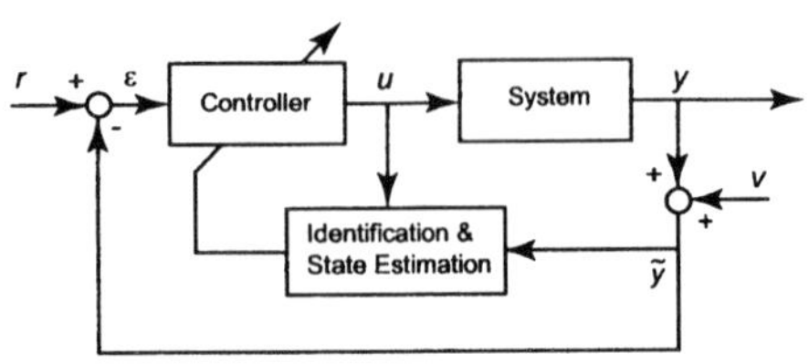

Fig. 1. Our control strategy

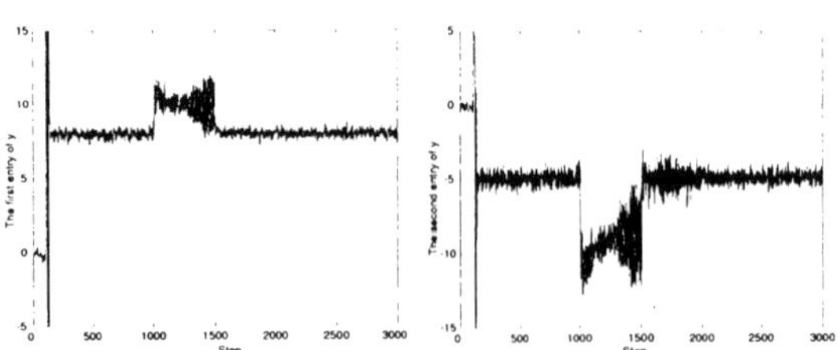

(a) First entry of the output y

(b) Second entry of the output y

Fig. 2. Simulation result

7. CONCLUSION

We have considered recursive state estimation of a MIMO model obtained by 4SID.

In the first place, we have introduced the recursive 4SID based on the matrix inversion lemma. The clearness of the structure of the algorithm has enabled us to analyze some properties of the 4SID schemes.

Then, we have proposed a scheme for the estimation of the state of models obtained by the recursive 4SID algorithm. To accomplish it, we have introduced a new kind of "*state*" which has maintained its meaning as a state over different update steps. We have adopted the equation (27) as such a "*state.*"

Finally, to illustrate the usefulness of our idea, a tracking control problem has been considered. Our simulation result has shown that both recursive identification and state estimation have worked well for the problem.

Our future works are to design an adaptive control law for the system by using the state estimation method in this paper, and to analyze the properties which the state estimation scheme owns, e.g., the convergence property of the estimated quasi-state and the covariance of the estimation error.

ACKNOWLEDGEMENT

We would like to thank Dr. Yasuaki Oishi of University of Tokyo for his helpful suggestions.

8. REFERENCES

Bierman, G. J. (1977). *Factorization Methods for Discrete Sequential Estimation.* Academic Press. New York.

Gustafsson, T., M. Lovera and M. Verhaegen (1998). A novel algorithm for recursive instrumental variable based subspace identification. In: *Proceedings of the 37th CDC.* Florida. pp. 3920–3925.

Ljung, L. (1999). *System Identification, 2nd Ed..* Prentice Hall. New Jersey.

Oku, H. and H. Kimura (1999). A recursive 4sid from the input-output point of view. *Asian Journal of Control* **1**(4), 258–269.

Van Overschee, P. and B. De Moor (1994). N4SID: Subspace algorithms for the identification of combined deterministic-stochastic systems. *Automatica* **30**(1), 75–93.

Van Overschee, P. and B. De Moor (1996). *Subspace Identification for Linear Systems.* Kluwer Academic Publishers. Massachusetts.

Verhaegen, M. (1993). Subspace model identification Part 3: Analysis of the ordinary output-error state-space model identification algorithm. *Int. J. Control* **58**(3), 555–586.

Verhaegen, M. (1994). Identification of the deterministic part of MIMO state space models given in innovations form from input-output data. *Automatica* **30**(1), 61–74.

Verhaegen, M. and E. Deprettere (1991). A fast, recursive MIMO state space model identification algorithm. In: *Proceedings of the 30th CDC.* Brighton. pp. 1349–1354.

Verhaegen, M. and P. Dewilde (1992*a*). Subspace model identification Part I: The output-error state space model identification class of algorithm. *Int. J. Control* **56**(5), 1187–1210.

Verhaegen, M. and P. Dewilde (1992*b*). Subspace model identification Part II: Analysis of the elementary output-error state-space model identification algorithm. *Int. J. Control* **56**(5), 1211–1241.

ON THE RANK DEFICIENCY OF THE LEAST SQUARES RESIDUALS IN SUBSPACE IDENTIFICATION

Tony Van Gestel *,1 **Bart De Moor** *,1
Peter Van Overschee **

* *Katholieke Universiteit Leuven, Dept. of Electrical Engineering ESAT-SISTA, Kard. Mercierlaan 94, B-3001 Leuven, Belgium. E-mail: {vangeste,demoor}@esat.kuleuven.ac.be*
** *ISMC N.V., Technologielaan 11-0101, B-3001 Heverlee, Belgium. E-mail: peter.vanoverschee@ismc.be*

Abstract: Subspace identification methods for linear time-invariant stochastic systems basically consist of two steps. In the first step, which is an orthogonal projection, followed by a Singular Value Decomposition (SVD), typically (non-steady state) Kalman filter states sequences are obtained directly from output data. In the second step, the system model matrices are identified in a least squares estimation step, which is asymptotically unbiased. For an infinite amount of data points, the measurement-process noise covariances are rank deficient, since the system is identified in innovation form. However, due to finite sample errors, this rank deficiency property is generically not satisfied when identifying on a finite number of measurement data. The required rank deficiency is achieved by replacing the least squares step by a total least squares estimate with exact row. *Copyright ©2000 IFAC*

Keywords: Subspace identification, Least squares estimate, Total least squares estimate, Kalman filter

1. INTRODUCTION

A linear time-invariant stochastic system in state space form is described by the equations[2] :

$$x_{k+1} = Ax_k + w_k, \quad (1)$$
$$y_k = Cx_k + v_k, \quad (2)$$

with:

$$\mathbf{E}(\begin{bmatrix} w_p \\ v_p \end{bmatrix} [\, w_q^T \; v_q^T \,]) = \begin{bmatrix} Q & S \\ S^T & R \end{bmatrix} \delta_{pq} \geq 0. \quad (3)$$

The vector $y_k \in \mathbb{R}^l$ contains the observed l outputs at the discrete time instant k. The vector $x_k \in \mathbb{R}^n$ is the state vector of the system at the time instant k. The (unobserved) process and measurement noise vectors at the time instant k denoted by $w_k \in \mathbb{R}^n$ and $v_k \in \mathbb{R}^l$ respectively. are independent of the state x_k. It is assumed that they are zero mean, stationary, white noise vector sequences and that v_k and w_k are independent of the state x_k. Their covariance matrices (3) are $Q \in \mathbb{R}^{n\times n}$, $S \in \mathbb{R}^{n\times l}$ and $R \in \mathbb{R}^{l\times l}$. The system matrix A and the output matrix C are real matrices of appropriate dimensions. The pair (A, C) is assumed to be observable and the pair $(A, Q^{\frac{1}{2}})$ to be controllable.

[1] T. Van Gestel is a Research Assistant with the Fund for Scientific Research-Flanders (FWO-Vlaanderen). B. De Moor is a Senior Research Associate with the FWO-Vlaanderen and Professor at the K.U.Leuven.

[2] **E** denotes the expected value operator and δ_{pq} the Kronecker delta. It's assumed that the data is ergodic: $\mathbf{E}(a_k b_k^T) = lim_{j\to\infty}[\frac{1}{j}\sum_{i=0}^{j} a_i b_i^T]$.

The stochastic process is assumed to be stationary: $\mathbf{E}(x_k) = 0$ and $\mathbf{E}(x_k x_k^T) = \Sigma = A\Sigma A^T + Q$. The output covariance matrices, defined as $\Lambda_i = \mathbf{E}(y_{k+i}y_k^T)$, can be calculated as $\Lambda_i = CA^{i-1}G$ for $i = 1, 2, \ldots$, with G defined as $G = A\Sigma C^T + S$. The energy of the output is measured by $\Lambda_0 = \mathbf{E}(y_k y_k^T) = C\Sigma C^T + R$.

Based on the state space equations (1)-(3), Kalman generated a recursive scheme for the linear filtering problem. Let $\hat{x}_{i|i-1} = \mathbf{E}(x_i|y_{0,i-1})$ denote the optimal predicted state at time step i, given the observations $y_{0,i-1} = y_0, \cdots, y_{i-1}$. The state estimate error is denoted by $\tilde{x}_i = x_i - \hat{x}_{i|i-1}$ and is orthogonal to $y_{0,i-1}$. Given the new observation y_i, we define the innovation of the process as: $e_i = y_i - C\hat{x}_{i|i-1}$. Since the innovation is also $e_i = C\tilde{x}_i + v_i$, this innovation is orthogonal to $y_{0,i-1}$ and $\hat{x}_{i|i-1}$. Let $R_i^e = \mathbf{E}(e_i e_i^T)$, then the next state Kalman filter prediction can be calculated as

$$\hat{x}_{i+1|i} = A\hat{x}_{i|i-1} + K_i(y_i - C\hat{x}_{i|i-1}), \qquad (4)$$

$$K_i = A\tilde{P}_i C^T(R + C\tilde{P}_i C^T)^{-1}. \qquad (5)$$

By calculating the covariance of $\tilde{x}_{i+1} = A\tilde{x}_i + w_i - K_i e_i$, the following recursion formula is obtained for the state error covariance matrix $\tilde{P}_i = \mathbf{E}(\tilde{x}_i\tilde{x}_i^T)$:

$$\tilde{P}_{i+1} = A\tilde{P}_i A^T + Q - (A\tilde{P}_i C^T + S)$$
$$(C\tilde{P}_i C^T + R)^{-1}(A\tilde{P}_i C^T + S)^T. \qquad (6)$$

These recursions are started with initial values $\hat{x}_{0|-1}$ and $\tilde{P}_0$, which are assumed to be available. For finite i, above equations are called non-steady state Kalman filter equations. They are called steady state Kalman filter equations when $i \to \infty$, (6) then becomes an algebraic Riccati equation and the system based on the steady state Kalman filter gain is called the (forward) innovation model: $x_{k+1}^f = Ax_k^f + Ke_k$, $y_k = Cx_k^f + e_k$, with $\mathbf{E}(e_k e_k^T) = C\tilde{P}_\infty C^T + R$. The forward innovation model has the same second order properties (the same autocorrelation function Λ_i, for $i = 0, 1, \ldots$) as the stochastic model (1)-(3) (see e.g. (Faurre *et al.*, 1978)).

2. SUBSPACE IDENTIFICATION

In system identification, we are concerned with the inverse problem: *Given* a large number of observations of the output y_k generated by the unknown system (1)-(3). *Determine* the system order n of the unknown system (1)-(3) and the estimates $\hat{A}$, $\hat{C}$ of the system matrices A, C up to within a similarity transformation and the estimates $\hat{Q}$, $\hat{S}$ and $\hat{R}$ of the noise covariances Q, S, R so that the second order statistics of the output of the model *and* of the given output are equal (or equivalent in the sense of Faurre (Faurre *et al.*, 1978)).

Below, a brief description of the subspace algorithms for the above identification problem is given, full details can be found in (Van Overschee and De Moor, 1996). One of the basic ideas in subspace identification methods is that the Kalman filter states $\hat{x}_{i|i-1}$ can be estimated directly from output data, without knowing the system matrices A, C and covariances Q, S, R:

Theorem 1. Given the initial state estimate $\hat{x}_0 = 0$, with covariance of the state $\tilde{P}_0 = \Sigma$, and the measurements $y_0, \cdots, y_{k-1}$, then the non-steady state Kalman filter state estimate $\hat{x}_{k|k-1}$ can be explicitly written as:

$$\hat{x}_{k|k-1} = \Delta_k^c L_k^{-1}\left[y_0^T\ y_1^T\ \ldots\ y_{k-1}^T\right]^T,$$

with the reversed extended observability matrix $\Delta_k = \left[A^{k-1}G\ A^{k-2}G\ \ldots\ AG\ G\right]$.

Let the output data block Hankel matrix $Y_{0|i-1}$ be constructed from the observations $y_0, y_1, \cdots, y_{i+j-2}$ as follows:

$$Y_{0|i-1} = \begin{bmatrix} y_0 & y_1 & \cdots & y_{j-1} \\ y_1 & y_2 & \cdots & y_j \\ \cdots & \cdots & \cdots & \cdots \\ y_{i-1} & y_i & \cdots & y_{i+j-2} \end{bmatrix}, \qquad (7)$$

where i is a user defined dimension with typically $i << j$ and the subscript $0|i-1$ denotes the time indices of the outputs in the first column of $Y_{0|i-1}$. In the section, we assume $j \to \infty$. The past and future output block Hankel matrices are defined as follows: $Y_p = Y_{0|i-1}$ and $Y_f = Y_{i|2i-1}$. From these past and future block Hankel matrices a projection matrix $\mathcal{O}_i$ is calculated as the orthogonal projection of Y_f onto Y_p:

$$\mathcal{O}_i = Y_f/Y_p = Y_f Y_p^T(Y_p Y_p^T)^{-1}Y_p. \qquad (8)$$

If we now arrange the Kalman filter state estimates in

$$\hat{X}_i = [\,\hat{x}_i\ \hat{x}_{i+1}\ \ldots\ \hat{x}_{i+j-1}\,],$$

with $\hat{x}_{i+q}$ the non-steady state Kalman filter state estimate based on the observations $y_q, \cdots, y_{i+q-1}$ and the same initial conditions as in Theorem 1, the extended observability matrix Γ_i and the Kalman filter state sequence $\hat{X}_i$ are calculated as $\Gamma_i = US_d^{\frac{1}{2}}$ and $\hat{X}_i = S_d^{\frac{1}{2}}V^T$, with $\mathcal{O}_i = US_dV^T$ (see (Van Overschee and De Moor, 1996)). In an analogous way, the next Kalman filter states sequence $\hat{X}_{i+1}$ is estimated, leading to the following relation:

$$\begin{bmatrix} \hat{X}_{i+1} \\ Y_{i|i} \end{bmatrix} = \begin{bmatrix} \hat{A} \\ \hat{C} \end{bmatrix}\hat{X}_i + \begin{bmatrix} K_i\hat{E}_i \\ \hat{E}_i \end{bmatrix}, \qquad (9)$$

with $\hat{E}_i = [\hat{e}_i, \ \hat{e}_{i+1}, \ \hat{e}_{i+2}, \ldots]$. In the right hand side of the last equality, K_i is the non-steady state Kalman filter gain (5) and $\hat{E}_i$ is the vector containing the non-steady state Kalman filter innovations. Since we know that $\mathbf{E}(\hat{e}_k \hat{x}_k^T) = 0$, the system matrices A, C are asymptotically unbiased estimated by the least squares solution of (9). When also $i \to \infty$, the system in forward innovation form is obtained.

For finite i, the covariance matrices $\hat{Q}$, $\hat{S}$ and $\hat{R}$ are estimated either based on deterministic realization theory or approximated by $\hat{Q}_i$, $\hat{S}_i$ (Algorithms 1, 2 in (Van Overschee and De Moor, 1996)) and $\hat{R}_i$ (Algorithm 3 in (Van Overschee and De Moor, 1996)). For both $j \to \infty$ and $i \to \infty$, subspace algorithms yield an unbiased estimate. For $j \to \infty$ and i finite, subspace algorithms may fail to estimate a positive real system because of a wrong system order estimate. For both j and i finite, subspace algorithms may fail to generate stable and/or positive real systems and fail to have rank-deficient residuals in the least-squares estimate. This papers concerns are the required rank deficiency.

3. RANK DEFICIENT RESIDUALS

When the number of observations becomes infinite ($j \to \infty$), we know from (9) that[3] $\mathbf{rank}\left([K_i\hat{E}_i; \ \hat{E}_i]\right) = l$. Hence, the rank of the non-steady state noise covariance matrix is l:

$$\mathbf{rank}\left(\begin{bmatrix} \hat{Q}_i & \hat{S}_i \\ \hat{S}_i^T & \hat{R}_i \end{bmatrix}\right) = l.$$

This corresponds to the following theorem.

Theorem 2. When the number of data points becomes infinite ($j \to \infty$), the following rank properties hold for the subspaces of the next and present Kalman filter states $\hat{X}_{i+1}$, $\hat{X}_i$ (resp.) and the outputs $Y_{i|i}$:

$$\mathbf{rank}\left(\begin{bmatrix} \hat{X}_{i+1} \\ Y_{i|i} \\ \hat{X}_i \end{bmatrix}\right) = n + l, \tag{10}$$

$$\mathbf{rank}\left(\begin{bmatrix} Y_{i|i} \\ \hat{X}_i \end{bmatrix}\right) = n + l, \tag{11}$$

$$\mathbf{rank}\left(\begin{bmatrix} \hat{X}_{i+1} \\ Y_{i|i} \end{bmatrix}\right) = n + l - z, \tag{12}$$

where z is the number of transmission zero's at the origin.

[3] The Matlab notation $[X; \ Y] = [X^T \ Y^T]^T$ is used.

Proof The proof is based on the optimal state prediction property of the Kalman filter and the corresponding orthogonality of the innovations onto the states: $\mathbf{E}(\hat{x}_i \hat{e}_i^T) = 0$. Therefor, the innovations $\hat{E}_i$ are orthogonal onto the regressors $\hat{X}_i$. The row space of the innovations $\hat{E}_i$ has full rank. The rest of the proof is based on the following equation:

$$\begin{bmatrix} \hat{X}_{i+1} \\ Y_{i|i} \\ \hat{X}_i \end{bmatrix} = \begin{bmatrix} \hat{A}^{n\times n} & \hat{K}_i^{n\times l} \\ \hat{C}^{l\times n} & I_l \\ I_n & 0_{n\times l} \end{bmatrix} \cdot \begin{bmatrix} \hat{X}_i \\ \hat{E}_i \end{bmatrix}. \tag{13}$$

The proof of (11) then becomes trivial since the row space of $\hat{X}_i$ is of full rank since the states are persistently excited and since the innovation sequence $\hat{E}_i$ is contained in the null space of $\hat{X}_i$. Hence, $[\hat{X}_i; E_i]$ has full row rank. This also proves (10).

The proof of (12) is based on the following Lemma, which basically follows from the definition of transmission zeros:

Lemma 3. The rank of a (square) linear transformation matrix

$$\begin{bmatrix} A^{n\times n} & B^{n\times l} \\ C^{l\times n} & I^{l\times l} \end{bmatrix}$$

is equal to $n + l - z$, with z the number of transmission zeros of

$$\begin{bmatrix} A & B \\ C & I \end{bmatrix}$$

in the origin.

The row space of $[\hat{X}_{i+1}; \ Y_{i|i}]$ is a linear combination of the row space of $[\hat{X}_i; \ \hat{E}_i]$. Since, this row space has full row rank, the row space is equal to the rank of the linear combination $[\hat{A} \ \hat{K}_i; \ \hat{C} \ I]$: this rank is equal to the system order n minus the number of transmission zeros of $[\hat{A} \ \hat{K}_i; \ \hat{C} \ I]$ in the origin z. □

4. FINITE J

When dealing with the practical case of a finite number of observations (finite j), the projection matrix $\mathcal{O}_i$ typically has full rank, due to finite sample effects and the noise. The system order n is estimated by calculating the SVD of $\mathcal{O}_i = US_dV^T$ and inspecting the $l \cdot i$ singular values in S_d. The smallest $l \cdot i - n$ singular values in S_d are put to zero and the extended observability matrix[4] $\bar{\Gamma}_i$

[4] A $\bar{X}$ is used to denote the Kalman filter state sequence estimates based on a finite number of observations (finite j). A $\bar{M}$ denotes an estimate of a system parameter M based on these Kalman filter state sequences. The modified Kalman filter state estimates will be denoted by $\tilde{X}$ and $\tilde{M}$

and the present state sequence $\bar{X}_i$ are estimated as $\bar{\Gamma}_i = US_d^{\frac{1}{2}}$ and $\bar{X}_i = S_d^{\frac{1}{2}}V^T$. The next state sequence $\bar{X}_{i+1}$ is computed in an analogous way. Then the model matrices can be estimated by the least squares solution of:

$$\begin{bmatrix} \bar{X}_{i+1} \\ Y_{i|i} \end{bmatrix} = \begin{bmatrix} \bar{A} \\ \bar{C} \end{bmatrix} \bar{X}_i + \begin{bmatrix} \bar{W}_i \\ \bar{V}_i \end{bmatrix}. \tag{14}$$

Due to the finite sample effects, the Kalman filter state estimates are perturbed and the estimate of the noise covariance matrices is *not* rank deficient. In general, one has:

$$\textbf{rank}\left(\begin{bmatrix} \bar{W}_i \\ \bar{V}_i \end{bmatrix}\right) = l + q, \quad \text{with} \quad 0 \leq q \leq n,$$
$$\textbf{rank}\left(\begin{bmatrix} \hat{X}_i \\ Y_{i|i} \\ \hat{X}_{i+1} \end{bmatrix}\right) = n + l + q. \tag{15}$$

Since the Kalman filter state estimates are perturbed by noise, the correction for rank deficiency is done by modifying $\bar{X}_i \to \tilde{X}_i$ and $\bar{X}_{i+1} \to \tilde{X}_{i+1}$, which corresponds to the following optimization problem:

$$\min_{\tilde{X}_{i+1}, \tilde{X}_i, T_x, T_y} \alpha||\bar{X}_i - \tilde{X}_i||_F^2 + \beta||\bar{X}_{i+1} - \tilde{X}_{i+1}||_F^2 \tag{16}$$
$$s.t. \quad \tilde{X}_{i+1} = T_x\tilde{X}_i + T_yY_{i|i}. \tag{17}$$

The weights α and β take the relative accuracy of $\bar{X}_i$ and $\bar{X}_{i+1}$ into account. By use of the method of Lagrange multipliers, it can be shown that the optimization problem (16)-(17) corresponds to a (weighted) Total Least Squares (TLS) problem with exact rows.

The solutions $\tilde{X}_{i+1}$ and $\tilde{X}_i$ of (16) and (17) are obtained by first calculating the orthogonal complements

$$\bar{X}_{i+1}^{\perp} = \bar{X}_{i+1} - \bar{X}_{i+1}/Y_{i|i}$$
$$\bar{X}_i^{\perp} = \bar{X}_{i+1} - \bar{X}_{i+1}/Y_{i|i}$$

of respectively $\bar{X}_{i+1}$ and $\bar{X}_i$ onto $Y_{i|i}$. The subspace $[\bar{X}_{i+1}^{\perp};\ \bar{X}_i^{\perp}]$, orthogonal onto $Y_{i|i}$, generically has rank $n + q$. By calculating the SVD of $[\sqrt{\beta}\bar{X}_{i+1}^{\perp};\ \ \sqrt{\alpha}\bar{X}_i^{\perp}]$ and putting the n smallest singular values to zero, $[\sqrt{\beta}\tilde{X}_{i+1}^{\perp};\ \ \sqrt{\alpha}\tilde{X}_i^{\perp}]$ are obtained. The solutions $\tilde{X}_{i+1}$ and $\tilde{X}_i$ are obtained after recomposition with the parts that were projected on $Y_{i|i}$:

$$\tilde{X}_{i+1} = \tilde{X}_{i+1}^{\perp} + \bar{X}_{i+1}/Y_{i|i}$$
$$\tilde{X}_i = \tilde{X}_i^{\perp} + \bar{X}_i/Y_{i|i}.$$

---

will denote the corresponding system parameter. Note that for $j \to \infty$, we have $\tilde{X} \to \bar{X} \to \hat{X}$ and correspondingly $\tilde{M} \to \bar{M} \to \hat{M}$.

The system matrices are obtained as the solution of the least squares step:

$$\begin{bmatrix} \tilde{X}_{i+1} \\ Y_{i|i} \end{bmatrix} = \begin{bmatrix} \tilde{A} \\ \tilde{C} \end{bmatrix} \tilde{X}_i + \begin{bmatrix} \tilde{K}_i\tilde{E}_i \\ \tilde{E}_i \end{bmatrix}. \tag{18}$$

For different choices of the weights α and β in (16)-(17), different solutions for $\tilde{X}_{i+1}$ and $\tilde{X}_i$ are obtained. In the three following paragraphs, we discuss the solutions for both the limiting values and intermediate values of the weights:

1. **Modification of $\bar{X}_{i+1}$: least squares (LS):** In this case it is assumed that the subspace $\bar{X}_i$ is correct and hence $\alpha \to \infty$ or $\beta/\alpha \to 0$ in (16). Since $[\bar{X}_i; Y_{i|i}]$ has full rank $n + l$ (11) it can be shown , by means of the use of Lagrange multipliers, that the optimal solution $\tilde{X}_{i+1}$ is obtained by the orthogonal projection of $\bar{X}_{i+1}$ onto $[\bar{X}_i; Y_{i|i}]$: $\tilde{X}_{i+1} = \bar{X}_{i+1}/[\bar{X}_i; Y_{i|i}]$. Basically this solution corresponds to a two step least squares solution. The first least squares step corrects the subspace $\bar{X}_{i+1}$ of (14) such that the rank deficiency property of the residuals holds. The estimated process noise $\bar{W}_i$ of (14) can be decomposed into a first component $\bar{W}_{i1}$ perpendicular onto $[\bar{X}_i; Y_{i|i}]$ and a second component $\bar{W}_{i2}$ which is contained in the subspace $[\bar{X}_i; Y_{i|i}]$ and which is perpendicular onto $\bar{X}_i$. The first projection of $\bar{X}_{i+1}$ onto $[\bar{X}_i; Y_{i|i}]$ puts $\bar{W}_{i1} = 0$. The estimates of the model matrices, $\tilde{A}$ and $\tilde{C}$, obtained by the second least squares step (18), and the covariances S_i and R_i remain unaltered. When looking at Algorithm 3 in (Van Overschee and De Moor, 1996), it follows that putting $\bar{W}_{i1} = 0$ results in reduction of $\bar{\Lambda}_0$, while keeping the model positive real. In other words, without the rank deficiency modification, the estimate $\bar{\Lambda}_0$ is unnecessarily shifted upwards, resulting in a "more" positive real system. The LS modification is depicted in Figure 1 for $n = l = 1$.
2. **Modification of $\bar{X}_i$: data least squares (DLS):** It is assumed that the subspace $\bar{X}_{i+1}$ is correctly estimated, corresponding to $\beta = \infty$ or $\frac{\alpha}{\beta} \to 0$ in (16). It is assumed that the model $[A\ \ K_i;\ \ C\ \ I]$ has no transmission zeros in the origin, i.e. $z = 0$ in (12). This is equivalent with saying that $[\hat{X}_{i+1}; Y_{i|i}]$ has full row rank by means of (12) of Theorem 2. The optimal solution $\tilde{X}_i = \bar{X}_i/[\bar{X}_{i+1}; Y_{i|i}]$ is found as the projection of $\bar{X}_i$ onto $[\bar{X}_{i+1}; Y_{i|i}]$. Since the estimates of the system matrices depend on the corrected $\tilde{X}_i$ (by the least squares estimate (18)), generically the estimates of all parameters are modified. Since this least squares correction only modifies the data matrix A of the standard least squares problem $Ax = b + e$ in the unknowns x and e, this modification is

called a data least squares modification[5].

3. **Modification of $\bar{X}_i$ and $\bar{X}_{i+1}$: total least squares (TLS):** In this case it is assumed that there are perturbations on both $\bar{X}_i$ and $\bar{X}_{i+1}$. Then (15) and (11) hold and both $\bar{X}_i$ and $\bar{X}_{i+1}$ are modified in TLS sense as explained above.

4.1 *Numerical Implementation*

The three modifications can be implemented using a QR-decomposition and/or an SVD:

1. **LS: modification of $\bar{X}_{i+1}$:** reorder the outputs $Y_{i|i}$ and the next and present Kalman filter state sequences $\bar{X}_{i+1}, \bar{X}_i$ in a big matrix and calculate the QR-decomposition:

$$\begin{bmatrix} Y_{i|i} \\ \bar{X}_i \\ \bar{X}_{i+1} \end{bmatrix} = \begin{bmatrix} R_{11}^{l\times l} & 0 & 0 \\ R_{12}^{n\times l} & R_{22}^{n\times n} & 0 \\ R_{13}^{n\times l} & R_{23}^{n\times n} & R_{33}^{n\times n} \end{bmatrix} \cdot \begin{bmatrix} Q_1^T \\ Q_2^T \\ Q_3^T \end{bmatrix}.$$

The dimension of the R-blocks are written in superscript. The modified $\tilde{X}_{i+1}$ is obtained by putting $R_{33} = 0$. This is equivalent to putting $\bar{W}_{i1}$ to zero.

2. **DLS: modification of $\bar{X}_i$:** this implementation is analogous to the above paragraph. The solution is obtained by formally replacing $[Y_{i|i}; \bar{X}_{i+1}; \bar{X}_i]$ by $[Y_{i|i}; \bar{X}_i; \bar{X}_{i+1}]$ in the previous paragraph. The solution $\tilde{X}_i$ is obtained by putting $R_{33} = 0$.

3. **TLS: modification of $\bar{X}_i$ and $\bar{X}_{i+1}$:** The TLS solution is obtained by calculating the SVD $U_R S_R V_R^T$ of $[\sqrt{\alpha}R_{22}\ 0;\ \ \sqrt{\beta}R_{23}\ \ \sqrt{\beta}R_{33}]$ and putting the lowest n singular values in S_R equal to zero.

4.2 *Stability*

Since all corrections are minimum norm corrections (16), the modifications $\Delta\bar{X}_i = \tilde{X}_i - \bar{X}_i$ and $\Delta\bar{X}_{i+1} = \tilde{X}_{i+1} - \bar{X}_{i+1}$ are contained in the null space of $[\bar{X}_i;\ \bar{X}_{i+1}]$. Hence we obtain:

$$\bar{X}_{i+1} + \Delta\bar{X}_{i+1} = \tilde{A}(\bar{X}_i + \Delta\bar{X}_i) + \tilde{W}_i + \Delta\tilde{W}_i, \quad (19)$$

$$\bar{X}_{i+1} = \bar{A}\bar{X}_i + \bar{W}_i, \quad (20)$$

with $(\Delta\tilde{W}_i + \tilde{W}_i) \perp (\bar{X}_i + \Delta\bar{X}_i)$ and $\bar{W}_i \perp \bar{X}_i$. By multiplication of (19) and (20) to the right

[5] The standard Data Least Squares (DLS) solution was introduced in (Groat and Dowling, 1993), modifying the data matrix A into A_d such that there exists a vector x for which $A_d x = b$.

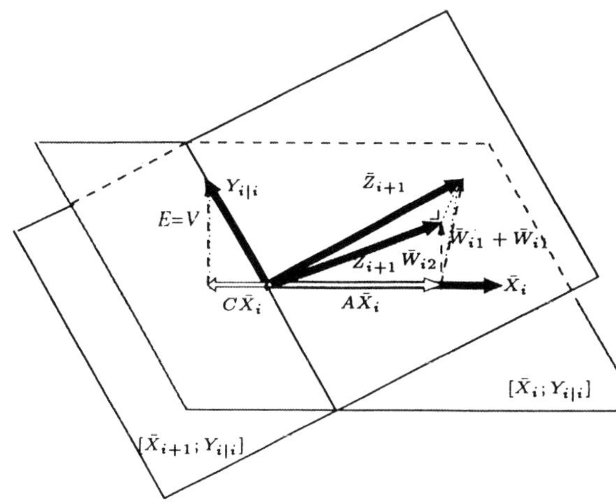

Fig. 1. Least Squares Modification $\bar{X}_{i+1} \rightarrow \tilde{X}_{i+1}$ for $n = l = 1$.

with $\bar{X}_i^T$ and $\tilde{X}_i^T$ respectively and substituting $\bar{X}_{i+1}\bar{X}_i^T$ in the first equation, the following expression is obtained:

$$\tilde{A}\Sigma_{\tilde{X}_i} - \bar{A}(\Sigma_{\bar{X}_i} + \Sigma_{\Delta\bar{X}_i}) + \Delta\bar{X}_{i+1}\Delta\bar{X}_i^T = 0 \quad (21)$$

with $\Sigma_{\tilde{X}_i} = \tilde{X}_i\tilde{X}_i^T$ and $\Sigma_{\Delta\bar{X}_i} = \Delta\bar{X}_i\Delta\bar{X}_i^T$. The stability analysis of the three modifications then becomes:

1. **LS: modification of $\bar{X}_{i+1}$:** as already explained above, we have $\tilde{A} = \bar{A}$.

2. **DLS: modification of $\bar{X}_i$:** from (21) and $\Delta\bar{X}_i = 0$, the relation between $\tilde{A}$ and $\bar{A}$ is given by:

$$\tilde{A}\Sigma_{\tilde{X}_i} = \bar{A}(\Sigma_{\bar{X}_i} + \Sigma_{\bar{X}_i}). \quad (22)$$

In (Van Gestel *et al.*, 1999) it is shown that the spectral radius of $\bar{A}$ can be limited by applying regularization (Sjöberg *et al.*, 1995). When comparing (22) with the results of (Van Gestel *et al.*, 1999), it follows that the DLS is the opposite of applying regularization. In other words, the system matrix $\tilde{A}$ in the data least squares can be expected to be relatively less stable than the least squares estimate $\bar{A}$. However, it cannot be proven that $\rho(\tilde{A}) < \rho(\bar{A})$ and for small corrections, numerical counter examples can be easily found.

3. **TLS: modification of $\bar{X}_i$ and $\bar{X}_{i+1}$:** the relative stability of $\tilde{A}$ with respect to $\bar{A}$ can be analyzed using (21). Due to the term $\Delta\bar{X}_{i+1}\Delta\bar{X}_i^T$, and the SVD needed for the solution, the influence on stability could not be stated explicitly.

5. SIMULATION RESULTS

The practical use of the three solutions is verified by Monte Carlo simulations. For each of the three

	A3	LS	DLS	TLS
Mean	6.96	6.76	419	8.52
SD	0.32	0.26	394	0.63
Mean	2.39	2.37	10.64	2.53
SD	0.10	0.09	1.93	0.10

Table 1. Comparison of the 3 solutions. See text for details. Top rows: 25 data points, bottom rows 100 data points.

algorithms, the stochastic system with system matrices:

$$A = \begin{bmatrix} 0.3 & 0 & 0 \\ 0 & 0.4 & 0 \\ 0 & 0 & 0.5 \end{bmatrix} \quad K = \begin{bmatrix} 1 \\ 0.1 \\ 0.01 \end{bmatrix}$$

$$C = \begin{bmatrix} 1 & 1 & 10 \end{bmatrix} \qquad R = 1$$

was simulated 1000 times for 25 and 100 data points. The Kalman filter state sequences were obtained with the CVA weights (Van Overschee and De Moor, 1996). Algorithm 3 of (Van Overschee and De Moor, 1996) was used. The results are compared on the difference of the first 10 points in the Autocorrelation Function (ACF) of the model ($\text{ACF}_{\text{est}}(0) = \Lambda_0$, $\text{ACF}_{\text{est}}(i) = \Lambda_i, i > 0$) and the data ($\text{ACF}_{\text{data}}(i) = \frac{1}{k}\sum_{k=0}^{k-i} y_k y_{k+i}^T$). The mean and standard deviation (SD) on the mean are shown in Table 1 for Algorithm 3 without modification (A3), for the LS, DLS and TLS (with $\alpha = \beta$) modification. The simulation results include only estimates where both estimates resulted in a stable system. From the simulation it follows that the LS-modification of $\hat{X}_{i+1} \rightarrow \tilde{X}_{i+1}$ has the best performance. This corresponds to a high β in (16) and can be explained by the fact that the noise in $\mathcal{O}_{i-1}$ is not totally reduced in the subspace algorithm.

6. CONCLUSIONS

When the number of data points in subspace identification goes to ∞, the residuals of the second least squares for the identification of the system matrices are rank deficient. However this rank deficiency is not satisfied in the practical case of a finite number of data points. In this paper, it is shown that the rank deficiency of the residuals corresponds to a rank deficiency of the row spaces formed by the next and present Kalman filter states and the observations. For a finite number of data points, the modifications for this rank deficiency are obtained as the solutions of a weighted total least squares problem with exact row, with least squares and data least squares solution as limiting cases for special choices of the weights. Both implementation and the influence on stability were discussed. In simulations, the least squares correction is the most appropriate, since noise is already cancelled out on the data matrix by the first SVD. The modifications have zero influence when the number of data points becomes infinite.

7. ACKNOWLEDGEMENTS

This work is supported by several institutions:

(1) the Flemish Government: Research Council K.U.Leuven: GOA-MEFISTO-666; the FWO (Fund for Scientific Research - Flanders) projects G.0240.99 and G.0256.97; the FWO Research Communities: ICCoS (Identification and Control of Complex Systems) and ANMMM (Advanced Numerical Methods for Mathematical Modelling); IWT projects EUREKA 1562 (SINOPSYS), EUREKA 2063 (IMPACT), STWW

(2) the Belgian State, Prime Minister's Office - Federal Office for Scientific, Technical and Cultural Affairs: Interuniversity Poles of Attraction Programme IUAP P4-02 (1997-2001) and IUAP P4-24 (1997-2001); Project MD/01/24 (1997-2000);

(3) the European Commission: TMR Networks: (ALAPEDES); System Identification; Brite/Euram Thematic Network : NICONET;

(4) Industrial Contract Research : ISMC, Electrabel, Laborelec, Verhaert, Europay.

The scientific responsibility is assumed by its authors.

8. REFERENCES

Faurre, P., M. Clerget and F. German (1978). *Opérateurs rationnels positifs, application à l'hyperstabilité et aux processus aléatoires.* Dunod.

Groat, R.D. De and E.M. Dowling (1993). The data least squares problem and channel equalization. *IEEE Trans. on Signal Processing* **41**, 407–411.

Sjöberg, J., Q. Zhang, L. Ljung, A. Benveniste, B. Deylon, P. Glorennec, H. Hjalmarsson and A. Juditsky (1995). Nonlinear black-box modelling in system identification: a unified overview. *Automatica* **31**, 1691–1724.

Van Gestel, T., J. Suykens, P. Van Dooren and B. De Moor (1999). Identification of stable models in subspace identification by using regularization. Technical Report TR99-81. ESAT-SISTA, K.U.Leuven. Submitted.

Van Overschee, P. and B. De Moor (1996). *Subspace Identification for Linear Systems: Theory, Implementation, Applications.* Kluwer Academic Publishers, Boston/London/Dordrecht.

LINEAR MATERIAL FLOW MODELS

Andreas Gleiß* Manfred Deistler* Thomas Matyus*

* *Institute of Econometrics, Operations Research and System Theory, Technical University, Vienna*

Abstract: This paper presents static and dynamic linear models for material flow systems. Based on the a priori knowledge given e.g. by mass balances these models are highly structured in a natural way. In the static case this structure is exploited for the reconciliation of the flow measurements, the estimation of the parameters and for simulation. In the dynamic case graph theoretic conditions are provided for stability, reachability and identifiability of three types of stock dynamics. *Copyright ©2000 IFAC*

Keywords: Linear Systems, Subsystems, Uncertainty, Environmental Engineering, Material Systems, Directed Graphs, Identifiability, Parameter Identification

1. INTRODUCTION

Material Flow Analysis (MFA) has become an important instrument in environmental science and pollution research, see (Baccini and Brunner, 1991), (Van der Voet *et al.*, 1995). MFA is a method for capturing and describing the metabolism of selected parts of the anthroposphere. Such parts may be regions or production sites in which flows of selected materials, i.e. chemical elements or their compounds, or of goods are measured. The usual procedure in MFA is to represent a system by a *flow sheet*, which is a block diagram showing the decomposition of the system into subsystems and the flows between the subsystems, additionally containing the flow measurements. From the flow sheet a *system graph* is directly deduced, where each vertex represents a subsystem and each (directed) edge a flow. This fact suggests the use of graph theoretic concepts for answering system theoretic questions.

Here the main objective is the identification of the system from the a priori information and the data. The special features of this system identification problem are: The decomposition into subsystems, the law of the conservation of mass within each subsystem and the positivity of the variables involved, i.e. of the flows and the stock levels.

In most applications the stock levels, which often are of great practical interest, are not directly observed. However, there are measurements of imports, i.e. physical inputs into the system, exports, i.e. physical outputs from the system, and internal flows, but in general in all three flow categories there are unobserved flows as well. Typically, there is only one measurement per balancing period (mostly one year). In the static case the *balance equations*, which equate the sum of the inputs of a subsystem with the sum of its outputs according to the law of the conservation of mass, provide important a priori information.

The problems considered for material flow systems are the reconciliation of the flow measurements, data driven modeling and policy simulation. In the following linear models for the static and the dynamic case are presented, for further details and proofs see (Bauer *et al.*, 1997), (Gleiß *et al.*, 1998) and (Gleiß, 1999).

2. THE STATIC CASE

2.1 *Model Class*

Let $\boldsymbol{f}$ be the vector of all "true" flows within the considered material flow system. Then the linear, static, symmetric model is given by

$$\boldsymbol{M f} = 0 \tag{1}$$

where in the matrix

$$\boldsymbol{M} = \begin{pmatrix} \boldsymbol{A} \\ \boldsymbol{T} \\ \boldsymbol{R} \end{pmatrix}$$

the submatrix $\boldsymbol{A}$ describes the *balance equations*, the submatrix $\boldsymbol{T}$ the *transfer coefficient equations* and the submatrix $\boldsymbol{R}$ the linearly independent *additional linear relations*, which do not satisfy a balance equation and which are linearly independent of the previous equations. Note that the relations described by (1) are assumed to be deterministic. Each transfer coefficient equation describes an output of a subsystem as a linear combination of its inputs, where the weights are called *transfer coefficients*.

In general neither the matrix $\boldsymbol{M}$ nor all true flow values are known a priori. For the formulation of the flow measurement model a partitioning of $\boldsymbol{f}$ is assumed as $\boldsymbol{f} = (\boldsymbol{z}'\boldsymbol{x}'\boldsymbol{y}')'$ where $\boldsymbol{z}$ denotes the subvector of flows with exact (i.e. noisefree) measurements, $\boldsymbol{x}$ the subvector of flows with noisy measurements $\boldsymbol{x}^*$ and $\boldsymbol{y}$ the subvector of *unmeasured flows*. For the noisy measurements it is assumed that the measurement error $\boldsymbol{x}^* - \boldsymbol{x}$ is normally distributed with zero mean and known variance covariance matrix $\boldsymbol{\Sigma}$ which in most cases will be assumed to be diagonal. Note that the normality assumption raises two problems: First, the positivity constraints are neglected and secondly, a symmetric error distribution is thus assumed which is not corroborated by many actual observations.

In the same way transfer coefficients, which are elements of $\boldsymbol{T}$, can be measured with or without noise or be unobserved. Thus the vector $\boldsymbol{t}$ consisting of all transfer coefficients is partitioned into exactly known, noisily measured and unmeasured coefficients $\boldsymbol{t}_e, \boldsymbol{t}_m$ and $\boldsymbol{t}_u$, respectively. For noisy measurements $t^*_{m,j}$ the measurement error $t^*_{m,j} - t_{m,j}$ for each such transfer coefficient is assumed to be normally distributed with zero mean and known variance. In addition the errors associated with transfer coefficients are assumed to be independent of each other (so that $\boldsymbol{\Omega} = \mathrm{Var}(\boldsymbol{t}^*_m)$ is diagonal) and of the flow measurements. The additional linear relations are assumed to be a priori known.

The a priori information which can be exploited in order to improve the data precision consists of the balance equations, the additional linear relations and those transfer coefficient equations which contain only observed coefficients. Now remove the rows from $\boldsymbol{T}$ which contain unknown elements. If the transfer coefficient equations associated with all outputs of a subsystem are known then in addition remove one of them since all these rows sum up to the corresponding balance equation in $\boldsymbol{A}$. Denote the resulting submatrix of $\boldsymbol{T}$ by $\check{\boldsymbol{T}}$ and let $\check{\boldsymbol{M}} = (\boldsymbol{A}'\check{\boldsymbol{T}}'\boldsymbol{R}')$. Under fairly general conditions (which require a connected system graph with links to the surroundings and the absence of ***traps***, i.e. of strong components which have no output to the rest of the system nor to the surroundings) this matrix has full row rank.

2.2 *Reconciliation and Estimation*

The flow and transfer coefficient measurements are reconciliated by solving the following minimization problem:

$$\min_{\hat{\boldsymbol{x}},\hat{\boldsymbol{y}},\hat{\boldsymbol{t}}_m} \; (\hat{\boldsymbol{x}} - \boldsymbol{x}^*)'\boldsymbol{\Sigma}^{-1}(\hat{\boldsymbol{x}} - \boldsymbol{x}^*) + (\hat{\boldsymbol{t}}_m - \boldsymbol{t}^*_m)'\boldsymbol{\Omega}^{-1}(\hat{\boldsymbol{t}}_m - \boldsymbol{t}^*_m) \tag{2}$$

$$\text{s.t.} \qquad \check{\boldsymbol{M}}\hat{\boldsymbol{f}} = 0 \tag{3}$$

where $\hat{\boldsymbol{f}} = \left(\hat{\boldsymbol{z}}'\hat{\boldsymbol{x}}'\hat{\boldsymbol{y}}'\right)'$ and $\hat{\boldsymbol{z}} = \boldsymbol{z}$.

Since the constraints (3) contain also variables which do not occur in the objective function (2) the following procedure is proposed: Transform $\check{\boldsymbol{M}}$ by a non-singular row transformation $\boldsymbol{S}$ (which leaves the right kernel unchanged) in such a way that the north-eastern zero block in

$$\begin{aligned} \tilde{\boldsymbol{M}} &= \boldsymbol{S}\check{\boldsymbol{M}} \\ &= \begin{pmatrix} \tilde{\boldsymbol{M}}_{1z} & \tilde{\boldsymbol{M}}_{1x} & 0 \\ \tilde{\boldsymbol{M}}_{2z} & \tilde{\boldsymbol{M}}_{2x} & \tilde{\boldsymbol{M}}_{2y} \end{pmatrix} \end{aligned} \tag{4}$$

has a maximal number of rows (The horizontal partitioning in (4) corresponds to that of $\boldsymbol{f}$). For one such $\boldsymbol{S}$ the first block row corresponds to the following equations: First, there are those equations which originally did not involve unmeasured flows. Secondly, those subsystems, which are incident with unmeasured flows, can be grouped in such a way that (i) any two subsystems within a group can be connected by a path of unmeasured flows (not regarding the direction) and (ii) these groups are connected with each other only by measured or exactly known flows. The flows within these groups contain no information relevant to the reconciliation of the measurements. For each group the sum of the corresponding balance equations occurs in the first block row of (4). This corresponds to the contraction of each of these groups to a single subsystem in the so called *Reduced Balance Scheme*, cf. (Mah *et al.*, 1976). Similarly, the first block row contains the coefficients of transformed transfer coefficient equations and additional linear relations.

Theorem 1. Assume that $\boldsymbol{\Sigma}$ is positive definite and exclude the case of transfer coefficients measured with noise. Then the unique solution $\hat{\boldsymbol{x}}$ of (2), (3) for the measured flows is given by

$$\boldsymbol{x}^* - \boldsymbol{\Sigma}\tilde{\boldsymbol{M}}'_{1x}(\tilde{\boldsymbol{M}}_{1x}\boldsymbol{\Sigma}\tilde{\boldsymbol{M}}'_{1x})^{-1}(\tilde{\boldsymbol{M}}_{1x}\boldsymbol{x}^* + \tilde{\boldsymbol{M}}_{1z}\boldsymbol{z})$$

which is an efficient estimate of $\boldsymbol{x}$. Given $\hat{\boldsymbol{x}}$ the estimates $\hat{\boldsymbol{y}}$ obtained from the second block row of (4) are unique if and only if the matrix $\tilde{\boldsymbol{M}}_{2y}$ has full column rank, in which case they are also efficient.

Remark 2. After a suitable reordering of the rows and columns in (4) the matrix $\tilde{\boldsymbol{M}}_{2y}$ is block diagonal so that the identifiability criterion given in theorem 1 can be applied blockwise. For blocks which do not contain coefficients of a priori known transfer coefficient equations and additional linear relations the criterion reduces to the absence of undirected cycles of unmeasured flows.

A common assumption in MFA is that the outputs of a subsystem only depend on the sum of the inputs. This allows to estimate the transfer coefficients as the quotients of the respective flow estimates.

Noisy transfer coefficient measurements can be taken into account by a two step procedure. In the first step only exactly known transfer coefficient equations are taken into account and the above procedure is applied. In the second step the transfer coefficient equations containing noisily measured transfer coefficients are added. For this purpose a suitably chosen linear combination of the respective estimate from the first step and the measurement value is taken as constant. Finally, the reconciliation and identification procedure is repeated.

2.3 *Analysis and Simulation*

Based on the identified model a tool called *origins analysis* has been developed which enables one to find sets of flows which give a complete physical explanation of a considered flow and which do not influence each other physically. Based on a *control analysis*, which gives the influence of flow modifications on other flows taking into account additional linear relations and upstream dependencies, simulations of policy scenarios (including both flow and transfer coefficient modifications) for the considered material flow system can be performed, see (Gleiß *et al.*, 1998).

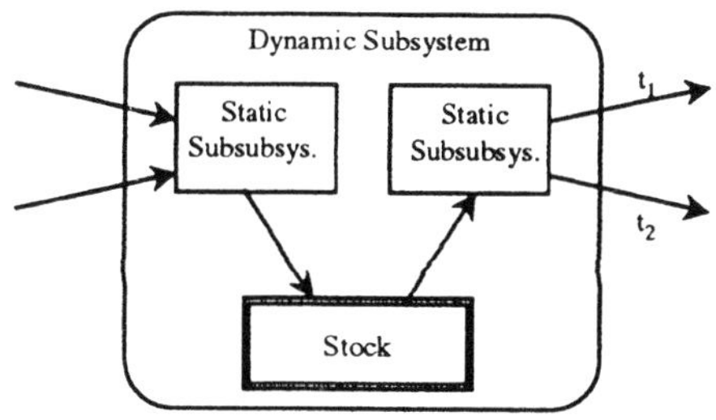

Fig. 1. A dynamic subsystem.

3. THE DYNAMIC CASE

3.1 *Model Class*

In Material Flow Analysis dynamics comes in via stocks, where a portion of the material under consideration enters a subsystem and does not leave it within the same balancing period.

A subsystem of a dynamic material flow system has the inner structure shown in fig. 1. For the "inner" static subsubsystems balance equations hold. Note that the stock behaviour describes a dynamic relationship between the sum of the inputs and the sum of the outputs of the dynamic subsystem. The transfer function for subsystem j (i.e. the j-th subsystem) is thus SISO and in general given by

$$k_j(z) = \sum_{l \geq 0} w_l(\boldsymbol{p}_j)z^{-l}$$

where the weights w_l describe a stock behavior depending on a *stock parameter* $\boldsymbol{p}_j$ and where z^{-1} is the backward shift operator. Here the law of the conservation of mass implies that the weights sum up to a value less than or equal to one. The output of the stock is divided by the right-hand static subsubsystem to the different outputs of the subsystem. Throughout the paper all subsystems are assumed to be properly dynamic in the sense that k_j is not constant.

The stock parameters $\boldsymbol{p}_j$ might be known a priori or have to be estimated from measurement data. The set of parameters to be estimated thus consists of both the a priori unknown stock parameters and the a priori unknown transfer coefficients of the right-hand static subsubsystem in each subsystem.

Remark 3. Note that, if the real phenomenon is properly dynamic but modelled by a static system, then the stock accumulation flows have to be modeled as exports and the stock depletion flows as imports (the stock shown in the figure thus lies outside the system boundary in this case), so that the intertemporal relation is neglected.

(a) The simplest kind of a stock pattern is the case when the outflow during period t is equal to the inflow during period $t - \tau_j$, i.e.

$$w_l(\tau_j) = \begin{cases} 1 \text{ for } l = \tau_j \\ 0 \text{ else} \end{cases}$$

(b) A more flexible stock depletion pattern is given by weights which describe a *discretized truncated normal distribution* and depend on the distribution parameters μ_j and σ_j via

$$w_l(\mu_j, \sigma_j) = \begin{cases} \dfrac{\Phi\left(\frac{l-\mu_j}{\sigma_j}\right)}{\sum_{i=a_j}^{b_j} \Phi\left(\frac{i-\mu_j}{\sigma_j}\right)}, & l = a_j, ..., b_j \\ 0 & \text{else} \end{cases} \quad (5)$$

where $\Phi(x)$ denotes the normal distribution function evaluated at x.

(c) Finally, a *geometrically declining pattern*, which will be suitable for modeling most environmental stocks, is easily described by adding a *loop* (i.e. a flow where the source and the target subsystem are the same) at the considered subsystem. The parameter determining the stock behavior in this case thus is the transfer coefficient associated with this loop.

A (discrete-time) state space model for the material flow system is of the form (neglecting measurement errors)

$$\boldsymbol{x}_{t+1} = \boldsymbol{F}\boldsymbol{x}_t + \boldsymbol{G}\boldsymbol{u}_t$$
$$\boldsymbol{y}_t = \boldsymbol{H}\boldsymbol{x}_t + \boldsymbol{K}\boldsymbol{u}_t$$

Here the vector $\boldsymbol{u}_t$ denotes the imports into the whole system, the vector $\boldsymbol{y}_t$ consists of the observed internal and export flows and the state vector $\boldsymbol{x}_t$ comprises the stock levels of the dynamic subsystems. Thus here the state dimension is given a priori by the number of subsystems m.

Since it is assumed that all subsystems are properly dynamic, in case (a) $\boldsymbol{K}$ is a zero matrix and each non-zero element of the system matrices contains the transfer coefficient from an import or an output of a dynamic subsystem to an observed internal or export flow or to an input of a dynamic subsystem. A subsystem j with $\tau_j > 0$ is modeled by a chain of τ_j subsystems each with stock parameter equal to one, giving an *extended system graph*.

The extended system graph for stock type (b) with $a_j > 1\ \forall j$ is obtained by representing each subsystem j by a system of $b_j - a_j + 2$ subsystems of type (a) (see fig. 2). The third stock type differs from the first only in the occurence of loops, i.e. non-zero diagonal elements in $\boldsymbol{F}$. Thus the first stock pattern can be seen as a prototype. Most of

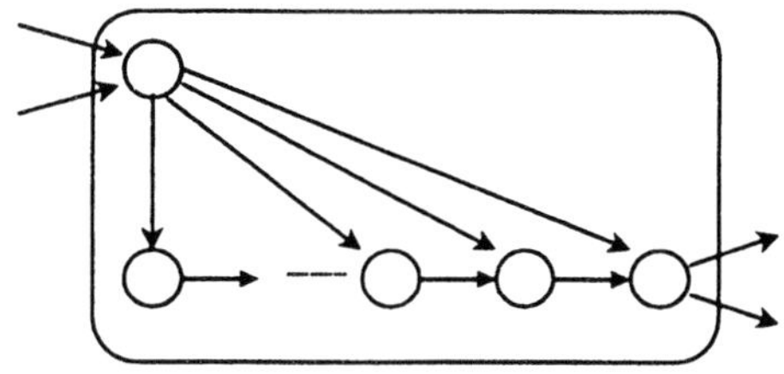

Fig. 2. A subsystem of type (b) with $a_j > 1$ as a system of $b_j - a_j + 2$ subsystems of type (a).

the results for the other cases follow on the basis of some additional assumptions and with only slight modifications. In any case the graph theoretic conditions described below apply directly to the non-extended system graph. In the following a *system of the second (third) stock type* denotes a system which contains subsystems with normally distributed (geometrically declining) stock pattern, but also some of type (a).

3.2 *Stability*

Beside some technical assumptions concerning the system graph it is assumed throughout this section that no stock levels, but at least one internal or export flow and all import flows are observed. Moreover the system theoretic inputs $\boldsymbol{u}_t$ are the imports and all flows are measured or unmeasured in all periods of observation.

Note that the (a priori) non-zero elements of the system matrices, which are the weights of the (extended) system graph, do not vary independently because of the balance equations, the restrictions given by the description of the stock behavior and because of the fact that the transfer coefficient associated with an observed internal flow occurs in $\boldsymbol{H}$ as well as in $\boldsymbol{F}$.

Definition 4. Let $\boldsymbol{\theta} : P \rightarrow N$ be a continuous map from a set P of free parameters to the set $N \subseteq \mathbb{R}^r$ of the r non-zero elements of the system matrices. Then a property that may be assigned to the system matrices is said to hold *$\boldsymbol{\theta}$-generically* if there exists a proper algebraic set $U \subset P$ such that the property holds for all values in N except on a set contained in $\boldsymbol{\theta}(U)$. A *system structure* $[\boldsymbol{F}, \boldsymbol{G}, \boldsymbol{H}, \boldsymbol{\theta}]$ denotes the model class defined by a classification of the elements of the system matrices into non-zero elements, which are parameterized by $\boldsymbol{\theta}$, and zero elements.

By the balance equations, the set of free parameters, P_1 say, for the first stock type is given as follows: For each subsystem k the (non-negative) transfer coefficients associated with its outflows can be described by $n^+(k) - 1$ (non-negative) free parameters ($n^+(k)$ denotes the number of successors of k) the sum of which has to be less than or equal to one. Let the according map be denoted by $\theta_1 : P_1 \to N$. For a system of the second stock type also the restrictions (5) have to be taken into account.

On the basis of the theorems of Perron and Frobenius, of Geršgorin and of Taussky the following is straightforward to prove:

Theorem 5. The system structure $[\boldsymbol{F}, \theta_1]$ is θ_1-generically asymptotically stable if and only if the system graph contains no traps.

The same result can be shown to hold for the normally distributed and the geometrically declining stock patterns, see (Gleiß, 1999).

3.3 *Reachability and Observability*

If in definition 4 the set P is identical to N and θ is the identity θ_0 then for simplicity the considered property is said to hold *generically*. The following well-known characterization of generic reachability dates back to (Shields and Pearson, 1976).

Lemma 6. A system structure $[\boldsymbol{F}, \boldsymbol{G}, \theta_0]$ is generically reachable if and only if the following conditions are satisfied:

A1 The system graph is *input-connectable* (i.e. to each subsystem there exists a directed path starting with some import).
A2 On the system graph there exists a family of disjoint directed cycles and *stems* (i.e. directed paths starting with an import) such that all vertices in the graph are covered.

Showing that θ_1-generic reachability is equivalent to generic reachability leads to the following

Theorem 7. $[\boldsymbol{F}, \boldsymbol{G}, \theta_1]$ is θ_1-generically reachable if and only if conditions A1 and A2 are both satisfied.

Whereas the necessity of the conditions A1 and A2 is obvious also for systems of the second stock type, up to now their sufficiency is not clear. For systems of the third type note that the subsystems with geometrically declining pattern are covered by their respective loops.

It can be shown that the dual analogons to A1 and A2 are necessary and sufficient for θ_1-generic observability.

Let K denote the set of subsystems with more than one output and at least one output which is not an export and $\tilde{V}$ denote the set of subsystems with exactly one output which is not an export. Then A2 can be shown to be equivalent to

A2' For all $S \subseteq K$ there exist $m - |K| + |S|$ different subsystems each of which is either the target of an import or the successor of a subsystem in $S \cup \tilde{V}$.

3.4 *Identifiability*

The following graph theoretic description (Audoly and D'Angio, 1983) of the transfer function $\boldsymbol{H}(z\boldsymbol{I} - \boldsymbol{F})^{-1}\boldsymbol{G}$ is, for simplicity, formulated for the case where $\tau_j = 1 \; \forall j = 1, \ldots, m$. For this purpose let $\varphi^{(l)}$ denote the sum of the weights of all disjoint unions of directed cycles the lengths of which sum up to l. Here the weight of a cycle involving the flows $i_1, \ldots, i_l$ with transfer coefficients $t_{i_1}, \ldots, t_{i_l}$ is defined as $(-1)^{l+1} t_{i_1} t_{i_2} \cdots t_{i_l}$. Analogously define $\rho_{s,t}^{(l)}$ as the sum of the weights of all disjoint unions of a path from t to s and directed cycles the lengths of which sum up to l so that $\rho_{s,t}^{(0)}$ equals one in the case $s = t$ and zero else.

Proposition 8. Let t be the target of the k-th import and s be the source of the j-th observed internal or export flow (with transfer coefficient t_j). Then the (j,k)-element of the nominal transfer function is given by

$$\frac{t_j \sum_{l=0}^{m-1} (-1)^{m-l} \rho_{s,t}^{(m-l-1)} z^l}{z^m + \sum_{l=0}^{m-1} (-1)^{m-l} \varphi^{(m-l)} z^l} \tag{6}$$

Remark 9. In the nominal transfer function per definition all denominators equal the determinant. For simplicity it is assumed that the transfer function is identical with the nominal transfer function. It can be shown that this implies that all subsystems are input- or output-unconnectable and that either $\varphi^{(m)} \neq 0$ or $\rho_{s,t}^{(m-1)} \neq 0$ for some target t of an import and some source s of an observed internal or export flow. The question whether this assumption implies further restrictions on the system graph is still open.

In the following all parameters are assumed to be a priori unknown. Based on the analogous result for general but a priori known τ_j the idenfiability problem is solved in two steps: First, the equations, which describe the $\varphi^{(\cdot)}$ and the $\rho_{\cdot,\cdot}^{(\cdot)}$ as the respective sums of products of transfer coefficients, are solved. These sums have only one summand under the assumption that there are no cycles with the same length and, for each pair (t, s) of the target t of an import and the source s of an observed internal or export flow, no paths of the same length from t to s. Thus taking logarithms leads to a linear system of equations, the solution of which is not unique (unless $V^* = E_0 = \emptyset$, see below). In the second step the balance equations are applied in order to solve for the free parameters.

For the formulation of the identifiability result denote by E_0 the set of flows which are contained in no directed cycle and no path from the target of an import to the target of an observed internal or export flow. Let V^* be the set of subsystems k such that there is a flow incident to k which is not an element of E_0 and there is no observed flow entering k. Finally, let V_1^* and V_1^- denote the subsets of V^* and of the set of the predecessors of V^*, respectively, the elements of which have no outflows in E_0, and set $V_2^* = V^* \setminus V_1^*$.

Theorem 10. The system $[\boldsymbol{F}, \boldsymbol{G}, \boldsymbol{H}, \boldsymbol{\theta}_1]$ is $\boldsymbol{\theta}_1$-generically identifiable if and only if
(i) No two flows contained in E_0 have the same source subsystems
and either
(ii) $V_2^* = \emptyset$ holds
or
(iii) There exists a set of vertex-disjoint paths from V_1^- to V_2^* either of length one or via V_1^* which covers all subsystems in V_2^*.

It can be shown that similar conditions guarantee identifiability in systems of the second stock type if the occurence of subsystems of type (b) are restricted in a certain manner. For systems of the third type the conditions of theorem 10 are equivalent only to local identifiability, see (Gleiß, 1999).

4. CONCLUSIONS

For the static material flow model a rather complete treatment of data reconciliation has been presented, procedures for parameter estimation and policy simulation have been sketched. The proposed methods take into account all available a priori information as well as the measurement data. Their applicability, however, is limited to the cases where the assumptions of both linearity of the input-output relations and normality of the flow measurements are justified. Useful extensions could be a tool for the detection of systematic measurement errors and leaks and a reconciliation procedure which implicitly takes such errors into account, see e.g. (Crowe, 1988). Another question of practical interest relates to the improvement (in a minimal variance sense) of the estimate of a given flow by more accurate or additional measurements of other flows in the system. In this context, but also in the context of control analysis and simulation, links with cost efficiency and evaluation considerations could be of interest.

Also in the dynamic case, it is possible to check important system theoretic properties of certain classes of material flow systems simply by inspecting the flow sheet of the system under consideration. It remains to elaborate methods for the estimation of the transfer coefficients and the dynamic parameters from flow measurement data. Furthermore it has yet to be examined whether the model classes considered so far are general enough to be suitable for being applied to actual data and how robust the presented results prove against violation of the underlying assumptions. Finally, appropriate simulation and prediction tools have to be developed.

ACKNOWLEDGEMENTS

Support by the Austrian "Fonds zur Förderung der wissenschaftlichen Forschung" (project no. P 12926-MAT) and the EU project MAcTEmPo is gratefully acknowledged.

5. REFERENCES

Audoly, S. and L. D'Angio (1983). On the identifiability of linear compartmental systems: A revisited transfer function approach based on topological properties. *Mathematical Biosciences* **66**, 201–228.

Baccini, P. and P. H. Brunner (1991). *The Metabolism of the Anthroposphere.* Springer.

Bauer, G., M. Deistler, A. Gleiß, E. Glenck and T. Matyus (1997). Identification of material flow systems. *Envir. Sci. and Pollut. Res.* **4**(2), 105–112.

Crowe, C.M. (1988). Recursive identification of gross errors in linear data reconciliation. *AIChE Journal* **34**(4), 541–549.

Gleiß, A. (1999). Identification of Linear Material Flow Systems - A Graph Theoretic Approach. PhD thesis. Technial University, Vienna.

Gleiß, A., T. Matyus, G. Bauer, M. Deistler, E. Glenck and C. Lampert (1998). Identification of material flow systems - extensions and case study. *Envir. Sci. and Pollut. Res.* **5**(4), 245–258.

Mah, R.S., G.M. Stanley and D.M. Downing (1976). Reconciliation and rectification of process flow and inventory data. *Ind. Eng. Chem., Process Des. Dev.* **15**(1), 175–183.

Shields, R.W. and J.B. Pearson (1976). Structural controllability of multiinput linear systems. *IEEE Trans. Automatic Control* **21**(2), 203–212.

Van der Voet, E., R. Kleijn, L. Van Oers, R. Heijungs, R. Huele and P. Mulder (1995). Substance flows through the economy and environment of a region - part 1: Systems definition. *Envir. Sci. and Pollut. Res.* **2**(2), 90–96.

TIME VARIABLE AND STATE DEPENDENT PARAMETER ESTIMATION

Peter C. Young [1]

Centre for Research on Environmental Systems and Statistics, Lancaster University, Lancaster LA1 4YQ, U.K

Abstract: Over the last forty years, much attention has been given to the estimation of time variable parameters in models of dynamic systems. This paper outlines how improved estimates of time variable parameters can be obtained using recursive fixed interval smoothing techniques. It then shows how similar techniques can be exploited to estimate state dependent parameter variations and so identify a widely applicable class of nonlinear stochastic systems, which includes chaotic processes.

Keywords: Nonlinear, Stochastic, Identification, Non-parametric, Estimation

1. INTRODUCTION

Previous publications (e.g. Young, 1978, 1984, 1993a,b 1998a,b, 1999a,b; Young and Beven, 1994; Young and Pedregal, 1998, 1999; Young *et al.*, 1989, 1999) have discussed an approach to nonstationary and nonlinear signal processing based on the identification and estimation of stochastic models with time variable (TVP) or, latterly, state dependent (SDP) parameters. Here the term 'nonstationarity' is assumed to mean that the statistical properties of the signal, as defined by the parameters in an associated stochastic model, are changing over time at a rate which is 'slow' in relation to the rates of change of the stochastic state variables in the system under study. Although such nonstationary systems exhibit nonlinear behaviour, this can often be approximated well by TVP (or piece-wise linear) models, the parameters of which can be estimated using recursive methods of estimation in which the parameters are assumed to evolve in a simple stochastic manner. On the other hand, if the changes in the parameters are functions of the state or input variables (i.e. they actually constitute stochastic state variables), then the system is truly nonlinear and likely to exhibit severe nonlinear behaviour. Normally, this cannot be approximated in a simple TVP manner; in which case, recourse must be made to the alternative, and more powerful SDP modelling methods that are the main topic of this paper.

The first identification stage in this SDP modelling process exploits recursive *Fixed Interval Smoothing* (FIS) algorithms, combined with special data re-ordering and 'back-fitting' procedures, to obtain estimates of any state dependent parameter variations. These state dependencies are estimated in the form of non-parametric relationships (graphs) between the estimated rapid parameter variation and the associated state or input variable(s). Parameterization of these non-parametric relationships can be accomplished in various ways: from simple curve fitting based on weighted least squares methods, to the use of neural networks or radial basis functions. Having identified a structural form for the nonlinear model of the system based on the parameterized nonlinear relationships, this model is converted into a stochastic state space form. The final estimation phase of the nonlinear modelling then exploits *Maximum Likelihood* (ML) methods of estimation, based on Gaussian Assumptions for the stochastic distur-

[1] Also Centre for Resource and Environmental Studies, Australian National University, Canberra, ACT 2600, Australia

bances and the application of a technique known as *Prediction Error Decomposition*.

In order to illustrate the practical application and utility of the SDP estimation approach, the paper contains a simulation example and a practical study involving a re-analysis of the famous Nicholson blowfly data. Other simulation and practical examples cited in the references cover a variety of application areas from the environment through engineering to economics.

2. STATE DEPENDENT PARAMETER ESTIMATION

Optimal TVP estimation works very well in situations where the parameters are slowly varying when compared to the observed temporal variation in the measured system inputs and outputs. Although such TVP models are nonlinear systems, since the same inputs, injected at different times, will elicit quite different output responses, the resultant nonlinearity is fairly mild. It is only when the parameters are varying at a rate commensurate with that of the system variables themselves that the model behaves in a heavily nonlinear or even chaotic manner. The paper will show how such *State Dependent Parameter* (SDP) models can be used to represent a wide variety of nonlinear stochastic systems and time series.

2.1 *SDARX Estimation*

A typical SDP model[2] is the *State Dependent Auto-Regressive eXogenous variables* (SDARX) model, which takes the following general form,

$$y_t = \mathbf{z}_t^T \mathbf{p}_t + e_t \qquad e_t = \mathrm{N}(0, \sigma^2), \qquad (1)$$

where now

$$\mathbf{z}_t^T = [-y_{t-1} \; -y_{t-2} \; \cdots \; -y_{t-n} \; u_{t-\delta} \; \cdots \; u_{t-\delta-m}]$$
$$\mathbf{p}_t = [a_1(\boldsymbol{\chi}_t) \; a_2(\boldsymbol{\chi}_t) \; \cdots \; a_n(\boldsymbol{\chi}_t) \; b_0(\boldsymbol{\chi}_t) \; \cdots \; b_m(\boldsymbol{\chi}_t)]^T$$

while $a_i(\boldsymbol{\chi}_t)$, $i = 1, 2, \ldots, n$ and $b_j(\boldsymbol{\chi}_t)$, $j = 0, 2, \ldots, m$ are the state dependent parameters, which are assumed to be functions of a non-minimal state vector $\boldsymbol{\chi}_t^T = [\mathbf{z}_t^T \; \mathbf{U}_t^T]$. Here $\mathbf{U}_t = [U_{1,t} \; U_{2,t} \; \cdots \; U_{r,t}]^T$ is a vector of other variables, not necessarily direct functions of y_t and u_t that may affect the relationship between these two primary variables.

The simpler signal topology for this 'affine' model means that FIS estimation of the SDPs can be based on the standard recursive least squares (RLS) filtering/smoothing equations (e.g. Bryson and Ho, 1970; Young, 1984). Although these recursive algorithms are nominally limited to linear systems, their ability to estimate time variable and state dependent parameters, as discussed below, means that they can be applied to time series data generated by nonlinear stochastic systems.

1. Forward Pass Recursive LS Equations

Prediction:

$$\hat{\mathbf{x}}_{t|t-1} = \mathbf{F}\hat{\mathbf{x}}_{t-1}$$
$$\hat{\mathbf{P}}_{t|t-1} = \mathbf{F}\hat{\mathbf{P}}_{t-1}\mathbf{F}^T + \mathbf{G}\mathbf{Q}_r\mathbf{G}^T.$$

Correction:

$$\hat{\mathbf{x}}_t = \hat{\mathbf{x}}_{t|t-1} + \boldsymbol{\Gamma}_t \left\{ y_t - \mathbf{H}_t \hat{\mathbf{x}}_{t|t-1} \right\}$$
$$\mathbf{P}_t = \mathbf{P}_{t|t-1} - \boldsymbol{\Gamma}_t \mathbf{H}_t \mathbf{P}_{t|t-1}$$
$$\boldsymbol{\Gamma}_t = \mathbf{P}_{t|t-1}\mathbf{H}_t^T \left[1 - \mathbf{H}_t \mathbf{P}_{t|t-1}\mathbf{H}_t^T\right]^{-1}$$

2. Backward Pass IV Smoothing Equations (single pass)

$$\hat{\mathbf{x}}_{t|N} = \mathbf{F}^{-1}\left[\hat{\mathbf{x}}_{t+1|N} + \mathbf{G}\mathbf{Q}_r\mathbf{G}^T\mathbf{L}_t\right]$$
$$\mathbf{L}_t = \left[\mathbf{I} - \mathbf{P}_{t+1}\mathbf{H}_{t+1}^T\mathbf{H}_{t+1}\right]^T\left[\mathbf{F}^T\mathbf{L}_{t+1} - \mathbf{H}_{t+1}^T\epsilon_t\right]$$
$$\epsilon_t = y_{t+1} - \mathbf{H}_{t+1}\hat{\mathbf{x}}_{t+1}$$
$$\mathbf{P}_{t|N} = \mathbf{P}_t + \mathbf{P}_t\mathbf{F}^T\mathbf{P}_{t+1|N}^{-1}\boldsymbol{\Delta}\mathbf{P}_{t+1|t}^{-1}\mathbf{F}\mathbf{P}_t$$
$$\boldsymbol{\Delta} = \left[\mathbf{P}_{t+1|N} - \mathbf{P}_{t-1|t}\right]$$
$$\mathbf{L}_N = 0.$$

where the state dependent parameters are elements of the state vector $\mathbf{x}_t$, the form of which depends on the assumed stochastic model for the parameter variations. Since the parameter vector $\mathbf{p}_t$ is potentially state-dependent, however, $\mathbf{x}_t$ may vary at a rate commensurate with the temporal variations in y_t, u_t and $U_{i,t}$, and so it cannot be assumed that the simple random walk-type models for the parameter variations, of the kind used in conventional TVP estimation, are appropriate to describe the parametric variation over time. In particular, it would appear, at first sight, that the stochastic state space model should include prior information on the nature of the parameter variation if the TVP estimation methodology is to work satisfactorily. Fortunately, it is possible to remove this requirement if we resort to the rather unusual procedure of sorting the data in a *non-temporal order*. Then, if the ordering is chosen so that the SDP variations associated with the *sorted* series are smoother and less rapid, it is more likely that a simple random walk-type process can be utilized to describe their evolution.

---

[2] for others, see Young (1999b). Note that the related SDAR model is handled in the same manner as the SDARX model: see later simulation example.

For example, if the time series are sorted in some common 'ascending order of magnitude' manner (i.e. the *sort* operation in Matlab), then the rapid natural variations in y_t and u_t are effectively eliminated from the data and replaced, in the sorted data space, by much smoother and less rapid variations. And if the SDP's are, indeed, related to these variables, then they will be similarly affected by the sorting. Following FIS estimation, however, these SDP estimates can be 'unsorted' (a trivial *unsort* operation to reverse Matlab's *sort*) and their true, rapid variation will become apparent. Of course, the nature of the sorting will affect the estimation and it seems likely that there will be an optimum sorting which results in minimum variance estimates. However, this will naturally depend upon the nature of the state dependency and its definition would require some sort of iterative estimation procedure. This is not considered here.

One obvious requirement with this new approach to SDP estimation is that the sorting of data, prior to FIS estimation, must be *common to all of the variables in the relationship* (1). If an ascending order strategy is selected, therefore, it is necessary to decide upon which variable in the model the sorting should be based. The simplest strategy is to sort according to the ascending order of the 'dependent' variable y_t. Depending upon the nature of each SDP in the vector $\mathbf{p}_t$, however, a single variable sorting strategy, such of this, may not produce satisfactory results. If this is the case, then a more complicated, but still straightforward, 'backfitting' procedure can be exploited, as outlined below.

Since the SDP estimates resulting from this backfitting algorithm are themselves time series, it will be noted that the algorithm constitutes a special form of non-parametric estimation and, as such, can be compared with other non-parametric estimation methods. However, in both conceptual and algorithmic terms, the SDP approach described here is significantly different from other approaches and seems more appropriate to the estimation of nonlinear, stochastic, dynamic models. Moreover, the recursive methodology, on which SDP estimation is based, is couched in optimal maximum likelihood terms that seem more elegant and flexible than the 'scatter-plot smoothing' procedures used in more conventional non-parametric estimation procedures.

Let $\mathbf{p}_t$ be defined as follows,

$$\mathbf{p}_t = [p_{1,t} \; p_{2,t} \; \cdots \; p_{n+m+1,t}]^T, \qquad (2)$$

with the $p_{i,t}$, $i = 1, 2, \ldots, n + m + 1$, relating to the SDARX model parameters $a_{i,t}$ and $b_{j,t}$ through (1). The backfitting algorithm then takes the following form:

- Assume that FIS estimation has yielded prior TVP estimates $\hat{p}^0_{i,t|N}, i = 1, 2, \cdots, m + n + 1$ of the SDPs.[3]
- Iterate: $i = 1, 2, \cdots, m+n+1; k = 1, 2, \cdots, k_c$
  (i) form the partial residual

$$y^i_t = y_t - \sum_{j \neq i} z_{j,t}.\hat{p}^k_{j,t} \qquad (3)$$

  (ii) sort[4] both y^i_t and $z_{i,t}$ according to the ascending order of $z_{i,t}$
  (iii) obtain an FIS estimate $\hat{p}^k_{i,t}$ of $p_{i,t}$ in the partial residual relationship $y^i_t = p_{i,t}.z_{i,t}$
- Continue 2. (each time forming the partial residual and then sorting according to the current right hand side variable $z_{i,t}$, prior to FIS estimation), until iteration k_c, when the individual SDPs (which are each time-series of length N) have not changed significantly according to some chosen criterion. The smoothing hyper-parameters required for FIS estimation at each stage are optimized by Maximum Likelihood (ML) using prediction error decomposition (see Schweppe, 1965).

Note that the ML optimization can be carried out in various ways: after every complete iteration (each involving $m+n+1$ FIS operations) until convergence is achieved; only at the initial complete iteration, with the hyper-parameters maintained at these values for the rest of the backfitting; or just on the first two iterations. The latter choice seems most satisfactory in practice, since very little improvement in convergence occurs if optimization is continued after this stage. Normally, convergence is completed after only a few iterations.

2.2 *SDAR Simulation Example: Identification and Estimation of the Cosine Map Model*

The cosine map model takes the form:

$$y_t = \cos(2.8 y_{t-1}) + 0.3 y_{t-2} + e_t$$
$$e_t = \mathrm{N}(0, 0.01)$$

This a typical stochastic model that exhibits underlying chaotic response characteristics shown in Figure 1.

[3] As a default, these can be simply the constant least squares parameter estimates, since the convergence of the backfitting procedure is not too sensitive to the prior estimates.

[4] depending on the nature of the state dependency, sorting may need to be with respect to another variable associated with this state dependency

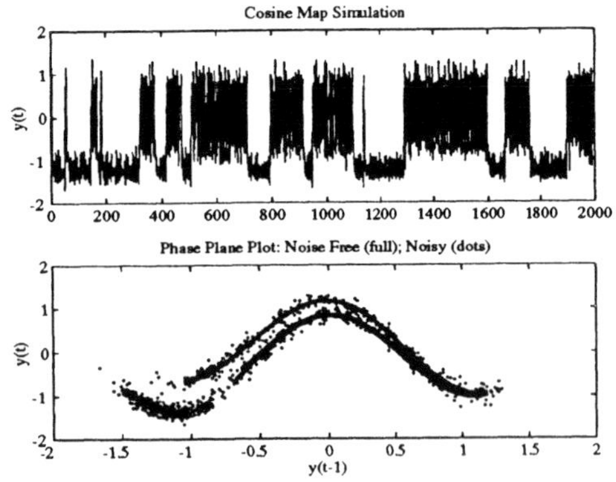

Fig. 1. Typical cosine map simulation

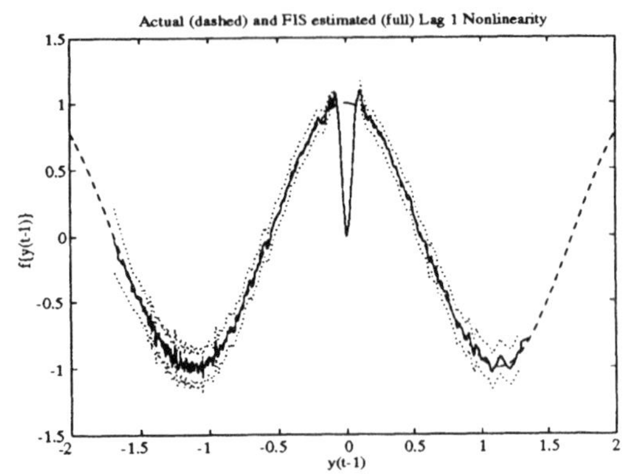

Fig. 2. Comparison of FIS estimated (full line) and actual cosine nonlinearity (dashed). The standard error bands shown as dotted lines.

This provides a testing example for the SDP approach, however, because it is not in the assumed affine form: in particular, the SDP term $a_1(y_{t-1}).\, y_{t-1}$ in the most appropriate SDAR model,

$$y_t = a_1(y_{t-1}).\, y_{t-1} + a_2(y_{t-2}).\, y_{t-2} + e_t$$
$$e_t = \mathrm{N}(0, \sigma^2),$$

is not able to represent the equivalent term $\cos(2.8y_{t-1})$ exactly, since $\cos(2.8y_{t-1})/y_{t-1}$ has a singularity at $y_{t-1} = 0$. Despite this difficulty, SDAR estimation yields excellent results, as illustrated in Figures 2 and 3.

These were obtained with an ML optimized Noise Variance Ratio (NVR) matrix $\mathbf{Q}_r = \text{diag}[0.0057\; 1.28 \times 10^{-7}]$. Figure 2 shows the FIS estimate of the cosine nonlinearity subsequent to the convergence of the backfitting procedure, which took 6 iterations in this case. Except for the region around the singularity at $y_{t-1} = 0$, the estimation is very good. The associated $\hat{a}_2(y_{t-2} \mid N) = 0.291(0.0038)$, for all t, is estimated as being time-invariant since the estimated NVR matrix element is small in this case (1.28×10^{-7}). Figure 3 compares the actual, simulated phase plane plot for the data used in the estimation (left hand panel) with a similar plot based on data from a typical random realization of the SDAR model (right hand panel).

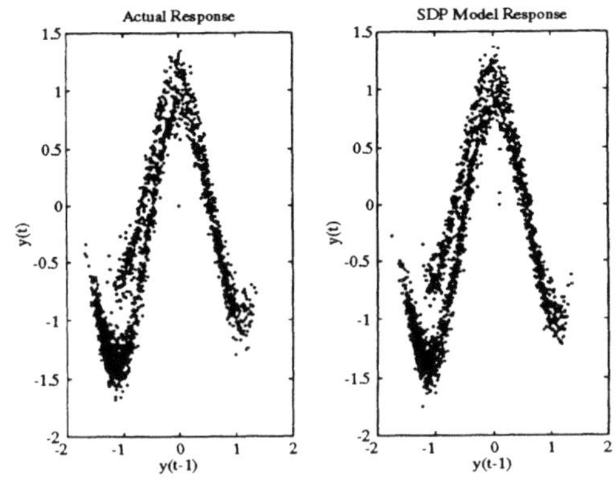

Fig. 3. Comparison of the phase-plane plot for the data used in the estimation (left panel) with that obtained from a random realization of the estimated SDAR model

Finally, on the basis of the above SDP results, the form of the nonlinear equation is identified correctly as:

$$y_t = \alpha \cos(\beta y_{t-1}) + \gamma y_{t-2} + e_t, \tag{4}$$

and the optimized ML estimates of the, now constant, parameters α, β and γ in this model are $\hat{\alpha} = 0.998(0.004)$, $\hat{\beta} = 2.797(0.004)$ and $\hat{\gamma} = 0.303(0.003)$.

2.3 *Real Example: The Nicholson Blowfly Data Revisited*

This example is concerned with the analysis of experimental data collected by Nicholson in his investigation of the Australian sheep blowfly *Lucilia Cuprina* (Nicholson, 1954). In this particular experiment, the food (liver) supplied to the blowflies was limited to $f_d = 0.5$g per day, so that the adult blowfly population y_t and the eggs laid per day by the blowflies u_t vary in an apparently systematic fashion that is redolent of nonlinear, limit cycle behaviour. Not surprisingly, these data have received much attention in the scientific literature (e.g. May, 1973; Gurney *et al.*, and the references therein).

In contrast to the analysis described in the previous references on this topic, however, the SDP modelling approach used here makes no *a priori* assumptions about the nature of the blowfly system but starts with a relatively non-prejudicial analysis of the data, recognising only that the eggs and blowfly series are causally related in some manner. The dynamic relationship between eggs laid per day and the adult blowfly population is identified as linear. Standard Transfer Function (TF) estimation yields a linear, first order TF

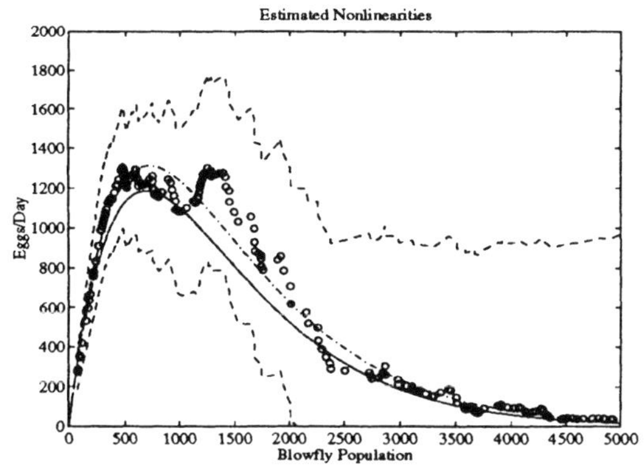

Fig. 4. FIS estimated feedback nonlinearity (circle points) compared with the weighted least squares estimate (dash-dot line) and final ML estimate (full line) of a suitable exponential parameterization of the nonlinear function (see text). The standard error bands on the FIS estimate are shown as dashed lines

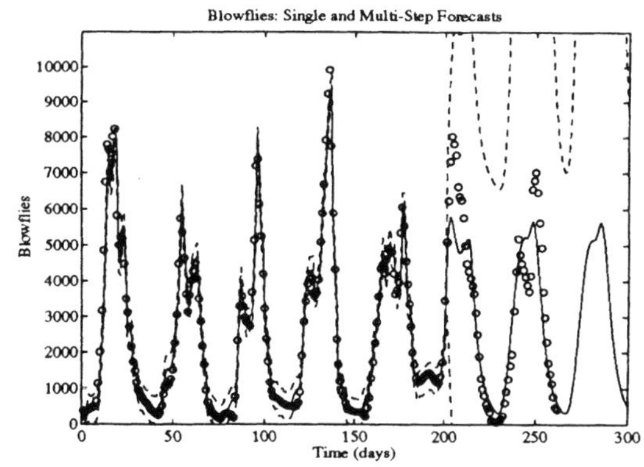

Fig. 5. Kalman filter forecasting results: forecast (full line) compared with the data (circle points), with the standard error band shown as dashed lines. Up to day 200, the forecasts are one-step-ahead; after this, the forecast is a true *ex-ante*, multi-step-ahead forecast, based only on the first 200 days of data

model with a pure time delay of 15 days that allows for the larval stage in the blowfly life-cycle. But the modelling story does not end with this linear model of the 'forward-path' dynamics: of much more interest is the nature of the 'feedback-path' dynamics, namely the mechanism by which the blowflies produce their eggs. In contrast to the forward-path, this mechanism is obviously nonlinear, with the blowflies producing eggs only when their numbers are low and the food supply per blowfly is plentiful. SDP estimation yields the estimate of the nonlinearity shown in Figure 4 as a function of y_t, with the standard error bounds shown dashed. Also shown as a dash-dot line is the WLS estimate of the nonlinearity based on an exponential-type parameterization (see below).

Having used SDP estimation to identify a suitable nonlinear model for the blowfly data, it is now possible to move on to the final ML estimation stage in the analysis. In this case, the nature of the coloured noise on the data requires that the model is enhanced to include stochastic elements. ML optimization of the resulting stochastic model yields the following result:

$$\begin{aligned} x_t &= 0.818x_{t-1} + u_{t-15} + \xi_t \\ u_t &= 4.63x_t.\exp\left(\frac{-1}{1392\frac{fd}{x_t}}\right) \\ \xi_t &= 1.137\xi_{t-1} - 0.491\xi_{t-2} + \varepsilon_t \\ y_t &= x_t + e_t \qquad e_t = \mathrm{N}(0,\sigma^2), \end{aligned} \tag{5}$$

where $\sigma^2 = 1.52 \times 10^5$ and $\varepsilon_t = \mathrm{N}(0, 44.5\sigma^2)$. In this ML optimization stage, the model (5) is formulated as the associated 17th order, discrete-time, stochastic state equation (15 orders account for the pure 'larval' time delay and 2 represent the noise dynamics). The optimization is then performed using prediction error decomposition based on a state dependent parameter implementation of the Kalman filter *that does not, therefore, require linearization*. The nonlinearity associated with this optimized model is shown as the full line in Figure 4 and we see that it is similar to the WLS estimate from the non-parametric estimation stage in the identification analysis, laying well within the standard error bounds.

This ML optimized model performs well in forecasting and validation tests. Figure 5, for exam-' provides typical forecasting results, with one s. ahead forecasts up to the 200th day and t. multi-step *ex-ante* forecasts thereafter. Also, explicit inclusion of fd in (5) is useful because .. allows us to evaluate the ecological realism of the model still further by examining its prediction of what should happen if the food supply is modified. In particular, Nicholson (1954) showed experimentally that, when the food supply was reduced from 0.5g per day to 0.1g per day, the average adult population dropped from 2520 to 527. In the case of the model (5) the average populations in these two same situations are 2657 and 548 respectively; a remarkable level of agreement in the circumstances (since the average values here are based on the deterministic limit cycle data produced by (5) when the stochastic inputs are removed, rather than the actual data). Certainly this result provides a suitable initial validation of the model.

3. CONCLUSIONS

The SDP approach to the identification of nonlinearities in stochastic systems appears to hold great promise. In contrast to other approaches, such as neural networks and NARMAX models, for example, it attempts to identify the type of nonlinearity and, therefore, the form of the nonlinear model, prior to the estimation of the parameters in the finally identified model. This helps to ensure that the final nonlinear model is efficiently parameterized (parsimonious) and it should avoid the over-parameterization that normally accompanies neural network and, to a lesser extent, NARMAX models.

This SDP approach also provides a non-parametric model that can be useful in its own right. For example, the non-parametric SDP model can be simulated easily in programs such as Simulink, thus removing the need for the final parametric estimation in some applications, such as simulation, forecasting and automatic control. In the latter case, for instance, it is clearly possible to develop state estimation and control system design methods based on this new class of nonlinear models, in parametric or non-parametric form (the latter providing a completely new way of considering control and estimation system design). Research on such developments is continuing and has so far led to encouraging initial results.

ACKNOWLEDGEMENTS

The author is grateful for the support of the UK Engineering and Physical Sciences and the Biotechnology and Biological Research Councils under grants GR/G 53866, GR/J 10136 and EO6813. The paper was completed while the author was at the Isaac Newton Institute for Mathematical Sciences, Cambridge, England.

REFERENCES

Bryson, A. E. and Y. C. Ho (1969). *Applied Optimal Control.* Blaisdel: Waltham, Mass..

Gurney, W. S. C., P. C. Blythe and R. M. Nisbet (1980). Nicholson's blowflies revisited. *Nature,* **287**, 17-21.

May, R. M. (1973) *Stability and Complexity in Model Ecosystems.* Princeton University Press: Princeton.

Ng, C. N. and P. C. Young (1990). Recursive estimation and forecasting of nonstationary time-series. *Jnl. of Forecasting,* **9**, 173-204.

Nicholson, A. J. (1954). An outline of the dynamics of animal populations'. *Australian Zoological Jnl.,* **2**, 9-65.

Schweppe, F. (1965). Evaluation of likelihood function for Gaussian signals. *IEEE Transactions on Information Theory,* **11**, 61-70.

Young, P. C. (1978). A general theory of modeling for badly defined dynamic systems. In: *Modeling, Identification and Control in Environmental Systems* (G.C. Vansteenkiste, ed.). North Holland: Amsterdam, 103-135.

Young, P. C. (1984). *Recursive Estimation and Time-Series Analysis.* Springer-Verlag: Berlin.

Young, P. C. (1993a). Time variable and state dependent modelling of nonstationary and nonlinear time series. In: *Developments in Time Series Analysis* (T. Subba Rao. ed.). Chapman and Hall: London, 374-413.

Young, P. C. (1993b). *Concise Encyclopedia of Environmental Systems,* Pergamon Press: Oxford.

Young, P. C. (1998a). Data-based mechanistic modelling of environmental, ecological, economic and engineering systems. *Environmental Modelling and Software,* **13**, 105-122.

Young, P. C. (1998b). Data-based mechanistic modelling of engineering systems. *Journal of Vibration and Control,* **4**, 5-28.

Young, P. C. (1999a). Nonstationary time series analysis and forecasting. *Progress in Environmental Science,* **1**, 3-48.

Young, P. C. (1999a). Nonstationary time series analysis and forecasting. *Progress in Environmental Science,* **1**, 3-48.

Young, P. C. (1999b). Stochastic, dynamic modelling and signal processing: time variable and state dependent parameter estimation. In: *Nonstationary and Nonlinear Signal Processing* (W. J. Fitzgerald, A. Walden, R. Smith and P. C. Young, eds.). Cambridge University Press: Cambridge, in press.

Young, P. C. and K. J. Beven (1994). Data-based mechanistic modelling and the rainfall-flow nonlinearity. *Environmetrics,* **5**, 335-363.

Young, P. C. and D. J. Pedregal (1998). Recursive and en-bloc approaches to signal extraction. *Journal of Applied Statistics,* **26**, 103-128.

Young, P. C. and D. J. Pedregal (1999). Macro-economic relativity: government spending, private investment and unemployment in the USA 1948-1998. *Jnl. Structural Change and Economic Dynamics* **10**, 359-380.

Young, P. C., C. N. Ng and P. Armitage (1989). A systems approach to economic forecasting and seasonal adjustment. *International Journal on Computers and Mathematics with Applications,* **18**, 481-501.

Young, P.C., D. J. Pedregal and W. Tych (1999). Dynamic harmonic regression. *Journal of Forecasting,* **18** , 369-394.

SYSTEM IDENTIFICATION OF AN OPEN WATER CHANNEL

Erik Weyer

Department of Electrical and Electronic Engineering
University of Melbourne, Parkville VIC 3010
Email: e.weyer@ee.mu.oz.au

Abstract: In this paper we derive models of the water level in an irrigation channel from system identification experiments. We present the complete system identification procedure from experiment design to model validation, taking into account prior physical information and that the intended use of the models is prediction and control. It is shown that a first order linear model captures the main trends in the data well, and together with prior information this model is probably sufficient for design of standard controllers. It is also shown that a higher order nonlinear model gives remarkably accurate predictions. *Copyright © 2000 IFAC*

Keywords: System identification, irrigation channels, environmental systems.

1. INTRODUCTION

Water is becoming an increasingly scarce resource throughout many parts of the world today, and it is therefore important to manage the water resources well and minimise the losses. This is particularly important in networks of irrigation channels, where the losses can be large.

As the level of instrumentation and automation in channel networks increase there is a large potential for reducing the losses via better prediction and control of the irrigation channels. Control of irrigation channels is attracting increased attention as is evident from the special issue of Journal of Irrigation and Drainage Engineering on canal automation [1], and the special session at the IEEE Systems, Man and Cybernetics conference in 1998 on modelling and control of water channels [2]. In most irrigation channels the water levels are the controlled variables, and the gate positions the manipulated variables. An alternative is volume control, but as volumetric and flow measurements are rarely available throughout a channel system, water level control remains the most common type of control, and besides the volume of water in a channel between two gates is roughly proportional to the water level.

Prediction and control of irrigation channels require models, and open channels have traditionally been modelled by the St. Venant equations which are nonlinear hyperbolic partial differential equations (see e.g. Cunge et al (1980)). The equation are derived from mass and momentum balances and are given by

$$\frac{\partial A}{\partial t} + \frac{\partial Q}{\partial x} = 0$$

$$\frac{\partial Q}{\partial t} + \left(\frac{gA}{B} - \frac{Q^2}{A^2}\right)\frac{\partial A}{\partial x} + \frac{2Q}{A}\frac{\partial Q}{\partial x} + gA(S_f - \bar{S}) = 0$$

where A is the cross sectional area of the channel, B is the width of the water surface, Q is the flow (discharge), g is the gravity, S_f is the friction slop, and $\bar{S}$ is the mean bed slope. The friction slope is a nonlinear function of the channel geometry. In the above equation it is assumed that the lateral inflow is zero. The St. Venant equations are known to be accurate from laboratory experiments (Brutsaert(1971)). The real world is however different from an ideal laboratory environment, and in practice channel geometry and friction coefficients are not exactly known, and they vary in both space and time. In addition the St. Venant equations are partial differential equations and can be too computationally demanding for prediction and control purposes, and moreover they do not describe the flow across a gate.

Nevertheless, despite their unknown accuracy in a real world environment, the St. Venant equations together with some equations for the flow across the gates are usually taken as starting point for developing models for control design. For example, for the two test cases for canal control algorithms developed by ASCE (American Society of Civil Engineers) Task Committee for on Canal Automation (Clemmens et al (1998)) it is specified that a simulation model of the St. Venant equations should be used.

As far as the author is aware of, very few, if any of the models used for control design have been put to the test against real data, and hence the accuracy and suitability of these models remain largely unknown.

From a control and prediction point of view it is important to have models that describe the real system adequately, and since operational data are widely available, irrigation channels are ideally suited for system identification. Using system identification we are able to obtain models which are in agreement with the physical reality and useful for prediction and control. In this paper we present the complete system identification procedure for irrigation channels from experiment design to model validation, taking into account prior physical information. This work is part of an ongoing research project between the University of Melbourne and Rubicon Systems, Australia on system identification and control of networks of irrigation channels.

The paper is organised as follows. In the next section, the requirements to the model and the prior information are discussed. The following sections follow the standard system identification procedure shown in Figure 1 (see e.g. Ljung (1999) or Söderström and Stoica (1988)) and present experiment design, model structure selection, parameter estimation and model validation. In section 6 more prior information is taken into account, and the system identification procedure is restarted from the model structure selection box using a more complex model. This illustrates the iterative nature of system identification, and also the grey box aspect of this particular application in the sense that we incorporate prior physical knowledge into our models. Discussions and conclusions are given in section 7.

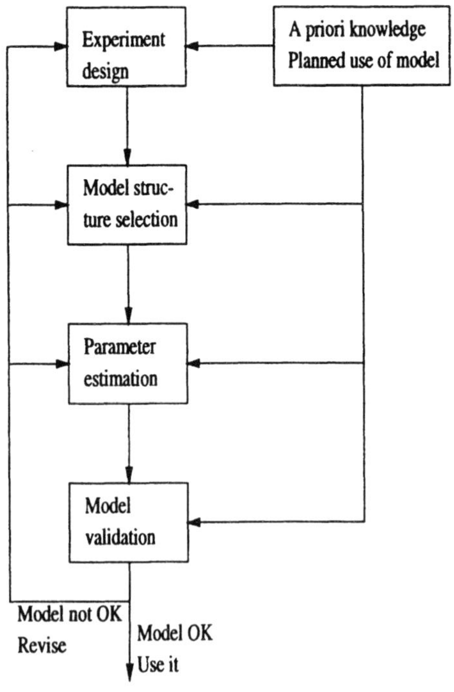

Fig. 1. The system identification procedure

2. PRIOR INFORMATION

The data used in this paper is from the Haughton Main Channel (HMC) in Queensland, Australia. The water levels in the HMC are controlled by overshot gates located along the channel as sketched in Figure 2. The measurements we have available are the water levels upstream and downstream of each gate and the gate position. The height of water above the gate is called the head over the gate, and it can be computed from the upstream water level and the gate position. The stretch of a channel between two gates is referred to as a pool.

As the water level is the controlled variable, and the gate position the manipulated variable, it is clear that a model for control should have water level as output and gate positions as input.

During the experiments all gates were in free flow, meaning that the top of the gate was above the down stream water level. In the literature it is suggested that an overshot gate behaves approximately as a sharp crested weir, and the flow over the gate can be written as (Bos (1978))

$$Q(t) = ch^{3/2}(t) \tag{1}$$

where Q is the flow, h the head over the gate and c an unknown parameter. c incorporates the effect of the discharge coefficients.

As a starting point for deriving a model structure we consider a simplified mass balance, and in addition to (1), we assume that the volume in the pool is proportional to the water level.

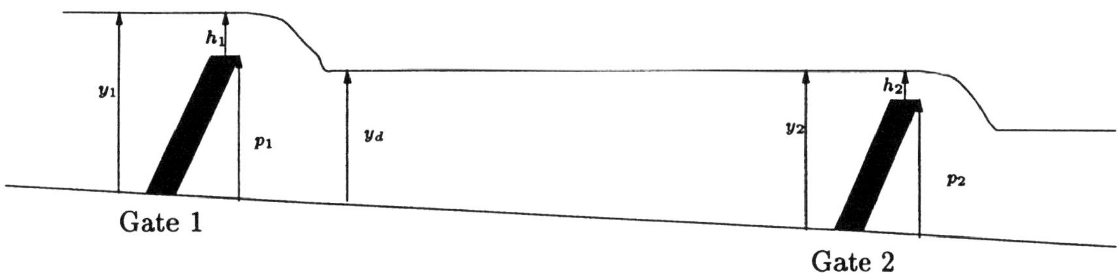

Fig. 2. Sketch of irrigation channel with gates.

$$\dot{y}_2(t) = \tilde{c}_1 h_1^{3/2}(t) - \tilde{c}_2 h_2^{3/2}(t) = \tilde{c}_1 h_1^{3/2}(t) - \tilde{c}_2 (y_2(t) - p_2(t))^{3/2} \quad (2)$$

where referring to Figure 2, y_2 is the water level upstream of gate 2, h_i the head over gate i, and p_2 the position of gate 2. It is clear that there is a time delay before the water flowing over gate 1 reaches the point where y_2 is measured. Taking this into account, and introducing an Euler approximation for the derivative we obtain

$$y_2(t+1) = y_2(t) + c_1 h_1^{3/2}(t-\tau) + c_2 (y_2(t) - p_2(t))^{3/2} \quad (3)$$

By just observing the channel during the experiments it was clear that waves were present. However, (3) can not describe waves and more complex models are needed for this, see section 6.

The HMC is mainly operated in downstream control, meaning that there is a feedback loop from $y_2(t)$ to $p_1(t)$ on Figure 2. From a control point of view it is therefore important to know how movements of gate 1 influence the water level upstream of gate 2.

An alternative is upstream control where there is a feedback loop from $y_2(t)$ to $p_2(t)$. The choice between downstream and upstream control is to a large extend dictated by the design of the channel, and often it is not a variable the control engineer can play with. It is expected that improved control can be achieved by using multivariable control techniques, and this is a current research area.

3. EXPERIMENT DESIGN AND NON PARAMETRIC IDENTIFICATION

The data and results reported in this paper is from a short pool of length 900m. The regulator structures at the upstream and downstream end of the pool are identical and consist of two identical overshot gates aligned in parallel. The water levels and gate positions are uniformly sampled with period 1 minute. The gate position is given in meters, and a fully shut gate has position 0 meters, and a positive value means that it is open. With reference to Figure 2, the measured gate position is $p_{\max} - p$ where p is shown in Figure 2. The water levels are given in mAHD (meters Australian Height Datum), and hence a correction term is included when the head over the gates are computed.

In this paper we only give results for one pool, although the pools upstream and downstream were also subject to similar system identification experiments at the same time.

3.1 *Step test*

Because of the simple form of (2) it was speculated that a first order linear model with time delay could provide an adequate description for control purposes, and the first experiment was therefore a step test. The gate at the downstream end was kept in a fixed position, and the head over the gate at the upstream end was stepped. With the downstream gate in a fixed position, the derivative of the water level is equal to the derivative of the head over the downstream gate. By replacing the terms raised to the power of 3/2 in (2) with linear terms, we obtain a transfer function of the form $Ke^{-\tau s}(1 + Ts)^{-1}$ between upstream and downstream head over gate.

The purpose of this experiment was twofold. Firstly to get rough estimates of the time delay, time constant and static gain in the linear first order plus time delay model of the channel, and then to use this information to design a more informative experiment where the gate positions follow binary signals. The results are shown in Figure 3. We see that the response looks very much like the response of a first order system with time delay. The presence of waves is evident from the repeated pattern of sharp rises and plateaus in the step response. For this reason drawing a tangent to determine the time constant is not going to give an accurate estimate. It is better to base estimates of the time constant on the time it takes for the water level to reach its final value (5 time constants) or 63% of its final value (1 time constant). From the figure we find that the time delay $\tau \approx 4.5$ minutes, and the time constant is $T \approx 20$ minutes. The static gain can not be accurately determined since the system is not in steady state before the step. However, due to the facts that the linear model represents the transfer

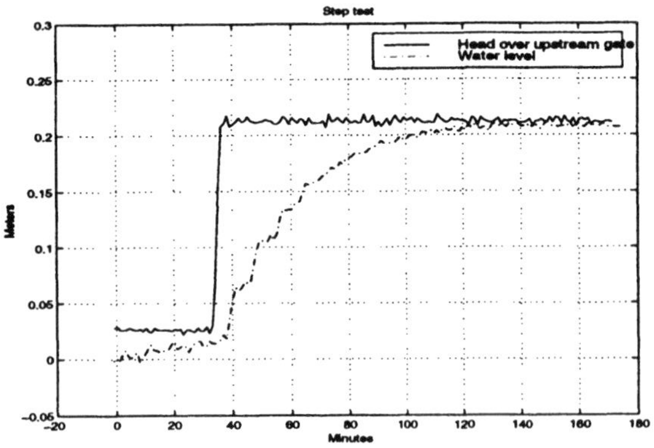

Fig. 3. Step test

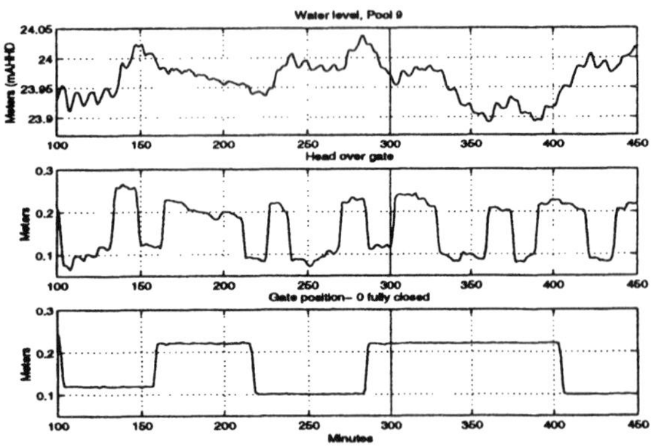

Fig. 4. Water level, head over upstream gate and downstream gate position.

function from head over the upstream gate to head over the downstream gate, and the two gates are identical, we know that the gain must be close to 1 since the gates pass all the flow.

3.2 *Experiment with binary signal*

In order to get more informative data, an experiment with binary signals was performed. According to linear theory the input signal should have most of its energy around the frequencies were there are "bends" in the transfer function to be maximally informative. From the step test we found that $T \approx 20$ minutes, and we decided to let the "period" of the upstream gate position to be 15 minutes, meaning that the gate was in a fixed position for a multiple of 15 minutes. By a similar consideration for the pool downstream, the "period" of the downstream gate position was set to 1 hour. (The pool downstream is about three and a half times as long as the one we are presenting results for.) The data from the experiment is shown in Figure 4. The data to the left of the vertical line were used for estimation and those to the right were used for model validation.

4. MODEL STRUCTURE SELECTION AND PARAMETER ESTIMATION

4.1 *Model structure selection*

The obvious initial choice of model structure based on the prior information is (3). This model structure has 3 unknown parameters, c_1, c_2 and the time delay τ. We let $\theta = [c_1, c_2]^T$. As control is one of the the intended uses of the model, a linear model of the form

$$y_2(t+1) = y_2(t) + c_1 h_1(t-\tau) + c_2(y_2(t) - p_2(t)) \quad (4)$$

was also considered, due to the control design methodologies available for linear systems. This model structure is also of interest since the step response of the system resembles the step response of a first order system with time delay. Depending on whether we assume only process noise or measurement noise present, we obtain the predictors

$$\hat{y}_2(t+1,\theta,\tau) = y_2(t) + c_1 h_1^{3/2}(t-\tau) + c_2(y_2(t) - p_2(t))^{3/2} \quad (5)$$

or

$$\hat{y}_2(t+1,\theta,\tau) = \hat{y}_2(t,\theta,\tau) + c_1 h_1^{3/2}(t-\tau) + c_2(\hat{y}_2(t,\theta,\tau) - p_2(t))^{3/2} \quad (6)$$

for the nonlinear model structure. The difference between (5) and (6) is that (5) uses the observed water level at time t to make a prediction of the level at time $t+1$ while (6) makes use of the predicted water level at time t. In the linear case (5) is the optimal one step ahead predictor for an ARX model, while (6) is the predictor for an OE model. From linear theory (e.g. Ljung (1999) p. 268) it is known that an OE model gives a good description of the low frequency properties, while an ARX model gives a better description of the high frequency properties. The waves in the pool are of relatively high frequency compared to the other interesting dynamics, and this may cause some problems for the ARX models. As we are mainly interested in control and prediction, the low frequency range is the more important one, and we therefore only report results using (6) and its linear counterpart.

4.2 *Parameter estimation*

The parameters $\theta = [c_1, c_2]^T$ were estimated for a number of time delays. From the step test we obtained an estimate of the time delay of around 4.5 minutes, and we therefore let τ vary between 3 and 10 to account for the uncertainty in the

Model structure	Time delay	$\hat{c}_1$	$\hat{c}_2$
Nonlinear	3	0.069	-0.063
Linear	3	0.043	-0.041

Table 1. Parameter estimates

Model	Averaged squared prediction error Est set $(\cdot 10^{-4})$	Averaged squared prediction error Val set $(\cdot 10^{-4})$
Nonlinear	0.734	0.690
Linear	0.746	1.141

Table 2. Prediction errors

estimate from the step test. The data points used were the ones to the left of the vertical line on figure 4. Due to some small variations in the sampling period the data were linearly interpolated in order to get uniform sampling.

The parameters were estimated using a prediction error method with a quadratic criterion, i.e.

$$\hat{\theta}_\tau = \arg\min_\theta \frac{1}{200} \sum_{t=1}^{200} (y_2(t) - \hat{y}_2(t, \theta, \tau))^2$$

where $\hat{y}_2(t, \theta, \tau))$ is given by (6) in the nonlinear case and by

$$\hat{y}_2(t+1, \theta, \tau) = \hat{y}_2(t, \theta, \tau) + c_1 h_1(t - \tau) + c_2(\hat{y}_2(t, \theta, \tau) - p_2(t)) \quad (7)$$

for the linear models. One model was obtained for every value of the time delay, and the final model was chosen as the one which gave the smallest error on the validation set. The time delay could also have been estimated using the estimation set, and for the data used here it would not have made any difference. $\tau = 3$ gave the best results on both data set, and the results are shown in Table 1 and 2.

5. DISCUSSION AND MODEL VALIDATION

There is good agreement between the results from the step test and the parameter estimates for the linear model. The time delay $\tau = 3$, is a bit shorter in the parametric model ($\tau \approx 4.5$ for the step test), while the time constant $T = 1/0.0408 \approx 25$ minutes is a bit longer ($T \approx 20$ minutes for the step test).

The nonlinear model gives about the same averaged squared prediction error on both the estimation and validation set, which is an indication that we are not trying to estimate too many parameters from too few data points. For the linear model we have a somewhat larger average squared prediction error on the validation set. This is due to that for a relatively large time interval the water level on the validation set is below the levels on estimation set, and the linear model is not

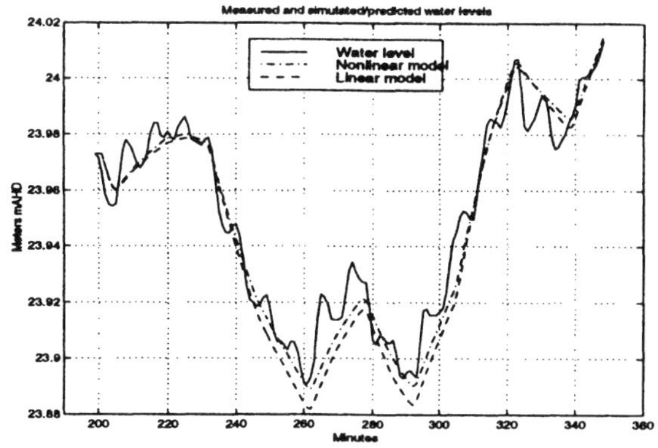

Fig. 5. Measured water level and simulated water levels on the validation set.

as accurate in that range. As we can see from Table 2 there is not much difference between the linear and nonlinear model on the estimation set. However, from Figure 5, we see that the nonlinear model more accurately simulates the water level on the validation set.

Notice that since we have used the predictors (6) and (7) the simulation errors and prediction errors are equal. The data used to compute the simulated levels in Figure 5 were the upstream and downstream gate positions from time 201 to 350, and the initial water level at time 201, no other values of the water level were used. The models are able to predict the main trends in the water level well over the whole time span of the validation set (two and a half hours). A feedback controller should be robust to uncertainties in the model, and the linear model may be sufficiently accurate for control design, although it has been observed that the time constant varies with the flow conditions. From Figure 5 the effects of the waves on the water level can be clearly seen, and it is equally clear that the linear and nonlinear models we have considered so far cannot represent wave motion. In the next section we investigate if better predictions can be obtained by using a more complex model structure able to represent waves.

6. HIGHER ORDER MODELS

In order to capture the wave dynamics we tried out a linear and nonlinear model of higher order. Table 3 gives the averaged squared prediction errors on the estimation and validation set. The prediction errors are much smaller in this case, but follow the same pattern as for the first order models. The nonlinear model actually has a smaller error on the validation set than on the estimation set, while the error increases with more than 100% for the linear model. The performance on the validation set is shown in Figure 6.

The nonlinear model has a remarkable predictive performance, particularly when we keep in mind

Model structure	Averaged squared prediction error Est set $(\cdot 10^{-4})$	Averaged squared prediction error Val set $(\cdot 10^{-4})$
Nonlinear	0.209	0.153
Linear	0.244	0.555

Table 3. Average prediction errors

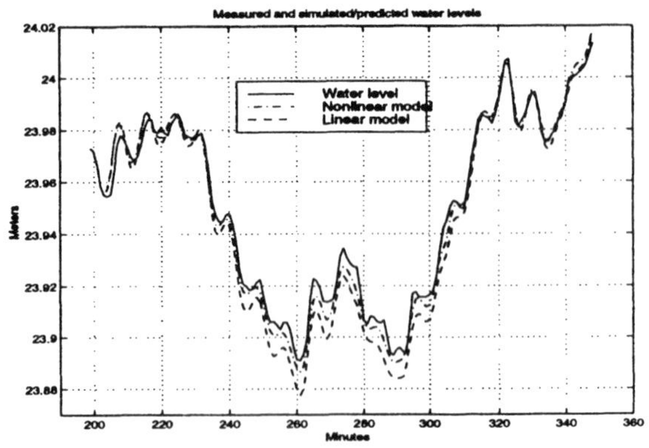

Fig. 6. Measured water level and simulated water levels on the validation set.

that the validation set was not used for parameter estimation, and that the predictor only had access to the initial values of the water level. The model is able to accurately predict both the main trends and the waves two and a half hours into the future. The linear model also shows good predictive performance, although it is not as accurate when the water level is low.

The results presented in this section are socalled "typical results" meaning they are the very best results obtained. For longer pools we did still obtain good results using a higher order nonlinear model, but they were not as good as those presented here. One reason for this is that waves with frequencies at the higher harmonics are also present, and their effect is more pronounced in longer pools.

7. DISCUSSIONS AND CONCLUSIONS

In this paper we have considered system identification of irrigation channels. We have presented the complete system identification procedure from experiment design to model validation, and prior information has been taken into account. It was found that a higher order nonlinear model gave very accurate predictions. It was even able to predict the waves. While the first order models were not able to capture waves, they represented the main trends in the data well.

The results obtained are very promising. However, system identification models are only valid under the operating conditions they were derived. The data set used here was characterised by that the flow was relatively low (up to 20% of the channel capacity), the gates were in free flow, and the pool was short. In order to obtain a system identification model (or models) valid for all conditions encountered during normal operation of an irrigation channel, further experiments are required. Preliminary results do indicate that the nonlinear models remain valid under much larger flows, but that for longer pools several delayed values of the head over the gates should be included. Another important issue which has to be addressed is closed loop identification. It is not often one is allowed to perform experiments as those presented here on an irrigation channel.

Acknowledgement. The author would like to thank Matthew Ryan and Bradley Ward at Rubicon Systems in Queensland for invaluable assistance in carrying out the experiments. He would also like to thank David Aughton, John Fenton, Iven Mareels and Carl Wendt for many fruitful discussions on operation of irrigation networks, hydraulics and system identification. This work was supported by Rubicon Systems Pty Ltd under the auspices of an AusIndustry Grant.

8. REFERENCES

[1] *Journal of Irrigation and Drainage Engineering*, Vol. 124, No. 1, 1998, pp. 1-62

[2] *Proceedings of IEEE Conference on System, Man and Cybernetics* San Diego, 1998, pp. 3850-3920.

[3] Bos, M.G. (Ed.) (1978). *Discharge measurement structures.* International Institute for Land Reclamation and Improvement/ILRI, Waageningen, The Netherlands.

[4] Brutsaert. W (1971). "De Saint-Venant equations experimentally verified" *J. Hyd. Div. ASCE* Vol. 97, pp. 1387-1401

[5] Clemmens, A.J., T.F. Kacerek, B. Grawitz, and W. Schuurmans (1998). "Test cases for canal control algorithms" *Journal of Irrigation and Drainage Engineering*, Vol. 124, No. 1, pp. 23-30.

[6] Cunge, J.A., F.M. Holly, and A. Verwey (1980). *Practical aspects of Computational River Hydraulics*

[7] Ljung, L. (1999). *System Identification - Theory for the User.* 2nd Edition, Prentice Hall.

[8] Söderström, T. and P. Stoica (1988). *System Identification.* Prentice Hall.

ESTIMATION OF POLLUTANT SOURCE SPECIFICATIONS USING A CONDITIONAL DECONVOLUTION METHOD.

G . Delmaire, M. Benjelloun

Laboratoire d'Analyse des Systèmes du Littoral,
195 Avenue du Pasteur Martin Luther King ,
62228 CALAIS BP 649
France
Email : **gilles.delmaire@lasl.univ-littoral.fr**
Phone: (33) 03-21-46-36-99

Abstract : In this paper, the general problem of the one dimensional pollutant scattering in the air or in water described by a parabolic partial differential equation is addressed. Two sources are considered, one of whom is operating. Source specifications can be reached through a deconvolution step and an hypothesis testing method. To prevent misdetections, an algorithm which thresholds the sensor measures and each conditional impulse response is proposed. This approach enables to minimize the number of parameters to be estimated and makes the detection reliable. The main interest lies in the joint estimation of the position and the source temporal flow. Copyright © 2000 IFAC

Keywords : deconvolution, decision making, convolution integral, condition numbers, scattering problems.

1. INTRODUCTION

Here we consider the generic framework of the joint off line estimation of the position and the temporal emission of a pollutant source from a partial differential model and a finite set of measurements. In most signal processing applications, the measured signal can be viewed as y(t) = h(t) * u(t) + b(t) where * stands for the convolution product. This is the case here when we consider a sensor located at a specific place and a fixed punctual source. The restoration of the emission u(t) then consists in a deconvolution of the sensor measures (Demoment, 89; Groetsch, 93; Tikhonov, and Arsenim, 77).

In the case when only a finite set of positions ($a \in A$) are possible, the task becomes a conditional deconvolution problem which can be formulated in the following way :

$$\underline{\hat{u}}_a(t) = \underset{[a \in A;\, u \in R^m]}{\arg\min} \; J_a(\underline{u}) \qquad (1)$$

where $J_a(\underline{u}) = Q_a(\underline{u}) + \lambda_a\,\Omega_a(\underline{u})$. $Q_a(\underline{u})$ measures the distance of u from the given data, $\Omega_a(\underline{u})$ measures the smoothness of u and λ_a stands for a set of control parameters. Some authors (Djafari, 97) propose to solve this mixed estimation decision problem in a bayesian framework. The major inconvenient is to assign prior probability laws for all the unknown parameters. This problem can also be split into a conditional input estimation phase followed by a detection step of the position into the set A. However, the parameters relative fitting is a difficult task especially when the number of hypotheses is high. As a result, it seems important to apply a previous processing so as to enlarge the regularization parameters range.

For this paper, we restrict ourselves firstly to the case of two possible hypotheses for the position and, secondly, to sensor measures corrupted with gaussian zero mean noise with known variance σ^2. Classical detection methods (Basseville,88) (for example the generalized likelihood ratio test) in these conditions amounts to compare the conditional error energy of each solution. Unfortunately, this detection stage can be disturbed by the result of the deconvolution since an excessive regularization can lead to misdetections (by increasing its error energy).

However, the reduction of the input vector size is a positive aspect for the precision of the estimation. In such a monitoring problem, measures are close to zero when no source is operating. An appropriate threshold enables to keep the significant part of the measures. Since the nature of the impulse responses appears to be time limited, a suitable off line threshold is performed jointly with each hypothesis so as not to change the relative size of the conditional energy. This leads to a decrease of significant inputs to be considered thanks to a temporal consistency analysis.

Finally, simulation results are proposed in order to confirm the interest for the method. It seems that large SNR (>50dB) yield good results in each stage while lower SNR can give misdetections.

2. MODELIZATION

We consider a pollutant propagation framework described by a parabolic one dimensional partial differential equation :

$$\frac{\partial C(x,t)}{\partial t} + V_x \cdot \frac{\partial C(x,t)}{\partial x} = K_x \frac{\partial^2 C(x,t)}{\partial x^2} + S(x,t) \quad (2)$$

where C(x,t) is the pollutant concentration (g/m^3), S(x,t) is the pollutant flow ($g/[s.m^3]$), V_x the wind velocity and K_x the diffusion parameter.

Solving this equation requires the knowledge of boundary conditions (Khapalov, 94) such as :

$C(x,0) = 0 \ \forall \ x \in \Omega$

$C(\Sigma,t) = 0 \ \forall \ t \in T$ (time range) where Σ represents the boundary of the geographical space Ω.

Namely, this last condition implies that the origin of the pollution belongs to the space Ω previously defined.

Using a Fourier expansion, it can be shown (Cornuel,98) that, in the case of a punctual source, the concentration is a 1D convolution product :

$$C(L,t) = \int_0^t h(L-a, t-\tau).s(\tau).d\tau \quad (3)$$

where s(.) [$g.m^{-2}.s^{-1}$] stands for the punctual source flow emission and h is the impulse response between a source located at x=a and a sensor at x=L.

For detailed expressions of h, the reader is invited to refer to Cornuel (98) or Groetsch (93) for similar examples. An example of an impulse response is given on figure 1. It appears that h can be approximated by a FIR with a significant time delay. The choice of the filter order will be discussed later on. From now, the measures and the input flow will be respectively called $\underline{y}$ and $\underline{u}$.

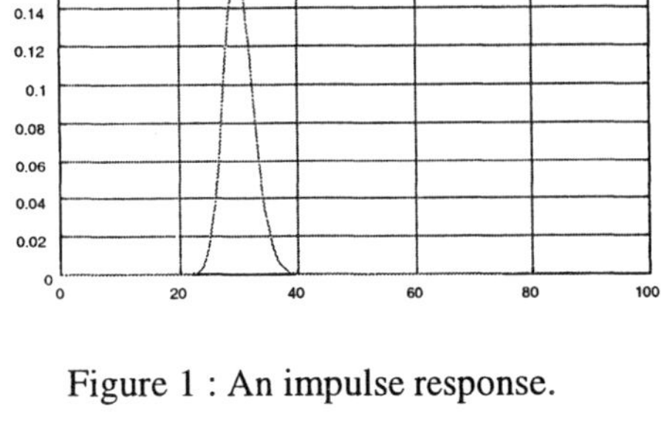

Figure 1 : An impulse response.

Referring to the convolution integral equation, recovering the input flow amounts to use a simple deconvolution technique (Groetsch, 93; Demoment, 89).

3 . USUAL DECONVOLUTION TECHNIQUES

Deconvolution is used to recover the system input from a set of measurements obtained from a linear time invariant system :

$$\underline{y} = H\,\underline{u} + \underline{b} \quad (4)$$

where H stands for a Toeplitz matrix made with the components of the previous impulse response, $\underline{b}$ is a random vector representing the noise, $\underline{u}$ is the temporal input flow.

Under these hypotheses, the deconvolution amounts to filter the measures by an appropriate filter. Two difficulties can affect the estimation quality. The first one comes from the nature of the impulse response which is generally a smooth function, this makes the deconvolution an ill-posed problem. In a discrete case, this inconvenient appears on the condition number of H which is relatively high (Philips, 62; Twomey, 65; Ekstrom, 73).

Secondly, measures are added with noise. Classical inverse filters (high pass filters) greatly amplify the effect of noise into the estimated input. As a result, standard inverse filtering is not suitable, however, it seems interesting to estimate the bounds of the least square error. For this purpose, Commenges (84) proposes an inequality for the Relative Root Mean Square error $RRMS(\underline{e}) = \frac{E(\|\underline{e}\|)}{\|\underline{u}\|}$ with $\underline{e} = \underline{u} - \underline{\hat{u}}_{LS}$, ($\underline{\hat{u}}_{LS}$ the least square estimate). It involves the condition number of the matrix H defined by $c = \frac{\lambda_{max}}{\lambda_{min}}$ and the ratio $\rho = \frac{n}{m}$ (λ denotes the singular values of H and n stands for the number of measures, m is the number of input parameters). Commenges (84) shows that this quantity is bounded :

$$\frac{1}{c.\rho.r^2} \le RRMS(\underline{e})^2 \le \frac{c}{\rho.r^2} \quad (5)$$

where r denotes the noise to signal ratio.

Under a fixed SNR, the estimation quality is increased if the ratio ρ is reduced. It seems interesting to take into account only the significant inputs to lower the quantity ρ.

Generally, discontinuities are badly recovered by classical deconvolution methods. Weak singular values are responsible for the damped oscillations in the estimated signal. Usually, this instability requires the use of techniques which include some extra a priori information on the solution. Some regularizing techniques (Demoment, 89;Hunt, 77;Tikhonov, and Arsenim, 77) apply a generalized inversion ($H^t.H + \alpha I$), which enables to shift the whole singular values of a value α, and as a result, to grow up the weakest singular values. Other stochastic methods also amount to increase the weak singular values, especially the Wiener filtering or some MAP estimators (Kailath et al , 74; Trussel and Hunt, 79). Indeed, such a technique use the inverse of ($H^t.H + R^t.R$). Weyl inequalities (Franchini and Jacquens, 96) enable to understand which changes are performed with the additional term $R^t.R$.

Let $\lambda_i(A)$ i=1..m be the singular values of A ordered as follow : $\lambda_1 < \lambda_2 \ldots\ldots < \lambda_m$. Weyl inequalities show that : $\forall k \in \{1 \ldots m\}$ (6)

$$\lambda_k(H^t.H)+\lambda_1(R^t.R) \le \lambda_k(H^t.H+R^t.R) \le \lambda_k(H^t.H)+\lambda_m(R^t.R)$$

Since $R^t.R$ is a definite quadratic form, all its singular values are strictly positive, especially $\lambda_1(R^t.R)$. Unfortunately, high singular values are also affected by these changes leading to a necessary trade off in the choice of the matrix R.

Another class of techniques involves truncated singular values decomposition (TSVD). They aim at keeping only the significant singular values of H in order to reduce the whole condition number. The major inconvenient lies in the choice of the threshold which yields an inverse approximation (Van Der Veen et al., 93) whose rank is equal to the remaining singular values (different from 0).

A third solution is to threshold the impulse response in advance in order to obtain a good conditioning. Weyl inequalities (Franchini and Jacquens, 96) show that weak singular values are generally increased for different thresholds. From a temporal point of view, this threshold enables to lower the number of input parameters and particularly improves the precision of the deconvolution step on the remaining parameters.

Nevertheless, all these methods present the inconvenient to increase the size of the error energy. In a multi-hypotheses framework, the necessary increase of the conditional error energy should not be performed independently. It is to be done relatively with each hypothesis so as not to change the relative size of the conditional energy.

4. SOURCE IDENTIFIABILITY PROBLEM

Now, we consider the problem of recovering the emission of one source among two possible ones. For this, we shall suppose that only one source is active during the simulation. A unique sensor gives sampled measurements on a time range T=[0 ; θ]. θ is chosen so as to obtain a configuration where measurements are zero at the beginning and at the end of the trial. Propagation causes a temporal delay between the time range of the active source and its effects on the sensor. We display a few signals evolving with time on figure 2.

The convolution filter is a finite IR from the time $(m_1\text{-}1)Te$ up to $(N_1\text{-}1)Te$ where Te is the sampling period. Moreover, the source is supposed to be operating only from $(n_1\text{-}1)Te$ to $(N_1\text{-}1)Te$. The resulting measure is also time limited from the time $(N_m\text{-}1)Te$ to $(N\text{-}1)Te$. These geometrical properties are useful for recovering the input. The convolution equation is organized with a m_1 samples shifting (due to the nature of the convolution filter). Here, the activity of the input is supposed to start at time 0 (while unknown).

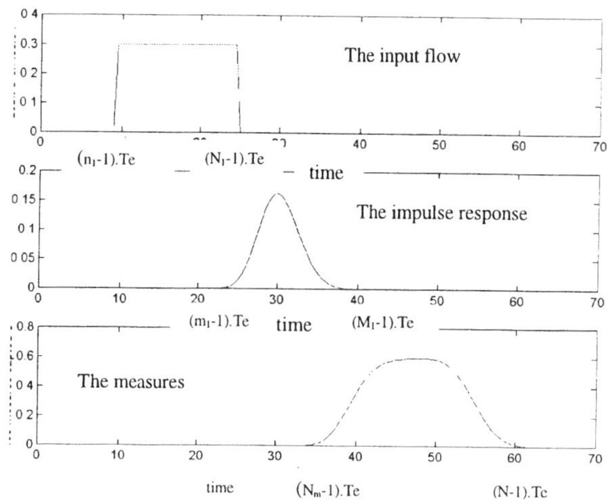

Figure 2 : evolution of some signals

The convolution matrix is a Toepliz type one and is called H_i for source number i.

$$\begin{bmatrix} y_{m_1-1} \\ \vdots \\ y_{N_m-2} \\ y_{N_m-1} \\ \vdots \\ \vdots \\ y_{N-1} \\ y_N \\ \vdots \\ y_{M-1} \end{bmatrix} = \begin{bmatrix} h_{m_1-1} & 0 & \cdots & 0 & \cdots\cdots & 0 \\ & \ddots & & & & \vdots \\ & & \ddots & 0 & & \vdots \\ h_{M_1-1} & & & \ddots & & \vdots \\ 0 & \ddots & & & \ddots & \vdots \\ \vdots & & \ddots & & 0 & \vdots \\ \vdots & & & \ddots & & \vdots \\ \vdots & & 0 & \ddots & & \vdots \\ & & & \ddots & \ddots & 0 \\ 0 & \cdots & 0 & h_{M_1-1} & \cdots & h_{m_1-1} \end{bmatrix} * \begin{bmatrix} u_0 \\ \vdots \\ u_{N_m-m_1-1} \\ u_{N_m-m_1} \\ \vdots \\ \vdots \\ u_{N-m_1} \\ u_{N-m_1+1} \\ \vdots \\ u_{M-m_1} \end{bmatrix} \quad (7)$$

First, we can look for the geometrical properties of the solution $\underline{u}_1$.

Proof :

We know that $\underline{y} \in Y_{Nm:N}$ according to the hypotheses with $Y_{Nm:N} = \{\underline{y} \in R^M \,/\, y(Nm:N) \neq 0 \text{ else } y(i) = 0\}$

Using the definition of the convolution product, it can be seen that $\underline{y} \in Y_{m1+n1-1:M1+N1-1}$. Equaling these quantities yields :

$$\begin{cases} m_1 - 1 + n_1 - 1 = Nm - 1 \\ M_1 - 1 + N_1 - 1 = N - 1 \end{cases} \Leftrightarrow \begin{cases} n_1 = Nm - m_1 + 1 \\ N_1 = N - M_1 + 1 \end{cases} \quad (8)$$

It means that the solution $\underline{u}_1$ belongs to $U_{Nm-m1+1:\,N-M1+1}$, which defines the bounds of its activity. The number of parameters is equal to $N\text{-}N_m\text{-}M_1\text{+}m_1\text{+}1$. The convolution matrix is supposed to be full column rank, then the system is overdetermined.

Using these hypotheses, some rows and columns can be cancelled into the matrix H_1. Let's now imagine a second source whose convolution matrix H_2 is also a Toeplitz one, with associated parameters m_2 and M_2. The main question is : are we able to solve the identifiability problem, namely to make the difference between a pollution coming from source 1 and from source 2 ?

The solution $\underline{u}_2$ belongs to $U_{Nm-m2+1\,:\,N-M2+1}$, according to the same previous arguments. Suppose that M_1-m_1<M_2-m2, then the number of parameters to be estimated is greater for the first solution than for second one.

Two conditional pseudo-inverse solutions can be found if H_i (i=1 and i=2) are full column rank ($H_i^t.H_i$ is full rank). A least square solution can be computed for each hypothesis. In the noiseless case, the measures are supposed to belong to Span(H_1) (so that H_1 is true). The projections of the measures on Span$(H_1)^{\perp}$ yields the null vector whereas the projection onto Span$(H_2)^{\perp}$ does not give the same results. This shows that the distance to Span(H_1) is zero under the right hypothesis, as a result, is minimum among all the hypotheses. Conditional distances are expressed below assuming that the measures are added with noise : $\underline{y}=H_1\,\underline{u}_1+\underline{b}$

$$d(\underline{y}, Sp(H_1))=\underline{b}^t.[I - H_1(H_1^t.H_1)^{-1}.H_1^t]^2.\underline{b} \qquad (9)$$
$$d(\underline{y}, Sp(H_2))=(H_1\,\underline{u}_1+\underline{b})^t.[I - H_2.(H_2^t.H_2)^{-1}.H_2^t]^2.(H_1\,\underline{u}_1+\underline{b})$$

Given the second order noise specifications, it appears that the true solution does not necessarily present a minimum distance with respect to the other hypotheses. The threshold and the regularization variables are a set of control parameters enabling to check the minimum distance condition.

5. TEMPORAL CONSISTENCY ANALYSIS

The expression (9) has shown that the noise can make the distance to Span(H_1) non minimal in the right hypothesis.

This is partly due to an ill-conditioning of the matrix H_1 relatively with matrix H_2 Indeed, the maximum error highly depends on the condition number $\left(\dfrac{\|\delta\underline{u}\|}{\|\underline{u}\|} \le \mathrm{cond}(H_i).\dfrac{\|\underline{b}\|}{\|\underline{y}\|}\right)$ and noise specifications. A threshold applied to the matrixes H_i (i =1,2) presents two major advantages, the first one is to slightly change the original matrix. As a result, only the low singular values which are responsible for a bad conditioning are increased. The idea proposed here is to choose a conditional threshold so that the resulting matrices H_i have got similar (and lower) conditions numbers. The maximum errors become in fact quite similar and smaller in each case.

The second advantage is to lower the order of each impulse response leading to a precise value of the delay brought by the system.

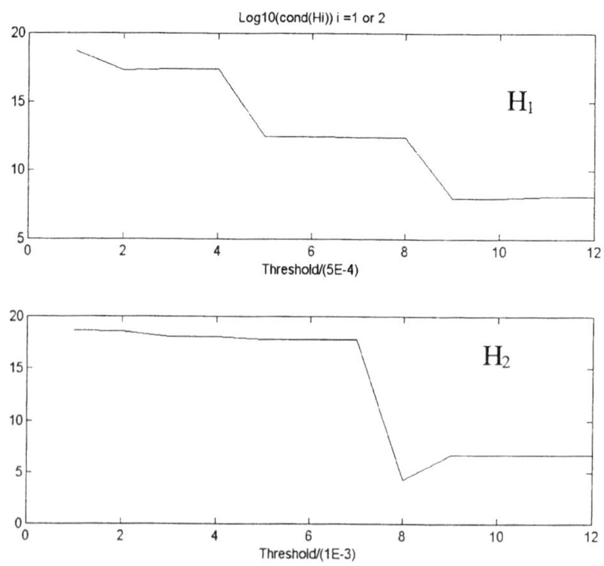

Fig 3 : The condition number for different thresholds

Figure 3 shows an example of the evolution of the condition number in each case for different thresholds. An appropriate choice of the threshold is 5 e-3 for H_1 and 8 e-3 for H_2, which leads to similar condition numbers.

Another processing can be applied to the measures in order to extract their active part : the measurement analysis in such a monitoring problem (quite zero at the beginning and at the end of the trial) leads to detect changes in the measurements. They represent the bounds of the source activity. These features can be obtained through a suitable threshold or through a change detection method (Basseville, 88).

We only consider here the thresholding case. To compute the starting time, we look for the 2 first successive times whose measures are greater than 3σ (under no source case and gaussian noise, Prob ($|y_k|>3\,\sigma$) = 5%). We apply the same principle for the ending time. This procedure seems to be sufficiently robust but is able to yield errors due to the random noise process.

These two steps can lead to a conditional consistency analysis of the input. As shown in part 3, this consistency analysis reduces the number of input parameters enabling to lower the errors caused by the restoration method. Another advantage lies in the indirect elimination of the input main changes which are badly recovered by a deconvolution procedure. Some authors propose to change the usual criteria and add an edge preservation term (Sanchez-Avila, 94; Geman and Reynolds, 92). In all cases, this introduces an extra control parameter, hardly tractable in a multi-hypotheses framework. The choice for the thresholding has 2 positive aspects : the source deconvolution is improved in the same time as the detection step.

6. ESTIMATION DETECTION ALGORITHM

The different sequential steps are outlined :

- The thresholding of H_1 and H_2 so as to obtain a similar and minimum condition number.
- Measures are threshold in order to compute the temporal conditional activity of each source. This enables to extract a sub-matrix from H_i (i=1,2) suitable for the situation (once the corresponding rows and columns are cancelled).
- Next, the computation of conditional solutions according to a chosen deconvolution method.
- Finally, a detection phase (which computes the ratio for the conditional error energy) to choose the most relevant solution.

Now, it's interesting to compare the performances of the algorithm and also understand the reasons for its limitations.

7 . SIMULATION RESULTS

To evaluate the performances of the whole method, we consider an example where the pollutant propagation is supposed to be unidirectional. The wind velocity and the diffusion parameters are known and equal to : U= 10m/s and K_x= 10m^2/s. The two possible sources are located respectively at x=1m and x=151m The remote sensor is situated at x = 301m.

We assume that only the source 1 is active. This case is the most difficult one since the filter which links the input source 1 to the output is even more low pass band, then the generalized inversion is getting more involved. This source is operating on a time limited range starting from t=10s to 25s. The measures are collected during 100 seconds with a sampling period equal to 0.5 second. Given these specific conditions, they are rigorously zero at the beginning and the end of the sequence. Moreover, they are corrupted with a zero mean gaussian noise with variance matrix σ^2 I with σ = 1E-3.

Conditional (source position =1 m) estimation of the input.

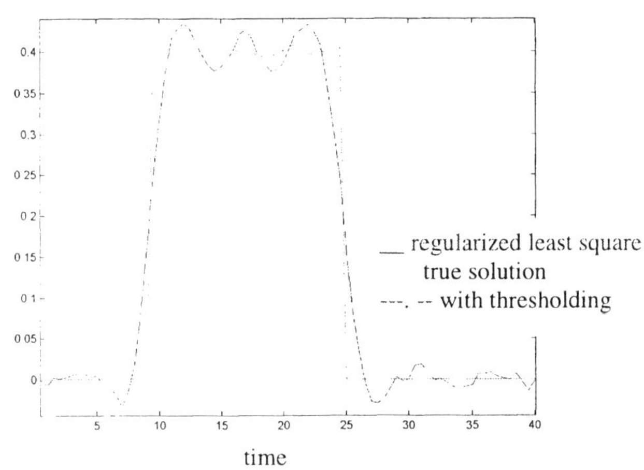

Fig 4 : conditional estimation of the input.

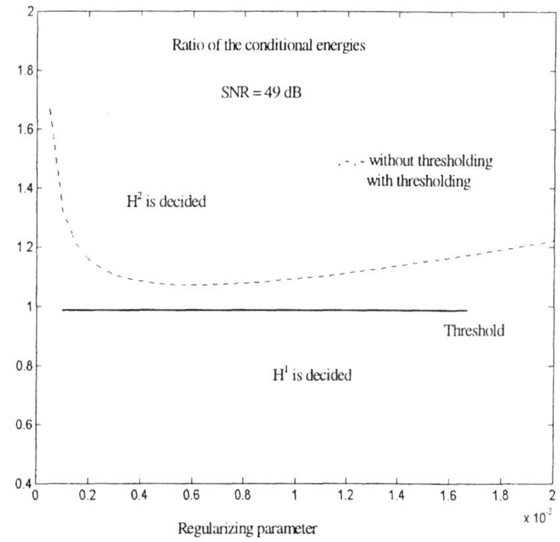

Fig 5 :The detection step quality for different smoothing parameters.

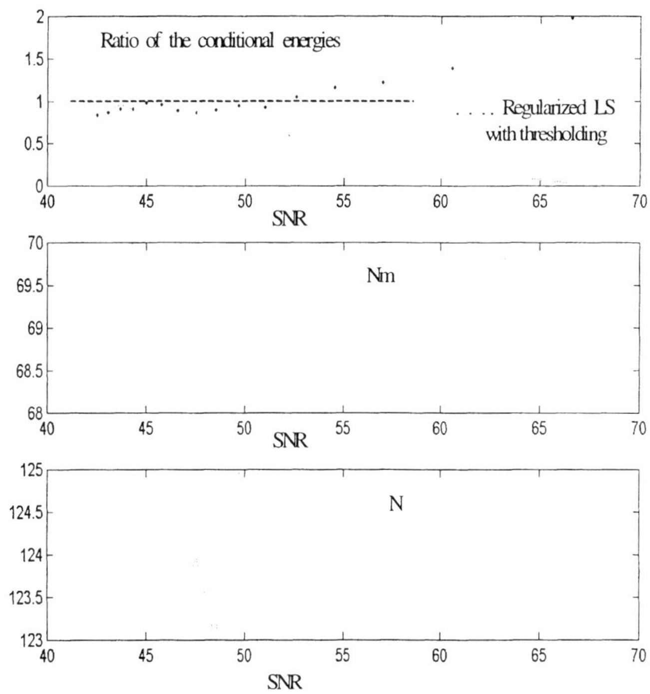

Fig 6: Evolution of the detection step with respect to the noise.

Their evolution is quite similar to the figure 2 since a small noise is added to the signal.

We show on figure 4 an example of conditional estimation obtained with and without threshold. Figure 5 illustrates that considering a fixed SNR (=49 dB), the smoothing parameters range leading to a right hypothesis is enlarged in case of a thresholding. Figure 6 is devoted to the comparison of the efficiency of the detection step for different SNR (the

smoothing parameter is identical for each method and is taken proportional to the variance).
The off line analysis of the measures and the impulse response appears to be an important previous processing to start a mixed estimation-decision problem. Once this processing is made, any other deconvolution method should be applied, stochastic or deterministic (Trussel and Hunt, 79; Demoment, 89; Tikhonov and Arsenim, 77). The advantage for this thresholding lies in the cancellation of the input signal changes which are hardly recovered in a deconvolution phase. To this end, we mention that a defective thresholding of the measures makes the detection step faulty. As a matter of fact, it could be interesting to use a change detection method such as a Generalized Likelihood Ratio test able to find the time changes without any error. Moreover, the importance of such a step encourage us to extend this framework to the multi-sensor case. Fusing the activity range of the sources obtained with each sensor enables to expect a more robust estimation.

8. CONCLUSION

In this paper, we describe the problem of both estimating the pollutant flow and choosing between two 2 possible sources from a set of measurements obtained by a unique remote sensor. This problem can be seen as a deconvolution stage followed by a detection technique. In these conditions, it appears that this last step can be affected by an excessive regularization leading to false detections. For this reason, we propose a procedure which lowers the number of input parameters to be estimated by a specific analysis of the measures and the impulse responses. This technique presents two major advantages : the deconvolution step is greatly improved by eliminating the main discontinuities of the input signal, second, the temporal consistency analysis enables to make the detection more reliable. The interest for this method is largely illustrated when SNR is greater than 50 dB but it can lead to false detection in the case of lower SNR. Consequently, it seems interesting for the future to apply a change detection method to the measures instead of a thresholding. Moreover, the use of extra distributed sensors would enable to fuse the temporal activities of the conditional input obtained by each sensor.

REFERENCES:

Basseville M. (1988). Detecting Changes in Signals and Systems. A survey, *Automatica*, **vol 24. n°3**, pp 309-326.

Commenges D (1984). The Deconvolution Problem : Fast Algorithms Including the Preconditioned Conjugate Gradient to Compute a MAP Estimator. *IEEE Trans. A.C.*, **Vol A-C 29, n° 3**, pp 229-243.

Cornuel H. (1998), *Estimation de caractéristiques d'une source sur la base d'informations hétérogènes*, DEA de l'Université de Lille 1, Juillet 1998.

Demoment G. (1989), Image Reconstruction : Overview of Common Estimation Structures and Problems, *IEEE Trans. On Acoustics Speech and Signal Processing*, **Vol. 37, n° 12**,.pp2024-2036.

Djafari A.M. (1997), Une Méthode Bayesienne pour la Localisation et la Segmentation de Sources de formes connues, *in Proc. GRETSI*, Grenoble, France, pp 136-138.

Ekstrom M.P.(1973), A spectral Characterization of the Ill-Conditioning in Numerical Deconvolution, *IEEE Trans. Audio-Electroacoust.*, **Vol. AV-21**, pp 344-348.

Franchini J. and J. C. Jacquens (1996). *Algèbre, Seconde Année Mathématiques Spéciales*. Les inégalités de Weyl , pp 182-185, collection Ellipses.

Geman D.and G. Reynolds (1992). Constrained Restoration and the Recovery of Discontinuities, *IEEE Trans. PAMI*, **Vol 14, n° 3**, pp 367-383.

Groetsch C.W. (1993), *Inverse Problems in the Mathematical Sciences*, Vieweg 1993.

Hunt B. R. (1977), Bayesian methods in digital image restoration, *IEEE Trans on Comp.*, **Vol C-26, n°3**, pp 219-229.

Kailath T., M. Morf and G.S. Sidhu (1974), Some New Algorithms for Recursive Estimation in Constant Linear Discrete-time Systems, *IEEE Trans. Aut. Cont.* **Vol AC n°19**, pp 315-323.

Khapalov A.V. (1994), Localization of Unknown Sources for Parabolic Systems on the Basis of Available Observations, *Int. Journal Systems Science*, **Vol 25, n° 8**, pp 1305-1322.

Philips D.L (1962), A technique for the Numerical Solution of Certain Integral Equations of the First Kind, *J. Ass. Comput. Mach.*, **Vol 9**, pp 97-101.

Sanchez-Avila C. (1994), "An adptive regularized method for Deconvolution of Signal with Edges by Convex projections". *IEEE Trans. S.P.* , **Vol 42, n° 7**, pp 1849-1851.

Tikhonov A. and V. Arsenim (1977). *Solutions of Ill Posed Problems*. Washington D.L. , Winston.

Trussel H.J. and B.R. Hunt (1979). Improved Methods of Maximum A Posteriori Restoration. *IEEE Trans. On Computers*, **Vol C-27**, n° 1, pp 57-62.

Twoney S. (1965), "The application of Numerical Filtering to the Solution of Integral Equation of the First Kind encountered in Indirect Sensing Measurements" – *J. Franklin Inst.*, **Vol 279**, pp 95-109.

Van Der Veen A., Deprettere Ed. F., Lee, Swindlehurst, A. (1993),"Subspace based Signal Analysis Using Singular Value Decomposition", *in Proc. IEEE*, **Vol 81**, pp 1277-1308.

IDENTIFICATION FOR A GENERAL CLASS OF LPV MODELS

Bassam Bamieh [*,1] **Laura Giarré** [**]

[*] *Mechanical and Environmental Engineering*
University of California, Santa Barbara, CA 93106, USA.
bamieh@engineering.ucsb.edu *(805)-893-4490*
[**] *Dipartimento di Ingegneria Automatica e Informatica*
Università di Palermo, Palermo, Italy.
giarre@ias.unipa.it

Abstract: In this paper we consider the problem of identifying discrete-time Linear Parameter Varying (LPV) models of non-linear or time-varying systems. LPV models are considered for their connection with the industrial practice of gain-scheduling. We assume that inputs, outputs and the scheduling parameters are measured, and a form of the functional dependence of the coefficients on the parameters is known. We show how the identification problem can be reduced to a linear regression so that a Least Mean Square identification algorithm can be reformulated. Conditions on the persistency of excitation in terms of the inputs and parameter trajectories are given to ensure the consistency of the algorithms when the functional dependence is of polynomial type. Simulation example are given for both periodic and non periodic trajectory of the scheduling parameters. *Copyright ©2000 IFAC*

Keywords: Identification, Linear Parameter Varying (LPV) models, Gain Scheduling, Persistence of Excitation

1. INTRODUCTION

In this paper, new techniques are developed for the identification of nonlinear systems, linearized along a varying trajectory. Let us assume that the system generating the data is described by the (possibly nonlinear) relationship

$$y = f(u, p, \eta) \tag{1}$$

where y, u and p are the measured experimental data, and η represents the unknown signals entering the system. In particular, y is the output, u is the input and p is the gain-scheduling parameter, a variable parameter that can be thought as determining the "set-point" of the system. In many process and aerospace control systems, it is a realistic assumption that the set point variable is measurable.

The aim of any identification procedure is to find conditions under which it is possible to find a parametric model of a possibly nonlinear, time-varying system to a desired accuracy.

As a *model* we consider a special class of discrete-time linear parameter-varying (LPV) models. A motivation for this type of model comes from the fact that in control applications the *true* system is not linear, yet LTI or LPV models are often satisfactory in industrial control design. In particular, we consider LPV models rather than generic Linear Time-Varying (LTV) models because of their connection to the industrial practice

[1] Research supported by NSF under CAREER award ECS-96-24152, and AFOSR-PRET 442530-25380

of gain-scheduling ((Rugh, 1991),(Shamma and Xiong, 1999)).

More recently, many methods to infer stability or construct stabilizing feedback controller for LPV systems have been also studied extensively throughout the literature ((Packard, 1994), (Blanchini, 1999),(Shamma and Athans, 1991), (Shamma and Xiong, 1999) and references therein).

The problem of identifying a linear parameter-varying systems has been studied in literature. We refer the reader to (Campi, 1994), where a survey of the literature has been presented for LTI models, and the exponentially weighted Least Squares algorithm and its stochastic properties have been studied.

More closely related to our approach is the work of (Lee and Poolla, 1999), which addresses the problem of identifying a discrete LPV model

$$y_k = P(p, \theta)u_k + v_k$$

where p is a measured time-varying parameter, and it reduces the problem to one of nonlinear programming.

In (M. Nemani and Bamieh, 1995), a special case of the LPV problem with state measurements and one parameter was shown to be equivalent to a linear regression, and certain conservative conditions for persistency of excitation were given.

In (Previdi and Lovera, 1999), LPV models are identified with an hybrid linear/nonlinear procedure. The nonlinear part is identified through neural network and the linear parametric part trough Least Square (LS) algorithm.

In this paper, we will present the solution to the problem of identifying an input-output LPV model via lincar regression and the Least Mean Square (LMS). Moreover we will consider the consistency of the algorithms in terms of persistency of excitation. For the solution to the problem via Recursive Least Square (RLS) see (Bamieh and Giarré, 1999).

The techniques presented here have also the following advantage: in order to identify the multiple LTI models linearized along the trajectory as usually done in gain scheduling oriented identification scheme, the experiment here can be carried on in one shot, without requiring multiple experiment to collect data and without requiring that the set-point vary slowly.

2. PROBLEM FORMULATION

Our special class of discrete-time LPV models is parameterized as follows

$$M(\delta, p) = \frac{B(\delta, p)}{A(\delta, p)}, \qquad (2)$$

where δ is the delay operator i.e. $(\delta\, y)_k := y_{(k-1)}$, and

$$\begin{aligned} B(\delta, p) &:= b_0(p) + b_1(p)\delta + \ldots + b_{n_b}(p)\delta^{n_b} \\ A(\delta, p) &:= 1 + a_1(p)\delta + \ldots + a_{n_a}(p)\delta^{n_a} \end{aligned} \qquad (3)$$

$n = n_a + n_b + 1$ is the number of parametric functions to be identified. Moreover, we assume that the varying parameter p is a function of discrete time ($p = p(k)$). The fractional representation in (2) is short hand for the operational notation

$$A(\delta, p)\ y(k) \ = \ B(\delta, p)\ u(k), \qquad (4)$$

where y is the output and u is the input.

We assume that the parameters $\{a_i\}$, $\{b_j\}$ in (3) are in general known function $x(p)$ of the parameter p of the form:

$$x(p) = c_1 f_1(p) + \ldots c_N f_N(p) \qquad (5)$$

where the constants $c_l, l = 1, \ldots, N$ are, respectively, $c_l = a_i^l, i = 1, \ldots, n_a$ for $a_i(p)$ and $c_l = b_j^l, j = 0, \ldots, n_b$ for $b_j(p)$.

Thus any particular model in our class is completely characterized by the real numbers $\{a_i^k\}$ and $\{b_j^l\}$.

For this general framework, many choices are possible for the functions $f_l(p)$. In particular if we consider a polynomial dependence, then the functions $f_l(p)$ are simply powers of p

$$f_l(p) = p^{l-1} \quad l = 1, \ldots, N$$

In this case the parameters are given by:

$$\begin{aligned} a_i(p) &= a_i^1 + a_i^2 p + \ldots + a_i^N p^{N-1} \\ b_j(p) &= b_j^1 + b_j^2 p + \ldots + b_j^N p^{N-1}. \end{aligned} \qquad (6)$$

An other possible choice is the following:

$$f_l(p) = sin(l\omega_0 p) \quad l = 1, \ldots, N$$

This representation would be useful for problems where the coefficients and the parameters are cyclical (e.g. angles of rotation). If the frequency ω_0 is known then the above form for $x(p)$ would be the first N harmonics of the representation of a general function of p. An example to which this is applicable, it is the so called non-circular machining, (see (Hanson and Tsao, 1998)).

An other possible choice is the one of the neural-network, where the constants c_l represent the weights and the function f_l are of the sigmoidal type.

Remark: In this paper we consider only the case of dependence on a single parameter. Most of our results in this paper are readily generalizable to the case where the functions $\{a_i\}$, $\{b_j\}$ are functions of several parameters. These results will be reported elsewhere.

We now observe that the input-output model in (4), together with (5) and (3) can be put into a nice linear regression form as follows. Let us define an $n \times N$ matrix Θ which contains all the coefficients to be identified:

$$\Theta := \begin{bmatrix} a_1^1 & \dots & a_1^N \\ a_2^1 & \dots & a_2^N \\ \vdots & \vdots & \vdots \\ a_{na}^1 & \dots & a_{na}^N \\ b_0^1 & \dots & b_0^N \\ \vdots & \vdots & \vdots \\ b_{nb}^1 & \dots & b_{nb}^N \end{bmatrix}. \qquad (7)$$

We also define the *extended regressor* Ψ, which will be made up of past i/o data and parameter trajectories

$$\Psi_k := \phi_k \pi_k :=$$

$$\begin{bmatrix} -y_{k-1} \\ \vdots \\ -y_{k-n_a} \\ u_k \\ \vdots \\ u_{k-n_b} \end{bmatrix} \left[f_1(p_k) \; f_2(p_k) \; \dots \; f_N(p_k) \right], \qquad (8)$$

where for notational simplicity we use a subscript for the time index (e.g. $p_k := p(k)$).

With the above definitions, it is easy to verify that the model (3)-(5) can now be rewritten as

$$y_k = \langle \Theta, \Psi_k \rangle, \qquad (9)$$

where the inner product is the matrix inner product defined as $< A, B >:= \text{trace}(A^* B) = \text{trace}(BA^*)$.

The advantage of rewriting the model in the inner product form (9) is that the usual RLS and LMS algorithms can be easily generalized to this case, (Bamieh and Giarré, 1999). To do so, one needs the correct notion of outer products (which is given by the tensor product), and to appropriately handle the issues of measurement noise and persistency of excitation.

We note that if $N = 1$, in the polynomial case $f_1(p_k) = 1$ and $\pi = 1$, then $\Theta = \theta \in \Re^{n \times 1}$, and $< \Theta, \Psi >=< \theta, \phi >= \text{trace}(\phi^* \theta) = \phi^T \theta$. This is the standard product among the vector of parameters θ and the regressor vector ϕ of any classical identification scheme, (Södeström and Stoica, 1989).

Disturbance considerations

The uncertainty model we consider in this paper is the following

$$y_k = \langle \Theta, \Psi_k \rangle + \eta_k. \qquad (10)$$

In this paper, for the sake of simplicity we consider the case when no unmodeled dynamics are present, i.e. there exists a *true* parameter matrix Θ_o. We will also assume the measurement noise to be white and uncorrelated with the input. We list our assumptions for reference:

A1 No unmodeled dynamics.
A2 η_k is a white noise sequences.
A3 η_k is uncorrelated with u_k

The relaxation of these assumptions is the topic of current research and a solution is partially reported in (Bamieh and Giarré, 1999).

3. IDENTIFICATION OF THE LPV SYSTEM

Let us consider the following *prediction error* ε defined as

$$\varepsilon(k, \Theta) = y_k - < \Theta, \Psi_k > . \qquad (11)$$

Let $\hat{\Theta}_k$ be the matrix of estimated parameters at time k.

Then, it is possible to derive the following LMS algorithm, see (Bamieh and Giarré, 1999) for details:

LMS Algorithm: *For the model in (10), under assumptions A.1, A.2 and A.3, the Least Mean Square identification algorithm is given by:*

$$\Psi_k = \phi_k \pi_k$$
$$\epsilon_k = y_k - \left\langle \hat{\Theta}_k, \Psi_k \right\rangle = y_k - trace\left(\hat{\Theta}_k^T \Psi_k\right)$$
$$\hat{\Theta}_{k+1} = \hat{\Theta}_k + \alpha \, \epsilon_k \, \Psi_k,$$

where the extended regressor Ψ_k is defined by (8).

We note that all that is needed for the implementation of this LMS algorithm is the formation of the extended regressor, and taking the matrix inner product involved in forming ε_k.

4. INPUT CHOICE AND PERSISTENCY OF EXCITATION

Forthe algorithm a *Persistency of Excitation* condition has to be satisfied in order to guarantee the convergence of the algorithm. We state the condition for the case of polynomial dependence of the coefficients on the parameters (i.e. when

$f_l(p) = p^{l-1}$). More general conditions will be presented in the final version of this paper.

Let Θ_o be the true value, then the identification algorithm converges if

$$\lim_{k\to\infty} \hat{\Theta}_k = \Theta_o. \tag{12}$$

Definition The extended regressor $\{\Psi_k\}$ *is said to be* persistently exciting (PE) *if*

$$\lim_{T\to\infty} \frac{1}{T} \sum_{k=0}^{T} E\{\Psi_k \otimes \Psi_k\} > 0. \tag{13}$$

We will need the following result, which is a generalization of a similar result for vectors, and an immediate consequence of the tensor product definition.

Lemma

Ψ_k *is PE if and only if, for any non-zero* $V \in \Re^{n\times N}$,

$$\lim_{T\to\infty} \frac{1}{T} \sum_{k=0}^{T} E\left\{\langle V, \Psi_k\rangle^2\right\} > 0.$$

Definition

Let $\mathcal{I}$ *be a subsequence of the natural numbers. The sequence* $\{\phi_k\}$ *is said to be* persistently exciting (PE) relative to $\mathcal{I}$ *if*

$$\lim_{T\to\infty} \frac{1}{T} \sum_{k\in\mathcal{I}}^{T} E\{\phi_k \phi_k^*\} > 0. \tag{14}$$

An example of such a sequence is a white noise vector process $\{\phi_k\}$, where the subsequence $\mathcal{I}$ goes to infinity not slower than its index, i.e.

$$\mathcal{I} := \{s_1, s_2, \ldots\}, \qquad \lim_{i\to\infty} \frac{i}{s_i} > 0.$$

The key to the PE condition for this special class of LPV systems is a natural interpolation type condition which we now state.

Theorem

Let $f_l(p) = p^{l-1}$. *The extended regressor* $\{\Psi_k\}$ *is PE if and only if the following two conditions hold*

1) *There exists* N *distinct points* $\bar{p}_i \in \Re$, $i = 1, \ldots, N$, *that are limit points of* $p(k)$. *Let* $\mathcal{I}_1, \ldots, \mathcal{I}_N$ *be subsequences corresponding to these limit points, i.e.*

$$\lim_{k\in\mathcal{I}_i} p(k) = \bar{p}_i.$$

2) *The sequence* $\{\phi_k\}$ *is PE relative to all subsequences* $\mathcal{I}_1, \ldots, \mathcal{I}_N$ *simultaneously, i.e. for all* $i = 1, \ldots, N$,

$$\lim_{T\to\infty} \frac{1}{T} \sum_{k\in\mathcal{I}_i}^{T} E\{\phi_k \phi_k^*\} = Q > 0. \tag{15}$$

Proof

We first use Lemma 4 to recast the PE condition for Ψ_k. We recall that Ψ_k is PE iff for any nonzero $V \in \Re^{n\times N}$,

$$\lim_{T\to\infty} \frac{1}{T} \sum_{k=0}^{T} E\left\{\langle V, \Psi_k\rangle^2\right\} > 0.$$

Using the definition of the inner product, we compute

$$\begin{aligned} E\left\{\langle V, \Psi_k\rangle^2\right\} &= E\left\{\text{trace}\,(\Psi_k V^*)^2\right\} \\ &= E\left\{\text{trace}\,(\phi_k \pi_k V^*)^2\right\} \\ &= E\left\{\text{trace}\,(\pi_k V^* \phi_k)^2\right\} \quad (16) \\ &= E\left\{\pi_k V^* \phi_k \phi_k^* V \pi_k^*\right\}, \quad (17) \end{aligned}$$

where in (16) we used the property trace(AB) = trace(BA), and (17) follows because $\pi_k V^* \phi_k$ is a scalar.

The above implies that the PE condition on $\{\Psi_k\}$ now becomes

$$\lim_{T\to\infty} \frac{1}{T} \sum_{k=0}^{T} E\{\pi_k V^* \phi_k \; \phi_k^* V \pi_k^*\} > 0. \tag{18}$$

Now using the existence of the limit points and the PE condition on ϕ_k relative to them, we compute

$$\begin{aligned} &\lim_{T\to\infty} \frac{1}{T} \sum_{k=0}^{T} E\{\pi_k V^* \phi_k \phi_k^* V \pi_k^*\} && (19) \\ &\geq \lim_{T\to\infty} \sum_{i=1}^{N} \frac{1}{T} \sum_{k\in\mathcal{I}_i}^{T} E\{\pi_k V^* \phi_k \phi_k^* V \pi_k^*\} && (20) \\ &= \sum_{i=1}^{N} \lim_{T\to\infty} \frac{1}{T} \sum_{k\in\mathcal{I}_i}^{T} E\{\bar{\pi}_i V^* \phi_k \phi_k^* V \bar{\pi}_i^*\} && (21) \\ &= \sum_{i=1}^{N} \bar{\pi}_i V^* Q_i V \bar{\pi}_i^*, && (22) \end{aligned}$$

where

$$\bar{\pi}_i := \begin{bmatrix} 1 \; \bar{p}_i \ldots \bar{p}_i^{N-1} \end{bmatrix},$$

and $\bar{p}_i$ are the limit points. Now the expression in (22) can be rewritten as

$$\sum_{i=1}^{N} \bar{\pi}_i V^* Q_i V \bar{\pi}_i^* \tag{23}$$

$$\geq \sum_{i=1}^{N} \bar{\pi}_i \bar{\pi}_i^* \lambda_{min}(V^* Q_i V) \tag{24}$$

$$\geq (\sum_{i=1}^{N} \bar{\pi}\bar{\pi}^*)(\min_i \lambda_{min}(V^* Q_i V)) \quad (25)$$

$$\geq \text{trace}\left(\bar{\Pi}\bar{\Pi}^*\right)\left(\min_i \lambda_{min}(V^* Q_i V)\right), \quad (26)$$

where the matrix $\bar{\Pi}$ is constructed from the limit points by

$$\bar{\Pi} := \begin{bmatrix} 1 & \bar{p}_1 & \dots & \bar{p}_1^{N-1} \\ \vdots & & & \vdots \\ 1 & \bar{p}_N & \dots & \bar{p}_N^{N-1} \end{bmatrix}.$$

Note that $\bar{\Pi}$ is a Vandermonde matrix, and is non-singular since the points p_i are distinct. The non-singularity of $\bar{\Pi}$, the fact that V is non-zero, together with the positive definiteness of Q_i implies that the right hand side of (26) is strictly positive. This in turn implies that condition (18) is satisfied, and the theorem is proved. □

Remark: Though the conditions of theorem 4 may seem a little technical, they can be intuitively understood. Assuming that the input is sufficiently rich to insure that ϕ_k is PE in the above sense, what is needed is that the trajectory of p_k "visit" N distinct points infinitely many times. Condition (15) means that the rate at which p_k revisits each of these limit points should *not slow down*. Moreover, each of these revisits should be sufficient to ergodically extract the correlation data of ϕ_k.

The above remark can be further elaborated as follows. The LPV model is essentially a parametrized family of LTI models. If the parameter determines a process set point, a possible identification scheme is to fix each set point, and collect enough small signal data to identify the LTI model at that set point. The identified LTI coefficients can then be used as interpolation points to find the coefficients as polynomial functions of the parameter p. This procedure would produce a similar model to ours, and would require at least N LTI models at distinct set points. The identification schemes we present in this paper essentially carry this process out in "one shot". The condition (15) insures that enough data is gathered around the N interpolation points. The advantage of our schemes is that the set point does not have to vary slowly. For example, periodic parameter trajectories of any period (as long as N points are covered) will be sufficient for identification. A case of non-periodic trajectories that are still sufficient for identification is illustrated in the example.

5. EXAMPLE

We illustrate our algorithms using a simple example involving the following LPV model:

$$y_k + a_1(p)y_{k-1} + a_2(p)y_{k-2} = b_0(p)u_{k-1} + b_1 u_{k-2} + \eta_k$$

where the functional dependence in p of the parameters is given by polynomials of order $N = 3$:

$$a_1(p) = 1 - 0.5p + 0.2p^2$$
$$a_2(p) = 1 - 0.7p - 0.1p^2$$
$$b_0(p) = 0.5 - 0.4p + 0.01p^2$$
$$b_1(p) = 0.2 - 0.3p + 0.02p^2$$

and η_k was chosen to be a random white noise sequences with variance 10^{-3}.

The *true* parameter matrix is then:

$$\Theta_o := \begin{bmatrix} 1 & -0.5 & 0.2 \\ 1 & -0.7 & -0.1 \\ 0.5 & -0.4 & 0.01 \\ 0.2 & -0.3 & 0.02 \end{bmatrix}. \quad (27)$$

In this case the *extended regressor* Ψ is:

$$\Psi_k := \phi_k \pi_k := \begin{bmatrix} -y_{k-1} \\ -y_{k-2} \\ u_{k-1} \\ u_{k-2} \end{bmatrix} \begin{bmatrix} 1 & p_k & p_k^2 \end{bmatrix},$$

A simulation was carried out with $T = 4500$.

The input u_k is chosen as random noise with uniform distribution in the interval $[-1, 1]$.

The sequence $p_k = \sin(\frac{\pi}{3}k)$ and, respectively, $p_k = \sin(k)$ was used for the parameter trajectory,

For $p_k = \sin(\frac{\pi}{3}k)$ the conditions of theorem 4 are clearly satisfied, visiting it infinitely time three point in the range $[-1, 1]$ at a linear rate.

We consider also $p_k = \sin(k)$, because since 1 is not a rational multiple of the transcendental number π, $\sin(k)$ is not a periodic function of k, but its range of values is dense in $[-1, 1]$. Moreover, given any neighborhood of a point in $[-1, 1]$, $\sin(k)$ we guess that it will revisit this neighborhood at a rate of order k. Thus the conditions of theorem 4 are also satisfied.

The convergence of the LMS algorithm with $\alpha = 0.025$ was somewhat slower. The estimated matrix at time $k = T$ for $p_k = \sin(\frac{\pi}{3}k)$ was

$$\hat{\Theta}_k = \begin{bmatrix} 1.00 & -0.5 & 0.2 \\ 1.0 & -0.7 & -0.1 \\ 0.5 & -0.4 & 0.004 \\ 0.2 & -0.3 & 0.02 \end{bmatrix}, \quad (28)$$

and the convergence of the parameters is shown in figure 1.

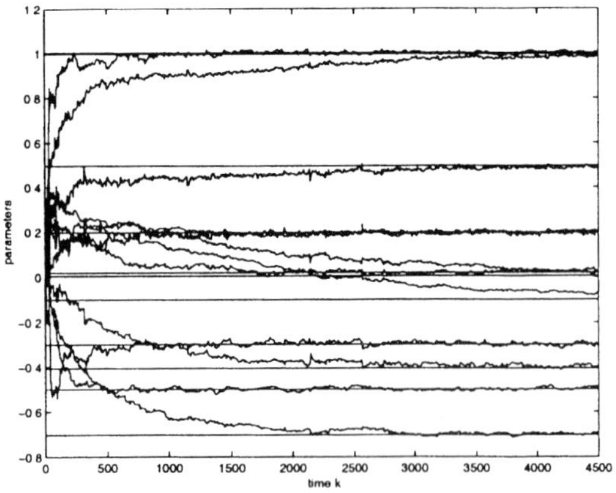

Fig. 1. LMS algorithm: parameters' convergence

The estimated matrix at time $k = T$ for $p_k = \sin(k)$ was

$$\hat{\Theta}_k = \begin{bmatrix} 1.00 & -0.5 & 0.2 \\ 0.99 & -0.7 & -0.1 \\ 0.49 & -0.4 & 0.007 \\ 0.2 & -0.3 & 0.02 \end{bmatrix}, \qquad (29)$$

and the convergence of the parameters is shown in figure 2.

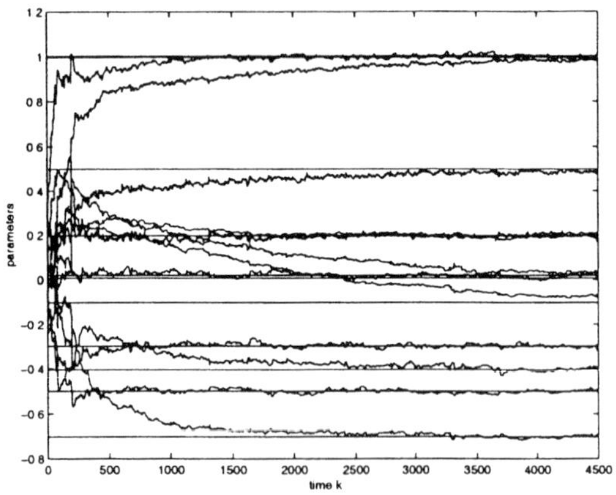

Fig. 2. LMS algorithm: parameters' convergence

6. CONCLUSIONS

We have considered Least Squares type identification problems for LPV systems with a general dependence on the parameters. For this particular class, we have shown that the problem can be solved by linear regression. By representing the regression as a matrix inner product, we derived simple representation of the LMS algorithms, and conditions for persistency of excitation.

The persistency of excitation conditions have a natural interpretation in terms of function interpolation. The algorithms converge with arbitrary changes of the parameters representing the set points, as long as certain "interpolating" values are visited regularly. This removes the typical restrictions of having to vary the set points slowly while collecting enough data to identify local LTI models.

7. REFERENCES

Bamieh, B. and L. Giarré (1999). Identification for linear parameter varying models. *In Proc. of the 38^{t}h IEEE Control and Decision Conference.*

Blanchini, F. (1999). The gain scheduling and the robust state feedback stabilization problems. *Proc. of the IEEE Conf. on Decision and Control.*

Campi, M.C. (1994). Exponentially weighted least squares identification of time-varying systems with white dusturbances. *IEEE Trans. on Signal Processing* **42**, 2906–2914.

Hanson, R.D. and Tsu-Chin Tsao (1998). Reducing cutting force induced bore cylindricity errors by learning control and variable depth of cut machining. *Trans. ASME, J. Manuf. Sci. Eng.* **120**(3), 547–54.

Lee, L. and K. Poolla (1999). Identification of linear parameter-varying systems using nonlinear programming. *ASME Journal of Dynamic Systems, Measurement, and Control* **121**, 71–78.

M. Nemani, R. Ravikanth and B. Bamieh (1995). Identification of linear parametrically varying systems. *Proc. of the 34'th IEEE Control and Decision Conference.*

Packard, A. (1994). Gain scheduling via linear fractional transformations. *System and Control Letters* **22**, 79–92.

Previdi, F. and M. Lovera (1999). Identification of a class of linear models with nonlinearly varying parameters. *Proc. of the V European Control Conference.*

Rugh, W.J. (1991). Analytical framework for gain-scheduling. *IEEE Control Systems Magazine* **11**, 79–84.

Shamma, J. and D. Xiong (1999). Set-valued methods for linear parameter varying systems. *Automatica* **35**, 1081–1089.

Shamma, J. and M. Athans (1991). Guaranteed properties of gain scheduled control for linear parameter varying plants. *Automatica* **27**(3), 559–564.

Södeström, S. and Stoica (1989). *System Identification.* Prentice Hall International.

IDENTIFICATION OF ARX MODELS WITH TIME-VARYING BOUNDED PARAMETERS: A SEMIDEFINITE PROGRAMMING APPROACH

Laurent El Ghaoui *,[1] **Giuseppe Calafiore** **

* *Department of Electrical Engineering and Computer Sciences, University of California at Berkeley, Berkeley, CA, 94720, USA. Internet: elghaoui@eecs.berkeley.edu*
** *DAUIN - Politecnico di Torino, Corso Duca degli Abruzzi 24, 10129 Torino, Italy. Internet: calafiore@polito.it*

Abstract: In this paper, we develop a new identification procedure for ARX models with uncertain, unknown-but-bounded time-varying coefficients. The method seeks the smallest ellipsoid containing the coefficients, such that the resulting model in unfalsified by the observed data. The problem is formulated as a semidefinite programming program. We interpret our deterministic method in a probabilistic setting, and show how this interpretation can be used to effectively discard outliers. We generalize the method to multivariate models with matrix unstructured uncertainty. The resulting model is directly amenable to worst-case simulation and robust control techniques.

Keywords: Identification, non-stationary systems, model validation, semidefinite programming.

1. INTRODUCTION

A discrepancy has emerged in recent years between robust control techniques (Zhou *et al.*, 1995), which require a detailed description of model uncertainty, and the traditional identification schemes (Ljung, 1987), which extract a unique model from the data. Ideally, robust control requires to build an uncertainty model from the data, for example upper and lower bounds on the coefficients of an ARX model. Unfortunately, these hard bounds are difficult to obtain from classical identification methods, which are mostly based on stochastic frameworks.

A more suitable paradigm for identification in a robust control context is that of model validation. A set of possible models is postulated, and we say that the model set is *unfalsified* by the observed data, if that data could actually have been produced by one member in the model set. The purpose of model validation is thus to asses whether a given class of models, with a-priori given bounds on the uncertainty, could have produced the observed data. Model validation has been the subject of several studies, in the framework of LTI systems affected by dynamic structured uncertainty in Linear Fractional form (LFT). The solution to the time domain model validation problem for structured dynamic uncertainty has been proven to be NP-hard, (Toker and Chen, 1998) in the general case, while it is tractable in the case of additive structured uncertainty, (Chen and Wang, 1995; Rangan and Poolla, 1998). A frequency domain version of the model validation problem is also treated in (Rangan and Poolla, 1998; Chen and Wang, 1996).

One of the main difficulties encoutered with model validation approaches stems from their high sensitivity to outliers in the data. Indeed, a determin-

[1] Partially supported by DARPA SEC under a grant managed by AFRL 33615-98-C-3614.

istic, worst-case point of view basically puts equal weight on all observations.

In this paper, we will consider ARX model structures with time-varying coefficients which are postulated to belong to an ellipsoidal set. This model set is suitable for modeling phenomena whose underlying data generation mechanism is uncertain and time-varying, including non-linear behaviors, multiplicative noise, and non-stationary uncertainty. Based on time-domain data, we seek the smallest ellipsoidal set in the space of the coefficients of the model, such that the resulting model is unfalsified by the data. Our main result is that the problem can be solved exactly in polynomial-time, using (convex) semidefinite programming (Vandenberghe and Boyd, 1996). Our approach can be termed "optimal model unfalsification" (OMU) in that the bounds on the coefficients of the model (the ellipsoid that is identified) are not assumed *a priori*. We also address the issue of outlier detection, which is crucial in the context of bounded noise approaches to estimation, by expressing the problem in a stochastic framework. We extend our results to a class of multivariable systems with unstructured uncertainty.

The identification scheme results in a model with time-varying, norm-bounded coefficients that is directly amenable to worst-case simulation (El Ghaoui and Calafiore, 1999) and robust control techniques (Boyd *et al.*, 1994) for systems with *unstructured uncertainty.* The absence of structure in the uncertainty makes the simulation and control problems very simple computationally.

The proposed approach is quite different from the set-membership estimation framework. In this latter case, the mechanism generating the data is assumed to be a linear time invariant system subject to bounded additive disturbance, and the identification procedure is aimed at determining the set of all *constant* parameters that are consistent with observed data. Extensive references to the set-membership identification approach may be found in (Milanese *et al.*, 1996).

Notation

For a square matrix X, $X \succ 0$ (resp. $X \succeq 0$) means X is symmetric, and positive-definite (resp. semidefinite). For $\Theta \in \mathbf{R}^{n \times n}$, with $\Theta \succ 0$, and $\hat{\theta} \in \mathbf{R}^n$, the notation $\mathbf{E}(\Theta, \hat{\theta})$ denotes the ellipsoid

$$\mathbf{E}(\Theta, \hat{\theta}) = \left\{ \xi \;\middle|\; \Theta \succeq (\hat{\theta} - \xi)(\hat{\theta} - \xi)^T \right\}.$$

Note that the above expression allows for singular Θ, that is, "flat" ellipsoids, for example singletons. The matrix Θ determines the "shape" of the ellipsoid (its eigenvalues are the squared semi-axis lengths), while the vector $\hat{\theta}$ is the center. The size of the ellipsoid is chosen to be measured by the sum of the squared semi-axis lengths, which is equal to $\mathbf{Tr}\,\Theta$ (the trace of Θ).

2. ARX MODELS

We consider discrete-time, time-varying AR model of the form

$$y(k) = \sum_{i=1}^{n} \theta_i(k) y(k-i) + e(k) \qquad (1)$$

where $y(k)$ is the measured output, and the vector of coefficients $\theta(k) \in \mathbf{R}^n$ is time-varying. The additive noise term is deterministic, bounded by $|e(k)| \leq \gamma$, where $\gamma \geq 0$ is given.

The *Deterministic Optimal Model Unfalsification* (DOMU) problem is as follows: Given a time horizon $T > 0$, a data sequence $y(0), \ldots, y(n+T-1)$, find the smallest (in the sense defined above) ellipsoid $\mathbf{E}(\Theta, \hat{\theta})$ in ithe space of parameters such that the model defined by (1) is unfalsified by the data, *i.e.*,

$$\begin{array}{l} \forall\, k,\ n \leq k \leq n+T-1,\ \exists\, \theta(k) \in \mathbf{E}(\Theta, \hat{\theta}), \\ \left| y(k) - \sum_{i=1}^{n} \theta_i(k) y(k-i) \right| \leq \gamma. \end{array} \qquad (2)$$

The proposed methodology amounts to assume that the observed signal has been generated by a linear time-varying system with deterministic, multiplicative *and* additive perturbation; we seek the "smallest" multiplicative perturbation that makes the system consistent, for a given level of additive perturbation.

2.1 *SDP solution*

We seek an ellipsoid $\mathbf{E}(\Theta, \hat{\theta})$ of minimal size $\mathbf{Tr}\,\Theta$, such that for every k, $n \leq k \leq n+T-1$, (1) holds for some $\theta(k) \in \mathbf{E}(\Theta, \hat{\theta})$ and $|e(k)| \leq \gamma$. For every k, and given $e(k)$, $|e(k)| \leq \gamma$, the hyperplane defined by (1) intersects the ellipsoid $\mathbf{E}(\Theta, \hat{\theta})$ if and only if

$$x(k)^T \Theta x(k) \geq (x(k)^T \hat{\theta} - y(k) - e(k))^2,$$

or, equivalently,

$$\begin{array}{l} \begin{bmatrix} x(k)^T \Theta x(k) & m(k) \\ m(k) & 1 \end{bmatrix} \succeq 0, \text{ where} \\ m(k) = x(k)^T \hat{\theta} - y(k) - e(k). \end{array} \qquad (3)$$

This yields the following result.

Theorem 1. A deterministic optimal unfalsified model is obtained by solving the following semidefinite programming (SDP) optimization problem in variables Θ, $\hat{\theta}$ and $e(k)$, $1 \le k \le T$

$$\begin{array}{c} \text{minimize } \mathbf{Tr}\,\Theta \text{ subject to} \\ \Theta \succeq 0, \begin{bmatrix} x(k)^T \Theta x(k) & m(k) \\ m(k) & 1 \end{bmatrix} \succeq 0, \\ |y(k) - x(k)^T \hat{\theta} - m(k)| \le \gamma, \\ n \le k \le n + T - 1. \end{array} \quad (4)$$

The convex optimization problem (4) is a semidefinite program, that is, it can be expressed as the minimization of a linear objective under linear matrix inequality (LMI) constraints:

$$\begin{array}{c} \text{minimize } c^T z \text{ subject to} \\ F(z) := F_0 + \sum_{i=1}^{m} z_i F_i \succeq 0, \end{array} \quad (5)$$

where $m = n(n+3)/2$ is the number of independent variables in $\Theta, \hat{\theta}$, z is a vector containing those variables, $c \in \mathbf{R}^m$ is an objective vector, and the F_i's are given symmetric, block-diagonal matrices consisting of T blocks of order 2, and one block of order n.

The SDP (4) can be solved very efficiently in a time polynomial in n (the order of the system) and T (the horizon of analysis). The precise complexity result is: using a *general-purpose* interior-point method such as the one described in (Vandenberghe and Boyd, 1996), the number of iterations needed to compute an ϵ-optimal solution grows as $O(\sqrt{n})$ in the worst-case, and is constant (between 50 and 200, depending on the method) in practice. Each iteration requires solving a least-squares problem, the complexity of which grows as $O(T^2 n^6)$. With no doubt, it is possible to exploit the particular structure of the above problem for greater efficiency; this is left for future research.

Remark 1. If we postulate that the optimal shape matrix Θ is positive-definite at the optimum, then we can ignore the inactive constraint $\Theta \succeq 0$ in the above SDP. The problem then reduces to a simple convex quadratically constrained problem, which can be solved in $O(n^3)$.

Remark 2. It is possible to develop a recursive version of the algorithm, by which the ellipsoid is updated with new each measurement. For details, we refer to the forthcoming paper (Ghaoui and Calafiore, 2000).

2.2 *A stochastic interpretation*

The above result can be interpreted in a stochastic context, using ideas from large deviations theory. This interpretation will prove useful in the context of outlier detection, which is discussed in the next section.

We assume now that the coefficient vector $\theta(k)$ is random, and obeys to a *time-varying* distribution, with *constant* mean $\hat{\theta}$ and covariance matrix Θ. We are given a confidence level η ($0 < \eta < 1$), and a parameter $\gamma > 0$. We now pose the *Stochastic Optimal Model Unfalsification* (SOMU) problem: Given a time horizon $T > n$, a data sequence $y(0), \ldots, y(T-1)$, find the smallest (in the sense of the trace) covariance matrix Θ, and a mean vector $\hat{\theta}$, such that, for every k, $1 \le k \le T$, there exist a distribution on the coefficient vector $\theta(k)$ having mean $\hat{\theta}$ and covariance matrix Θ that satisfies

$$\mathbf{Prob}\left\{ \left| y(k) - \sum_{i=1}^{n} \theta_i(k) y(k-i) \right| \le \gamma \right\} \ge \eta.$$

In the above problem, the parameter process $\theta(k)$ is stationary, in the sense its first and second moment are constant in time. However, the resulting data-generation process $y(k)$ is not, in general, stationary.

Define

$$p_k(\hat{\theta}, \Theta) = \sup \mathbf{Prob}\left\{ \left| y(k) - \sum_{i=1}^{n} \theta_i(k) y(k-i) \right| \le \gamma \right\},$$

where the sup is taken over all distributions with mean $\hat{\theta}$ and covariance matrix Θ. Our problem is expressed as

$$\begin{array}{c} \text{minimize } \mathbf{Tr}\,\Theta \text{ subject to} \\ \Theta \succeq 0, \;\; p_k(\hat{\theta}, \Theta) \ge \eta, \;\; n \le k \le n + T - 1. \end{array} \quad (6)$$

The following is a direct application of very general results reported in (Bertsimas and Sethuraman, 2000). We have

$$\sup \mathbf{Prob}\{|\xi| \le \gamma\} = \frac{1}{1 + d^2},$$

where the sup is taken over all distributions of a scalar, random variable ξ with given mean μ and variance $\sigma > 0$, and where

$$d^2 := \inf_{|e| \le \gamma} \frac{(e - \mu)^2}{\sigma^2}.$$

We conclude that the sup above is greater than η for some distribution with mean μ and variance $\sigma \ge 0$ if and only if there exists a scalar e such that

$$|e| \le \gamma, \;\; (e - \mu)^2 \le \frac{1 - \eta}{\eta} \sigma^2.$$

(Note that the result is trivial when $\sigma = 0$, and follows from the above otherwise.)

Let us now return to our problem, and assume that $\theta(k)$ obeys to a time-varying distribution with mean $\hat{\theta}$ and covariance matrix Θ. For every k, the random variable $y(k) - x(k)^T\theta(k)$ has mean $y(k) - x(k)^T\hat{\theta}$ and squared variance $x(k)^T\Theta x(k)$. Thus, $p_k(\hat{\theta}, \Theta) \geq \eta$ if and only if there exists a variable $e(k)$ such that $|e(k)| \leq \gamma$ and

$$(y(k) - x(k)^T\hat{\theta} - e(k))^2 \leq \frac{1-\eta}{\eta} x(k)^T\Theta x(k).$$

This yields the following result.

Theorem 2. The optimization problem (6) is equivalent to the semidefinite programming problem in variables $\hat{\theta}, \Theta$ and $e(k)$, $1 \leq k \leq T$:

$$\begin{array}{c} \text{minimize } \mathbf{Tr}\,\Theta \text{ subject to} \\ \Theta \succeq 0, \begin{bmatrix} \frac{1-\eta}{\eta} x(k)^T\Theta x(k) & m(k) \\ m(k) & 1 \end{bmatrix} \succeq 0, \\ |y(k) - x(k)^T\hat{\theta} - m(k)| \leq \gamma, \\ n \leq k \leq n+T-1. \end{array}$$

We observe the close connection between the above result and the one derived in a purely deterministic manner in theorem 1. Indeed, by setting $\eta = 1/2$ in the LMI conditions above, we recover those obtained in the latter theorem. In words, the deterministic OMU problem is equivalent to searching the "smallest" covariance matrix such that there exist a distribution that "explains" the data, up to a confidence level of 1/2.

We also observe that requiring a confidence level $\eta = 1$ requires that, for every k, $1 \leq k \leq T$, we have

$$|x(k)^T\hat{\theta} - y(k)| \leq \gamma.$$

In this case, there exists a *constant* parameter $\hat{\theta}$ such that the corresponding AR model explains the data. This condition holds if and only if the deterministic OMU problem has a solution with $\Theta = 0$ (that is, the ellipsoid $\mathbf{E}(\Theta, \hat{\theta})$ is reduced to a singleton).

Likewise, setting $\eta \rightarrow 0$ makes the constraints in the stochastic OMU problem feasible for any $\Theta \succ 0$. This should be expected: relaxing our level of confidence makes any covariance matrix feasible.

Finally, we observe that a solution $(\hat{\theta}, \Theta)$ to the deterministic OMU problem provides an immediate solution $(\hat{\theta}_\eta, \Theta_\eta)$ to the stochastic OMU problem for an arbitrary value of η: it suffices to set

$$\hat{\theta}_\eta = \hat{\theta}, \quad \Theta_\eta = \frac{\eta}{1-\eta}\Theta.$$

This means that, by scaling the ellipsoid obtained in a deterministic setting, by a factor $\frac{\eta}{1-\eta}$, we obtain the *family* of ellipsoids that solve the stochastic OMU problems with confidence level η.

2.3 *Outlier detection*

Detecting outliers is especially important in the context of bounded noise uncertainty, since every measurement is taken into account. The parallel between the stochastic and deterministic approach allows us to develop an efficient heuristic for outlier detection.

Consider a batch of data, for which we solve the deterministic OMU problem, based on the conditions of theorem 1, for a time period $n \leq k \leq n+T-1$. This result in an ellipsoid $\mathbf{E}(\Theta, \hat{\theta})$. Now consider a new data $y(n+T)$. We can easily test if $y(n+T)$ falsifies our model or not: it suffices to check if there exist $e(n+T)$ such that

$$\begin{array}{l} x(n+T)^T\Theta x(n+T) \geq \\ (x(n+T)^T\hat{\theta} - y(n+T) - e(n+T))^2, \\ |e(n+T)| \leq \gamma. \end{array} \quad (7)$$

(Note that $x(n+T)$ contains values of y up to time $n+T-1$ only, and is postulated to be correct.) What do we do if $y(n+T)$ does not satisfy the above condition?

We face two possibilites. Either we declare the model falsified, that is, the ellipsoid $\mathbf{E}(\Theta, \hat{\theta})$ should be recomputed, using the new measurement (that is, with (7) as an additional constraint). Another possibility is to declare the data that causes problem to be an outlier that should be discarded.

To decide upon this issue, we set an outlier threshold level η_{th}, with $0 < \eta_{\text{th}} < 1/2$. Define $\eta(n+T)$ to be the largest η such that the condition

$$\begin{array}{l} \frac{1-\eta}{\eta} x(n+T)^T\Theta x(n+T) \geq \\ (x(n+T)^T\hat{\theta} - y(n+T) - e(n+T))^2, \\ |e(n+T)| \leq \gamma \end{array}$$

holds for some $e(n+T)$. (Computing $\eta(n+T)$ is a simple exercise left to the reader.) If $\eta(n+T) < \eta_{\text{th}}$, we declare the new measurement $y(n+T)$ to be an outlier, and we discard it. Otherwise, we decide that the data falsifies the model and that a new ellipsoid $\mathbf{E}(\Theta, \hat{\theta})$ should be computed.

If we are to discard the new data $y(n+T)$, we must replace it with an estimate. Clearly, our best estimate obtains when $\theta(n+T)$ is set to the center of the ellipsoid $\mathbf{E}(\Theta, \hat{\theta})$, and $e(T) = 0$. The estimate is

$$\hat{y}(n+T) = x(n+T)^T\hat{\theta}.$$

2.4 *One-step ahead prediction*

Assuming that the parameter vector $\theta(k)$ lies in $\mathbf{E}(\Theta, \hat{\theta})$ for $k = n, \ldots, n+T-1$, we can compute the predicted bounds for the next output iterate $y(T+n)$. The possible values for this output lie in an interval $[y^{+}(n+T)\ y^{-}(n+T)]$, where

$$y^{\pm}(n+T) = \hat{\theta}^T x(n+T) \pm \left(\gamma + \sqrt{x(n+T)^T \Theta x(n+T)}\right), \quad (8)$$

Thus, the one-step ahead prediction error is $\gamma + \sqrt{x(n+T)^T \Theta x(n+T)}$.

If the purpose of the identification is prediction in the above sense, it is clear from the above result that we could modify the objective in the optimization problem (4) in order to optimize the prediction error. This is done by replacing the objective $\mathbf{Tr}\,\Theta$ in the SDP by the squared prediction error $x(n+T)^T \Theta x(n+T)$. The resulting problem is still an SDP, with objective of the form $\mathbf{Tr}\,\Theta W$, where $W = x(n+T)x(n+T)^T$.

In our experiments, we have found it is better to account for the prediction errors at every step, and include some forgetting factor. The corresponding objective is of the form $\mathbf{Tr}\,\Theta W$, with

$$W = \sum_{k=0}^{N} \eta^k x(n+T-k)x(n+T-k)^T,$$

where $\eta > 0$ is the forgetting factor, and $N < n+T$ is the window length.

3. MULTIVARIATE MODELS

The simple ARX model in (1) is a special case of multivariate models of the form

$$\begin{aligned} y(k+1) &= (A + L\Delta(k)R)\, x(k) + e(k), \\ \|z(k)\|_2 &\leq 1, \end{aligned} \quad (9)$$

where $x(k) = (y(k-1), \ldots, y(k-n))$ is the state, and A, L, R are appropriate matrices that depend on the parameters $\hat{\theta}, \Theta$ computed by the identification procedure. The above model can be interpreted as one with norm-bounded, unstructured uncertainty (in the sense classical in robust control).

We may generalize the previous approach to arbitrary models of the form above. For simplicity, we set $e(k) = 0$, and seek the "bset" model parameters A, L, R that are consistent with data values $y(k), x(k)$. Let us introduce new variables,

$$P := LL^T \in \mathbf{R}^{n_y \times n_y}, \quad Q := R^T R \in \mathbf{R}^{n_u \times n_u},$$

We would like to minimize the "size" of the set $\mathbf{U} := \{A + L\Delta R \mid \|\Delta\| \leq 1\}$, subject to the unfalsification conditions. Note that $\mathbf{U}$ can be interpreted as an "ellipsoid" in a matrix space equipped with the largest singular value norm. One obvious measure of the size is $\mathbf{Tr}\,P + \mathbf{Tr}\,Q$. By homogeneity, we may without loss of generality set $\mathbf{Tr}\,P = 1$. We obtain the following result.

Theorem 3. An optimal unfalsified model for (9) is obtained by solving the following SDP optimization problem in variables A, P, Q:

$$\begin{aligned} &\text{minimize } \mathbf{Tr}\,Q \text{ subject to} \\ &Q \succeq 0, \ \ \mathbf{Tr}\,P = 1, \\ &\begin{bmatrix} P & m(k) \\ m(k)^T & x(k)^T Q x(k) \end{bmatrix} \succeq 0, \\ &|y(k) - Ax(k) - m(k)| \leq \gamma, \\ &n \leq k \leq n+T-1. \end{aligned} \quad (10)$$

At the optimum, an unfalsified model is given by (9), with L a full column rank matrix such that $LL^T = P$, and R a full row matrix such that $Q = R^T R$.

The above result reduces to the basic theorem 1 when we impose an additional linear constraint on the variable A, that A should be in upper companion form (the coefficients of which have the interpretation of the variable $\hat{\theta}$ used in the previous section).

4. NUMERICAL EXAMPLES

We have generated data by a time-varying ARX(3,2) process with a prevalence of multiplicative noise over additive noise.

We applied the OMU paradigm to identify an uncertainty model of the form (1), from the input-output observations. We assumed the correct order of the model (i.e. $n = 3$ and $m = 2$) in the identification step, and then computed robust bounds on one-step ahead predictions, using the technique described in Section 2.4. The predictions are computed using a new set of validation data, different from the one used for identification. We also compared the so obtained results with classical probabilistic bounds obtained identifying a standard stochastic ARX(3,2) model on the same data. The results are displayed in Figure 1, where the shaded region represents 98% confidence bounds on the predictions, computed according to the standard stochastic approach.

It is worth to notice that the 98% bounds computed from the stochastic ARX model are actually conservative with respect to the robust ones. This is to be ascribed to the fact that the standard ARX tries to explain the data using only additive noise, while the observed process has a prevalence of multiplicative noise. A plot of the type of that

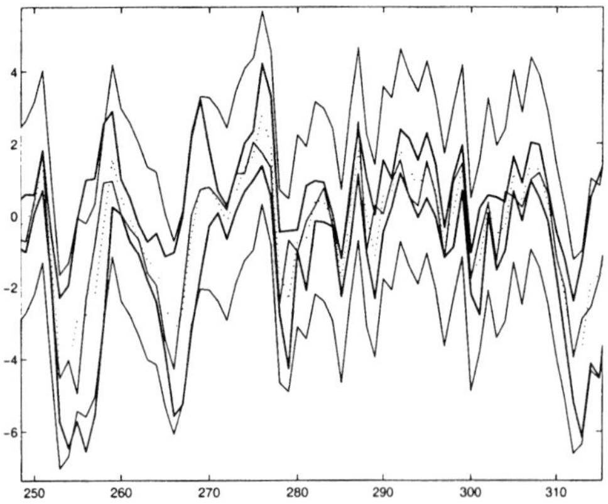

Fig. 1. Comparison of robust predictions: thick lines represent the robust bounds obtained using the OMU method; the shaded region represent the 98% confidence region resulting from standard stochastic ARX identification. The solid line is the actual data, and the dotted line is the central prediction from the OMU method.

in Figure 1 may therefore be revealing for a better understanding of the uncertainty phenomena affecting the data.

We can easily compute a trade-off curve between the level of additive uncertainty (measured by the parameter γ bounding the additive noise $e(k)$ in (1)), and that of multiplicative uncertainty (measured by the square root of the trace of the optimal matrix Θ). Using the data used previsously, we obtained the curve shown in Figure 2.

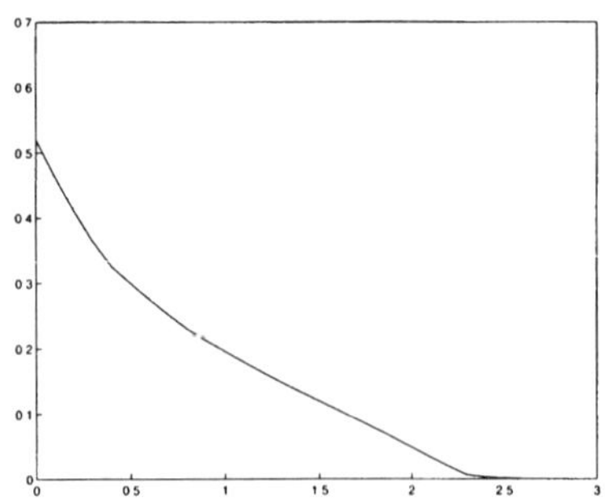

Fig. 2. A point in the shaded region represents a level of additive uncertainty γ, and a level of multiplicative uncertainty (measured as the square root of the trace of matrix Θ) that permit to explain the observed data in the unfalsification sense.

5. CONCLUDING REMARKS

In this paper, we outlined a method that for modeling non-stationary times using ARX models with time-varying, bounded coefficients, and additive, bounded uncertainty. We have derived an efficient numerical solution based on convex, semidefinite programming. We have shown how we can interpret our results in a stochastic framework and used this interpretation for outlier detection purposes. Finally, we have shown how to generalize our results to a multivariate, norm-bound model; the basic assumption is that the whole state should be measured. The resulting model is directly amenable to robust control (Zhou *et al.*, 1995) and exhaustive ("worst-case") simulation techniques (El Ghaoui and Calafiore, 1999). Further research will elaborate on this connection.

6. REFERENCES

Bertsimas, D. and J. Sethuraman (2000). Moment problems and semidefinite optimization. In: *Semidefinite Programming and Applications* (R. Saigal, L. Vandenberghe and H. Wolkowicz, Eds.). Kluwer Academic.

Boyd, S., L. El Ghaoui, E. Feron and V. Balakrishnan (1994). *Linear Matrix Inequalities in System and Control Theory*. Vol. 15 of *Studies in Applied Mathematics*. SIAM. Philadelphia, PA.

Chen, J. and S. Wang (1995). Validation of linear fractional uncertain models: solutions via matrix inequalities. In: *Proc. American Control Conf.*

Chen, J. and S. Wang (1996). Validation of linear fractional uncertain models: solutions via matrix inequalities. *IEEE Trans. Aut. Control* **41**(6), 844–849.

El Ghaoui, L. and G. Calafiore (1999). Worst-case simulation of uncertain systems. In: *Robustness in Identification and Control* (A. Tesi A. Garulli and A. Vicino, Eds.). Vol. 245 of *Lecture Notes in Control and Information Sciences*. Springer.

Ghaoui, L. El and G. Calafiore (2000). Identification of time-varying systems with norm-bounded coefficients: A semidefinite programming approach. In preparation.

Ljung, L. (1987). *System Identification: Theory for the User*. Prentice-Hall.

Milanese, Mario, Norton, John, Piet-Lahanier, Hélène and Walter, Eric, Eds.) (1996). *Bounding Approaches to System Identification*. Plenum Press. New York, NY, USA.

Rangan, S. and K. Poolla (1998). Model validation for structured uncertainty models. In: *Proc. American Control Conf.*. pp. 629–633.

Toker, O. and J. Chen (1998). On computational complexity of invalidating structured uncertainty models. *Syst. Control Letters* **33**(3), 199–207.

Vandenberghe, Lieven and Stephen Boyd (1996). Semidefinite programming. *SIAM Review* **38**(1), 49–95.

Zhou, K., J. Doyle and K. Glover (1995). *Robust and Optimal Control*. Prentice Hall, New Jersey.

H^∞ PARAMETER IDENTIFICATION FOR INFLIGHT DETECTION OF AIRCRAFT ICING: THE TIME-VARYING CASE ⋆

James W. Melody * **Thomas Hillbrand** ** **Tamer Başar** * **William R. Perkins** *

* *Coordinated Science Laboratory, University of Illinois; 1308 West Main Street; Urbana, IL 61801-2307 / USA*
** *Darmstadt University of Technology, Institute of Automatic Control, Laboratory of Control Systems and Process Automation, Landgraf-Georg-Strasse 4, D-64283 Darmstadt, Germany*

Abstract: Inflight parameter identification of aircraft flight dynamics is considered in this paper in the context of an ice management system. In particular, an H^∞ parameter identification algorithm is evaluated in terms of detecting an aircraft icing event when only a noisy state measurement is available. While previous studies have addressed identification during a pilot-induced maneuver, taking advantage of the excitation due to input, this paper addresses identification during cruise, where excitation is provided only by unknown disturbances due to turbulence. Simulation results show that for moderate turbulence levels, the H^∞ algorithm provides a timely and unambiguous icing indication. *Copyright © 2000 IFAC*

Keywords: Aircraft Control, H^∞, Identification, Time-Varying Systems, Robustness

1. INTRODUCTION

Aircraft icing accidents result from the effect of ice accretion on the performance, stability, and controllability of the aircraft. Small to moderate amounts of icing will generally cause a reduction in aircraft performance in terms of climb rates, range, endurance, and maximum speed and acceleration. This regime of icing effects is known as performance events. As icing increases, separation of air flow from the flight surfaces can cause loss of pilot control and even wildly unstable behavior. These more severe icing events, known as handling events, are often precipitated by a change in the aircraft configuration (*e.g.*, flap retraction) or an aircraft maneuver effected by a pilot unaware of the flight dynamics degradation. These handling events can be classified into either tailplane stall, where the aircraft pitches forward, or asymmetric wing effects causing a roll upset. Two aspects of icing accidents are particularly of interest: 1) icing is of concern insofar as it affects the aircraft flight dynamics, and 2) icing events can be exacerbated when the pilot takes action while unaware of the severity of these effects.

The new approach to aircraft icing safety, advanced in Bragg et al. (1998), is to provide the pilot with a near real-time indication of the effect of ice on the aircraft performance, stability, and controllability. Current aircraft inflight ice protection systems (IPS) consist of de-icing and anti-icing systems that respectively remove ice or inhibit accretion, as well as rudimentary stall protection systems that seek to restrict the aircraft to some presumed safe region of operation, or "flight envelope," by limiting pilot input. In addition to these existing ice protection systems, this approach adds an Ice Management System (IMS) that provides three functions: 1) sensing of the presence of ice accretion and its effect on the flight dynamics; 2) activation and management of the IPS while providing the pilot with a characterization of the icing effects; and 3) modification of the flight envelope in order to maintain safe operation. As an additional level of safety, control

⋆ This work was supported in part by NASA, and in part by the University of Illinois at Urbana-Champaign under the Critical Research Initiative Program.

adaptation could be added to maintain acceptable flying qualities during a performance icing event and in the worst contingency, to stabilize the aircraft after the onset of a handling event.

The IMS incorporates parameter identification of the flight dynamics, as well as external icing sensors, steady-state characterization of drag and lift using force-balance relations, and control hinge moment measurements (Bragg, 1996). Although all of these techniques can provide indications of the presence of icing, only parameter identification provides an explicit characterization of the type and severity of the icing effects on a parametrization of the flight dynamics, since it identifies the flight dynamics directly. In the icing context, the identification algorithms must estimate aircraft performance, stability, and control parameters over time, based on measurements of the aircraft state variables and control input. Furthermore, the algorithms must do so without the introduction of test excitation input since this input could itself induce a handling event. Excitation due to normal operational maneuvers must be sufficient. Since icing safety problems can occur at any time from takeoff to landing, effective ice detection and characterization must function in all phases of flight. Melody et al. (1999) have established the utility of H^{∞} parameter identification under noisy full-state measurements during maneuver, where excitation due to pilot input provides parameter convergence that results in unambiguous icing indication in a matter of seconds. In this flight regime, fast icing indication is necessary in order to provide the pilot with an icing indication before the possible onset of a handling event. In the case where there is negligible excitation, steady-state characterization via force-balance equations as discussed in Bragg et al. (2000) can be used to provide an icing indication. This paper addresses identification during steady, level flight, with moderate turbulence, attempting to fill the gap between ID during maneuver, and steady-state characterization for steady, level flight without turbulence. In this flight regime, a handling event is unlikely except in the most severe icing conditions, and hence a slower icing indication is acceptable.

2. FLIGHT DYNAMICS MODEL

In actuality, all flight dynamics are nonlinear, and in this case the flight dynamics are time-varying due to the effect of ice accretion on the aerodynamics of the flight surfaces. An ideal flight dynamics model would capture both the nonlinear nature and the time variability. A typical simplification is to linearize the flight dynamics about a "trim condition" via Taylor series expansion, and to determine the values of the first-order partial derivatives via numerical simulation, wind tunnel measurements, or flight tests (Roskam, 1982). This linearization is valid for small excursions of the state from the operating condition. Since this flight dynamics model is based on the tailplane icing research of the NASA Twin Otter reported by Ratvasky et al. (1999), only the linearized longitudinal flight dynamics are considered, given by Roskam (1982) as

$$\dot{u} = -g\theta\cos(\Theta_\circ) + (X_u + X_{T_u})u + X_\alpha\alpha + X_{\delta_E}\delta_E \quad (1)$$

$$\dot{w} = -g\theta\sin(\Theta_\circ) + Z_u u + Z_\alpha\alpha + Z_{\dot{\alpha}}\dot{\alpha} + Z_{\delta_E}\delta_E + (Z_q + U_\circ)q \quad (2)$$

$$\dot{q} = (M_u + M_{T_u})u + (M_\alpha + M_{T_\alpha})\alpha + M_{\dot{\alpha}}\dot{\alpha} + M_q q + M_{\delta_E}\delta_E \quad (3)$$

where u is the forward velocity, w is the downward velocity, q is the pitch rate, θ is the pitch angle, α is the angle of attack, and δ_E is the elevator deflection angle, the input to the system. $U_\circ$ and $\Theta_\circ$ represent the trim values of forward velocity and pitch angle, respectively, and M_*, Z_*, and X_*, are known as "stability" or "control" derivatives, as appropriate. The effects of icing are manifested in these stability and control derivatives. During an icing encounter, the stability and control derivatives will change slowly over a period of several minutes. In particular, based on the NASA Twin Otter tailplane icing flight data reported in Ratvasky and Ranaudo (1993), it has been determined that the only significant variations occur in the M_α, M_q, Z_α, Z_q, X_α and $(X_u + X_{T_u})$ stability derivatives, and in the M_{δ_E} and Z_{δ_E} control derivatives. Whereas in Melody et al. (1999) these stability and control derivatives were assumed to be constant over a period of 10 s, this paper addresses identification over longer periods of time, and hence the stability and control derivatives are now allowed to be time varying.

Table 1. Twin Otter S/C derivatives

	Clean	Iced ($\eta = 1$)
M_α	-7.86	-7.08
M_{δ_E}	-10.44	-9.40
M_q	-3.055	-2.948
Z_α	-378.7	-342.7
Z_{δ_E}	-40.30	-36.45
Z_q	-19.70	-19.43
X_α	13.71	13.90
$X_u + X_{T_u}$	-0.018	-0.020

The values for the clean and "fully" iced parameters (corresponding to a TIP simulated ice shape) are given in Table 1. These are the only two conditions for which test data exists. In order to capture a continuous variation in the stability and control derivatives, an icing severity factor η_{ice} is used. This factor represents the degree to which the aircraft has accreted tailplane ice, with $\eta_{\text{ice}} = 0$ representing a clean aircraft, and $\eta_{\text{ice}} = 1$ representing tailplane icing corresponding to the TIP simulated ice shape. A linear variation of the parameters with respect to this icing severity factor is assumed:

$$C_* = C_*^{\text{clean}}(1 + K_{C_*}\eta_{\text{ice}}) \quad (4)$$

where C_* represents any of the stability or control derivatives listed in Table 1, C_*^{clean} is its clean value, and K_{C_*} is the associated slope, determined from the parameter values corresponding to $\eta_{\text{ice}} = 1$ in Table 1.

With this model, continuous accretion of ice over time is captured by allowing η_{ice} to vary with time. Since the ice accretion rate is a function of both atmospheric conditions and the amount of ice already accreted, and furthermore since no ice accretes without proper atmospheric conditions, the differential equation

$$\dot{\eta} = \aleph_1(1 + \aleph_2\eta)d_\eta \tag{5}$$

is used as a model for ice accretion, where d_η represents the conduciveness of the atmosphere to icing. Although d_η may be measured in some sense via atmospheric sensors on board the aircraft, in this study it is assumed that d_η is unknown. The constants $\aleph_1$ and $\aleph_2$ depend on the ice collection efficiency and aerodynamics of the flight surfaces as well as the physics of ice accretion in a complicated fashion. The values of $\aleph_1$ and $\aleph_2$ are determined from an assumed icing severity profile characterized by the duration of the icing encounter, denoted by T_{cld}, and the final and middle values of the icing severity factor, $\eta_{ice}(T_{cld})$ and $\eta_{ice}(T_{cld}/2)$, respectively. (5) is a nonlinear integrator, where $\aleph_2$ represents the degree of the nonlinearity. $\aleph_2 > 0$ results in increased ice accretion rate as ice accretion increases, and $\aleph_2 < 0$ provides the reverse. Consistent with ice accretion physics, the system is driftless in the sense that the state is fixed when the input is zero. Furthermore, this simplified model precludes discontinuities in the icing severity, as produced for example by ice shedding due to activation of the IPS.

The scenario modeled is a period of steady, level flight with turbulence through a "cloud" of icing conditions. The icing encounter is specified by its duration T_{cld} and the icing severity values at T_{cld} and at $T_{cld}/2$. In the following simulations, d_η is assumed to be a raised cosine

$$d_\eta(t) = \tfrac{1}{2}(1 - \cos(2\pi t/T_{cld})).$$

For this shape, the constants $\aleph_1$ and $\aleph_2$ in (5) as functions of the assumed values of T_{cld}, $\eta_{ice}(T_{cld})$ and $\eta_{ice}(T_{cld}/2)$ are determined via integration, which yields the relations

$$\aleph_1 = \tfrac{2}{\aleph_2 T_{cld}} \log[1 + \aleph_2\eta_{ice}(T_{cld})] \tag{6}$$

$$\aleph_2 = \frac{\eta_{ice}(T_{cld}) - 2\eta_{ice}(T_{cld}/2)}{\eta_{ice}^2(T_{cld}/2)} \tag{7}$$

Two icing encounter scenarios, depicted in Figure 1, will be investigated. The first is a moderate icing encounter that has $T_{cld} = 600$ s, $\eta_{ice}(T_{cld}) = 1$, and $\eta_{ice}(T_{cld}/2) = 0.5$. The second is a rapid and severe icing encounter that is characterized by $T_{cld} = 300$ s, $\eta_{ice}(T_{cld}) = 1.5$, and $\eta_{ice}(T_{cld}/2) = 1$. For each scenario, the parameter estimates are evaluated in terms of reliability and timeliness. By reliable, it is meant that the parameter estimates for clean and iced aircraft are distinguishable. Hence both clean and icing scenarios will be considered. Timeliness means roughly that the pilot should have an unambiguous indication of at least some level of icing before the icing severity becomes critical. This timeliness is measured in terms of the delay between the time when η_{ice} obtains some level, and the time when the estimate of η_{ice} reaches that level.

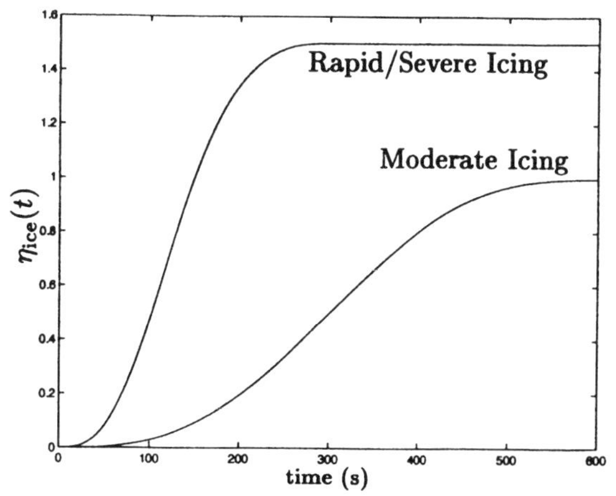

Fig. 1. Icing severity factor, $\eta_{ice}(t)$, for moderate and severe icing scenarios.

3. H^∞ PARAMETER ID ALGORITHM

In Melody et al. (1999) the linearized flight dynamics model has been converted into the general parameter identification framework considered by Didinsky et al. (1995) for time-invariant parameter systems

$$\dot{x} = A(x, v)\chi + b(x, v) + d_p \tag{8}$$

$$y = x + d_m, \tag{9}$$

via the assignments $x = \text{col}(q, \theta, \alpha, u)$, $\chi = \text{col}(M_\alpha, M_{\delta_E}, M_q, Z_\alpha, Z_{\delta_E}, Z_q, X_\alpha, X_u + X_{T_u})$, and $v = \delta_E$. Here $x \in \mathbb{R}^n$ is the system state, $y \in \mathbb{R}^n$ is the measured output, $v \in \mathbb{R}^m$ is the input, $d_p \in \mathbb{R}^n$ is the (unknown) process disturbance, $d_m \in \mathbb{R}^n$ is the measurement noise, and $\chi \in \mathbb{R}^r$ is the parameter vector to be estimated. In the case of systems with time-varying parameters, the formulation incorporates a linear model of the parameter evolution

$$\dot{\chi} = H\chi + Kd_\chi, \quad \chi(0) = \chi_\circ \tag{10}$$

where $d_\chi \in \mathbb{R}^s$ is unknown and represents uncertainty in the parameters. The goal is to obtain estimates of the parameters from the available information, that will be accurate in the presence of disturbances and measurement noise.

As reported in Melody et al. (1999), it has been determined that only the pitching moment derivatives M_α, M_{δ_E}, and M_q provide useful estimates. The Z parameters generally converge too slowly, and the X parameters are too sensitive to noise. During steady-level flight, the M_{δ_E} parameter cannot be identified, since it is the coefficient of the input δ_E that is zero. Furthermore, although the M_q estimate is useful for icing detection in the absence of measurement noise, it is not useful when measurement noise is added. Hence only the M_α estimate will be used for icing detection in the forthcoming simulations.

In order to obtain (10) for the icing problem, a linearization of (5) is performed with the composition of the equations (4) as an output equation about the expected operating condition, $\eta_{\text{ice}} = 0$ and $d_\eta = 0$. Since the system is driftless, $H = 0$, and thus $K = \aleph_1 \aleph_2 [K_*]$, where again $*$ varies over the stability and control derivatives in Table 1.

The H^∞ approach to identification treats noise signals as deterministic but unknown quantities. For the icing identification problem, it is assumed that only noisy measurements of the state are available. Didinsky et al. (1995) develop an H^∞ identification algorithm for this information state, known as the noise-perturbed full-state information (NPFSI) case, which achieves a guaranteed disturbance attenuation level γ between the unknown quantities and the parameter estimate error. Specifically, for any $\gamma \geq \gamma^*$

$$\|\chi - \hat{\chi}\|^2_{Q(x,v)} \leq \gamma^2 \left(\|d_p\|^2 + \|d_m\|^2 + \|d_\chi\|^2 + |\chi - \hat{\chi}_\circ|^2_{Q_\circ} + |x_\circ - \hat{x}_\circ|^2_{P_\circ}\right)$$

where $\hat{\chi}_\circ$ is an *a priori* estimate of the parameter vector χ, $\|\cdot\|_Q$ is an L_2 norm with a chosen weighting function $Q(x,v) \geq 0$, $|\cdot|_{Q_\circ}$ is a generalized Euclidean norm with a chosen weighting matrix $Q_\circ > 0$, similarly, $P_\circ > 0$ is a chosen weighting matrix, and γ^* is the (problem dependent) best achievable level of disturbance attenuation. In general, γ^* is not known *a priori*, and it may even be infinite (meaning that no guaranteed disturbance attenuation level can be achieved). Here it will turn out that γ^* is finite for a particular choice of the weighting matrix Q.

The NPFSI algorithm for $\gamma \geq \gamma^*$ is described by

$$\begin{bmatrix} \dot{\hat{x}} \\ \dot{\hat{\chi}} \end{bmatrix} = \begin{bmatrix} 0 & A(y,v) \\ 0 & H \end{bmatrix} \begin{bmatrix} \hat{x} \\ \hat{\chi} \end{bmatrix} + \begin{bmatrix} b \\ 0 \end{bmatrix} + \Sigma^{-1} \begin{bmatrix} I \\ 0 \end{bmatrix} (z - \hat{x}), \quad (11)$$

$$\dot{\Sigma} = -\Sigma \begin{bmatrix} 0 & A(y,v) \\ 0 & H \end{bmatrix} - \begin{bmatrix} 0 & 0 \\ A(y,v)^T & H^T \end{bmatrix} \Sigma + \begin{bmatrix} I & 0 \\ 0 & -\gamma^{-2} Q(y,v) \end{bmatrix} - \Sigma \begin{bmatrix} I & 0 \\ 0 & KK^T \end{bmatrix} \Sigma \quad (12)$$

with initial conditions $\hat{\chi}(0) = \hat{\chi}_\circ$ and $\Sigma(0) = \text{diag}(P_\circ, Q_\circ)$, and where $\hat{\chi}(t)$ is the estimate of parameter $\chi(t)$ and $\Sigma(t) \in I\!R^{(n+r)\times(n+r)}$. Design of this algorithm entails the choices of the cost terms $Q(x,v)$, $Q_\circ$, and $P_\circ$, as well as the choice of some $\gamma > \gamma^*$. Notice that for the identification algorithm, y is used in place of x as input to the functions $A(\cdot, v)$ and $Q(\cdot, v)$, since x is not available. In fact, Q is allowed to be a function of not only the current value of y, but also its past values. Typically, Q is chosen so that γ^*, or at least some upper bound on γ^*, can be determined analytically. To arrive at a choice of Q which yields a known γ^*, the result of Didinsky et al. (1995) is used which states that for a given γ, if $\Sigma(t) > 0$ for all possible v, d_p, d_m, d_χ, $\hat{x}_\circ$, and $\hat{\chi}_\circ$, then $\gamma^* \leq \gamma$. Conversely, if for a particular γ, there is some time t_1 and some $(v, d_p, d_m, d_\chi, \hat{x}_\circ, \hat{\chi}_\circ)$ for which $\Sigma(t_1) \not> 0$, then $\gamma^* > \gamma$.

This result can be used to specify an NPFSI algorithm with $\gamma^* \leq 1$. Using the partition

$$\Sigma =: \begin{bmatrix} \Sigma_1 & \Sigma_2 \\ \Sigma_2^T & \Sigma_3 \end{bmatrix}$$

with $\Sigma_1 \in I\!R^{n\times n}$, the Schur test for positivity states that $\Sigma(t) > 0$ if and only if $\Sigma_1(t) > 0$ and $\Pi(t) > 0$ where $\Pi := \Sigma_3 - \Sigma_2^T \Sigma_1^{-1} \Sigma_2$. Following an argument similar to that of Didinsky et al. (1995), from (12), which in this case has $H = 0$, it can be shown that

$$\dot{\Sigma}_1 = I - \Sigma_1 \Sigma_1 - \Sigma_2 KK^T \Sigma_2^T, \quad \Sigma_1(0) = P_\circ \quad (13)$$

$$\dot{\Sigma}_2 = -\Sigma_1 A - \Sigma_1 \Sigma_2 - \Sigma_2 KK^T \Sigma_3, \quad \Sigma_2(0) = 0 \quad (14)$$

$$\dot{\Pi} = -\Pi KK^T \Pi - \Sigma_2^T \Sigma_1^{-2} \Sigma_2 - \gamma^{-2} Q, \quad \Pi(0) = Q_\circ. \quad (15)$$

Choosing $Q = \Sigma_2^T \Sigma_1^{-2} \Sigma_2 \geq 0$ and $\gamma = 1$, (15) becomes

$$\dot{\Pi} = -\Pi KK^T \Pi. \quad (16)$$

It is well known that the solution to (16) with $\Pi(0) = Q_\circ > 0$ is positive definite for all $t < \infty$.[1]

Furthermore, for any $\gamma > 1$,

$$\dot{\Pi} = -\Pi KK^T \Pi + (1 - \gamma^{-2}) Q \geq -\Pi KK^T \Pi,$$

and $\Pi(t)$ is again positive definite for all positive time. Hence, $\gamma^* \leq 1$ so long as $\Sigma_1(t) > 0$ for $\gamma \geq 1$ and for all (finite) time. Using $\Omega_1(t) := \Sigma_1^{-1}(t)$ it is clear that $\Sigma_1(t) > 0$ so long as $\Omega_1(t)$ remains finite. From (13)

$$\dot{\Omega}_1 = I - \Omega_1 \Omega_1 + \Omega_1 \Sigma_2 KK^T \Sigma_2^T \Omega_1 \quad (17)$$

with $\Omega_1(0) = P_\circ^{-1} > 0$. Finiteness of the solution to (17) for any finite time $T > 0$ is demonstrated by showing that the last term is bounded by $\kappa \Omega_1$ for some $\kappa \in I\!R$. In order to see this, first notice that

$$\Omega_1 \Sigma_2 KK^T \Sigma_2^T \Omega_1 \leq \rho(KK^T) \Omega_1 \Sigma_2 \Sigma_2^T \Omega_1, \quad (18)$$

where $\rho(\cdot)$ denotes the spectral radius. Furthermore, using the unique positive definite Cholesky decomposition of Ω_1, denoted $\Omega_1^{1/2}$

$$\begin{aligned} \Omega_1 \Sigma_2 \Sigma_2^T \Omega_1 &= \Omega_1^{1/2} \Omega_1^{1/2} \Sigma_2 \Sigma_2^T \Omega_1^{1/2} \Omega_1^{1/2} \\ &\leq \rho(\Omega_1^{1/2} \Sigma_2 \Sigma_2^T \Omega_1^{1/2}) \Omega_1 \\ &= \rho(\Sigma_2^T \Omega_1 \Sigma_2) \Omega_1 \end{aligned}$$

Substituting back into (17) yields

$$\begin{aligned} \dot{\Omega}_1 &\leq I - \Omega_1 \Omega_1 + \rho(KK^T) \rho(\Sigma_2 \Sigma_1^{-1} \Sigma_2) \Omega_1 \\ &= I - \Omega_1 \Omega_1 + \rho(KK^T) \rho(\Sigma_3 - \Pi) \Omega_1 \\ &\leq I - \Omega_1 \Omega_1 + \rho(KK^T) \rho(\Sigma_3) \Omega_1 \end{aligned}$$

If $\rho(\Sigma_3(t))$ is bounded on $[0, T]$, then the comparison lemma (Khalil, 1996) demonstrates that Ω_1 is also bounded on $[0, T]$. Keeping in mind the evolution of $\Sigma_3(t)$,

$$\dot{\Sigma}_3 = -\Sigma_2^T A(t) - A(t)^T \Sigma_2 - \Sigma_2^T \Sigma_2 - \Sigma_3 KK^T \Sigma_3 - \Sigma_2^T \Sigma_1^{-2} \Sigma_2, \quad (19)$$

[1] That this is true is easily obtained by noticing that $S(t) := \Pi^{-1}(t)$ is the solution to the differential equation $\dot{S} = KK^T$, with $S(0) = Q_\circ^{-1}$.

rearrangement of $(\Sigma_2 + A(t))^T(\Sigma_2 + A(t)) \geq 0$ yields

$$-\Sigma_2^T\Sigma_2 - A(t)^T\Sigma_2 - \Sigma_2^T A(t) \leq A(t)^T A(t).$$

Furthermore, since $A(t)$ is piecewise continuous, [2] there exists some matrix $M > 0$ with $A(t)^T A(t) \leq M$ for all $t \in [0, T]$. With this consideration, it is easily shown that for all $\gamma \geq 1$

$$\dot{\Sigma}_3 \leq M - \Sigma_3 K K^T \Sigma_3 - \Sigma_2^T \Sigma_1^{-2} \Sigma_2. \quad (20)$$

Application of the comparison lemma provides the bound $\Sigma_3(t) \leq MT$ on $[0, T]$. Thus $\Sigma_1(t) \geq 0$ for all $t < \infty$ and $\gamma^* \leq 1$ is established. [3]

4. SIMULATION RESULTS

Simulations of the H^∞ NPFSI parameter ID algorithm were performed in order to evaluate the accuracy and timeliness of the icing indication. The two icing scenarios depicted in Figure 1 are considered. In each scenario, excitation is provided only by turbulence. Both iced-aircraft and clean-aircraft simulations were performed. The iced-aircraft simulations investigate a correct indication of icing, whereas the clean-aircraft simulations investigate an incorrect icing indication, or false-alarm. In the iced case, the question is whether the parameter estimates can track the actual parameters in a timely fashion. The accuracy of the iced-aircraft estimates must be considered in light of the clean-aircraft estimate error.

Table 2. Measurement noise intensities

σ_q	σ_θ	σ_α	σ_u
0.0167°/s	0.0293°	0.003°	0.039 m/s

The unknown exogenous signals, d_m and d_p are determined from a sensor noise model, and a turbulence model, respectively. It is assumed that both d_p an d_m are zero-mean, bandlimited white Gaussian noise, with a bandlimit of 10 Hz. Furthermore, both the measurement noise and process noise are assumed to have uncorrelated elements, and hence are completely characterized by their covariance matrix diagonals. For the measurement noise, these intensities are the NASA Twin Otter instrument resolution specifications as reported by Ratvasky and Ranaudo (1993), and are given in Table 2. The process noise is assumed to be turbulence, which is modeled as independent vertical and horizontal acceleration perturbations (Roskam, 1982). Since the vertical velocity is represented in the state x by α according to $w = U_\circ \alpha$, turbulence enters the dynamics through α as well as u. A noise intensity corresponding to $0.2g$ of acceleration was used for both α and u, and all other elements of d_p are assumed to be zero.

The algorithm simulated used $Q = \Sigma_2^T \Sigma_1^{-2} \Sigma_2$ so that $\gamma^* \leq 1$. A value of γ slightly larger than 1 was chosen, in this case $\gamma = 1.01$, since $\gamma = \gamma^*$ often results in numerically sensitive

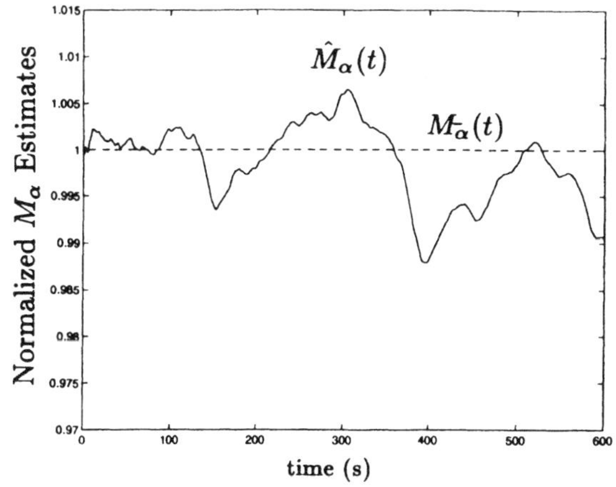

Fig. 2. NPFSI identification of M_α for clean aircraft. The dotted line is M_α for $\eta_{ice} = 0.125$.

algorithms. Finally, some experimentation was used to determine that $Q_\circ = (1 \times 10^{-5})I$ and $P_\circ = I$ provided a good combination of noise sensitivity and convergence for the assumed noise.

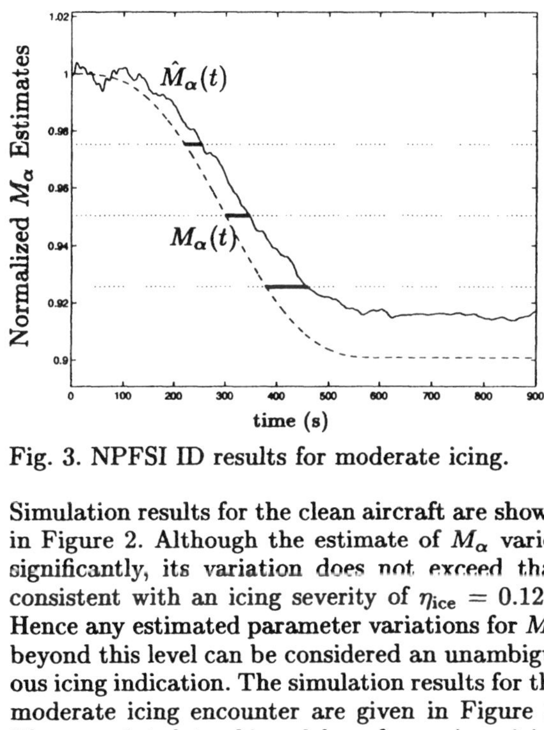

Fig. 3. NPFSI ID results for moderate icing.

Simulation results for the clean aircraft are shown in Figure 2. Although the estimate of M_α varies significantly, its variation does not exceed that consistent with an icing severity of $\eta_{ice} = 0.125$. Hence any estimated parameter variations for M_α beyond this level can be considered an unambiguous icing indication. The simulation results for the moderate icing encounter are given in Figure 3. The associated tracking delays for various icing severity levels are listed in Table 3. The important feature of these delays, as can be discerned from Figure 3, is that the pilot is aware of each icing severity level before the next severity level is actually reached. Finally, the simulation results for the severe icing encounter are given in Figure 4 and Table 4. In this case, the delay is more substantial in that the pilot does not receive notification for a particular icing severity level before the next level is reached. However, even in this most extreme case, the pilot will receive an unambiguous indication before the icing becomes severe. Put more precisely, the M_α estimate goes below the $\eta_{ice} = 0.125$ value before η_{ice} reaches one. This should be sufficient, since there is little danger of

[2] This assumes that the control input $v(t)$ is piecewise continuous as well.

[3] In fact, slight modification of the above argument demonstrates that $\gamma^* = 1$.

inducing a handling event without pilot input for $\eta_{ice} < 1$. If the pilot makes a maneuver during this period, then the previous analysis in Melody et al. (1999) applies, and an icing indication would be available in a matter of seconds.

Table 3. Icing classification delay: moderate icing

Ice accretion level		
$\eta = 0.25$	$\eta = 0.5$	$\eta = 0.75$
33.3 s	47.3 s	79.7 s

It may seem curious that a completely unknown exogenous signal which on the one hand produces parameter estimate error, can on the other hand provide excitation that results in parameter convergence. This fact is the benefit of the state-space formulation of the problem, namely that the state-space formulation, as opposed to a frequency-domain input-output formulation, explicitly contains the effect of the history of the state on the present value of the state. Speaking theoretically, this is in fact the definition of the state, namely, all information about the history of the system pertinent to the future evolution of the system. At any time instant, the unknown present value of the turbulence is a source of error. However, past turbulence has caused the state to move about, and the effect of this past state excitation on the present value of the state is a function of the parameter, χ. Hence measurements, or in this case estimates, of the state provide a means of identifying the parameters. In the case of flight dynamics, this effect is elucidated by (3) which contains M_α. Assuming that the state is measured, α and u on the right hand side of (3) can be thought of as input to the system represented by (3) alone. Turbulence directly affects only α and u, so if the state is measured perfectly, the turbulence provides input excitation to the system of (3) alone in a manner similar to δ_E. In this case, the state is not measured perfectly and a state estimate is used in place of the measured state.

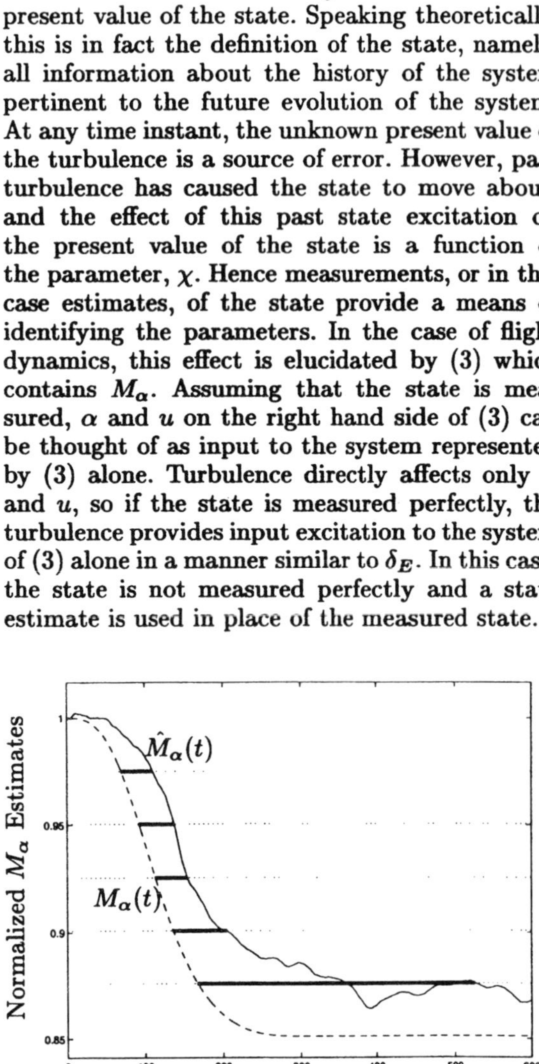

Fig. 4. NPFSI ID results for severe icing.

Table 4. Icing classification delay: severe icing

Ice accretion level				
$\eta = 0.25$	$\eta = 0.5$	$\eta = 0.75$	$\eta = 1$	$\eta = 1.25$
41.1 s	45.3 s	41.1 s	68.4 s	359.3 s

5. CONCLUSION

This study has investigated the performance of an H^∞ NPFSI ID algorithm in the context of icing detection during steady, level flight with turbulence. Ice accretion was modeled as a continuous process, where the tailplane icing severity was captured by a severity factor, η_{ice}. The relation between aircraft stability and control derivatives and the icing severity factor was determined by linearly interpolating between iced and clean aircraft models based on the NASA Twin Otter inflight icing research aircraft. The typical linearization about a trim condition was used, and only longitudinal dynamics were considered. Simulations for the iced and clean aircraft were performed in order to investigate both accurate icing indications and false alarms. The simulation results demonstrate that for both moderate and severe icing encounters, reliable indications of nonnegligible levels of icing severity are available prior to the onset of severe icing. These promising results justify further evaluation of the algorithms with the full nonlinear flight dynamics. Ultimately the parameter identification algorithms and ice detection may be coupled with control adaptation.

References

M. Bragg. Aircraft aerodynamic effects due to large droplet ice accretions. In *Proc. 34th AIAA ASM*, number AIAA-96-0932, January 1996.

M. Bragg, T. Hutchison, R. Oltman, D. Pokhariyal, and J. Merritt. Effect of ice accretion on aircraft flight dynamics. In *Proc. 38th AIAA ASM*, number AIAA-2000-0360, January 2000.

M. Bragg, W. Perkins, N. Sarter, T. Başar, P. Voulgaris, H. Gurbacki, J. Melody, and S. McCray. An interdisciplinary approach to inflight aircraft icing safety. In *Proc. 36th AIAA ASM*, number AIAA-98-0095, January 1998.

G. Didinsky, Z. Pan, and T. Başar. Parameter identification of uncertain plants using H^∞ methods. *Automatica*, 31(9):1227–1250, September 1995.

H. Khalil. *Nonlinear Systems*. Prentice-Hall, Upper Saddle River, NJ, 2nd edition, 1996.

J. Melody, T. Başar, W. Perkins, and P. Voulgaris. Parameter identification for inflight detection of aircraft icing. Submitted, August 1999.

T. Ratvasky and R. Ranaudo. Icing effects on aircraft stability and control determined from flight data. In *Proc. 31st AIAA ASM*, number AIAA-93-0398, January 1993.

T. Ratvasky, J. Van Zante, and J. Riley. NASA/FAA tailplane icing program overview. In *Proc. 37th AIAA ASM*, number AIAA-99-0370, January 1999.

J. Roskam. *Airplane Flight Dynamics and Automatic Flight Controls*. Roskam Aviation and Engineering Corporation, Ottawa, KS, 2nd edition, 1982.

Impulse Response Model Identification in the Case of a Multi-Period ZOH Input

ZI-JIANG YANG and YOSHIHIRO KODAMA

Faculty of Computer Engineering and System Science
Kyushu Institute of Technology
Iizuka, Fukuoka, 820-8502 Japan
TEL+81-948-29-7672, FAX+81-948-29-7651
E-Mail yang@cse.kyutech.ac.jp

Abstract— This paper considers the problem of impulse response identification for a linear sampled-data system where the input signal is held constantly within a multiple of the sampling period of the output signal. To improve identification accuracy, this paper proposes a new identification approach by employing the Haar scaling and wavelet functions. At first, we point out that the discrete-time impulse response of a linear system with zero-order hold (ZOH) input is actually the piecewise-constant approximation of the continuous-time impulse response. Based on the close relation between piecewise-constant approximation with Haar scaling and wavelet functions, a hierarchical identification procedure is proposed which identifies the system impulse response from a coarse resolution level to a fine resolution level successively. At each resolution level, the BIC is utilized to determine the length of the decomposed impulse response in the corresponding subspace, so that some redundant parameters in the high frequency-domain which are sensitive to the noise effects are discarded. Since the identified impulse response model is not smooth when it is represented by some Haar scaling functions of different widths, we can replace each Haar scaling function in the preidentified impulse response model by a Gaussian basis function with corresponding position and width. Then an improved identification method is also proposed to achieve smooth continuous-time impulse response model from sampled data. It is shown through simulation study that the proposed methods yield accurate estimate of the impulse response even in the ill-conditioned cases.

Keywords: identification, impulse response, ill-conditioned problem, multiresolution analysis, Haar wavelet, multirate sampling.

1. INTRODUCTION

An important benefit of impulse response identification is that no *a priori* knowledge about system order and dead time of the process is necessary but only a rough estimate of the settling time. Therefore, besides the parameteric models, identification and controller design based on impulse response models have aslo been studied by many researchers [5, 7, 8, 12].

In this paper, we consider the problem of impulse response model identification for a sampled-data system where the input signal is held constantly within a multiple of the sampling period of the output signal. In practice, such systems are frequently encountered. For example, in a multirate digital control system controlled by a multi-period ZOH input, it is necessary to investigate the system dynamics with a faster sampling rate. And in some practical situations, to avoid fluctuating the system roughly, the amplitude of the input signal for identification should not be changed in a fast rate even if an M-sequence or a white noise is applied. In these cases, compared to the output sampling frequency, the test input is concentrated relatively in a low frequency-domain. And the identification problem is usually ill-conditioned, if the system under study is discretized with a much shorter sampling period of the output signal [12].

To improve identification accuracy under ill-condition, this paper proposes a new identification approach by employing the Haar scaling and wavelet functions. At first, we point out that the discrete-time impulse response of a linear system with ZOH input is actually the piecewise-constant approximation of its continuous-time counterpart. Then based on the close relation between piecewise-constant approximation with Haar scaling and wavelet functions, we show theoretically that it is possible to identify the impulse response from a coarse resolution level to a fine resolution level successively. And then a hierarchical identification algorithm is proposed. At each resolution level, the BIC is utilized to determine the length of the decomposed impulse response, so that some redundant parameters in the high frequency-domain which are sensitive to the noise effects are discarded. This appraoch is especially effective for systems with fast or slowly decaying poles. Usually, the high frequency components in the impulse response require a small sampling period which may tend to cause the ill-conditioned problem, whereas a large sampling period may cause some information loss of the high frequency components of the impulse response. Therefore, multiresolution approach may be a good candidate to treat identification problem of such systems [4].

Since the identified impulse response model is not smooth when it is represented by some Haar scaling functions of different widths, we can replace each Haar scaling function in the preidentified impulse response model by a Gaussian basis function with corresponding position and width. Then an improved identification method is also proposed to achieve smooth continuous-time impulse response model from sampled data, based on the designed Gaussian basis functions. It is shown through simulation study that the proposed methods yield accurate estimate of the impulse response even in the ill-conditioned cases.

2. STATEMENT OF THE PROBLEM

Consider a linear time-invariant continuous system

$$x(t) = \int_0^t g(\tau)u(t-\tau)\, d\tau \tag{1}$$

where $u(t)$ and $x(t)$ are the input and output signals respectively.

As mentioned in the previous section, in this study, it is assumed that a ZOH is utilized in the input side

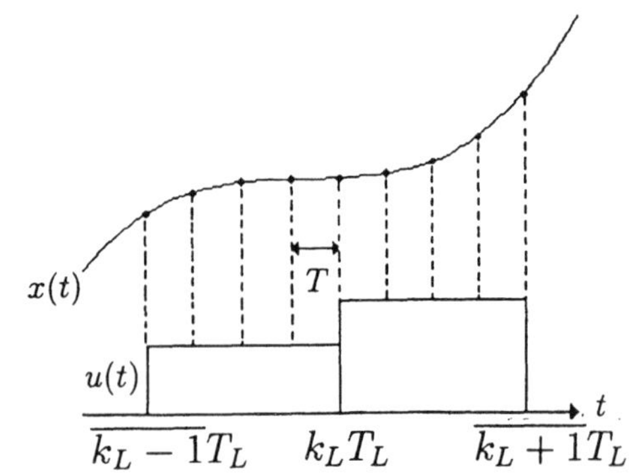

Figure 1: Input and output signals of a sampled-data system where the input signal is the output of a multi-period ZOH.

such that

$$u(t) = u(k_L T_L), \qquad k_L T_L \le t < \overline{k_L + 1} T_L \tag{2}$$

That is, the input is held with a ZOH within a frame period T_L as the following.

$$G_{h0}(s) = \frac{1 - e^{-sT_L}}{T_L s} \tag{3}$$

And the frequency response of the ZOH is

$$G_{h0}(j\omega) = \frac{\sin(\omega T_L/2)}{\omega T_L/2} e^{-j\omega T_L/2} \tag{4}$$

As shown in figure 1, to identify the system dynamics beyond the dominant frequency band $[0, \pi/T_L)$ of the input signal, we sample the output signal $x(t)$ via a finer sampling period as $T = T_L/2^L$. According to the theory of multiresolution analysis, T_L/T should be a power of 2, i.e., 2^L for $L = 0, 1, 2, \ldots$

It is obvious here that the spectra of the input signal decay significantly beyond the frequency band $[0, \pi/T_L)$ so that the Noise/Signal Ratio (NSR) of the output decreases in the high frequency-domain.

3. CONVENTIONAL METHOD OF IMPULSE RESPONSE IDENTIFICATION

Considering the sampling instants $kT(k = 1, 2, \cdots)$, we can rewrite equation (1) as

$$x(kT) = \int_0^{kT} g(\tau) u(kT - \tau)\, d\tau \tag{5}$$

In practice, the measured output is corrupted by a measurement noise such that

$$y(kT) = x(kT) + e(kT) \tag{6}$$

where $e(kT)$ is independent of $u(kT)$, zero-mean, Gaussian and white.

Then we have the finite impulse response model.

$$\begin{aligned} y(kT) &= \sum_{i=1}^{k} u(\overline{k-i}T) \int_{\overline{i-1}T}^{iT} g(\tau)\, d\tau + e(kT) \\ &= \sum_{i=1}^{n} h_{0,i} u(\overline{k-i}T) + e(kT) \end{aligned} \tag{7}$$

where n is a natural number such that nT is sufficiently longer than the setting time and it is assumed here that $n/2^L$ is an integer number.

Equation (7) can be rewritten in vector form:

$$\begin{aligned} y(kT) &= \boldsymbol{p}^T(kT) \boldsymbol{h}_0 + e(kT) \\ \boldsymbol{p}^T(kT) &= \left[u(\overline{k-1}T), u(\overline{k-2}T) \cdots, u(\overline{k-n}T)\right] \\ \boldsymbol{h}_0^T &= [h_{0,1}, h_{0,2}, \cdots, h_{0,n}] \end{aligned} \tag{8}$$

And vector $\boldsymbol{h}$ of the discrete-time impulse response can be estimated by the LS method provided that the inverse matrix of

$$\mathcal{R} = \left[\frac{1}{N+1} \sum_{k=n}^{n+N} \boldsymbol{p}(kT) \boldsymbol{p}^T(kT)\right]$$

exists. However, in the case where a fast sampling period is chosen and the spectra of the input signal are concentrated in the low frequency-domain relatively, the impulse response identification problem often becomes ill-conditioned.

4. MULTIRESOLUTION ANALYSIS OF $g(\tau)$

Consider a sequence $V_i (i \in Z)$ of closed subspaces of $L^2(R)$. Based on the multiresolution analysis theory, space V_0 can be decomposed as

$$\begin{aligned} V_0 &= W_1 \oplus V_1 \\ &= W_1 \oplus W_2 \oplus V_2 \\ &\ \vdots \\ &= W_1 \oplus W_2 \oplus \cdots \oplus W_L \oplus V_L \end{aligned} \tag{9}$$

where $W_1, W_2, \cdots, W_L, V_L$ are mutually orthogonal subspaces:

$$\begin{aligned} V_m &= span\{2^{-\frac{m}{2}} \phi(2^{-m}\tau - iT)\} \\ W_m &= span\{2^{-\frac{m}{2}} \psi(2^{-m}\tau - iT)\} \end{aligned} \tag{10}$$

and $\phi(\tau)$ と $\psi(\tau)$ are scaling and wavelet functions respectively.

So far various scaling and associated wavelet functions are proposed. The Haar scaling and wavelet functions while not being smooth, are well localized in time-domain and are easy to implement in digital control systems with a ZOH input[10, 2, 3, 6, 1, 9], due to close relation bewteen the Haar scaling and wavelet functions with piecewise-constant ZOH signals. Keeping the ZOH input shown in equation (2) in mind, we here employ the Haar scaling and wavelet functions shown as the following.

$$\phi(\tau) = \begin{cases} \dfrac{1}{\sqrt{T}} & 0 \le \tau < T \\ 0 & \tau \ge T \end{cases} \tag{11}$$

$$\psi(\tau) = \begin{cases} \dfrac{1}{\sqrt{T}} & 0 \le \tau < \dfrac{T}{2} \\ -\dfrac{1}{\sqrt{T}} & \dfrac{T}{2} \le \tau < T \end{cases} \tag{12}$$

The continuous-time impulse $g(\tau)$ can be approximated in space V_0 by the Haar scaling functions for a sufficiently small samping period T as

$$g_v^0(\tau) = \sum_{i=0}^{n-1} c_{0,i} \phi(\tau - iT) \tag{13}$$

where

$$c_{0,i-1} = \frac{1}{\sqrt{T}} \int_{\overline{i-1}T}^{iT} g(\tau)\, d\tau = \frac{h_{0,i}}{\sqrt{T}}, \quad i = 1, \cdots, n \tag{14}$$

Instituting equation (13) into equation (5), we have

$$\begin{aligned} x(kT) &= \int_0^{kT} g_v^0(\tau) u(kT - \tau)\, d\tau \\ &= \sum_{i=0}^{n-1} c_{0,i} \sqrt{T} u(\overline{k-i-1}T) = \sum_{i=1}^{n} h_{0,i} u(\overline{k-i-1}T) \end{aligned} \tag{15}$$

This implies that approximation of $g(\tau)$ via the Haar scaling functions is equivalent to $\boldsymbol{h}_0$ in the sense of piecewise-constant approximation. In the continuous time-domain, $g_v^0(\tau)$ is a piecewise-constant approximation of $g(\tau)$ with interval T.

Decompose $g_v^0(\tau) \in V_0$ successively such that

$$\begin{aligned} g_v^0(\tau) &= g_w^1(\tau) + g_v^1(\tau) \\ &= g_w^1(\tau) + g_w^2(\tau) + g_v^2(\tau) \\ &\vdots \\ &= g_w^1(\tau) + g_w^2(\tau) + \cdots + g_w^L(\tau) + g_v^L(\tau) \end{aligned} \tag{16}$$

where $g_w^m(\tau) \in W_m$ $(m = 1, 2, \cdots, L)$, $g_v^L(\tau) \in V_L$.

Although the localization properties are not sufficient in frequency-domain, V_L is the space where the dominant spectra are concentrated in $[0, \pi/(2^L T))$, while W_m is the space where the peak of the spectra is located in $[\pi/(2^m T), \pi/(2^{m-1}T))$ [10].

Using the scaling and wavelet functions, we have

$$\begin{aligned} g_v^0(\tau) &= \sum_{i=0}^{\frac{n}{2}-1} d_{1,i}\ \psi_{1,i}(\tau) + \sum_{i=0}^{\frac{n}{2}-1} c_{1,i}\ \phi_{1,i}(\tau) \\ &= \sum_{m=1}^{2} \sum_{i=0}^{\frac{n}{2^m}-1} d_{m,i}\ \psi_{m,i}(\tau) + \sum_{i=0}^{\frac{n}{2^2}-1} c_{2,i}\ \phi_{2,i}(\tau) \\ &\vdots \\ &= \sum_{m=1}^{L} \sum_{i=0}^{\frac{n}{2^m}-1} d_{m,i}\ \psi_{m,i}(\tau) + \sum_{i=0}^{\frac{n}{2^L}-1} c_{L,i}\ \phi_{L,i}(\tau) \end{aligned} \tag{17}$$

where

$$\begin{aligned} \phi_{m,i}(\tau) &= 2^{-\frac{m}{2}}\phi\left(2^{-m}\tau - iT\right) \\ \psi_{m,i}(\tau) &= 2^{-\frac{m}{2}}\psi\left(2^{-m}\tau - iT\right) \end{aligned} \tag{18}$$

$$c_{m,i} = \int_{-\infty}^{\infty} \phi_{m,i}(t)g(\tau)\ dt, \quad d_{m,i} = \int_{-\infty}^{\infty} \psi_{m,i}(t)g(\tau)\ dt \tag{19}$$

and for the Haar scaling and wavelet functions, we have [4].

$$c_{m,i} = \frac{c_{m-1,2i} + c_{m-1,2i+1}}{\sqrt{2}}, \quad d_{m,i} = \frac{c_{m-1,2i} - c_{m-1,2i+1}}{\sqrt{2}} \tag{20}$$

Based on the above discussions, the system output can be described as

$$\begin{aligned} x(kT) &= \int_0^{kT} g_v^0(\tau)u(kT-\tau)\ d\tau \\ &= \int_0^{kT} g_v^L(\tau)u(t-\tau)\ d\tau \\ &+ \int_0^{kT} \left[g_w^1(\tau) + g_w^2(\tau) + \cdots g_w^L(\tau)\right] u(t-\tau)\ d\tau \end{aligned} \tag{21}$$

Before providing the concrete identification algorithm, some results on the sampled-data system with multi-period ZOH input are given here. Our purpose here is to show that it is possible to identify the impulse response from a coarse resolution level to a fine resolution level successively. At first, based on the satndard discretization procedure of a continuous-time system with a ZOH input, we have

Lemma 1 *Consider a linear time-invariant continuous system whose input-output relationship is described by its continuous-time impulse response $g(\tau)$ as*

$$x(t) = \int_0^t g(\tau)u(t-\tau)\ d\tau$$

Suppose that the above system is excited by a multi-period ZOH input of frame period $T_L = 2^L T$, and let $T_m = 2^m T$ for $m = 0, 1, \cdots, L$, then the following results hold:

$$\begin{aligned} x(k_m T_m) &= \int_0^{k_m T_m} g(\tau)u(k_m T_m - \tau)\ d\tau \\ &= \int_0^{k_m T_m} g_v^m(\tau)u(k_m T_m - \tau)\ d\tau \\ &= \sum_{i=1}^{\frac{n}{2^m}} h_{m,i}u(\overline{k_m - i}T_m) \\ &= \sum_{i=0}^{\frac{n}{2^m}-1} c_{m,i}\ \sqrt{T_m}u(\overline{k_m - i - 1}T_m) \end{aligned}$$

where n is a natural number such that nT represents the length of the finite impulse response, and

$$k_m = \frac{n}{2^m}, \frac{n}{2^m}+1, \frac{n}{2^m}+2, \cdots$$

Lemma 1 implies that the discrete-time impulse response $\widehat{\boldsymbol{h}}_m = [\widehat{h}_{m,1}, \widehat{h}_{m,2}, \cdots, \widehat{h}_{m,n/2^m}]$ of the system can be estimated by sampling the input-output data with sampling period $T_m = 2^m T$. And the corresponding piecewise-constant $\widehat{g}_v^m(\tau) \in V_m$ can be readily obtained from the estimated $\widehat{\boldsymbol{h}}_m$. However, since $\widehat{\boldsymbol{h}}_m$ or $\widehat{g}_v^m(\tau)$ is estimated with sampling period T_m, Only the system dynamics that is dominant in $[-\pi/T_m, \pi/T_m]$ is identified. To identify the system dynamics in the higher frequency-domain, we should sample the input-output data with a finer sampling period $T_{m-1} = 2^{m-1}T$. Also from lemma 1, we have

$$x(k_{m-1}T_{m-1}) = \int_0^{k_{m-1}T_{m-1}} g_v^{m-1}(\tau)u(k_{m-1}T_{m-1} - \tau)d\tau \tag{22}$$

Furthermore, since we can decompose $g_v^{m-1}(\tau)$ into $g_v^m(\tau) + g_w^m(\tau)$, then we have

$$\eta_m(k_{m-1}T_{m-1}) = \int_0^{k_{m-1}T_{m-1}} g_w^m(\tau)u(k_{m-1}T_{m-1} - \tau)\ d\tau \tag{23}$$

$$\begin{aligned} \eta_m(k_{m-1}T_{m-1}) &= x(k_{m-1}T_{m-1}) \\ &- \int_0^{k_{m-1}T_{m-1}} g_v^m(\tau)u(k_{m-1}T_{m-1} - \tau)\ d\tau \end{aligned} \tag{24}$$

where

$$\begin{aligned} &\int_0^{k_{m-1}T_{m-1}} g_v^m(\tau)u(k_{m-1}T_{m-1} - \tau)\ d\tau \\ &= \sum_{i=0}^{\frac{n}{2^m}-1} c_{m,i} \int_0^{k_{m-1}T_{m-1}} \phi_{m,i}(\tau)u(k_{m-1}T_{m-1} - \tau)\ d\tau \\ &= \sum_{i=0}^{\frac{n}{2^m}-1} c_{m,i}\frac{\sqrt{T_m}}{2} \\ &\times \left[u(\overline{k_{m-1} - 2i - 1}T_{m-1}) + u(\overline{k_{m-1} - 2i - 2}T_{m-1})\right] \end{aligned} \tag{25}$$

When $g_v^m(\tau)$ is identified, $\eta_m(k_{m-1}T_{m-1})$ can be calculated by equation (24), and hence $g_w^m(\tau)$ can be identified based on the LS method. Finally, the estimate of $g_v^{m-1}(\tau) \in V_{m-1}$ can be synthesized as $\widehat{g}_v^m(\tau) + \widehat{g}_w^m(\tau) = \widehat{g}_v^{m-1}(\tau)$. To identify the parameters associated with $g_w^m(\tau)$, we have lemma 2 from equation (23), according to the properties of the Haar scaling and wavelet functions.

Lemma 2 *For the same system defined in lemma 1, $\eta_m(k_{m-1}T_{m-1})$ can be expressed as*

$$\begin{aligned} \eta_m(k_{m-1}T_{m-1}) &= \sum_{i=0}^{\frac{n}{2^m}-1} d_{m,i}\frac{\sqrt{T_m}}{2} \\ &\times \left[u(\overline{k_{m-1} - 2i - 1}T_{m-1}) - u(\overline{k_{m-1} - 2i - 2}T_{m-1})\right] \end{aligned}$$

Moreover, for

$$k_{m-1} = \frac{n}{2^{m-1}} + 2l, \qquad l = 0, 1, \cdots$$

we have

$$\eta_m(k_{m-1}T_{m-1}) = 0$$

An example of the synthesized $\widehat{g}_v^{m-1}(\tau) \in V_{m-1}$ is shown in figure 2. Since the length of $\widehat{g}_w^m(\tau)$ is determined by the BIC as discussed in the next section, the synthesized $\widehat{g}_m^{m-1}(\tau)$ is composed of Haar scaling functions of different widths, i.e., figure 2 represents a multiresolution approximation of $g(\tau)$. The begining part corresponds to a discrete-time impulse response of sampling period T_{m-1}, while the ending part corresponds to a discrete-time impulse response of sampling period T_m. In frequency-domain, the begining part contains frequency property of the system that is dominant in $[-\pi/T_{m-1}, \pi/T_{m-1}]$, while the ending part contains that in $[-\pi/T_m, \pi/T_m]$.

5. IDENTIFICATION ALGORITHM

Based on the results of lemmas 1 and 2, we show the hierarchical identification algorithm which identifies the impulse response of the system successively in order of $g_v^L(\tau)$, $g_v^{L-1}(\tau)$, $\cdots$, $g_v^1(\tau)$, $g_v^0(\tau)$ (from coarse resolutions to fine resolutions).

Step 1.
Excite the system by a multi-period ZOH input as

$$u(t) = u(k_L T_L), \qquad k_L T_L \le t < \overline{k_L + 1}T_L \tag{26}$$

and record the input-output data $\{u(kT), y(kT)\}(k = 0, 1, \cdots, n + N)$ with a smaller sampling period T.

Step 2.
Identify $g_v^L(\tau)$ by the LS method based on the following model.

$$y(k_L T_L) = \sum_{i=0}^{M_{vL}-1} c_{L,i}\sqrt{T_L}u(\overline{k_L - i - 1}T_L) + e(k_L T_L) \tag{27}$$

where

$$k_L = \frac{n}{2^L}, \frac{n}{2^L} + 1, \cdots, \frac{n+N}{2^L} \tag{28}$$

and $0 \le M_{vL} \le n/2^L$ is chosen such that the following BIC is minimized. If M_{vL} chosen as $M_{vL} = 0$, then none of the parameters are included.

$$\mathrm{BIC}(M_{vL}) = \left(\frac{N}{2^L} + 1\right)\ln\sigma_{vL}^2 + M_{vL}\ln\left(\frac{N}{2^L} + 1\right) \tag{29}$$

$$\sigma_{vL}^2 = \frac{1}{\frac{N}{2^L} + 1}\sum_{k_L = \frac{n}{2^L}}^{\frac{n+N}{2^L}} (y(k_L T_L) - \widehat{x}(k_L T_L))^2 \tag{30}$$

$$\widehat{x}(k_L T_L) = \sum_{i=0}^{M_{vL}-1} \widehat{c}_{L,i}\sqrt{T_L}u(\overline{k_L - i - 1}T_L) \tag{31}$$

Therefore, the estimated parameter vector $\widehat{c}_L \in R^{\frac{n}{2^L}}$ associated with $\widehat{g}_v^L(\tau)$ is obtained as

$$\widehat{c}_L^T = [\widehat{c}_{L,0}, \widehat{c}_{L,1}, \cdots, \widehat{c}_{L,M_{vL}-1}, 0, \cdots, 0] \tag{32}$$

and $\widehat{g}_v^L(\tau) \in V_L$ is obtained as

$$\widehat{g}_v^L(\tau) = \sum_{i=0}^{\frac{n}{2^L}-1} \widehat{c}_{L,i}\,\phi_{L,i}(\tau) \tag{33}$$

Finally, let $m = L$ and go to the next step.

Step 3.
Identify $g_w^m(\tau)$ based on the following model.

$$\eta_m(k_{m-1}T_{m-1}) = e(k_{m-1}T_{m-1}) + \sum_{i=0}^{M_{wm}-1} d_{m,i}\frac{\sqrt{T_m}}{2} \times \left[u(\overline{k_{m-1} - 2i - 1}T_{m-1}) - u(\overline{k_{m-1} - 2i - 2}T_{m-1})\right] \tag{34}$$

where

$$k_{m-1} = \frac{n}{2^{m-1}} + 2l + 1, \qquad l = 0, 1, \cdots, \frac{N}{2^m} - 1 \tag{35}$$

$$\bar{\eta}_m(k_{m-1}T_{m-1}) = y(k_{m-1}T_{m-1}) - \sum_{i=0}^{M_{wm}-1} \widehat{c}_{m,i}\frac{\sqrt{T_m}}{2} \tag{36}$$

and $0 \le M_{wm} \le n/2^m$ is chosen such that the following BIC is minimized. If M_{wm} chosen as $M_{wm} = 0$, then none of the parameters are included.

$$\mathrm{BIC}(M_{wm}) = \frac{N}{2^m}\ln\sigma_{wm}^2 + M_{wm}\ln\left(\frac{N}{2^m}\right) \tag{37}$$

$$\sigma_{wm}^2 = \frac{1}{\frac{N}{2^m}}\sum_{l=0}^{\frac{N}{2^m}-1}\left[\bar{\eta}_m\left(\overline{\frac{n}{2^{m-1}} + 2l + 1}T_{m-1}\right) - \widehat{\eta}_m\left(\overline{\frac{n}{2^{m-1}} + 2l + 1}T_{m-1}\right)\right]^2 \tag{38}$$

$$\widehat{\eta}_m(k_{m-1}T_{m-1}) = \sum_{i=0}^{M_{wm}-1} \widehat{d}_{m,i}\frac{\sqrt{T_m}}{2} \times \left[u(\overline{k_{m-1} - 2i - 1}T_{m-1}) - u(\overline{k_{m-1} - 2i - 2}T_{m-1})\right] \tag{39}$$

Therefore, the estimated parameter vector $\widehat{\boldsymbol{d}}_L \in R^{\frac{n}{2^m}}$ associated with $\widehat{g}_w^m(\tau)$ is obtained as

$$\widehat{\boldsymbol{d}}_m^T = \left[\widehat{d}_{m,0}, \widehat{d}_{m,1}, \cdots, \widehat{d}_{L,M_{wm}-1}, 0, \cdots, 0\right] \tag{40}$$

Step 4.
Synthesize $\widehat{\boldsymbol{c}}_{m-1}$ from $\widehat{\boldsymbol{c}}_m$ and $\widehat{\boldsymbol{d}}_m$ as

$$\widehat{c}_{m-1,2i} = \frac{\widehat{c}_{m,i} + \widehat{d}_{m,i}}{\sqrt{2}}, \quad \widehat{c}_{m-1,2i+1} = \frac{\widehat{c}_{m,i} - \widehat{d}_{m,i}}{\sqrt{2}} \tag{41}$$

And $\widehat{g}_v^{m-1}(\tau) \in V_{m-1}$ is obtained as

$$\widehat{g}_v^{m-1}(\tau) = \sum_{i=0}^{\frac{n}{2^{m-1}}-1} \widehat{c}_{m-1,i}\,\phi_{L,i}(\tau) \tag{42}$$

Step 5.
Let $m = m - 1$ and go to Step 3 until $m = 0$.

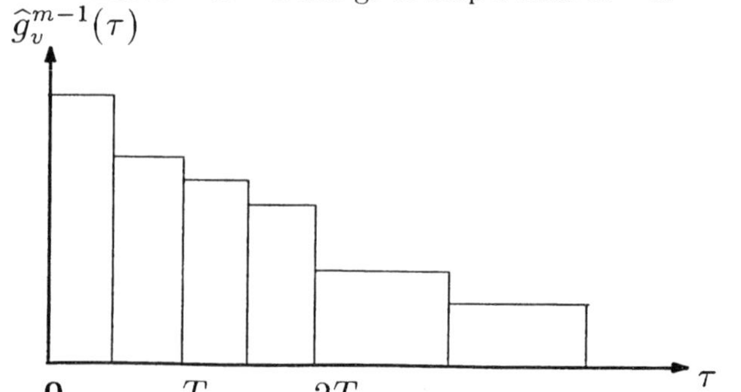

Figure 2: An example of synthesized $\widehat{g}_v^{m-1}(\tau)$.

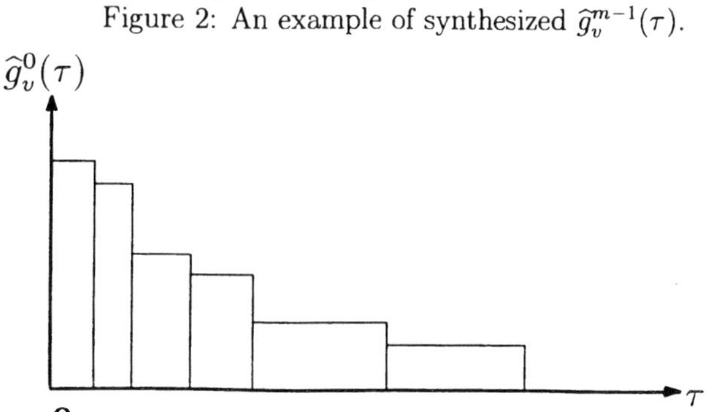

Figure 3: An example of synthesized $\widehat{g}_v^0(\tau)$.

6. IMPROVED IDENTIFICATION USING THE GAUSSIAN BASIS FUNCTIONS

Consider $\widehat{g}_v^0(\tau) \in V_0$ obtained by the proposed hierarchical identification algorithm:

$$\widehat{g}_v^0(\tau) = \sum_{m=1}^{L}\sum_{i=0}^{M_{wm}-1} \widehat{d}_{m,i}\ \psi_{m,i}(\tau) + \sum_{i=0}^{M_{wL}-1} \widehat{c}_{L,i}\ \phi_{L,i}(\tau) \quad (43)$$

An example of $\widehat{g}_v^0(\tau)$ is shown figure 3. Although the Haar scaling and wavelet functions are easy to treat, as shown in figurc 3, the synthesiezed $\widehat{g}_v^0(\tau)$ is composed of Haar scalings functions with various widths and is hence not smooth.

To achieve smooth estimate of the impulse response, an *ad hoc* improved identification procedure is proposed here by replacing each of the Haar scaling function as shown in figure 3 by a Gaussian basis function with a corresponding essential width. That is, instead of $\widehat{g}_v^0(\tau)$ expressed by equation (43), we approximate $g(\tau)$ by the following RBF (Radial Basis Function) network.

$$g_s(\tau) = \sum_{j=1}^{J} b_j\ r\left(\frac{\tau - c_j}{\sigma_j}\right) \quad (44)$$

where

$$r\left(\frac{\tau - c_j}{\sigma_j}\right) = \frac{1}{\sigma_j\sqrt{2\pi}}\exp\left[-\frac{1}{2}\left(\frac{\tau - c_j}{\sigma_j}\right)^2\right] \quad (45)$$

and J is the number of the included Haar scaling functions as shown in figure 3. c_j which coincises with the center position of the corresponding Haar scaling function is the center of jth Gaussian function.

As claimed in section 4, a Haar scaling function of width T_m corresponds to the discrete-time impulse response of sampling period T_m and hence contains the system property that is dominant in $[-\pi/T_m, \pi/T_m]$. Therefore, when we replace a Haar scaling function of width T_m by a Gaussian basis function, it is recommendable to determine parameter σ_j so that the Fourier transformation of the basis function decays significantly beyond $[-\pi/T_m, \pi/T_m]$ [13].

The Fourier transformation of $r(\tau/\sigma_j)$ is given as

$$\begin{aligned} R(\omega) &= \int_{-\infty}^{\infty} \frac{1}{\sigma_j\sqrt{2\pi}}\exp\left[-\frac{1}{2}\left(\frac{\tau}{\sigma_j}\right)^2\right] e^{-j\omega\tau}\ d\tau \\ &= \exp\left[-\frac{{\sigma_j}^2\omega^2}{2}\right] \end{aligned} \quad (46)$$

Let the cut-off frequency be $\omega_c = \pi/T_m$, then we can determine σ_j such that

$$R(\omega_c) = \exp\left[-\frac{{\sigma_j}^2{\omega_c}^2}{2}\right] = \exp(-0.3466) \approx \frac{\sqrt{2}}{2} \quad (47)$$

And hence we have

$$\sigma_j = 0.83\frac{T_m}{\pi} \quad (48)$$

The parameters of the RBF network can be estimated by the LS method based on the following model.

$$\begin{aligned} y(kT) &= \sum_{i=1}^{n} u(\overline{k - i}T)\int_{(i-1)T}^{iT} g(\tau)\ d\tau + e(kT) \\ &= \sum_{j=1}^{J} b_j\ \xi_j(kT) + e(kT) \end{aligned} \quad (49)$$

where

$$\xi_j(kT) = \sum_{i=1}^{n} u(\overline{k - i}T)\int_{(i-1)T}^{iT} r\left(\frac{\tau - c_j}{\sigma_j}\right)\ d\tau \quad (50)$$

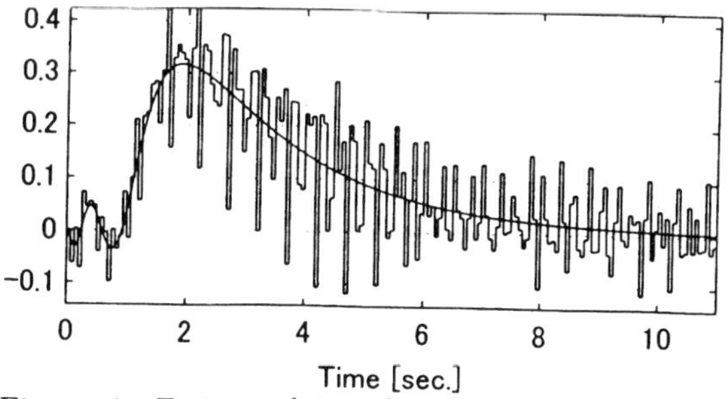

Figure 4: Estimated impulse response (conventional method).

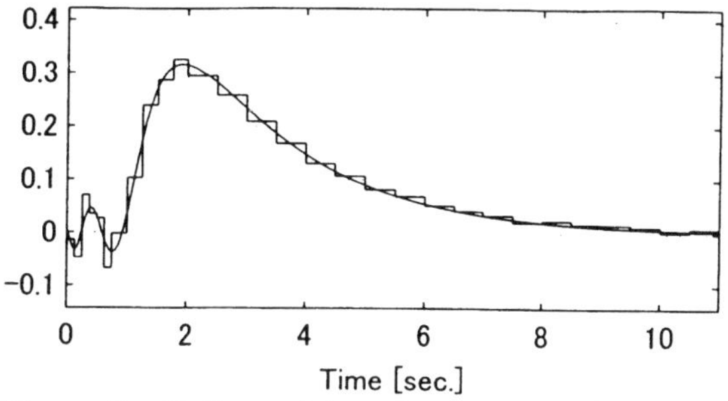

Figure 5: Estimated impulse response (hierarchical identification algorithm).

And the identified continuous-time impulse response $\widehat{g}_s(\tau)$ is obtained as

$$\widehat{g}_s(\tau) = \sum_{j=1}^{J} \widehat{b}_j\ r\left(\frac{\tau - c_j}{\sigma_j}\right) \quad (51)$$

Finally, the discrete-time impulse response of sampling period T_m can be obtained by discretizing $\widehat{g}_s(\tau)$ as the following, if required.

$$\widehat{h}_{m,i} = \int_{(i-1)T_m}^{iT_m} \sum_{j=1}^{J} \widehat{b}_j\ r\left(\frac{\tau - c_j}{\sigma_j}\right)\ d\tau \quad (52)$$

where $i = 1, \cdots, n/2^m$.

7. SIMULATION EXAMPLE

To verify the efficiency of our proposed identification methods, consider a continuous-time system governed by the following transfer function model.

$$\frac{-20s^3 + 180s^2 - 840s + 1680}{s^6 + 22.5s^5 + 231s^4 + 1310s^3 + 3960s^2 + 5040s + 1680} \quad (53)$$

This is a nonminimum phase system with fast and slowly decaying poles located at $-4.2076 \pm 5.3148i$, $-5.7924 \pm 1.7345i$, -2.0000, -0.5000. The input signal is a white noise umiformly distributed in $[-0.5, 0.5]$ and is applied to the system with a ZOH of frame period $T_L = 0.5$[sec]. The output signal is sampled with a fast sampling period $T = T_L/2^3 = 0.0625$[Sec.]. The data length is $N = 2^9$ and NSR= 5%. The order of the impulse response is chosen *a priori* as $n = 176$.

At first, the conventional impulse response identification method mentioned in section 3 is applied to obtain $\boldsymbol{h}_0$. And the corresponding piecewise-constant representation $\widehat{g}_v^0(\tau)$ (see lemma 1) together with the true continuous-time impulse response $g(\tau)$ is shown in figure 4. The result by the hierarchical identification algorithm is shown in figure 5, and the result of the improved identification method using the Gaussian basis functions is shown in figure 6. It can be verified

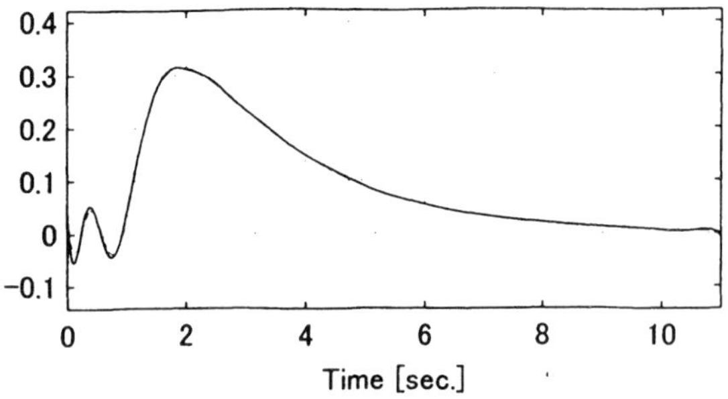

Figure 6: Estimated continuous-time impulse response (improved identification method).

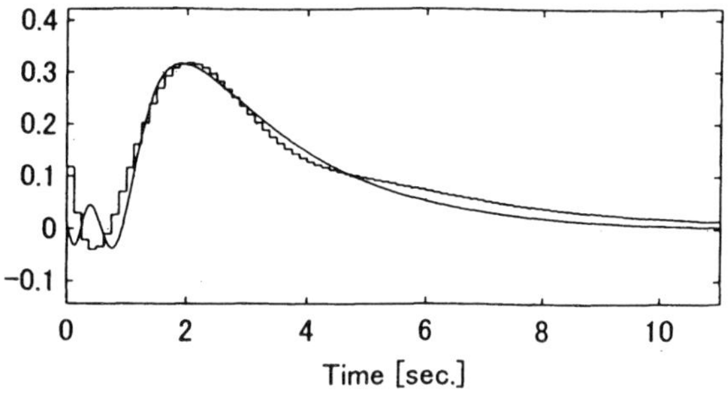

Figure 8: Estimated impulse response (N4SID method, $T_1 = 0.125$[Sec.]).

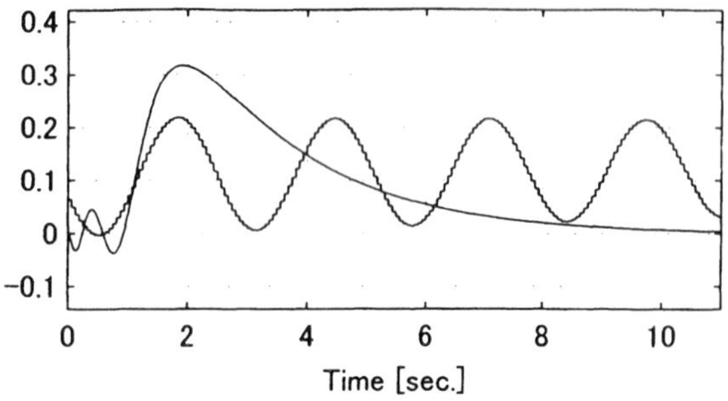

Figure 7: Estimated impulse response (N4SID method, $T = 0.0625$[Sec.]).

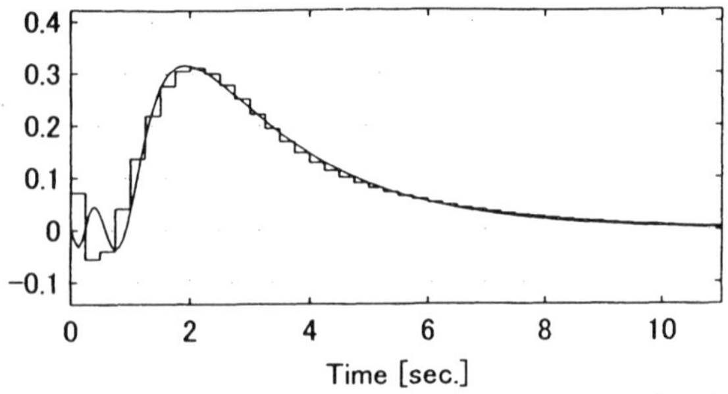

Figure 9: Estimated impulse response (N4SID method, $T_2 = 0.25$[Sec.]).

that the results of figures 5 and 6 are much better than the result by the conventional method.

Finally, for comparison, the results by the N4SID method in matlab are also shown in figures 7~9, for $T = 0.0625$[Sec.], $T_1 = 0.125$[Sec.] and $T_2 = 0.25$[Sec.]. It can be verified that in our example, for the N4SID method, too small a sampling period leads to unacceptable result, whereas a large sampling period leads information loss of the begining part of the impulse response.

8. CONCLUSIONS

In this paper, we treated the problem of impulse response identification for a linear sampled-data system where the input signal is held constantly within a multiple of the sampling period of the output signal. To improve identification accuracy, we proposed a hierarchical identification algorithm by employing the Haar scaling and wavelet functions. Furthermore, an improved identification method is also proposed to achieve smooth continuous-time impulse response model from sampled data by using the Gaussian basis functions. The efficiency of the proposed methods is verified through simulation results.

References

[1] C. F. Chen and C. H. Hsiao: Haar wavelet method for solving lumped and distributed-parameter systems, IEE Proc.-Control Theory Appl., **144**-1 87/94(1997).

[2] D. Clancy and U. Ozguner: Multiresolutional controller design: sensitivity minimization, Proceedings of the 35th Conference on Decision and Control, 3218/3223(1996).

[3] D. Clancy and U. Ozguner: Multiresolutional controller for sensitivity minimization, The 13th IFAC World Congress, 91/95(1996).

[4] I. Daubechies: Ten Lectures on Wavelets, Society for Industrial and Applied Mathematics, Philadelphia, Pennsylvania(1992).

[5] B. S. Dayal and J. F. MacGewgor: Identification of finite impulse response models: methods and robustness issues, Ind. Eng. Chem. Res., **35**, 4078/4090(1994).

[6] M. Doroslovacki, H. Fan and L. Yao: Wavelet-based identification of linear discrete-time systems: robustness issue, Automatica, **34**-12, 1637/1640(1998).

[7] K. Furuta, T. T. Hiep, T. Hojo and H.Sone: Design of response-based servo-mechanism and its application to simulator of distiliation column, 11th IFAC/IFORS Sympo. on Identification and System Parameter Estimation, 589/594(1997).

[8] K. Furuta, M. Wongsaisuwan and S. Umeki: Discrete-time LQG dynamic controller dersign using plant Markov parameters, Automatica, **31**-9, 1317/1324(1995).

[9] J. S. Gu and W. S. Jiang: The Haar wavelets operational matrix of integration, International Journal of Systems Science, **27**-7 623/628(1996).

[10] J. H. Lee, Y. Chikkula, Z. Yu and J. C. Kantor: Improving computational efficiency of model predictive control algorithm using wavelet transformation, International Journal of Control, **61**-4, 859/883(1995).

[11] L. Ljung: System Identification: Theory for the User, Prentice-Hall(1987).

[12] A. Sano and H. Tsuji: Optimal sampling rate for system identification based on decimation and interpolation. The 12th IFAC World Congress, Sydney, **Vol.IV**, 219/222(1993).

[13] R. M. Sanner and J. J. E. Slotine: Gausian networks for direct adaptive control. I.E.E.E. Transactions on Neural Networks, **NN-3**-6, 837/863 (1992)

A NEARLY INTERPOLATORY ALGORITHM FOR $\mathcal{H}_\infty$ IDENTIFICATION WITH MIXED TIME AND FREQUENCY RESPONSE DATA *

Guoxiang Gu[§,1] and Jie Chen[†]

[§] Department of Electrical and Computer Engineering
Louisiana State University, Baton Rouge, LA 70803-5901, U.S.A.
[†] College of Engineering, University of California, Riverside, CA 92521, U.S.A.

Abstract: A novel algorithm is proposed for $\mathcal{H}_\infty$ identification with mixed time and frequency response data. This algorithm shares similar properties to the interpolatory algorithm and is nearly interpolatory. Meanwhile it greatly reduces the computational complexity of the interpolatory algorithm. Worst-case identification error bounds are also obtained for this nearly interpolatory algorithm. A numerical example is used to illustrate the proposed identification algorithm. *Copyright © 2000 IFAC*

Key Words: $\mathcal{H}_\infty$ identification, robust convergence, interpolatory algorithm, mixed time and frequency response data, worst-case identification error bound.

1 Introduction

Since the work of Helmicki, Jacobson, and Nett (1991), the research area of $\mathcal{H}_\infty$ identification has made much progress. Although the initial work has focused on the two-stage nonlinear algorithm (Helmicki, *et al.*, 1991; Gu and Khargonekar, 1992; Partington, 1991) with frequency response data, interpolatory algorithms (Chen, *et al.*, 1994; Chen, *et al.*, 1995; Gu, *et al.*, 1993; Parrilo, *et al.*, 1998) gradually gain more recognition due to its essentially optimal identification error (within a factor of two). An interpolatory algorithm was also developed in (Chen and Nett, 1995) for $\mathcal{H}_\infty$ identification with time response data.

In this paper we consider $\mathcal{H}_\infty$ identification with mixed time and frequency response data, which was initially tackled in (Parrilo, *et al.*, 1998). An interpolatory algorithm was proposed in (Parrilo, *et al.*, 1998) to seek the identified model. Similar to (Chen, *et al.*, 1994; Chen, *et al.*, 1995; Gu, *et al.*, 1993), the interpolatory algorithm consists of two steps: first solve a data consistency problem which determines whether or not the *a posteriori* data are consistent with the *a priori* information; and second construct an interpolatory model, if the answer to the first one is positive. Again the data consistency problem is an equivalent LMI feasibility problem, and there are standard solution methods and software packages available (Boyd, *et al.*, 1994; Gahinet, *et al.*, 1995). The interpolatory model can be constructed using the procedure as outlined in (Ball, *et al.*, 1990), which in special cases reduces to Nevanlinna-Pik or Carathéodory-Fejér interpolation — the main tool for the interpolatory algorithm with frequency or time response data respectively.

While the interpolatory algorithm presents a viable approach to $\mathcal{H}_\infty$ identification, and is essentially optimal within a factor of two (Chen, *et al.*, 1995) for its associated identification error, the corresponding computational complexity is high for both the data consistency problem, and the construction of the interpolatory model. This is the main motivation for the suboptimal algorithm proposed in (Gu, 1994) as an alternative to the interpolatory algorithm in (Chen, *et al.*, 1995; Gu, *et al.*, 1993), where only frequency response data are used. See also (Chen, *et al.*, 1996). Our goal in this paper is to generalize the results in (Gu, 1994) to the case of mixed time and frequency response data so that the computationally prohibitive interpolatory algorithm can be replaced by a much simpler one with much lower computational complexity. This can indeed be achieved. To accomplish this goal, we propose a nearly interpolatory algorithm, which produces an identified model of FIR (finite impulse response) form with a small quantifiable increment in the worst-case identification error. In fact this nearly interpolatory algorithm converges asymptotically to the interpolatory algorithm as the order of the identified model increases. Moreover the associated computation complexity is comparable to that in (Gu, 1994; Chen, *et al.*, 1996). Hence our results complement the development of interpolatory algorithms for $\mathcal{H}_\infty$ identifica-

*This work was supported in part by AFOSR, ARO and NSF. [1] The corresponding author with email: ggu@ee.lsu.edu, and fax: (225) 388-5200.

tion. For simplicity, we consider only single-input and single-output discrete-time systems. The generalization of our results to multivariable systems is straightforward, and thus omitted.

2 Problem Formulation and Notions of IBC

The systems under consideration are causal, stable, linear, and shift-invariant. They can be represented by transfer function

$$\hat{h}(\lambda) = \sum_{k=0}^{\infty} h(k)\lambda^k, \quad \lambda = z^{-1}. \tag{1}$$

Hence stability implies analyticity of $\hat{h}$ on the open unit disc. The set of all such transfer functions such that

$$\|\hat{h}\|_\infty = \sup_{|\lambda|<1} |\hat{h}(\lambda)| < \infty \tag{2}$$

is denoted as $\mathcal{H}_\infty$. A subset of $\mathcal{H}_\infty$ is

$$\mathcal{H}_{\rho,M} = \left\{ \hat{h}(\lambda) : \sup_{|\lambda|<\rho} |\hat{h}(\lambda)| < M \right\} \tag{3}$$

where $\rho > 1$ and $M > 0$. Roughly speaking, ρ represents the relative stability of the system, and M represents the overall gain to weighted sinusoidal inputs in worst-case.

Our identification problem assumes that the system to be identified admits a transfer function $\hat{h}(\lambda) \in \mathcal{H}_{\rho,M}$. The experimental data consist of

$$\begin{aligned} y^t &= E_{N_t}(\hat{h}, u, v) & (4)\\ &:= \left\{ y(k) = \sum_{i=0}^{k} h(k-i)u(i) + v(k) \right\}_{k=0}^{N_t-1}, \\ y^f &= E_{N_f}(\hat{h}, \eta) & (5)\\ &:= \left\{ y_k = \hat{h}(e^{j\omega_k}) + \eta_k \right\}_{k=0}^{N_f-1}, \end{aligned}$$

where $\{u(i)\}$ and $\{y(i)\}$ are the sequences of observed input and output data in time-domain respectively, and $\{v(k)\}$ and $\{\eta_k\}$ are the sequences of measurement noises in time-domain and frequency-domain respectively. It is assumed that $v = \{v(k)\}$ is a real sequence while $\eta = \{\eta_k\}$ is a complex sequence, and

$$\begin{aligned} v \in \mathcal{N}_R(\epsilon^t) &:= \left\{ v \in \mathbb{R}^{N_t} : |v(k)| \le \epsilon_k^t \right\}_{k=0}^{N_t-1}, \\ \eta \in \mathcal{N}_C(\epsilon^f) &:= \left\{ \eta \in \mathbb{C}^{N_f} : |\eta_k| \le \epsilon_k^f \right\}_{k=0}^{N_f-1}. \end{aligned}$$

Hence $(\epsilon^t, \epsilon^f, \rho, M)$ represents the *a priori* information, and (y^t, y^f) the *a posteriori* data. The problem of $\mathcal{H}_\infty$ identification is to search for an identified model $\hat{h}_{id}(\lambda)$ such that the worst-case identification error

$$e(\epsilon^t, \epsilon^f, \rho, M, u) := \sup_{\substack{\hat{h} \in \mathcal{H}_{\rho,M} \\ v \in \mathcal{N}_R(\epsilon^t) \\ \eta \in \mathcal{N}_C(\epsilon^f)}} \|\hat{h}_{id} - \hat{h}\|_\infty \tag{6}$$

is quantified, which can be used, together with the identified model, for $\mathcal{H}_\infty$-based robust feedback control system design. The arguments of the worst-case identification error indicate its dependence on the *a priori* information of the system and noise levels, as well as the time-domain input sequence. If the worst-case identification error corresponding to an identification algorithm A_N converges to zero as $N \to \infty$ and the noise levels go to zero, then A_N is said to be convergent. If the convergence is independent of the *a priori* information, then the algorithm A_N is said to be robustly convergent.

Since the *a priori* information is partial and the *a posteriori* data are incomplete and corrupted, there will generally exist a set of systems that are indistinguishable from the true plant, or systems that cannot be falsified by the given information. Let the set of systems indistinguishable by y^t be denoted by $\mathbb{Q}_t(y^t)$. Then it is defined as

$$\begin{aligned} \mathbb{Q}_t(y^t) &:= \left\{ \hat{h} : y^t = E_{N_t}(\hat{h}, u, v), \right. & (7)\\ &\qquad \left. \text{for some } \hat{h} \in \mathcal{H}_{\rho,M},\ v \in \mathcal{N}_R(\epsilon^t) \right\}. \end{aligned}$$

Let the set of systems indistinguishable by y^f be denoted by $\mathbb{Q}_f(y^f)$. Then it is defined as

$$\begin{aligned} \mathbb{Q}_f(y^f) &:= \left\{ \hat{h} : y^f = E_{N_f}(\hat{h}, \eta), \right. & (8)\\ &\qquad \left. \text{for some } \hat{h} \in \mathcal{H}_{\rho,M},\ \eta \in \mathcal{N}_C(\epsilon^f) \right\}. \end{aligned}$$

It follows that the set $\mathbb{Q}_t(y^t) \cap \mathbb{Q}_f(y^f)$ contains all systems indistinguishable by (y^t, y^f). Let $\mathbf{Y}_t$ be the set of all possible data generated from the experiment E_{N_t}, and $\mathbf{Y}_f$ from E_{N_f}. Then the set $\mathbf{Y}_t \times \mathbf{Y}_f$ contains all possible time-domain and frequency-domain data which may be generated by E_{N_t} and E_{N_f} jointly. In the normed space $\mathcal{H}_\infty$, define the radius and diameter for $\mathbb{Q}(y) := \mathbb{Q}_t(y^t) \cap \mathbb{Q}_f(y^f)$ by

$$\begin{aligned} r(\mathbb{Q}(y)) &:= \inf_{\hat{g} \in \mathcal{H}_\infty} \sup_{\hat{h} \in \mathbb{Q}(y)} \|\hat{h} - \hat{g}\|_\infty, \\ d(\mathbb{Q}(y))) &:= \sup_{\hat{g}, \hat{h} \in \mathbb{Q}(y)} \|\hat{h} - \hat{g}\|_\infty, \end{aligned}$$

respectively. Then in the framework of IBC (information based complexity) (Traub, *et al.*, 1988), $r(\mathbb{Q}_t(y^t) \cap \mathbb{Q}_f(y^f))$ and $d(\mathbb{Q}_t(y^t) \cap \mathbb{Q}_f(y^f)))$ are called local radius of information and local diameter of information, respectively. The global radius of information and global diameter of information are defined by

$$\begin{aligned} r^* &= \sup_{y^t \in \mathbf{Y}_t, y^f \in \mathbf{Y}_f} r\left(\mathbb{Q}_t(y^t) \cap \mathbb{Q}_f(y^f)\right), \\ d^* &= \sup_{y^t \in \mathbf{Y}_t, y^f \in \mathbf{Y}_f} d\left(\mathbb{Q}_t(y^t) \cap \mathbb{Q}_f(y^f)\right), \end{aligned}$$

respectively. If $\hat{h}_{id}$ is an identified model based on local data (y^t, y^f), then its associated local identification error is defined as

$$e(\epsilon^t, \epsilon^f, \rho, M, u, y^t, y^f) = \sup_{\hat{h} \in \mathbb{Q}_t(y^t) \cap \mathbb{Q}_f(y^f)} \|\hat{h} - \hat{h}_{id}\|_\infty.$$

The global identification error coincides with the worst-case identification error as defined by:

$$e(\epsilon^t, \epsilon^f, \rho, M, u) = \sup_{y^t \in \mathbf{Y}_t, y^f \in \mathbf{Y}_f} e(\epsilon^t, \epsilon^f, \rho, M, u, y^t, y^f).$$

The above discussions introduce the notion of optimality regarding to the identification error.

Definition 2.1 *The optimal local identification error with a data record* (y^t, y^f) *is defined as*

$$e^*(\epsilon^t, \epsilon^f, \rho, M, u, y^t, y^f) := \inf_{\{A_N : \hat{h}_{id} \in \mathcal{H}_\infty\}} e(\epsilon^t, \epsilon^f, \rho, M, u, y^t, y^f).$$

An algorithm A_N^* *is said to be locally optimal at* (y^t, y^f) *if it achieves optimal local identification error. Similarly, the optimal global identification error is defined by*

$$e^*(\epsilon^t, \epsilon^f, \rho, M, u) := \inf_{\{A_N : \hat{h}_{id} \in \mathcal{H}_\infty\}} e(\epsilon^t, \epsilon^f, \rho, M, u),$$

and A_N^* *is said to be globally optimal if it achieves optimal global identification error.*

The following lemma follows directly from IBC, and thus its proof is omitted.

Lemma 2.2 *The following statements are true.*

(i) *For any* $y^t \in \mathbf{Y}_t$ *and* $y^f \in \mathbf{Y}_f$,

$$e^*(\epsilon^t, \epsilon^f, \rho, M, u, y^t, y^f) = r\left(\mathbb{Q}_t(y^t) \cap \mathbb{Q}_f(y^f)\right).$$

(ii) *For any* $y^t \in \mathbf{Y}_t$ *and* $y^f \in \mathbf{Y}_f$,

$$\frac{1}{2} d\left(\mathbb{Q}_t(y^t) \cap \mathbb{Q}_f(y^f)\right) \leq e^*(\epsilon^t, \epsilon^f, \rho, M, u, y^t, y^f) \leq d\left(\mathbb{Q}_t(y^t) \cap \mathbb{Q}_f(y^f)\right).$$

(iii) $r^* = e^*(\epsilon^t, \epsilon^f, \rho, M, u)$, *and*

$$\frac{1}{2} d^* \leq e^*(\epsilon^t, \epsilon^f, \rho, M, u) \leq d^*.$$

(iv) *Let* $\hat{h}_{id} = \hat{A}_N(y^t, y^f) \in \mathbb{Q}_t(y^t) \cap \mathbb{Q}_f(y^f)$ *be the identified model. Then for any* $(y^t, y^f) \in \mathbf{Y}_t \times \mathbf{Y}_f$,

$$e^*(\epsilon^t, \epsilon^f, \rho, M, u, y^t, y^f) \leq e(\epsilon^t, \epsilon^f, \rho, M, u, y^t, y^f) \leq 2e^*(\epsilon^t, \epsilon^f, \rho, M, u, y^t, y^f).$$

A *central algorithm* is the one which picks the center of $\mathbb{Q}_t(y^t) \cap \mathbb{Q}_f(y^f)$ as the identified model. Thus it is locally optimal at every data record (y^t, y^f). Although the central algorithm is most desirable, the task of finding the center is rather formidable, and in rare cases central algorithm is manageable computationally. On the other hand, an *interpolatory* algorithm picks an element in $\mathbb{Q}_t(y^t) \cap \mathbb{Q}_f(y^f)$, and thus is relatively easier to implement. More important, the corresponding identification error is essentially optimal within a factor of two in light of (iv) of Lemma 2.2. The next section will focus on the interpolatory algorithm and its associated identification error bounds.

3 An Interpolatory Algorithm

An interpolatory algorithm for the identification problem posed in this paper was studied in (Parrilo, *et al.*, 1998). However the results in (Parrilo, *et al.*, 1998) are incomplete in the sense that time-domain data include only impulse response, and no explicit upper bounds are derived for the corresponding identification error. Both will be addressed in this section.

Denote the Toeplitz matrix associated with sequence $\{s_k\}_{k=0}^{n-1}$ as

$$T_s = \begin{bmatrix} s_0 & 0 & \cdots & 0 \\ s_1 & s_0 & \ddots & \vdots \\ \vdots & \ddots & \ddots & 0 \\ s_{n-1} & s_{n-2} & \cdots & s_0 \end{bmatrix}$$

Then T_u, T_{y^t}, and T_v are Toeplitz matrices associated with input, output, and noise sequences respectively. Denote D_s as the diagonal matrix associated with sequence $\{s_k\}_{k=0}^{n-1}$. Then D_{y^f} and D_η are diagonal matrices associated with frequency response data y^f and noise η, respectively. Define further

$$Z := \begin{bmatrix} 1 & z_0 & \cdots & z_0^{N_t-1} \\ 1 & z_1 & \cdots & z_1^{N_t-1} \\ \vdots & \vdots & \cdots & \vdots \\ 1 & z_{N_f-1} & \cdots & z_{N_f-1}^{N_t-1} \end{bmatrix},$$

$$R := \text{diag}\left(1, \rho, \cdots, \rho^{N_t-1}\right),$$

$$Q := \left[\frac{1}{1-\rho^{-2} z_i \bar{z}_j}\right],$$

where $z_k = e^{j\omega_k}$ for $0 \leq k \leq N_f - 1$. The following result gives a necessary and sufficient condition under which $\mathbb{Q}_t(y^t) \cap \mathbb{Q}_f(y^f) \neq \emptyset$, or equivalently the *a priori* information is consistent with the *a posteriori* data. For completeness, its proof is included.

Theorem 3.1 *The experimental data set* $(y^t, y^f) \in \mathbb{R}^{N_t} \times \mathbb{C}^{N_f}$ *is consistent with* $(\epsilon^t, \epsilon^f, \rho, M)$, *if and only if there exists a* $(v, \eta) \in \mathcal{N}_R(\epsilon^t) \times \mathcal{N}_C(\epsilon^f)$ *such that*

$$A(v, \eta) = \begin{bmatrix} A_{11} & A_{12} \\ A_{21} & A_{22} \end{bmatrix} \geq 0$$

where

$$A_{11} = M^2 \begin{bmatrix} T_u^T R^2 T_u & T_u^T Z^H \\ Z T_u & Q \end{bmatrix},$$

$$A_{12} = A_{21}^H = \text{diag}\left\{T_{y^t} - T_v, D_{y^f} - D_\eta\right\},$$

$$A_{22} = \begin{bmatrix} W & -W Z^H Q^{-1} \\ -Q^{-1} Z W & (Q - Z R^{-2} Z^H)^{-1} \end{bmatrix}$$

and $W = (R^2 - Z^H Q^{-1} Z)^{-1}$.

Theorem 3.1 converted the data consistency problem into that of LMI (linear matrix inequality) for which there are ready software tools. We would like to comment that if $N_t = 0$, then the data consistency problem is equivalent to $\mathbb{Q}_f(y^f) \neq \emptyset$ for which $A(v, \eta) \geq 0$ reduces to (Chen, *et al.*, 1995; Gu, *et al.*, 1993)

$$A(\eta) := \begin{bmatrix} Q & D_{y^f} - D_\eta \\ (D_{y^f} - D_\eta)^H & M^2 Q^{-1} \end{bmatrix} \geq 0. \quad (9)$$

If $N_f = 0$, then the data consistency problem is equivalent to $\mathbb{Q}_t(y^t) \neq \emptyset$ for which $A(v, \eta) \geq 0$ reduces to (Chen and Nett, 1995)

$$A(v) := \begin{bmatrix} T_u^T R^2 T_u & (T_{y^t} - T_v)^T \\ (T_{y^t} - T_v) & M^2 R^{-2} \end{bmatrix} \geq 0. \quad (10)$$

Both (9) and (10) are special cases of $A(v, \eta) \geq 0$ (Parrilo, *et al.*, 1998).

As mentioned earlier, the interpolatory algorithm consists of two steps. For our identification problem, the first step searches (v^*, η^*) such that $A(v^*, \eta^*) \geq 0$, and the second step constructs the identified model which interpolates the new data set $(y^t - v^*, y^f - \eta^*)$. Define

$$\gamma^* := \sup_{\hat{h} \in \mathbb{Q}_t(0) \cap \mathbb{Q}_f(0)} \|\hat{h}\|_\infty. \quad (11)$$

Then the worst-case identification error is characterized as follows.

Lemma 3.2 *For any $\rho > 1$, $M > 0$, the following inequalities hold.*

(i) $\gamma^* \leq r^* \leq d^* = 2\gamma^*$.

(ii) $\gamma^* = r(\mathbb{Q}(0)) \leq \min\{r(\mathbb{Q}_t(0)), r(\mathbb{Q}_f(0))\}$ *where* $\mathbb{Q}(0) = \mathbb{Q}_t(0) \cap \mathbb{Q}_f(0)$.

(iii) *The identification error associated with the interpolatory algorithm satisfies*

$$e(\epsilon^t, \epsilon^f, \rho, M, u) \leq 2\gamma^*.$$

The following result gives an explicit upper bound to γ^*, and thus the upper bound for the worst-case identification error associated with the interpolatory algorithm.

Theorem 3.3 *For the given frequency points $\omega_0 < \omega_1 < \cdots < \omega_{N_f - 1}$, define*

$$\bar{\delta} := \max_{1 \leq k \leq N_f - 1} (\omega_k - \omega_{k-1}).$$

Moreover, assume that $u(0) \neq 0$, and define

$$\begin{aligned} \epsilon : &= M^2 \frac{\epsilon_f + (\epsilon_t / |u_0|)}{M^2 + \epsilon_f(\epsilon_t/|u_0|)}, \\ \epsilon_t &:= \max_k \{\epsilon_k^t\}, \\ \epsilon_f &:= \max_k \{\epsilon_k^f\}, \\ \phi(\delta) &:= \frac{2\rho \sin(\delta/2)}{\sqrt{(\rho^2-1)^2 + 4\rho^2 \sin^2(\delta/2)}}, \\ \psi(\delta) &:= \frac{\epsilon + (M/\rho)\phi(\delta)}{1 + (\epsilon\rho/M)\phi(\delta)}. \end{aligned}$$

Then there holds

$$\gamma^* \leq M^2 \frac{(\epsilon_t/|u_0|) + \psi(\bar{\delta}/2)}{M^2 + (\epsilon_t/|u_0|)\psi(\bar{\delta}/2)}. \quad (12)$$

The worst-case identification error bound for the special case of $N_t = 0$ or $N_f = 0$ is characterized as follows.

Theorem 3.4 *Suppose that $N_t = 0$, $N_f > 0$, and the frequency response samples are uniformly distributed. Then the worst-case identification error bound satisfies*

$$\begin{aligned} e(\epsilon^t, \epsilon^f, \rho, M, u) &= e(\epsilon^f, \rho, M) \quad (13) \\ &\leq 2\left(\sqrt{2}\epsilon_f + \frac{(1+\sqrt{2})M}{\rho^{1+N_f/2}}\right). \end{aligned}$$

Suppose that $N_f = 0$, $N_t > 0$, and $u(0) \neq 0$, then the worst-case identification error bound satisfies

$$\begin{aligned} e(\epsilon^t, \epsilon^f, \rho, M, u) &= e(\epsilon^t, \rho, M, u) \quad (14) \\ &\leq \sum_{k=0}^{n-1} \left(1 - \rho^{2(k-n)}\right)\gamma_k + M\rho^{-n}, \end{aligned}$$

where $\gamma_k = \min\left\{\sum_{i=0}^{k} |w_i| \epsilon_{k-i}^t, M\rho^{-k}\right\}$, *and* $n \leq N_t$.

If both $N_t > 0$, $N_f > 0$, and $u(0) \neq 0$, then the worst-case identification error bound satisfies

$$e(\epsilon^t, \epsilon^f, \rho, M, u) \leq \min\left\{e(\epsilon^t, \rho, M, u), e(\epsilon^f, \rho, M)\right\}. \quad (15)$$

4 A Nearly Interpolatory Algorithm

The LMI problem associated with data consistency problem in the previous section involves a complex hermitian matrix of size $2N \times 2N$ with $N = N_t + N_f$. Equivalently it is an LMI involving a real symmetric matrix of size $4N \times 4N$. It is noted that even for $N_t = N_f = 64$, we have a real symmetric matrix of sized 512×512 for the associated LMI problem which would be extremely difficult to solve numerically. A natural question is thus whether or not we can reduce its computational complexity through searching an approximate solution to the data consistency problem, and meanwhile eliminate the construction of the interpolatory model which is again ill-conditioned numerically. The answer is affirmative.

We are motivated by (Gu, 1994), and seek an identified model of the form:

$$\hat{h}_{id}(\lambda) = \sum_{k=0}^{n-1} \left(1 - \rho^{2(k-n)}\right) p_k \lambda^k \quad (16)$$

such that it interpolates the *a posteriori* data approximately. In fact $\hat{h}_{id}$ in (16) is an n-width approximant to some interpolatory model. In light of (Gu, 1994;

Chen, *et al.*, 1996), it is necessary that $I - H_{N_t}^T H_{N_t} \geq 0$ where

$$H_{N_t} = \begin{bmatrix} h_0 & 0 & \cdots & 0 \\ h_1 & h_0 & \ddots & \vdots \\ \vdots & \ddots & \ddots & 0 \\ h_{N_t-1} & \cdots & h_1 & h_0 \end{bmatrix} \tag{17}$$
$$= \frac{1}{M} R(T_y - T_v) T_u^{-1} R^{-1},$$

and $h_k = p_k \rho^k / M$ for $0 \leq k \leq n-1$. It can be easily shown that $I - H_{N_t}^T H_{N_t} \geq 0$ is equivalent to $A(v) \geq 0$ where $A(v)$ has the expression as in (10), which is exactly the same condition for the existence of the interpolatory model for the case $N_f = 0$. Thus our next goal is to approximately interpolate the frequency response data with $\hat{h}_{id}$.

Denote p as a column vector consisting of $p_0, p_1, \cdots, p_{n-1}$ which parameterize the identified model $\hat{h}_{id}$ in (16). Then there holds:

$$p = (T_y - T_v) T_u^{-1} e_n, \quad e_n = \begin{bmatrix} 1 & 0 & \cdots & 0 \end{bmatrix}^T \in \mathbb{R}^n.$$

Hence we can write

$$\hat{h}_{id}(e^{j\omega_i}) = \Omega_i p = \Omega_i (T_y - T_v) T_u^{-1} e_n,$$
$$\Omega_i = \left[(1 - \rho^{2(k-n)}) e^{jk\omega_i} \right]_{k=0}^{n-1} \in \mathbb{C}^n.$$

Assume that $\epsilon_k^f = \epsilon_f$ for $k = 0, 1, \cdots, N_f - 1$. Then approximate interpolation in frequency-domain can be achieved through minimization of

$$J_\infty := \max_i |\Omega_i (T_y - T_v) T_u^{-1} e_n - y_i| \tag{18}$$

subject to the constraint $I - H_n^T H_n \geq 0$, or equivalently $A(v) \geq 0$. This results in the mixed suboptimal algorithm proposed in this paper. For simplicity we consider only the case $n \leq N_t$. The other case is similar and thus omitted. It should be clear that for $n \leq N_t$, the constraint $I - H_n^T H_n \geq 0$ is implied by $I - H_{N_t}^T H_{N_t} \geq 0$. Hence the following algorithm is developed.

Mixed Suboptimal Algorithm:

- Step 1: Seek a noise sequence $\{v_k\}_{k=0}^{N_t-1}$ which minimizes (18), subject to the constraints $|v_k| \leq \epsilon_t$ for $0 \leq k \leq n-1$, and $A(v) \geq 0$.
- Step 2: If such a minimizer v^* exists, set $p = (T_y - T_{v^*}) T_u^{-1} e_n$, and the identified model as in (16).

End.

Clearly the constrained minimization problem in Step 1 can be converted into an equivalent LMI problem. If ϵ_k^t's are not uniform, $|v_k| \leq \epsilon_k^t$ can be used for $0 \leq k \leq N_t - 1$. If ϵ_k^f's are not uniform, then (18) can be replaced by

$$J_\infty := \max_i \ \epsilon_f |\Omega_i (T_y - T_v) T_u^{-1} e_n - y_i| / \epsilon_i^f.$$

These changes do not alter the nature of Step 1 as an LMI problem. Hence a MATLAB program can be easily compiled to implement the mixed suboptimal algorithm. The questions in hands are the existence conditions for the minimizer in Step 1, and the worst-case identification error bound associated with the mixed suboptimal algorithm. These are answered as follows.

Theorem 4.1 *If the a posteriori data* (y_t, y_f) *and a priori information* $(\epsilon_t, \epsilon_f, M, \rho)$ *are consistent, then there exists a feasible solution to* $J_\infty \leq \epsilon_f + M\rho^{-n}$, *and* $A(v) \geq 0$ *for* $|v_k| \leq \epsilon_t$ *where* $0 \leq k \leq n-1$. *Conversely if Step 1 of the mixed suboptimal algorithm is solvable for* $J_\infty \leq \epsilon_f + M\rho^{-n}$ *as* $n \to \infty$, *then the a posteriori data* (y_t, y_f) *and the a priori information* $(\epsilon_t, \epsilon_f, M, \rho)$ *are consistent.*

Theorem 4.2 *Suppose that the a posteriori data* (y_t, y_f) *and a priori information* $(\epsilon^t, \epsilon^f, M, \rho)$ *are consistent, and* $n \leq N_t$. *Then the worst-case identification error for the mixed suboptimal algorithm* A_N *is no larger than*

$$2 \min \left\{ e^*(\epsilon_f + M\rho^{-n}, \rho, M), e^*(\epsilon_t, \rho, M, u) \right\} + M\rho^{-n}.$$

Explicit upper bounds for the worst-case identification error can be easily obtained by combining Theorem 4.2, and the results in the previous section. The details are omitted.

5 An Example and Concluding Remarks

A simple example is used to illustrate the proposed algorithm. The system to be identified has a transfer function

$$\hat{h} = \frac{1}{1 - 2r\cos(\theta)\lambda + r^2\lambda^2}, \qquad r = 0.88, \quad \theta = \pi/6.$$

It can be verified that $\hat{h} \in \mathcal{H}_{\rho,M}$ with $M = 44$, and $\rho = 1.1$. A set of $N_f = 64$ uniformly distributed samples of frequency response data is obtained with $\epsilon_f = 0.8$ which is approximately one-tenth of $\|\hat{h}\|_\infty$. After solving the data consistency problem $A(\eta) \geq 0$, Nevanlinna-Pick interpolation is used to calculate the interpolation model. The magnitude error response in frequency-domain is plotted in Figure 1.

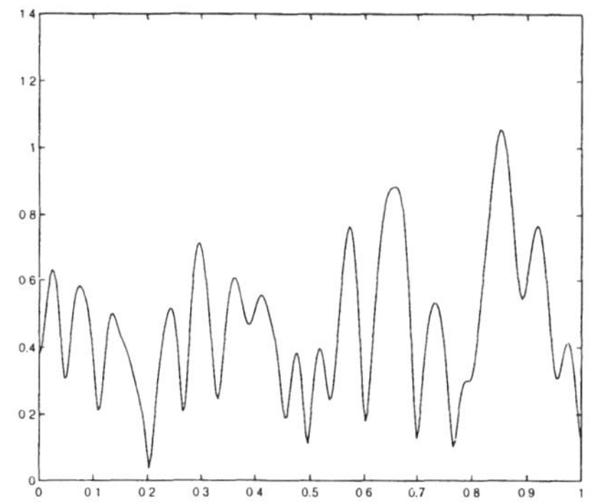

Fig. 1: Error response for frequency response data

A set of $N_t = 64$ samples of unit step response is also obtained with $\epsilon_t = 0.2$. The noise sequence v is computed such that $A(v) \geq 0$ with an LMI software package. The Carathéodory-Fejér procedure is employed to compute the interpolatory model. The magnitude frequency response error is plotted in Figure 2. We note that the error response in Figure 2 is worse than that in Figure 1.

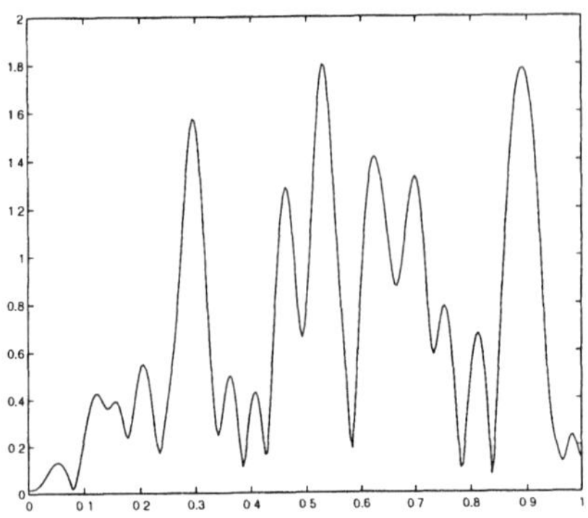

Fig. 2: Error response for time response data

If both time and frequency response data are used, then we need compute "tuned" sequences (v^*, η^*) which solve $A(v^*, \eta^*) \geq 0$ jointly. Since $N_t = N_f = 64$, the size of the corresponding LMI is too large to compute numerically. Hence we used our proposed nearly interpolatory algorithm with $n = N_t/2 = N_f/2 = 32$. The identified model is obtained, and its error response in frequency domain is plotted as in Figure 3 with a maximum error 0.916. Even though the identified model has a lower order, non-interpolatory, and the same data samples are used, the actual error is about 20% smaller than those in Figure 1 and Figure 2 with much lower computation complexity.

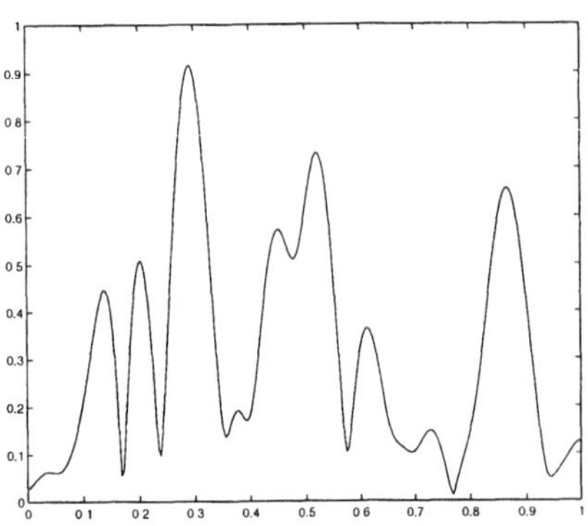

Figure 3: Error response for mixed data

To conclude this paper, we would like to mention that the "tuned" noise sequence η^* for Nevanlinna-Pick interpolation, and "tuned" noise sequence v^* for Carathéodory-Fejér interpolation do not solve the joint time and frequency data consistency problem in general. The contribution of our paper is the much simplified nearly interpolatory algorithm. It is worth to mention also that the associated worst-case identification error is close to that of the true interpolatory algorithm, in light of Theorem 4.2.

References:

Ball, J.A., I. Gohberg, and L. Rodman (1990). *Interpolation of Rational Matrix Functions*, Operator Theory: Advances and Applications, **45**, Birkhäuser, Basel.

Boyd, S., L. El Ghaoui, E. Feron, and V. Balakrishnan (1994). *Linear Matrix Inequalities in System and Control Theory*, SIAM, Philadelphia, PA.

Chen, J., G. Gu, and C.N. Nett (1994). Worst case identification of continuous-time systems via interpolation. *Automatica*, **30**, pp. 1825–1837.

Chen., J., C. N. Nett, and M. K. H. Fan (1995). Worst case identification in $\mathcal{H}_\infty$: validation of *a priori* information, essentially optimal algorithms, and error bounds. *IEEE Trans. Automat. Contr.*, **40**, pp. 1260–1265.

Chen, J. and C.N. Nett (1995). The Carathédory-Fejér problem and $\mathcal{H}_\infty/\ell^1$ identification – A time domain approach. *IEEE Trans. Automat. Contr.*, **40**, pp. 729–735.

J. Chen, J.A. Farrell, C.N. Nett, and K. Zhou (1996). $\mathcal{H}_\infty$ identification of multivariable systems by Tangential interpolation methods. *IEEE Trans. Automat. Contr.*, **41**, pp. 1822–1828.

Gahinet, P., A. Nemirovski, A. Laub, and M. Chilali (1995). *The LMI Control Toolbox*, The MathWorks, Inc., Natick, MA.

Gu, G. (1994). Suboptimal algorithms for worst case identification and model validation. *IEEE Trans. on Automat. Contr.*, **39**, pp. 1657–1661.

Gu, G. and P. P. Khargonekar (1992). A class of algorithms for identification in $\mathcal{H}_\infty$. *Automatica*, **28**, pp. 199–312.

Gu, G., D. Xiong, and K. Zhou (1993). Identification in H_∞ using Pick's interpolation. *Syst. and Contr. Lett.*, **20**, pp. 263–272.

Helmicki, A.J., C.A. Jacobson, and C.N. Nett (1991). Control oriented system identification: a worst-case/deterministic approach in $\mathcal{H}_\infty$. *IEEE Trans. Automat. Contr.*, **36**, pp. 1163–1176.

Parrilo, P.A., M. Sznaier, and R.S. Sánchez Peña (1998). Mixed time/frequency-domain based robust identification. *Automatica*, **34**, pp. 1375–1389.

Partington, J. (1991). Robust identification and interpolation in $\mathcal{H}_\infty$. *Int. J. Contr.*, **54**, pp. 1281–1290.

Pinkus, A. (1985). *n-Widths in Approximation Theory*, Springer-Verlag, Berlin.

Traub, J.F., G.W. Wasilkowski, and H. Wozniakowski (1988). *Information-Based Complexity*, Academic Press, New York, NY.

Use of Error Criteria in Identification for Control

Yucai Zhu
Faculty of Electrical Engineering
Eindhoven University of Technology
P.O. Box 513, 5600 MB Eindhoven, The Netherlands
email: y.zhu@tue.nl

Also with: Tai-Ji Control
Hageheldlaan 62, 5641 GP Eindhoven, The Netherlands
email: y.zhu@tip.nl

Abstract

This work discusses the use of error criteria in parameter estimation and in model order selection where the purpose of identification is control. It will be shown that in general a model estimated using output-error criterion is not optimal according to output-error criterion. Prediction-error criterion will perform better in parameter estimation. This fact implies that an output-error model is not optimal for simulation. For model order selection, on the other hand, prediction-error criterion is not control relevant; and output-error will do a better job. Therefore, for control relevant identification, it is suggested that prediction error criterion is used for parameter estimation and output error criterion for model order selection. Simulations will be used to illustrate the ideas. *Copyright ©2000 IFAC*

Keywords: Identification for control, error criteria, parameter estimation, order selection

1 Introduction and Problem Statement

The choice of error criterion in parameter estimation and in model order selection is a key issue in system identification. When the identified model is used for control, often a model that minimizes output error (simulation error) criterion for a given class of input can be desired. This is because an accurate input-output model is important for control. Some researchers and users of identification believe that a model estimated using output error criterion will be optimal according output error criterion and therefore optimal for control.

On the other hand, when prediction error method is used for parameter estimation, the same criterion is often also used for model order selection following the work of Akaike (1974, 1981). In this notes, we will first shown that output error model is NOT optimal for output error criterion when evaluated using fresh data set and when the process is not noise-free. Then we will argue that prediction error criterion is not the best one for order selection in identification for control. Based on these observations, for control relevant identification, we suggest a cross-criterion approach: use prediction error method, or maximum likelihood method for parameters estimation, and output error criterion for model order selection.

Given a linear time-invariant discrete-time process

$$\begin{aligned} y(t) &= G^o(z^{-1})u(t) + v(t) \\ v(t) &= H^o(z^{-1})e(t) \end{aligned} \qquad (1)$$

where $u(t)$ is the process input at time t, $y(t)$ is the process output, $v(t)$ is the disturbance at the output, $\{e(t)\}$ is white noise with zero mean, $G^o(z^{-1})$ and $H^o(z^{-1})$ are stable transfer functions of order n. For the discussion, assume $H^o(z^{-1}) \neq 1$, which implies that the disturbance $v(t)$ is NOT white noise.

Assume that an open loop identification test has been carried out. This implies that the input $\{u(t)\}$ and white noise $\{e(t)\}$ are mutually independent. Denote the collected input/output data as

$$\{u(1), y(1), u(2), y(2), \ldots\ldots, u(N), y(N)\} \qquad (2)$$

where N is the number of samples.

The model to be identified is

$$y(t) = G(z^{-1})u(t) + H(z^{-1})e(t) \qquad (3)$$

where $G(z^{-1})$ is a stable transfer function, $H(z^{-1})$ is stable and minimumphase and $\{e(t)\}$ is white noise with zero mean.

Suppose that the control requirement for model identification has been translated into the minimization of the following loss function

$$V_{OE} = \frac{1}{2\pi}\int_{-\pi}^{\pi} |G^o(e^{i\omega}) - \hat{G}(e^{i\omega})|^2 W(\omega) d\omega \quad (4)$$

where $G^o(e^{i\omega})$ and $\hat{G}(e^{i\omega})$ are the frequency responses of the process and the model respectively, $W(\omega)$ is a weighting function that is continuous over frequency ω. Assume that an open loop identification test is performed and the test input signal $u(t)$ is chosen such that its power spectrum $\Phi_u(\omega)$ is equal to $W(\omega)$, namely,

$$\Phi_u(\omega) = W(\omega) \quad (5)$$

Under these conditions, two questions will be answered in this work:

1) What error criterion should be used in order to minimize V_{OE} in (4)?
2) What criterion should be used for order selection in order to minimize V_{OE}?

Section 2 will discuss criterion for parameter estimation; Section 3 is concerned with criterion for order selection; Section 4 is the conclusion.

2 Criterion for Parameter Estimation

The discussion below compares the model quality between an output-error model and a prediction error model which does not depend on specific model parametrization. However, for the readability for control engineers, we will use the following specific model structures. Parametrize the output-error model as

$$\begin{aligned} y(t) &= G(z^{-1},\theta)u(t) + e(t) \\ &= \frac{b_0 + b_1 z^{-1} + ... + b_n z^{-n}}{1 + a_1 z^{-1} + ... + a_n z^{-n}} + e(t) \end{aligned} \quad (6)$$

where $\theta = [b_o, b_1, ...b_n, a_1, ...a_n]^T$ is the vector of process model parameters. Parametrize the prediction error model as a Box-Jenkins model

$$\begin{aligned} y(t) &= G(z^{-1},\theta)u(t) + H(z^{-1},\alpha)e(t) \\ &= \frac{b_0 + b_1 z^{-1} + ... + b_n z^{-n}}{1 + a_1 z^{-1} + ... + a_n z^{-n}} u(t) \\ &+ \frac{1 + c_1 z^{-1} + ... + c_n z^{-n}}{1 + d_1 z^{-1} + ... + d_n z^{-n}} e(t) \end{aligned} \quad (7)$$

where $\theta = [b_o, b_1, ...b_n, a_1, ...a_n]^T$ is the parameter vector of process model and $\alpha = [c_1, ...c_n, d_1, ...d_n]^T$ is the parameter vector of disturbance model. Note that the key difference between a prediction error model and an output error model is on the use of the disturbance model $H(z^{-1})$.

Suppose that one will use the identified model to simulate the process output using an input $u^*(t)$ with spectrum $\Phi_u^*(\omega) = W(\omega)$. Perform an open-loop identification test using a input signal $u(t)$ that has the same spectrum $\Phi_u(\omega) = W(\omega)$. Denote $\hat{G}(z^{-1}, \hat{\theta}^{BJ})$ as the Box-Jenkins model estimated using the test data and $\hat{G}(z^{-1}, \hat{\theta}^{OE})$ as that of the output-error model. Note that $u^*(t)$ has NOT been used in identification. Denote

$$y^o(t) = G^o(z^{-1})u^*(t)$$

as the true noise-free output. When simulating the process using $u^*(t)$, the error of the Box-Jenkins model is

$$\tilde{y}(t, \hat{\theta}^{BJ}) = [\hat{G}(z^{-1}, \hat{\theta}^{BJ}) - G^o(z^{-1})]u^*(t) \quad (8)$$

with the variance, or, the output-error criterion when $N \to \infty$ as

$$\begin{aligned} & V_{OE}(\hat{\theta}^{BJ}) \\ = & \frac{1}{2\pi}\int_{-\pi}^{\pi} |G^o(e^{i\omega}) - \hat{G}(e^{i\omega}, \hat{\theta}^{BJ})|^2 \Phi_u^*(\omega) d\omega \end{aligned} \quad (9)$$

similarly, the output-error criterion for the output error model is when $N \to \infty$

$$\begin{aligned} & V_{OE}^V(\hat{\theta}^{OE}) \\ = & \frac{1}{2\pi}\int_{-\pi}^{\pi} |G^o(e^{i\omega}) - \hat{G}(e^{i\omega}, \hat{\theta}^{OE})|^2 \Phi_u^*(\omega) d\omega \end{aligned} \quad (10)$$

Because the output disturbance $v(t)$ is assumed to be a realization of a stochastic process, the estimated models $\hat{G}(e^{i\omega}, \hat{\theta}^{BJ})$ and $\hat{G}(e^{i\omega}, \hat{\theta}^{OE})$ are random variables; see Ljung (1987). Then, taking expectations with respect to $\hat{G}(e^{-i\omega}, \theta^{BJ})$, we obtain the average output-error criterion of the Box-Jenkins model

$$EV_{OE,1}^V(\hat{\theta}^{BJ}) = \frac{1}{2\pi}\int_{-\pi}^{\pi} \mathrm{var}[\hat{G}(e^{i\omega}, \hat{\theta}^{BJ})]\Phi_u^*(\omega) d\omega \quad (11)$$

and the average output-error criterion of the output-error model

$$EV_{OE,1}^V(\hat{\theta}^{OE}) = \frac{1}{2\pi}\int_{-\pi}^{\pi} \mathrm{var}[\hat{G}(e^{i\omega}, \hat{\theta}^{OE})]\Phi_u^*(\omega) d\omega \quad (12)$$

Theorem 2.1. *Given a linear process as in (1). Assume that: 1) Output disturbance is colored noise; 2) an open-loop test has been performed using a input signal with a continuous power spectrum; 3) the true process model order is n which is used for estimating a Box-Jenkins model $\hat{G}(z^{-1}, \hat{\theta}^{BJ})$ and an output-error* *model* $\hat{G}(z^{-1}, \hat{\theta}^{BJ})$*; 4) the numerical search routines have converged to their global minima. Then*

$$EV^{V}_{OE,1}(\hat{\theta}^{BJ}) < EV^{V}_{OE,1}(\hat{\theta}^{OE}) \text{ as } N \to \infty \qquad (13)$$

See Appendix 1 for the proof.

This result is somewhat against intuition. Some researchers believed that an output-error model is optimal for output-error criterion. This is true when data is noise-free and bias is the only error source, also true when the identification data are used to evaluate the error criterion. However, when the process $G^o(z^{-1})$ is in the model set $\{G(z^{-1})\}$ and when the data are noisy, the Box-Jenkins or a prediction error model will have a smaller output-error criterion than the output-error model when evaluated on a fresh data set. Note that in (13) the result is given in expectations, meaning that exceptions may exist.

Simulation Example 1. Check the result for finite data sequence. Given the process

$$y(t) = \frac{z^{-1} + 0.5z^{-2}}{1 - 1.5z^{-1} + 0.7z^{-2}} u(t) + v(t)$$

with

$$v(t) = \frac{\alpha}{1 - 0.9q^{-1}} e(t)$$

The input signal is a GBN (generalized binary noise) with average switch time of 10 samples. The test time is 1000 samples. The disturbance power is 40% of that of the noise-free output. In total 100 open-loop simulations have been carried out. The data of each run are used to estimate a 2nd order Box-Jenkins model and a 2nd order output-error model using Matlab(R) System Identification Toolbox of MathWorks Inc. The following noise-free output-error criterion is used to compared the quality of the two models

$$V_{OE} = \frac{1}{1000} \sum_{t=1}^{1000} [G^o(z^{-1})u(t) - \hat{G}(z^{-1})u(t)]^2$$

The results are shown in Figure 2.1. One can see that Box-Jenkins mode indeed has much smaller average output-error criterion. On some occasions, the output-error model has slightly smaller error criteria.

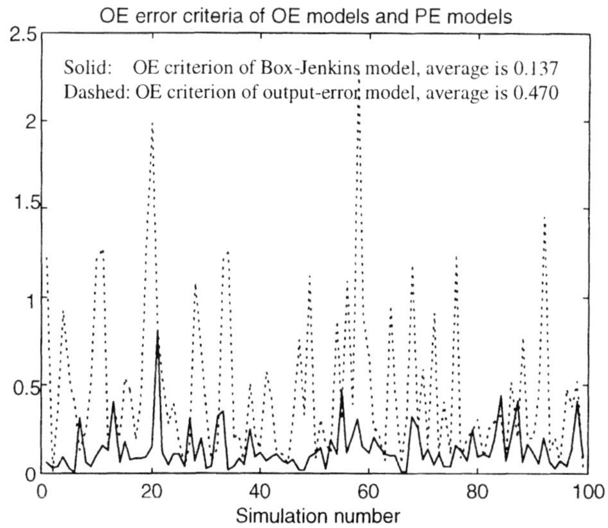

Figure 2.1 Compare Box-Jenkins model with output-error model

From numerical point of view, it is sometimes advantageous to first estimate a high order model and then perform a model reduction on the high order model; see, e.g., Wahlberg (1989), Bayard (1994), Zhu and Backx (1993) and Zhu (1998). Recently result shows that this approach can reduce the variance when compared with direct estimation of a low order model; see Tjärnström and Ljung (1999). Basically, two types of models are used for the high order model: ARX model and output error model. According to Theorem 2.1, an ARX model will be more accurate than an output error model with the same order, because a high order ARX model is practically unbiased and it is a prediction error model. The following example shows this fact.

Simulation Example 2. The process is

$$y(t) = \frac{1 + .5q^{-1} + .4q^{-2} + 2q^{-3}}{1 + .5q^{-1} - .4q^{-2} - .26q^{-3} - .03q^{-4}} u(t) + v(t)$$

where

$$v(t) = \frac{\alpha}{1 - 0.9q^{-1}} e(t)$$

$e(t)$ is a white Gaussian noise and α is a constant that is used to obtain a signal-to-noise ratio of 20% in power at the output. The input signal used is GBN signal with average switch time of 10 samples; the number of samples is 1000. In total 100 open-loop simulations are carried out. For each data set, a 10th order ARX model and a 10th order OE model are estimated using the Identification Toolbox of Matlab (R). Noise-free output-error criterion is used to compared the quality of the two models. Figure 2.2 shows the results. It is

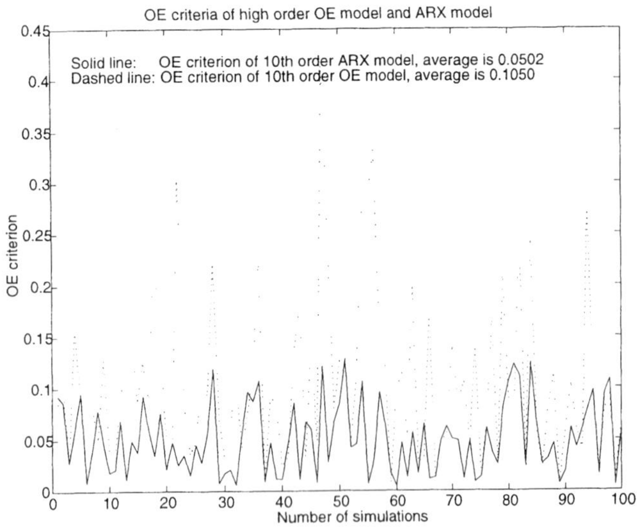

Figure 2.2 Compare high order OE model and high order ARX model

interesting to note that the ARX models are much simpler numerically and are more accurate than the output error models. If one uses an ARX model as initial value for high order output error model estimation, the iteration will in fact increase the true output error criterion. This is a situation where one gets more accuracy for less effort.

It follows to analyze the bias error. Assume that the disturbance is zero and the model order is lower than the true order, then it can be shown that (Wahlberg and Ljung, 1986) the limiting estimate of a OE model is

$$\theta_*^{OE} = \arg\min_\theta \int_{-\pi}^{\pi} |G^o(e^{i\omega}) - G(e^{i\omega}, \theta)|^2 \Phi_u^*(\omega) d\omega \tag{14}$$

and the limiting estimate of a Box-Jenkins model is

$$\theta_*^{BJ} = \arg\min_\theta \int_{-\pi}^{\pi} |G^o(e^{i\omega}) - G(e^{i\omega}, \theta)|^2 \Phi_u^*(\omega) \frac{D(e^{i\omega})}{C(e^{i\omega})} d\omega \tag{15}$$

These two expressions show that when only bias error exists, an output error is optimal for output error criterion. This is in contrary to variance error situation.

In practice, when low order model is estimated directly, both variance error and bias error can exist. The question is what argument should be used for selecting the estimation criterion, variance or bias. An answer to the question is based on the following result (Guo and Ljung, 1994).

Theorem 2.2. *Denote the bias error of the model as*

$$\widetilde{B}(e^{i\omega}) = G^o(e^{i\omega}) - E[\hat{G}(e^{i\omega})] \tag{16}$$

and variance error as

$$\widetilde{V}^2(e^{i\omega}) = E\{G^o(e^{i\omega}) - E[\hat{G}(e^{i\omega})]\}^2 \tag{17}$$

Then the total mean square error is

$$\widetilde{M}^2(e^{i\omega}) = E[G^o(e^{i\omega}) - \hat{G}(e^{i\omega})]^2 = \widetilde{B}^2(e^{i\omega}) + \widetilde{V}^2(e^{i\omega}) \tag{18}$$

Assume that, when model order increases, the decrease of bias error $\widetilde{B}^2(e^{i\omega})$ is faster then the increase of variance error $\widetilde{V}^2(e^{i\omega})$, then for the optimal model (order) that minimizes the total error $\widetilde{M}^2(e^{i\omega})$, the following relation holds

$$\widetilde{B}^2(e^{i\omega}) \leqslant \widetilde{V}^2(e^{-i\omega}) \tag{19}$$

that is, the bias error is dominated by the variance error.

Therefore, in low order model estimation, it is logical to consider variance error for criterion selection when the total error is to be minimized, and a prediction error model is preferred to an output error model.

3 Criterion for Order Selection

Basically two choices need to be made in order selection: 1) error criterion, and 2) data set on which the criterion is evaluated. From identification literature, often prediction error criterion is used in model order selection and model validation. For the purpose of control, the most important requirement is to obtain a model between process input and output. This requires that the simulation error, or, output error of the model should be small. Many practical examples has shown that a small prediction error do not always lead to a small output error and sometimes the result based on prediction error can be totally misleading; see, e.g., Söderström (1974). Therefore, output error criterion is much more suitable for order selection for control.

The second problem is the choice of data set on which the chosen criterion will be evaluated. The most reliable way is to use a fresh data set which is often called *validation data set.* The data set used for model estimation is then called *estimation data set.* This approach is called order selection by *cross-validation*; see e.g., Ljung (1987). The use of validation data set is a natural one, because if the estimation data and estimation criterion are used for order selection, higher order will be always better due to overfit problem. The disadvantage of cross-validation is its high cost, because additional test is required to collect validation data. In order to avoid the additional test, many researchers have proposed to use estimation data in a modified criterion that compensate for the effect of overfit. One of the well known criterion is the Akaike's *final prediction-error criterion* (FPE)

$$FPE = \frac{N+d}{N-d} V_{pe} \tag{20}$$

where N is the number of data, d is the number of parameters and V_{pe} is the prediction-error criterion used for parameter estimation evaluated on the estimation data set. Here the factor $\frac{N+d}{N-d}$ is used to correct for the overfit effect. As mentioned before, a small prediction-error criterion does not always lead to a small output-error criterion, which means that FPE may not give reliable results in practical identification.

Based on this observation and motivated by saving test time, we will propose the following *final output error criterion* (FOE)

$$\begin{aligned} FOE(m) &= \frac{N+d}{N-d} V_{oe} \\ &= \frac{N+d}{N-d} \frac{1}{N} \sum_{t=1}^{N} [y(t) - \hat{G}(q)u(t)]^2 \end{aligned} \tag{21}$$

where V_{oe} is the output-error (or simulation error) criterion evaluated on the estimation data set. Here, the factor $\frac{N+d}{N-d}$ is used to correct for the overfit and V_{oe} is used to reflect the model purpose, namely, control. Note that V_{oe} has not been used in model estimation, therefore, intuitively, one expect less overfit problem. We may call this way of using different criteria for model estimation and order selection *cross-criterion validation*.

Example 3.1. Simulated process:

$$y(t) = \frac{B(z^{-1})}{A(z^{-1})} u(t) + v(t)$$

where

$$\begin{aligned} A(z^{-1}) &= 1.0 - 1.15z^{-1} \quad 0.295z^{-2} + 0.7525z^{-3} \\ &\quad -0.02575z^{-4} - 0.14555z^{-5} - 0.00005z^{-6} \\ &\quad +0.002465z^{-7} - 0.000105z^{-8} \end{aligned}$$

$$B(z^{-1}) = 0 + 1.0z^{-1} + 1.0z^{-2} + 0.65z^{-3} + 2.20z^{-4} + 1.0z^{-5}$$

$$v(t) = \frac{\alpha}{1 - 0.9z^{-1}} e(t)$$

$e(t)$ is a white Gaussian noise and α is a constant that is used to obtain a signal-to-noise ratio of 10% in power at the output.

The input signal used is GBN signal with average switch time of 10 samples; the number of samples is 1000. In total 100 open-loop simulations are carried out. For each simulation, Box-Jenkins models of orders between 2 and 12 are estimated. FOE, Akaike's FPE and *noise-free output error criterion* are calculated for each model. The average of the three criteria over the 100 simulations are plotted in Figure 3.1.

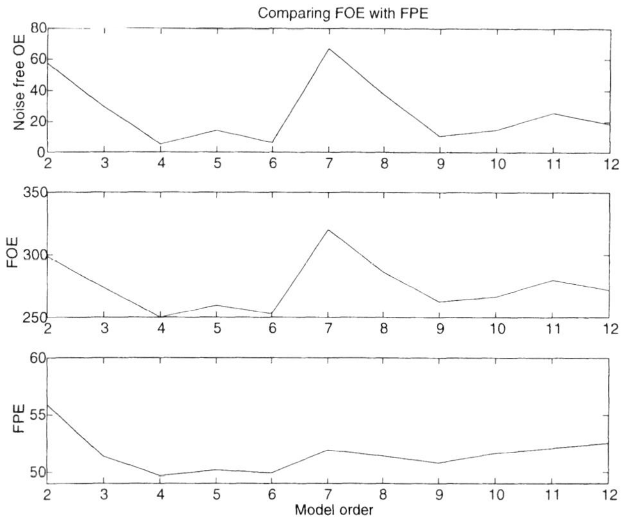

Figure 3.1 The averages of FOE, FPE and the true OE criteria over the 100 simulations.

One can see that, for this example, the shape of FOE is closer to noise free OE than FPE is. ARMAX models have also been estimated using the data from the same simulations. For ARMAX models, the shapes of FOE and FPE are different from that of noise free OE, but it is hard to say which criterion is better.

4 Conclusion

The use of criteria for control-relevant identification is studied. In parameter estimation, we have shown that an output-error model is not optimal for its criterion. In general, a prediction-error model has smaller output-error criterion. Therefore, a prediction-error model, is recommended for parameter estimation for minimizing an output-error criterion or a frequency weighted L_2 norm for control purpose. For model order selection, on the other hand, one should still use output-error criterion. Final output error (FOE) criterion is proposed. This leads to a so called *cross-criterion* approach.

References

[1] Akaike, H. (1974). A new look at the statistical model identification. *IEEE Trans. Autom. Control*, Vol. AC-19, pp. 716-723.

[2] Akaike, H. (1981). Modern development of statistical methods. In P. Eykhoff, ed.; *Trends and Progress*

in System Identification, Pergamon Press, Elmsford, New York.

[3] Guo, L. and L. Ljung (1994). The role of model validation for assessing the size of the unmodelled dynamics. *Proceedings of the 33rd Conference on Decision and Control*, Lake Buena Vista, Florida, pp. 3894-3899.

[4] Ljung, L. (1987). *System Identification: Theory for the User*. Prentice-Hall, Englewood Cliffs, N.J.

[5] Söderström, T. (1974). Convergence properties of the generalised least squares identification method. *Automatica*, Vol. 10, pp. 617-626.

[6] Söderström, T. and Stoica (1982). Some properties of the output error method. *Automatica*, Vol. 18, No. 1, pp. 93-99.

[7] Wahlberg, B. (1989). Model reduction of high-order estimated models: the asymptotic ML approach. *Int. J. Control*, Vol. 49, No. 1, pp. 169-192.

[8] Tjärnström and Ljung (1999). L_2 model reduction and variance reduction. Report No. LiTH-ISY-R-2158, Department of Electrical Engineering, Linköping University, S-581 83 Linköping, Sweden.

[9] Wahlberg, B. and L. Ljung (1986). Design variables for bias distribution in transfer function estimation. *IEEE Trans. Autom. Control*, Vol. AC-31, pp.134–144.

[10] Zhu, Y.C. (1998). Multivariable process identification for MPC: the asymptotic method and its applications. *Journal of Process Control*, Vol. 8, No. 2, pp. 101-115.

[11] Zhu, Y.C. and A.C.P.M. Backx (1993). *Identification of Multivariable Industrial Processes: for Simulation, Diagnosis and Control.* Springer-Verlag, London.

Appendix 1. Proof of Theorem 2.1.

Denote $\theta^o = [b^o_o, b^o_1, ...b^o_n, a^o_1, ...a^o_n]^T$ as the vector of true process model parameters. Under above assumptions, according to Ljung (1987), we have

$$\hat{\theta}^{BJ} \to \theta^o \text{ with probability one as } N \to \infty \qquad (22)$$

$$\hat{\theta}^{OE} \to \theta^o \text{ with probability one as } N \to \infty \qquad (23)$$

and

$$\sqrt{N}(\hat{\theta}^{BJ} - \theta^o) \in AsN(0, \text{cov}(\hat{\theta}^{BJ})) \text{ as } N \to \infty \quad (24)$$

$$\sqrt{N}(\hat{\theta}^{OE} - \theta^o) \in AsN(0, \text{cov}(\hat{\theta}^{OE})) \text{ as } N \to \infty \quad (25)$$

where $\text{cov}[\hat{\theta}]$ is the covariance matrix of the parameter vector; see Ljung (1987) for the expression.

According to Söderström and Stoica (1982, 1989), under conditions 1) to 4), we have

$$\text{cov}(\hat{\theta}^{BJ}) < \text{cov}(\hat{\theta}^{OE}) \qquad (26)$$

Note that here the inequality holds strictly because of condition 2) and 3), namely, the input is not multiple sinusoids and the disturbance is not white noise.

By Taylor's expansion

$$\begin{aligned} & \hat{G}(e^{i\omega}, \hat{\theta}^{BJ}) - G^o(e^{i\omega}, \theta^o) \\ = \; & [G'_\theta(e^{i\omega}, \theta^o)]^T(\hat{\theta}^{BJ} - \theta^o) + O(|\hat{\theta}^{BJ} - \theta^o|) \end{aligned} \quad (27)$$

where $G'_\theta(e^{i\omega}, \theta^o)$ is a column vector, the derivative of $G(e^{i\omega}, \theta)$ with respect to θ. Then, according to (24) we obtain the asymptotic distribution in the frequency domain

$$\begin{aligned} & \sqrt{N}(\hat{G}(e^{i\omega}, \hat{\theta}^{BJ}) - G^o(e^{i\omega}, \theta^o)) \\ \in \; & AsN(0, \text{var}[\hat{G}(e^{i\omega}, \hat{\theta}^{BJ})]) \text{ as } N \to \infty \end{aligned} \quad (28)$$

with

$$\text{var}\hat{G}(e^{i\omega}, \hat{\theta}^{BJ}) = [G'_\theta(e^{i\omega}, \theta^o)]^T \text{cov}(\hat{\theta}^{BJ}) G'_\theta(e^{i\omega}, \theta^o) \quad (29)$$

Similarly, we have for the output error model

$$\begin{aligned} & \sqrt{N}(\hat{G}(e^{i\omega}, \hat{\theta}^{OE}) - G^o(e^{i\omega}, \theta^o)) \\ \in \; & AsN(0, \text{var}[\hat{G}(e^{i\omega}, \hat{\theta}^{OE})]) \text{ as } N \to \infty \end{aligned} \quad (30)$$

with

$$\text{var}\hat{G}(e^{i\omega}, \hat{\theta}^{OE}) = [G'_\theta(e^{i\omega}, \theta^o)]^T \text{cov}(\hat{\theta}^{OE}) G'_\theta(e^{i\omega}, \theta^o) \quad (31)$$

Now

$$\begin{aligned} & EV^V_{OE,1}(\hat{\theta}^{BJ}) - EV^V_{OE,1}(\hat{\theta}^{OE}) \\ = \; & \frac{1}{2\pi}\int_{-\pi}^{\pi} \{\text{var}\hat{G}(e^{i\omega}, \hat{\theta}^{BJ}) \\ & \qquad - \text{var}\hat{G}(e^{i\omega}, \hat{\theta}^{OE})\}\Phi_u^*(\omega)d\omega \\ = \; & \frac{1}{2\pi}\int_{-\pi}^{\pi} \{[G'_\theta(e^{i\omega}, \theta^o)]^T[\text{cov}(\hat{\theta}^{BJ}) \\ & \qquad - \text{cov}(\hat{\theta}^{OE})]G'_\theta(e^{i\omega}, \theta^o)\}\Phi_u^*(\omega)d\omega \\ < \; & 0 \end{aligned} \quad (32)$$

which proves the theorem. Note that the last inequality is due to (26).

APPROACHES TO CONTROL-ORIENTED H_∞ IDENTIFICATION

Mehrzad Namvar and Alina Besançon-Voda [1]

Laboratoire d'Automatique de Grenoble
Bp 46, 38402-St.Martin d'Heres, France
saba@lag.ensieg.inpg.fr

Abstract: This paper presents deterministic approaches to control oriented H_∞ identification of linear systems using time domain measurements. In the open-loop approach, we identify the models with the minimum additive uncertainty in H_∞ norm and then constant or frequency dependent upper-bounds for the modeling error, in the same optimization procedure, are obtained. In the closed-loop approach, the models are identified so that the multiplicative error between real and nominal closed-loop sensitivity functions is minimized. The identified models will then suitable for iterative identification-control design. Finally the applicability of the proposed methods is illustrated by a numerical example. *Copyright ©2000 IFAC*

Keywords: Identification for control, Model Validation, Linear Matrix Inequality, Convex optimization, Model uncertainty, Robust control

1. INTRODUCTION

Identification for control or control oriented identification can be viewed as choosing a suitable model for control design, from a given model set, using finite corrupted data. The corruption of data may be due to disturbance signal belonging to some class of signals or to unmodeled dynamics.

In general two branches for control based identification can be considered. In the first branch, which we call open-loop approach, the research is mainly concentrated on identifying nominal models with the minimum possible uncertainty preferably expressed in a norm compatible with that used in robust control. For example, in the context of set-membership identification (Milanese and Vicino 1991), the best nominal model belonging to the feasible model sets, is identified so that a worst-case modeling error is minimized. Following an identification procedure and usually in a *separate* stage, worst-case bounds on the additive modeling error are found for a *finite* number of frequencies. This stage necessitates solving an optimization problem in each separate frequency (Giarre *et al.* 1997). This is also the case for the deterministic uncertainty bounding procedure proposed in (Hakvoort 1994).

On the other hand, in (Kosut 1995), knowing the frequency distribution of the uncertainty and for ARX model structure, the optimization constraints turn to be affine in model parameters and the feasible parameter sets take the form of convex ellipsoids or hyperboloids (Kosut *et al.* 1992).

In the second branch for the control based identification introduced in (Van den Hof and Schrama 1995) and (Hjalmarsson *et al.* 1996), the objective is to identify the nominal models minimizing a common criteria with control design in an iterative *identification-control design* scheme. We call this approach as closed-loop identification.

Recent results in model validation (Poolla *et al.* 1994), (Newlin and Smith 1998) and the techniques in solving convex optimization problems (Boyd *et al.* 1994) provide powerful tools to unify model validation with system identification in a single optimization problem and in a deterministic environment. Formulation of the problem by LMI's allows to consider different convex model and disturbance signals sets with various properties. See

[1] The authors would like to thank Professor I.D. Landau for his useful comments and suggestions.

(Venkatesh and Dahleh 1997) and (Kosut 1995) for some examples. This approach, for the case of general LFT structures and with frequency domain data samples (H_∞ case) is also studied in (Smith 1992) and the problem is solved via μ calculations.

In the present paper, adopting the specific uncertain model structures introduced in (Kosut 1995), and assuming that the norm of the disturbance signal is bounded by a known value, we state and solve system identification and additive uncertainty bounding problems in H_∞ norm via convex optimization algorithms in the form of LMIs. Moreover, at the same procedure, we obtain constant or frequency dependent upper-bounds for the unmodeled dynamics. The results are then extended to closed-loop identification. Our study is restricted in time domain (see (Namvar *et al.* 1999) for frequency domain approach).

The paper is organized as follows: In section 2, notations and some definitions are given, then in section 3, the open-loop approach to the identification and additive uncertainty bounding is studied. In section 4, the results for closed-loop identification are presented and in section 5 a numerical example is presented.

2. NOTATIONS AND DEFINITIONS

Unit delay operator denotes z^{-1}. Throughout the paper the nominal model is expressed as

$$\hat{G}(z^{-1}) = \frac{B(z^{-1})}{A(z^{-1})} \tag{1}$$

with $B(z^{-1}) = b_0 + b_1 z^{-1} + ... + b_m z^{-m}$ and $A(z^{-1}) = a_0 + a_1 z^{-1} + ... + a_n z^{-n}, (a_0 > 0)$

If G is the plant, the actual additive uncertainty is defined as $\Delta G = G - \hat{G}$. The measured input-output data u and y are produced as $y(t) = Gu(t) + d(t)$ where $d(t)$ is an unknown bounded disturbance.

Throughout the paper we call Lower triangular Toeplitz matrix as LT matrix. If $x = [x(0), x(1), \cdots x(N)]$ is a real time domain sequence, $T_x \in \mathcal{R}^{N+1}$ is a LT matrix constructed from $x(t)$s. l_2 denotes the space of real sequences p with bounded energy and with norm $||p||_2 = \sum_{t=0}^{\infty} p^2(t) < \infty$. l_2^N is defined as $l_2^N = \{p(t)|p(t) \in \mathcal{R}, \sum_{t=0}^{N} p^2(t) < \infty\}$. Let $\mathcal{D}$ denote the unit circle, then the operator $\Delta \in \mathcal{H}_\infty$ if $||\Delta||_\infty = \sup_{\omega \in \mathcal{R}} |\Delta(j\omega)| < \infty$ and Δ has no poles outside and on $\mathcal{D}$.

3. OPEN-LOOP APPROACH TO IDENTIFICATION AND UNCERTAINTY BOUNDING

Consider the uncertain model structure introduced in (Kosut 1995)

$$y(t) = (\hat{G}(\theta) + \hat{H}(\theta)\Delta)u(t) + \hat{H}(\theta)w(t) \tag{2}$$

where Δ is a linear causal operator representing the uncertainty block and $w(t)$ is a disturbance signal belonging to a known convex signal set $\mathcal{W}$. $u(t)$ and $y(t)$ are the given measured time domain input and output of the system.

In the context of model validation, given the model $\hat{G} \in \mathcal{G}$, the uncertainty set $\mathbf{\Delta}$ and the disturbance signal set $\mathcal{W}$, the problem in general is imposed as whether the uncertain structure is consistent with the given input-output measurements $u(t)$ and $y(t)$. See (Poolla *et al.* 1994) (Newlin and Smith 1998) for complete discussion.

Problem 1: Consider the uncertain model structure (2). Given the sets $\mathcal{W}$, $\mathcal{G}$ and $\mathbf{\Delta}$, assume that the unknown model $\hat{G}(\theta)$ is parameterized by parameter vector θ, the objective is to solve the following optimization problem which identifies the unknown parameter vector θ

$$\min_{\theta, w} ||\hat{H}\Delta||_\infty$$
$$w \in \mathcal{W}, \quad \hat{G} \in \mathcal{G}, \quad \Delta \in \mathbf{\Delta}$$
$$y(t) = (\hat{G}(\theta) + \hat{H}(\theta)\Delta)u(t) + \hat{H}(\theta)w(t)$$

Moreover, the ***uncertainty bounding problem*** is defined as finding a constant or frequency dependent upper-bounds for the additive uncertainty $\Delta G = \hat{H}\Delta$ in the above optimization problem. If Δ depends only on the model parameter vector θ (through $\hat{G}$), the above optimization can give only a *constant* upper-bound for the uncertainty, but if Δ is independently parameterized by a parameter vector ρ, then a frequency dependent-upper bound for the uncertainty, through solving the above optimization problem, can also be obtained. □

The properties of the model set $\mathcal{G}$ can be characterized by a bounded linear operator $\mathcal{A}$ as $\mathcal{A}[\hat{G}(\theta)] \leq B$.

In general the optimization problem with uncertain model structure (2) is not convex, but for some uncertain model structures, the problem reduces to convex optimizations. Consider first the case of the ARX model structure

$$Ay(t) = (B + \Delta)u(t) + w(t) \tag{3}$$

We state at first the following result

Theorem 3.1. (Poolla *et al.* 1994) Given $N+1$ samples of time domain sequences p, q as $p^N, q^N \in {l_2}^N$, there exists a stable, causal, LTI operator Δ with $||\Delta||_\infty < \gamma$ such that $p = \Delta q$ iff:

$$T_p' T_p < \gamma^2 T_q' T_q \tag{4}$$

where $T_p \in \mathcal{R}^{(N+1(N+1)}$ is a LT matrix constructed from p^N. T_q is defined similarly.

A version of the following result which is derived in (Kosut 1995), solves the identification problem via a convex optimization

Theorem 3.2. Given $N + 1$ samples of time domain sequences $u^N, y^N \in l_2^N$, and the ARX model structure (3), the following convex optimization problem identifies a model $\hat{G}$ such that $||\Delta||_\infty \leq \gamma$ is minimized

$$\min_{T_A, T_B, w} \gamma^2$$
$$w \in \mathcal{W}, \quad \hat{G} \in \mathcal{G}$$
$$\begin{bmatrix} \gamma^2 T_u' T_u & * \\ T_y T_A - T_u T_B - T_w & I \end{bmatrix} > 0$$

where $T_A \in \mathcal{R}^{(N+1)(N+1)}$ is a LT matrix constructed from $[a_0, a_1, ..., a_n, 0, ..., 0] \in \mathcal{R}^{N+1}$.

T_B is defined similarly. $T_u, T_y, T_w \in \mathcal{R}^{(N+1)(N+1)}$ are LT matrices constructed from u^N, y^N and $w^N = [w_0, ..., w_N]$ respectively.

3.1 *Additive uncertainty minimization*

According to the ARX model structure (3), the additive uncertainty is expressed as $\Delta G = \Delta \frac{1}{A}$ with $\hat{H} = \frac{1}{A}$. Therefore, the minimization of $||\Delta||_\infty$ by Theorem 3.2 may not lead to the minimization of $||\Delta G||_\infty$ because the coefficients of $A(\theta)$ are themselves optimization variables. In order to treat the problem, an iterative algorithm is proposed where in each iteration an optimization problem as in Theorem 3.2 is solved.

In the i'th iteration of the algorithm we consider the following model structure

$$A_i y(t) = \Delta \underbrace{A_{i-1}u(t)} + B_i u(t) + \underbrace{A_{i-1}w} \tag{5}$$

It can be verified that the additive uncertainty is now $\Delta G = \frac{A_{i-1}}{A_i}\Delta$. The presence of the term A_{i-1} in series with Δ is to eliminate the effect of term $\frac{1}{A_i}$. Note that in iteration i, A_{i-1} is known from the previous iteration and A_i is unknown.

Suppose that in iteration i, the optimal value of γ_i^o is achieved ($||\Delta||_\infty < \gamma_i^o$). When A_i approaches A_{i-1} in the sense that

$$\sup_\omega \left| \frac{A_{i-1}(j\omega) - A_i(j\omega)}{A_i(j\omega)} \right| \leq \epsilon_i \leq 1 \Longrightarrow$$

$$||\Delta||_\infty (1 - \epsilon_i) \leq ||\Delta G||_\infty \tag{6}$$

$$\leq ||\Delta||_\infty (1 + \epsilon_i) \leq \gamma_i^o (1 + \epsilon_i) \tag{7}$$

It is seen that the minimization of $||\Delta||_\infty$ through minimization of γ_i in iteration i, results in the minimization of $||\Delta G||_\infty$ with ϵ_i degree of accuracy.

In summary in order to minimize the H_∞ norm of the additive uncertainty, the following iterative algorithm is proposed

- *Algorithm* 1, Minimization of $||\Delta G||_\infty$:

Step 1 (Initialization) Put $i = 0$ and solve the optimization problem as in Theorem 3.2 and find $A_0(z^{-1})$ and $B_0(z^{-1})$. Then put $i = 1$.

Step 2 Construct the filtered signal $u_f(t) = A_{i-1}u(t)$

Step 3 Solve the following optimization problem

$$\min_{T_A, T_B, w} \gamma^2$$
$$w \in \mathcal{W}, \quad \hat{G} \in \mathcal{G}$$
$$\begin{bmatrix} \gamma^2 T_{u_f}' T_{u_f} & * \\ T_y T_{A_i} - T_u T_{B_i} - T_w T_{A_{i-1}} & I \end{bmatrix} > 0$$
$$\begin{bmatrix} d_{i-1}^2 & * \\ T_{A_i} - T_{A_{i-1}} & I \end{bmatrix} > 0$$

with $d_i = ||T_{A_i} - T_{A_{i-1}}||$.

Step 4 When $d_i < d_p$ and $\epsilon_i \leq 1$ stop, else $i = i + 1$ and go to step 2 (d_p is a given small positive real number).

Note that according to Theorem 3.3 and Corollary ??, $d_i = ||A_i - A_{i-1}||_\infty$ for large N. The second inequality in step 3 which is equivalent to $d_i < d_{i-1}$ guarantees the convergence of the algorithm.

3.2 *Frequency dependent uncertainty bounding*

An important problem in control design is to know the frequency distribution of the uncertainty. This section deals with solving H_∞ identification when considering a FIR structure for the uncertain block Δ and simultaneously identifying the plant G together with the parameters of Δ. This leads to obtain a frequency dependent uncertainty overbound *continuous* in frequency.

Both of these solutions are presented for the case of ARX model structure (3) for minimizing $||\Delta||_\infty$. Obviously Algorithm 1 in section 3.1 can be applied in each case for minimizing $||\Delta G||_\infty$.

Asymptotic solution

The following theorem shows how a lower approximation to the H_∞ norm of a polynomial can be obtained.

Theorem 3.3. (Grenander and Szego 1958) Let $\Delta(z^{-1}) = d_0 + \sum_{k=1}^n d_k z^{-k}$ be analytic in $|z| > 1$. We denote $\bar{\sigma}_n$ the largest singular value of T_Δ where T_Δ is a LT matrix constructed from $[d_0, ..., d_n]$, then

i) $\lim_{n\to\infty} \bar{\sigma}_n = ||\Delta||_\infty \geq \bar{\sigma}_n$

ii) $\bar{\sigma}_n$'s form a non-decreasing sequence

The theorem states that the sequence of $\bar{\sigma}_n$'s tend asymptotically and monotonically to the infinite norm of Δ. The monotonic convergence of $\bar{\sigma}_n$'s, allows us to approach $||\Delta||_\infty$ from below with a desired degree of accuracy when the order of $\Delta(z^{-1})$ is increased.

We define T_{Δ_1} as a $(n_1+1)(n_1+1)$ LT matrix constructed from $[d_0, d_1, ..., d_n, 0, ..., 0]'_{n_1+1}$, with $n_1 > n$. clearly $\bar{\sigma}(T_{\Delta_1})$ is nearer to $||\Delta||_\infty$ than $\bar{\sigma}(T_\Delta)$.

In the sequel we assume that $N \leq N_1 < 2N + 2$. Suppose that $u, y \in l_2^N$ are known. We construct the extended sequences $u^e, y^e \in l_2^{N_1}$ as $y^e = [y' \quad \hat{y}']'$ and $u^e = [u' \quad \hat{u}']'$ with $\hat{u} = 0$. Since only y is known, we consider $\hat{y}$ as the unknown output. Let $T_y^e \in \mathcal{R}^{(N_1+1)(N_1+1)}$ be a LT matrix constructed from y^e and be partitioned as

$$T_y^e = \begin{bmatrix} T_y & 0 \\ T_{21}(y^e) & T_{22}(y) \end{bmatrix} \tag{8}$$

with $T_{21}(y^e) = T_{21}(\hat{y}) + T_{21}(y)$.

$T_y \in \mathcal{R}^{(N+1)(N+1)}$ is a LT matrix constructed from y. T_u is defined similarly.

$T_{21}(\hat{y})$ depends only on $\hat{y}$ and $T_{21}(y)$ depends on y. It can be verified that by the assumption $N_1 < 2N + 2$, $T_{22}(y)$ doesn't depend on the unknown output vector $\hat{y}$.

T_u^e, $T_{21}(u^e)$, $T_{21}(\hat{u})$, $T_{21}(u)$ and $T_{22}(u)$ are defined similarly. Note that $T_{21}(\hat{u}) = 0$ by assumption.

Similarly T_A^e is a LT matrix constructed from

$$[a_0, ..., a_n, 0, ..., 0] \in \mathcal{R}^{N_1+1}$$

and is partitioned as

$$T_A^e = \begin{bmatrix} T_A & 0 \\ T_{21}(A) & T_{22}(A) \end{bmatrix} \quad (9)$$

$T_A \in \mathcal{R}^{(N+1)(N+1)}$ is a LT matrix constructed from $[a_0, ..., a_n, 0, ..., 0] \in \mathcal{R}^{N+1}$. The same definitions hold for T_B^e.

Theorem 3.4. Consider the ARX model structure in (3). Given time domain input-output measurements $u, y \in l_2^N$ with $u(0) \neq 0$, suppose that the uncertainty block Δ is represented as a FIR filter of degree N (as in Theorem 3.3). Then the H_∞ identification and uncertainty bounding problem can be solved via the following optimization problem

$$\min_{T_A, T_B, T_N, w} \gamma^2$$
$$w \in \mathcal{W}, \quad \hat{G} \in \mathcal{G}$$
$$\begin{bmatrix} \gamma^2 T_u T_u & T_u T_{21}'(u) & T_y T_A - T_u T_B - T_w & 0 \\ * & \mathcal{T}_u & \mathcal{T}_R & \mathcal{T}_S \\ * & * & I & 0 \\ * & * & * & I \end{bmatrix} > 0$$

with $\mathcal{T}_u = \gamma^2 (T_{21}(u) T_{21}'(u) + T_{22}(u) T_{22}'(u))$

$$\mathcal{T}_R = T_N + T_{21}(y) T_A - T_{21}(u) T_B + T_{22}(y) T_{21}(A)$$
$$- T_{22}(u) T_{21}(u) - T_{21}(w)$$
$$\mathcal{T}_S = T_{22}(y) T_{22}(A) - T_{22}(u) T_{22}(B) - T_{22}(w)$$

The coefficients of the optimum FIR filter representing the uncertainty $\Delta(z^{-1})$ are calculated as

$$T_\Delta^o = T_u^{-1}(T_y T_A^o - T_w^o) - T_B^o \quad (10)$$

The theorem is stated when the degree of $\Delta(z^{-1})$ is taken as its maximal value N.

Remark 1. When the optimal value of the slack variable T_N^o is obtained from the optimization, the unknown output vector $\hat{y}$ can be calculated by $T_{21}^o(\hat{y}) = T_N^o T_A^{o^{-1}}$. In fact $\hat{y}$ can be seen as the prediction of the output signal y in the interval $N < t \leq N_1 + 1$ based on the measured values of the output signal in $0 \leq t \leq N$. The assumption $N_1 < 2N + 2$ limits the prediction to be in a maximum interval of $N + 2$ samples.

Remark 2. By Theorem 3.4, only a finite samples of the impulse response of Δ are identified. If we define $\Delta^N(z^{-1}) = \sum_{k=N+1}^{\infty} d_k z^{-k}$ and assume $\|\Delta^N(z^{-1})\|_\infty < \delta_t$, then

$$\lim_{N \to \infty} \|\Delta\|_\infty \leq \gamma^o + \delta_t$$

Note that in Theorem 3.2 no such assumption on the tail of Δ was necessary and γ^o was exactly an upper-bound for the infinite norm of Δ.

Non-Asymptotic Solution

In this case we apply bounded real lemma and propose an algorithm which solves the identification and uncertainty bounding problem in a non-asymptotic way. This means that with a finite order FIR filter corresponding to $\Delta(z^{-1})$, we can exactly make $\|\Delta\|_\infty < \gamma^o$.

Lemma 3. A necessary and sufficient condition for a LTI system Δ with realization $[A_d, B_d, C_d, D_d]$ to have H_∞ norm smaller than γ is that

$$\begin{bmatrix} Q & A_d'Q & C_d' & 0 \\ * & Q & 0 & QB_d \\ * & * & I & D_d \\ * & * & * & \gamma^2 I \end{bmatrix} > 0$$

for some symmetric positive definite matrix Q.

Theorem 3.5. Consider the ARX model structure (3) and suppose that $\Delta(z^{-1})$ is represented as a FIR filter of degree N. Using data samples $u, y \in l_2^N$ with $u(0) \neq 0$, the non-asymptotic solution to the problem is

$$\min_{\theta_a, \theta_b, Q>0, w} \gamma^2$$
$$w \in \mathcal{W}, \quad \hat{G} \in \mathcal{G}$$
$$\begin{bmatrix} T_u^N Q T_u'^N & T_u^N A_d' Q & \mathcal{T} & 0 \\ * & Q & 0 & Q e_1 u(0) \\ * & * & 1 & \mathcal{S} \\ * & * & * & \gamma^2 u^2(0) I \end{bmatrix} > 0$$
$$\mathcal{S} = a_0 y(0) - b_0 u(0) - w(0)$$
$$\mathcal{T} = T_1 \theta_a - T_2 \theta_b - T_3 w$$

Where $T_u \in \mathcal{R}^{(N+1)(N+1)}$ is a LT matrix constructed from u and is partitioned as $T_u = \begin{bmatrix} u(0) & 0 \\ u_N & T_u^N \end{bmatrix}$ with $u_N = [u(1), ..., u(N)]'$ and $T_u^N \in \mathcal{R}^{N \times N}$ is also a LT matrix. y_N and T_y^N are defined similarly.

$$T_1 = [y_N - \frac{y_0}{u_0} u_N \quad T_y^N] \in \mathcal{R}^{N(N+1)}$$
$$T_2 = [0 \quad T_u^N] \in \mathcal{R}^{N(N+1)}$$
$$T_3 = [\frac{-1}{u_0} u_N \quad I] \in \mathcal{R}^{N(N+1)}$$
$$A_d = \begin{bmatrix} 0 & 0 \\ I & 0 \end{bmatrix} \in \mathcal{R}^{N \times N}$$
$$e_1 = [1, 0, ..., 0]' \in \mathcal{R}^N$$
$$w = [w_0, ..., w_N]'$$
$$\theta_a = [a_1, ..., a_n, 0, ...0]' \in \mathcal{R}^{N+1}$$
$$\theta_b = [b_1, ..., b_m, 0, ...0]' \in \mathcal{R}^{N+1}$$

Remark 4. The number of optimization variables in the asymptotic solution is $\mathcal{N}(asym) = 2N_1 - N + n + m + 4$ and the dimension of the inequality $\mathcal{D}(asym) = 2(N_1 + 1)$. With the assumption $N \leq N_1 \leq 2(N+1)$, $\max \mathcal{N}(asym) = 3N + n + m + 8$ and $\max \mathcal{D}(asym) = 4N + 6$

On the other hand, in non-asymptotic solution, the number of variables is $\mathcal{N}(nasym) = \frac{N(N+1)}{2} + N + n + m + 4$ and the dimension of the inequality is $\mathcal{D}(nasym) = 2(N+1)$. Then

$$\max \mathcal{N}(asym) < \mathcal{N}(nasym) \quad \text{for} \quad N > 4$$
$$\max \mathcal{D}(asym) > \mathcal{D}(nasym)$$

The choice between the asymptotic and non-asymptotic solutions will be then a compromise between the dimension of the inequality, the number of the optimization variables and the degree of convergence of γ^o to $\|\Delta\|_\infty$ in the asymptotic solution.

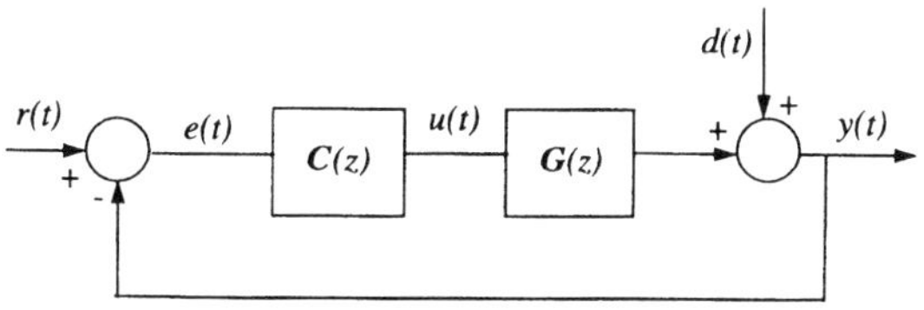

Fig. 1. Feedback connection

4. CLOSED-LOOP IDENTIFICATION

While in section 3 we concentrated on the identification of the nominal models with the minimum additive uncertainty in $\mathcal{H}_\infty$ norm and the characterization of the uncertainty, in the present section our focus is on the identification of nominal models which are suitable for iterative identification-control design. This means that we identify the models which are more accurate in the frequencies which are important for the control design. We present the results for the case where the control objective is to minimize the effect of output disturbance $d(t)$ on the input of the system $u(t)$ or equivalently to minimize the H_∞ norm of the real input sensitivity function (see Figure 1). The objective can also be regarded as increasing the robustness of the closed-loop against the additive perturbations to the nominal model of the system.

In what follows the real input sensitivity function is $S_u = \frac{C}{1+GC}$ and the nominal input sensitivity is $\hat{S}_u = \frac{C}{1+\hat{G}C}$. The additive uncertainty in the input sensitivity function is called $\Delta S_u = S_u - \hat{S}_u$. The output sensitivity functions is equally defined as $S = \frac{1}{1+GC}$.

Problem 2: Given $N+1$ samples of the measured signals $u(t), y(t)$ and $r(t) \in l_N^2$ obtained from the experimental setup in Figure 1 and the model set $\mathcal{G}$, disturbance set $\mathcal{W}$, and the model structure

$$u(t) = (\Delta_s + 1)\hat{S}_u(\theta)r(t) + \hat{H}(\theta)w(t) \qquad (11)$$

identify a nominal model $\hat{G}(\theta)$ such that $\|\frac{\Delta S_u}{\hat{S}_u}\|_\infty = \|\Delta_s\|_\infty$ is minimized. □

Note that $\frac{\Delta S_u}{\hat{S}_u} = \frac{\Delta S}{\hat{S}}$ It can be verified that

$$\|S_u\|_\infty \leq \|\hat{S}_u\|_\infty(1 + \|\Delta_s\|_\infty)$$

Therefore, the minimization of $\|\Delta_s\|_\infty$ makes the nominal input sensitivity approach the real one. This can be treated as the control based identification step in a consecutive *identification-control design* scheme discussed in (Hof and Schrama 1995) and (Hjalmarsson *et al.* 1996). Moreover, note that $\Delta_s = S_u(\Delta G)$, this means that the minimization of $\|\Delta_s\|_\infty$ minimizes the additive uncertainty weighted by the real input sensitivity function. Therefore, the identification in closed-loop gives a model which is a good approximation of the true plant in the frequencies where the real sensitivity in large.

As in section 3, in order to solve the problem we begin by considering the following model structure affine in the parameters

$$Ay(t) = \Delta r(t) + Bu(t) + w(t) \qquad (12)$$

It can be verified that (12) is equivalent to

$$u(t) = (1 + \frac{\Delta}{A})\hat{S}_u r(t) - \frac{\hat{S}_u}{A}w(t) \qquad (13)$$

Comparing (13) with (11) shows that $\Delta_s = \frac{1}{A}\Delta$ and $\hat{H} = -\frac{1}{A}\hat{S}_u$.

Theorem 4.1. Given the same conditions in Problem 2 and taking (12) as the model structure, the following optimization problem minimizes $\|\Delta\|_\infty \leq \gamma$

$$\min_{T_A, T_B, w} \gamma^2$$
$$w \in \mathcal{W}, \quad \hat{G} \in \mathcal{G}$$
$$\begin{bmatrix} \gamma^2 T_r' T_r & * \\ T_y T_A + T_u T_B - T_w & I \end{bmatrix} > 0$$

Note that the knowledge of the controller is not essential in the optimization. This is significant specially when the controller is unstable or non-minimum phase.

To eliminate the effect of the term $\frac{1}{A}$ and make $\|\Delta\|_\infty \longrightarrow \|\Delta_s\|_\infty$, as in section 3.1 , we take the following model structure in the iteration i of an iterative algorithm

$$A_i y(t) = \Delta r_f(t) + B_i u(t) + A_{i-1}w(t)$$

with $r_f(t) = A_{i-1}r(t)$ and we replace the following inequality in Theorem 4.1

$$\begin{bmatrix} \gamma^2 T_{r_f}' T_{r_f} & * \\ T_y T_{A_i} + T_u T_{B_i} - T_{A_{i-1}} T_w & I \end{bmatrix} > 0$$

Note that the convergence condition as in Algorithm 1 can also be considered here.

5. EXAMPLE

In this example we apply the open-loop and closed-loop approaches to the identification studied in sections 3 and 4. Our plant G is a high degree model (10th order) of a pneumatic actuator system. The model $\hat{G}$ is considered to be of order 4 ($n = m = 4$).

Open-loop identification:

In a simulation experiment we produced the input-output signals as $y(t) = Gu(t) + d(t)$ where $d(t)$ is a disturbance signal with $\|d\|_2 = 0.065$ such that $\frac{\|d\|_2}{\|y\|_2} = 3.5\%$ and the input signal is $u(t) = u_0(t) - 2u_0(t - 5T) + u_0(t - 10T)$ with $u_0(t)$ as the unit step. Here $T = 0.01$ sec. is the sampling time. The number of data samples is $N = 60$.

The asymptotic solution to the frequency dependent uncertainty bounding problem in Theorem 3.4 is applied iteratively in Algorithm 1. $\Delta(z^{-1})$ is considered to have FIR structure of order N. We took $N_1 = N = 60$ in solving the optimization in Theorem 3.4 in each iteration. w is assumed to be energy bounded $\|w\|_2 < 0.04$ (note that the bound is intentionally taken to be different from the energy of the real disturbance signal). Application of Algorithm 1 makes the estimated uncertainty $\Delta(z^{-1})$ approach the actual uncertainty ΔG. After $i = 3$ iterations when $d_3 = 2.1 \times 10^{-5}$, $\epsilon_3 = 2.3 \times 10^{-4}$ and $\gamma_3^o = 0.075$. Figure 2 shows the difference between $|\Delta(j\omega)|$ and $|\Delta G(j\omega)|$. Note that $\gamma_3^o < \|\Delta\|_\infty$, this is because of the asymptotic nature of the solution in Theorem 3.4

Closed-loop identification:

Using $\hat{G}_{ol}(z)$ and the uncertainty distribution $\Delta(j\omega)$, an optimal controller $C(z)$ is designed. Then in an experimental set-up shown in Figure 1 with $\|d\|_2 = 0.07$, closed-loop data sequences r, u, and y are produced.

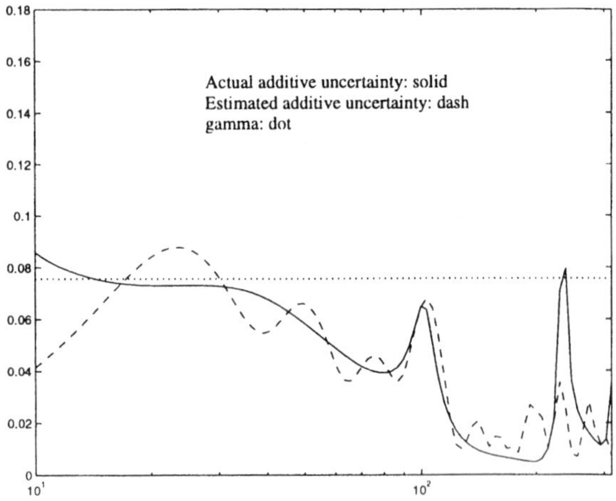

Fig. 2. Actual additive uncertainty $|\Delta G(j\omega)|$ (solid) versus estimated additive uncertainty $|\Delta(j\omega)|$ (dash) in iteration $i = 3$

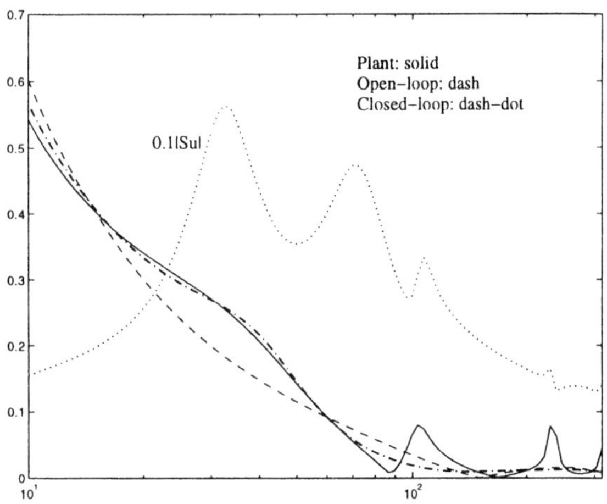

Fig. 3. Plant $|G(j\omega)|$ (solid) versus $|\hat{G}_{ol}(j\omega)|$ (dash) and $|\hat{G}_{cl}(j\omega)|$ (dash-dot)

As discussed in section 4, the optimization problem in Theorem 4.1 is solved iteratively in order to minimize $||\Delta_s||_\infty < \gamma_i^o$. In iteration $i = 3$, when $d_3 = 0.015$, $\epsilon_3 = 0.02$ and $\gamma_3^o = 0.051$. Figure 3 shows the comparison between the identified models in open-loop and closed-loop. It is clearly seen that $\hat{G}_{cl}(z)$ approximates the plant much better than $\hat{G}_{ol}(z)$ in the frequencies where the real sensitivity function S_u is large.

6. CONCLUSION

In this paper we studied two deterministic approaches to control oriented H_∞ system identification. In the open-loop approach it was shown that for some given convex model sets and disturbance sets, we can identify the models with the minimum H_∞ uncertainty in additive modeling error and then quantify the model uncertainty by constant or frequency dependent upper-bounds at the same optimization procedure. The results then were extended for closed-loop identification which weights the additive modeling error by the real closed-loop sensitivity function. The formulation of these problems by LMIs results in the capability of considering different objectives in the identification.

The simulation examples show that the proposed algorithms are implementable, nevertheless unlike the least-squared based algorithms which are often solved recursively, the proposed optimizations are solved non-recursively and may become computationally complex. This problem together with the choice of the best inputs or reference signals can be the subject of further research.

7. REFERENCES

Boyd, S., L. El Ghaoui and V. Balakrishnan (1994). *Linear matrix inequality in system and control theory.* Siam studies in applied mathematics.

Giarre, L., M. Milanese and M. Taragna (1997). H_∞ identification and model quality evaluation. *IEEE Transactions on Automatic control* **42**(2), 188–199.

Grenander, U. and G. Szego (1958). *Toeplitz forms and their applications.* Unv. of California Press. Berkeley and Los Angeles.

Hakvoort, R. (1994). System identification for robust process control: Nominal models and error bounds. *PhD dissertation, Delf Univ. Tech.*

Hjalmarsson, H., M. Gevers and F. De Bruyne (1996). For model-based control design, cloosed-loop identification gives better performance. *Automatica* **32**(12), 1659–1673.

Hof, P. Van-Den and R. Schrama (1995). Identification and control, closed-loop issues. *Automatica* **31**(12), 1751–1770.

Kosut, R. (1995). Uncertainty model unfalsification: A system identification paradigm compatible with robust control design. *Proceeding of 34th IEEE CDC, New Orleans, LA* pp. 3492–3497.

Kosut, R., M. Lau and S. Boyd (1992). Set-membership identification of systems with parametric and nonparametric uncertainty. *IEEE Transactions on Automatic control* **37**, 929–941.

Milanese, M. and A. Vicino (1991). Optimal estimation theory for dynamic systems with set membership uncertainty: An overview. *Automatica* **27**(6), 997–1009.

Namvar, M., A. Voda and I. Landau (1999). Approaches to identification and model validation for robust control design. *Proceedings of 14th IFAC World Congress, China* pp. 307–312.

Newlin, M.P. and R.S. Smith (1998). A genelalization of the structured singular value and its application to model validation. *IEEE Transactions on Automatic control* **43**(7), 901–907.

Poolla, K., P. Khargonekar, A. Tikku, J. Krause and K. Nagpal (1994). A time domain approach to model validation. *IEEE Transactions on Automatic Control* **39**(5), 951–959.

Smith, R. (1992). Model validation and parameter identification in H_∞ and l_1. *Proc. of American Control Conference* pp. 2852–2856.

Venkatesh, S. and M. Dahleh (1997). Identification in the presence of classes of unmodeled dynamics and noise. *IEEE Transactions on Automatic control* **42**(12), 1620–1635.

CONTROLLER VALIDATION FOR STABILITY AND PERFORMANCE BASED ON AN UNCERTAINTY REGION DESIGNED FROM AN IDENTIFIED MODEL [1]

X. Bombois* M. Gevers* G. Scorletti**
B.D.O. Anderson***

* *CESAME, Université Catholique de Louvain*
Bâtiment EULER, 4 av. Georges Lemaitre, B-1348 Louvain-la-Neuve, Belgium
Tel: +32 10 472596, Fax: +32 10 472180,
Email: {Bombois, Gevers}@csam.ucl.ac.be
** *LAP ISMRA Caen, France*
*** *RSISE, Australian National University*

Abstract: This paper focuses on the validation (for stability and for performance) of a controller that has been designed from an unbiased model of the true system, identified either in open-loop or in closed-loop using a prediction error framework. A controller is said to be *validated for stability* if it stabilizes all models defined by an ellipsoidal parametric uncertainty set containing the true system with some prescribed probability. The same controller is said to be validated for performance if the worst case performance achieved by this controller over the plants in the uncertainty region is better than some threshold value. *Copyright © 2000 IFAC*

Keywords: identification, robust stability, robust performance

1. INTRODUCTION

This paper is part of our continuing investigation of identification for control and controller validation based on uncertainty regions defined by prediction error identification methods (Bombois *et al.*, 1999a). Here we consider the case where a nominal model G_{mod} has been identified, together with an uncertainty set $\mathcal{D}$ to which the true system G_0 is known to belong with some prescribed probability. This uncertainty set $\mathcal{D}$ is defined as a set of parametrized rational transfer functions whose parameter vector lies in an ellipsoidal confidence region. This is clearly a nonstandard uncertainty set in robust control analysis and design. We focus on the validation of a controller C, designed from G_{mod}, both for robust stability and for robust performance. We present a validation procedure for stability which ensures that the controller C stabilizes all systems in this nonstandard uncertainty set $\mathcal{D}$. We also present a validation procedure for performance in which we compute the worst case performance over all closed loop systems made up of the controller C and all plants in $\mathcal{D}$.

Uncertainty region. Prediction error identification theory (see e.g. (Ljung, 1999)) delivers an estimated model G_{mod} for the true plant G_0 and provides us with tools for the estimation of an uncertainty region. If the parametric structure is

[1] The first three authors acknowledge the Belgian Programme on Inter-university Poles of Attraction, initiated by the Belgian State, Prime Minister's Office for Science, Technology and Culture. The last author acknowledges support of Office of Naval Research, Washington and U.S. Army Research Office Far East, Tokyo. The scientific responsibility rests with its authors.

sufficiently complex to represent the true system, then G_{mod} is asymptotically unbiased and the uncertainty is described by the covariance matrix of the identified model G_{mod}. This covariance matrix allows one to construct a parametric uncertainty region U containing the parameters of the true system G_0 at a certain probability level that we can fix at, say, 95 %. The uncertainty region U in the parameter space defines an equivalent uncertainty region $\mathcal{D}$ in the space of transfer functions. This uncertainty region $\mathcal{D}$ can be obtained for both open-loop identification and indirect closed-loop identification.

Controller validation for stability. Robust stability theory developed in e.g. (Zhou *et al.*, 1995; Hinrichsen and Pritchard, 1988) provides necessary and sufficient conditions for the stabilization, by some given controller C, of all plants in an uncertainty region, provided this uncertainty region is defined in the general LFT (linear fractional transformation) framework for robust stability analysis. Our contribution in the proposed procedure is to show that one can rewrite the closed-loop connection of the controller C and all plants in the uncertainty region $\mathcal{D}$ obtained from both types of identification (open-loop and indirect closed-loop identification) into a particular LFT that takes into account the parametric description of $\mathcal{D}$ (i.e. the uncertainty part of the obtained LFT is a real vector) and whose (real) stability radius is exactly computable, using a result presented in (Hinrichsen and Pritchard, 1988; Rantzer, 1992). The problem of robust stability of all plants in the domain $\mathcal{D}$ of parametrized transfer functions was already addressed in (Bombois *et al.*, 1999a). The solution presented there was to embed the uncertainty region $\mathcal{D}$ into a larger coprime factor uncertainty region, leading therefore only to a sufficient robust stability condition. In the new approach of this paper, we obtain a necessary and sufficient condition, because our new stability results apply directly to the parametrized set $\mathcal{D}$ resulting from the identification step, thereby avoiding the conservativeness resulting from the overbounding of $\mathcal{D}$ by a coprime factor uncertainty set. In the case of open-loop identification of an ARMA structure, the structure of $\mathcal{D}$ has a simpler expression. In (Rantzer, 1992), it is shown that this simpler structure can be expressed as an LFT. In this paper, we give a general formulation of this LFT for all model sets (not just ARMA) and for both open-loop and indirect closed-loop identification. In the case of open-loop identification and an ARX structure, an approach similar to ours and to that presented in (Rantzer, 1992) can also be found in (Kosut and Anderson, 1994).

Controller validation for performance. Our procedure for controller validation for performance is based on the computation of the worst case performance over all closed-loops made up of the controller C and all systems in the uncertainty region $\mathcal{D}$. The performance for a closed loop (see (Zhou *et al.*, 1995)) is often defined via the modulus of the frequency response of one (or several) of the four closed-loop transfer functions. The worst case performance, for each of these four transfer functions, is defined as the maximum of their modulus computed over all plants in the uncertainty region $\mathcal{D}$. These maxima over all plants in $\mathcal{D}$, for a given controller C, define four templates, which are used as performance indicators. A number of standard performance indicators (such as disturbance rejection properties, resonance peak, ...) can be derived from these four indicators. Our contribution is to show that the computation of the worst case performance can be formulated as an LMI-based optimization problem. In fact, we give a general LMI-based optimization problem which allows the computation of the worst case performance for the four different closed-loop transfer functions by an appropriate choice of the weights in the general LMI problem. The LMI formulation of the problem uses the fact that the parametric uncertainty appears linearly in the expression of both the numerator and the denominator of the systems in the uncertainty region $\mathcal{D}$ and, as a consequence, also appears linearly in the expression of the four closed-loop transfer functions.

Paper outline. In Section 2, the general expression of the parametric uncertainty regions delivered by prediction error identification is presented. In Section 3, we show how the closed-loop connections of the systems in the uncertainty region $\mathcal{D}$ and the "to-be-validated controller" can be expressed in the general LFT framework for robust stability analysis. A necessary and sufficient condition for the robust stabilization of all plants in $\mathcal{D}$ is then derived from classical robust stability theory. In Section 4, the worst case performance is defined and the LMI-based optimization problem developed for its computation is given. The procedures for validation for stability and for performance are illustrated by an example in Section 5. Finally, some conclusions are given in the last section.

2. UNCERTAINTY REGION DELIVERED BY PREDICTION ERROR IDENTIFICATION

In this section, we give the general expression of the uncertainty regions delivered by classical prediction error identification, assuming that unbiased model structures are used. This general expression, valid for both open-loop and indirect closed-loop identification, is summarized in the following proposition, where we assume that the true *open-loop* system is linear and time-invariant, with a rational input-output transfer function G_0 such that $y = G_0 u + v$, where v is additive noise. See (Bombois *et al.*, 1999a; Bombois *et al.*, 1999b) for more details.

Proposition 1. Consider $\delta \in \mathbf{R}^{k\times 1}$, the real parameter vector of the parametrized transfer function set; $G_0 = G(\delta_0)$, the true open-loop system, and $G(\hat{\delta})$, the full order identified model obtained either "directly" by open-loop identification or "indirectly" by indirect closed-loop identification. The uncertainty region $\mathcal{D}$ containing $G(\delta_0)$ at a certain probability level has the following general form.

$$\mathcal{D} = \left\{ \begin{array}{l} G(\delta) \mid G(\delta) = \dfrac{e + Z_N\delta}{1 + Z_D\delta} \text{ and} \\ \delta \in U = \{\delta \mid (\delta - \hat{\delta})^T R(\delta - \hat{\delta}) < 1\} \end{array} \right\} \quad (1)$$

where $R \in \mathbf{R}^{k\times k}$ is proportional to the inverse of the covariance matrix of $\hat{\delta}$ and has been scaled so as to obtain 1 on the right hand side, $Z_N(z)$ and $Z_D(z)$ are known row vectors of transfer functions of size k and $e(z)$ is either a known transfer function or is equal to 0.

3. CONTROLLER VALIDATION FOR STABILITY

Consider an identified model $G_{mod} = G(\hat{\delta})$ and the corresponding uncertainty region $\mathcal{D}$ of (1) containing G_0 at some probability level. We say that a controller C, designed from G_{mod}, is *validated for stability* if it stabilizes all models in this uncertainty region $\mathcal{D}$ (and therefore also the true system G_0). Our contribution in this section is to show that the uncertainty region $\mathcal{D}$ is amenable to classical robust stability analysis. Indeed, we present a way to describe the set of closed-loop connections of all plants in $\mathcal{D}$ with the "to be validated controller" C as a set of loops $[M_{\mathcal{D}}(z)\ \phi]$ where the uncertainty part ϕ is a real vector. We also show that the (real) stability radius linked with the set of loops $[M_{\mathcal{D}}(z)\ \phi]$ can be computed exactly and efficiently, using the result presented in (Hinrichsen and Pritchard, 1988; Rantzer, 1992). First, we recall an important result of robust stability analysis (Rantzer, 1992; Hinrichsen and Pritchard, 1988) in the case when the uncertainty is assumed to be a real vector.

3.1 *Robust stability for a real vector uncertainty*

We consider here a set of loops $[M(z)\ \beta]$ that obey the following system of equations:

$$\left\{ \begin{array}{l} p = \beta q \\ q = M(z)p \end{array} \right. \quad (2)$$

In this set of loops, it is assumed that $M(z) \in H_\infty$ is a known *fixed* row vector of size b and that the uncertainty part β is a real vector $\in \mathbf{R}^{b\times 1}$ that varies in the following uncertainty domain : $|\beta|_2 < 1$. $|\beta|_2$ represents the 2-norm of the vector β. The robust stability theorem for the set of loops $[M(z)\ \beta]$ is summarized in the following proposition.

Proposition 2. If $M(z) \in H_\infty$ and $\beta \in \mathbf{R}^{b\times 1}$, then the loops $[M(z)\ \beta]$ given in (2) are internally stable for all $\beta \in \mathbf{R}^{b\times 1}$ such that $|\beta|_2 < 1$ if and only if

$$\max_{\Omega} \mu_\beta(M(e^{j\Omega})) \leq 1 \quad (3)$$

The value $\mu_\beta(M(e^{j\Omega}))$ is called the stability radius of the loop $[M(z)\ \beta]$ at the frequency Ω and is defined below.

Definition 3. For M a known complex matrix $\in \mathbf{C}^{1\times b}$ and $\beta \in \mathbf{R}^{b\times 1}$, the stability radius $\mu_\beta(M)$ is defined as follows if $Im(M) \neq 0$:

$$\mu_\beta(M) = \sqrt{|Re(M)|_2^2 - \frac{(Re(M)Im(M)^T)^2}{|Im(M)|_2^2}} \quad (4)$$

and $\mu_\beta(M) = |M|_2$ if $Im(M) = 0$.

Note that the stability radius is discontinuous only at the frequencies where M is real (Qiu *et al.*, 1995).

3.2 *LFT framework for $\mathcal{D}$ and C*

In order to apply Proposition 2 to check the stabilization of all plants in $\mathcal{D}$ by some controller C, we must show that the closed-loop connections of all plants in $\mathcal{D}$ with C can be described as a particular set of loops $[M(z)\ \beta]$. This first step can be achieved using the following theorem.

Theorem 4. Consider an uncertainty region $\mathcal{D}$ of plant transfer functions given by (1) and a controller $C(z) = X(z)/Y(z)$. The set of closed-loop connections $[G(\delta)\ C]$ for all $G(\delta) \in \mathcal{D}$ are equivalent to the set of loops $[M_{\mathcal{D}}\ \phi]$ which obey the system of equations defined in (2), where the uncertainty part ϕ is a real column vector of size k such that $|\phi|_2 < 1$, and where $M_{\mathcal{D}}(z)$ is a row vector of size k defined as :

$$M_{\mathcal{D}}(z) = \frac{-(Z_D + \frac{X(Z_N - eZ_D)}{Y + eX})T^{-1}}{1 + (Z_D + \frac{X(Z_N - eZ_D)}{Y + eX})\hat{\delta}}, \quad (5)$$

with T a square root of the matrix R defining U in (1) : $R = T^T T$.

Proof. The closed-loop connection of C and a particular plant $G(\delta) = (e + Z_N\delta)/(1 + Z_D\delta)$ in $\mathcal{D}$ (see (1)) can be rewritten as follows by introducing two new signals q and p_1.

$$\begin{cases} \begin{pmatrix} q \\ y \end{pmatrix} = \begin{pmatrix} -Z_D & 1 \\ Z_N - eZ_D & e \end{pmatrix} \begin{pmatrix} p_1 \\ u \end{pmatrix} \\ p_1 = \delta q \\ u = -Cy \end{cases} \quad (6)$$

The variables y and u are now eliminated from (6), yielding the following system of equations:

$$\begin{cases} p_1 = \delta q \\ q = \overbrace{(-Z_D - \frac{C(Z_N - eZ_D)}{1 + eC})}^{M_1(z)} p_1 \end{cases} \quad (7)$$

The system (7) is equivalent with the closed-loop connection of a particular $G(\delta)$ in $\mathcal{D}$ with the controller C. In order to consider the closed-loop connections for all plants in $\mathcal{D}$, we have to consider all $\delta \in \mathbf{R}^{k \times 1}$ lying in the ellipsoid U defined in (1). In order to apply Proposition 2 to the set $[M_1(z)\ \delta]$, $\delta \in U$, we still need a normalization step, as follows. Using $R = T^T T$, define the real vector $\phi \in \mathbf{R}^{k \times 1}$: $\phi \triangleq T(\delta - \hat{\delta})$. Then $\delta \in U \Leftrightarrow |\phi|_2 < 1$. To replace δ by ϕ in (7), we first denote $p \triangleq \phi q$. Since $\delta = \hat{\delta} + T^{-1}\phi$, (7) is equivalent with:

$$\begin{cases} p = \phi q \\ q = \frac{M_1 T^{-1}}{1 - M_1 \hat{\delta}} p = M_{\mathcal{D}}(z) p \end{cases} \quad (8)$$

The result then follows from Proposition 2. □

3.3 *Robust stability condition for $\mathcal{D}$*

Using Proposition 2 and Theorem 4, we can now formulate our main stability theorem.

Theorem 5. Consider an uncertainty region $\mathcal{D}$ of plant transfer functions having the general form given in (1) and let C be a controller that stabilizes the nominal model $G(\hat{\delta})$. All plants in the uncertainty region $\mathcal{D}$ are stabilized by the controller C if and only if

$$\max_{\Omega} \mu_\phi(M_{\mathcal{D}}(e^{j\Omega})) \leq 1 \quad (9)$$

where the stability radius μ and $M_{\mathcal{D}}(z)$ are defined in Definition 3 and in (5), respectively.

Proof. $M_{\mathcal{D}}(z)$ lies in H_∞ since its denominator is the denominator of the sensitivity function of the closed loop $[G(\hat{\delta})\ C]$ which is stable by assumption. Therefore, this theorem is a direct consequence of Proposition 2 and Theorem 4. □

4. CONTROLLER VALIDATION FOR PERFORMANCE

In this section, we show that we can evaluate the worst case performance in the uncertainty region $\mathcal{D}$, i.e. the worst level of performance of a closed loop made up of the connection of the considered controller and any plant in $\mathcal{D}$. The worst case performance in $\mathcal{D}$ is of course a lower bound for the closed-loop performance achieved with the true system. We then say that a controller is validated for performance if this worst case performance in $\mathcal{D}$ remains below some threshold. There is no unique way of defining the performance of a closed-loop system. However most commonly used performance criteria can be derived from some norm of a frequency weighted version of the stability matrix $H(G, C)$ of the closed-loop system $[C\ G]$ made up of G in feedback with the controller C.

Definition 6. Given a plant G and a stabilizing controller C, the stability matrix $H(G, C)$ of the closed loop $[C\ G]$ is given by:

$$H(G, C) = \begin{pmatrix} \frac{GC}{1+GC} & \frac{G}{1+GC} \\ \frac{C}{1+GC} & \frac{1}{1+GC} \end{pmatrix}. \quad (10)$$

4.1 *The general criterion measuring the worst case performance*

The worst case performance criterion over all plants in an uncertainty region $\mathcal{D}$ will be similarly defined as the worst possible norm, over all plants in $\mathcal{D}$, of a frequency weighted version of the stability matrix $H(G(\delta), C)$, where $G(\delta)$ is any plant in $\mathcal{D}$ and C is the "to-be-validated" controller C.

General Criterion. Consider an uncertainty region $\mathcal{D}$ given by (1) and containing all systems $G(\delta) = G(z, \delta)$ with $\delta \in U$. Consider also a controller $C(z)$ that is validated for stability. The general criterion J_{WC} measuring the worst case performance level is defined at a frequency Ω as follows:

$$J_{WC}(\mathcal{D}, C, W_l, W_r, \Omega) = \max_{G(z,\delta) \in \mathcal{D}} \sigma_1 \left(W_l H(G(e^{j\Omega}, \delta), C(e^{j\Omega})) W_r \right) \quad (11)$$

where $W_l(z) = diag(W_{l1}, W_{l2})$ and $W_r(z) = diag(W_{r1}, W_{r2})$ are diagonal weights that allow one to define specific worst case performance levels and where $\sigma_1(A)$ denotes the largest singular value of A. Note that J_{WC} is a frequency function : it defines a template.

4.2 *Computation of the general criterion*

We now present a procedure for the computation of the general criterion $J_{WC}(\mathcal{D}, C, W_l, W_r, \Omega)$ at a given frequency Ω.

Theorem 7. Consider an uncertainty region $\mathcal{D}$ defined in (1) and a controller $C(z) = X(z)/Y(z)$. The general criterion J_{WC} defined in (11) is equal to $\sqrt{\gamma_{opt}}$, where γ_{opt} is the optimal value of γ for *the following standard convex optimization problem involving LMI constraints evaluated at the frequency Ω:*

$$\begin{array}{ll} \textit{minimize } \gamma & \\ \textit{over} \quad \gamma, \tau & (12) \\ \textit{subject to } \tau \geq 0 \textit{ and} & \end{array}$$

$$\begin{pmatrix} Re(a_{11}) & Re(a_{12}) \\ Re(a_{12}^*) & Re(a_{22}) \end{pmatrix} - \tau \begin{pmatrix} R & -R\hat{\delta} \\ (-R\hat{\delta})^T & \hat{\delta}^T R\hat{\delta} - 1 \end{pmatrix} < 0$$

where $a_{11} = (Z_N^* W_{l1}^* W_{l1} Z_N + Z_D^* W_{l2}^* W_{l2} Z_D) - \gamma(Q Z_1^* Z_1)$, $a_{12} = Z_N^* W_{l1}^* W_{l1} e + W_{l2}^* W_{l2} Z_D^* - \gamma(Q Z_1^*(Y + eX))$, $a_{22} = e^* W_{l1}^* W_{l1} e + W_{l2}^* W_{l2} - \gamma(Q(Y + eX)^*(Y + eX))$, $Z_1 = XZ_N + YZ_D$ *and* $Q = 1/(X^* W_{r1}^* W_{r1} X + Y^* W_{r2}^* W_{r2} Y)$.

Proof. Proving this theorem is equivalent to proving that the solution γ_{opt} of the LMI problem (12), evaluated at Ω, is such that: $\gamma_{opt} = \max_{\delta \in U} \lambda_1(H_w(e^{j\Omega}, \delta)^* H_w(e^{j\Omega}, \delta))$, where $H_w(z, \delta) = W_l H(G(\delta), C) W_r$ and $\lambda_1(A)$ denotes the largest eigenvalue of A. An equivalent and convenient way of restating the problem of computing $\max_{\delta \in U} \lambda_1(H_w(e^{j\Omega}, \delta)^* H_w(e^{j\Omega}, \delta))$ is as follows:

minimize γ *such that*
$\lambda_1(H_w(e^{j\Omega}, \delta)^* H_w(e^{j\Omega}, \delta)) - \gamma \leq 0 \; \forall \delta \in U$.

Using the fact $H_w(e^{j\Omega}, \delta)$ has rank one, we have that $\lambda_1(H_w(e^{j\Omega}, \delta)^* H_w(e^{j\Omega}, \delta)) - \gamma \leq 0$ is equivalent with

$$\begin{pmatrix} W_{l1}(e + Z_N \delta) \\ W_{l2}(1 + Z_D \delta) \\ Y + eX + Z_1 \delta \end{pmatrix}^* A \begin{pmatrix} W_{l1}(e + Z_N \delta) \\ W_{l2}(1 + Z_D \delta) \\ Y + eX + Z_1 \delta \end{pmatrix} \leq 0$$

where $A = diag(I_2, -\gamma Q)$ and Q, Z_1 are defined in (12). This last expression is equivalent to the following constraint on the *real* vector δ:

$$\overbrace{\begin{pmatrix} \delta \\ 1 \end{pmatrix}^T \begin{pmatrix} Re(a_{11}) & Re(a_{12}) \\ Re(a_{12}^*) & Re(a_{22}) \end{pmatrix} \begin{pmatrix} \delta \\ 1 \end{pmatrix}}^{\alpha(\delta)} \leq 0$$

with a_{11}, a_{12} and a_{22} as defined in (12). This last expression is equivalent to stating that $\lambda_1(H_w(e^{j\Omega}, \delta)^* H_w(e^{j\Omega}, \delta)) - \gamma \leq 0$ for a particular δ in U. However, this must be true for all $\delta \in U$. Therefore the last expression must be true for all δ such that

$$\overbrace{\begin{pmatrix} \delta \\ 1 \end{pmatrix}^T \begin{pmatrix} R & -R\hat{\delta} \\ (-R\hat{\delta})^T & \hat{\delta}^T R\hat{\delta} - 1 \end{pmatrix} \begin{pmatrix} \delta \\ 1 \end{pmatrix}}^{\rho(\delta)} < 0$$

which is equivalent to the statement "$\delta \in U$".

Let us now recapitulate. Computing $\max_{\delta \in U} \lambda_1(H_w(e^{j\Omega}, \delta)^* H_w(e^{j\Omega}, \delta))$ is equivalent to finding the smallest γ such that $\alpha(\delta) \leq 0$ for all δ for which $\rho(\delta) \leq 0$. By the $\mathcal{S}$ procedure (Boyd *et al.*, 1994), this problem is equivalent to finding the smallest γ and a positive scalar τ such that $\alpha(\delta) - \tau\rho(\delta) \leq 0$, for all $\delta \in \mathbf{R}^{k \times 1}$, which is precisely (12). □

5. EXAMPLE

To illustrate our results, we present an example of controller validation for a model identified in closed-loop. Let us consider the following true system G_0 with an Output Error structure:

$$y = \overbrace{\frac{0.1047z^{-1} + 0.0872z^{-2}}{1 - 1.5578z^{-1} + 0.5769z^{-2}}}^{G_0} u + e,$$

where e is a unit-variance white noise. The sampling time is 0.05 second. We perform an indirect closed-loop identification of an unbiased closed-loop transfer function $T(\hat{\xi})$ by collecting 1000 reference and output data on the true system in closed loop with an output-feedback controller $u = 3(r - y)$ (see (Bombois *et al.*, 1999a) for details). This controller stabilizes G_0. The open-loop model $G_{mod} = G(\hat{\xi}) = T(\hat{\xi})/(K(1 - T(\hat{\xi})))$ corresponding to $T(\hat{\xi})$ is equal to

$$G_{mod} = \frac{0.1060z^{-1} + 0.0928z^{-2}}{1 - 1.5308z^{-1} + 0.5467z^{-2}}$$

Control design. From the model G_{mod}, we have designed a controller with a phase advance : $C(z) = (1.8464 - 1.3647z^{-1})/(1 - 0.4545z^{-1})$. With this controller, the designed closed-loop $[G_{mod} \; C]$ has a stability margin of 57 degrees and a gain margin of 10dB. The cut-off frequency Ω_c is equal to 0.5 which corresponds to a *real* frequency $\omega_c = 11 \; rad/s$. Before applying this controller $C(z)$ to the true system, we verify whether it achieves satisfactory behaviour with all plants in the uncertainty region $\mathcal{D}_{CL}$. The uncertainty region $\mathcal{D}_{CL}$ containing the true system G_0 at a probability level equal to 0.95 is given by

$$\mathcal{D}_{CL} = \{G(\xi) \mid G(\xi) = \frac{T(\xi)}{K(1 - T(\xi))} \; and \; \xi \in U_{CL}\}$$

where $U_{CL} = \{\xi \mid (\xi - \hat{\xi})^T P_{\xi}^{-1}(\xi - \hat{\xi}) < 12.6\}$ and P_{ξ} is the covariance matrix of $\hat{\xi}$. As shown in (Bombois *et al.*, 1999a; Bombois *et al.*, 1999b), it is easy to prove that $\mathcal{D}_{CL}$ has the general structure (1).

Validation of C for stability. Using the procedure presented in Section 3, we check whether C stabilizes all plants in $\mathcal{D}_{CL}$. For this purpose, we construct the row vector $M_{\mathcal{D}_{CL}}(z)$ defined in Theorem 4 and we compute the corresponding stability radius $\mu_{\phi}(M_{\mathcal{D}_{CL}}(e^{j\Omega}))$ at all frequencies in $[0\ \pi]$. The maximum over these frequencies is 0.1313. Since this maximum is smaller than 1, we conclude that $C(z)$ stabilizes all plants in $\mathcal{D}_{CL}$ and therefore also the true system G_0.

Validation of C for performance. In order to verify that C gives satisfactory performance with all plants in $\mathcal{D}_{CL}$, we compute at each frequency the worst case modulus $t_{\mathcal{D}_{CL}}(\Omega, S)$ of the sensitivity function "S" achieved by C over all plants in $\mathcal{D}_{CL}$. This can be done by computing $J_{WC}(\mathcal{D}_{CL}, C, W_l, W_r, \Omega)$ using Theorem 7 with the particular weights $W_l = W_r = diag(0, 1)$. The worst case modulus of the sensitivity function over all models in $\mathcal{D}_{CL}$ is represented in Figure 1. In this figure, the worst case performance level $t_{\mathcal{D}_{CL}}(\Omega, S)$ is compared with the sensitivity functions of the designed closed loop $[G_{mod}\ C]$ and of the achieved closed loop $[G_0\ C]$. From $t_{\mathcal{D}_{CL}}(\Omega, S)$, we can find that the worst case static error ($=t_{\mathcal{D}_{CL}}(0, S)$) resulting from a constant disturbance of unit amplitude is equal to 0.1692, whereas this static error is 0.0834 in the designed closed-loop. The achieved static error is 0.1017. Using $t_{\mathcal{D}_{CL}}(\Omega, S)$, we can also see that the bandwidth of $\Omega_c = 0.5$ in the designed closed-loop is preserved for all closed loops with a plant in $\mathcal{D}_{CL}$, since $t_{\mathcal{D}_{CL}}(\Omega, S)$ is equal to 1 at $\Omega_c \simeq 0.5$. The difference between the resonance peak of the designed sensitivity function (i.e. $\max_\Omega \parallel S(G_{mod}, C) \parallel = 1.6184$) and the worst case reasonance peak achieved by a plant in $\mathcal{D}_{CL}$ (i.e. $\max_\Omega t_{\mathcal{D}_{CL}}(\Omega, S) = 1.7075$) also remains small. Note that the actually achieved resonance peak (i.e. $\max_\Omega \parallel S(G_0, C) \parallel$) is equal to 1.6229.

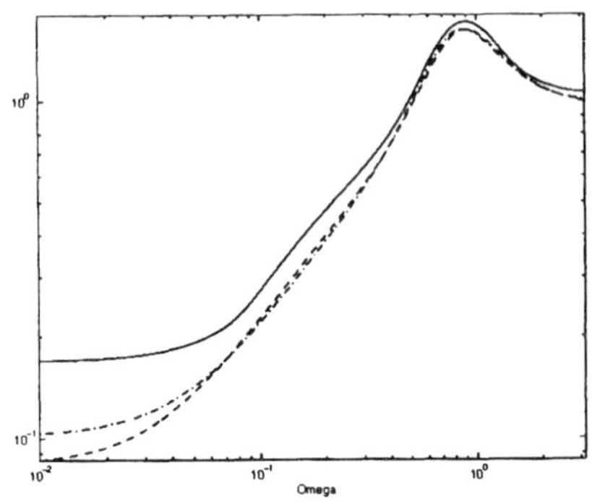

Fig. 1. $t_{\mathcal{D}_{CL}}(\Omega, S)$ (solid), $\parallel S(G_{mod}, C) \parallel$ (dashed) and $\parallel S(G_0, C) \parallel$ (dashdot)

We may therefore conclude that the controller C is validated for performance since the difference between the nominal and worst case performance level remains very small at every frequency. With such stability and performance analysis results, one would confidently apply the controller to the real system, assuming that the nominal performance is judged to be satisfactory.

6. CONCLUSIONS

We have developed tools for the robust stability and robust performance analysis of a controller designed from a nominal model, when the uncertainty set $\mathcal{D}$ containing the true system is described via ellipsoidal perturbations around the parameter vector of the nominal model. Such ellipsoidal parameter perturbations arise when the nominal model is the result of a prediction error identification procedure using an unbiased model structure. Our solution to the validation for stability problem is in the form of a necessary and sufficient condition for the stabilization of all models in this parametric uncertainty set $\mathcal{D}$ by a given controller C. Our solution to the validation for performance problem takes the form of the exact computation of the worst case performance of the controller C in closed loop with all models in the uncertainty set $\mathcal{D}$.

7. REFERENCES

Bombois, X., M. Gevers and G. Scorletti (1999a). Controller validation based on an identified model. In: *Proc. IEEE Conference on Decision and Control.* pp. 2816–2821.

Bombois, X., M. Gevers, G. Scorletti and B.D.O. Anderson (1999b). Controller validation for stability and performance based on an uncertainty region designed from an identified model. CESAME Technical Report 99.36. Submitted to Automatica.

Boyd, S., L. El Ghaoui, E. Feron and V. Balakrishnan (1994). *Linear Matrix Inequalities in Systems and Control Theory.* Vol. 15 of *Studies in Appl. Math..* SIAM. Philadelphia.

Hinrichsen, D. and A.J. Pritchard (1988). New robustness results for linear systems under real perturbations. In: *Proc. Conference on Decision and Control,.* pp. 1375–1378.

Kosut, R.L. and B.D.O. Anderson (1994). Least-squares parameter set estimation for robust control design. In: *Proc. American Control Conference.* pp. 3002–3006.

Ljung, L. (1999). *System Identification: Theory for the User (second edition).* Prentice-Hall. Upper Saddle, NJ.

Qiu, L., B. Bernhardsson, A. Rantzer, E.J. Davison, P.M. Young and J.C. Doyle (1995). A formula for computation of the real stability radius. *Automatica* **31**(6), 879–890.

Rantzer, A. (1992). Convex robustness specifications for real parametric uncertainty in linear systems. In: *Proc. American Control Conference.* pp. 583–585.

Zhou, K., J.C. Doyle and K. Glover (1995). *Robust and Optimal Control.* Prentice Hall, New Jersey.

BLIND ESTIMATION AND DECONVOLUTION OF COMMUNICATION CHANNELS WITH UNBALANCED NOISE

Paolo Castaldi * Roberto Diversi * Roberto P. Guidorzi *
Umberto Soverini *

* *Dipartimento di Elettronica, Informatica e Sistemistica*
Università di Bologna
Viale del Risorgimento 2, 40136 Bologna, Italy
tel. +39 051 209 3026; e-mail: pcastaldi@deis.unibo.it

Abstract: Blind identification is a significant problem in many communication systems. A new approach to blind estimation and input deconvolution is developed in this paper. By using the structural shift properties of the generalized Sylvester matrix of the system, an estimate of the covariance matrix of the unknown source is obtained and a suitable identification criterion for the estimation of the channel coefficients is introduced. A reconstruction of the input sequence is then obtained by means of a standard LS equalizer. *Copyright © 2000 IFAC*

Keywords: Blind estimation, blind deconvolution, linear systems, FIR systems, errors-in-variables models

1. INTRODUCTION

Recent years have seen an increasing interest in so called "blind identification" problems (Kalouptsidis, 1997; Abed-Meraim *et al.*, 1997a); these problems consider the estimation of channel parameters without knowing its input and/or the reconstruction of the input sequence without knowing the channel characteristics.

Problems of this type arise in many fields and are of particular interest in modern communications and signal processing. For example, the distorsion caused by multipath interference affects the transmission quality and efficiency of several communication systems. Another important example concerns one of the main limiting aspects in digital communication systems, i.e. intersymbol interference (ISI). This phenomenon is caused by a reduced bandwidth and/or by frequency selectivity in the transmission channel, as it happens in mobile communication systems. The result is that each transmitted symbol lasts over a period of time longer than its original duration.

In most cases the channel characteristics are identified by using input training sequences. However, in several situations where the channel characteristics are highly time-varying a training sequence is not always used and blind equalizers can be designed for identifying the communication channel and reconstructing the transmitted signal.

In the literature several blind identification schemes have been described (Van Der Veen, 1998; Tong and Perreau, 1998). Since communication channels are, in general, non-minimum phase, the second-order statistics of the stationary signals are not sufficient for their identification; for this reason, some blind identification approaches are based on higher-order statistics of the received signal (Tugnait, 1996).

In some applications, however, the underlying signals exhibit ciclostationarity (for example, when fractionally sampled data are considered). For these cases (Tong *et al.*, 1994) it has been shown that it is possible to estimate non-minimum phase channels using only second-order statistics. This has led to the development of several blind channel estimation algorithms (Xu *et al.*, 1995; Abed-Meraim *et al.*, 1997b; Moulines *et al.*, 1995).

In this paper a new blind identification method is presented, that can be considered as the development of the techniques proposed for identifiying errors–in–variables models (Beghelli *et al.*, 1990; Diversi *et al.*, 1999). This method is based on the structural shift properties of time–invariant dynamic systems and can be considered as a generalization of the eigenvector method proposed by Gürelli and Nikias, (1995). Unlike several other blind identification approaches, this method can be used successfully also when the output measurements are affected by unbalanced amounts of noise.

The paper is organized as follows. Section 2 defines the blind identification problem for multiple FIR channels. In section 3 a new method for estimating the channel parameters and deconvolving the unknown input sequence is proposed. Section 4 reports the results obtained in testing the procedure on a multichannel FIR model already described in the literature. Some concluding remarks are finally reported in section 5.

2. STATEMENT OF THE PROBLEM

Let us consider a linear, discrete, time–invariant single–input multi–output (SIMO) finite impulse response (FIR) system, whose noiseless outputs $x_i(t)$ $(i = 1, \ldots, M)$ are linked to the scalar input $u(t)$ by the following convolutional model

$$x_i(t) = \sum_{k=0}^{L} h_i(k)\, u(t-k), \qquad (1)$$

where $h_1(k), h_2(k), \cdots, h_M(k)$ $(k = 0, \cdots, L)$ are M FIR channels with degree L, which is assumed as known. This model is useful to describe the case of a single unknown source and multiple spatially or/and temporally distributed sensors. The outputs $x_i(t)$ $(i = 1, \ldots, M)$ are assumed as not directly accessible; only the noisy signals

$$\begin{aligned} y_1(t) &= x_1(t) + n_1(t) \\ y_2(t) &= x_2(t) + n_2(t) \\ &\vdots \\ y_M(t) &= x_M(t) + n_M(t), \end{aligned} \qquad (2)$$

are available. Moreover, for the sake of generality, all the variables in (1) and (2) are assumed as complex. It has been proved (Hua and Wax, 1996) that the identifiability of system (1) is ensured (up to a scalar factor) under these sufficient conditions:

Assumption 1

1a) the input $u(t)$ is persistently exciting of order $2L + 1$;

1b) the transfer functions of the channels

$$\mathbf{h}_i(z) = h_i(0) + h_i(1)\, z^{-1} + \cdots + h_i(L)\, z^{-L} \quad (i = 1, \ldots, M) \qquad (3)$$

do not share any common zero;

1c) the number of available samples of every channel N, satisfies the condition $N \geq 3L+1$.

The following additional hyphoteses concern the measurement noises:

Assumption 2

2a) The additive noises $n_i(t)$ $(i = 1, \ldots, M)$ are zero mean, mutually uncorrelated, circular white noises with unknown variances $\sigma_1^*, \sigma_2^*, \ldots, \sigma_M^*$;

2b) the input $u(t)$ is a zero mean stationary stochastic process uncorrelated with $n_i(t)$ $(i = 1, \ldots, M)$.

The purpose of blind channel identification and equalization can be stated as follows.

Problem 1 – Given N observations $y_i(0), y_i(1), \ldots, y_i(N-1)$ $(i = 1, \ldots, M)$ of each channel, estimate the coefficients $h_i(k)$ $(k = 0, \ldots, L)$ $(i = 1, \ldots, M)$ and the input sequence $u(\cdot)$.

The procedure described in this paper can be applied only for the case $M = 2$ (see Figure 1); its extension to an arbitrary number of channels will be studied in the future.

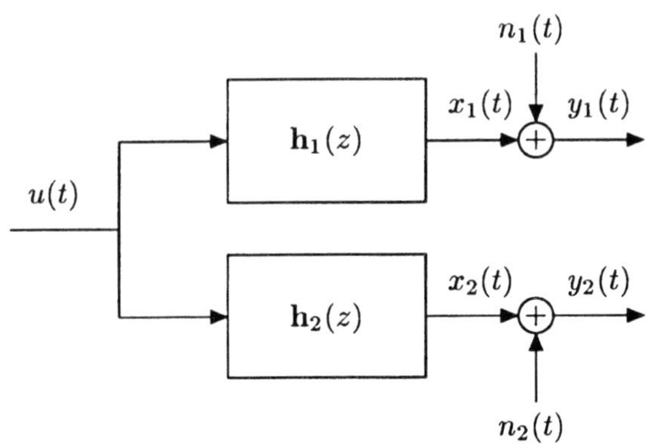

Fig. 1. Block structure of the considered system

For the subsequent analysis, it is convenient to rewrite model (1)-(2) by using the matrix notation

$$\mathbf{y}(t) = \mathbf{x}(t) + \mathbf{n}(t) = \mathcal{H}_L(w)\,\mathbf{u}(t) + \mathbf{n}(t) \qquad t \geq 0, \qquad (4)$$

where $0 \leq w \leq N-L-1$ in an integer "window parameter" which determines the length of the vectors in (4) as follows

$$\mathbf{y}(t) = [y_1(t), \ldots, y_1(t+L+w), \\ y_2(t), \ldots, y_2(t+L+w)]^T \quad (5)$$

$$\mathbf{x}(t) = [x_1(t), \ldots, x_1(t+L+w), \\ x_2(t), \ldots, x_2(t+L+w)]^T \quad (6)$$

$$\mathbf{u}(t) = [u(t-L), \ldots, u(t+L+w)]^T \quad (7)$$

$$\mathbf{n}(t) = [n_1(t), \ldots, n_1(t+L+w), \\ n_2(t), \ldots, n_2(t+L+w)], \quad (8)$$

and $\mathcal{H}_L(w)$ is the following generalized Sylvester matrix with dimension $2(L+w+1) \times (2L+w+1)$

$$\mathcal{H}_L(w) = \begin{bmatrix} \mathcal{H}_L^{(1)}(w) \\ \mathcal{H}_L^{(2)}(w) \end{bmatrix}, \quad (9)$$

where $\mathcal{H}_L^{(i)}(w)$ $(i = 1, 2)$ is the $(L+w+1) \times (2L+w+1)$ Sylvester matrix of the i-th channel

$$\mathcal{H}_L^{(i)}(w) = \begin{bmatrix} h_i(L) & \cdots & h_i(0) & \cdots & 0 \\ \vdots & \ddots & \vdots & \ddots & \vdots \\ 0 & \cdots & h_i(L) & \cdots & h_i(0) \end{bmatrix}. \quad (10)$$

3. BLIND ESTIMATION AND DECONVOLUTION

For the sake of simplicity, we will use the following short notation for the case $w = 0$

$$\mathbf{H}_L \equiv \mathcal{H}_L(0) \quad \mathbf{H}_L^{(i)} \equiv \mathcal{H}_L^{(i)}(0) \quad i = 1, 2. \quad (11)$$

From the structural properties of $\mathbf{H}_L$ it is easy to verify that

$$\mathbf{c}_L^H \mathbf{H}_L = 0, \quad (12)$$

where

$$\mathbf{c}_L = \left[\mathbf{c}_2^T, -\mathbf{c}_1^T\right]^T = \\ [h_2(L), \ldots, h_2(0), -h_1(L), \ldots, -h_1(0)]^T. \quad (13)$$

We can construct now the $2(L+1) \times 2(L+1)$ sample covariance matrix $\hat{\mathbf{\Sigma}}_L$ of $\mathbf{x}(t)$

$$\hat{\mathbf{\Sigma}}_L = \frac{1}{N-L} \sum_{t=0}^{N-L-1} \mathbf{x}(t)\mathbf{x}^H(t) = \mathbf{H}_L \, \mathbf{\Sigma}_L^u \, \mathbf{H}_L^H, \quad (14)$$

where

$$\mathbf{\Sigma}_L^u = \frac{1}{N-L} \sum_{t=0}^{N-L-1} \mathbf{u}(t) \, \mathbf{u}^H(t). \quad (15)$$

is the $(2L+1) \times (2L+1)$ covariance matrix of the unknown input.

When Assumption 1 is fulfilled, $\mathbf{\Sigma}_L^u$ has full rank, $\hat{\mathbf{\Sigma}}_L$ has rank $2L+1$ and from (12) we obtain

$$\hat{\mathbf{\Sigma}}_L \mathbf{c}_L = 0. \quad (16)$$

This relation allows to determine $\mathbf{c}_L$ if $\hat{\mathbf{\Sigma}}_L$ is known. In the noisy case only the sample covariance matrix $\mathbf{\Sigma}_L$ of $\mathbf{y}(t)$ is available

$$\mathbf{\Sigma}_L = \frac{1}{N-L} \sum_{t=0}^{N-L-1} \mathbf{y}(t)\mathbf{y}^H(t). \quad (17)$$

Since matrix $\mathbf{\Sigma}_L$ is positive definite, the coefficient vector $\mathbf{c}_L$ can no longer be extracted from a relation of type (16).

However, when Assumption 2 is satisfied and $N \to \infty$ the following relation holds

$$\mathbf{\Sigma}_L = \hat{\mathbf{\Sigma}}_L + \tilde{\mathbf{\Sigma}}_L^*, \quad (18)$$

where

$$\tilde{\mathbf{\Sigma}}_L^* = \text{diag}[\,\sigma_1^* \, I_{L+1}, \, \sigma_2^* \, I_{L+1}\,] \quad (19)$$

is the unknown covariance matrix of the noise vector $\mathbf{n}(t)$.

In this case, the solution of Problem 1 can be obtained as follows.

Let us consider the family of non-negative definite diagonal matrices $\tilde{\mathbf{\Sigma}}_\ell = \text{diag}[\sigma_1 \, I_{\ell+1}, \, \sigma_2 \, I_{\ell+1}]$ such that

$$\hat{\mathbf{\Sigma}}_\ell = \mathbf{\Sigma}_\ell - \tilde{\mathbf{\Sigma}}_\ell \geq 0. \quad (20)$$

It has been proved (Beghelli *et al.*, 1990) that, for every $\ell > 0$, the matrices $\tilde{\mathbf{\Sigma}}_\ell$ satisfying relation (20) are defined by the points of a convex curve, belonging to the first quadrant of the noise plane (σ_1, σ_2). Every point $P = (\sigma_1, \sigma_2)$ of the curve of order ℓ is related to FIR models $\mathbf{c}_1(P), \mathbf{c}_2(P)$ given by

$$\hat{\mathbf{\Sigma}}_\ell(P) \begin{bmatrix} \mathbf{c}_2(P) \\ -\mathbf{c}_1(P) \end{bmatrix} = 0, \quad (21)$$

where

$$\hat{\mathbf{\Sigma}}_\ell(P) = \mathbf{\Sigma}_\ell - \text{diag}\left[\sigma_1 \, I_{\ell+1}, \, \sigma_2 \, I_{\ell+1}\right] \geq 0. \quad (22)$$

In particular, the point $P^* = (\sigma_1^*, \sigma_2^*)$, corresponding to the actual variances of the noises, belongs to all curves corresponding to orders $\ell \geq L$. The FIR models associated with P^* are characterized by the actual coefficients (up to a scalar factor) $\mathbf{c}_1(P^*)$, $\mathbf{c}_2(P^*)$. The search for the solution of Problem 1 may thus start from the determination, in the noise space, of the common point P^*.

For this purpose it can be verified that, in correspondence of the point P^*, the following relation is satisfied for $\ell = (L+1)$

$$\hat{\Sigma}_{L+1}(P^*) = \Sigma_{L+1} - \tilde{\Sigma}_{L+1}(P^*) = \bar{\mathbf{H}}_{L+1}(P^*)\Sigma^u_{L+1}(P^*)\bar{\mathbf{H}}^H_{L+1}(P^*), \quad (23)$$

where

$$\bar{\mathbf{H}}_{L+1}(P^*) = \begin{bmatrix} \bar{\mathbf{H}}^{(1)}_{L+1}(P^*) \\ \bar{\mathbf{H}}^{(2)}_{L+1}(P^*) \end{bmatrix}, \quad (24)$$

and $\bar{\mathbf{H}}^{(i)}_{L+1}(P^*)$ is the Sylvester matrix with dimension $(L+2) \times (2L+3)$ built with the coefficients of the i-th channel of order L as follows

$$\bar{\mathbf{H}}^{(i)}_{L+1}(P^*) = \left[\begin{array}{c|c|c} \mathbf{0}_{(L+1)\times 1} & \mathbf{H}^{(i)}_L(P^*) & \mathbf{0}_{(L+1)\times 1} \\ \hline 0 & 0 \ldots 0 \; h_i(L) \ldots h_i(1) & h_i(0) \end{array}\right]. \quad (25)$$

From (23) it is possible to determine the covariance matrix of the input sequence Σ^u_{L+1} up to a scalar factor

$$\Sigma^u_{L+1}(P^*) = \left(\bar{\mathbf{H}}_{L+1}(P^*)\right)^+ \hat{\Sigma}_{L+1}(P^*) \left(\bar{\mathbf{H}}^H_{L+1}(P^*)\right)^+, \quad (26)$$

where + denotes pseudoinversion. Note that in correspondence of P^* we have

$$\Sigma^u_{L+1}(P^*) = \left[\begin{array}{c|c|c} r_u(0) & \ldots & r_u(2L+2) \\ \hline \vdots & \Sigma^u_L(P^*) & \vdots \\ \hline r_u(2L+2) & \ldots & r_u(0) \end{array}\right], \quad (27)$$

where $r_u(\cdot)$ is, up to a scalar factor, the autocorrelation function of the actual input $u(\cdot)$.

The following procedure can now be stated for the determination of the common point P^*.

Let us consider two points $P = (\sigma_1, \sigma_2)$ and $\bar{P} = (\bar{\sigma}_1, \bar{\sigma}_2)$ on the curves of order L and $L+1$, respectively, linked by the following relation

$$\frac{\sigma_1}{\sigma_2} = \frac{\bar{\sigma}_1}{\bar{\sigma}_2}, \quad (28)$$

i.e. P and $\bar{P}$ are points belonging to the straight line with slope–rate (28) departing from the origin. For every point P on the curve of order L, the covariance matrix of the input can be estimated from relation (14). In fact we have

$$\Sigma^u_L(P) = (\mathbf{H}_L(P))^+ \hat{\Sigma}_L(P) \left(\mathbf{H}^H_L(P)\right)^+, \quad (29)$$

where $\mathbf{H}_L(P)$ is built with the coefficients of $\mathbf{c}_L(P)$. In an analogous way, from relation (26) it is now possible to compute, for every pont $\bar{P}$ on the curve of order $L+1$, the matrix

$$\Sigma^u_{L+1}(\bar{P}) = \left(\bar{\mathbf{H}}_{L+1}(P)\right)^+ \hat{\Sigma}_{L+1}(\bar{P}) \left(\bar{\mathbf{H}}^H_{L+1}(P)\right)^+. \quad (30)$$

By using the structural shift properties described in (25) and (27), we can introduce the cost function

$$J(P, \bar{P}) = \left\|\Sigma^u_L(P) - \Sigma^u_L(\bar{P})\right\|_F, \quad (31)$$

where $\Sigma^u_L(\bar{P})$ is obtained partitioning $\Sigma^u_{L+1}(\bar{P})$ as described in (27) and $\|\cdot\|_F$ denotes Frobenius norm. From condition $J(P^*) = 0$, it is then possible to univocally determine the point P^*. When Assumption 2 is not satisfied and/or the length N of the sequences is finite, the curves of order L and $L+1$ do not present any common point P^* so that $J(P) > 0 \quad \forall P$. However, an estimate of the FIR models can be obtained by minimizing cost function (31).

Remark 1 – It is possible to prove (Guidorzi and Pierantoni, 1995) that, when relation (28) is satisfied, the coordinates of $\bar{P}$ are linked to point P by

$$\bar{\sigma}_1 = \frac{\sigma_1}{\lambda_M} \qquad \bar{\sigma}_2 = \frac{\sigma_2}{\lambda_M}, \quad (32)$$

where

$$\lambda_M = \max\mathrm{eig}\left(\Sigma^{-1}_{L+1}\,\mathrm{diag}[\sigma_1 I_{L+2}, \sigma_2 I_{L+2}]\right). \quad (33)$$

On the basis of previous considerations, the following algorithm can be devised for real cases.

Algorithm 1

1) Start from a generic point P on the curve of order L and compute the model $\mathbf{c}_L(P) = [\mathbf{c}^T_2(P), -\mathbf{c}^T_1(P)]^T$ by means of (21).
2) Compute the corresponding point $\bar{P}$ on the curve of order order $L+1$ using (32)-(33).
3) Compute the matrices $\Sigma^u_L(P)$ and $\Sigma^u_{L+1}(\bar{P})$ by means of (29)-(30).
4) Compute the value of cost function (31).
5) Use a search procedure on the curve of order L to obtain the point associated with the minimum of cost function (31).

Once the noise variances and the channels coefficients have been estimated, it is possible to address the blind deconvolution problem, in order to reconstruct the unknown input sequence $u(\cdot)$. When the whole set of data is considered, i.e when $t = 0$ and $w = w_{\max} = N - L - 1$ in relations (4)-(5), the system model can be rewritten as

$$\mathbf{y} = \mathbf{H}\,\mathbf{u} + \mathbf{n}, \quad (34)$$

where $\mathbf{H} = \mathcal{H}_L(w_{\max})$ and $\mathbf{y} = \mathbf{y}(0)$, $\mathbf{u} = \mathbf{u}(0)$, $\mathbf{n} = \mathbf{n}(0)$ are the vectors containing the whole sequences. Since, in general, $\sigma^*_1 \neq \sigma^*_2 \neq 0$, it is possible to scale the signal of the i–th channel output by the reciprocal of the corresponding noise variance, $1/\sigma_i^*$, and compute the least squares estimate of the input sequence

$$\mathbf{u}_{LS} = (\mathbf{H}^H \mathbf{H})^{-1} \mathbf{H}^H \mathbf{y} \ . \qquad (35)$$

4. NUMERICAL RESULTS

In this section the results of a Monte Carlo simulation are presented in order to evaluate the performance of the proposed algorithm in a digital communication situation. The source symbols where drawn from a 16–QAM signal constellation with an uniform distribution. The following two–FIR–channel model, already proposed in the literature (Abed-Meraim *et al.*, 1997b) has been considered

$$\mathbf{h}_1(z) = -1.1836 + 0.4906\,z^{-1} - 0.3093\,z^{-2} + \\ +0.4011\,z^{-3} + 0.1269\,z^{-4} - 1.8522\,z^{-5}$$
$$\mathbf{h}_2(z) = 1.2965 + 0.0525\,z^{-1} + 0.3410\,z^{-2} - \\ -0.0260\,z^{-3} + 0.3991\,z^{-4} + 0.8817\,z^{-5}.$$

Note that $\mathbf{h}_1(z)$ has 4 nonminimum phase zeros. The output sequences $x_1(t)$ and $x_2(t)$ have been corrupted by adding circular white noises $n_1(t)$ and $n_2(t)$ corresponding to signal–to–noise ratios (SNR) ranging from 5 dB to 35 dB, where

$$\text{SNR} = 20 \log \frac{\text{std}(x_i)}{\text{std}(n_i)} \ (\text{dB}) \qquad i = 1, 2 \ . \quad (36)$$

The undisturbed signals are characterized by the standard deviations $\text{std}(x_1) = 7.71$ and $\text{std}(x_2) = 5.10$ so that the same SNR actually corresponds to an unbalance of 3.5 dB on the actual amounts of noise on the channels. Figure 2 reports the plots of 1000 output symbols of the unequalized channels.

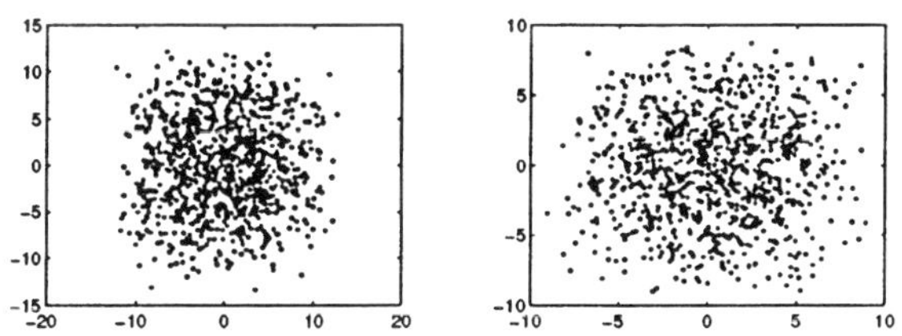

Fig. 2. Outputs y_1, y_2 of the unequalized channels; SNR=30dB

For the identification of the channel parameters, a sequence of $N = 100$ output samples has been used. The performance of the estimation procedure is summarized in Figure 3 where, for a single realization with an amount of noise equal to 30 dB, 1000 symbols of the input reconstructed by means of (35) have been plotted. The figure shows that the channel is properly equalized.

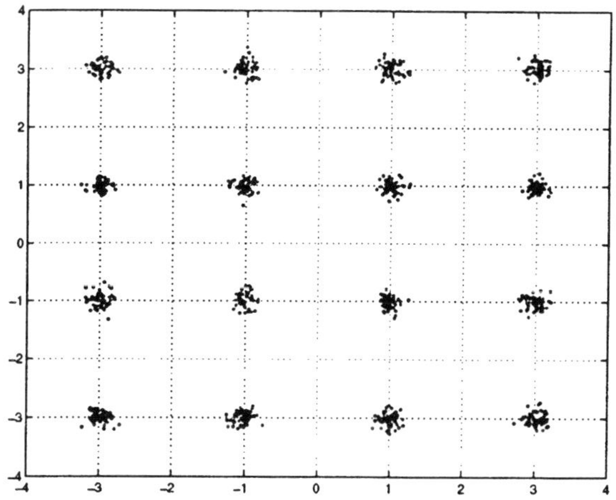

Fig. 3. Reconstructed input symbols. 1000 symbols are plotted while 100 symbols have been used used for the estimation. SNR=30dB

A Monte Carlo simulation of 100 independent trials has been performed in the same simulation scenario. In this case the performance of the algorithm has been measured by means of the normalized root–mean–square–error (NRMSE) defined as

$$\text{NRMSE} = \frac{1}{||\mathbf{c}_L||} \sqrt{\frac{1}{R} \sum_{i=1}^{R} ||\hat{\mathbf{c}}_L^i - \mathbf{c}_L||^2} \quad (37)$$

where R is the number of runs and $\hat{\mathbf{c}}_L^i$ is the estimate, from the the i-th trial, of the coefficient vector $\mathbf{c}_L = \left[\mathbf{c}_2^T \ -\mathbf{c}_1^T\right]^T$.

Figure 4 shows the NRMSE versus the signal–to–noise ratio (SNR). 100 runs with 100 input symbols have been performed in every Monte Carlo simulation.

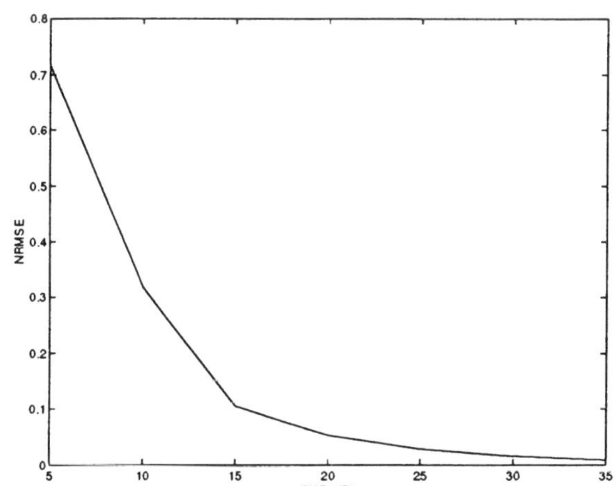

Fig. 4. NRMSE versus SNR. 100 runs with 100 input symbols have been performed in every Monte Carlo simulation.

The obtained results show good estimates of the FIRs coefficients even with low signal to noise ratios.

As a further meaningful performance index, the symbol error rate (SER) has been used; it is defined as the frequency of errors averaged over the number of Monte Carlo runs. Figure 5 plots the SER against the SNR under the same conditions of Figure 4.

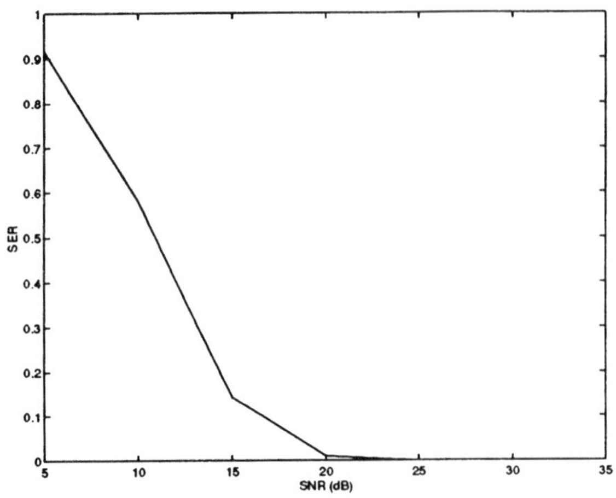

Fig. 5. Symbol-error-rate versus SNR. 100 runs with 100 input symbols have been performed in every Monte Carlo simulation.

It can be observerd that the proposed procedure exhibits excellent performances for SNR up to 20 dB.

5. CONCLUSION

Blind equalization is a significant problem in many communication systems. A new approach to blind identification and input deconvolution has been developed in this paper. By using the structural shift properties of the generalized Sylvester matrix of the system, an estimate of the covariance matrix of the unknown source has been obtained and a suitable identification criterion for the estimation of the channel coefficients introduced. A reconstruction of the input sequence is then obtained by means of a standard LS equalizer.

This method, limited at present to the case of two channels, allows to deal with output measurements affected by different amounts of noise on the channels. This is one of the main properties of the procedure, not shared by other procedures presently available in the literature. This approach has been tested on a simulated system taken from the literature. The numerical results show that the method can be advantageously used especially in presence of poor signal-to-noise ratios.

REFERENCES

Abed-Meraim, K., W. Qiu, and Y. Hua (1997a). Blind system identification. *Proceedings of the IEEE*, **85**, 1310–1322.

Abed-Meraim, K., E. Moulines and P. Loubaton (1997b). Prediction error method for second-order blind identification. *IEEE Transactions on Signal Processing*, **45**, 694–705.

Beghelli, S., R.P. Guidorzi and U. Soverini (1990). The Frisch scheme in dynamic system identification. *Automatica*, **26**, 171–176.

Diversi, R., R. P. Guidorzi, U. Soverini and P. Castaldi (1999). Blind identification of SIMO FIR systems. In: *Recent Advances in Signal Processing and Communications*, (N.E. Mastorakis, Ed.), pp. 60–64. World Scientific & Engineering Society Press.

Guidorzi, R.P. and M. Pierantoni (1995). A new parametrization of Frisch scheme solutions. In: *Proc. of the XII Int. Conf. on Systems Science*, Wroclaw, Poland, 114–120.

Gürelli, M.I. and C.L. Nikias (1995). EVAM: an eigenvector–based algorithm for multichannel blind deconvolution of input colored signals. *IEEE Transactions on Signal Processing*, **43**, 134–149.

Hua, Y. and M. Wax (1996). Strict identifiability of multiple FIR channels driven by unknown arbitrary sequence. *IEEE Transactions on Signal Processing*, **44**, 756–758.

Kalouptsidis, N. (1997). *Signal Processing Systems: Theory and Design.* John Wiley & Sons.

Moulines, E., P. Duhamel, J.F. Cardoso and S. Mayrargue (1995). Subspace methods for the blind identification of multichannel FIR filters. *IEEE Transactions on Signal Processing*, **43**, 516–525.

Tong, L., G. Xu and T. Kailath (1994). Blind identification and equalization based on second-order statistics: a time domain approach. *IEEE Transactions on Information Theory*, **40**, 340–349.

Tong, L. and S. Perreau (1998). Multichannel blind identification: from subspace to maximum likelihood methods. *Proceedings of the IEEE* , **86**, 1951–1968.

Tugnait, J.K. (1996). Blind equalization and estimation of FIR communications channels using fractional sampling. *IEEE Transactions on Communications* , **44**, 324–336.

Van Der Veen, A. (1998). Algebraic methods for deterministic blind beamforming. *Proceedings of IEEE*, **86**, 2082–2089.

Xu, G., H. Liu, L. Tong and T. Kailath (1995). A least–squares approach to blind channel identification. *IEEE Transactions on Signal Processing*, **43**, 2982–2993.

ADAPTIVE IDENTIFICATION AND EQUALIZATION FOR TIME-VARYING CHANNEL

Takayuki Naito, Hiromitsu Ohmori, Akira Sano * Kohichi Hidaka **

* *Keio University, 3-14-1 Hiyoshi, Yokohama 223-8522, Japan*
sano@sano.elec.keio.ac.jp
** *Tokyo Metropolitan College of Aeronautical Engineering, 8-52-1 Minami-Senju, Arakawa-ku, Tokyo 116-0003, Japan*

Abstract. A new deterministic adaptive identification and equalization approach is investigated based on an internal model principle to take account of time-varying dynamics of the channel parameters. The proposed adaptive algorithm incorporates an internal model for compensating the dynamics of the time-varying parameters of the FIR channel model. The stability of the adaptation can be assured by designing the parameter adjusting dynamics in the adaptive algorithm by the help of the small gain theorem and passivity theorem. Its effectiveness of the proposed algorithms is examined and compared to ordinary LMS approaches in numerical simulation of adaptive identification and equalization of a fading channel with parameters changing with a first-order function of time. *Copyright © 2000 IFAC*

Keywords. Adaptive filter, time-varying systems, system identification, adaptive algorithm, adaptive equalization, LMS algorithm, wireless communication

1. INTRODUCTION

Adaptive equalization is needed to deal with uncertain changes in channel dynamics to recover transmitted symbols correctly using the received signals. The parameters of the equalizer are updated in an on-line manner by using the training symbols to minimize the detection errors. Adaptive filtering approach plays a significant role of adjusting the equalizer parameters, and becomes now one of important tools in area of digital signal processing and communication systems (Widrow and Stern, 1985)(Haykin, 1996). Furthermore, when the channel dynamics and transmitted symbols are both uncertain, blind channel equalization approaches are also very useful to recover the transmitted symbol using only received signal. Recently such blind adaptive approaches are also very important in high rate band-limited digital communication systems in which the training symbols are not needed and equalization can start its self-adaptation without transmitting the training symbols (Sato, 1990)(Godard, 1972)(Liu and Tong, 1998).

However, almost all adaptive filtering algorithms have been derived on an assumption that the channel parameters are unknown but constant. In order to treat with time-varying situations, selection of a step size or forgetting factor play a very important role in the LMS algorithm or RLS algorithm respectively. However, the choice of the step size are given by a compromise between the tracking ability and steady-state error performance. The step size is limited also by the stability constraint.

Therefore, in this paper we propose a new deterministic adaptive approach in which an internal model is introduced to compensate the dynamics of the time-varying parameters of an FIR channel. The stability of the adaptation can be assured by designing the parameter adjusting dynamics in the adaptive algorithm by the help of the small gain theorem and the passivity theorem (Anderson, 1972)(Desoer and Vidyasager, 1975)(Åström and Wittenmark, 1995).

In wireless communication systems, channels are influenced by fading and multipath effects and then it should be treated as a time-varing system. Therefore, the proposed adaptive algorithm will also contribute to the adaptive equalization for such a time-varying channel. The effectiveness of the proposed algorithms is examined and compared to ordinary LMS and RLS approaches in numerical simulations

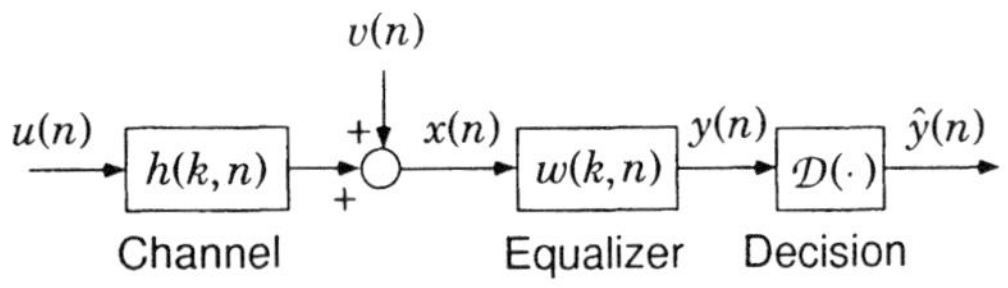

Fig. 1. Input-output expression of channel and equalizer.

of adaptive identification and equalization of a channel with parameters changing with a first-order function of time.

2. PROBLEM STATEMENT

Consider a simplified digital communication system in Fig.1, given by the input-output relation as

$$x(n) = \sum_{k=0}^{M} u(n-k)h(k,n) + v(n) \\ = \boldsymbol{u}^T(n)\boldsymbol{h}(n) + v(n) \quad (1)$$

where $u(n)$ is a transmitted symbol, $x(n)$ is a received signal and $v(n)$ is a random noise which is statistically independent of the symbols. $h(k,n)$ is a time-varying impulse response including the transmitter filter, channel and receiver filter, and the channel dynamics change with time as stated later. $\boldsymbol{h}(n)$ and $\boldsymbol{u}(n)$ are the impulse response vector and the transmitted symbol vector respectively, as defined below:

$$\boldsymbol{h}(n) = [h(0,n),\ h(1,n),\ \cdots, h(M,n)]^T$$
$$\boldsymbol{u}(n) = [u(n),\ u(n-1),\ \cdots, u(n-M)]^T$$

It is assumed in the following discussion that

(A1) The impulse response length of the channel is assumed to be known as $M+1$.
(A2) The coefficients $\{h(i,n)\}$ of the FIR channel change as piecewise polynomial functions of time, where the maximum order is known as q and the time-varying coefficients are unknown but bounded.
(A3) The noise $v(n)$ is a bounded white noise with zero mean and $|v(n)| < c$, and is statistically independent of the transmitted symbols $\{u(n)\}$.
(A4) The transmitted symbol $u(n)$ is a binary signal with ± 1 or a QPSK signal.

Fig.1 shows an FIR equalizer with adjustable weights $w(k,n)$ which are updated by an adaptive equalization algorithm, as given later. In a simple adaptive equalizer, for instance, the equalizer output $\hat{y}(n)$ is given by

$$y(n) = \sum_{k=0}^{N} x(n-k)w(k,n) = \boldsymbol{x}^T(n)\boldsymbol{w}(n) \quad (2a)$$
$$\hat{y}(n) = \mathcal{D}[y(n)] \quad (2b)$$

where $\boldsymbol{x}(n)$ and $\boldsymbol{w}(n)$ are the received signal vector and the equalizer weight vector respectively defined by

$$\boldsymbol{x}(n) = [x(n),\ x(n-1),\ \cdots,\ x(n-N)]^T$$

$$\boldsymbol{w}(n) = [w_0(n),\ w_1(n),\ \cdots,\ w_N(n)]^T$$

where $N+1$ is the number of the equalizer weights, and $\mathcal{D}(\cdot)$ is a memoryless nonlinear decision slicer for the symbol detection.

A new adaptive identification algorithm for a time-varying channel is investigated in Section 3, and its application to adaptive equalization for time-varying fading channels is discussed in Section 4.

3. ADAPTIVE IDENTIFICATION OF TIME-VARYING SYSTEM

3.1 *Interpretation of Ordinary LMS Algorithm from Intrnal Model Principle*

We first revisit the ordinary LMS adaptive algorithm from a new viewpoint of the internal model principle, when the parameters of FIR channel are unknown but constant.

Let the output error be denoted by

$$e(n) = \hat{x}(n) - x(n) \quad (3)$$

where $\hat{x}(n)$ is the output of the FIR channel model, which is given by

$$\hat{x}(n) = \boldsymbol{u}(n)^T\hat{\boldsymbol{h}}(n) \quad (4)$$

where $\hat{\boldsymbol{h}}(n) = (\hat{h}(0,n), \hat{h}(1,n), \cdots, \hat{h}(M,n))^T$ is an updated parameter vector of the FIR channel model, and $\boldsymbol{u}(n)$ is the transmitted training symbol vector for the identification of the channel model.

One of the ordinary LMS adaptive algorithm to update the parameter $\hat{\boldsymbol{h}}(n)$ is described by

$$\text{LMS1: } \hat{\boldsymbol{h}}(n) = \hat{\boldsymbol{h}}(n-1) - \mu e(n)\boldsymbol{u}(n) \\ = -\frac{\mu}{1-z^{-1}}e(n)\boldsymbol{u}(n) \quad (5)$$

where $\mu > 0$ is a step size and z^{-1} is a backward shift oprator. Let the error be rewritten by

$$e(n) = \boldsymbol{u}^T(n)\tilde{\boldsymbol{h}}(n) - v(n) \quad (6)$$

where $\tilde{\boldsymbol{h}}(n) \equiv \hat{\boldsymbol{h}}(n) - \boldsymbol{h}$.

The above relation shows that the adaptive algorithm is a closed-loop system consisting of the parameter adjusting dynamics (5) and the error system (6), as illustrated in Fig.2, in which $v(n)$ is omitted for the simplicity. Thus, it can be seen in Fig.2 that the parameters $\boldsymbol{h}$ can be regarded as an uncertain external input to the closed-loop system. In the time-invariant case, since $\boldsymbol{h}$ is a constant input which satisfies $(1-z^{-1})\boldsymbol{h} = 0$, then $1/(1-z^{-1})$ plays a role of the internal model in the closed-loop system. Thus the parameter adjusting dynamics $\mu/(1-z^{-1})$ can compensate for the unknown constant input $\boldsymbol{h}$ to assure the boundedness of the error (Hidaka *et al.*, 1999). Furthermore, since the upper subsystem $\boldsymbol{H}_1$ satisfies the strict passivity and the lower subsystem $\boldsymbol{H}_2$ is passive

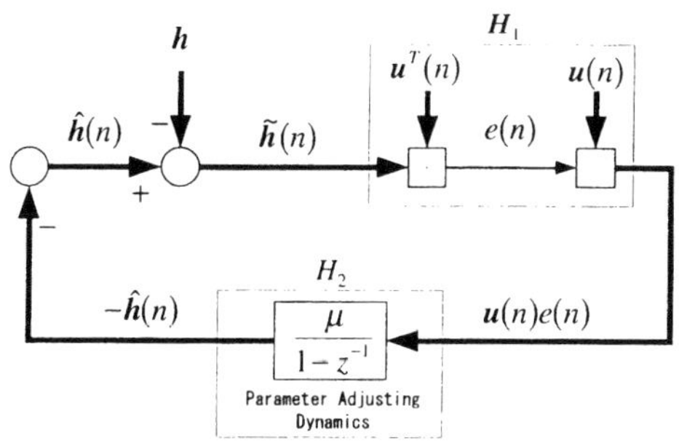

Fig. 2. Structure of ordinary LMS adaptive filtering.

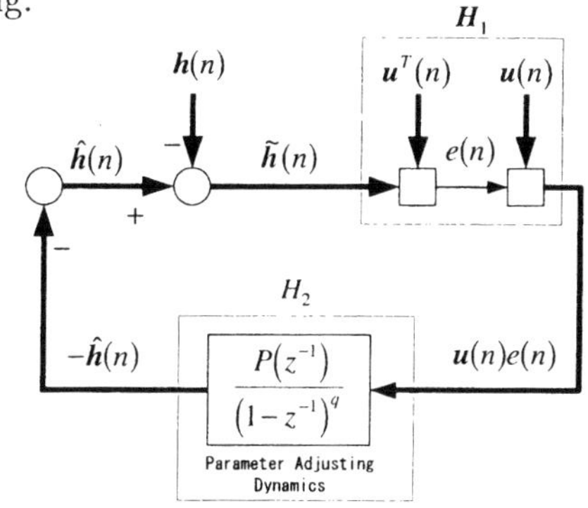

Fig. 3. New stucture of time-varying LMS algorithm

(positive real), then it follows from the passivity theorem (Åström and Wittenmark, 1995)(Anderson, 1972) that the closed-loop system of the total adaptive algorithm is BIBO stable.

Finally, it should be remarked that since (5) for updating $\hat{\boldsymbol{h}}(n)$ includes $\hat{\boldsymbol{h}}(n)$ itself through $e(n)$ in (6), the expression of $e(n)$ should be rewritten by substituting $\hat{\boldsymbol{h}}(n)$ in (6) into (5), as

$$e(n) = \frac{\boldsymbol{u}^T(n)\hat{\boldsymbol{h}}(n-1) - x(n)}{1 + \mu \boldsymbol{u}^T(n)\boldsymbol{u}(n)} \tag{7}$$

An alternative LMS algorithm which is simple and widely used is described by

$$\text{LMS2:}\ \hat{\boldsymbol{h}}(n) = \hat{\boldsymbol{h}}(n-1) - \mu e(n-1)\boldsymbol{u}(n-1) \tag{8}$$

In the latter case, the stability condition for the step size μ is needed (Widrow and Stern, 1985) (Haykin, 1996).

3.2 *New LMS Algorithm for Time-varying System*

We will investigate an extension of the above interpretation of the LMS algorithm to a general case with time-varying channels. In Fig.3 showing the error system, we can regard the unknown time-varying channel $\boldsymbol{h}(n)$ as an external input added into the closed-loop system, and adopt the corresponding internal model into the parameter adjusting dynamics. If the true channel parameters change piecewise as a polynomial function of time, for instance, a first-order function of time, second-order function of time, or so, then the corresponding internal model is incorporated into the closed-loop, such as $1/(1-z^{-1})^2$, $1/(1-z^{-1})^3$, respectively. However, even if the above internal model is directly put into the parameter adjusting dynamics in Fig.2, the stability of the closed-loop system cannot be assured. Therefore, the important issue is how to attain the stability of the adaptive algorithm in a case when the above internal model is employed to compensate for the time-varying parameters, as given in Fig.3.

As stated in the time-invariant case in Fig.2, the stability of the closed-loop system consisting of two blocks is closely related with the passivity of the two subsystems. If the error system $\boldsymbol{H}_1$ satisfies the strictly output passivity and the parameter adjusting dynamics satisfies the passivity, the closed-loop system can be stable (Åström and Wittenmark, 1995). However, unlike the time-invariant case, the adjusting dynamics $H_2(z^{-1})$ including $1/(1-z^{-1})^q$ with $q \geq 2$ is no longer passive, so we have to introduce a new adjusting dynamics as in Fig.3 as

$$H_2(z^{-1}) = \frac{P(z^{-1})}{(1-z^{-1})^q} \tag{9}$$

where $P(z^{-1})$ is the numerator polynomial described by

$$P(z^{-1}) = p_0 + p_1 z^{-1} + \ \cdots \ + p_r z^{-r}, \qquad r \leq q, \quad p_0 \neq 0 \tag{10}$$

where $r \leq q$ and $p_0 \neq 0$, and these coefficients should be determined so that the closed-loop can be stabilized, as described later.

Thus the new LMS type of adaptive algorithm for updating $\hat{\boldsymbol{h}}(n)$ with the time-varying parameters in the $(q-1)$-th order of polynomial function of time can be summarized as follows:

$$\hat{\boldsymbol{h}}(n) = -\frac{P(z^{-1})}{(1-z^{-1})^q}[\boldsymbol{u}(n)e(n)] \tag{11}$$

$$e(n) = \boldsymbol{u}^T(n)\hat{\boldsymbol{h}}(n) - x(n) \tag{12}$$

The above two representations are realized in Fig.3. The error $e(n)$ should also be rewritten by substituting (11) into (12) to make the calculation causal, as

$$e(n) = \frac{\boldsymbol{u}^T(n)(1-(1-z^{-1})^q)[\hat{\boldsymbol{h}}(n)]}{1 + p_0 \boldsymbol{u}^T(n)\boldsymbol{u}(n)} - \frac{\boldsymbol{u}(n)(P(z^{-1}) - p_0)[\boldsymbol{u}(n)e(n)] - x(n)}{1 + p_0 \boldsymbol{u}^T(n)\boldsymbol{u}(n)} \tag{13}$$

For instance, in a case with $q = 2$, the proposed time-varying LMS algorithm can be described from (11) by

$$\begin{aligned}\hat{\boldsymbol{h}}(n) &= -\frac{P(z^{-1})}{(1-z^{-1})^2}[\boldsymbol{u}(n)e(n)] \\ &= 2\hat{\boldsymbol{h}}(n-1) - \hat{\boldsymbol{h}}(n-2) \\ &\quad -[p_0\boldsymbol{u}(n)e(n) + p_1\boldsymbol{u}(n-1)e(n-1)]\end{aligned} \tag{14}$$

where

$$e(n) = \frac{\boldsymbol{u}^T(n)}{1 + p_0 \boldsymbol{u}^T(n)\boldsymbol{u}(n)} \Big[2\hat{\boldsymbol{h}}(n-1)$$

$$-\hat{\boldsymbol{h}}(n-2) - p_1 \boldsymbol{u}(n-1)e(n-1) - x(n) \Big]$$

Thus it is noticed that the algorithm is causal because $e(n)$ does not include $\hat{\boldsymbol{h}}(n)$.

3.3 *Stability Considerarion*

In order to investigate the stability of the adaptive algorithm, first we perform the equivalent transformations on the operator $\boldsymbol{H}_1$ to obtain the new operator $\boldsymbol{S}_1$, and similarly on $\boldsymbol{H}_1$ to obtain S_2, as $\boldsymbol{S}_1 = [\boldsymbol{H}_1 - \boldsymbol{I}][\boldsymbol{H}_1 + \boldsymbol{I}]^{-1}$ and $S_2 = (H_2 - 1)(H_2 - 1)\boldsymbol{I}$, respectively. It is generally known from the pasisvity theorem (Anderson, 1972)(Åström and Wittenmark, 1995) that if and only if the operator $\boldsymbol{H}$ is strictly input passive, then $\gamma(\boldsymbol{S}) < 1$, and if and only if the operator $\boldsymbol{H}$ is passive, then $\gamma(\boldsymbol{S}) \leq 1$. Thus, in the time-invarinat case in Fig.2, it follows from the above result the BIBO stability in (5) and (6) can be assured since $\gamma(\boldsymbol{S}_1 \cdot \boldsymbol{S}_2) \leq \gamma(\boldsymbol{S}_1)\gamma(\boldsymbol{S}_2) < 1$. $\gamma(\boldsymbol{S})$ denotes a gain defined by

$$\gamma(\boldsymbol{S}) \equiv \sup_{v \in X_e} \frac{\| \boldsymbol{S}(\boldsymbol{v}) \|}{\| \boldsymbol{v} \|} \qquad (15)$$

where $X_e \equiv \{x(n) \in X,\ x_T(n) \in X,\ \exists T > 0,\ \|x_T\| < \infty\}$ and the norm is arbitrary.

Unlike the above time-invariant case, the passivity theorem is not available, since H_2 is not passive (not positive real) in Fig.3 in the time-variant case, although $\boldsymbol{H}_1$ is strictly input passive since the input signal is binary random then the persistently exciting (PE) condition is satisfied. However, since $\boldsymbol{H}_1$ is strictly input passive, then there exists $\varepsilon > 0$ such that $\gamma(\boldsymbol{S}_1) = 1 - \varepsilon$. Therefore, if $P(z^{-1})$ in $H_2(z^{-1})$ in Fig.3 can be determined so that

$$\gamma(\boldsymbol{S}_2) < 1/\gamma(\boldsymbol{S}_1) = 1/(1-\varepsilon) \qquad (16)$$

then the BIBO stability of the adaptive algorithm can be guaranteed. Thus, the design problem is now to obtain $P(z^{-1})$ so that the operator gain $\gamma(\boldsymbol{S}_2)$ should satisfy

$$\gamma(\boldsymbol{S}_2) = \left\| \frac{P(z^{-1}) - (1 - z^{-1})^q}{P(z^{-1}) + (1 - z^{-1})^q} \right\|_\infty < \frac{1}{1-\epsilon} \quad (17)$$

and that the denominator polynomial of (17) should be stable. It is difficult to explicitly determine the stability margin ε in $\gamma(\boldsymbol{S}_1) = 1 - \varepsilon$ prior to the equalization, but we can estimate $\gamma(\boldsymbol{S}_1)$ by using the input-output signal data. In practical cases with little prior information, we first may set ε small so long as (17) is satisfied.

3.4 *Numerical Simulation Results*

The proposed LMS algorithm given in (11) and (13) is applied to identify the time-varying channel impulse response, and is compared to the two ordinary adaptive algorithms LMS1 in (5) and LMS2 in (8).

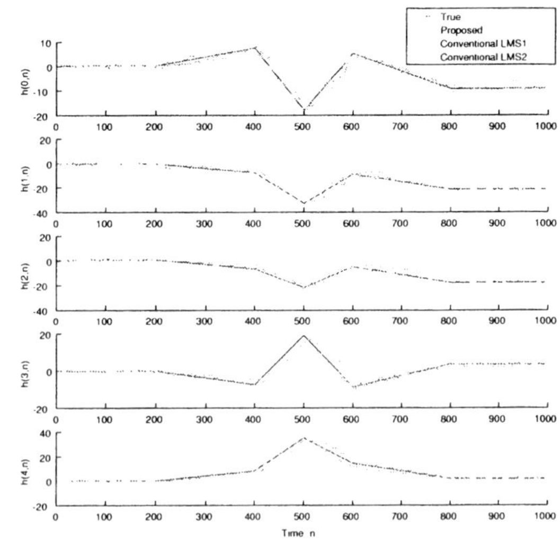

Fig. 4. True and identified coefficients of time-varying FIR channel obtained by proposed method and ordinary LMS1 and LMS2

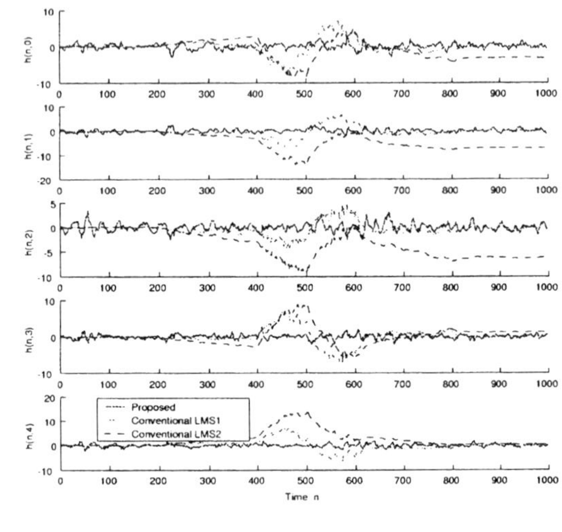

Fig. 5. Comparison of identification error in time-varying parameters between proposed method and ordinary LMS1 and LMS2

The transmitted symbols which are binary random signals taking ± 1 are used as input signals for the identification. The length of the impulse response is $M = 4$. The number of the observed symbols are 1000, and the SNR is 25(dB). The setup of the proposed algorithm is given as: $q = 2$, $P(z^{-1}) = 0.5 - 0.49z^{-1}$, $\epsilon = 0.12$. As for the conventional LMS algorithms, the step size is chosen as $\mu = 2$ for LMS1, and $\mu = 0.2$ for LMS2. Fig.4 shows the comparison of the proposed algorithm with the ordinary LMS1 and LMS2. The identified impulse response obtained by the proposed method can almost perfectly track the true one given by the solid line even if the channel dynamics changed from the time instant 200 to 800, while the identified results obtained by the two ordinary methods have some large delay from the true ones from the time instant 400 to 600. Fig.5 also shows more clearly the better tracking performance of the proposed algorithm compared to LMS1 and LMS2.

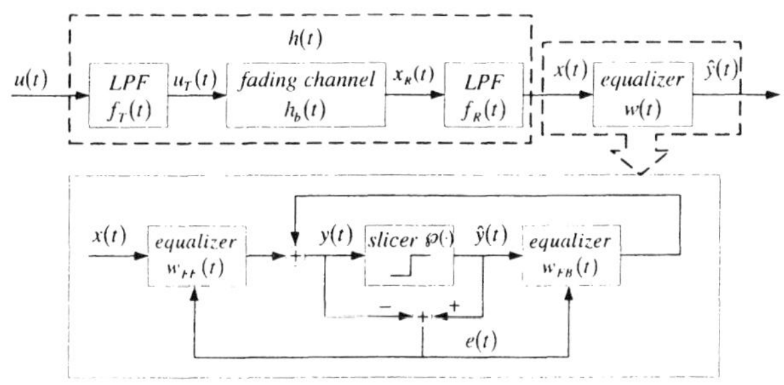

Fig. 6. Structure of adaptive equalization and RLS-DFE algorithm

4. RLS ADAPTIVE EQUALIZATION OF FADING CHANNEL

4.1 *Time-varying RLS Adaptive Equalization*

We apply the proposed time-varying adaptive algorithm to the RLS type of adaptive decision feedback equalization. Fig.6 gives the total structure of the adaptive DFE, in which $f_T(t)$ and $f_R(t)$ are the transmitter and receiver filters, and $h_b(t)$ is a fading channel. $h(t)$ expresses a total impulse response. The output of the DFE is given by the feedforward and feedback filters as

$$y(n) = \sum_{k=0}^{M} \hat{w}_{FF}(n,k)x(n-k) + \sum_{k=1}^{MM} \hat{w}_{FB}(n,k)\hat{y}(n-k)\boldsymbol{\xi}^T(n)\hat{\boldsymbol{w}}(n) \quad (18)$$

$$\hat{y}(n) = \mathcal{D}(y(n)) \quad (19)$$

$$\hat{\boldsymbol{w}}(n) = [\hat{\boldsymbol{w}}_{FF}^T(n)\ \hat{\boldsymbol{w}}_{FB}^T(n)]^T$$
$$\hat{\boldsymbol{w}}_{FF}(n) = [\hat{w}_{FF}(n,0)\cdots\hat{w}_{FF}(n,M)]^T$$
$$\hat{\boldsymbol{w}}_{FB}(n) = [\hat{w}_{FB}(n,1)\cdots\hat{w}_{FB}(n,MM)]^T$$
$$\boldsymbol{\xi}(n) = [x(n)\ x(n-1)\ \cdots\ x(n-M)\ \hat{y}(n-1)\ \hat{y}(n-2)\ \cdots\ \hat{y}(n-MM)]^T$$

where $\hat{\boldsymbol{w}}_{FF}(n)$ and $\hat{\boldsymbol{w}}_{FB}(n)$ are the feedforward and feedback taps respectively, $\hat{y}(n)$ is the output of the decision slicer which is input to the feedback taps. The known input symbols $u(n)$ are used in the training phase, and the detected symbols $\hat{y}(n)$ are used in the transmitting phase. The purpose of the equalization is to give the recovery of the transmitted symbols as $\hat{y}(n) = u(n-d)$, where d is an adequate delay time. $\mathcal{D}(\cdot)$ is a decision slicer, and its output is given, for instance, in the QPSK case, by

$$\hat{y}(n) = \mathcal{D}(y(n)) = \hat{y}_I(n) + j\cdot\hat{y}_Q(n) = \mathrm{sgn}[Re(y(n))] + j\cdot\mathrm{sgn}[Im(y(n))] \quad (20)$$

where

$$\mathrm{sgn}(c) = \begin{cases} 1 & (c \geq 0) \\ -1 & (c < 0) \end{cases} \quad (21)$$

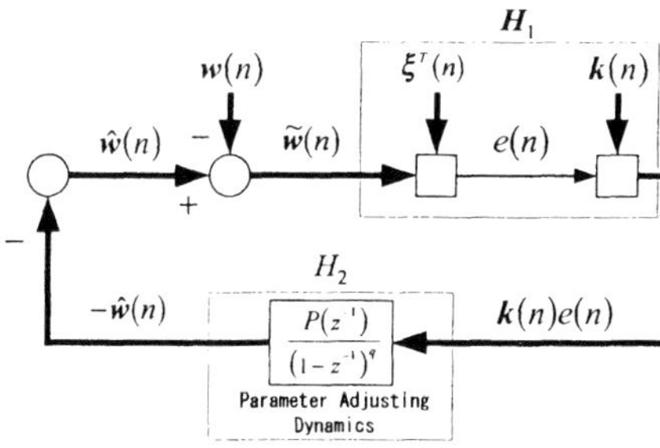

Fig. 7. Structure of proposed adaptive equalization algorithm in time-varying case

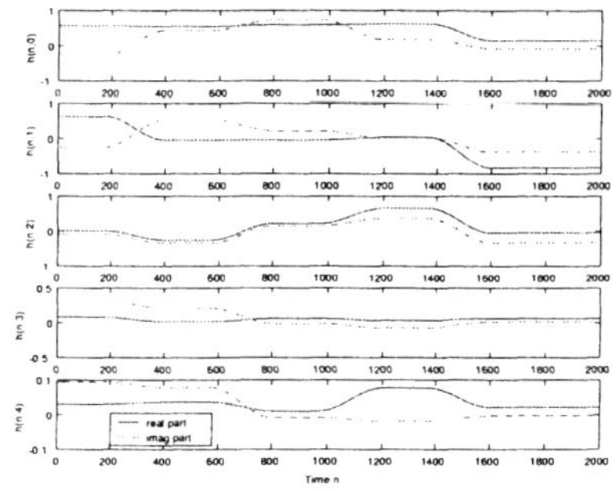

Fig. 8. Time-varying FIR channel on mobile wireless communivation (mobile speed:100km/h)

On the above assumptions, the new RLS type of adaptive equalizer based on the proposed internal model principle is summarized as follows:

$$y(n) = \boldsymbol{\xi}^T(n)\hat{\boldsymbol{w}}(n) \quad (22)$$

$$e(n) = y(n) - \hat{y}(n) = \boldsymbol{\xi}^T(n)\tilde{\boldsymbol{w}}(n) \quad (23)$$

$$\boldsymbol{k}(n) = \frac{\boldsymbol{C}(n-1)\boldsymbol{\xi}(n)}{\lambda + \boldsymbol{\xi}^H(n)\boldsymbol{C}(n-1)\boldsymbol{\xi}(n)} \quad (24)$$

$$\boldsymbol{C}(n) = \frac{1}{\lambda}[\boldsymbol{C}(n-1) - \boldsymbol{k}(n)\boldsymbol{\xi}^H(n)\boldsymbol{C}(n-1)] \quad (25)$$

$$\hat{\boldsymbol{w}}(n) = -\frac{P(z^{-1})}{(1-z^{-1})^q}\boldsymbol{k}(n)e(n) \quad (26)$$

$$P(z^{-1}) = p_0 + p_1 z^{-1} + \cdots + p_r z^r, \quad p_0 = 0, \quad r \geq q \quad (27)$$

where the equalization error $e(n)$ can be rewritten like (13), as

$$e(n) = \frac{Q(n) - \mathcal{D}(Q(n))}{1 + \boldsymbol{\xi}^T(n)\boldsymbol{k}(n)} \quad (28)$$

$$Q(n) = \boldsymbol{\xi}^T(n)(1-(1-z^{-1})^q)[\hat{\boldsymbol{w}}(n)] - \boldsymbol{\xi}^T(n)(P(z^{-1}) - p_0)[\boldsymbol{k}(n)e(n)] \quad (29)$$

4.2 *Numerical Simulation Results*

The proposed RLS adaptive equalization scheme given in (24)~(29) is applied to a frequency-selective fading channel in mobile communication. The setup conditions are summarized as follows: The transmitted symbols are 4-QPSK random signals, its bit rate is 56(kbps) and the carrier frequency is

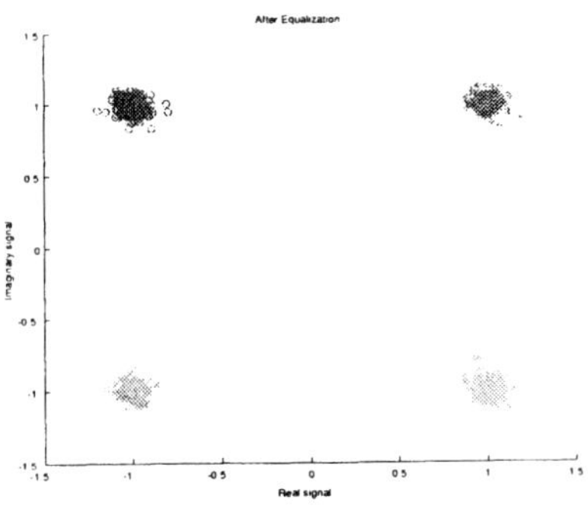

Fig. 9. Recovered symbols obtained by proposed RLS-DFE algorithm (mobile speed: 100km/h)

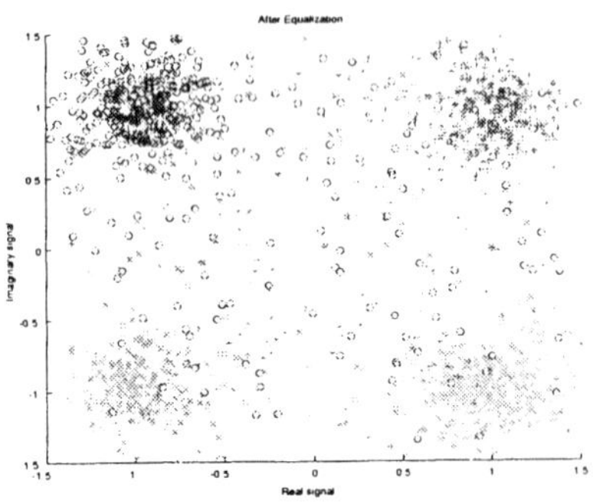

Fig. 10. Recovered symbols by conventional RLS-DFE algorithm (mobile speed: 100km/h)

1.5(GHz). The SNR is 25(dB), the transmitter and receiver filters are the cosine raised rolled-off filters, and the numbers of the feedforward taps and feedback taps are six and five respectively. The number of multipath is six, and their delay profiles are 0, $0.8T$, $1.4T$, $1.9T$, $2.1T$ and $3.3T$. The fading is characterized by the Rayleigh distribution and the maximum Doppler shift is 140(Hz). Thus, Fig.8 shows the time profiles of the real and imaginary parts of the time-varying fading channel in a case of mobile speed 100(km/h). Now the proposed RLS-DFE adaptive equalization is designed by setting $q = 2$, $P(z^{-1}) = 0.5 - 0.49z^{-1}$, and the stability margin is $\varepsilon = 0.07$.

Fig.9 shows the recovered symbols obtained by the proposed RLS-DFE adaptive equalization, while Fig. 10 gives the recovery symbols by the ordinary RLS-DFE equalization. The proposed algorithm can give excellent performance in the equalization results even if the channel varies largely, as shown in Fig.8. Fig.11 and Fig.12 shows the comparison in the real part of the convolution of the time-varying channel and the adaptive equalizer. The proposed method gives very clear peak at the delay of three, *i.e.* $y(n) \cong u(n-3)$, and the convolution result is almost zero at other delay, so the zero-forcing can be almost perfectly attained even in the presence of the time-varying effect. On the other hand, the peak obtained by the ordinary method is much more influenced by the time-varying channel.

5. CONCLUSION

The new adaptive identification and equalization algorithms have been presented to compensate the

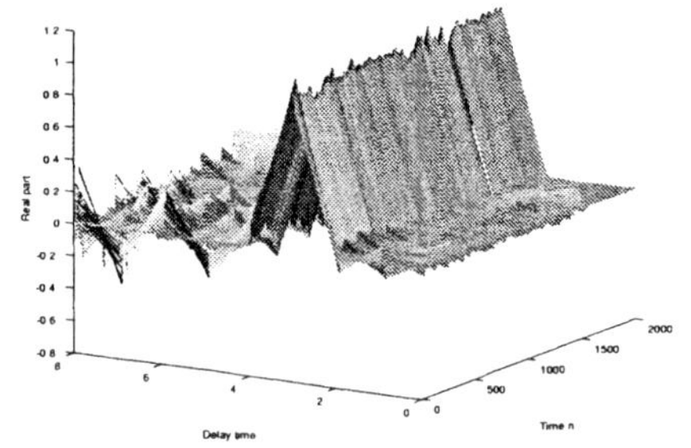

Fig. 11. Convolution of channel and proposed equalizer (real-part)(mobile speed:100km/h)

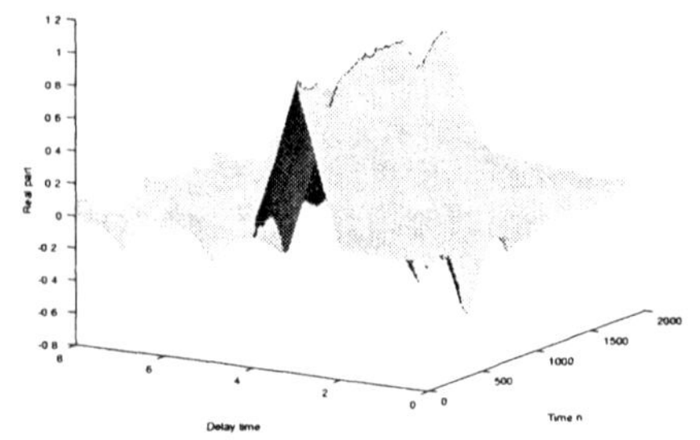

Fig. 12. Convolution of channel and ordinary equalizer (real-part)(mobile speed:100km/h)

time-varying dynamics of the impulse response parameters expressed by polynomial functions of time. The stability and convergence is assured by applying the passivity and small gain theorems to the closed-loop system of the adaptive algorithm. The effectiveness has been validated in the simulation of identification and equalization for time-varying channels.

REFERENCES

Anderson, B.D.O. (1972). The small-gain theorem, the passivity theorem and their equivalence. *Journal of the Franklin Institute* **298**, 105-115.

Åström, K.J. and B. Wittenmark (1995). *Adaptive Control*, Addison-Wesley Publishing Company.

Desoer, C.A. and M.Vidyasager (1975). *Feedback Systems: Input-Output Properties*,, Academic Press NY.

Godard, D.N. (1980). Self-recovering equalization and carrier tracking in tow-dimensional data communication systems. *IEEE Trans. Communications* **COM-28**, 1867-1875.

Haykin, S.(1996). *Adaptive Filter Theory*. Prentice-Hall, Englewood Cliffs, NJ.

Hidaka, K., H. Ohmori and A. Sano (1999). Adaptive control design for linear time-varying system based on internal model principle. *IEICE Trans. Fundamentals* **E82-A**, 1047-1054.

Liu R. and L. Tong (1980). Special Issue on Blind System Identification and Estimation. *Proc.IEEE* **86**, 1901-2116.

Sato, Y. (1990). *Linear Equalization THeory.* Maruzen.

Widrow, B. and S.D. Stearns (1985). *Adaptive Signal Processing*. Prentice-Hall, Englewood Cliffs, NJ.

BLIND LOCALIZATION BY SUBSPACE METHOD FOR A SCATTERING MODEL

E.Ternisien, G. Roussel, M. Benjelloun

Laboratoire d'Analyse des Systèmes du Littoral
195 rue du Pasteur Martin Luther King
B.P 649 62228 Calais CEDEX, FRANCE
Eric.Ternisien@lasl-gw.univ-littoral.fr
Tel : (+33) 3.21.46.06.81

Abstract: This paper deals with the parameters estimation of a scattering environment modeled by a state-space equation, and the localization of a source emitting in it. Measurements are performed thanks to a multisensor network located in the monitored area. The problem is then to localize the source and to identify the propagation parameters using only the sensors signals. The blind aspect of this procedure requires the use of specific methods such as subspace one. This method is combined with the *a priori* information brought by the specific form of the state-space matrices to reduce the number of estimated parameters. *Copyright © 2000 IFAC*

Keywords: Blind identification, Impulse responses, Scattering problems, State-space models, Subspace methods

1. INTRODUCTION

This paper deals with the physical models of dispersion. Their study is realized by means of system theory. Indeed, the source-multisensor approach permits to apply blind identification methods, which are very used in digital communications, to physical systems. Let's consider a region monitored by a number of sensors geographically distributed. These sensors are used to evaluate the concentration of a gas or matter emitted by an unknown source. The problem is to localize this unknown source and the dispersion system on the basis of measurement. A mathematical solution have already been proposed in (Khapalov, 1994). Another approach is to consider each source-sensor system as an ARMA filter having an impulse response which is approximated by a FIR channel.

Generally, the localization of source is studied from a **D**irections **O**f **A**rrival point of view (DOA) with the formulation (*Méthodes Haute Résolution*, 1999).

$$x = \mathbf{A}\underline{s} + b \qquad (1)$$

where $\mathbf{A} = [\underline{a}(\theta_1), \cdots, \underline{a}(\theta_k)]$ is the matrix containing the source vectors $\underline{a}(\theta_k)$. $\underline{a}(\theta_k)$ essentially depends on the DOA, thus the estimation of $\mathbf{A}$ gives the localization. The present approach is quite different because of the problem formulation into state-space model. Indeed, in this contribution, localization is performed by estimating the control matrix which caracterizes the source position.

The modelling of the dispersion channel by a FIR leads to use subspace technics (Tong, *et al.*, 1991; Moulines, *et al.*, 1995). This method, quite adapted to the multisensor case, operates with the second order statistics of the observed signals. Its use with the state-space representation of the model reduces the channel parameters determination and the source localization to a functionnal minimization which provides the impulse response of the source-sensor system. This impulse response is no more determined under the unity norm constraint as in (Moulines, *et al.*, 1995), but thanks to *a priori* information brought by the form of the state-space matrices. Another advantage is that the impulse response is no more estimated up to a scale factor.

This paper is organized as follows : In section 2, the model and its state-space formulation will be presented. The theory of the subspace method will be reminded in section 3 and the relation between the two previous sections will appear in section 4. A new criterion will be defined in section 5 and the results will be shown in section 6.

2. PROBLEM FORMULATION

The model used to describe the dispersion phenomenon had to be deliberately distant of the highly complex physical or chemical models. Therefore, the very classical **Navier-Stokes** differential equation, limited to the bidimensionnal case, was chosen.

$$\begin{aligned}\frac{\partial v(x,y,t)}{\partial t} &+ U_x(t)\frac{\partial v(x,y,t)}{\partial x} + U_y(t)\frac{\partial v(x,y,t)}{\partial y} \\ &= \frac{\partial}{\partial x}(K_x(t)\frac{\partial v(x,y,t)}{\partial x}) + \frac{\partial}{\partial y}(K_y(t)\frac{\partial v(x,y,t)}{\partial y}) \\ &\quad + S(x,y,t) + R(t) \qquad (2)\end{aligned}$$

where $v(x,y,t)$ is the gas concentration. Indeed, this equation takes into account the two phenomena called advection and diffusion who are fundamental for gas transport. A such model may be applied to the propagation of the pollution.

Generally, some simplifications are allowed :

- The chemical recombinations are not considered $\Rightarrow R(t) = 0$.
- The diffusion coefficients are supposed to be temporally and spatially invariant. This involves that the solution has a gaussian spatial behaviour.
- the wind (defined by its cartesian vector coordinates U_x and U_y) is supposed to be constant with respect to the observations range time.

The simplified formulation is then :

$$\begin{aligned}\frac{\partial v(x,y,t)}{\partial t} &+ U_x\frac{\partial v(x,y,t)}{\partial x} + U_y\frac{\partial v(x,y,t)}{\partial y} \\ &= K_x\frac{\partial^2 v(x,y,t)}{\partial x^2} + K_y\frac{\partial^2 v(x,y,t)}{\partial y^2} + S(x,y,t) \qquad (3)\end{aligned}$$

The numeric discretization is provided by the finite difference method and more precisely by the so-called "upwind" method (Euvrard, 1994). This one avoids averaging effect of the spatial derivative by calculating it on the upwind space of the concerned point instead of on its all neighbourhood.

$$\begin{aligned}\Upsilon_{i,j}^{n+1} &= m_1\Upsilon_{i+1,j}^{n} + m_2\Upsilon_{i,j}^{n} + m_3\Upsilon_{i-1,j}^{n} \\ &\quad + m_4\Upsilon_{i,j+1}^{n} + m_5\Upsilon_{i,j-1}^{n} + m_6 S_{i,j}^{n} \qquad (4)\end{aligned}$$

where $\Upsilon_{i,j}^{n}$ is the discrete form at position (i,j) and time n of the continuous concentration $v(x,y,t)$ and

$$\begin{aligned}m_1 &= \frac{pK_x}{h^2} \qquad (5)\\ m_2 &= 1 - \frac{p|U_x|}{h} - \frac{p|U_y|}{k} - \frac{2pK_x}{h^2} - \frac{2pK_y}{k^2}\\ m_3 &= \frac{p|U_x|}{h} + \frac{pK_x}{h^2}\\ m_4 &= \frac{pK_y}{k^2}\\ m_5 &= \frac{p|U_y|}{k} + \frac{pK_y}{k^2}\\ m_6 &= p\end{aligned}$$

h and k are the spatial steps respectively along the x-axis and the y-axis. p is the sample period. As the source is ponctual, $S_{i,j}^{n} = 0 \ \forall (i,j) \neq (x_s, y_s)$. (x_s, y_s) are the cartesian coordinates of the source in the monitored area. The stability has been studied by means of the **Gerschgörin-Hadamard** theorem (Lascaux and Theodor, 1993). The stability of the numerical model is ensured if :

$$p \leq \frac{h^2k^2}{2k^2K_x + 2h^2K_y + hk^2|U_x| + kh^2|U_y|} \qquad (6)$$

The equation (4) without the source term can be formulated as $\underline{\Upsilon}^{n+1} = \Phi\underline{\Upsilon}^{n}$ (where Φ is the transition matrix). This new formulation looks like the evolution of the system during the free response. So, a state-space representation of the model is possible by integrating the source term $m_6 S_{i,j}^{n}$. Let $(L \times l)$ be the size of the monitored area (Figure 1). This area can be seen as a matrix whose each element is the concentration at the point (i,j). So, the state vector can be constructed by stacking the L rows one after the other. $\underline{\Upsilon}$ is then a $(L \times l) \times 1$ vector defined as $\underline{\Upsilon} = [\underline{\Upsilon}_l^{(1)^T}, \underline{\Upsilon}_l^{(2)^T}, \cdots, \underline{\Upsilon}_l^{(L)^T}]^T$ where $\underline{\Upsilon}_l^{(i)} = [\Upsilon_{i,1}, \Upsilon_{i,2}, \cdots, \Upsilon_{i,l}]^T$. Then, the state-space model can be defined as :

$$\begin{cases}\underline{\Upsilon}^{n+1} &= \mathbf{A}\underline{\Upsilon}^{n} + m_6\mathbf{B}S^{n}\\ \underline{y}^{n} &= \mathbf{C}\underline{\Upsilon}^{n}\end{cases} \qquad (7)$$

$\underline{y}^{n}$ represents the observations of the system performed by a distributed sensors network located in the monitored area.

The particular form of the state vector explains the one of the state-space matrices. Indeed, evolution matrix $\mathbf{A}$ has a block-tridiagonal form

$$\mathbf{A} = \begin{pmatrix} \mathbf{A}_{1,1} & \mathbf{A}_{1,2} & 0 & \cdots & & \cdots & 0\\ \mathbf{A}_{2,1} & \mathbf{A}_{2,2} & \mathbf{A}_{2,3} & \ddots & & & \vdots\\ 0 & \ddots & \ddots & \ddots & & \ddots & \vdots\\ \vdots & \ddots & \ddots & \ddots & & \ddots & 0\\ \vdots & & \ddots & \mathbf{A}_{L-1,L-2} & & \mathbf{A}_{L-1,L-1} & \mathbf{A}_{L-1,L}\\ 0 & \cdots & \cdots & 0 & & \mathbf{A}_{L-1,L} & \mathbf{A}_{L,L}\end{pmatrix}$$

where each block has the form

$$\mathbf{A}_{k,k} = \begin{bmatrix} m_2 & m_1 & & & \\ m_3 & m_2 & m_1 & & \\ & \ddots & \ddots & \ddots & \\ & & m_3 & m_2 & m_1\\ & & & m_3 & m_2\end{bmatrix},$$

$$\mathbf{A}_{k,k+1} = \begin{bmatrix} m_4 & & \\ & \ddots & \\ & & m_4\end{bmatrix}, \mathbf{A}_{k-1,k} = \begin{bmatrix} m_5 & & \\ & \ddots & \\ & & m_5\end{bmatrix}.$$

Coding of matrices **B** and **C** respectively depends on source and sensors positions in the observed area, but the principle is the same than for the state vector. Indeed, if $(L \times l)$ is the size of the area, then **B** will be a full zero $(L \times l) \times 1$ matrix with a 1 corresponding to the position of the source in the area *i.e* the $(((y_s - 1)l) + x_s)^{th}$ element. The matrix **C** is a full zero $n_c \times (L \times l)$ matrix (where n_c is the number of sensors) with a 1 on each row corresponding to a sensor position in the area.

$$\mathbf{B} = \begin{pmatrix} 0 \\ \vdots \\ 0 \\ 1 \\ 0 \\ \vdots \\ 0 \end{pmatrix} \quad \mathbf{C} = \begin{pmatrix} 0 & \cdots & 0 & 1 & 0 & \cdots & 0 \\ \vdots & & & & & & \vdots \\ 0 & 1 & 0 & \cdots & \cdots & \cdots & 0 \end{pmatrix}$$

So, the problem of blind identification and localization amounts to estimate the matrix **B** for the localization and the m_i parameters of **A** for the evolution system. To achieve this identification, the classical subspace method is used.

3. PRESENTATION OF THE SUBSPACE METHOD

The subspace method is generally dedicated to a signal processing domain called equalization. The equalization is very used in digital communication to guarantee a good transmission between a source and a sensor. Blind equalization methods are more interesting because they don't require a training sequence to calculate the channel coefficients. Most of these methods directly perform the channel equalization (Godart, 1980; Sato, 1975). Other ones propose to identify the channel before equalizing it. In 1991, Tong and al. (1991) have presented a method for identifying the channel coefficients using only second order statistics, and in 1995 an extension of this method was proposed by Moulines (1995).

Let $s(n)$ be a discrete source emitting through a channel $h(n)$

$$y(n) = \sum_{\Delta=0}^{\infty} h(n)s(n - \Delta T) + b(n) \qquad (8)$$

Where T is the sample period and $b(n)$ is an additive noise independant of $s(n)$ with variance σ^2. Stocking N successive samples of the received signal gives

$$\underline{Y}_n = \mathcal{H}_N \underline{S}_n + \underline{B}_n \qquad (9)$$

with $\underline{Y}_n = [y_n, \cdots, y_{n-N+1}]^T$, $\underline{B}_n = [b_n, \cdots, b_{n-N+1}]^T$ and $\mathcal{H}_N$ is the $N \times (M+N)$ filtering matrix associated with the channel

$$\mathcal{H}_N = \begin{pmatrix} h_0 & \cdots & h_M & 0 & \cdots & 0 \\ 0 & \ddots & & \ddots & \ddots & \vdots \\ \vdots & \ddots & \ddots & & \ddots & 0 \\ 0 & \cdots & 0 & h_0 & \cdots & h_M \end{pmatrix}$$

The generalization with n_c sensors gives

$$\begin{bmatrix} \underline{Y}_n^{(0)} \\ \vdots \\ \underline{Y}_n^{(n_c-1)} \end{bmatrix} = \begin{bmatrix} \mathcal{H}_N^{(0)} \\ \vdots \\ \mathcal{H}_N^{(n_c-1)} \end{bmatrix} \underline{S}_n + \begin{bmatrix} \underline{B}_n^{(0)} \\ \vdots \\ \underline{B}_n^{(n_c-1)} \end{bmatrix} \qquad (10)$$

The identifiability conditions are very important :

- The zeros of the filters $h^{(i)}(n)$ $(0 \le i \le n_c)$ have to be different. That means that the spatio-temporal diversity is guaranteed. A sufficient distance between the sensors verifies this condition.
- N has to be greater than the maximum degree M of all the filters $h^{(i)}(n)$.
- At least one filter has degree M

These three conditions involve that $\mathcal{H}_N$ is rank $M+N$. If the hypothesis are verified, then the blind identification is possible using the autocorrelation matrix of the observations :

$$\mathbf{R}_y = E[\underline{Y}_n \underline{Y}_n^T] \qquad (11)$$

The independance between the noise and the source signal leads to:

$$\mathbf{R}_y = \mathcal{H}_N \mathbf{R}_s \mathcal{H}_N^T + \mathbf{R}_b \qquad (12)$$

The next step is to compute the eigenvalue decomposition of $\mathbf{R}_y$. If we assumed that $\mathbf{R}_b = \sigma^2 I$ and that $\mathcal{H}_N$ is rank $M + N$ (consequently the signal part of $\mathbf{R}_y$ too), the subspace decomposition gives :

$$\lambda_i > \sigma^2 \quad \text{for} \quad 0 < i < M + N - 1$$
$$\lambda_i = \sigma^2 \quad \text{for} \quad M + N < i < n_c N - 1$$

and the associated eigenvectors :

$$\underline{S}_S = [\underline{S}_{S_0}, \cdots, \underline{S}_{S_{M+N-1}}] \qquad \text{signal subpace}$$
$$\underline{N}_S = [\underline{N}_{S_0}, \cdots, \underline{N}_{S_{n_c N-M-N-1}}] \qquad \text{noise subpace}$$

The order M is then given by the size of the signal subspace.

In pratice, only a finite number of observations are available. So, the autocorrelation matrix is an estimated one. This implies that the eigenvalues are different one by one, so the boundary between the signal and the noise subspace is not easily detectable. Thus, for the detection of the order M, selection order methods like the use of a decreasing model of eigenvalues have to be performed (*Méthodes Haute Résolution*, 1999). On the other hand the subspace method is not robust to an overdetermination of M. So, others methods have been developed to solve this problem (Deneire, 1998; Meraim, 1995). In this paper, the order M is supposed to be known.

Once the two subpaces are separated, as they are orthogonal, a criterion $J(.)$ can be defined in order to find the impulse responses. The signal subspace being the space spanned by the columns of the filtering matrix $\mathcal{H}_N$, $J(.)$ has the form :

$$J(\mathcal{H}_N) = \sum_{i=0}^{n_c N-M-N-1} |\underline{N}_{S_i}^T \mathcal{H}_N|^2 \qquad (13)$$

In (Moulines, *et al.*, 1995) minimizing this criterion is the same as reaching the solution of :

$$J(\underline{H}) = \underline{H}^T \mathbf{Q} \underline{H} \tag{14}$$

under the constraint $|\underline{H}| = 1$, where $\mathbf{Q} = \sum_{i=0}^{n_c N-M-N-1} \mathcal{N}_{S_i} \mathcal{N}_{S_i}^T$ and $\mathcal{N}_{S_i}$ is the filtering matrix associated with the eigenvector $\underline{N}_{S_i}$. $\underline{H}$ is then the vector containing the impulse responses $\underline{H} = [\underline{H}^{(0)^T}, \cdots, \underline{H}^{(n_c-1)^T}]^T$.

Now, instead of using the criterion in its original form, the state-space model will be integrated in it.

4. RELATION BETWEEN STATE-SPACE MODEL AND SUBSPACE METHOD

In the following, the relation between state equation and vector $\underline{H}$ defined in equation (14) is shown.

Let $\mathcal{O}$ be the observability matrix and $\mathcal{C}$ the commandability matrix of the system :

$$\mathcal{O} = \begin{pmatrix} \mathbf{C} \\ \mathbf{CA} \\ \vdots \\ \mathbf{CA}^{M-1} \end{pmatrix} \qquad \mathcal{C} = \begin{pmatrix} \mathbf{B} & \mathbf{AB} & \cdots & \mathbf{A}^{M-1}\mathbf{B} \end{pmatrix}$$

We know that the matrix $\mathbb{H}$, containing all the impulse responses, can be defined by these two matrices :

$$\mathbb{H} = \mathcal{O}\mathcal{C} \tag{15}$$

$\mathbb{H}$ is a block-Hankel matrix of the form :

$$\mathbb{H} = \begin{pmatrix} \mathbb{H}_0 & \mathbb{H}_1 & \cdots & \cdots & \mathbb{H}_{\frac{M}{2}-1} \\ \mathbb{H}_1 & \mathbb{H}_2 & & & \\ \vdots & & \ddots & & \\ \mathbb{H}_{\frac{M}{2}-1} & \cdots & \cdots & \cdots & \mathbb{H}_{M-1} \end{pmatrix}$$

with $\mathbb{H}_i = [h_i^{(0)}, \cdots, h_i^{(n_c-1)}]^T$. Each row of $\mathbb{H}_i$ corresponds to one of the n_c sensor, so to one row of the matrix C :

$$\mathbb{H}_i = \begin{bmatrix} \mathbf{C}^{(0)}\mathbf{A}^{i-1}\mathbf{B} \\ \mathbf{C}^{(1)}\mathbf{A}^{i-1}\mathbf{B} \\ \vdots \\ \mathbf{C}^{(n_c-1)}\mathbf{A}^{i-1}\mathbf{B} \end{bmatrix}$$

Hence, we can formulate $\underline{H}$ as in (Moulines, *et al.*, 1995) :

$$\underline{H} = \begin{pmatrix} \begin{bmatrix} \mathbf{C}^{(0)}\mathbf{B} \\ \mathbf{C}^{(0)}\mathbf{AB} \\ \vdots \\ \mathbf{C}^{(0)}\mathbf{A}^{M-1}\mathbf{B} \end{bmatrix} \\ \vdots \\ \begin{bmatrix} \mathbf{C}^{(n_c-1)}\mathbf{B} \\ \mathbf{C}^{(n_c-1)}\mathbf{AB} \\ \vdots \\ \mathbf{C}^{(n_c-1)}\mathbf{A}^{M-1}\mathbf{B} \end{bmatrix} \end{pmatrix} \tag{16}$$

5. NEW FORM OF THE CRITERION

5.1 *Reduction of the problem*

As shown previously, the criterion (14) depends on matrices $\mathbf{A}$ and $\mathbf{B}$. Furthermore, the form of these matrices introduce *a priori* information. Indeed, $\mathbf{A}$ is build with the m_i parameters, and $\mathbf{B}$ is a column vector with only one non-zero element. Let be a few assumptions :

- The steps h and k of the observed area and the sample period p are known.
- The order M of the largest filter is assumed to be known or determined by an order selection method.
- U_x and U_y standing for the advection velocity (*i.e* the wind) are supposed known and constant.

These hypothesis lead to the following expressions of m_i :

$$\begin{aligned} m_1 &= \frac{pK_x}{h^2} \\ m_2 &= 1 - m_1 - m_3 - m_4 - m_5 \\ m_3 &= \alpha_1 + m_1 \\ m_4 &= \frac{pK_y}{k^2} \\ m_5 &= \alpha_2 + m_4 \end{aligned} \tag{17}$$

α_1 and α_2 depend on the wind, and hence are known. So the estimation of $\mathbf{A}$ is resumed to the estimation of the couple (m_1, m_4). If $\mathbf{B}$ is known, the criterion (14) can be wrote as :

$$\min_{m_1, m_4} \underline{H}(m_1, m_4)^T \mathbf{Q} \underline{H}(m_1, m_4) \tag{18}$$

without constraint.

5.2 *Localization problem*

Localization is performed with the intercorrelation information of the observed signals. Let $h_i(n)$ and $h_j(n)$ be a couple of differents impulse responses. Now, let's define $\underline{Z}_n^i$ and $\underline{Z}_n^j$ as

$$\underline{Z}_n^i = \mathcal{H}_N^i \underline{S}_n \,,\; \underline{Z}_n^j = \mathcal{H}_N^j \underline{S}_n \quad i, j = 1, .., n_c, i \neq j \tag{19}$$

with any signal $\underline{S}(n)$. The computation of the intercorrelation matrix $\mathbf{R}_{\underline{Z}^i \underline{Z}^j}$ gives a maximum value which is not on the main diagonal but on a subdiagonal.

$$\begin{aligned} \mathbf{R}_{\underline{Z}^i \underline{Z}^j} &= E[\underline{Z}^i \underline{Z}^{j^T}] \\ &= \mathcal{H}_N^i E[\underline{S}_n \underline{S}_n^T] \mathcal{H}_N^{j^T} \\ &= \underbrace{\mathcal{H}_N^i \mathbf{R}_s \mathcal{H}_N^{j^T}}_{\mathbf{R}_{\mathcal{H}_N^i \mathcal{H}_N^j}} \end{aligned} \tag{20}$$

As $\mathbf{R}_s$ is a symetric matrix, the maximum of the product $\mathcal{H}_N^i \mathbf{R}_s \mathcal{H}_N^{j^T}$ is on the same subdiagonal than the $\mathbf{R}_{\underline{Z}^i \underline{Z}^j}$ one in the noiseless case. If we call τ the shifting between the main diagonal and the one containing the maximum value of the intercorrelation matrix, then :

$$\tau_{\mathbf{R}_{\underline{Z}^i \underline{Z}^j}} = \tau_{\mathbf{R}_{\mathcal{H}_N^i \mathcal{H}_N^j}} \tag{21}$$

Furthermore, if $\mathbf{R}_s$ is the identity matrix, then $\tau_{\mathbf{R}_{\mathcal{H}_N^i \mathcal{H}_N^j}}$ corresponds to the shifting of the matrix $\mathbf{R}_{h_i h_j}$ based on the impulse responses. So, if we have noisy observations $\underline{Y}_n^i$, the computation of $\tau_{\mathbf{R}_{\underline{Y}^i \underline{Y}^j}}$ brings information about the localization since it has to be equal to $\tau_{\mathbf{R}_{h_i h_j}}$. Indeed, the position of the maximum of intercorrelation essentially depends on the shifting between the two impulse responses who depends itself on the source-sensors distance. Nevertheless, this is not a sufficient information. In fact, if all the possible source positions are take into account, several of them give the same value of τ (a better discrimination can be realized if the number of sensors is increased). In order to solve this problem, the following method has been established.

5.3 *Estimation method*

The estimation method is made up of two steps. The first one is the minimization of a joint criterion for all the admissible source positions.

$$\begin{cases} \min\limits_{m_1, m_4, \mathbf{B}} \underline{H}(m_1, m_4, \mathbf{B})^T \mathbf{Q} \underline{H}(m_1, m_4, \mathbf{B}) \\ \min\limits_{m_1, m_4, \mathbf{B}} \tau_{\mathbf{R}_{h_i h_j}} - \tau_{\mathbf{R}_{\underline{Y}^i \underline{Y}^j}} \quad i, j = 1, .., n_c, i \neq j \end{cases} \tag{22}$$

The optimization method used is a nonlinear least square method which minimizes a system of equations. At least, there are $\frac{n_c!}{2!(n_c-2)!} + 1$ equations. The **Levenberg-Marquardt** algorithm is used to achieve the minimization. So, for each iteration on $\mathbf{B}$, a vector containing the optimization results of the equation system is memorized. As said previously, there is several positions who cancel the second term. So, a selection criterion (23) is used to choose among these positions. This one makes the summation of all the terms in each result vector to discriminate the solution who gives the right localization $\mathbf{B}^*$ and the best estimation $(m_1(\mathbf{B}^*), m_4(\mathbf{B}^*))$.

$$\mathbf{B}^* = Arg \min_{\mathbf{B}} (\underline{H}(m_1(\mathbf{B}), m_4(\mathbf{B}))^T \mathbf{Q} \underline{H}(m_1(\mathbf{B}), m_4(\mathbf{B})) + \sum_{i,j=1}^{n_c} \tau_{\mathbf{R}_{h_i h_j}}(\mathbf{B}) - \tau_{\mathbf{R}_{\underline{Y}^i \underline{Y}^j}}(\mathbf{B})) \tag{23}$$

Then, the solution provides both the second term of (23) equal to zero and the minimum of the first term.

6. RESULTS

The simulation has been realized with the following assumptions :

- The wind U is equal to $(14.14 m/s, \frac{\pi}{4})$.
- The size of the monitored area is $(10, 10)$. The sensors are located respectively at $(7, 4)$, $(3, 7)$, $(8, 8)$ and the source at $(1, 1)$. In order to accelerate the computation, a subarea has been defined with all the admissible positions of source (Figure 1).
- The largest filter order is $M = 40$.

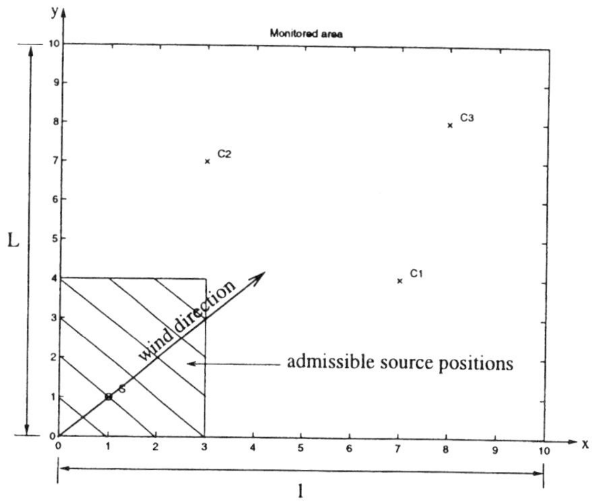

Figure 1. Monitored area

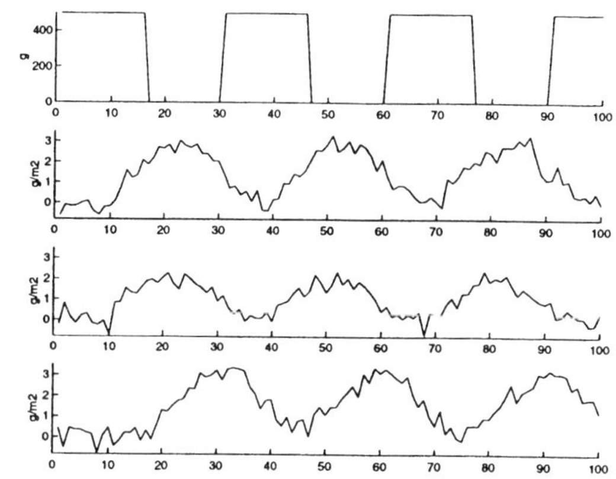

Figure 2. Source and observed signals

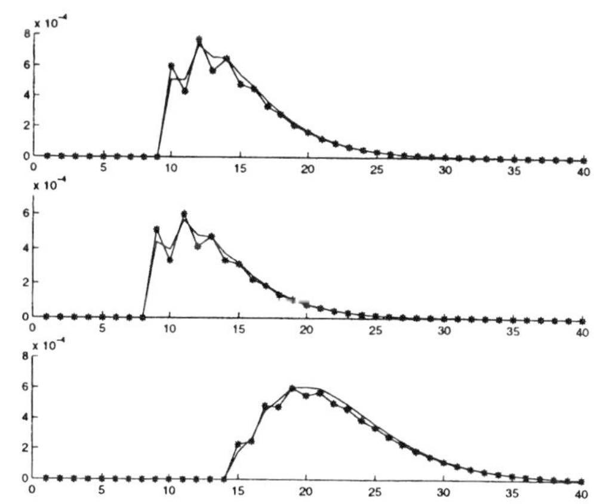

Figure 3. Impulse response (originals (-) and estimated (-*-) responses)

- The noise variance is $\sigma^2 = 0.09$.

The figure 2 shows the emmited source and the signals observed by the three sensors. These last have a SNR close to 20 dB. The source is well localized after the computation (Figure 4). Parameters m_1 and m_4 are estimated with respectively 10% and 11% of relative error. This error is attenuated by the fitting of the others parameters (m_2, m_3, m_5) who depend on (m_1, m_4). We can see on figure 3 that the impulse

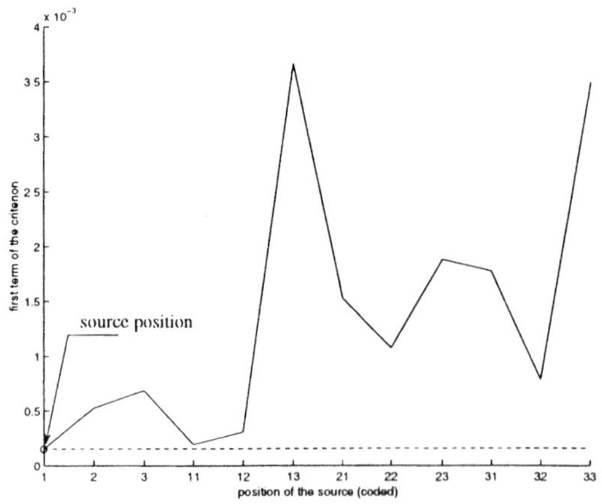

Figure 4. Evolution of the first term of the criterion (23) for each source position

responses are well approximated. Indeed, the error is about 8% on the estimation. So, identification provides a system quite close to the real one and the source is localized.

Error depends essentially on the noise. Experiments have shown that a good localization is achieved with a near 40 dB SNR, but for good estimation of (m_1, m_4), a smaller one is needed. Obviously, the noise can't be controlled, so one of the criterion (23) terms is always most favorized than the other. Nevertheless, the method gives good results with a SNR going up to 20 dB.

7. CONCLUSION

This contribution introduces a subspace-based approach for blind identification and source localization in the case of scattering model. This method is very interesting because it identifies the real value of the coefficients whereas most of blind methods gives the coefficients up to a scale factor. This is very useful for deconvolution. Indeed, the deconvolution step is then able to restore not only the form of the signal but also its amplitude (very important in the case of pollution for example). Simulations have performed good results but the method needs a long computation time because of the iterations on the matrix $\mathbf{B}$, *i.e* on all the possible source positions. Nevertheless, it is possible to reduce the search to a wideless area by *a priori* information or by methods like elliptic models (Vannoorenberghe, *et al.*, 1998). Then the computation time is greatly improved.

8. REFERENCES

Deneire, Luc (1998). Estimation aveugle de canal et accès multiple par répartition spatiale. PhD thesis. ENST.

Euvrard, D. (1994). *Résolution numérique des équations aux dérivées partielles*. MASSON.

Godart, D. (1980). Self-recovering equalization and carrier-tracking in two-dimensional data communication systems. *IEEE Trans. Commun.* **28**, 1867–1875.

Khapalov, A.Y. (1994). Localization of unknown sources for parabolic systems on the basis of available observations. *INT. J. Systems Sci.* **25(8)**, 1305–1322.

Lascaux, P. and R. Theodor (1993). *Analyse numérique matricielle appliquée à l'art de l'ingénieur, Tome I : Méthodes directes*. MASSON.

Meraim, Karim Abed (1995). Identification/Egalisation autodidacte au second ordre : Algorithmes et performances. PhD thesis. ENST.

Méthodes Haute Résolution (1999). HERMES.

Moulines, E., P. Duhamel, J-F. Cardoso and S. Mayrargue (1995). Subspace methods for the blind identification of multichannel FIR filters. *IEEE Trans on Signal Processing* **43**, 516–525.

Sato, Y. (1975). A method of self-recovering equalization for multi-level amplitude modulation. *IEEE Trans. Commun.* **COM-6**, 679–682.

Tong, L., G. Xu and T Kailath (1991). A new approach to blind identification and equalization of multipath channels. In: *Proc. 25th Asilomar Conf.*. pp. 856–860.

Vannoorenberghe, P., R. Lherbier, G. Roussel and G. Delmaire (1998). Possibilistic identification of polluant source by multisensor data fusion. In: *Proc APMS'98*. pp. 767–768.

Blind System Identification and Equalization

Erwei Bai
Department of Electrical and Computer Engineering
University of Iowa, Iowa City, IA 52242 USA
Email: erwei@icaen.uiowa.edu

Abstract

In this paper, we presents some new results on blind system identification and equalization, especially blind equalization for a class of time-varying channels. A decision feedback structure is proposed which is simple and convergent. *Copyright © 2000 IFAC*

1. Introduction

The blind system identification (BSI) and blind channel equalization (BCE) problems addressed in this paper can be formulated as follows: A sequence of input signal $u[kh_i]$ is transmitted at sampling rate $f_i = 1/h_i$ to a continuous time system via an impulse generator or a Zero Order Hold. The received output signal $y[nh_o]$ is sampled at the rate $f_o = 1/h_o$. The BSI and BCE problems are to identify from $y[nh_o]$ both the system transfer function and the input $u[kh_i]$. These problems have received a lot of attention in recent years for applications in digital image processing, speech coding and mobile communications such as cellular and cordless telephony where communication channels drift constantly.

In this paper, we present some new approachs to the BSI and BCE problems.

2. Problem Statement and Preliminary

In this paper, the sampled data system consists of an ideal impulse generator. The input to the continuous time system is a sequence of δ-functions whose magnitudes vary according to the input sequence $u[mh_i]$ with the sampling interval h_i. The output $y[kh_o]$ is also a discrete sequence with the sampling interval h_o. The sampling intervals h_i and h_o are usually different. It is assumed in the paper that $h_o = h_i/p$ for some positive integer $p \geq 1$, referred to as the over-sampling ratio. The continuous time system is assumed to be represented by an unknown linear time-invariant nth order state space equation

$$\begin{aligned} \dot{x}(t) &= Ax(t) + bu(t), \qquad u, y \in R, \quad A \in R^{n\times n} \\ y &= cx(t). \end{aligned} \tag{2.1}$$

The transfer function from the input u to the output y, considered at the output sampling period h_o, is given by an nth order rational transfer function:

$$G(z) = c(zI - F)^{-1}\Gamma = \frac{b(z)}{a(z)} = \frac{b_1 z^{-1} + \dots + b_n z^{-n}}{1 - a_1 z^{-1} - \dots - a_n z^{-n}} \tag{2.2}$$

for some constants a_i and b_i.

From the derivation of the discrete time transfer function $G(z)$, we see that the order of the sampled data system is n which is the same as the order of the continuous time system and is independent of the over-sampling ratio p, except in some pathological cases for which pole-zero cancellation happens. This is one of the advantages to model the channel as an IIR system. In contrast, if an FIR model is used to approximate an IIR channel, the number of parameters in the model needs to increase as the over-sampling ratio p increases.

Our goal is to identify the parameters in (2.2) as well as the input signal $u[mh_i]$ based on the output measurements $y[kh_o]$ with as little knowledge as possible on the unknown input sequence. Let $Y(z)$ and $U(z)$ be the z-transforms of output and input sequences, respectively. Then

$$Y(z) = G(z)U(z) = (\alpha G(z))(\frac{1}{\alpha}U(z))$$

for any non-zero constant α. Therefore, the best we can do is to identify $G(z)$ and $U(z)$ up to a scaling constant.

Let $Y(z)$ and $U(z)$ denote the z-transforms of $y[kh_o]$ and $u[kh_o]$ respectively. For $h_i/h_o = p$, note that the input to the system is non-zero only once for every p samples, i.e.,

$$Y(z) = G(z)U(z) = G(z)U_0(z^p). \tag{2.3}$$

3. Identification of All-Pole Systems

We first consider a simple case of IIR systems, namely an all-pole filter, to convey the idea.

Let the over-sampling ratio $p = 2$, i.e., $h_i = 2h_o$. Suppose the resultant sampled data system is an all-pole system, in terms of the output sampling interval h_o, i.e.,

$$G(z) = \frac{b_1}{1 - a_1 z^{-1} - \dots - a_n z^{-n}}. \tag{3.1}$$

We have dropped the delay z^{-1} in the numerator for notational simplicity. The result in this section remains valid with some minor notational changes if z^{-1} is present.

The time domain expression of (3.1) is

$$y[kh_o] = \sum_{i=1}^{n} a_i y[(k-i)h_o] + b_1 u[kh_o].$$

Observe that $p = 2$ and the input $u[kh_o]$ to the impulse generator is zero for odd k. It follows that

$$y[kh_o] = \sum_{i=1}^{n} a_i y[(k-i)h_o] \tag{3.2}$$

for $k = 2l+1$, $l = 0, 1, 2, \cdots$ In other words, the coefficients a_i's are the solution of the following linear equation

$$\Phi[a_1 \;\; a_2 \;\; \cdots \;\; a_n]' = q \tag{3.3}$$

where

$$q = (\,y[(2l+1)h_o], y[(2l+3)h_o], y[(2l+5)h_o], \ldots)' \tag{3.4}$$

and

$$\Phi = \begin{pmatrix} y[(2l)h_o] & y[(2l-1)h_o] & \ldots & y[(2l+1-n)h_o] \\ y[(2l+2)h_o] & y[(2l+1)h_o] & \ldots & y[(2l+3-n)h_o] \\ y[(2l+4)h_o] & y[(2l+3)h_o] & \ldots & y[(2l+5-n)h_o] \\ \vdots & \vdots & \vdots & \vdots \end{pmatrix}. \tag{3.5}$$

Thus, a_i's can be uniquely solved by many standard methods if the matrix Φ has full column rank. In fact, a_i's can also be calculated on-line recursively by employing recursive least squares or other recursive algorithms. In this case, to guarantee the asymptotical convergence of the estimates $\hat{a}_i$'s to the true but unknown a_i's, some persistent excitation (PE) condition is required. Define

$$\phi[l] = [y[(2l+n)h_o] \;\; y[(2l+n-1)h_o] \;\; \cdots \;\; y[(2l+1)h_o]]'. \tag{3.6}$$

The PE condition means that the condition below holds uniformly in l_0 for some constants α, $m > 0$:

$$\sum_{l=l_0+1}^{l_0+m} \phi[l]\phi'[l] \geq \alpha I > 0 \tag{3.7}$$

Theorem 3.1 *Consider the system with* $p = 2$ *and* $G(z)$ *given by (3.1). Then*

1. $\phi[l]$ *is PE if the spectral measure of the input* $u[kh_i]$ *(in* h_i *not in* h_o*, a weaker condition) is not concentrated on* $k < n$ *points.*

2. *The recursive least squares estimates* $\hat{a}_i$*'s of equation (3.3) converge asymptotically to the true value* a_i*'s, if the PE conditiōn in 1 above is satisfied.*

To illustrate the result of Theorem 3.1, we give a simulation example here.

Simulation Example 1. Suppose the (unknown) sampled data system is represented by a third order transfer function

$$G(z) = \frac{1}{(1-0.5z^{-1})(1+0.2z^{-1})(1-0.3z^{-1})} = \frac{1}{1-\alpha_1 z^{-1} - \alpha_2 z^{-2} - \alpha_3 z^{-3}},$$

where

$$\alpha_1 = 0.6, \;\; \alpha_2 = 0.01, \;\; \alpha_3 = -0.03.$$

To obtain a performance measure of the blind channel identification, we have tried 100 Monte Carlo trials. 50 symbols were used in each trial to estimate $(\alpha_1, \alpha_2, \alpha_3)'$. The input signal is estimated by feeding the received output signal into the inverse of the estimated transfer function.

For the simulation, both the input and the noise signals are assumed to be independent random variables in $[-1, 1]$. Figure 1 shows the 100 estimates of the input signal (SNR=25db). Figure 2 shows the actual input signal (solid line) and the sample mean of its 100 estimates (dash-dot line) with SNR=25db. We see that the sample mean is very close to the actual input signal.

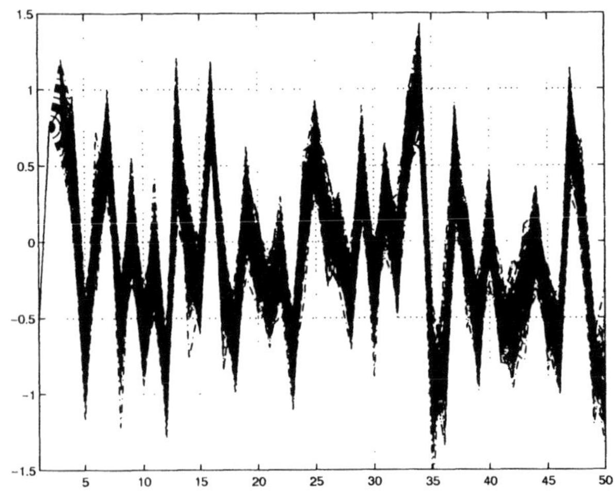

Figure 1: 100 estimates of the input signal (SNR=25db).

4. Identification of IIR Systems

We now consider arbitrary IIR systems characterized by (2.2). The algorithm requires $p = 2(n+1)$ and identifies a set of parameters α_i and β_i which are equivalent to a_i and b_i, respectively. The identification of α_i and β_i will be done separately with the advantage that estimation errors of α_i and β_i are independent.

Algorithm 4.1.

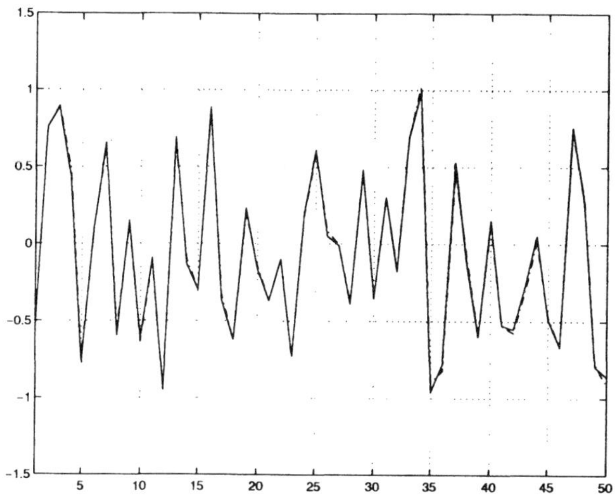

Figure 2: The actual input (solid) and the sample mean of 100 estimates (dash-dot) (SNR=25db).

Let $p = 2(n+1)$. The input $U(z)$ is given by

$$U(z) = U_0(z^{2(n+1)}). \tag{4.1}$$

Decompose the output

$$Y(z) = Y_0(z^2) + z^{-1}Y_1(z^2)$$

and let $G_0(z)$ and $G_1(z)$ be respectively the transfer functions from $U_0(z^{n+1})$ to $Y_0(z)$ and $Y_1(z)$, i.e.,

$$Y_0(z) = G_0(z)U_0(z^{n+1}); \quad Y_1(z) = G_1(z)U_0(z^{n+1}). \tag{4.2}$$

Then, we have

$$Y(z) = G_0(z^2)U_0(z^{2(n+1)}) + z^{-1}G_1(z^2)U_0(z^{2(n+1)}).$$

On the other hand,

$$Y(z) = G(z)U_0(z^{2(n+1)}) = \frac{b(z)a(-z)}{a(z)a(-z)}U_0(z^{2(n+1)}) \tag{4.3}$$

where $a(z)a(-z)$ is an even polynomial. Denote

$$\alpha(z^2) = a(z)a(-z) \tag{4.4}$$

where $\alpha(z) = 1 - \alpha_1 z^{-1} - ... - \alpha_n z^{-n}$, and decompose

$$b(z)a(-z) = \beta_0(z^2) + z^{-1}\beta_1(z^2) \tag{4.5}$$

with

$$\beta_0(z) = \beta_{01}z^{-1}+...+\beta_{0n}z^{-n}, \ \beta_1(z) = \beta_{11}+...+\beta_{1n}z^{-n-1}.$$

We obtain the following equation

$$Y(z) = \frac{\beta_0(z^2)}{\alpha(z^2)}U_0(z^{2(n+1)}) + z^{-1}\frac{\beta_1(z^2)}{\alpha(z^2)}U_0(z^{2(n+1)}).$$

Hence,

$$G_0(z) = \frac{\beta_0(z)}{\alpha(z)}; \quad G_1(z) = \frac{\beta_1(z)}{\alpha(z)}. \tag{4.6}$$

Now because $G_0(z)$ and $G_1(z)$ are the transfer functions from $U_0(z^{(n+1)})$ to the outputs $Y_0(z)$ and $Y_1(z)$ respectively in terms of the sampling period $h = 2h_o$, $G_0(z)$ and $G_1(z)$ share the common nth order denominator. Also, as far as equalization is concerned, either $G_0(z)$ or $G_1(z)$ is sufficient to recover the input because

$$U(z^{(n+1)}) = G_0^{-1}(z)Y_0(z) = G_1^{-1}(z)Y_1(z).$$

Note that the equations in (4.2) are expressed in terms of the sampling interval $h = 2h_o$. Its time domain representation is given by

$$y_j[kh] = \sum_{i=1}^{n} \alpha_i y[(k-i)h] + \sum_{i=1}^{n} \beta_{ji}u[(k-i)h]. \tag{4.7}$$

The input sequence $u[kh]$ is non-zero only if $k = l(n+1)$, i.e.,

$$u[(l(n+1)-1)h] = u[(l(n+1)-2)h] = \cdots = u[(l(n+1)-n)h] = 0$$

for all l. Now consider equation (4.7) at the sampling interval h for $k = l(n+1)$,

$$y_j[l(n+1)h] = \sum_{i=1}^{n} \alpha_i y_j[(l(n+1)-i)h].$$

As this holds for $j = 0, 1$, we can write more generally that

$$y[l(n+1)h] = \sum_{i=1}^{n} \alpha_i y[(l(n+1)-i)h]. \tag{4.8}$$

This equation is linear in α_i and can be written as follows (similar to (3.3)):

$$\Phi[\alpha_1 \ \ \alpha_2 \ \ \cdots \ \ \alpha_n]' = q \tag{4.9}$$

where

$$q = (\, y[l(n+1)h], y[(l+1)(n+1)h], y[(l+2)(n+1)h], \dots)' \tag{4.10}$$

$$\Phi = \begin{pmatrix} y[(l(n+1)-1)h] & \dots & y[(l(n+1)-n)h] \\ y[((l+1)(n+1)-1)h] & \dots & y[((l+1)(n+1)-n)h] \\ y[((l+2)(n+1)-1)h] & \dots & y[((l+2)(n+1)-n)h] \\ \vdots & \vdots & \vdots \end{pmatrix}. \tag{4.11}$$

Again any standard identification method (e.g., least squares) can be used to estimate α_i's.

To identify $\beta_0(z)$ and $\beta_1(z)$, we have by substituting (4.6) into (4.2) that

$$\beta_1(z)Y_0(z) - \beta_0(z)Y_1(z) = 0. \tag{4.12}$$

From equations that $\beta_0(z) = \beta_{01}z^{-1} + \cdots + \beta_{0n}z^{-n}$ and $\beta_1(z) = \beta_{11} + \cdots + \beta_{1n}z^{-n-1}$, we have in time domain

$$\bar{\Phi}(\beta_{11} \ \ \dots \ \ \beta_{1n} \ \ \beta_{01} \ \ \cdots \ \ \beta_{0n})' = 0 \tag{4.13}$$

where

$$\bar{\Phi} = \begin{pmatrix} y[kh] & \dots & -y[(k-n)h+h_o] \\ y[(k+1)h] & \dots & -y[(k-n+1)h+h_o] \\ y[(k+2)h] & \dots & -y[(k-n+2)h+h_o] \\ \vdots & \vdots & \vdots \end{pmatrix}. \quad (4.14)$$

Again, β_{0i} and β_{1i} appear linearly in (4.13). The difference between this equation and equation (4.9) is that the solution to this equation is unique only up to a scaling constant, provided that $\bar{\Phi}$ has full column rank. This is consistent with our goal of blind system identification, i.e., to find $G(z)$ up to a scaling constant. In fact, any vector in the null space of $\bar{\Phi}$ is a solution. There are many ways to solve this equation. The simpliest one is probably to normalize one of the components of $\beta_0(z)$ or $\beta_1(z)$. For instance, let $\beta_{01} = 1$, then we have

$$\phi'[kh](\beta_{11} \quad \cdots \quad \beta_{1n} \ \beta_{02} \quad \cdots \quad \beta_{0n})' = y[(k-1)h+h_o] \quad (4.15)$$

$$\phi'[kh] = (\,y[kh], ..., -y[(k-n)h+h_o]\,)$$

or in a compact form

$$\tilde{\Phi}(\beta_{11} \quad \cdots \quad \beta_{1n} \ \beta_{02} \quad \cdots \quad \beta_{0n})' = q \quad (4.16)$$

where

$$q = (\,y[(k-1)h+h_o], y[kh+h_o], y[(k+1)h+h_o], \dots)' \quad (4.17)$$

$$\tilde{\Phi} = \begin{pmatrix} y[kh] & \dots & -y[(k-n)h+h_o] \\ y[(k+1)h] & \dots & -y[(k-n+1)h+h_o] \\ y[(k+2)h] & \dots & -y[(k-n+2)h+h_o] \\ \vdots & \vdots & \vdots \end{pmatrix}. \quad (4.18)$$

Thus, β_{0i} and β_{1i} can be solved in a similar way as for equation (4.9). Moreover, the full-rankness of $\tilde{\Phi}$ can be established in a similar way as for all-pole systems. We summarize the results for blind system identification of the IIR system in the following theorem.

Theorem 4.1 *Consider the sampled data system with* $p = 2(n+1)$. *Let* $G_0(z)$ *and* $G_1(z)$ *be the transfer functions described above. Then,*

1. *The recursive least squares estimates* $\hat{\alpha}_i$*'s from equation (4.9) converge asymptotically to* α_i*'s if the spectral measure of the input* $u[kh_i]$ *is not concentrated on* $k < n$ *points.*

2. *The recursive least squares estimate* $(\hat{\beta}_{01}, \dots, \hat{\beta}_{0n})$ *and* $(\hat{\beta}_{11}, \dots, \hat{\beta}_{1n})$ *from equation (4.16) converges asymptotically to* $(\beta_{01}, \dots, \beta_{0n})$ *and* $(\beta_{11}, \dots, \beta_{1n})$ *up to a scaling constant, provided that the spectral measure of the input* $u[kh_i]$ *is not concentrated on* $k < 2n-1$ *points and that the numerator of the transfer functions* $G(z)$ *has coprime even and odd components.*

5. Time varying channels

As shown in the literature [2], for a large class of time-varying fading channels, the received signal samples from oversampling and/or multiple antenna sensors can be represented by

$$\vec{y}(k) = \sum_{m=0}^{L} \sum_{i=1}^{P} \vec{c}_i(m) e^{j\omega_i k} s(k-m) + w(k), \qquad s_k \in \mathcal{A}, \quad (5.1)$$

where $\mathcal{A}$ is the input signal set of the user from which channel input sequence $\{s_k\}$ is typically i.i.d. The noise $w(k)$ is stationary, white, and independent of channel input sequence s_k, but not necessarily Gaussian. $\vec{y}(k)$ and $\vec{c}_i(k)$ are $M-$dimensional vectors obtained from oversampling and a multiple sensor array.

In blind *equalization*, the desired objective is to recover each channel input, without necessarily needing to estimate the channel responses.

Denote

$$\vec{h}_i[m] \doteq c_i[m] e^{j\omega_i m}$$

and

$$s_i[k-m] \doteq s[k-m] e^{j\omega_i (k-m)}.$$

Then it follows that the received signal vector can be written as

$$\vec{y}[k] = \sum_{m=0}^{L} \sum_{i=1}^{P} \vec{h}_i[m] s_i[k-m] + w[k]$$

$$= \begin{pmatrix} h_1[0] \\ \vdots \\ h_1[L] \\ \vdots \\ h_P[0] \\ \vdots \\ h_P[L] \end{pmatrix} \begin{pmatrix} s_1[k] \\ \vdots \\ s_1[k-L] \\ s_2[k] \\ \vdots \\ s_P[k] \\ \vdots \\ s_P[k-L] \end{pmatrix} + w[k].$$

For simplicity, we first consider the development for noiseless channels. Collecting the received signal over K multiple period results in

$$Y[k] \doteq \begin{pmatrix} \vec{y}[k] \\ \vec{y}[k-1] \\ \vdots \\ \vec{y}[k-K+1] \end{pmatrix} = \mathbf{H} \underbrace{\begin{pmatrix} s_1[k] \\ \vdots \\ s_1[k-K-L+1] \\ \vdots \\ s_P[k]\vdots \\ s_P[k-K-L+1] \end{pmatrix}}_{\mathbf{s}[k]}.$$

for some $\mathbf{H}$. Notice that $\mathbf{H}$ is an $MK \times P(L+K)$ block Toeplitz matrix. So long as the channel diversity condition $M > P$ is satisfied, i.e., when the number of sub-channels exceeds the number of Doppler shifts, a sufficiently large

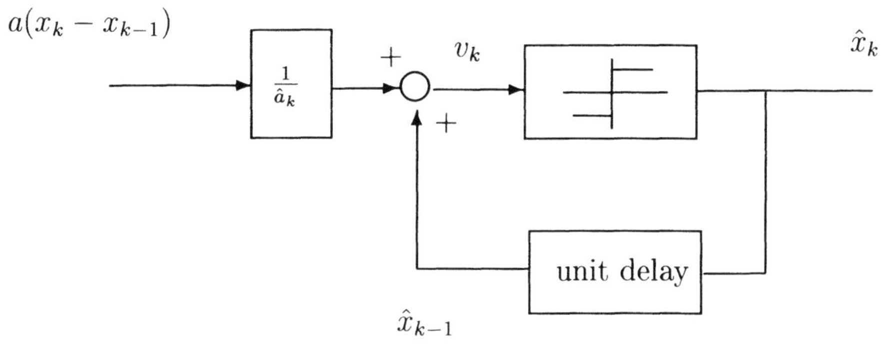

Figure 3: A simple decision feedback equalizer.

K will allow $\mathbf{H}$ to be over-determined. Throughout this paper, we assume that

- $\mathbf{H}$ has full column rank after removing all zero columns;
- $s[k]$ is i.i.d. with zero mean.

To identify $\mathbf{H}$ from $Y[k]$ without the assistance of training, observe that every entry in the outer-product $\mathbf{s}[k]\mathbf{s}[k]^H$ is of the form

$$s_i[k-m]s_q^*[k-n] = s[k-m]s^*[k-n]e^{j[\omega_i(k-m)-\omega_q(k-n)]}. \tag{5.2}$$

Thus under the ergodic condition

$$\lim_{N\to\infty} \frac{1}{N} \sum_{k=k_0+1}^{k_0+N} s_i[k-m]s_q^*[k-n] \tag{5.3}$$

$$= E\{|s(k)|^2\}\delta[m-n]\delta[i-q].$$

This implies that

$$R_Y = \mathbf{H}\left(\frac{1}{N}\sum_{k=k_0+1}^{k_0+N} \mathbf{s}[k]\mathbf{s}[k]^H\right)\mathbf{H}^H \to E\{|s_k|^2\}\mathbf{H}\mathbf{H}^H. \tag{5.4}$$

For simplicity, we write $s_k = s[k]$.

Without loss of generality, we assume that $E\{|s_k|^2\} = 1$. Let the singular value decomposition of $\mathbf{H}$ be

$$\mathbf{H} = U\begin{pmatrix}\Sigma & \\ & 0\end{pmatrix}V^H$$

where $U = [U_1\ U_2]$ and V are unitary matrices and Σ is a (full rank) diagonal matrix of singular values. Thus, the eigen-decomposition of R_Y is

$$R_Y = \mathbf{H}\mathbf{H}^H = U\Sigma^2U^H. \tag{5.5}$$

This allows U and Σ to be determined from R_Y. We can then form an output signal

$$\begin{aligned}\hat{\mathbf{s}}[k] &\doteq \Sigma^{-1}U_1^H Y[k] \\ &= V^H\mathbf{s}[k].\end{aligned} \tag{5.6}$$

Since V is unitary,

$$\hat{\mathbf{s}}^H[k]\hat{\mathbf{s}}[k-1] = \mathbf{s}^H[k]\mathbf{s}[k-1] \tag{5.7}$$

$$= \left(\sum_{i=1}^{P} e^{-j\omega_i}\right)\left[s_k^*s_{k-1} + \ldots s_{k-K-L+1}^*s_{k-K-L}\right].$$

Denoting the unknown scalar $a = \left(\sum_{i=1}^{P} e^{-j\omega_i}\right)$, we can generate a new signal

$$z_k = \hat{\mathbf{s}}^H[k]\hat{\mathbf{s}}[k-1] - \hat{\mathbf{s}}^H[k-1]\hat{\mathbf{s}}[k-2] \tag{5.8}$$

$$= a\left(s_k^*s_{k-1} - s_{k-K-L}^*s_{k-K-L-1}\right).$$

Given a constellation $\mathcal{A}$, we can estimate a from $\max Re\{z_k\}$, $\max Im\{z_k\}$, $\min Re\{z_k\}$, and $\min Im\{z_k\}$. Alternatively, we can use an adaptive decision feedback equalizer as described below.

6. Decision Feedback

It is apparent that to recover $s_k^*s_{k-1}$ from $z_k = a(s_k^*s_{k-1} - s_{k-K-L}^*s_{k-K-L-1})$ is generally impossible. However, we can simplify the task by recovering $s_k^*s_{k-1}$ from z_k. With the additional assumption that the input signal s_k is differentially encoded PSK, the input symbol information can be exactly retrieved from the successive phases of $s_k^*s_{k-1}$. For illustrate the concept, we will further assume, in this paper, that x_k is a data stream of i.i.d. equi-probable QPSK symbols randomly taking value $\pm1 \pm j$. This is only for simplicity and in fact a similar analysis can be carried out for many other constellations with non-equi-probable distributions.

To simplify notation, we will solve an equivalent problem, i.e, to determine x_k from $a(x_k - x_{k-1})$. The approach we will use is the decision feedback shown in Figure 3, where $a(x_k - x_{k-1})$, $a \neq 0$, is the received signal. a is possibly unknown.

The estimate $\hat{x}_k$ of x_k at time k is defined as

$$\hat{x}_k = sgnRe(v_k) + jsgnIm(v_k) \tag{6.1}$$

where sgn is defined as usual, namely

$$sgn(x) = \begin{cases} 1 & x \geq 0 \\ -1 & x < 0 \end{cases}.$$

If a is unknown, its estimate $\hat{a}_k$ can be obtained by the following adaptive algorithm

$$\hat{a}_{k+1} = \hat{a}_k + \epsilon(\hat{x}_k - \hat{x}_{k-1})^* [a(x_k - x_{k-1}) - \hat{a}_k(\hat{x}_k - \hat{x}_{k-1})] \tag{6.2}$$

where $\epsilon > 0$ is a small positive constant. The convergence analysis of the above decision feedback is divided into two parts: (1) known a and (2) unknown a.

6.1. Known a

If a is known, $\hat{a}_k$ can be set identically to a for all k and this implies

$$v_k = x_k - x_{k-1} + \hat{x}_{k-1}. \tag{6.3}$$

Define the error

$$\tilde{x}_k = x_k - \hat{x}_k \tag{6.4}$$

it follows that

$$\hat{x}_k = sgnRe(x_k - \tilde{x}_{k-1}) + jsgnIm(x_k - \tilde{x}_{k-1}). \tag{6.5}$$

Theorem 6.1 *Consider the decision feedback equalizer shown in Fig 1 and the adaptive algorithm (6.5). Suppose $\hat{a}_k \equiv a$ for all k. Then, with probability one as $k \to \infty$, $\hat{x}_k$ converges to x_k globally.*

6.2. Unknown a

If a is unknown, it is easily verified that all 4 points

$$\{\hat{a}_k = ae^{jn0.5\pi},\ \hat{x}_k = x_k e^{-jn0.5\pi}\}, \quad n = 0,\ 1,\ 2,\ 3, \tag{6.6}$$

are equilibrium points of the decision feedback equalizer shown in Fig 1. In other words, they are all stationary points of the following two equations

$$\hat{x}_k = sgnRe(v_k) + jsgnIm(v_k) \tag{6.7}$$

$$\hat{a}_{k+1} = \hat{a}_k + \epsilon(\hat{x}_k - \hat{x}_{k-1})^*[a(x_k - x_{k-1}) - \hat{a}_k(\hat{x}_k - \hat{x}_{k-1})]. \tag{6.8}$$

Clearly, in the case that a is unknown, the global convergence of $\hat{a}_k \to a$ is not possible. If it can be established that the these four points are the only possible equilibria, then the decision feedback equalizer will simply have

$$\hat{x}_k = x_k e^{-jn0.5\pi}, \quad n = 0,\ 1,\ 2,\ 3.$$

Since $x_k \in \{\pm 1 \pm j\}$, the rotated sequence $\hat{x}_k$ will have exactly the same statistical feature as x_k. Thus, this constant phase rotation is an intrinsic ambiguity in blind equalization systems. Nonetheless, because of differential encoding, the transmitted information is contained in the phase difference of successive symbols x_k. Thus, with differential encoding, this phase rotational ambiguity does not pose any problem to channel input signal recovery.

Thus, the desirable result is to show that these 4 equilibria are the only ones reachable under certain conditions. It is with this important observation in mind that we present the convergence results of $\hat{a}_k$.

Lemma 6.1 *Consider the decision feedback equalizer (6.1) and (6.2). Then,*

1. *$\hat{a}_k$ is bounded for any bounded initial $\hat{a}_0$.*
2. *If $|\hat{a}_k| \leq \frac{2\epsilon|a|}{1-4\epsilon}$, then $|\hat{a}_{k+1}| \geq |\hat{a}_k|$.*

The meaning of the lemma is that the estimate $\hat{a}_k$ is bounded and will not diverge. Moreover, there is a lower bound on $\hat{a}_k$ which roughly translates into $\hat{a}_k \not\to 0$ as $k \to \infty$.

Lemma 6.2 *Assume that $a \neq 0$. Define the set*

$$Q = \{\hat{a}\ ;\ |\hat{a}| < \sqrt{2}|a|\}. \tag{6.9}$$

Then, $\hat{a}_0 \in Q \Longrightarrow \hat{a}_k \in Q$ for all k.

The result says that if the estimate $\hat{a}_{k_0}$ lies inside the circle Q centered at the origin with radius $\sqrt{2}|a|$, then the estimate $\hat{a}_{k_0+k}$ will never leave the circle Q for any random sequence $\{x_k\}_0^\infty$. We now in a position to state the main result of the paper.

Theorem 6.2 *Consider the decision feedback equalizer and the adaptive algorithms (6.1) and (6.2). Assume $a \neq 0$ and the initial $\hat{a}_0 \in Q$. Then, with probability one as $k \to \infty$,*

$$(\hat{a}_k, \hat{x}_k) \to (ae^{j\theta}, x_k e^{-j\theta})$$

for some $\theta = 0,\ \frac{\pi}{2},\ \pi,\ \frac{3}{2}\pi$.

7. Concluding Remarks

In the paper, a novel approach has been proposed to blindly identify an IIR system. The key is to sample the output faster. We feel that the work reported in the paper is just a preliminary result on the problem of blind identification of an arbitrary IIR system, which we believe is largely an untouched field and deserves more work.

This paper also presents a new algorithm for the blind equalization of a class of cyclostationary time-varying wireless linear channels. Previous works require that the cyclic frequencies of the channel models be known or accurately estimated.

References

[1] E.W. Bai and M. Fu, "Blind system identification and channel equalization of IIR systems", IEEE Trans. on SP, July, 1999

[2] E.W. Bai and Z. Ding, "Blind Decision Feedback Equalization of Time-Varying Channels with DPSK Inputs", Technical reports, Dept. of ECE, University of Iowa, 1999

DIRECTION ESTIMATION OF COHERENT SIGNALS WITHOUT EIGENDECOMPOSITION AND SPATIAL SMOOTHING

Jingmin XIN [†] and **Akira SANO** [§]

[†] *YRP Mobile Telecommunications Key Technology Research Laboratories
3-4 Hikari-no-oka, Yokosuka 239-0847, Japan*
[§] *Department of System Design Engineering, Keio University
3-14-1 Hiyoshi, Kohoku-ku, Yokohama 223-8522, Japan*

Abstract: By exploiting the inherently temporal properties exhibited by most communication signals, in this paper, we propose a new cyclic approach for estimating the directions of coherent narrow-band signals without eigendecomposition and spatial smoothing processes. In which, the cyclostationarity and subarray scheme are combined to decorrelate the signal coherency and to suppress the noise and interfering signals, while the noise subspace is estimated from the resulting cyclic correlation matrix by a linear operation. As a result, the proposed approach has the advantages of computational simplicity and robust detection capability. The performance of the presented approach is verified through numerical examples. *Copyright © 2000 IFAC*

Keywords: Array processors, signal detection, subspace methods, communication environments, singular value decomposition

1. INTRODUCTION

In recent years, subspace-based methods for estimating the direction-of-arrival (DOA) of the signals impinge on an array of sensors have received a lot of attention due to their high-resolution abilities, such as MUSIC, ESPRIT, and minimum-norm methods. Most of these methods and their variants require either the eigenvalue decomposition (EVD) of a covariance matrix or the singular value decomposition (SVD) of a data matrix to estimate the signal or noise subspace. Unfortunately, these decomposition processes are computationally intensive and cumbersome, and the applications of these algorithms are limited in many practical situations. To alleviate this difficulty, many computationally more efficient techniques have been studied. The data adaptive eigenvector estimation algorithms were presented (Yang and Kaveh, 1988; Comon and Golub, 1990; Bischof and Shroff, 1992). The QR decomposition and fast subspace decomposition were used to replace the EVD or SVD for significantly computational reduction (Comon and Golub, 1990; Bischof and Shroff, 1992; Reilly, 1987; Prasad and Chandna, 1991). Several approximation methods were proposed to get signal or noise subspace without eigendecomposition, e.g. power method and Prony-Lanczos method (Tufts and Melissinos, 1986), eigenvalue filtering method (Kay and Shaw, 1988), discrete Fourier transform (DFT) based MUSIC method (Karhunen and Joutsensalo, 1992) and DFT-based minimum-norm method (Shaw and Xia, 1994). Recently some new subspace-based DOA estimation methods have been proposed to remove the need for computation of the EVD or SVD by using linear operations, such as BEWE (Yeh, 1987), propagation method (PM) (Marcos, *et al.*, 1994), SWEDE (Eriksson, *et al.*, 1994), and prediction filter method (Davila and Azmoodeh, 1994). In the BEWE, SWEDE and PM algorithms, the signal or noise subspace is estimated from a partition of the array covariance matrix or the array data by using linear operations, while in the prediction filter method, the signal subspace is obtained from the regularized linear prediction solutions to array data.

Most of subspace-based methods for DOA estimation almost always assume that the impinging signals are uncorrelated. They usually suffer serious degradation when the incident signals are mutually coherent in the practical scenario due to multipath propagation, which is often encountered in a variety of communication systems because of various reflections. For decorrelating the signal coherency, the popular spatial smoothing (SS) preprocessing is used to improve the estimation performance of subspace-based methods (Shan, *et al.*, 1985). Motivated by the PM (Marcos, *et al.*, 1994), a simpler algorithm was proposed for direction finding of coherent signals without eigendecomposition and spatial smoothing (Li, *et al.*, 1996). Although the computational load is reduced in this method, a separately placed single sensor is required to eliminate the effects of the additive noise. The BEWE can be used to resolve the coherent sources without spatial smoothing (Yeh, 1987), but the evaluation of array covariance matrix is needed.

Therefore the purpose of this paper is to investigate a new approach for DOA estimation of the coherent narrow-band signals in communication systems without eigendecomposition and spatial smoothing. In communications, most different types of modulated signals can have highly distinct cyclic correlation functions with a known cycle frequency due to the underlying periodicity arising from carrier frequencies or baud rates, while stationary noise and interference exhibit no cyclostationarity with the same cycle frequency (Gardner, 1991). By exploiting this inherently temporal characteristics exhibited by most communication signals, we have proposed a cyclic MUSIC algorithm (Xin, *et al.*, 1998) and a linear prediction method (Xin, *et al.*, 1999) to improve the resolution capability of DOA estimation of coherent signals, where the SVD is required to estimate subspace. In this paper, a new cyclic approach is developed for the localization of coherent signals in a more effective and less burdensome way. In which, the cyclostationarity and subarray scheme are combined to decorrelate the signal coherency and to suppress the noise and interfering signals, then the noise subspace is estimated from the resulting cyclic correlation matrix by considering the linear dependence between the rows of array response matrix. As a result, the better capability is attained to discriminate in favor of the desired coherent signals against noise and interference signals, and the computation load is efficiently reduced. The effectiveness is demonstrated and compared with the ordinary methods through numerical examples.

2. PROBLEM FORMULATION

2.1 Data Model

We consider q far-field narrow-band zero-mean signals impinge on an equally spaced linear array composed of M sensors with adjacent spacing d from the distinct directions $\theta_1,\cdots,\theta_q$ measured relative to the normal of array. The signal received at the ith sensor can be written as

$$x_i(n) = \sum_{k=1}^{q} s_k(n) e^{j\omega_0 (i-1)\tau(\theta_k)} + w_i(n) \qquad (1)$$

where $w_i(n)$ is the additive noise, $\omega_0 = 2\pi f_c$, $\tau(\theta_k) = (d/c)\sin\theta_k$, f_c and c are the carrier frequency and speed of propagation respectively.

From (1), the noisy measurements $\{x_i(n)\}$ can be expressed compactly

$$\boldsymbol{x}(n) = \boldsymbol{A}(\theta)\boldsymbol{s}(n) + \boldsymbol{w}(n) \qquad (2)$$

where $\boldsymbol{x}(n) = [x_1(n), x_2(n),\cdots,x_M(n)]^T$, $\boldsymbol{s}(n) = [s_1(n), s_2(n),\cdots,s_q(n)]^T$, $\boldsymbol{w}(n) = [w_1(n), w_2(n),\cdots,w_M(n)]^T$, $\boldsymbol{A}(\theta) = [\boldsymbol{a}(\theta_1), \boldsymbol{a}(\theta_2),\cdots,\boldsymbol{a}(\theta_q)]$, and $\boldsymbol{a}(\theta_k) = [1, e^{j\omega_0\tau(\theta_k)},\cdots, e^{j\omega_0(M-1)\tau(\theta_k)}]^T$.

In this paper, we assume that the number of coherent signals is q_α ($q_\alpha \le q$ and $q_\alpha < M$), then the coherent impinging signals can be expressed by

$$s_k(n) = \beta_k s_1(n), \qquad \text{for } k = 1,2,\cdots,q_\alpha \qquad (3)$$

where β_k is the multipath coefficient which represents the complex attenuation of the kth signal with respect to the first one $s_1(n)$ with $\beta_k \ne 0$ and $\beta_1 = 1$. The signal $s_1(n)$ exhibits cyclostationarity with the cycle frequency α, which can be determined from the carrier frequency and baud rate (Gardner, 1991), and we assume that the cycle frequency α is known. The additive noises $\{w_i(n)\}$ are assumed to be cyclically uncorrelated with themselves and with the signals $\{s_k(n)\}$ at the cycle frequency α.

In addition, the detection of number of signals is not discussed in this paper, we assume that it is known or it has been estimated. The objective of this paper is to estimate the directions $\{\theta_k\}$ of the coherent signals from the noisy array data $\{\boldsymbol{x}(n)\}_{n=0}^{N-1}$.

2.2 Cyclic Correlation of Array Data

It follows from (1) and (2) that the noise-corrupted signal $x_i(n)$ can be rewritten as

$$x_i(n) = \boldsymbol{b}_i^T(\theta)\boldsymbol{s}(n) + w_i(n) \qquad (4)$$

where $\boldsymbol{b}_i(\theta) = [e^{j\omega_0(i-1)\tau(\theta_1)}, e^{j\omega_0(i-1)\tau(\theta_2)},\cdots, e^{j\omega_0(i-1)\tau(\theta_q)}]^T$.

From the definition of the cyclic correlation (Gardner, 1991), and under the assumptions, we have the cyclic correlation $r^\alpha_{x_i,x_k}(\tau)$ between the noisy signals $x_i(n)$ and $x_k(n)$ in (1)

$$\begin{aligned} r^\alpha_{x_i,x_k}(\tau) &= \langle x_i(n) x_k^*(n+\tau) e^{-j2\pi\alpha n}\rangle \\ &= \langle \boldsymbol{b}_i^T(\theta)\boldsymbol{s}(n)\boldsymbol{s}^H(n+\tau)\boldsymbol{b}_k^*(\theta) e^{-j2\pi\alpha n}\rangle \\ &= \boldsymbol{b}_i^T(\theta)\boldsymbol{R}_s^\alpha(\tau)\boldsymbol{b}_k^*(\theta) \end{aligned} \qquad (5)$$

for $i,k = 1,2,\cdots,M$, where $\langle\cdot\rangle = \lim_{N\to\infty} 1/N \sum_{n=0}^{N-1}(\cdot)$, τ is the lag parameter, $\boldsymbol{R}_s^\alpha(\tau)$ is the cyclic source covariance matrix given by $\boldsymbol{R}_s^\alpha(\tau) = \langle \boldsymbol{s}(n)\boldsymbol{s}^H(n+\tau) \cdot e^{-j2\pi\alpha n}\rangle$, asterisk and H denote complex conjugate and conjugate transposition operations. Clearly the contributions of the arbitrary (not necessarily stationary and/or spatial white) noise and interference disappear in (5) if the cycle frequency α is appropriately selected, then the signal detection capability can be achieved.

Because the q_α signals $\{s_k(n)\}$ are coherent, like the source covariance matrix, the cyclic source covariance matrix $\boldsymbol{R}_s^\alpha(\tau)$ in (5) becomes singular, which is given

$$\begin{aligned} \boldsymbol{R}_s^\alpha(\tau) &= \langle \boldsymbol{\beta} s_1(n) s_1^*(n+\tau)\boldsymbol{\beta}^H e^{-j2\pi\alpha n}\rangle \\ &= r_s^\alpha(\tau)\boldsymbol{\beta}\boldsymbol{\beta}^H \end{aligned} \qquad (6)$$

where $\boldsymbol{\beta} = [\beta_1,\cdots,\beta_{q_\alpha},\beta_{q_\alpha+1},\cdots,\beta_q]$ with $\beta_{q_\alpha+1} = \cdots = \beta_q = 0$, $r_s^\alpha(\tau)$ is the cyclic auto-correlation of the signal $s_1(n)$ defined by $r_s^\alpha(\tau) = \langle s_1(n)s_1^*(n+\tau)e^{-j2\pi\alpha n}\rangle$. Thus the directions of the impinging signals cannot be estimated from the subspaces of the array covariance matrix or cyclic array covariance matrix directly.

For restoring the dimension of signal subspace, some modification schemes have been proposed such as spatial smoothing (Shan, *et al.*, 1985), which is a spatial averaging preprocessing that groups the total array into overlapping subarrays and then forms the average of the subarray covariance matrices to decorrelate the coherent signals. For estimating the directions of the coherent signals in a more effective and less burdensome way, in this paper, we consider the decorrelation of coherent signals by using the cyclostationary property of impinging signals and avoiding the computations of eigenvector and spatial smoothing.

3. DIRECTION ESTIMATION OF COHERENT SIGNALS

3.1 Decorrelation without Spatial Smoothing

By dividing the total array into L overlapping subarrays with m sensors ($m \geq q_\alpha$), *i.e.* the lth forward subarray comprises of sensors $l, l+1, \cdots, l+m-1$, where $L = M - m + 1$ and $L > m$, then the received signal vector $\boldsymbol{x}_{fl}(n)$ can be expressed by (Shan, *et al.*, 1985)

$$\boldsymbol{x}_{fl}(n) = \boldsymbol{A}_1(\theta)\boldsymbol{D}^{l-1}\boldsymbol{s}(n) + \boldsymbol{w}_{fl}(n) \tag{7}$$

where $\boldsymbol{x}_{fl}(n) = [x_l(n), x_{l+1}(n), \cdots, x_{l+m-1}(n)]^T$, $\boldsymbol{D} = \mathrm{diag}(e^{j\omega_0\tau(\theta_1)}, e^{j\omega_0\tau(\theta_2)}, \cdots, e^{j\omega_0\tau(\theta_q)})$, $\boldsymbol{w}_{fl}(n) = [w_l(n), w_{l+1}(n), \cdots, w_{l+m-1}(n)]^T$, and $\boldsymbol{A}_1(\theta)$ is the matrix consisting of the first m rows of the array response matrix $\boldsymbol{A}(\theta)$ in (2).

From (5)-(7), we obtain the cyclic correlation $\boldsymbol{\varphi}_{fl}(\tau)$ between the signal vector $\boldsymbol{x}_{fl}(n)$ and signal $x_M(n)$

$$\begin{aligned}\boldsymbol{\varphi}_{fl}(\tau) &= \left\langle \boldsymbol{x}_{fl}(n)x_M^*(n+\tau)e^{-j2\pi\alpha n}\right\rangle \\ &= \boldsymbol{A}_1(\theta)\boldsymbol{D}^{l-1}\left\langle \boldsymbol{s}(n)\boldsymbol{s}^H(n+\tau)\boldsymbol{b}_M^*(\theta)e^{-j2\pi\alpha n}\right\rangle \\ &= r_s^\alpha(\tau)\boldsymbol{A}_1(\theta)\boldsymbol{D}^{l-1}\boldsymbol{\beta}\boldsymbol{\beta}^H\boldsymbol{b}_M^*(\theta) \\ &= \rho_M r_s^\alpha(\tau)\boldsymbol{A}_1(\theta)\boldsymbol{D}^{l-1}\boldsymbol{\beta}\end{aligned} \tag{8}$$

where $\rho_M = \boldsymbol{\beta}^H\boldsymbol{b}_M^*(\theta)$.

By letting $l = 1$ to L, from (8), we have a matrix to express the cyclic correlations between the signal vectors $\{\boldsymbol{x}_{fl}(n)\}$ of L subarrays and signal $x_M(n)$ as

$$\boldsymbol{\Phi}_f = [\boldsymbol{\varphi}_{f1}(\tau), \boldsymbol{\varphi}_{f2}(\tau), \cdots, \boldsymbol{\varphi}_{fL}(\tau)]^T \tag{9}$$

For estimating the directions of the coherent signals from the cyclic matrix $\boldsymbol{\Phi}_f$, we have to verify the dimension of the signal subspace of $\boldsymbol{\Phi}_f$. The identifiable condition is given as follows.

Theorem: *If the numbers of the coherent signals is not larger than* $(M+1)/2$, *then the matrix* $\boldsymbol{\Phi}_f$ *will be of full rank.*

Proof: By substituting (8) into (9) and some manipulations, the matrix $\boldsymbol{\Phi}_f$ can be rewritten as

$$\begin{aligned}\boldsymbol{\Phi}_f &= \rho_M r_s^\alpha(\tau)\begin{bmatrix}\boldsymbol{\beta}^T\boldsymbol{A}_1^T(\theta) \\ \boldsymbol{\beta}^T\boldsymbol{D}\boldsymbol{A}_1^T(\theta) \\ \vdots \\ \boldsymbol{\beta}^T\boldsymbol{D}^{L-1}\boldsymbol{A}_1^T(\theta)\end{bmatrix} \\ &= \rho_M r_s^\alpha(\tau)\boldsymbol{A}_s(\theta)\boldsymbol{B}\boldsymbol{A}_1^T(\theta)\end{aligned} \tag{10}$$

where $\boldsymbol{B} = \mathrm{diag}(\beta_1, \beta_2, \cdots, \beta_q)$, and $\boldsymbol{A}_s(\theta)$ is the matrix consisting of the first L rows of the array response matrix $\boldsymbol{A}(\theta)$ in (2).

Because it has been assumed $\beta_k \neq 0$ for $k = 1,2,\cdots,q_\alpha$, then we have $\mathrm{rank}(\boldsymbol{B}) = q_\alpha$. As $\boldsymbol{A}_1(\theta)$ and $\boldsymbol{A}_s(\theta)$ are submatrices of the Vandermonde matrix $\boldsymbol{A}(\theta)$, their ranks are given by $\mathrm{rank}(\boldsymbol{A}_1(\theta)) = \min(m,q) \geq q_\alpha$ and $\mathrm{rank}(\boldsymbol{A}_s(\theta)) = \min(L,q)$. Hence if $L \geq q_\alpha$, we can obtain that $\mathrm{rank}(\boldsymbol{A}_s(\theta)) \geq q_\alpha$, and it follows that $\mathrm{rank}(\boldsymbol{\Phi}_f) = q_\alpha$. Since the number of forward subarrays is given by $L = M - m + 1$ and $m \geq q_\alpha$, the condition of full rank can be rewritten as $M - m + 1 \geq q_\alpha$. Therefore in order to ensure the full rank of matrix $\boldsymbol{\Phi}_f$, the number of coherent signals should satisfy the inequality $q_\alpha \leq (M+1)/2$. ■

As the dimension of signal subspace of matrix $\boldsymbol{\Phi}_f$ is q_α when $q_\alpha \leq (M+1)/2$, it is possible to estimate the DOA of coherent signal from (9). In the sequel, we will study the direction estimation by avoiding eigendecomposition process.

3.2 Estimation of Noise Subspace without Eigendecomposition

By adapting the idea of the linear operator suggested in PM (Marcos, *et al.*, 1994), we consider the following partition of the $L \times m$ matrix $\boldsymbol{\Phi}_f$ in (9)

$$\boldsymbol{\Phi}_f = \begin{bmatrix}\boldsymbol{\Phi}_{f1} \\ \boldsymbol{\Phi}_{f2}\end{bmatrix}\begin{matrix}\}\ m \\ \}\ L-m\end{matrix} \tag{11}$$

From the definition of the matrices $\boldsymbol{A}_1(\theta)$ and $\boldsymbol{A}_s(\theta)$, the $L \times q$ matrix $\boldsymbol{A}_s(\theta)$ can be partitioned similarly as

$$\boldsymbol{A}_s(\theta) = \begin{bmatrix}\boldsymbol{A}_1(\theta) \\ \boldsymbol{A}_2(\theta)\end{bmatrix}\begin{matrix}\}\ m \\ \}\ L-m\end{matrix} \tag{12}$$

Then from (10)-(12), the expressions of the submatrices $\boldsymbol{\Phi}_{f1}$ and $\boldsymbol{\Phi}_{f2}$ in (11) can be obtained

$$\boldsymbol{\Phi}_{f1} = \rho_M r_s^\alpha(\tau)\boldsymbol{A}_1(\theta)\boldsymbol{B}\boldsymbol{A}_1^T(\theta) \tag{13}$$

$$\boldsymbol{\Phi}_{f2} = \rho_M r_s^\alpha(\tau)\boldsymbol{A}_2(\theta)\boldsymbol{B}\boldsymbol{A}_1^T(\theta) \tag{14}$$

Hence we can find that there exists a $m \times (L-m)$ linear operator $\boldsymbol{P}$ between the matrices $\boldsymbol{\Phi}_{f1}$ and $\boldsymbol{\Phi}_{f2}$ as (Marcos, *et al.*, 1994)

$$\boldsymbol{P}^H\boldsymbol{\Phi}_{f1} = \boldsymbol{\Phi}_{f2} \tag{15}$$

By defining $\boldsymbol{Q} = [\boldsymbol{P}^T, -\boldsymbol{I}_{L-m}]^T$, (15) can be rewritten as

$$\boldsymbol{Q}^H\boldsymbol{\Phi}_f = \boldsymbol{O}_{(L-m)\times m} \tag{16}$$

where $\boldsymbol{I}_{L-m}$ and $\boldsymbol{O}_{(L-m)\times m}$ denote a $(L-m)\times(L-m)$ identity matrix and a $(L-m)\times m$ null matrix. Under the assumptions, from (10) and (16), we can obtain the following orthogonal relation

$$\boldsymbol{Q}^H\boldsymbol{A}_s(\theta) = \boldsymbol{O}_{(L-m)\times q} \tag{17}$$

Since the $L \times (L-m)$ matrix $\boldsymbol{Q}$ has rank equal to $L-m$, then (17) implies that the space spanned by the columns of the matrix $\boldsymbol{Q}$ is orthogonal to the array response vector $\boldsymbol{a}_s(\theta_k)$ given by $\boldsymbol{a}_s(\theta_k) = [1, e^{j\omega_0\tau(\theta_k)}, \cdots, e^{j\omega_0(L-1)\tau(\theta_k)}]^T$. In other words, a basis of the noise subspace can be estimated readily from the cyclic matrix $\boldsymbol{\Phi}_f$ without using eigendecomposition, and this result can be used to estimate the directions of coherent signals in a manner similar to MUSIC methods because $\boldsymbol{Q}^H\boldsymbol{a}_s(\theta_k) = \boldsymbol{0}_{(L-m)\times 1}$ for $k = 1,2,\cdots,q_\alpha$, where $\boldsymbol{0}_{(L-m)\times 1}$ is a $(L-m)\times 1$ null vector.

3.3 Direction Estimation with Forward and Back Subarrays

The total array can be also partitioned properly to

generate L backward subarrays, where the conjugate signal vector of the lth backward subarray is given by (Pillar and Kwon, 1989)

$$\boldsymbol{x}_{bl}(n) = \boldsymbol{A}_1(\theta)\boldsymbol{D}^{l-1}(\boldsymbol{D}^{M-1}\boldsymbol{s}(n))^* + \boldsymbol{w}_{bl}(n) \tag{18}$$

where $\boldsymbol{x}_{bl}(n) = [x^*_{M-l+1}(n), x^*_{M-l}(n), \cdots, x^*_{L-l+1}(n)]^T$, and $\boldsymbol{w}_{bl}(n) = [w^*_{M-l+1}(n), w^*_{M-l}(n), \cdots, w^*_{L-l+1}(n)]^T$.

Then we obtain the cyclic matrix $\boldsymbol{\Phi}_b$ between the signal $x_1(n)$ and the L signal vectors $\{\boldsymbol{x}_{bl}(n)\}$ of backward subarrays

$$\boldsymbol{\Phi}_b = [\boldsymbol{\varphi}_{b1}(\tau), \boldsymbol{\varphi}_{b2}(\tau), \cdots, \boldsymbol{\varphi}_{bL}(\tau)]^T$$

$$= \rho_1^* r_s^\alpha(\tau)\boldsymbol{A}_s(\theta)\boldsymbol{B}^*\boldsymbol{D}^{-(M-1)}\boldsymbol{A}_1^T(\theta) \tag{19}$$

where $\boldsymbol{\varphi}_{bl}(\tau) = \langle x_1(n)\boldsymbol{x}_{bl}(n+\tau)e^{-j2\pi\alpha n}\rangle$, and $\rho_1 = \boldsymbol{\beta}^H\boldsymbol{b}_1^*(\theta)$. We can prove that the rank of the matrix $\boldsymbol{\Phi}_b$ is given by $\mathrm{rank}(\boldsymbol{\Phi}_b) = q_\alpha$ if $q_\alpha \le (M+1)/2$.

Therefore by partitioning the matrix $\boldsymbol{\Phi}_b$ into two submatrices $\boldsymbol{\Phi}_{b1}$ and $\boldsymbol{\Phi}_{b2}$ as (11), from (10), (12) and (19), and by some manipulations, we have

$$[\boldsymbol{\Phi}_f, \boldsymbol{\Phi}_b]$$

$$= \boldsymbol{A}_s(\theta)[\rho_M r_s^\alpha(\tau)\boldsymbol{B}\boldsymbol{A}_1^T(\theta), \rho_1^* r_s^\alpha(\tau)\boldsymbol{B}^*\boldsymbol{D}^{-(M-1)}\boldsymbol{A}_1^T(\theta)]$$

$$= \begin{bmatrix} \boldsymbol{A}_1(\theta) \\ \boldsymbol{A}_2(\theta) \end{bmatrix} [\rho_M r_s^\alpha(\tau)\boldsymbol{B}\boldsymbol{A}_1^T(\theta), \rho_1^* r_s^\alpha(\tau)\boldsymbol{B}^*\boldsymbol{D}^{-(M-1)}\boldsymbol{A}_1^T(\theta)]$$

$$= \begin{bmatrix} \boldsymbol{\Phi}_{f1} & \boldsymbol{\Phi}_{b1} \\ \boldsymbol{\Phi}_{f2} & \boldsymbol{\Phi}_{b2} \end{bmatrix} = \begin{bmatrix} \boldsymbol{\Phi}_1 \\ \boldsymbol{\Phi}_2 \end{bmatrix} \tag{20}$$

Then we can estimate the $m \times (L-m)$ linear operator $\hat{\boldsymbol{P}}$ as

$$\hat{\boldsymbol{P}} = (\boldsymbol{\Phi}_1\boldsymbol{\Phi}_1^H)^{-1}(\boldsymbol{\Phi}_1\boldsymbol{\Phi}_2^H) \tag{21}$$

Since the columns of the projector $\boldsymbol{Q}$ in (17) which form the basis of noise subspace are not orthogonal, then the directions of the coherent signals can be estimated from the positions of the peaks of the spectrum given by (Marcos, *et al.*, 1994)

$$P(\theta) = \frac{1}{\boldsymbol{a}^H(\theta)\hat{\boldsymbol{Q}}_o\hat{\boldsymbol{Q}}_o^H\boldsymbol{a}(\theta)} \tag{22}$$

where $\boldsymbol{a}(\theta) = [1, e^{j\omega_0\tau(\theta)}, \cdots, e^{j\omega_0(L-1)\tau(\theta)}]^T$, $\tau(\theta) = (d/c)\sin\theta$, and the estimated orthogonal projector $\hat{\boldsymbol{Q}}_o$ is given by $\hat{\boldsymbol{Q}}_o = \hat{\boldsymbol{Q}}(\hat{\boldsymbol{Q}}^H\hat{\boldsymbol{Q}})^{-1/2}$.

3.4 Direction Estimation Algorithm

The algorithm for estimating the directions of coherent signals from the finite array data $\{x_1(n), x_2(n), \cdots, x_M(n)\}_{n=0}^{N-1}$ is summarized as follows.

Step 1: Set the subarray size m to satisfy $m \ge q_\alpha$, $L \ge q_\alpha$ and $L > m$, where $L = M - m + 1$.

Step 2: Calculate the estimates of the cyclic correlation functions $r^\alpha_{x_i,x_M}(\tau)$ and $r^\alpha_{x_1,x_k}(\tau)$ from the finite N received signals $\{\boldsymbol{x}(n)\}_{n=0}^{N-1}$ as

$$\hat{r}^\alpha_{x_i,x_k}(\tau) = \frac{1}{N}\sum_{n=0}^{N-1-\tau} x_i(n)x_k^*(n+\tau)e^{-j2\pi\alpha n}, \text{ for } \tau \ge 0 \tag{23}$$

$$\hat{r}^\alpha_{x_i,x_k}(\tau) = \frac{1}{N}\sum_{n=-\tau}^{N-1} x_i(n)x_k^*(n+\tau)e^{-j2\pi\alpha n}, \text{ for } \tau < 0$$

(24)

where $i = 1, 2, \cdots, M$ and $k = M$ for $r^\alpha_{y_i,y_M}(\tau)$, while $k = 1, 2, \cdots, M$ and $i = 1$ for $r^\alpha_{y_1,y_k}(\tau)$.

Step 3: Form the estimated cyclic matrices of the matrices $\boldsymbol{\Phi}_f$ in (9) and $\boldsymbol{\Phi}_b$ in (19) by using (23) and (24) as

$$\hat{\boldsymbol{\Phi}}_f = \begin{bmatrix} \hat{r}^\alpha_{x_1,x_M}(\tau) & \hat{r}^\alpha_{x_2,x_M}(\tau) & \cdots & \hat{r}^\alpha_{x_m,x_M}(\tau) \\ \hat{r}^\alpha_{x_2,x_M}(\tau) & \hat{r}^\alpha_{x_3,x_M}(\tau) & \cdots & \hat{r}^\alpha_{x_{m+1},x_M}(\tau) \\ \vdots & \vdots & \ddots & \vdots \\ \hat{r}^\alpha_{x_{M-m+1},x_M}(\tau) & \hat{r}^\alpha_{x_{M-m+2},x_M}(\tau) & \cdots & \hat{r}^\alpha_{x_M,x_M}(\tau) \end{bmatrix}$$

$$= \begin{bmatrix} \hat{\boldsymbol{\Phi}}_{f1} \\ \hat{\boldsymbol{\Phi}}_{f2} \end{bmatrix} \begin{matrix} \} \, m \\ \} \, L-m \end{matrix} \tag{25}$$

$$\hat{\boldsymbol{\Phi}}_b = \begin{bmatrix} \hat{r}^\alpha_{x_1,x_M}(\tau) & \hat{r}^\alpha_{x_1,x_{M-1}}(\tau) & \cdots & \hat{r}^\alpha_{x_1,x_{M-m+1}}(\tau) \\ \hat{r}^\alpha_{x_1,x_{M-1}}(\tau) & \hat{r}^\alpha_{x_1,x_{M-2}}(\tau) & \cdots & \hat{r}^\alpha_{x_1,x_{M-m}}(\tau) \\ \vdots & \vdots & \ddots & \vdots \\ \hat{r}^\alpha_{x_1,x_m}(\tau) & \hat{r}^\alpha_{x_1,x_{m-1}}(\tau) & \cdots & \hat{r}^\alpha_{x_1,x_1}(\tau) \end{bmatrix}$$

$$= \begin{bmatrix} \hat{\boldsymbol{\Phi}}_{b1} \\ \hat{\boldsymbol{\Phi}}_{b2} \end{bmatrix} \begin{matrix} \} \, m \\ \} \, L-m \end{matrix} \tag{26}$$

Step 4: Form the estimates of the matrices $\boldsymbol{\Phi}_1$ and $\boldsymbol{\Phi}_2$ in (20) from $\hat{\boldsymbol{\Phi}}_f$ and $\hat{\boldsymbol{\Phi}}_b$ as

$$\hat{\boldsymbol{\Phi}}_1 = [\hat{\boldsymbol{\Phi}}_{f1}, \hat{\boldsymbol{\Phi}}_{b1}] \tag{27}$$

$$\hat{\boldsymbol{\Phi}}_2 = [\hat{\boldsymbol{\Phi}}_{f2}, \hat{\boldsymbol{\Phi}}_{b2}] \tag{28}$$

Step 5: Calculate the linear operator $\hat{\boldsymbol{P}}$ and form the orthogonal matrix $\hat{\boldsymbol{Q}}$ as

$$\hat{\boldsymbol{P}} = (\hat{\boldsymbol{\Phi}}_1\hat{\boldsymbol{\Phi}}_1^H)^{-1}(\hat{\boldsymbol{\Phi}}_1\hat{\boldsymbol{\Phi}}_2^H) \tag{29}$$

$$\hat{\boldsymbol{Q}} = \begin{bmatrix} \hat{\boldsymbol{P}} \\ -\boldsymbol{I}_{L-m} \end{bmatrix} \tag{30}$$

Step 6: Estimate the DOA of the coherent signals from the q_α highest peak locations of the spectrum given by

$$P(\theta) = \frac{1}{\boldsymbol{a}^H(\theta)\hat{\boldsymbol{Q}}(\hat{\boldsymbol{Q}}^H\hat{\boldsymbol{Q}})^{-1}\hat{\boldsymbol{Q}}^H\boldsymbol{a}(\theta)} \tag{31}$$

3.5 Computational Complexity

Here we define one operation as one multiplication plus an optional addition, and make no distinctions between real and complex numbers, e.g. the inner production of two $M \times 1$ vectors (either real or complex) will be said to require M operations, *i.e.* $O(M)$. The spatial smoothing based DOA estimation method (Shan, *et al.*, 1985) involves $O(NM^2)$ operations for the evaluation of array covariance matrix in decorrelation step, where N is the number of snapshots, while the proposed approach requires $O(2NM)$ operations to construct the cyclic matrix in which the evaluation of covariance matrix is avoided. For subspace estimation step, the eigendecomposition used in MUSIC usually require $O(m^3)$ operations, whereas the proposed approach needs $O(m(M-m+1)^2)$ approximately (Marcos, *et al.*, 1994). Hence the proposed approach is computationally simpler than the method with eigen-decomposition and spatial smoothing.

As analyzed for the linear operations based methods (Eriksson, *et al.*, 1994; Marcos, *et al.*, 1994), we should be noted that the proposed approach is computationally much more efficient than the SS-based (cyclic) MUSIC and its variants but only slightly less accurate than SS-based cyclic MUSIC from the statistical standpoint.

4. NUMERICAL EXAMPLES

The effectiveness of the proposed approach is studied through several numerical examples, where the desired coherent binary phase-shift keying (BPSK) signals have raised cosine shape with 50% excess bandwith. The uniform linear array has $M=10$ sensors with half-wavelength spacing. The sensor outputs are collected at the rate $f_s = 8\text{MHz}$, the speed of propagation is $c = 3\times10^6 \text{m/s}$, the number of snapshots is $N=32$, and the lag parameter is chosen as $\tau = 3$. The additive noises are assumed to be complex white noise with the same variance, and the signal-to-noise ratio (SNR) is defined as the ratio of the signal power to the noise power at each sensor.

Example 1: Performance versus SNR

We examine the estimation performance of the proposed approach in terms of SNR. Two coherent BPSK signals (*i.e.* BPSK 1) with equal power impinge on the array from $\theta_1 = -15^o$ and $\theta_2 = 9^o$ with 1.6MHz baud rate ($\alpha = 0.2$ normalized to the sampling rate), while an interfering BPSK 2 signal comes from with 2.0MHz baud rate ($\alpha = 0.25$). Here the numbers of impinging and coherent signals are $q=3$ and $q_\alpha = 2$. The SNR of the coherent signals is varied from 0 dB to 20dB, while that of the interfering signal is fixed at 10dB, and the number of snapshots is $N=256$.

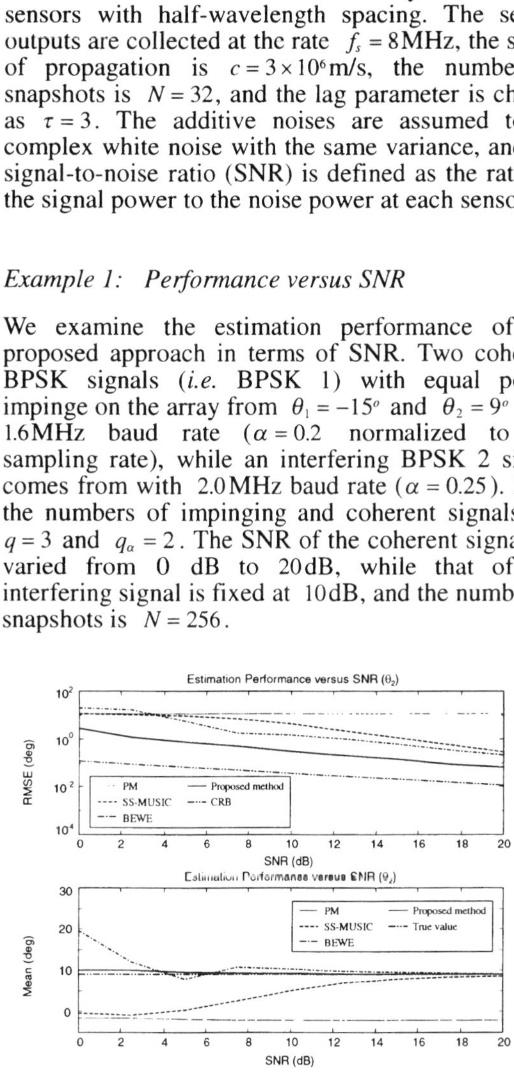

Fig. 1. RMSE and averaged estimates of $\theta_2 = 9^o$ versus SNR by using the PM, SS-MUSIC, BEWE and the proposed approach in Example 1.

We run 100 independent trials at each SNR, where the PM (Marcos, *et al.*, 1994), SS-MUSIC (Shan, *et al.*, 1985) and BEWE (Yeh, 1987) are also performed for comparison. In this example, the subarray size is $m=2$ for the proposed approach. The root mean-square-error (RMSE) and the averaged estimates of θ_2 are shown in Fig. 1, where the Cramer-Rao bounds (CRB's) (Stoica and Nehorai, 1989) and the true values are also plotted for reference. It is clear that the SS-MUSIC fails to give correct estimates even the optimal subarray size is chosen as $m_{opt} = 5$ (Gershman and Ermolaev, 1995) and the dimension of signal subspace is assumed as $q_\alpha = 2$, because it has no signal selectivity. Although the BEWE is computationally efficient than the SS-MUSIC to resolve the coherent signals, it cannot distinguish the desired coherent signals from the interfering signals as the SS-MUSIC. The PM has advantage of simpler computational complexity, but it cannot resolve the DOA estimation of coherent signals. However, since the inherently temporal property of the desired signals is exploited, the proposed cyclic approach outperforms the conventional subspace methods in suppression of additive noise and interference and has the reduced computational burden. The proposed approach can estimate the direction of coherent signals accurately (the results for $\theta_1 = -15^o$ are similar are omitted here).

Example 2: Performance versus Subarray Size

Here we assess the effect of the subarray size on the estimation performance of the proposed cyclic approach. The parameters for simulation are similar to those in Example 1, except that the subarray size is chosen as $m = 2, 3, 4$ and 5, *i.e.* the number of forward (backward) subarrays is $L = 9, 8, 7$ and 6.

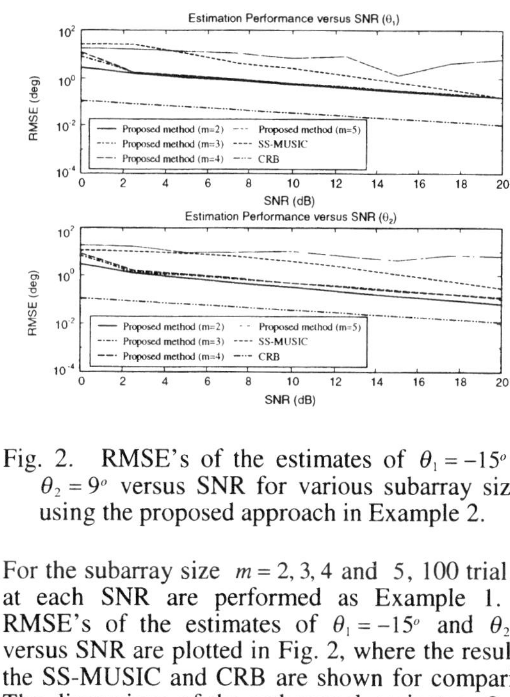

Fig. 2. RMSE's of the estimates of $\theta_1 = -15^o$ and $\theta_2 = 9^o$ versus SNR for various subarray size by using the proposed approach in Example 2.

For the subarray size $m = 2, 3, 4$ and 5, 100 trial runs at each SNR are performed as Example 1. The RMSE's of the estimates of $\theta_1 = -15^o$ and $\theta_2 = 9^o$ versus SNR are plotted in Fig. 2, where the results of the SS-MUSIC and CRB are shown for comparison. The dimensions of the orthogonal projector $\boldsymbol{Q}$ used in direction estimation can be rewritten as $(M-m+1)\times(M-2m+1)$. Thus for improving the estimation performance, the subarray size m ($m \geq q_\alpha$) should be chosen as small as possible in order to increase the effective aperture. We can find that the choice of subarray size m can significantly improve the estimation performance, and the best estimation is attained when the subarray size m equals the number of the desired coherent signals, *i.e.* $m = q_\alpha$.

Example 3: Performance versus Snapshot Number

Now we study the estimation performance of the proposed approach in terms of the number of snapshots N. The simulation conditions are similar to those in Example 1, except that the SNR of two coherent signals is fixed at 10 dB, and the number of snapshots is varied as $N = 16, 32, 64, 128, 256, 1024$ and 2048.

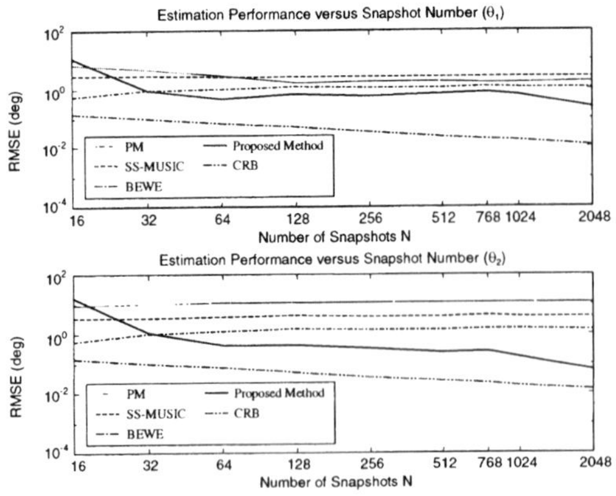

Fig. 3. RMSE's of the estimates of $\theta_1 = -15^o$ and $\theta_2 = 9^o$ versus number of snapshots by using the PM, SS-MUSIC, BEWE and the proposed approach in Example 3.

One hundred independent trials are carried out for each value of N. The RMSE's of the estimates of $\theta_1 = -15^o$ and $\theta_2 = 9^o$ versus the number of snapshots are shown in Fig. 3, where they are compared with that of the PM, SS-MUSIC and BEWE and the CRB is also plotted. As described in previous examples, the PM fail completely because it is unable to resolve the coherent signals, while the SS-MUSIC and BEWE cannot give correct estimates of the desired signals since the temporal property of the impinging signals is not exploited. It is shown that the proposed approach can estimate the directions of the two coherent signals accurately as the number of snapshots is increased.

5. CONCLUSIONS

In some applications of array processing such as mobile communication systems, computationally much more efficient algorithms for estimating the directions of the impinging signals are desired. In this paper, a new approach was developed for DOA estimation of narrow-band cyclostationary signals in in multipath propagation. In the proposed approach, the cyclostationarity of the modulated signals and the linear algebraic relations of array response matrix are exploited, and then the eigendecomposition and spatial smoothing processes are avoided to reduce the computational burden. The proposed approach has an advantage of the computational simplicity over the eigendecomposition based subspace methods and that of selective detection capability over the existing linear operations based subspace methods. The effectiveness of the presented approach was verified through numerical examples.

REFERENCES

Bischof, C.H. and G.M. Shroff (1992). On updating signal subspace. *IEEE Trans. Signal Processing*, **40/1**, 96-105.

Comon, P. and G.H. Golub (1990). Tracking a few extreme singular values and vectors in signal processing. *Proc. IEEE*, 78/8, 1327-1343.

Davila, C.E. and M. Azmoodeh (1994). Efficient estimation of the signal subspace without eigendecomposition. *IEEE Trans. Signal Processing*, **42/1**, 236-239.

Eriksson, A., P. Stoica and T. Söderström (1994). On-line subspace algorithms for tracking moving sources. *IEEE Trans. Signal Processing*, **42/9**, 2319-2330.

Gardner, W.A. (1991). Exploitation of spectral redundancy in cyclostationary signals. *IEEE Signal Processing Mag.*, **8**, 14-37, April.

Gershman, A.B. and V.T. Ermolaev (1995). Optimal subarray size for spatial smoothing. *IEEE Signal Processing Letters*, **2/2**, 28-30.

Karhunen, J.T. and J. Joutsensalo (1992). Sinusoidal frequency estimation by signal subspace approximation. *IEEE Trans. Signal Processing*, **40/12**, 2961-2972.

Kay, S.M. and A.K. Shaw (1988). Frequency estimation by principle component AR spectral estimation method without eigencomposition. *IEEE Trans. Acoust., Speech, Signal Processing*, **36/1**, 95-101.

Li, P., J. Sun and B. Yu (1996). Two-dimensional spatial-spectrum estimation of coherent signals without spatial smoothing and eigen-decomposition. *IEE Proc.-Radar, Sonar Navig.*, **143/5**, 295-299.

Marcos, S., A. Marsal and M. Benidir (1994). The propagator method for source bearing estimation. *Signal Processing*, **42/2**, 121-138.

Pillar, S.U. and B.H. Kwon (1989). Forward/backward spatial smoothing techniques for coherent signal identification. *IEEE Trans. Acoust., Speech, Signal Processing*, **37/1**, 8-15.

Prasad, S. and B. Chandna (1991). Direction-of-arrival estimation using rank revealing QR factorization. *IEEE Trans. Signal Processing*, **39/5**, 1224-1229.

Reilly, J.P. (1987). A real-time high resolution technique for angle-of-arrival estimation. *Proc. IEEE*, **75/12**, 1692-1694.

Shan, T.-J., M. Wax and T. Kailath (1985). On spatial smoothing for direction-of-arrival estimation of coherent signals. *IEEE Trans. Acoust., Speech, Signal Processing*, **33/4**, 806-811.

Stoica, P. and A. Nehorai (1989). MUSIC, maximum likelihook, and Cramer-Rao bound. *IEEE Trans. Acoust., Speech, Signal Processing*, **37/5**, 720-741.

Tufts, D.W. and C.D. Melissinos (1986). Simple, effective computation of principal eigenvectors and their eigenvalues and application to high-resolution estimation of frequencies. *IEEE Trans. Acoust., Speech, Signal Processing*, **34/5**, 1046-1053.

Yang, J.-F. and M. Kaveh (1988). Adaptive eigensubspace algorithms for direction or frequency estimation and tracking. *IEEE Trans. Acoust., Speech, Signal Processing*, **36/2**, 241-251.

Yeh, C.-C. (1987). Simple computation of projection matrix for bearing estimation. *IEE Proc., Pt. F*, **134/2**, 146-150.

Xin, J., H. Tsuji, Y. Hase and A. Sano (1998). Directions-of-arrival estimation of cyclo-stationary coherent signals in array processing. *IEICE Trans. Fundamentals*, **E81-A/8**, 1560-1569.

Xin, J., H. Tsuji, Y. Hase and A. Sano (1999). Localization of cyclostationary signals in multipath based on linear prediction. *Proc. IFAC 14th World Congress*, **I**, 13-18.

INFORMATION CRITERION FOR MODEL SELECTION WITH CONTROLLER DESIGN

Kouji Tsumura *

* *Department of Mathematical Engineering and Information Physics*
The University of Tokyo
Hongo 7-3-1, Bunkyo-ku, Tokyo 113-8656, Japan
tsumura@crux.t.u-tokyo.ac.jp

Abstract: In this paper, an information criterion based on AIC/TIC is proposed which addresses not only on the estimation and selection of model for plant but also on designing of controller. A key point is that the model is estimated on the space of control performance, and the resultant criterion also contains a bias term as AIC/TIC however it depends on the controller. Moreover, when the controller is also designed with stochastic data generated by the estimated model, the criterion is added by another bias term, which is proportional to the dimension of the control parameter. *Copyright © 2000 IFAC*

Keywords: information criterion, robust control, identification, complexity

1. INTRODUCTION

In the field of robust control and the corresponding identification, the necessity of probabilistic evaluation in some sense for the uncertainties of controlled systems and the performances of the closed loop systems has been actively discussed in the last several years (Dahleh *et al.*, 1993; Vidyasagar, 1997; Tempo *et al.*, 1996). One of the motivations of these researches is to avoid the conservatism of the uncertainties which are derived from identification, and another is the application of the Monte Carlo method for robust control problem, which is hard to be solved analytically. These researches in the field of identification and design have been advanced independently, however the modeling of systems and the attained performance of the closed loop system which is designed from the model, essentially have deep relations each other. For example, the complexity of model directly influences that of controller such as its order or the number of free parameters, and it is conjectured that the complexity of controller and the reliability of the attained performance of the resultant closed loop systems will conflict, when the controlled systems are evaluated in probablistic sense. These facts say that when we apply probablistic evaluation for the system uncertainties and the attained control performances, the evaluation should be to address both of them simultaneously.

On the other hand, in the field of statistics or coding, several information criteria for model estimation and selection, such as AIC/TIC, MDL and so on, are proposed in connection with information and complexity (Akaike, 1974; Takeuchi, 1976; Rissanen, 1984), where the criteria can make decision which models are appropriate for the objective systems by using given data. In particular, the well-known results, which show a trade-off between consistency of the estimated model for the given data and its complexity, are essence of model estimation and selection. However note that their focuses have been only on model estimation and selection problem. In controller designing, the most important point is the performance of closed loop systems, and when some model, which is selected by the information criterion, is used for the controller designing, the resultant closed loop system or its performance should be consistently evaluated by the same paradigm or the criterion. However the connection between the selected model and the resultant closed loop system has not been investigated in the fields of statistics or coding.

On these backgrounds, in this paper, an information criterion is proposed for model estimation and selection problem in consideration of controller designing. The criterion is a derivation of AIC/TIC basically, but it reflects the selection of controller or its class. A key point is the transformation from the modeling of the distribution of systems itself to that of performance indices which are consequently attained with the systems and a controller. Moreover, it is shown that when a controller is designed by using the estimated model from the data, the complexity of the controller, that is, the number of its parameter, influences on the estimation results of the model in a probablistic sense. These results reveal the indivisibility of the modeling of the objective systems and controller designing.

2. MEASURABLE SPACE AND VARIABLE TRANSFORMATION

When an uncertainty addressed in robust control is not given a priori, a possible process we can take may be identification. This means that a model set of systems is derived from a data set, and then a controller is designed from the model set. Here it should be noted that the final purpose of control is the performance of the closed loop system, the model set which has a consistency with the given data, should be modeled with respect to the resultant control performance (Tsumura and Shin, 1995; Tsumura and Shin, 1996; Tsumura and Shin, 1997).

With this discussion, here it is considered to transform the modeling of the distribution of systems in a sense of probability directly to that of objective performance indices which are attained in the closed loop system as a result. For example, let x be an index of system, X the set of x, $f(x)$ a probability density, and γ a performance index of a closed loop system composed of the system x and a controller ϕ. Note that γ is a function of x and ϕ, in other word, the controller ϕ is considered as a mapping $\phi : X \to \Gamma$. Corresponding to this mapping, the probability density $f(x)$ on a set of system index is transformed to that on a space of the performance index as $\phi : f(x) \to g(\gamma)$. And then it is natural to consider the modeling of $g(\gamma)$ substitute for that of $f(x)$. For the analysis along this discussion, several formulae are needed for the variable transformation on measurable spaces, and next show several basic notations (Lasota and Mackey, 1995) about variable transformation of probability density for the preparation of the following sections.

At first define two measurable spaces, $(\mathcal{A}, X, \mu)$, $(\mathcal{B}, Y, \nu)$, where X, Y are sets, $\mathcal{A}$, $\mathcal{B}$ are σ-algebras composed of all the subsets of X and Y, μ, ν are measures on X, Y, and f, g are probability densities on $(\mathcal{A}, X, \mu)$, $(\mathcal{B}, Y, \nu)$. Also define a transformation $S : X \to Y$. Here the following holds (Lasota and Mackey, 1995).

Proposition 1. For any nonsingular transformation S: $X \to Y$, there exists a unique operator $P : \mathcal{L}^1 \to \mathcal{L}^1$ for all $B \in \mathcal{B}$ such that

$$\int_{B \in \mathcal{B}} Pf(y)\nu(dy) = \int_{S^{-1}(B)} f(x)\mu(dx). \qquad (1)$$

Definition 2. Let $(\mathcal{A}, X, \mu)$, $(\mathcal{B}, Y, \nu)$ be measurable spaces. For a nonsingular measurable transformation $S : X \to Y$, the unique operator $P : \mathcal{L}^1 \to \mathcal{L}^1$ which satisfies (1) is called a Frobenius-Perron operator.

Proposition 3. For a nonsingular measurable transformation $S : X \to Y$, the following holds

$$\int_{S^{-1}(B)} f(x)\mu(dx) = \int_B f(S^{-1}(y))\mu S^{-1}(dy)$$
$$= \int_B f(S^{-1}(y))J^{-1}(y)\nu(dy),$$

where μS^{-1}, J^{-1} are defined as

$$\mu S^{-1}(B) = \int_B J^{-1}(y)\nu(dy), \ B \in \mathcal{B}.$$

From this proposition, the next representation of the Frobenius-Perron operator follows.

Proposition 4. For an invertible nonsingular measurable transformation $S : X \to Y$, let P be the corresponding Frobenius-Perron operator. Then, the following holds.

$$Pf(y) = f(S^{-1}(y))J^{-1}(y)$$

Moreover, when $S(y)^{-1}$ is differentiable, $J^{-1}(y)$ and $J(x)$ are given by the absolute values of the Jacobians of $S^{-1}(y)$ and $S(x)$ as

$$J^{-1}(y) = \left| \frac{dS^{-1}(y)}{dy} \right|, \ J(x) = \left| \frac{dS(x)}{dx} \right|.$$

In this paper, it is considered that μ, ν are Lebesgue measures.

3. INFORMATION CRITERION

In this section, an information criterion such as AIC or TIC will be introduced for model estimation and selection problem. At first define several notions. Let x be an index of systems and also a variable of stochastic process, $g(x)$ the true probability density of x and unknown, $f_\theta(x)$ a model of $g(x)$ which is parametrized by θ, $x^{(n)} := \{x_1, x_2, \ldots, x_n\}$ sample data of x, γ a performance index of a closed loop system.

Based on the Kullback-Leibler information:

$$\int g(x) \log \frac{g(x)}{f_\theta(x)} dx, \qquad (2)$$

which is considered to be appropriate for the distance of two probability densities (Kullback, 1959), the consistency of probability density $f_\theta(x)$ and the true density $g(x)$ is evaluated by the value of

$$\int g(x) \log f_\theta(x) dx, \qquad (3)$$

and the maximizer $f_\theta(x)$ of (3) is considered to be a good model for $g(x)$. However because the true density $g(x)$ is unknown, it should be substituted to the empirical distribution and consider the following maximization problem of the following likelihood function for (3).

$$\frac{1}{n}\sum_{i=1}^{n} \log f_\theta(x_i) \qquad (4)$$

This is called the maximum likelihood method (Kullback, 1959).

In usual, the modeling of $f_\theta(x)$, which is a distribution function of the system index x, is considered, but following the discussion in Section 2, it is natural to consider the modeling of the distribution of the performance index γ substitute for that of x from the point of control problem. Here note that γ is defined by a system x and a controller. When a controller is fixed, γ of the closed loop system composed of the controller and system x can be figured out, and then a mapping

$$S_\phi(x) = \gamma, \qquad (5)$$

from x to γ for a controller ϕ is derived. That is, a controller ϕ is regarded to decide a mapping from a system index to a performance index. In the following of this paper, S_ϕ is supposed to be an invertible non-singular measurable transformation. This condition seems to be somewhat restrictive and in some cases it does not hold. However in usual cases of control problem, by introducing appropriate deformation of the space of γ, this condition holds. Here omit the detail of such manipulation.

Corresponding the variable transformation (5), the probability density $f_\theta(x)$ is transformed to $P_\phi f_\theta(\gamma)$ where the Frobenius-Perron operator P_ϕ is defined by the mapping (5), and therefore the evaluation (3) is also transformed to

$$\int \hat{g}_\phi(\gamma) \log P_\phi f_\theta(\gamma) d\gamma, \qquad (6)$$

where $\hat{g}_\phi(\gamma)$ is the unknown true density of γ. And now consider the problem:

$$\max_\theta \sum_{i=1} L_{\phi,\theta}(\gamma_i) \qquad (7)$$

$$L_{\phi,\theta}(\gamma^{(n)}) := \frac{1}{n}\sum_{i=1}^{n} \log P_\phi f_\theta(\gamma_i), \qquad (8)$$

where $\gamma_i := S_\phi(x_i)$ and $\gamma^{(n)} := \{\gamma_1, \gamma_2, \ldots, \gamma_n\}$, substitute for the maximization of (4). Here the following two facts should be noted:

1. the constant value

$$\int g(x) \log g(x) dx \qquad (9)$$

   in the normal model estimation may be changed by the transformation of the variable x.
2. However, the Kullback-Leibler information (2) itself is constant for the transformation of the variables by a cancellation between the numerator and the denominator in (2).

These facts seemingly say that the maximization of (6) or (7) is unreasonable, however note that (6) can be decomposed as

$$\int \hat{g}_\phi(\gamma) \log P_\phi f_\theta(\gamma) d\gamma$$
$$= \int \hat{g}_\phi(\gamma) \log \hat{g}_\phi(\gamma) d\gamma - \int g(x) \log \frac{g(x)}{f_\theta(x)} dx. \qquad (10)$$

The first term is the minus of the entropy of the distribution of the resultant performance index and the second term is the minus of the normal Kullback-Leibler information between the true density $g(x)$ and the estimation $f_\theta(x)$. Therefore, the maximization of (10) can be interpreted as a simultaneous optimization to restrict the dispersion of the resultant performance and to approximate $g(x)$ by $f_\theta(x)$. In particular, to restrict the dispersion of the performance is significant for the robust control. From these reasons, the current problem is concluded to be sufficiently meaningful.

By employing the formulae on variable transformation introducing in Section 2, $L_{\phi,\theta}(\gamma^{(n)})$ is written as

$$L_{\phi,\theta}(\gamma^{(n)}) = \frac{1}{n}\sum_{i}^{n} \log P_\phi f_\theta(\gamma_i)$$
$$= \frac{1}{n}\sum_{i=1}^{n} \log f_\theta(x_i) + \frac{1}{n}\sum_{i=1}^{n} \log K_\phi(x_i), \qquad (11)$$
$$K_\phi(x) := J_\phi^{-1}(S_\phi(x)). \qquad (12)$$

The equations (7) and (11) show the followings. When ϕ does not depend on θ, that is, the controller ϕ is designed regardless of the system distribution $f_\theta(x)$, then, the maximization of (11) is identical to that of (4) and it is nothing but the ordinary modeling and estimation problem of system distribution by using data. Of course, it is usual case that the controller is designed with respect to the system distribution model $f_\theta(x)$, and the maximum likelihood of (7) is different to (4) at that time. In the following of this paper, it is considered that $\phi(\theta)$ is a function of θ.

Now define

$$\hat{\theta} := \arg\max_\theta \frac{1}{n}\sum_{i=1}^{n} \log P_{\phi(\theta)} f_\theta(\gamma_i) \qquad (13)$$

$$\theta_o := \arg\max_{\theta} \int \hat{g}_\phi(\gamma) \log P_{\phi(\theta)} f_\theta(\gamma_i) d\gamma \quad (14)$$

Next consider to evaluate the maximum likelihood model $f_{\hat{\theta}}$ in the sense of (6). The value of (6) at $\theta = \hat{\theta}$ can be reduced to that of

$$\int \left\{ \log f_{\hat{\theta}}(x) + \log K_{\phi(\hat{\theta})}(x) \right\} g(x) dx. \quad (15)$$

Of course (15) can not be calculated because $\hat{g}_\phi(\gamma)$ is unknown and it should be approximated. The Akaike's information criterion AIC (Akaike, 1974) or the Takeuchi's information criterion TIC (Takeuchi, 1976; Konishi and Kitagawa, 1996) is known as an approximation of (3) for the maximum likelihood model. By following the process of deriving AIC/TIC, here the following proposition is given.

Proposition 5. Define the followings,

$$\Psi(\hat{\theta}, \theta_o) := \frac{1}{n}\sum_{i=1}^{n} \left(\log f_{\hat{\theta}}(x_i) + \log K_{\phi(\hat{\theta})}(x_i) \right) - \frac{1}{n} \mathrm{tr}\, J(\phi, \theta_o)^{-1} I(\phi, \theta_o) \quad (16)$$

$$J(\phi, \theta) := -\int \frac{\partial^2}{\partial\theta\partial\theta'} \log\left\{ f_\theta(x) K_{\phi(\theta)}(x) \right\} \times g(x) dx$$

$$I(\phi, \theta) := \int \frac{\partial}{\partial\theta} \log\left\{ f_\theta(x) K_{\phi(\theta)}(x) \right\} \times \frac{\partial}{\partial\theta'} \log\left\{ f_\theta(x) K_{\phi(\theta)}(x) \right\} g(x) dx,$$

then, the next holds.

$$\mathbf{E}\left[\Psi(\hat{\theta}, \theta_o) - \int \left\{ \log f_{\hat{\theta}}(x) + \log K_{\phi(\hat{\theta})}(x) \right\} \times g(x) dx \right] = o\left(n^{-1}\right) \quad (17)$$

PROOF. (outline)

Now consider the Taylor expansion of (15) at $\theta = \theta_o$,

$$\begin{aligned} &\int \left\{ \log f_{\hat{\theta}}(x) + \log K_{\phi(\hat{\theta})}(x) \right\} g(x) dx \\ &= \int \left\{ \log f_{\theta_o}(x) + \log K_{\phi(\theta_o)}(x) \right\} g(x) dx \\ &\quad - \frac{1}{2}(\hat{\theta} - \theta_o)' J(\phi, \theta_o)(\hat{\theta} - \theta_o) + o(n^{-1}). \end{aligned} \quad (26)$$

Here the first term in the right hand side of the above equation can be rewritten as follows.

$$\begin{aligned} &\int \left\{ \log f_{\theta_o}(x) + \log K_{\phi(\theta_o)}(x) \right\} g(x) dx \\ &= \frac{1}{n}\sum_{i}^{n} \left(\log f_{\hat{\theta}}(x_i) + \log K_{\phi(\hat{\theta})}(x) \right) \\ &\quad - \frac{1}{2}(\theta_o - \hat{\theta})'(J(\phi, \theta_o) + o(1))(\theta_o - \hat{\theta}) \\ &\quad + M + o(n^{-1}), \end{aligned} \quad (27)$$

where

$$M := \int \left(\log f_{\theta_o}(x) + \log K_{\phi(\theta_o)}(x) \right) g(x) dx - \frac{1}{n}\sum_{i}^{n} \left\{ \log f_{\theta_o}(x_i) + \log K_{\phi(\theta_o)}(x_i) \right\} \quad (28)$$

Substituting (27) to (26), we get

$$\begin{aligned} &\int \left\{ \log f_{\hat{\theta}}(x) + \log K_{\phi(\hat{\theta})}(x) \right\} g(x) dx \\ &= \frac{1}{n}\sum_{i}^{n} \left(\log f_{\hat{\theta}}(x_i) + \log K_{\phi(\hat{\theta})}(x_i) \right) \\ &\quad - (\hat{\theta} - \theta_o)'(J(\phi, \theta_o) + o(1))(\hat{\theta} - \theta_o) \\ &\quad + M + o(n^{-1}). \end{aligned} \quad (29)$$

Moreover,

$$\int (\hat{\theta} - \theta_o)'(J(\phi, \theta_o) + o(1))(\hat{\theta} - \theta_o) g(x^{(n)}) dx^{(n)} = -\frac{1}{n} \mathrm{tr}\, J(\phi, \theta_o)^{-1} I(\phi, \theta_o) + o(n^{-1}) \quad (30)$$

$$\int M g(x^{(n)}) dx^{(n)} = 0, \quad (31)$$

where $g(x^{(n)})$ means the joint density of $x^{(n)}$, and we get

$$\begin{aligned} &\Psi(\hat{\theta}, \theta_o) - \int \left\{ \log f_{\hat{\theta}}(x) + \log K_{\phi(\hat{\theta})}(x) \right\} g(x) dx \\ &= -\frac{1}{n} \mathrm{tr}\, J^{-1}(\phi, \theta_o) I(\phi, \theta_o) + (\hat{\theta} - \theta_o)'(J(\phi, \theta_o) \\ &\quad + o(1))(\hat{\theta} - \theta_o) - M + o(n^{-1}). \end{aligned} \quad (32)$$

Finally take the average of (32) for $x^{(n)}$ and the proposition is given. □

The equation (17) shows the properness of using $\Psi(\hat{\theta}, \theta_o)$ as an estimation of (15). However at first θ_o is unknown and $\Psi(\hat{\theta}, \hat{\theta})$ is considered as a possible estimation. Moreover $J(\phi, \theta)$, $I(\phi, \theta)$ are also unknown and their average based on the empirical data is considerable to be used. Further approximation is also applicable but such discussion is not essential and here omit.

Note that $\Psi(\hat{\theta}, \hat{\theta})$ is identical to TIC except for the ϕ-depended terms, and moreover it is well-known that when $f_{\theta_o}(x) = g(x)$, then,

$$\frac{1}{n} \mathrm{tr}\, J(\phi, \theta_o)^{-1} I(\phi, \theta_o) = \frac{\dim\theta}{n}$$

and TIC is equal to AIC. Note that $\Psi(\hat{\theta}, \hat{\theta})$ depends not only on $f_\theta(x)$ but also the controller $S_\phi(x)$. This suggests that when the model is estimated or selected based on $\Psi(\hat{\theta}, \hat{\theta})$, the selected model or its estimation value will changes depending on the class of the selected controller and the design method.

4. SELECTION OF CONTROLLER DESIGNING

In the previous section, an information criterion is induced for model estimation and selection. In this section, at first define concrete controller designs and then extend the above derived criterion to new one. At first show one of considerable design method of controller $S_{\phi(\theta)}$.

Let $P_\phi f_\theta(\gamma)$ be a performance distribution of a closed loop system which depends on the model, then, a considerable evaluation of the controller ϕ is the average of the attained performance as

$$\int \gamma P_\phi f_\theta(\gamma) d\gamma. \tag{33}$$

The other evaluation is also considerable but this formula will be addressed to understand the following process. The optimal controller ϕ can be defined by

$$\phi_o(\theta) := \arg\min_\phi \int \gamma P_\phi f_\theta(\gamma) d\gamma. \tag{34}$$

Here the minimized function (33) is also written as

$$\int \gamma P_\phi f_\theta(\gamma) d\gamma = \int S_\phi(x) f_\theta(x) dx. \tag{35}$$

For example when γ is some operator norm of a closed loop system, then, (35) is an infinite dimensional optimization and it is very hard to be solved in general. A reasonable one is the Monte Carlo method, that is, at first prepare sample points $\hat{x}_i$ $(i = 1 \sim m)$ which obey probability density $f_\theta(x)$ and transform them to the performances on the sample points. Then the controller is selected as the optimizer of the sample points of the performance as follows.

$$\hat{\phi} := \arg\min_\phi \frac{1}{m}\sum_{i=1}^{m} \gamma_i = \arg\min_\phi \frac{1}{m}\sum_{i=1}^{m} S_\phi(\hat{x}_i) \tag{36}$$

Here assume that f_θ is nonsingular, and then the sample points $\hat{x}_i$ which obey f_θ are generated by using uniformly distributed sample data $\{z_i\}$ $(=: z^{(m)})$ on a sample space Z of the corresponding dimension and a variable transformation T_θ which is defined by f_θ as follow (Fushimi, 1994).

$$\hat{x}_i = T_\theta(z_i) \tag{37}$$

Therefore (36) is written as

$$\begin{aligned}\hat{\phi}(\theta, z^{(m)}) &= \arg\min_\phi \frac{1}{m}\sum_{i=1}^{m} S_\phi(T_\theta(z_i)) \\ &=: \arg\min_\phi \frac{1}{m}\sum_{i=1}^{m} U_{\phi,\theta}(z_i) \\ U_{\phi,\theta}(z) &:= S_\phi(T_\theta(z)). \end{aligned} \tag{38}$$

Corresponding to $\hat{\phi}(\theta, z^{(m)})$, $\phi_o(\theta)$ is also given by

$$\phi_o(\theta) = \arg\min_\phi \int h(z) U_{\phi,\theta}(z) dz \tag{39}$$

where $h(z)$ is a uniform distribution density. The important point is that $\hat{\phi}$ is a function of θ and a random variable $z^{(m)}$, and then $\hat{\phi}$ is also a random variable.

Because $\hat{\phi}$ is a random variable, next consider the average of the term

$$\frac{1}{n}\sum_{i=1}^{n} \log K_{\hat{\phi}(\hat{\theta}, z^{(m)})}(x_i)$$

in $\Psi(\hat{\theta}, \theta_o)$ at $\phi = \hat{\phi}$. At first the next is given.

$$\begin{aligned}&\int h(z^{(m)}) \frac{1}{n}\sum_{i=1}^{n} \log K_{\hat{\phi}(\hat{\theta}, z^{(m)})}(x_i) dz^{(m)} \\ &= \frac{1}{n}\sum_{i=1}^{n} \log K_{\phi_o(\hat{\theta})}(x_i) + \int h(z^{(m)}) \\ &\times \Bigg((\hat{\phi}(\hat{\theta}, z^{(m)}) - \phi_o)' \\ &\times \frac{1}{2n}\sum_{i=1}^{n} \frac{\partial^2}{\partial\phi\partial\phi'} \log K_{\phi_o(\hat{\theta})}(x_i) \\ &\times (\hat{\phi}(\hat{\theta}, z^{(m)}) - \phi_o) \Bigg) dz^{(m)} + o(m^{-1}) \end{aligned} \tag{40}$$

Moreover the first term in the right hand side of (40) is written as follows.

$$\begin{aligned}&\frac{1}{n}\sum_{i=1}^{n} \log K_{\phi_o(\hat{\theta})}(x_i) \\ &= \frac{1}{n}\sum_{i=1}^{n} \log K_{\hat{\phi}(\hat{\theta}, z^{(m)})}(x_i) \\ &+ \frac{1}{n}\sum_{i=1}^{n} \frac{\partial}{\partial\phi} \log K_{\hat{\phi}(\hat{\theta}, z^{(m)})}(x_i)(\phi_o - \hat{\phi}(\hat{\theta}, z^{(m)})) \\ &+ (\phi_o - \hat{\phi}(\hat{\theta}, z^{(m)}))' \\ &\times \frac{1}{2n}\sum_{i=1}^{n} \frac{\partial^2}{\partial\phi\partial\phi'} \log K_{\hat{\phi}(\hat{\theta}, z^{(m)})}(x_i) \\ &\times (\phi_o - \hat{\phi}(\hat{\theta}, z^{(m)})) + o(m^{-1}) \end{aligned} \tag{41}$$

Now define

$$V(\phi, \theta) := -\sum_{i=1}^{n} \frac{\partial^2}{\partial\phi\partial\phi'} \log K_{\phi(\theta)}(x_i)$$

$$W := \frac{1}{n}\sum_{i=1}^{n} \frac{\partial}{\partial\phi} \log K_{\hat{\phi}(\hat{\theta}, z^{(m)})}(x_i)(\phi_o - \hat{\phi}),$$

then (40) is written as

$$\begin{aligned}&\int h(z^{(m)}) \frac{1}{n}\sum_{i=1}^{n} \log K_{\hat{\phi}(\hat{\theta}, z^{(m)})}(x_i) dz^{(m)} \\ &= \frac{1}{n}\sum_{i=1}^{n} \log K_{\hat{\phi}(\hat{\theta}, z^{(m)})}(x_i) + W \end{aligned}$$

$$- \frac{1}{nm} \mathrm{tr}\, R(\phi_o, \hat{\theta})^{-1} Q(\phi_o, \hat{\theta}) R(\phi_o, \hat{\theta})^{-1} \times V(\phi_o, \hat{\theta}) + o\left(m^{-1}\right), \quad (42)$$

where

$$R(\phi, \theta) := - \int h(z) \frac{\partial^2}{\partial\phi\partial\phi'} U_{\phi,\theta}(z) dz$$
$$Q(\phi, \theta) := \int h(z) \frac{\partial}{\partial\phi} U_{\phi,\theta}(z) \frac{\partial}{\partial\phi'} U_{\phi,\theta}(z) dz.$$

By taking the average of $\Psi(\hat{\theta}, \theta_o)$ on $z^{(m)}$ and the following is given.

$$\begin{aligned} &\int h(z^{(m)}) \int \left\{ \log f_{\hat{\theta}}(x) + \log K_{\hat{\phi}(\hat{\theta},\, z^{(m)})}(x) \right\} \\ &\quad \times g(x) dx dz^{(m)} \\ &= \frac{1}{n} \sum_{i=1}^{n} \log K_{\hat{\phi}(\hat{\theta}, z^{(m)})}(x_i) \\ &\quad - \frac{1}{n} \mathrm{tr}\, J(\hat{\phi}, \theta_o)^{-1} I(\hat{\phi}, \theta_o) + W \\ &\quad - \frac{1}{nm} \mathrm{tr}\, R(\phi_o, \hat{\theta})^{-1} Q(\phi_o, \hat{\theta}) R(\phi_o, \hat{\theta})^{-1} \\ &\quad \times V(\phi_o, \hat{\theta}) + o\left(n^{-1}\right) + o\left(m^{-1}\right) \end{aligned} \quad (43)$$

In the above equation, the true value of W is unknown but its average is 0, therefore approximate it as 0 and also introduce an approximation of $R(\phi_o, \hat{\theta}) \rightarrow R(\hat{\phi}, \hat{\theta})$ and so on, then, an information criterion can be given for the estimation/selection of system model and also the selection of controller design method as follows.

$$\begin{aligned} &\Upsilon(\hat{\phi}, \hat{\theta}) \\ &= \frac{1}{n} \sum_{i=1}^{n} \log K_{\hat{\phi}(\hat{\theta}, z^{(m)})}(x_i) \\ &\quad - \frac{1}{n} \mathrm{tr}\, J(\hat{\phi}, \hat{\theta})^{-1} I(\hat{\phi}, \hat{\theta}) \\ &\quad - \frac{1}{nm} \mathrm{tr}\, R(\hat{\phi}, \hat{\theta})^{-1} Q(\hat{\phi}, \hat{\theta}) R(\hat{\phi}, \hat{\theta})^{-1} V(\hat{\phi}, \hat{\theta}) \end{aligned} \quad (44)$$

The formulae (43) and (44) show that Υ adds a new bias term to Ψ, which is proportional to the number of parameters of the controller. This suggests that when the structure of controller is complex then a controller which shows high performance for the given data will be designed. However, when the data x for the controller designing obeys a stochastic process, the complex controller tends to be specialized only for the given data. The new bias term can be interpreted as to avoid such complex controller.

5. CONCLUSION

In this paper, information criteria were proposed for model estimation and selection problem based on AIC/TIC with respect to the attained performance of closed loop systems. They are composed of entropy not only with a bias term which is proportional to the complexity of model but also a new term which is proportional to the complexity of controller.

The concrete methods of the controller designing were not referred and then the discussion was advanced under the assumption of their feasibility, however the real execution of solving controller designing as (36) is very hard and this problem is left for future work.

The same discussion based on MDL (Rissanen, 1984), which is related to the complexity of model and its estimation/selection problem, can be also applicable.

6. REFERENCES

Akaike, H. (1974). A new look at the statistical model identification. *IEEE Trans. Automat. Control* **AC-19-6**, 716–723.

Dahleh, M. A., T. V. Theodosopoulos and J. N. Tsitsiklis (1993). The sample complexity of worst–case identification of FIR linear systems. *Systems & Control Letters* **20–3**, 157–166.

Fushimi, M. (1994). *Stochastic method and simulation*. Iwanami Seminar, Applied Mathematics. Iwanami.

Konishi, S. and G. Kitagawa (1996). Generalised information criteria in model selection. *Biometrika* **83–4**, 875–890.

Kullback, S. (1959). *Information theory and statistics*. Mathematical statistics. John Wiley & Sons, Inc.. New York.

Lasota, A. and M. C. Mackey (1995). *Chaos, fractals, and noise; Stochastic aspects of dynamics, Second edition*. Vol. 97 of *Applied mathematical sciences*. Springer-Verlag. New York.

Rissanen, J. (1984). Universal coding, information, prediction and estimation. *IEEE Trans. Information Theory* **IT-30**, 629–636.

Takeuchi, K. (1976). Distribution of information statistics and criteria for adequacy of models. *Mathematical Sciences* **153**, 12–18.

Tempo, R., E. W. Bai and F. Dabbene (1996). Probabilistic robustness analysis: explicit bounds for the minimum number of samples. In: *Proceedings of the 35th CDC*. pp. 3424–3428.

Tsumura, K. and S. Shin (1995). Decision of nominal model and closed loop system design. In: *The 18th SICE Symposium on Dynamical System Theory*. pp. 285–288.

Tsumura, K. and S. Shin (1996). Robust control with distribution condition of system parameters. In: *The 25th SICE Symposium on Control Theory*. pp. 247–252.

Tsumura, K. and S. Shin (1997). Simultaneous modeling and data-distribution-dependent robust control system design. In: *Preprints of the 11th IFAC Symposium on System Identification*. Kitakyushu. pp. 129–134.

Vidyasagar, M. (1997). *A theory of learning and generalization*. Springer. London.

ON MODEL STRUCTURE SELECTION FOR NONPARAMETRIC PREDICTION METHODS

Anders Stenman *,[1]

* *Department of Electrical Engineering, Linköpings universitet, SE-581 83 Linköping, SWEDEN. Fax: +46 13 28 26 22*

Abstract: In this paper we continue to explore identification of nonlinear systems using the previously proposed concept of model-on-demand. The idea is to estimate the process dynamics locally and on-line using process data stored in a database, and has in earlier contributions proven to be capable to produce results comparable to (or better than) other nonlinear black-box approaches. The modeling part of the method is based on local polynomial modeling ideas. This has several implications on the choice of model structure, which is discussed at length in the paper. It is concluded that the NARX structure should be considered as the default choice in the local polynomial context. Furthermore, it is shown that the predictions in some situations can be enhanced by tuning other parameters that are special for the nonparametric case. The usefulness of the method is illustrated in numerical simulations. For the chosen application it is shown that the prediction errors are in order of magnitude directly comparable to more established modeling tools such as artificial neural nets.

Keywords: system identification, local polynomial modeling

1. INTRODUCTION

The problem of modeling dynamical systems from experimental data with poor *a priori* knowledge of the system is usually referred to as *black-box modeling* (Ljung, 1999). It has traditionally been solved using linear models of different sophistication. However, problems usually occur when encountering highly nonlinear systems that poorly allow themselves to be approximated by linear models. As a consequence of this, the interest for nonlinear modeling alternatives like neural networks and radial basis functions has been growing in recent years (Sjöberg *et al.*, 1995; Chen *et al.*, 1990). When the number of observations grows very large, however, this approach appears less attractive because of the difficulties in specifying model structure and the complexity of the associated optimization problem. Inspired by ideas from local modeling and database systems technology, we have taken a conceptually different point of view: We assume that all available data are stored in a database, and that local models are built "on demand" as the actual need arises.

It is clear that many possibilities for constructing the local model exist. In this contribution we will focus on different nonparametric methods derived in earlier contributions (Stenman *et al.*, 1996; Stenman, 1999), and will address some of the design considerations that have to be taken into account when using such methods for prediction and simulation.

The organization of the paper is as follows: Section 2 states the basic problem considered, whereby Section 3 shows how it can be solved using the model-on-demand concept. The different aspects of model structure selection associated with the method are discussed in Section 4, and are exemplified in Section 5. In Section 6, finally, the conclusions are drawn.

2. PRELIMINARIES

The class of systems considered here will be nonlinear, multi-input, single-output (MISO) systems in

[1] Email: stenman@isy.liu.se

discrete time, governed in input-output form by the dynamic relationship

$$y(t) = m(\varphi(t)) + v(t), \quad (1)$$

where

$$\varphi(t) = \varphi(Z^{t-1})$$

is a regression vector constructed from past measurements

$$Z^N \triangleq \{(u(1), y(1)), \ldots, (u(N), y(N))\}.$$

The goal of the modeling is to produce a good *predictor*, i.e., one that estimates

$$\hat{y}(t) = \mathrm{E}(y(t)|\varphi(t))$$

as well as possible, utilizing all the available information contained in $\varphi(t)$. This problem has traditionally been solved by parametric black-box approaches. In this contribution we shall instead utilize the recently proposed concept of model-on-demand for prediction and simulation purposes.

3. MODEL-ON-DEMAND

The basic principles behind the model-on-demand philosophy is as follows: Given a database of observed output-regressor pairs from the plant $\{(y(k), \varphi(k))\}$, a local prediction is formed by retrieving relevant data from the database and applying a suitable modeling operation.

For the modeling part we have adopted a "sliding window" approach. That is, for each operating point $\varphi(t)$, a local model is identified via the weighted regression problem; minimize

$$\sum_{k=1}^{N} \ell(y(k) - m(\varphi(k), \beta)) \cdot w_k(\varphi_t), \quad (2)$$

subject to β. Here $w_k(\cdot)$ are weights that control the locality and $\ell(\cdot)$ is a scalar-valued and positive norm function.

The weights can of course be selected in many ways. In the statistics they are often implicitly selected according to

$$w_k(\varphi(t)) = W\left(\frac{\|\varphi(k) - \varphi(t)\|_{\mathcal{M}}}{h}\right),$$

where h is a *bandwidth* parameter controlling the size of the local neighborhood, $\|u\|_{\mathcal{M}}$ is a scaled distance function on the regressor space, and $W(\cdot)$ is a window function (usually referred to as the *kernel*) assigning weights to each remote data point according to its distance from $\varphi(t)$. The kernel is typically a bell-shaped function with support on the interval [0, 1], see Figure 1, where some commonly used windows are depicted. In Stenman (1999) we also studied a direct optimization scheme that explicitly determines the optimal weights. It has been shown that this might result in better accuracy in some cases (Stenman *et al.*, 1997).

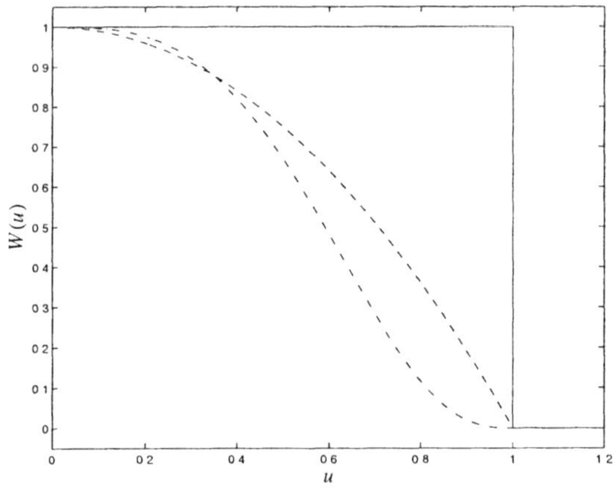

Fig. 1. Some commonly used window functions: Uniform (solid), Epanechnikov, $W(u) = (1 - u^2)_+$ (dashed), and tricube, $W(u) = (1 - u^3)_+^3$ (dash-dotted).

3.1 *Local Model Structures*

In principle, it is possible to use any model structure as local model in (2). However, if a quadratic norm, $\ell(\varepsilon) = \frac{1}{2}\varepsilon^2$, is used and the model is linear in the parameters, the estimate can be explicitly computed using ordinary weighted least squares. This idea of estimating β has a long tradition within the statistics and is usually referred to as *local polynomial modeling* (Fan and Gijbels, 1996).

For simplicity we assume a local linear model structure,

$$m(\varphi(k), \beta) = \beta_0 + \beta_1^T(\varphi(k) - \varphi(t)) \quad (3)$$

as the the default model in the sequel of the paper. Thus, if $\hat{\beta}_0$ and $\hat{\beta}_1$ denote the minimizers of (2) using the model (3), it is easy to realize that an estimate of $y(t)$ (i.e., the one-step-ahead prediction), is given by

$$\hat{y}(t) = m(\varphi(t), \hat{\beta}) = \hat{\beta}_0. \quad (4)$$

Note that each optimization of (2) produces a single local prediction $\hat{y}(t)$ corresponding to the current regression vector $\varphi(t)$. To obtain predictions at other locations in the regressor space, the weights change and new optimization problems have to be solved. This is in contrast to the global modeling approach where the model is fitted to data only once. Since no global description is obtained, estimators like (2) is often called nonparametric.

3.2 *Model Tuning*

It is well known that the bandwidth h has a critical impact on the resulting estimate, since it governs a trade-off between the bias and variance errors of the estimate. Methods that use the available data to produce a good bandwidth are usually referred to as *bandwidth selectors*, and have been thoroughly studied within the statistical literature, see for instance Fan and Gijbels (1996) and the references therein. They can roughly be divided into "classical" methods which are based on cross-validation ideas, and "plug-in" methods which rely on minimizing asymptotic MSE expressions. The majority of the bandwidth selectors proposed so far, though, have been of *global* type, i.e., they produce a single global value. However, adaptive (i.e., *local*) methods, which select local bandwidths on-line for each estimation point, have gained a significant interest in recent years, even though the development of them still seems to be an open and active research area. See Stenman (1999) for a detailed discussion.

The scaling matrix $\mathcal{M}$ that controls the distance function can be optimized in a similar way. The simplest choice, though, is to let it be the identity matrix. This corresponds to the usual Euclidean norm. A diagonal scaling matrix transforms the radially symmetric distance function into an axis parallel ellipse. By also allowing cross-terms in $\mathcal{M}$ it is possible to make the ellipse arbitrary oriented. Selecting the scaling matrix is very important, especially when the regressor components have very different magnitudes. An obvious default choice, which has been adopted here, is to make it proportional to the inverse covariance of the regressors.

3.3 *Extensions*

It is clear that several extensions of the local polynomial regression idea exist. One possibility is to replace (2) with a localized maximum likelihood criterion (Tibshirani and Hastie, 1987), although this will increase the computational burden even more. Recent work by Kulhavý and Ivanova (1999) has also shown how it is possible to incorporate *a priori* information to the local model through a localized Bayesian estimation approach.

4. DESIGN CONSIDERATIONS

In order to obtain a local model that is suitable for prediction or simulation, there are a number of design questions which have to be taken into consideration:

(1) Which quantities, constructed from past measurements, should be included in the regression vector $\varphi(t)$?
(2) Which model structure should be used?
(3) Which criteria should be used when determining which data that are relevant for the local fit?

The first question is the same for all modeling approaches (linear as well as nonlinear ones) and in most situations it turns out that the most useful choices of regression vectors are those who are built up by past inputs and outputs. The second question appears in parametric black-box modeling too, but then usually includes a wide range of different parameterizations. In a local modeling framework, we essentially just have to decide upon the polynomial degree and the norm for the fit. The third question is rather unique for the nonparametric estimation problem, because of its local properties. Nevertheless, it is an important one, since we can emphasize different features in data with different relevance measures.

4.1 *Model Structures for Nonparametric Methods*

For parametric models, it is well known that adding an extra regressor component enables an increased flexibility, but that this comes to the price of an increased variability of the estimated parameters. It is thus often desirable to keep the dimension of the regression vector as low as possible. This conclusion still holds and is perhaps even more important in the nonparametric case. It is a well-known fact that the rate of convergence typically gets slower in higher regressor dimensions. This is primarily due to the so-called *curse of dimensionality*, which follows from the sparseness of data in the multidimensional regressor space. Another, but yet equally important reason is the dataset searching problem associated with the local modeling. Certain storage structures, like *k-d trees* (Bentley, 1975), have proven to be most efficient for a moderate number of dimensions.

As we shall see later, a possible alternative to restrict the size of the regression vector, is to impose restrictions on the mapping $m(\cdot)$. One such possibility is to assume that the regression function is global in certain regressor directions. This is usually referred to as *conditionally parametric models*, and could be a useful weapon to combat the curse-of-dimensionality.

Let us here review different choices on model structures and their ability to be adapted to the local polynomial framework. The discussion will be made assuming single-input, single-output systems. Generalization to the multiple-input, single-output case is straightforward however; think of $u(t)$ as a vector of inputs.

4.1.1. *NFIR Structures* The simplest nonlinear dynamical model is the nonlinear finite impulse response model, NFIR, which only uses past inputs as regressors, i.e.,

$$y(t) = m(u(t-n_k), \ldots, u(t-n_b-n_k+1)) + e(t).$$

Here n_b and n_k denote the number of past inputs and the time delay, respectively. The corresponding predictor is obtained by deleting the noise term. That is,

$$\hat{y}(t) = m(u(t-n_k), \ldots, u(t-n_b-n_k+1)).$$

A major advantage with this structure is that it will always generate bounded predictions (provided, of course, that $m(\cdot)$ does not have any singularities), since it only depends on past inputs which are assumed to be bounded. However, in order to achieve accurate predictions, the NFIR structure normally requires quite large model orders n_b, which, as discussed above, is unsuitable for the local modeling approach.

4.1.2. *NARX Structures* Augmenting the NFIR model structure with an autoregressive part results in the nonlinear, autoregressive with exogenous input, NARX, model. As shown earlier by Ljung and Wahlberg (1992), the linear FIR and ARX models are capable of approximating any linear system provided arbritrary high model orders are allowed. It easy to believe that the NARX model will play a similar role for the nonlinear case. The most general NARX structure is

$$\begin{aligned} y(t) = m(&y(t-1), \dots, y(t-n_a), \\ &u(t-n_k), \dots, u(t-n_b-n_k+1)) + e(t), \end{aligned} \quad (5)$$

which was referred to as the NARX_1 structure by Sjöberg *et al.* (1995). Here n_a denotes the number of lagged outputs. Its predictor is

$$\begin{aligned} \hat{y}(t) = m(&y(t-1), \dots, y(t-n_a), \\ &u(t-n_k), \dots, u(t-n_b-n_k+1)) . \end{aligned}$$

The NARX model has several nice properties which make it suitable to work with and a default choice for many applications:

- It is general and can describe any nonlinear system well.
- It is not *recurrent*, i.e., the regressors are independent of previous model outputs.

Unfortunately, there are also drawbacks: There is no specific noise model built into the structure. Instead the noise dynamics are modeled along with the plant dynamics. As a consequence of this, a good approximation of both the system and the noise dynamics may require a high dimension of the regression vector $\varphi(t)$. A solution to these problems is to include separate noise dynamics in the model.

4.1.3. *Structures with Separate Noise Models: NARMAX, NOE and NBJ* Structures with separate noise models can be constructed by incorporating past prediction errors $\varepsilon(t-k) = y(t-k) - \hat{y}(t-k)$ into the regression vector. However, since it under normal circumstances is quite difficult to model the noise in a general nonlinear way, it seems natural to impose the restriction that the model is linear in the prediction errors. This corresponds to the NARMAX-type structure

$$y(t) = m(\varphi(t)) + C(q)e(t), \quad (6)$$

where $\varphi(t)$ is constructed as in the NARX case. The corresponding predictor is

$$\hat{y}(t) = m(\varphi(t)) + (C(q) - 1)\varepsilon(t),$$

which can be interpreted as a conditionally parametric model that is global in the $\varepsilon(t-k)$-directions of the regressor space.

A NARMAX model of the form (6) can be estimated by modifying the loss function in (2) according to

$$\arg\min_{C,\beta} \sum_k \left[C^{-1}(q)(y(k) - m(\varphi(k), \beta))\right]^2 \cdot w_k(\varphi(t))$$

However, since this model is not linear in the parameters, a numerical optimization routine must be used. Thus this model is not that suitable for the nonparametric setting. The same conclusion will of course also hold for the nonlinear Box-Jenkins (NBJ) and output error (NOE) structures.

4.1.4. *NFIR Systems with Colored Noise* So far we have assumed that the noise term $e(t)$ is white. Let us see what happens if we remove this assumption. Consider an NFIR model with an additive but not necessarily white noise term $v(t)$,

$$y(t) = m(u(t-1), \dots, u(t-n_b)) + v(t). \quad (7)$$

For the noise $v(t)$ we will in general only need a spectral description. It can thus be described by

$$v(t) = H(q)e(t)$$

where $\{e(t)\}$ is white noise. A predictor for (7) is

$$\hat{y}(t) = H^{-1}(q)m(\varphi_u(t)) + \left(1 - H^{-1}(q)\right) y(t)$$

In the first term, the linear filter $H^{-1}(q)$ can equally well be absorbed into the nonlinear mapping $m(\cdot)$. The structure (7) can thus be viewed as a NFIR structure, accomplished by a linear term consisting of past $y(t)$'s. That is,

$$\hat{y}(t) = (1 - A(q))y(t) + \bar{m}(\varphi_u(t)). \quad (8)$$

Since this structure has the same regressor configuration as the NARX_1 model in equation (5), it has been referred to as a NARX_2 model by Sjöberg *et al.* (1995). The fact that it is not recurrent makes it a very interesting and useful alternative to work with.

In the model-on-demand framework, the NARX_2 model can similarly to the NARMAX structure above be interpreted as a conditionally parametric local linear model which is global in the $y(t-k)$-directions of the regressor space. Estimation of such models can be performed by zeroing out the corresponding elements of the scaling matrix $\mathcal{M}$ that controls the distance function. Another alternative is to start by fitting a global linear model to the $y(t-k)$-regressors, and use the residuals for nonparametric modeling in the $u(t-k)$-directions of the regressor space.

The NARX_2 concept can be extended to the case where the predictor consists of a sum of two nonlinear terms which are independent of each other. That is

$$\hat{y}(t) = m_1(\varphi_y(t)) + m_2(\varphi_u(t)). \quad (9)$$

A model of this type is referred to as an *additive model* in the statistical literature (Hastie and Tibshirani, 1990), and can (although even more computational demanding) be estimated using the backfitting algorithm which alternates between estimating $m_1(\cdot)$ and $m_2(\cdot)$ from previous residuals. In the work by Sjöberg *et al.* (1995), a model of the form (9) was called a NARX_3 model, and can be interpreted as a FIR model of the type (7) where the noise is state dependent. That is,

$$v(t) = H(q, \varphi_u(t))e(t). \tag{10}$$

5. AN EXAMPLE

In this section we will illustrate some of the previously described concepts through simulations. The application has previously been investigated by Sjöberg *et al.* (1995) and Gunnarsson and Krus (1990) and has to some extent become a "benchmark" example for nonlinear black-box modeling.

The system in question is a hydraulic actuator that controls the position of a robot arm. The arm position is directly coupled to the oil pressure in the actuator which is determined by a valve that regulates the oil flow into the actuator. The modeling aim is to model the oil pressure as a function of the valve opening.

The dataset consists of 1024 samples in total and has been split into two equally sized parts – one for estimation and one for validation. Figure 2 shows the estimation part. The valve opening is here denoted $u(t)$ and the oil pressure $p(t)$. As observed from

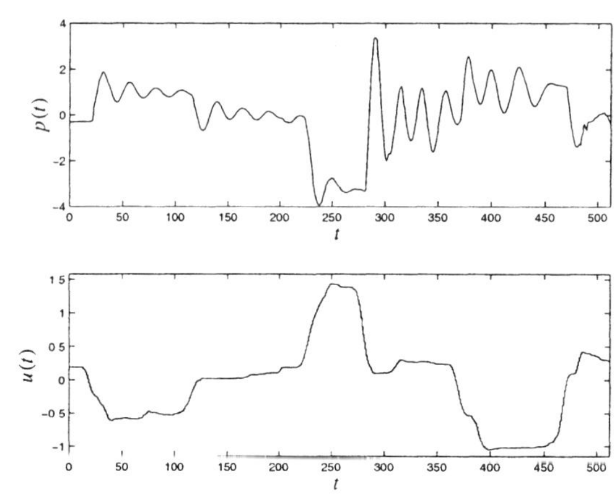

Fig. 2. Measured values of oil pressure $p(t)$ and valve position $u(t)$.

the figure, the oil pressure signal has very oscillative settling periods after step changes in the valve size. This is primarily due to mechanical resonances in the robot arm.

It seems reasonable to start by first considering simple linear black-box models. Experiments performed by Sjöberg *et al.* (1995) show that a model that predicts the pressure $p(t)$ using three old values of the pressure and two old values of the valve size signal (i.e., an ARX 321 model) is appropriate for this particular application. The result of a simulation on validation data using this configuration is shown in Figure 3 (a). The achieved root mean squared error, RMSE, for the simulated output is 0.925, but it is quite obvious from visual inspection that it would be possible to improve this rather poor result by considering more complex nonlinear models.

A severe problem with this application from a local modeling viewpoint is of course the sparseness of data (512 data points in a five-dimensional regressor space). However, it might anyway be interesting to see how far we can get with the a local polynomial approach. We found that in order to obtain reasonable results we had to increase the model order by one. That is, to use a NARX 431 predictor,

$$\hat{p}(t) = m(p(t-1), p(t-2), p(t-3), p(t-4), \\ u(t-1), u(t-2), u(t-3)).$$

Figure 3 (b) shows a simulation on validation data using the model-on-demand concept with a local linear NARX_1 model based on the above regressor configuration. The achieved RMSE is 0.499 and represents a significant enhancement compared to the linear ARX model. It turns out though that the result can be even further improved by considering a conditionally global NARX_2 model. This can be achieved by altering the scaling matrix according to

$$\mathcal{M} = \text{diag}\begin{pmatrix}0 & 0 & 0 & 0 & 1 & 1 & 1\end{pmatrix}$$

A simulation with this setup in the model-on-demand framework is shown in Figure 3 (c). The RMSE error is now 0.471. Trying the NARX_3 model and other choices of model parameters did not result in any further improvements. However, Sjöberg *et al.* (1995) obtained a best value of 0.328 using a complex neural network structure. The conclusion is therefore that this dataset is too small to produce results directly comparable to the parametric case. Nevertheless, we have demonstrated how local modeling techniques even for rather small datasets quite drastically can improve the quality of the simulations compared to traditional linear black-box modeling.

Table 1. The best achieved RMS errors.

Method	RMSE
ARX 321	0.926
NARX 431 (MOD)	0.471
NARX 431 (NN)	0.328

6. CONCLUSIONS

We have in this paper discussed several aspects of model structure selection associated with the local polynomial modeling concept, and have shown that the modeling approach is a useful for predicting and simulating nonlinear systems when sufficiently large datasets are available. A number of commonly used model structures have been evaluated from a model-on-demand perspective and we have concluded that the different NARX structures should be considered as the default choice in this context. Furthermore, it has been shown that the predictions in some situations

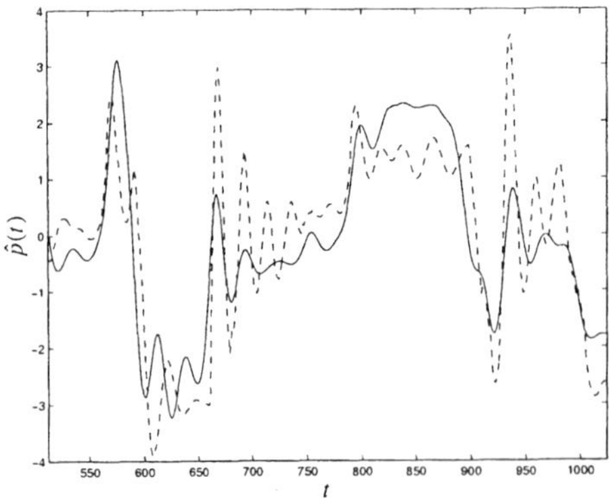

(a)

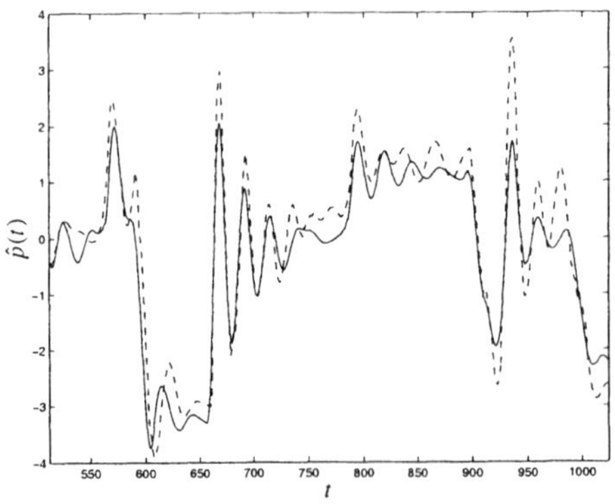

(b)

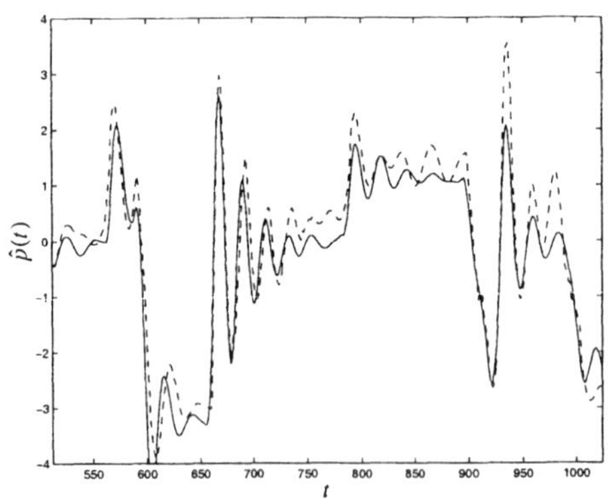

(c)

Fig. 3. Simulation of the crane behavior using validation data: (a) ARX model. (b) $NARX_1$ model. (c) $NARX_2$ model. The solid lines are the simulated output and the dashed lines are the true system output.

can be enhanced by tuning certain parameters like the the scaling matrix.

The usefulness of the model-on-demand concept has been illustrated in numerical simulations. For the chosen application we have shown that the prediction errors are in order of magnitude directly comparable to more established modeling tools such as artificial neural nets. However, we have at the same time shown that the demand for large datasets is more obvious in the model-on-demand case than for traditional parametric modeling approaches.

7. REFERENCES

Bentley, J.L. (1975). Multidimensional binary search trees used for associative searching. *Communications of the ACM* **18**(9), 509–517.

Chen, S., S.A. Billings, C.F.N. Cowan and P.M. Grant (1990). Non-linear system identification using radial basis functions. *International Journal of Systems Science* **21**(12), 2513–2539.

Fan, J. and I. Gijbels (1996). *Local Polynomial Modelling and Its Applications*. Chapman & Hall.

Gunnarsson, S. and P. Krus (1990). Modeling of a flexible mechanical system containing hydralic actuators. Technical report. Dept of EE. Linköping University, S-581 83, Linköping, Sweden.

Hastie, T. and R. Tibshirani (1990). *Generalized Additive Models*. Chapman & Hall.

Kulhavý, R. and P. Ivanova (1999). Memory-based prediction in control and optimisation. In: *Preprints of the 14th World Congress of IFAC* (H.F. Chen and B. Wahlberg, Eds.). Vol. H. Elsevier Science. pp. 289–294.

Ljung, L. (1999). *System Identification – Theory for the user*. 2nd ed.. Prentice-Hall, New Jersey.

Ljung, L. and B. Wahlberg (1992). Asymptotic properties of the least-squares method for estimating transfer functions and disturbance spectra. *Advances in Applied Probability* **24**, 412–440.

Sjöberg, J., Q. Zhang, L. Ljung, A. Benveniste, B. Deylon, P.-Y. Glorennec, H. Hjalmarsson and A. Juditsky (1995). Nonlinear black-box modeling in system identification: a unified overview. *Automatica* **31**, 1691–1724.

Stenman, A. (1999). Model on Demand: Algorithms, Analysis and Applications. PhD thesis. Dept of EE, Linköping University. SE-581 83, Linköping, Sweden.

Stenman, A., A.V. Nazin and F. Gustafsson (1997). Asymptotic properties of Just-in-Time models. In: *Preprints of the 11th IFAC Symposium on System Identification, Kitakyushu, Japan* (Y. Sawaragi and S. Sagara, Eds.). pp. 1249–1254.

Stenman, A., F. Gustafsson and L. Ljung (1996). Just in time models for dynamical systems. In: *Proceedings of the 35th IEEE Conference on Decision and Control, Kobe, Japan*. pp. 1115–1120.

Tibshirani, J.R. and T.J. Hastie (1987). Local likelihood estimation. *Journal of the American Statistical Association* **82**, 559–567.

MODEL ORDER SELECTION OF N-FIR MODELS BY THE ANALYSIS OF VARIANCE METHOD

Ingela Lind

Division of Automatic Control, Department of Electrical Engineering, Linköping University, SE-581 83 Linköping, Sweden. E-mail: ingela@isy.liu.se

Abstract: Identification of nonlinear finite impulse response (N-FIR) models is studied. In particular the selection of model structure, i.e., to find the contributing input time lags, is examined. A common method, exhaustive search among models with all possible combinations of the input time lags, has some undesired drawbacks, as a tendency that the minimization algorithm gets stuck in local minima and heavy computations. To avoid these drawbacks it is needed to know the model structure prior to identifying a model.
In this contribution it is shown by experiment that a statistical method, the analysis of variance, is a good alternative to exhaustive search in the identification of the structure of nonlinear FIR-models.
It is possible to reduce the risks of getting an erroneous model structure due to the non-convexity of the minimization problems, reduce the computation time needed and also get a good estimate of how far the fit of the desired model can be enhanced.

Keywords: Identification, Nonlinear models, Analysis of variance

1. PROBLEM DESCRIPTION

Assume that a nonlinear finite impulse response model (Sjöberg *et al.* 1995) describes the measurements y_t from a system with input u_t, that is,

$$y_t = g(u_t, u_{t-T}, u_{t-2T}, ..., u_{t-kT}) + e_t. \qquad (1)$$

The value of k is unknown in addition to which time lags of u_t that contribute to the value of y_t. Moreover g is an unknown static nonlinear function of up to $k+1$ variables. The problem is to identify from measurements the model order k, the contributing time lags u_{t+iT} and the function g.

A common method used to estimate the function g is to parameterize it as a nonlinear neural network. If we want to get a small network, with few parameters to estimate, and information about what inputs contribute to the measurements, this method has some drawbacks:

Since the model order is not known in advance, it is necessary to assume a largest k and estimate a network of each combination of the time lags of u_t under that assumption. Of these models, the one with the best performance on validation data is chosen. Each model takes time to estimate and depending on the assumption of k, there can be many models.

During the estimation of the neural network, the optimization algorithm can stop in a local minimum, which can lead to a model with wrong assumptions on relevant input time lags coming out best.

Both these drawbacks can be avoided if the structure of the model is known in advance, i.e., the contributing time lags and k are known. In order to obtain this knowledge, it is possible to use the Analysis of Variance method (ANOVA) from statistics (Montgomery 1991, Krishnaiah 1980). This way we only need to estimate one model, thereby reducing the computation time considerably, and we need not worry about getting an erroneous model structure because the optimization algorithm has been stuck in a local minimum.

2. EXHAUSTIVE SEARCH FOR FINDING MODEL STRUCTURE

The idea behind the exhaustive search method to find the model structure is to enumerate all possible combinations of the input time lags and estimate a model for each such combination. When all the models are estimated, their prediction performance are compared with some goodness measure, e.g., the root mean square error between measured and predicted output. The model that has the best performance is chosen as the model for the system. The goal is that this model has the same structure as the system, whose structure we want to identify.

This method does not distinguish between the task of finding the model structure and the task of finding a model, thereby tuning is done to improve models before we know if they are going to be used or not.

The models used to parameterize the system depend on our assumptions on how the system is working, i.e., if it is linear, nonlinear, dynamic or static etc. If we had been identifying a linear system there would not have been any big problems with using the exhaustive search method to find the correct structure. The identification methods for finding good linear models are efficient and reliable. Of course, we could still make the wrong choices sometimes, due to the non-deterministic nature of the system. In this contribution we try to identify the structure of nonlinear systems though, and that is where the problems start. There are no really efficient and completely reliable methods to estimate a nonlinear model of a nonlinear system. It is common to use sigmoidal neural networks to do it, but these suffer from the lack of search algorithms that guarantee that the global minimum is found. They also may take a considerable amount of computations to find a minimum at all. So, we cannot be sure that the global minimum is found for each of the models we compare and can thereby not be sure that we have found the model structure that describes the system best.

3. THE ANALYSIS OF VARIANCE

The statistical analysis method ANOVA (Montgomery 1991, Krishnaiah 1980) is a widely spread tool for finding out which factors contribute to given measurements. Though common in medicine and quality control applications, it does not seem to have been tried in system identification applications.

The method is based on hypothesis tests with F-distributed test variables computed from the residual quadratic sum. Below, one variant of the method is stated in a statistical framework for two factors. The complexity grows rapidly with the number of factors.

Assume that we have a batch of abn observations, corresponding to the ab treatment/level combinations of the factors A and B. A has a different levels, B has b different levels and the experiment is repeated n times. The measurements are made in random order to be sure to avoid effects of time dependency etc. The observations may be described by a linear statistical model,

$$y_{ijk} = \mu + \tau_i + \beta_j + (\tau\beta)_{ij} + \epsilon_{ijk}, \tag{2}$$

where $i = 1, \dots, a$, $j = 1, \dots, b$, $k = 1, \dots, n$, μ is the overall mean effect, τ_i is the effect of the ith level of the factor A, β_j is the effect of the jth level of the factor B, $(\tau\beta)_{ij}$ is the effect of the interaction between the ith level of factor A and the jth level of the factor B and ϵ_{ijk} is a random error component. This means that we assume that there are deterministic effects from the level of the factors, and that the stochastic effects are totally due to measurement noise. The treatment effects are defined to be fixed deviations from the overall mean, so $\sum_{i=1}^{a} \tau_i = 0$, $\sum_{j=1}^{b} \beta_j = 0$, $\sum_{i=1}^{a} (\tau\beta)_{ij} = 0, \forall\, j$ and $\sum_{j=1}^{b} (\tau\beta)_{ij} = 0, \forall\, i$.

We are interested in testing the following hypotheses regarding the treatment effects:

$$H_{0AB} : \ (\tau\beta)_{ij} = 0, \forall\, i, j, \tag{3}$$

that is, the measurements do not depend on the level combination of factors A and B (interaction effect), against

$$H_{1AB} : \ \text{at least one } (\tau\beta)_{ij} \neq 0, \tag{4}$$

that there is at least one significant interaction effect. If the null hypothesis is accepted, we are also interested in testing:

$$H_{0A} : \ \tau_1 = \tau_2 = \dots = \tau_a = 0, \tag{5}$$

$$H_{0B} : \ \beta_1 = \beta_2 = \dots = \beta_b = 0, \tag{6}$$

against, respectively,

$$H_{1A} : \ \text{at least one } \tau_i \neq 0, \tag{7}$$

$$H_{1B} : \ \text{at least one } \beta_j \neq 0. \tag{8}$$

If any of the null hypotheses is rejected, we assume that the factor involved does have some effect on the measurements. To do the hypothesis tests we use a two-factor analysis of variance. We compute the overall mean, the cell means and the means over the factor levels. Let

$$\overline{y}_{...} = \frac{1}{abn}\sum_{i=1}^{a}\sum_{j=1}^{b}\sum_{k=1}^{n} y_{ijk}, \tag{9}$$

$$\overline{y}_{i..} = \frac{1}{bn}\sum_{j=1}^{b}\sum_{k=1}^{n} y_{ijk}, \tag{10}$$

$$\overline{y}_{.j.} = \frac{1}{an}\sum_{i=1}^{a}\sum_{k=1}^{n} y_{ijk}, \tag{11}$$

$$\overline{y}_{ij.} = \frac{1}{n}\sum_{k=1}^{n} y_{ijk}. \tag{12}$$

The total residual quadratic sum can be written

$$\begin{aligned} SS_T &= \sum_{i=1}^{a}\sum_{j=1}^{b}\sum_{k=1}^{n}(y_{ijk}-\overline{y}_{...})^2 \\ &= \sum_{i=1}^{a} bn(\overline{y}_{i..}-\overline{y}_{...})^2 + \sum_{j=1}^{b} an(\overline{y}_{.j.}-\overline{y}_{...})^2 \\ &+ \sum_{i=1}^{a}\sum_{j=1}^{b} n(\overline{y}_{ij.}-\overline{y}_{i..}-\overline{y}_{.j.}+\overline{y}_{...})^2 \\ &+ \sum_{i=1}^{a}\sum_{j=1}^{b}\sum_{k=1}^{n}(y_{ijk}-\overline{y}_{ij.})^2 \\ &= SS_A + SS_B + SS_{AB} + SS_E, \end{aligned} \tag{13}$$

where

$$SS_A = \sum_{i=1}^{a} bn(\overline{y}_{i..}-\overline{y}_{...})^2, \tag{14}$$

$$SS_B = \sum_{j=1}^{b} an(\overline{y}_{.j.}-\overline{y}_{...})^2, \tag{15}$$

$$SS_{AB} = \sum_{i=1}^{a}\sum_{j=1}^{b} n(\overline{y}_{ij.}-\overline{y}_{i..}-\overline{y}_{.j.}+\overline{y}_{...})^2 \tag{16}$$

and

$$SS_E = \sum_{i=1}^{a}\sum_{j=1}^{b}\sum_{k=1}^{n}(y_{ijk}-\overline{y}_{ij.})^2. \tag{17}$$

From a theorem by Cochran (Montgomery 1991, page 59) it is possible to show that

- the stochastic variables SS_A, SS_B, SS_{AB} and SS_E are independent,
- the stochastic variable $\frac{1}{\sigma^2}\cdot SS_E \sim \chi^2(ab(n-1))$,
- if $\tau_1 = \ldots = \tau_a = 0$, then $\frac{1}{\sigma^2}\cdot SS_A \sim \chi^2(a-1)$,
- if $\beta_1 = \ldots = \beta_b = 0$, then $\frac{1}{\sigma^2}\cdot SS_B \sim \chi^2(b-1)$,
- if $(\tau\beta)_{ij} = 0, \forall i,j$, then $\frac{1}{\sigma^2}\cdot SS_{AB} \sim \chi^2((a-1)(b-1))$.

These observations are used to design test variables to test the interesting hypotheses. The test variable associated with factor A is chosen as

$$v_A = \frac{SS_A/(a-1)}{SS_E/(ab(n-1))}. \tag{18}$$

If H_{0A} is true, then $v_A \sim F(a-1, ab(n-1))$, i.e., v_A is F-distributed with $a-1$ and $ab(n-1)$ degrees of freedom. The null-hypothesis is rejected if we get a large value of v_A, that is, we reject H_{0A} if $v_A > c$, where c is taken from an $F_\alpha(a-1, ab(n-1))$-table and α denotes the level of significance (the probability to reject H_0 though H_0 is true). H_{0B} and H_{0AB} are tested analogously. We can also estimate the standard deviation, σ, associated with the random error component, as $\hat{\sigma}^2 = \frac{SS_E}{ab(n-1)}$. The degrees of freedom are $ab(n-1)$. Note that n needs to be larger than 1 to do the analysis with all interaction effects.

The results from the hypotheses testing can be used to determine which factors have effect on the measurements and if there are interaction effects between different factors.

The most important modelling simplifications made are the assumptions that the variance is constant through the batch and that the random error component is Gaussian distributed. The F-tests are quite robust against violations against both assumptions (Krishnaiah 1980, Chapter 7). If the levels of the input cannot be regarded as fully deterministic or if the effects of the input are partly stochastic it is better to use another variant of the analysis of variance method (Montgomery 1991, Scheffé 1959). There are also ways to treat missing data, see Hocking (1984).

Now, to use ANOVA in a system identification application, we can consider the nonlinear FIR-model

$$y_t = g(u_t, u_{t-1}) + e_t. \tag{19}$$

y_t is seen as the measurements, u_t as factor A and u_{t-1} as factor B. This structure gives that $a = b = m$, the number of levels we allow u_t to assume.

4. EXPERIMENTAL DESIGN

We want to design an experiment to determine which time lags have effect on the output of a system, assumed to be described by a nonlinear FIR model, $y_t = g(u_t, u_{t-1}, \ldots, u_{t-k}) + e_t$, where y is the output, g is an unknown function of the input and e is additive Gaussian white noise with variance σ^2. The result will both give us the important time lags and the main structure of the function g, e.g., with k assumed equal to 3,

$$g(u_t, u_{t-1}, u_{t-2}, u_{t-3}) = g_1(u_t, u_{t-2}) + \\ + g_2(u_t, u_{t-3}). \quad (20)$$

To do identification experiments with the described variant of ANOVA we need to construct a signal u_t that contains all possible combinations of levels of the time lags we want to examine. Assume that we have p time lags we want to examine, m levels of u_t and need n repetitions of each combination. The most straightforward way is to enumerate all possible sequences of length p, with different level combinations of the time lags, repeat them n times, and sort them in random order. This gives us a sequence u_t of length pnm^p. In this longer sequence, the short sequences will occur more than n times each, and we use only every pth measurement of the output from the system. This is a wasteful way to treat the measurements and there are better methods, see e.g. Godfrey (1993, page 41).

5. TEST SPECIFICATION AND RESULTS

To compare the different methods, we test their performance on a list of different functions that fits into

$$y_t = g(u_t, u_{t-1}, u_{t-2}) + e_t, \quad (21)$$

see Table 1. That is, we have assumed that $k = 2$ and will perform a three-way analysis of variance ($p = 3$). Assume that we have two sets of input/output data for the system we want to identify. Also assume that the datasets are suited for analysis with ANOVA, that is, all possible level combinations of the input are included. We put the second data set aside for validation use. From these data sets we want to find what input time lags are relevant, and if possible, a good model for the system.

The networks used have two layers with 5, 10 or 15 tansig neurons in the first layer and one purelin neuron in the second layer. We vary the number of input time lags between one and three and get seven different combinations of possible time lags. This means there are 16 to 76 parameters to estimate, depending on the number of input time lags included. The minimization algorithm used is Levenberg-Marquardt (Demuth and Beale 1998).

The only performance criterion used in this test is the root mean square error (RMS) between measured output and simulated output from the model.

Our own implementation of the ANOVA test used in this experiment does not employ all useful features of the theory. It is possible to get the test more efficient and to gain more information by using it.

We will compare the two structure identification methods in the following way.

(1) **Exhaustive search** Divide the first data set into estimation data and verification data. Construct neural networks for all possible combinations of time lags under the assumption of largest k. For each such combination, construct networks with different numbers of parameters (e.g. 5, 10 and 15 neurons in the first layer). Start with random network parameters and estimate the parameters on the estimation data. Start over 4 or 5 times with new random network parameters to try to avoid getting stuck in a local minimum. Of all these estimated networks, choose the one with the smallest RMS on the verification data.

(2) **ANOVA** Use all of the first data set and perform ANOVA. This gives a model (which input time lags contribute to measurements) and a model structure (the interaction pattern of the inputs). In this report, we do not use the information about the model structure to improve the estimation of the model, that is, only the information about the contributing input time lags is used. Construct neural networks with different numbers of neurons (5, 10 ,15) for the chosen model. Estimate as in the exhaustive search and choose the network with the smallest RMS on the verification data.
Note that one benefit of the ANOVA test is that we get a good estimate of the noise variance more or less automatically. This information can be used to determine when the minimization algorithm has failed to find a minimum close to the global minimum. If the RMS is much larger than the ANOVA estimated standard deviation we know that we have got a bad model. Try to restart the estimation process more times with new random network parameters to find a better local minimum. If this does not lead to smaller RMS try to get more estimation data or change to another type of network. In the following test, this feature is not used.

Finally, we compare the network chosen by exhaustive search and the network chosen by ANOVA on the validation data.

We are interested in how often the two methods can pinpoint the correct input time lags, and for ANOVA, also the correct model structure, for a specific function. We test this by running 100 Monte Carlo simulations on a list of test functions, see Table 1. We also include an additional test function, a random network function, which is the same type of network as the ones we try to estimate the function g with. We vary the signal

Function	$\sigma = 1,\ \alpha = 0.01$ Exhaustive search	ANOVA	$\sigma = 0.0001,\ \alpha = 0.0001$ Exhaustive search	ANOVA
$u_t - 0.03u_{t-2}$	10	6	94	100
$\ln\lvert u_t\rvert + u_{t-1} + e^{u_{t-2}}$	77	100	78	100
$u_{t-1} \cdot [u_t + \frac{1}{u_{t-2}}]$	100	100	100	100
$\mathrm{sgn}(u_{t-1})$	84	94	80	100
$\mathrm{sgn}(u_{t-1}) \cdot u_{t-2}$	93	96	92	100
$\mathrm{sgn}(u_{t-1}) \cdot u_t \cdot u_{t-2}$	100	100	100	100
$\ln\lvert u_{t-1} + u_{t-2}\rvert$	95	96	94	100
$\ln\lvert u_{t-1} \cdot u_{t-2}\rvert$	94	92	82	100
$u_{t-2} \cdot \ln\lvert u_{t-1}\rvert$	97	97	95	100
$u_{t-2}^3 \cdot \ln\lvert u_{t-1}\rvert$	50	95	56	100
$u_{t-2} \cdot (\ln\lvert u_{t-1}\rvert)^3$	93	95	91	99
$\lvert u_{t-2}\rvert \cdot e^{u_{t-1}}$	54	96	54	100
$u_{t-2} \cdot e^{u_{t-1}}$	49	94	49	100
$u_{t-2} \cdot e^{u_{t-1} - 0.03u_t}$	58	100	58	100
$\lvert u_t\rvert$	83	96	73	100
network $g(u_{t-1}, u_{t-2})$	73	88	94	100
TOTAL	75.6	90.3	80.6	99.9

Table 1. Results from Monte Carlo simulations to test structure identification methods. Stated are the number of correctly chosen models out of 100. $n = 4$.

to noise ratio and the level of significance for the ANOVA method and let the number of repeated measurements, n, equal 4. The possible levels of u_t are -2, 1, 3 and 5, that is, $m = 4$, so we get datasets of length $n * m^p = 192$. The results are collected in Table 1. It is also possible to compute the theoretical probability to find the correct model structure by the ANOVA method (Lind 1999).

From Table 1 we can draw the conclusion that the ANOVA method is much better at spotting what input time lags contribute to the output than the exhaustive search method. The results for the first function, $u_t - 0.03u_{t-2}$, shows that it is important to have not too small a signal to noise ratio (SNR). If the SNR is increased by a factor 4 the theoretical probability of finding the correct model structure by ANOVA increases from 4% to 89%.

The difference in performance between the two methods becomes more profound when the functions have a more nonlinear behaviour, e.g., exponential functions. This indicates that the network we have been working with does not handle this kind of functions very well, which we can confirm by looking at the RMS values for the validation data.

In Table 1, Column 5, the better performance for ANOVA as compared to Column 3 is mostly due to the increased significance level, except for the first function, where the decrease in noise variance is important to explain the better performance. We can also see that the decrease in noise variance does not affect the performance for the exhaustive search method either, except for the first function.

6. CONCLUSIONS

We can conclude from experiments that the analysis of variance method is a good alternative to exhaustive search for identifying what input contribute to the output in nonlinear finite impulse response (N-FIR) models.

The ANOVA method manages much better to identify the correct input signals (contributing time lags) and reduces the number of erroneous models from 19.4-24.4% to 0.1-9.7% when compared to the exhaustive search method. The main source of the differences is that the minimization algorithm gets stuck in local minima, due to the non-convexity of the identification problems. ANOVA is also computationally much more effective. The ANOVA method (including estimation of a model) is at least 14 times faster in finding appropriate input time lags and a corresponding model. When the complexity increases, due to more possible input time lags, it is an even greater advantage to use the ANOVA method.

7. REFERENCES

Demuth, Howard and Mark Beale (1998). *Neural Network Toolbox: Manual.* The MathWorks, Inc.. Natick, Mass.

Godfrey, Keith (1993). *Perturbation Signals for System Identification.* Prentice Hall. New York.

Hocking, Ronald R. (1984). *The Analysis of Linear Models.* Brooks/Cole. Monterey.

Krishnaiah, Paruchuri R., Ed.) (1980). *Handbook of Statistics.* Vol. 1. North-Holland. Amsterdam.

Lind, Ingela (1999). Use of the analysis of variance method for model order selection of N-

FIR models. Technical Report LiTH-ISY-R-2177. Department of Electrical Engineering, Linköping University.

Montgomery, Douglas C. (1991). *Design and Analysis of Experiments*. 3rd ed.. John Wiley & Sons. New York.

Scheffé, Henry (1959). *The Analysis of Variance*. John Wiley & Sons. New York.

Sjöberg, Jonas, Q Zhang, Lennart Ljung, A Benveniste, B Deylon, P-Y Glorennec, Håkan Hjalmarsson and A Juditsky (1995). Nonlinear black-box modeling in system identification: a unified overview. *Automatica* **31**, 1691–1724.

ALGORITHM FOR THREE-STEP ESTIMATION OF TRANSFER FUNCTION WITH UNKNOWN DELAY STEPS AND ORDER

Mitsuki MASHINO* , Fujio OHKAWA** ,
Kazushi NAKANO*** and Masayoshi TOMIZUKA****

* *Mechanics & Electronics Research Institute, Fukuoka Industrial Technology Center; 3-6-1 Norimatsu, Yahatanishi-ku, Kitakyushu 807-0831, JAPAN*
** *Dept. of Control Engineering and Science, Faculty of Computer & System Engineering, Kyushu Institute of Technology; Kawazu, Iizuka 820-8502, JAPAN*
*** *Dept. of Electronic Engineering, The University of Electro-Communicatons; 1-5-1 Chofu-ga-oka, Chofu, Tokyo 182-8585, JAPAN*
**** *Dept. of Mechanical Engineering, University of California at Berkeley; 5100 B Etcheverry Hall, Berkeley, CA 94720-1740, USA*

Abstract: This paper deals with the estimation of an unknown process transfer function in the presence of colored measurement noise. A three-step estimation procedure has been previously developed for transfer functions, the delay steps and the orders of which are known in advance. The procedure is extended to deal with transfer functions with unknown delay steps and orders. The auto-correlation function of the error between the process output and model output is utilized for evaluating the model fitness. The effectiveness of the proposed method is demonstrated by a simulation study using a sample set of data in MATLAB. *Copyright © 2000 IFAC*

Keywords: colored noise, I/O correlation information, BCLS estimates, parameter transformation, MSE

1. INTRODUCTION

For the problem of consistently estimating the transfer function model in the presence of colored measurement noise, several effective estimation methods and software packages have been already developed (Ljung,1987; Söderström and Stoica,1989).

We proposed a *three-step estimation procedure* (Mashino and Nakano,1992) with the following steps : 1) an initial parameter estimation by using the input-based instrumental variable (IV) method, 2) an estimation of noise auto-correlation functions, and 3) a final consistent estimation based on the optimal IV method. The procedure yielded a bias-compensated least squares (BCLS) estimate. It is important in the proposed method to set the order of the model in the first and second steps higher than the actual order so that the *input/output (I/O) correlation information* is expanded (Mashino and Nakano,1992,1997). Recently reconsidering the relation between the model order in the first and second steps and that in the third step from the viewpoint of *compressive parameter transformation* (Mashino and Nakano,1997), we presented a *compatibility condition* for satisfying both the BCLS estimates and the compressive parameter transformation (Mashino, Nakano and Tomizuka,1998; Nakano, Mashino and Tomizuka,2000). This condition is the consequence of the first and second steps in the previous three-step procedure.

The previous works on the three-step estimation procedure assumed that the delay steps and the order of the process are known. In this paper, we propose an algorithm for estimating the parameters as well as the delay steps and the order. For this purpose, we introduce a criterion in the correlation domain to evaluate the fitness between the actual process and its model (Mashino and Nakano,1999). We demonstrate the effectiveness of the proposing estimation algorithm by using the data provided in MATLAB (Ljung,1995).

Notations: An $(m+1)\times(n+1)$ Toeplitz-type matrix $\boldsymbol{\Phi}_{uy}(l)$, which becomes the Toeplitz matrix when $n=m$, is defined by

$$\boldsymbol{\Phi}_{uy}(l)=\begin{bmatrix}\phi_{uy}(l) & \phi_{uy}(l-1) & \cdots & \phi_{uy}(l-n)\\ \phi_{uy}(l+1) & \phi_{uy}(l) & \cdots & \phi_{uy}(l+1-n)\\ \vdots & \vdots & \vdots & \vdots\\ \phi_{uy}(l+m) & \phi_{uy}(l+m-1) & \cdots & \phi_{uy}(l+m-n)\end{bmatrix}$$

For notational simplicity, $\boldsymbol{\Phi}_{uy} = \boldsymbol{\Phi}_{uy}(0)$ and $\boldsymbol{\Phi}_u(l) = \boldsymbol{\Phi}_{uu}(l)$ ($\phi_u(l) = \phi_{uu}(l)$). An $(n+1) \times (n+1)$ symmetric Toeplitz matrix $\boldsymbol{\Phi}_u$ ($=\boldsymbol{\Phi}_u(0)$) is defined by

$$\boldsymbol{\Phi}_u = \begin{bmatrix} \phi_u(0) & \phi_u(1) & \cdots & \phi_u(n) \\ \phi_u(1) & \phi_u(0) & \cdots & \phi_u(n-1) \\ \vdots & \vdots & \vdots & \vdots \\ \phi_u(n) & \phi_u(n-1) & \cdots & \phi_u(0) \end{bmatrix}$$

An $(m+1)$-vector $\boldsymbol{\phi}_{uy}(l)$ is defined by

$$\boldsymbol{\phi}_{uy}(l) = [\phi_{uy}(l), \phi_{uy}(l+1), \cdots, \phi_{uy}(l+m)]^{\mathrm{T}}$$

where $\boldsymbol{\phi}_{uy} = \boldsymbol{\phi}_{uy}(0)$. In the same manner, $\boldsymbol{\phi}_y(l) = \boldsymbol{\phi}_{yy}(l)$ and $\boldsymbol{\phi}_y = \boldsymbol{\phi}_y(0)$.

2. PROBLEM STATEMENT

A single-input/single-output linear discrete process as configured in **Fig.1** is considered. The measurement y_k of the process output x_k is contaminated by the colored measurement noise v_k

$$y_k = x_k + v_k \tag{1}$$

The process input u_k and the output x_k satisfy the difference equation:

$$x_k = \sum_{i=1}^{\nu} \alpha_i x_{k-i} + \sum_{i=0}^{\mu} \beta_i u_{k-i-\delta} \tag{2}$$

The transfer function of this process with delay steps δ is given by

$$G(q^{-1}) = \frac{q^{-\delta} \cdot B(q^{-1})}{1 - A(q^{-1})} \tag{3}$$

where

$$1 - A(q^{-1}) = 1 - \sum_{i=1}^{\nu} \alpha_i q^{-i}, \quad B(q^{-1}) = \sum_{i=0}^{\mu} \beta_i q^{-i}$$

It is assumed that poles and zeros of this transfer function are all outside the unit circle centered at the origin of q^{-1}-plane. The degrees ν and μ are denoted as order (ν, μ). For notational simplicity, we will present the methodology for $\nu = \mu$, and will comment on cases where $\nu \neq \mu$.

The key assumption is that the colored noise v_k is independent of both u_k and x_k. Thus

$$\left.\begin{aligned} E[u_j v_k] = 0 \\ E[x_j v_k] = 0 \end{aligned}\right\} \quad (\text{ for all } j, k) \tag{4}$$

where $E[\,\cdot\,]$ denotes the expectation of $\cdot$.

The cross-correlation function $\phi_{uy}(l)$ and the auto-correlation function $\phi_y(l)$ are defined as

$$\left.\begin{aligned} \phi_{uy}(l) &= E[u_{k-l} y_k] \\ \phi_y(l) &= E[y_{k-l} y_k] \end{aligned}\right\} \tag{5}$$

They satisfy

$$\left.\begin{aligned} \phi_{uy}(l) &= \phi_{yu}(-l) \\ \phi_y(l) &= \phi_y(-l) \\ \boldsymbol{\Phi}_{uy}^T(l) &= \boldsymbol{\Phi}_{yu}(-l) \end{aligned}\right\} \tag{6}$$

where l is the correlation time (lag). From (4), the following relations hold:

$$\left.\begin{aligned} \phi_{uy}(l) &= \phi_{ux}(l) \\ \phi_y(l) &= \phi_x(l) + \phi_v(l) \end{aligned}\right\} \tag{7}$$

where $\phi_v(l)$ is the auto-correlation function of noise v_k.

Under the ergodic assumption, correlation functions are estimated by time average. For examples

$$\widehat{\phi}_{uy}(l) \simeq \frac{1}{N} \sum_{k=1}^{N} u_{k-l} y_k \tag{8}$$

where N is the number of data. We call $\widehat{\phi}_{uy}(l)$, $\widehat{\phi}_u(l)$ and $\widehat{\phi}_y(l)$ collectively the I/O correlation function, and a set consisting of all the correlation functions the I/O correlation information.

Our problem is to estimate the unknown parameters of $G(q^{-1})$ from the data $\{u_k,\ y_k\}$ $(k = 1, 2, \cdots, N)$. The purpose of this paper is to extend the three-step estimation procedure already proposed to the case where the process structure; i.e. the delay steps δ and the order ν, is unknown.

3. THREE-STEP ESTIMATION PROCEDURE FOR CONSISTENT ESTIMATES

For $\nu = \mu$, Eq.(2) becomes

$$x_k = \sum_{i=1}^{\nu} \alpha_i x_{k-i} + \sum_{i=0}^{\nu} \beta_i u_{k-i-\delta} \tag{2$'$}$$

The delay steps δ and the process order ν are denoted as structure $(\delta; \nu)$.

Let us consider Model-1 with structure $(d; n)$. Then the Model-1 output x_k^o is generated by

$$x_k^o = \sum_{i=1}^{n} a_i^o x_{k-i} + \sum_{i=0}^{n} b_i^o u_{k-i-d} \tag{9}$$

The regression vector is defined by

$$\begin{bmatrix} \boldsymbol{a}_n^o \\ \boldsymbol{b}_n^o \end{bmatrix} = [a_1^o,\ a_2^o,\ \cdots,\ a_n^o,\ b_0^o,\ b_1^o,\ \cdots,\ b_n^o]^{\mathrm{T}}$$

Model-2 is assumed to have a structure different from Model-1. When using the structure $(\delta;\nu)$, the Model-2 output is generated by

$$x_k^t = \sum_{i=1}^{\nu} \widehat{a}_i x_{k-i} + \sum_{i=0}^{\nu} \widehat{b}_i u_{k-i-\delta} \tag{10}$$

For the moment, we assume that δ and ν are known. The regression vector for Model-2 is defined by

$$\begin{bmatrix} \widehat{a}_\nu \\ \widehat{b}_\nu \end{bmatrix} = [\widehat{a}_1,\ \widehat{a}_2,\ \cdots,\ \widehat{a}_\nu,\ \widehat{b}_0,\ \widehat{b}_1,\ \cdots,\ \widehat{b}_\nu]^{\mathrm{T}}$$

Under the Model-1 structure $(d; n)$ and the Model-2 structure $(\delta; \nu)$, the three-step estimation procedure is described as follows:

1st-Step: Initial Estimation of Parameters

Assuming the Model-1 structure $(d; n)$, we obtain initial parameter estimates only from $\widehat{\phi}_u(l)$ and $\widehat{\phi}_{uy}(l)$ by using the input-based IV (instrumental variable) method (Wouters,1972) (which is sometimes called the correlation method)

$$\begin{bmatrix} a_n^o \\ b_n^o \end{bmatrix} = \begin{bmatrix} \widehat{\boldsymbol{\Phi}}_{uy}(d-1) & \widehat{\boldsymbol{\Phi}}_u \end{bmatrix}^{-1} \widehat{\phi}_{uy}(d) \quad (11)$$

where $\widehat{\boldsymbol{\Phi}}_{uy}(d-1)$ is a $(2n+1) \times n$ matrix, $\widehat{\boldsymbol{\Phi}}_u$ a $(2n+1) \times (n+1)$ matrix, and $\widehat{\phi}_{uy}(d)$ a $(2n+1)$-dimensional vector.

2nd-Step: Direct Estimation of Noise Auto-Correlation Functions

Assuming the Model-1 structure $(d; n)$, we obtain the estimates of noise auto-correlation functions

$$\widehat{\phi}_v = [\widehat{\phi}_v(0),\ \widehat{\phi}_v(1),\ \cdots,\ \widehat{\phi}_v(n)]^{\mathrm{T}}$$

from a_n^o and b_n^o in the first step (Mashino and Nakano,1992): i.e.

$$\widehat{\phi}_v = \widehat{\phi}_y + \boldsymbol{\Gamma}_{a^o}^{-1} \widehat{\boldsymbol{\Phi}}_{uy}^{\mathrm{T}}(d) b_n^o \quad (12)$$

where $\widehat{\boldsymbol{\Phi}}_{uy}(d)$ is an $(n+1) \times (n+1)$ matrix, $\widehat{\phi}_y$ an $(n+1)$-dimensional vector, $a_0^o = -1$, and

$$\boldsymbol{\Gamma}_{a^o} = \begin{bmatrix} a_0^o & o & \cdots & 0 \\ a_1^o & a_0^o & \cdots & 0 \\ \vdots & \vdots & \vdots & \vdots \\ a_{n-1}^o & a_{n-2}^o & \cdots & 0 \\ a_n^o & a_{n-1}^o & \cdots & a_0^o \end{bmatrix} + \begin{bmatrix} 0 & a_1^o & \cdots & a_{n-1}^o & a_n^o \\ 0 & a_2^o & \cdots & a_n^o & 0 \\ \vdots & \vdots & \vdots & \vdots & \vdots \\ 0 & a_n^o & \cdots & 0 & 0 \\ 0 & 0 & \cdots & 0 & 0 \end{bmatrix}$$

is an $(n+1) \times (n+1)$ matrix.

Notice that $\widehat{\phi}_v(l)(l = n+1, n+2, \cdots)$ may be obtained recursively.

3rd-Step: Final Estimation of Parameters

Assuming the Model-2 structure $(\delta; \nu)$, we obtain the final parameter estimates by using the BCLS estimation method (Sagara and Wada,1997; Zhdanov and Katsyuba,1979)

$$\begin{bmatrix} \widehat{a}_\nu \\ \widehat{b}_\nu \end{bmatrix} = \begin{bmatrix} \widehat{\boldsymbol{\Phi}}_y - \widehat{\boldsymbol{\Phi}}_v & \widehat{\boldsymbol{\Phi}}_{uy}^{\mathrm{T}}(\delta-1) \\ \widehat{\boldsymbol{\Phi}}_{uy}(\delta-1) & \widehat{\boldsymbol{\Phi}}_u \end{bmatrix}^{-1} \begin{bmatrix} \widehat{\phi}_y(1) - \widehat{\phi}_v(1) \\ \widehat{\phi}_{uy}(\delta) \end{bmatrix} \quad (13)$$

where $\widehat{\boldsymbol{\Phi}}_y$ and $\widehat{\boldsymbol{\Phi}}_v$ are $\nu \times \nu$ matrices, $\widehat{\boldsymbol{\Phi}}_{uy}(\delta-1)$ a $(\nu+1) \times \nu$ matrix, $\widehat{\boldsymbol{\Phi}}_u$ a $(\nu+1) \times (\nu+1)$ matrix, $\widehat{\phi}_y(1)$ and $\widehat{\phi}_v(1)$ ν-dimensional vectors, and $\widehat{\phi}_{uy}(\delta)$ a $(\nu+1)$-dimensional vector.

Eq.(13) is originally derived by the optimal IV method (Pandya,1974). It can also be derived by minimizing the following mean square error (MSE) between the process output x_k and the Model-2 output x_k^t:

$$J = E[(x_k^t - x_k)^2] \quad (14)$$

4. COMPATIBILITY BETWEEN PARAMETER TRANSFORMATION AND BCLS ESTIMATION

4.1 Parameter Transformation

We consider the transformation from Model-1 to Model-2 as shown in **Fig.1**. This is carried out so as to minimize the loss function J_m defined by the MSE between the model outputs

$$J_m = E[(x_k^t - x_k^o)^2] \quad (15)$$

By partially differentiating J_m with respect to $\widehat{a}_\nu$ and $\widehat{b}_\nu$, we obtain the following formula for parameter transformation [1] (Mashino and Nakano,1997; Nakano, Mashino and Tomizuka,2000):

$$\begin{bmatrix} \widehat{a}_\nu \\ \widehat{b}_\nu \end{bmatrix} = \begin{bmatrix} \underbrace{\begin{matrix} \widehat{\boldsymbol{\Phi}}_y - \widehat{\boldsymbol{\Phi}}_v \\ \widehat{\boldsymbol{\Phi}}_{uy}(\delta-1) \end{matrix}}_{\nu} & \underbrace{\begin{matrix} \widehat{\boldsymbol{\Phi}}_{uy}^{\mathrm{T}}(\delta-1) \\ \widehat{\boldsymbol{\Phi}}_u \end{matrix}}_{\nu+1} \end{bmatrix}^{-1}$$

$$\times \begin{bmatrix} \underbrace{\begin{matrix} \widehat{\boldsymbol{\Phi}}_y - \widehat{\boldsymbol{\Phi}}_v \\ \widehat{\boldsymbol{\Phi}}_{uy}(\delta-1) \end{matrix}}_{n} & \underbrace{\begin{matrix} \widehat{\boldsymbol{\Phi}}_{uy}^{\mathrm{T}}(d-1) \\ \widehat{\boldsymbol{\Phi}}_u(\delta-d) \end{matrix}}_{n+1} \end{bmatrix} \begin{bmatrix} a_n^o \\ b_n^o \end{bmatrix} \quad (16)$$

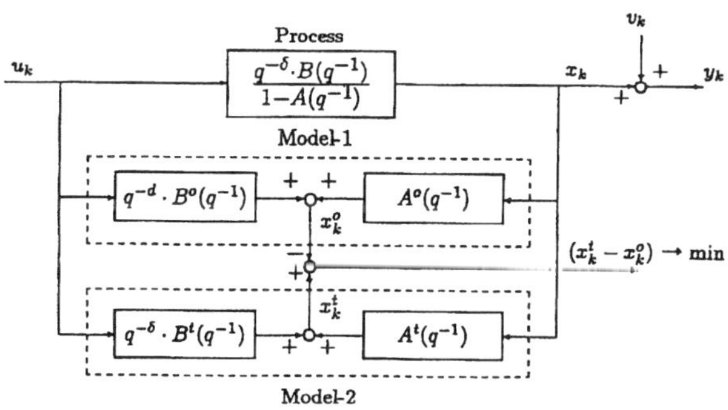

$$G^o(q^{-1}) = \frac{q^{-d} \cdot B^o(q^{-1})}{1 - A^o(q^{-1})} \ :\ A^o(q^{-1}) = \sum_{i=1}^{n} a_i^o q^{-i},\ B^o(q^{-1}) = \sum_{i=0}^{n} b_i^o q^{-i}$$

$$G^t(q^{-1}) = \frac{q^{-\delta} \cdot B^t(q^{-1})}{1 - A^t(q^{-1})} \ :\ A^t(q^{-1}) = \sum_{i=1}^{\nu} \widehat{a}_i q^{-i},\ B^t(q^{-1}) = \sum_{i=0}^{\nu} \widehat{b}_i q^{-i}$$

Fig.1. Transformation from Model-1 to Model-2

In the special case where $d = \delta$, the formula (16) is classified into the following three cases depending on the order of Model-1: (a) the over-parameterized case $(n > \nu)$, (b) the matched order case $(n = \nu)$, (c) the under-parameterized case

[1] Both x_k^t and x_k^o in the time domain are not necessary to calculate the formula (16) in the correlation domain.

$(n < \nu)$. The cases (b) and (c) reduce to the formula (16) with

$$\widehat{a}_i = a_i^o \ (i = 1, \cdots, n), \ \ \widehat{a}_i = 0 \ (i = n+1, \cdots, \nu)$$
$$\widehat{b}_i = b_i^o \ (i = 0, \cdots, n), \ \ \widehat{b}_i = 0 \ (i = n+1, \cdots, \nu)$$

This implies that Model-2 is neither better nor worse than Model-1. On the other hand, there is a possibility that the parameters of Model-2 represents an improvement for the case (a). In the over-parameterized case (a), the formula (16) reduces to the compressive parameter transformation

$$\begin{bmatrix} \widehat{\boldsymbol{a}}_\nu \\ \widehat{\boldsymbol{b}}_\nu \end{bmatrix} = \begin{bmatrix} \boldsymbol{a}_\nu^o \\ \boldsymbol{b}_\nu^o \end{bmatrix} + \begin{bmatrix} \widehat{\boldsymbol{\Phi}}_y - \widehat{\boldsymbol{\Phi}}_v & \widehat{\boldsymbol{\Phi}}_{uy}^{\mathrm{T}}(\delta-1) \\ \widehat{\boldsymbol{\Phi}}_{uy}(\delta-1) & \widehat{\boldsymbol{\Phi}}_u \end{bmatrix}^{-1}$$
$$\times \begin{bmatrix} \widehat{\boldsymbol{\Phi}}_y(-\nu) - \widehat{\boldsymbol{\Phi}}_v(-\nu) & \widehat{\boldsymbol{\Phi}}_{yu}(-\delta-\nu) \\ \widehat{\boldsymbol{\Phi}}_{uy}(\delta-\nu-1) & \widehat{\boldsymbol{\Phi}}_u(-\nu-1) \end{bmatrix} \begin{bmatrix} \boldsymbol{a}_{\nu+1,n}^o \\ \boldsymbol{b}_{\nu+1,n}^o \end{bmatrix} \quad (17)$$

where

$$\boldsymbol{a}_\nu^o = [a_1^o, a_2^o, \cdots, a_\nu^o]^{\mathrm{T}}, \ \boldsymbol{a}_{\nu+1,n}^o = [a_{\nu+1}^o, a_{\nu+2}^o, \cdots, a_n^o]^{\mathrm{T}}$$
$$\boldsymbol{b}_\nu^o = [b_0^o, b_1^o, \cdots, b_\nu^o]^{\mathrm{T}}, \ \boldsymbol{b}_{\nu+1,n}^o = [b_{\nu+1}^o, b_{\nu+2}^o, \cdots, b_n^o]^{\mathrm{T}}$$

and the dimensions of the two matrices are as shown below

$$\begin{matrix} \nu \{ \\ \nu+1 \{ \end{matrix} \begin{bmatrix} \underbrace{\begin{matrix} \widehat{\boldsymbol{\Phi}}_y - \widehat{\boldsymbol{\Phi}}_v \\ \widehat{\boldsymbol{\Phi}}_{uy}(\delta-1) \end{matrix}}_{\nu} & \underbrace{\begin{matrix} \widehat{\boldsymbol{\Phi}}_{uy}^{\mathrm{T}}(\delta-1) \\ \widehat{\boldsymbol{\Phi}}_u \end{matrix}}_{\nu+1} \end{bmatrix}$$

$$\begin{matrix} \nu \{ \\ \nu+1 \{ \end{matrix} \begin{bmatrix} \underbrace{\begin{matrix} \widehat{\boldsymbol{\Phi}}_y(-\nu) - \widehat{\boldsymbol{\Phi}}_v(-\nu) \\ \widehat{\boldsymbol{\Phi}}_{uy}(\delta-\nu-1) \end{matrix}}_{n-\nu} & \underbrace{\begin{matrix} \widehat{\boldsymbol{\Phi}}_{yu}(-\delta-\nu) \\ \widehat{\boldsymbol{\Phi}}_u(-\nu-1) \end{matrix}}_{n-\nu} \end{bmatrix}$$

The second term in the right hand side of Eq.(17) is a modifier.

4.2 Compatibility Condition

The parameter transformation (16) yields estimates identical with the BCLS estimates (13) under the following compatibility condition (Mashino, Nakano and Tomizuka,1998; Nakano, Mashino and Tomizuka,2000):

$$\begin{matrix} \nu \{ \\ \nu+1 \{ \end{matrix} \begin{bmatrix} \underbrace{\begin{matrix} \widehat{\boldsymbol{\Phi}}_y - \widehat{\boldsymbol{\Phi}}_v \\ \widehat{\boldsymbol{\Phi}}_{uy}(\delta-1) \end{matrix}}_{n} & \underbrace{\begin{matrix} \widehat{\boldsymbol{\Phi}}_{uy}^{\mathrm{T}}(d-1) \\ \widehat{\boldsymbol{\Phi}}_u(\delta-d) \end{matrix}}_{n+1} \end{bmatrix} \begin{bmatrix} \boldsymbol{a}_n^o \\ \boldsymbol{b}_n^o \end{bmatrix} = \begin{bmatrix} \widehat{\boldsymbol{\phi}}_y(1) - \widehat{\boldsymbol{\phi}}_v(1) \\ \widehat{\boldsymbol{\phi}}_{uy}(\delta) \end{bmatrix} \quad (18)$$

This is derived by picking out the elements of the equations in the first and second steps of the three-step estimation procedure under the order condition $((n \geq \nu)$ and $(2n + d \geq \nu + \delta))$ (see **Appendix A**). Under the compatibility condition, the use of Eq.(16) in the third step yields the BCLS estimates (13) (Mashino and Nakano,1997). Conversely, when applying the BCLS estimation method in the same third step, it is nothing but the compressive parameter transformation (16).

Therefore, the compatibility condition (18) makes it possible to minimize both the MSEs $E[(x_k^t - x_k^o)^2]$ and $E[(x_k^t - x_k)^2]$ [2]. Then, the final parameter estimates $[\widehat{\boldsymbol{a}}_\nu, \widehat{\boldsymbol{b}}_\mu]^{\mathrm{T}}$ are optimal in the sense of minimizing MSEs among x_k, x_k^o and x_k^t.

For a more general case of $\nu \neq \mu$ (instead of $\nu = \mu$), the Model-1 order corresponds to $(n, m)(n \neq m)$. In this case in *4.1*, possibilities of improvement of the transformed parameter estimates are for the over-parameterized case $(n \geq \nu, m \geq \mu)$ except $(n = \nu, m = \mu)$. Also in *4.2*, the compatibility condition is satisfied for the case ($(n \geq \nu)$ and $(n + m + d \geq \mu + \delta)$).

5. THREE-STEP ESTIMATION ALGORITHM CONSIDERING MODEL STRUCTURE

5.1 Evaluation for Model Fitting

In the previous section, we considered the loss function J given by (14) for the BCLS estimates. This criterion made it possible to minimize the MSE between the Model-2 output and the process output. In this section, let us consider the following new function (Mashino and Nakano,1999):

$$S_J = J^2(0) + J^2(1) + \cdots + J^2(\tau) \quad (19)$$

where $J(l)$ is the auto-correlation function of the output error as defined by

$$J(l) = E[(x_{k-l}^t - x_{k-l})(x_k^t - x_k)] \quad (20)$$

and τ is an integer which denotes the range of lag to be taken into account.

The minimization of the function S_J makes it possible to evaluate the degree of fitness between the process and its model in the correlation domain [3]. Now, setting $(\delta; \nu)$ as $(\widehat{\delta}; \widehat{\nu})$, we have

$$J(l) = \begin{bmatrix} \widehat{\boldsymbol{a}}_{0,\widehat{\nu}}^{\mathrm{T}} & \widehat{\boldsymbol{b}}_{\widehat{\nu}}^{\mathrm{T}} \end{bmatrix} \begin{bmatrix} \widehat{\boldsymbol{\Phi}}_y(l) - \widehat{\boldsymbol{\Phi}}_v(l) & \widehat{\boldsymbol{\Phi}}_{yu}(l-\widehat{\delta}) \\ \widehat{\boldsymbol{\Phi}}_{uy}(l+\widehat{\delta}) & \widehat{\boldsymbol{\Phi}}_u(l) \end{bmatrix} \begin{bmatrix} \widehat{\boldsymbol{a}}_{0,\widehat{\nu}} \\ \widehat{\boldsymbol{b}}_{\widehat{\nu}} \end{bmatrix} \quad (21)$$

where $\widehat{\boldsymbol{a}}_{0,\widehat{\nu}} = [\widehat{a}_0, \widehat{a}_1, \cdots, \widehat{a}_{\widehat{\nu}}]^{\mathrm{T}}$ with $\widehat{a}_0 = -1$ and $\widehat{\boldsymbol{b}}_{\widehat{\nu}} = [\widehat{b}_0, \widehat{b}_1, \cdots, \widehat{b}_{\widehat{\nu}}]^{\mathrm{T}}$. Moreover, the dimensions of all the correlation matrices $\widehat{\boldsymbol{\Phi}}_y(l)$, $\widehat{\boldsymbol{\Phi}}_v(l)$, $\widehat{\boldsymbol{\Phi}}_{yu}(l-\widehat{\delta})$, $\widehat{\boldsymbol{\Phi}}_{yu}(l+\widehat{\delta})$ and $\widehat{\boldsymbol{\Phi}}_u(l)$ are $(\widehat{\nu}+1) \times (\widehat{\nu}+1)$. $J(l)$ of Eq.(21) can be obtained in practice from the I/O correlation functions and the estimates $\widehat{\phi}_v$ in the second step.

5.2 Three-Step Estimation Algorithm and Its Application

---

[2] Although the model transformation based on $E[(x_k^t - x_k^o)^2]$ is independent of the actual process, the transformed model output x_k^t can be connected with the process output x_k via the compatibility condition.

[3] When $J(0)$ in eq.(21) is calculated using parameter estimates $\widehat{a}_{\widehat{\nu}}, \widehat{b}_{\widehat{\nu}}$ which are obtained under the MSE of J, $J(0)(= J)$ is minimum for any structure $(\widehat{\delta}; \widehat{\nu})$. Then, only use of $J(0)$ is inadequate for evaluating the structure $(\widehat{\delta}; \widehat{\nu})$.

We will explain the proposed three-step estimation algorithm according to the flow-chart in **Fig.2** (Mashino and Nakano,1999). The chart is made for cases where $n \neq m$ and $\widehat{\nu} \neq \widehat{\mu}$, and is explained below for case where $n = m$ and $\widehat{\nu} = \widehat{\mu}$:

First of all, we select a structure $(d; n)$ of Model-1 such that $d \leq \widehat{\delta}$ and $n > \widehat{\nu}$ for the structure $(\widehat{\delta}; \widehat{\nu})$. Then, the order n should be selected higher; i.e. be selected such that the I/O correlation information in the structure $(\widehat{\delta}; \widehat{\nu})$ is inclusive as shown in **Fig.3** in **Appendix-B**. Under the structure $(d; n)$, we obtain the parameter estimates $\boldsymbol{a}_n^o$, $\boldsymbol{b}_n^o$ in the first step, and estimates of noise auto-correlation functions $\widehat{\boldsymbol{\phi}}_v$ in the second step. Finally, through the parameter transformation, we obtain the BCLS estimates under the structure $(\widehat{\delta}; \widehat{\nu})$. From the combinations of the model structures $(\widehat{\delta}; \widehat{\nu})$, we first select the delay steps $\widehat{\delta}$ with stable poles and zeros so as to minimize the function S_J among $(\widehat{\delta}; \widehat{\nu})$. Secondly, making the delay steps d corresponding with $\widehat{\delta}$ and executing the three steps after returning to the first selected model, we obtain the parameter estimates $\widehat{\boldsymbol{a}}_{\widehat{\nu}}, \widehat{\boldsymbol{b}}_{\widehat{\nu}}$ to satisfy the parameter transformation. Finally, we obtain the optimal structure $(\delta^o; \nu^o)$ to attain the minimum S_J after correcting the combinations of n and $(\widehat{\delta}; \widehat{\nu})$.

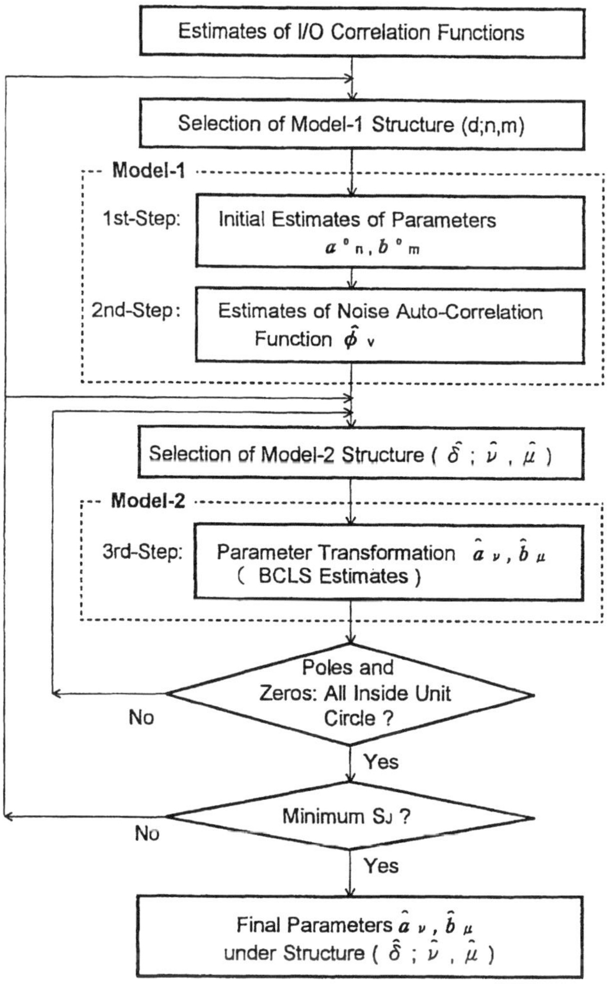

Fig.2. Three-Step Estimation Algorithm

Applications of the three-step estimation algorithm to the data (Dryer 2) provided in MATLAB resulted in a slightly better result in terms of the model fitness than the Box-Jenkins estimation.

6. CONCLUSIONS

First, we extended the previously proposed three-step estimation procedure to the time-delay process without correction of the original estimation method. Next, introducing a new criterion for evaluating the degree of fitness in the model structure, we proposed a three-step estimation algorithm for estimating both the model structure and parameters. The effectiveness of the algorithm has been demonstrated thought simulation using the data provided in MATLAB.

REFERENCES

Cadzow,J.A., B.Baseghi and T.Hsu(1983). Singular-Value Decomposition Approach to Time Series Modelling; *IEE Proc.*, Vol.130, No.3, pp.202-210

Ljung,L.(1987). *System Identification – Theory for the User*, Prentice Hall

Ljung,L.(1995). *System Identification Toolbox User's Guide*, The Math Works, Inc.

Mashino,M. and K.Nakano(1992). Estimation Procedure for Parameters of Transfer Function by Using Estimates of Autocorrelation Function of Measurement Noise, *Trans. of the Institute of Systems, Control and Information Engineers*, Vol.5, No.6, pp.221-228 (in Japanese)

Mashino,M. and K.Nakano(1997). A Three-Step Estimation Procedure for the Pulse Transfer Function from the Viewpoint of a Reductive Parameter Transformation, *Proc. of 11th IFAC Symp. on System Identification (SYSID'97)*, Vol.3, pp.1387-1392

Mashino,M., K.Nakano and M.Tomizuka (1998). A Three-Step Estimation Procedure for the Transfer Function Based on the Minimization of the Mean Square Error between Model Outputs, *Proceedings of the 3rd Asia-Pacific Conference on Control & Measurement*, pp.48-52

Mashino,M. and K.Nakano(1999). Three-Step Estimation Procedure for Transfer Function Considering Model Structure, *Preprints of the 28th SICE Symposium on Control Theory*, pp.111-116 (in Japanese)

Nakano,K., M.Mashino and M.Tomizuka(2000). Bias-Compensated Least-Squares Estimation and Parameter Transformation in Identification with Over-Parameterized Model, *Trans. of the Institute of Systems, Control and Information Engineers*, Vol.13, No.2, pp.72-79 (in Japanese)

Pandya,R.N. (1974). A Class of Bootstrap Estimators and Their Relationship to the Generalized Two Stage Least Squares Estimators, IEEE Trans. *Automat. Contr.*, Vol.AC-19, No.6, pp.831-835

Söderström,T. and P.Stoica(1989). *System Identification,* Prentice Hall
Sagara,S. and K.Wada(1977). On-line Modified Least-Squares Parameter Estimation of Linear Discrete Dynamical Systems, *Int.J.Control,* Vol.25, No.3, pp.329-345
Wouters,W.R.(1972). On-Line Identification in an Unknown Stochastic Environment, *IEEE Trans. Syst. Man Cybern.,* Vol.SMC-2, No.5, pp.666-668
Zhdanov,A.I. and O.A.Katsyuba(1979). Specifics of Applying the Method of Least Squares to Estimation of Linear Difference Operators in Identification Problems, *Avtom. Telemekh.,* Vol.8, No.8, pp.86-92

APPENDIXES

Appendix-A Compatibility Condition

Eqs.(11) and (12) in the first and second steps are described in the matrix form

$$\begin{matrix} n+1\,\{ \\ 2n+1\{ \end{matrix}\begin{bmatrix} \underbrace{\widehat{\boldsymbol{\Phi}}_y(-1)-\widehat{\boldsymbol{\Phi}}_v(-1) \\ \widehat{\boldsymbol{\Phi}}_{uy}(d-1)}_{n} & \underbrace{\widehat{\boldsymbol{\Phi}}^T_{uy}(d) \\ \widehat{\boldsymbol{\Phi}}_u}_{n+1} \end{bmatrix}\begin{bmatrix} a^o_n \\ b^o_n \end{bmatrix}=\begin{bmatrix} \widehat{\phi}_y-\widehat{\phi}_v \\ \widehat{\phi}_{uy}(d) \end{bmatrix} \qquad \text{(A1)}$$

where the lower block corresponds to Eq.(11) and the upper block to Eq.(12)(Mashino and Nakano,1992). By neglecting the first element in the above equality, we have

$$\begin{matrix} n\;\{ \\ 2n+1\{ \\ +d-\delta \end{matrix}\begin{bmatrix} \widehat{\boldsymbol{\Phi}}_y-\widehat{\boldsymbol{\Phi}}_v & \widehat{\boldsymbol{\Phi}}^T_{uy}(d-1) \\ \underbrace{\widehat{\boldsymbol{\Phi}}_{uy}(\delta-1)}_{n} & \underbrace{\widehat{\boldsymbol{\Phi}}_u(\delta-d)}_{n+1} \end{bmatrix}\begin{bmatrix} a^o_n \\ b^o_n \end{bmatrix}=\begin{bmatrix} \widehat{\phi}_y(1)-\widehat{\phi}_v(1) \\ \widehat{\phi}_{uy}(\delta) \end{bmatrix} \qquad \text{(A2)}$$

ν elements in the upper block and $(\nu+1)$ elements in the lower block of this equality are written as the compatibility condition (18). The general condition that Eq.(A1) covers Eq.(18) is $(n \geq \nu)$ and $(2n+d \geq \nu+\delta)$.

Appendix-B I/O Correlation Information

In the three-step estimation procedure, the final BCLS estimates depend only on the estimates of the noise auto-correlation $\widehat{\phi}_v$. Then, we consider how the estimates $\widehat{\phi}_v$ are determined from data sets of the I/O correlation information.

B.1 Information of Toeplitz-Type Matrix

For example, in the first step with the Model-1 structure $(d;n)$ of the three-step estimation procedure, let us consider the information of the Toeplitz-type matrix

$$\widehat{\boldsymbol{\Phi}}_{uy}(d-1)=\begin{bmatrix} \widehat{\phi}_{uy}(d-1) & \widehat{\phi}_{uy}(d-2) & \cdots & \widehat{\phi}_{uy}(d-n) \\ \widehat{\phi}_{uy}(d) & \widehat{\phi}_{uy}(d-1) & \cdots & \widehat{\phi}_{uy}(d-n+1) \\ \vdots & \vdots & \vdots & \vdots \\ \widehat{\phi}_{uy}(d+2n-1) & \widehat{\phi}_{uy}(d+2n-2) & \cdots & \widehat{\phi}_{uy}(d+n) \end{bmatrix}$$

in Eq.(11). As shown in **Fig.3**, we can characterize $\widehat{\boldsymbol{\Phi}}_{uy}(d-1)$ by not only the I/O correlation information as a set of $\{\widehat{\phi}_{uy}(d-n), \widehat{\phi}_{uy}(d-n+1),\cdots,\widehat{\phi}_{uy}(d+2n-1)\}$ but also the number (frequency) of the elements $\widehat{\phi}_{uy}(l)$ which appears diagonally in $\widehat{\boldsymbol{\Phi}}_{uy}(d-1)$. The I/O information and frequency are determined uniquely by the structure $(d;n)$ and vary with the range of lag l.

B.2 BCLS Estimates Based on Inclusive Expansion

Next, let us consider how the information is expanded by use of the order n instead of the true order ν, when delay steps d is fixed by $d=\delta$ in $d \leq \delta$ for simplicity. **Fig.3** shows the case of inclusive expansion which means to expand both the I/O information and the frequency including the original ones. This is obviously realized by selecting the over-parameterized case $n(>\nu)$. Also, the inclusive expansion is made for $\widehat{\boldsymbol{\Phi}}_u$ in the first step and $\widehat{\boldsymbol{\Phi}}^T_{uy}(d)$ in the second step. For these inclusive expansions, only the case of the I/O correlation information can be considered, since the frequency is necessarily inclusive when the I/O information is inclusive.

We consider that the estimates of both a^o_n, b^o_n and $\widehat{\phi}_v$ in the first and second steps are less sensitive to the estimation error of the I/O correlation information; i.e. both $\{\widehat{\phi}_u\}$ and $\{\widehat{\phi}_{uy}\}$ according to the assertion[4] which relates to the inclusive expansion. So, we insist that BCLS estimates in the final step become better in precision by using $\widehat{\phi}_v$ which have little effect of the estimation error of $\{\widehat{\phi}_{uy}\}$ and $\{\widehat{\phi}_u\}$.

For the general case of the order (ν,μ) (instead of $\nu=\mu$), the above description is satisfied under $(n \geq \nu, m \geq \mu)$ except $(n=\nu, m=\mu)$ when $d=\delta$.

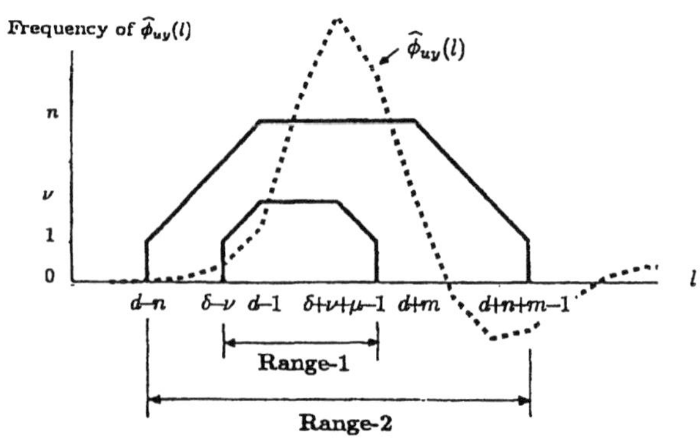

Fig.3. Frequency Functions for Lag l for the Case of $\widehat{\boldsymbol{\Phi}}_{uy}(d-1)$ in the First Step
(The vertical axis indicates the frequency of correlation functions. Range-1 depends on $(\delta;\nu,\mu)$ and Range-2 on $(d;n>\nu,m>\mu)$.)

[4]For the Toeplitz-type matrix composed of auto-correlation functions, it should be noted that, by taking this over-parameterized model, the resultant modeling will be less sensitive to the statistical quirks of the data being analysed than the case for the matched-order choice (Cadzow, Baseghi and Hsu,1983).

AUTOMATIC MODEL SELECTION FOR LINEAR TIME-INVARIANT SYSTEMS

Gyula Simon*, Johan Schoukens and Yves Rolain****

* Dept. Measurement and Information Systems, Budapest University of Technology and Economics, Hungary. e-mail: simon@mit.bme.hu
** Dept. ELEC, Vrije Universiteit, Brussel, Belgium.
e-mail: {Johan.Schoukens, Yves.Rolain}@vub.ac.be

Abstract: A completely automatic identification system is described which uses the latest results of the frequency domain identification approach to provide an easy-to-use and reliable tool to the inexperienced users. The proposed system accepts periodic measurement data and performs the whole identification procedure automatically from the data-preprocessing step to model and parameter selection. The validation phase provides information on the results at the level of the non-expert user. The performance of the system is illustrated by results obtained from real measurement data. *Copyright © 2000 IFAC*

Keywords: model approximation, system identification, frequency domain

1. INTRODUCTION

The theory of estimation of linear time-invariant (LTI) systems and systems with a dominant LTI behavior perturbed by nonlinear disturbances is well defined and available in the literature (Ljung 1999, Söderström *et al.*, 1989, Norton 1986, Schoukens *et al.*, 1991). The system identification process, however, is so complicated that the proper usage of the theoretical results (or even the usage of the available tools (Ljung 1986, Kollár 1994)) requires a solid theoretical background, signal processing skills, and know-how on the identification field. While some parts of the identification process can be (and have been) automated, some parts still require decisions of human experts. To allow users that are experienced in their own field to use identification – and hence extract high quality validated models – without having to become an identification expert, the process should be fully automated. The final goal is that users "drop in the measurement results, have a coffee, and get the final validated model" without further intervention.

Many new contributions have been published lately in the identification field (Ljung 1999, Söderström *et al.*, 1989, Norton 1986, Schoukens *et al.*, 1991, 1998, Rolain *et al.*, 1997), which can serve as fundamental elements in the construction of a fully automated identification process. Based upon these results in this paper an automatic model selection and validation system is described.

The flowchart of the system identification process can be seen in Fig. 1. After the measurement, that is set up to obtain repeated measurements (at least 4 repeated measurements or 4 periods of the signal) a non-parametric noise model is extracted out of the data by calculating the non-parametric variances (Schoukens *et al.*, 1997). This preprocessing step can easily and automatically be accomplished by using periodic excitation signals, as will be discussed in Section 2.

The solution of the complete identification problem is then further split into two parts: a model order estimation part and a parameter estimation part. The model selection phase estimates the order of the LTI model, which can adequately describe the linear dynamic behavior of the system. Note that many practical systems contain perturbations that avoid the system to be fully described by a LTI model. Hence, no exact model order will exist as the system falls out of the chosen model class. Other small perturbations, such as device non-linearities, unmodeled device dynamics, instrumentation errors or calibration residuals will further contribute to the lack of identifiability of "the

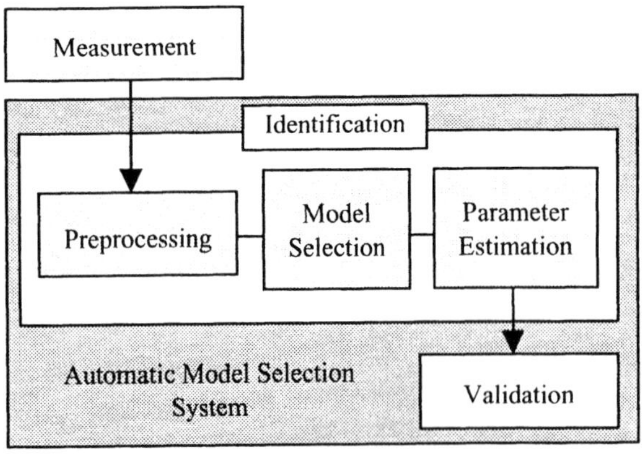

Fig. 1. The identification process and the tasks of the Automatic Model Selection System

exact model" even if it exists. The method proposed will hence deliver a model which describes the system dynamics adequately, and in this sense is "close" to this ideal model order. In any case, at some moment modeling errors may require a user decision to be made.

In some cases, there is a priori knowledge upon the system: the order determination may then be straight-forward through the analysis of the underlying physical phenomena. In most cases, however, the knowledge of the system is too idealized, and hence enough only to an approximate estimation of the correct model order. In many situations, there may not be any *a priory* information about the system's order. In these cases, the model order must be determined either by guess-and-trial methods (which are very commonly used), or by systematic order selection methods, which to the best knowledge of the authors, do not exist in the literature.

A systematic order selection method can use either bottom-up (from simple to complex) or top-down (from complex to simple) approach. A bottom-up approach starts the search from a simple model and tries to improve the fit by adding additional degrees of freedom until the desired (user defined) precision is reached. The main advantage of this approach is that the complexity of the model during the estimation process remains always below the final complexity, which is a useful property from the viewpoint of the numerical stability of many widely used algorithms. The top-down approach starts from a high-order system and the complexity is reduced while the fit is still satisfactory. Practical experience shows that this approach can avoid local minima with higher probability, and thus the top-down models can provide better estimations than their bottom-up counterparts, especially in the cases when the model order is high. Since top-down methods must be able to cope with high orders and common pole-zero pairs, the numerical stability is a key issue for the considered application.

The parameter estimation phase is a very deeply examined field and thus many theoretical results are available (Ljung 1999, Söderström *et al.*, 1989, Norton 1986, Schoukens *et al.*, 1991). The main problem of all the known algorithms is that they tend to find local minima of different quality, depending on the initial conditions. Therefore, these algorithms require a good initial guess in order to provide "optimal" results. The determination of the initial values of the algorithms and the parameterization of the sophisticated search methods to make the best of their capabilities requires a great deal of know-how from the users of these algorithms.

The heart of the described system is a top-down mixed order and model approximation algorithm. In an initialization phase, an appropriate starting value is determined, based upon Rolain *et al.* (1997). This initial order is first determined by a coarse order estimation step, and then using this as an initial guess, the final order and model are determined in an iterative way. To provide the necessary numerical stability orthogonal polynomials are used in the representation of the LTI model, and during the iterative estimation loop sophisticated initial values are provided for the model estimation algorithm.

Since the model estimation should serve the requirements of the user, the proposed system is able to provide a set of models to fulfill strict requirements with higher model orders and loosen the requirements with lower complexity as well. The high order models extract all linear dynamics from the data, while a low order model is assumed to have errors below a user defined value. Note that although the system gives reasonable models, it is possible that personal experience leads to a better decision. The full process will be described in Section 3.

The calculated models are useless unless the user knows their quality, thus the automatic validation and qualification of the models is as important for the user as the model itself. Since the model quality strongly depends on the goal of the user, the final decision is in his/her hands. The proposed system can give useful hints and tips in addition to the global qualification of the provided model, which is based on different numerical and statistical properties. The additional textual information helps the inexperienced user to understand and interpret the meaning of the validation tests, as described in Section 4.

The performance of the proposed system is illustrated by examples based on real measurement data in Section 5.

2. AUTOMATIC DATA PREPROCESSING

To make automatic processing easy, the measurement must be made with periodic excitation signals. The described approach uses frequency domain methods, so each period of the measured input and output signals is transformed to the frequency domain by the DFT. Using these input and output spectra, the sample mean $(\overline{U}(\omega_k), \overline{Y}(\omega_k))$, the sample variance $(\sigma_U^2(\omega_k), \sigma_Y^2(\omega_k))$, and the sample covariance $(\sigma_{YU}^2(\omega_k))$ of the input and output spectra are calculated (Schoukens *et al.*, 1997).

3. MODEL SELECTION AND PARAMETER ESTIMATION

The inputs of the automatic model and order selection algorithm are the preprocessed data with the non-parametric noise model (i.e. the mean value, the sample variances and covariance of the input and output). The process can be divided into four major components:

A. coarse order estimation,
B. parameter estimation and model validation,
C. order reduction #1, and
D. order reduction #2.

The coarse order estimation step gives a rough (and usually conservative) estimation of the model order alone. Step B determines the model parameters with the estimated order and tries to validate it. If the validation fails, the model order is increased until the validation is successful. The first order reduction step produces a validated good quality model with the lowest possible order, while the second order reduction step tries to further decrease the order at the expense of a minor diminution of the model quality. The components of the automatic model and parameter estimation system are explained below.

Step A: The coarse order detection algorithm is based upon the method proposed in Rolain *et al.* (1997), which is able to reduce the initial model order using stochastic methods, without any parameter estimation. Since this step does not require parameters to be estimated, it is faster than a full estimation run. Therefore, it is worthwhile to refine this step, to obtain maximal accuracy for the model order, as it will reduce the number of subsequent runs to get the final model. In the proposed system, the original coarse estimation method is extended in two ways: to provide possibly lower order estimates, and to avoid the necessity of an initial order, which is "high enough".

Since the coarse order reduction step is fast, it is repeated iteratively until no further decrease is gained. If the initial order is detected to be too low, the order is automatically increased by a constant factor (in the present system by 20 percent) until a satisfactory (or the maximum allowed) order is reached. See Simon *et al.* (1999) for details.

Step B: Because of the stochastic nature of this order estimation algorithm, it sometimes reduces the model order too much. Hence, the result must be verified before further processing. Using the estimated order, first the model parameters are estimated (Schoukens *et al.*, 1991, 1997). Then, the model is checked using correlation test on the residual error (Schoukens, *et al.*, 1998). If the correlation test fails, the model order is increased and the parameter estimation and correlation test steps are repeated until the correlation test succeeds, or the maximum allowed order is reached.

Note that the reason of the failure of the correlation test may be not only the low model order. A suboptimal solution (local minimum) of the used parameter estimation algorithm may lead to analogous results. In both cases, however, the right solution to the validation failure is to increase the model order until the parameter estimation algorithm can find an acceptable – even high order – solution, from which Step C is able to produce a good quality lower order model.

Step C: The validated model is then processed by a peeling algorithm based upon Rolain *et al.* (1997). The peeling process eliminates the roots (poles, zeros, or canceling pairs) which have no significant effect on the transfer function. Theoretically, the peeling algorithm is very simple:

Step 1 Choose some roots which can possibly be eliminated.
Step 2 Re-estimate the model and parameters with the decreased order.
Step 3 Check the result. If the reduced order model is validated then accept the decreased order and go to Step 1.
Step 4 Restore the previous order. If there are other roots to try, go to Step 1.

Apart from the very high computational requirements of this approach, several other questions arise concerning the practical realization of the peeling process. The most important practical issues are the following:

- how to decrease the computational complexity of the root elimination procedure (Step 2),
- how to decide which roots to eliminate (Step 1),
- is it possible to decrease the possibility of local minima in the parameter estimation phase using available extra information (Step 2),
- how to decide whether an estimation is acceptable or not (Step 3)?

The computational complexity can effectively be decreased by reducing the number of the parameter estimation steps. The greater the number of the eliminated roots in one parameter estimation step, the faster the algorithm. The used method is based upon the idea proposed in Rolain *et al.* (1997), which uses a simplified approach to give a conservative estimation on the cost function after eliminating some roots. (Note that in this paper the term "cost function" refers to the minimum description length cost function, Ljung 1999, Rissanen 1978.) This non-parametric estimation is much faster than the parameter estimation step (which is from now referred to as slow estimation), but since the remaining roots are not rearranged, good candidates may be refused in this step. However, each hit saves precious seconds or even minutes.

The proposed algorithm combines the fast and slow estimation methods as follows:

Step 1' Choose a (real or complex) zero, pole, and a pole-zero pair, which can possibly be eliminated. Rank the three candidates.
Step 2' Make a non-parametric estimation on the cost function by eliminating the best candidate which was not yet tried.

Step 3' Check the result. If the reduced order model is acceptable then accept the decreased order and go to Step 1.

Step 4' Restore the previous order. If there are other candidates to try, go to Step 2'.

Step 1'' Choose a (real or complex) zero, pole, and a pole-zero pair, which can possibly be eliminated. Rank the three candidates.

Step 2'' Re-estimate the model and parameters by eliminating the best candidate, which was not yet tried.

Step 3'' Check the result. If the reduced order model is acceptable then accept the decreased order and go to Step 1'.

Step 4'' Restore the previous order. If there are other candidates to try, go to Step 2''.

In both the fast (') and the slow ('') estimation steps 3 possible candidates are tested: the less significant pole, zero, and pole-zero pair. If the fast estimation step can no longer decrease the order, then it is tried by the slow estimation method. If any roots are eliminated in the slow step, the fast method is tried again with the new root placement. The algorithm terminates if no more pole, zero, or pole-zero pair can be eliminated. More details on the algorithm can be found in Simon *et al.* (1999).

The choice of the candidates in Step 1' is made by estimating the contribution of the roots to the transfer function. In the case of the poles and zeros, the distance from the measurement band is considered, while in the case of the canceling pairs the distance of roots from each other is considered (Rolain *et al.* 1997). The order of the trials is determined by ranking the results (see Simon *et al.*, 1999 for details).

In the present system the candidates in Step 1'' are the same as in Step 1', but their ranking is re-evaluated using the cost function estimations produced in the previous three (unsuccessful) fast estimation steps.

The parameter estimation in Step 2'' uses an iterative parameter estimation algorithm (Schoukens *et al.*, 1991, 1997), for which the initial values must be set. The convergence speed and the quality of the supplied model strongly depend on the initialization. It is obvious, that the elimination of an unnecessary root does not change significantly the transfer function. It can also be assumed that the elimination of such a root does not change too much the numerator and the denominator of the transfer function, either. Based upon this assumption the following initialization scheme is used:

If the estimated fit in Step 2' is 'not too bad' (which means that the candidate root possibly can be eliminated with a small rearrangement of the other roots) then use the reduced-order (non-parametric) numerator and denominator, otherwise (the candidate root can not be eliminated, or major rearrangement is necessary) use the previous good non-reduced-order (non-parametric) numerator and denominator, as initial values.

The decision in Step 3' is made upon the estimated cost function: if it is smaller than a maximal allowed value (Rolain *et al.*, 1997), then the reduced order is accepted otherwise it is rejected. In Step 3'' the quality test is combined: in addition to the cost-function test a correlation test is also made on the residuals (Schoukens, *et al.*, 1998).

Since the automatic model selection algorithm is a top-down approach, high model orders may appear during the root elimination phase. The numerical stability of the algorithms was ensured by using orthogonal polynomials instead of power polynomials (Rolain *et al.*, 1995). With this solution, orders of 100/100 were successfully handled.

Step D: The second peeling step produces a somewhat lower quality, lower order model starting from the verified model produced in Step D. (Note that a possible and straightforward further extension of the system is the dynamic order reduction, which enables the user to specify the requirements in terms of the allowed model errors.) In many cases, the loose model is also acceptable for the user, who may decide which model to choose.

The only difference between Step C and D is that the algorithm uses a different qualification method in Step 3''. While in the validated case the peeling algorithm uses the combined cost function and correlation test, in the loose model case only the cost function values are tested against a maximum value determined from the validated model's cost function.

4. QUALIFICATION OF MODELS

The two models provided by the model selection algorithm must be validated and qualified before being applied. The user may not have the necessary theoretical background and experience to use and interpret the conventional validation tests, such as the theoretical and the observed cost function, and the correlation test. Instead, the proposed system provides some additional easy-to-understand information on the model quality, the presence of the unmodeled dynamics, and the effect of nonlinear distortion.

Based upon the result of the correlation test (Schoukens, *et al.*, 1998) it can be decided whether the linear dynamic behavior is correctly modeled or if there are some unmodeled linear dynamics present. The correlation test also gives estimation on the presence of (unmodeled) nonlinear distortions. The ratio of the mean error power and the mean of the variance of the transfer function taken over the frequency gives the basis of the qualification of the overall model fit. Upon the provided textual information, even the inexperienced user is able to decide which of the provided models to choose (see Fig. 3).

5. EXPERIMENTAL ILLUSTRATIONS

In this section, three examples are presented to illustrate the performance of the automatic model selection system. The examples cover a wide range of difficult practical problems. The first example is a Brüel&Kjaer passive bandpass filter, where the SNR of the

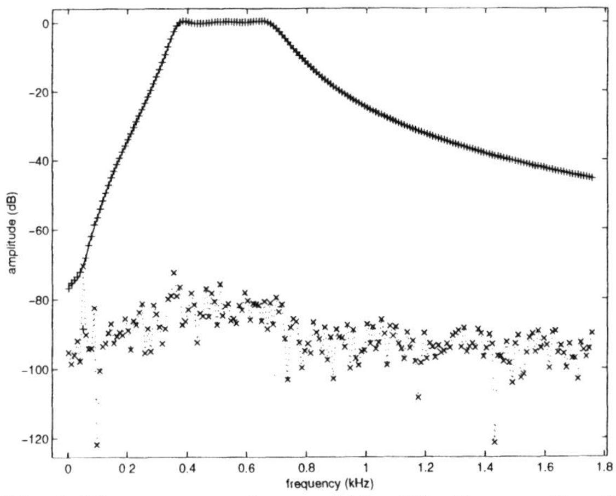

Fig. 2. Measurement data and identification results of the Brüel&Kjaer bandpass filter.

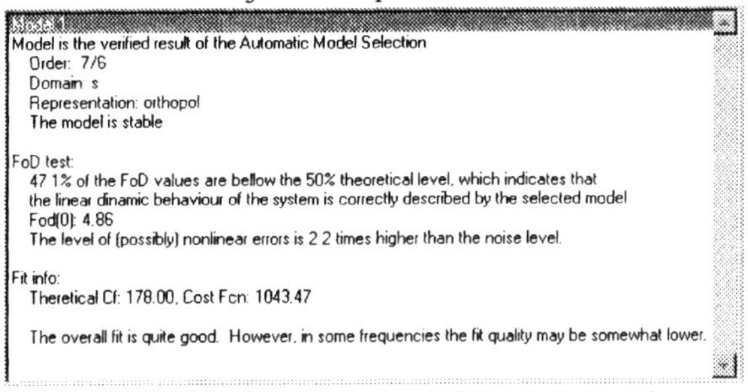

Fig. 3. The validation message of the Brüel&Kjaer bandpass filter.

measurement is very high, so the presence of small nonlinear distortions is disturbing. The second example is a second order mechanical system with poor SNR and high nonlinear distortion. The third example is the radial servo system of a CD player, which is a high complexity system. On the figures the following notation is used: measured transfer function (+), estimated transfer function (solid line), variance of the transfer function (dotted line), and the residual error (x with dotted line).

Brüel&Kjaer bandpass filter. The measured transfer function and the estimated variance can be seen in Fig. 2, along with the estimated model and the model error. No loose model is provided in this case, since no lower order model can be found to fulfill the requirements in Step D. The information on the validated model can be seen in Fig. 3.

Second order mechanical system. A mechanical resonating system (mass, viscous damping, nonlinear spring) was simulated with an electrical circuit. Although all the linear dynamics are modeled, the overall fit is very poor because of the unmodeled nonlinear behavior, as can be seen in Fig. 4. The automatic identification process produced the expected 0/2 order system, for which the qualification message can be seen in Fig. 5.

Radial servo system of a CD player. The radial servo system of a Philips CD320/00G was measured. Since open-loop measurement is impossible, the system was operated in closed loop while applying an external excitation signal and the input and output of the servo system was measured. The result of the automatic identification process is a 14/15 order verified model

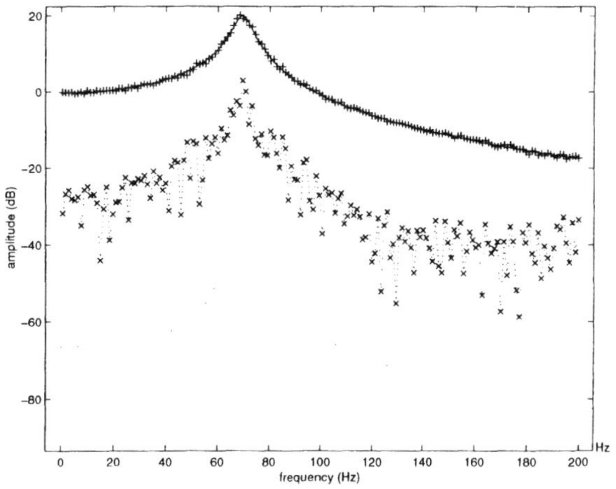

Fig. 4. Measurement data and identification results of the mechanical system.

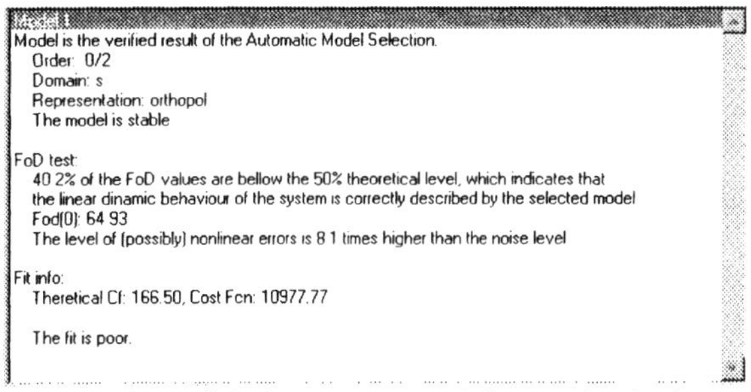

Fig. 5. The validation message of the mechanical system.

and a 7/8 order loose model, as can be seen in Figs. 6-9. Note, that in this case a human expert was able to provide somewhat better model quality by increasing the model order (the decrease of the cost function by 30% required a 24/24 order model).

6. CONCLUSION

The described automatic identification system is able to provide reasonable linear time invariant models for linear and almost linear systems from measurement data, without any human interaction. The input of the system is the measured data, and the output is one strictly validated, and one (somewhat lower quality) 'loose' model. The generated models are also qualified in a form, which is interpretable also by the inexperienced user.

The proposed system was build using advanced frequency domain identification algorithms, and a special care was taken to choose proper excitation signal to enable fully automatic processing, robust data representation to ensure numerical stability. The speed of the algorithm was increased by combining fast, but not precise techniques with slow and reliable methods.

The system after rough order estimation uses a top-down approach to select the proper model order and parameters. The main advantage of this approach is that the possible local minima of the parameter estimation algorithms can be avoided with greater chance, using appropriate initialization methods. The performance of the proposed system was also illustrated though examples with real measurement data.

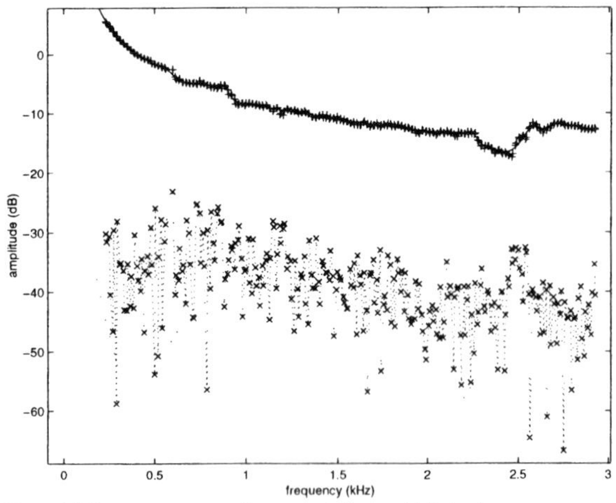

Fig. 6.Measurement data and the validated identification results of the CD player.

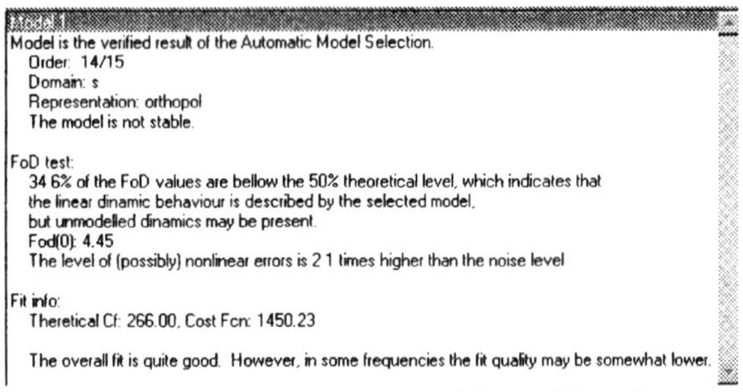

Fig. 7. The validation message of the validated identification results of the CD player.

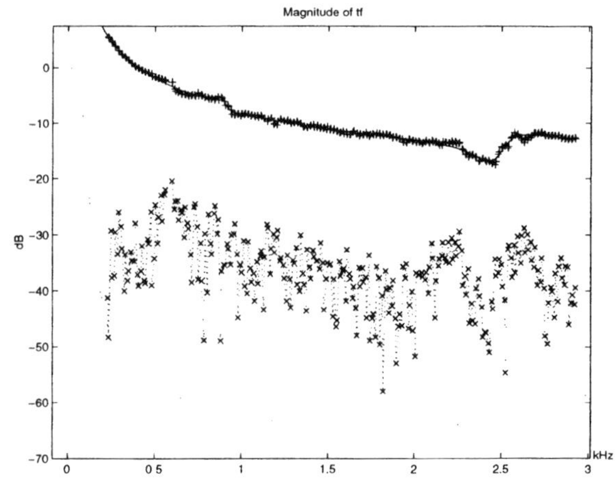

Fig. 8. Measurement data and the loose identification results of the CD player.

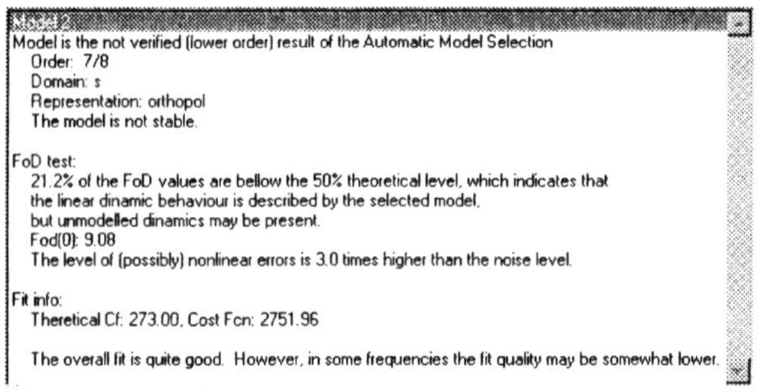

Fig. 9. The validation message of the loose identification results of the CD player.

ACKNOWLEDGEMENTS

This work was supported by the Hungarian National Scientific Research Fund (OTKA) under contact number F 033055, the Belgian National Fund for Scientific Research, the Flemish government (GOA-IMMI, and BIL99/18), and the Belgian government as a part of the Belgian program on Interuniversity Poles of attraction (IUAP4/2) initiated by the Belgian State, Prime Minister's Office, Science Policy Programming.

REFERENCES

Kollár, I. (1994). *Frequency-Domain System Identification Toolbox for Use with Matlab*, The Mathworks, Natick, M. A.

Ljung, L. (1986). *System Identification Toolbox for Use with Matlab*. The MathWorks, Inc., Sherborn, A.

Ljung, L. (1999). *System Identification :Theory for the Use*. Prentice Hall, Englewood Cliffs, NJ.

Norton, J. P. (1986). *An Introduction to Identification*, Academic, London, England.

Rissanen J., (1978). Modeling by shortest data description. *Automatica*, **Vol. 14**, pp. 465-471.

Rolain, Y., R. Pintelon, K. Q. Xu, and H. Vold (1995). Best conditioned parametric identification of transfer function models in the frequency domain. *IEEE Trans. on Automatic Control*, **Vol. 40**, no. 11, pp. 1954-1960.

Rolain, Y., J. Schoukens, and R. Pintelon (1997). Order estimation for linear time-invariant systems using frequency domain identification methods. *IEEE Trans. on Automatic Control*, **Vol. 42**, No. 10, pp. 1408-1417.

Schoukens, J. and R. Pintelon (1991). *Identification of Linear Systems: A Practical Guideling to Accurate Modeling*. Pergamom, Oxford.

Schoukens, J., R. Pintelon, G. Vandersteen and P. Guillaume (1997). Frequency-domain system identification using non-parametric noise models estimated from a small number of data sets. *Automatica*, **Vol. 33**, no.6, pp. 1073-1086.

Schoukens, J., T. Dobrowiecki, and R. Pintelon (1998). Identification of linear systems in the presence of nonlinear distortions. A frequency domain approach. *IEEE Trans. on Automatic Control*, **Vol. 43**, No.2, pp. 176-190.

Simon, Gy, J. Schoukens, and Yves Rolain (1999). Automatic Model Selection for Linear Time-Invariant Systems – Practical Issues. *Electronic publication*: http://www.mit.bme.hu/~simon/publications/automodel.ps.zip.

Söderström, T. and P. Stoica (1989). *System Identification*. Englewood Cliffs, NJ: Prentice-Hall, pp. 256.

GENERAL FORMS OF ADAPTIVE NONLINEAR $\mathcal{H}^{\infty}$ CONTROL FOR PROCESSES WITH BOUNDED VARIATIONS OF PARAMETERS

Yoshihiko Miyasato*

**The Institute of Statistical Mathematics*
4-6-7 Minami-Azabu, Minato-ku, Tokyo 106-8569, JAPAN

Abstract : A new class of adaptive nonlinear $\mathcal{H}^{\infty}$ control systems for processes with bounded variations of time-varying parameters, is proposed in this manuscript. Those control schemes are derived as solutions of particular nonlinear $\mathcal{H}^{\infty}$ control problems, where unknown system parameters are regarded as exogenous disturbances to the processes, and thus, in the resulting control systems, the $\mathcal{L}^2$ gains from system parameters to generalized outputs are made less than the prescribed $\gamma(> 0)$. It is shown that the proposed control schemes are sub-optimal to some $\mathcal{H}^{\infty}$ cost functionals, when the high-frequency gains are time-invariant. *Copyright ©2000 IFAC*

Keywords: adaptive control; nonlinear systems; H-infinity control; time-varying systems; optimal control

1. INTRODUCTION

In the study of adaptive control, the main topics have been an asymptotic stability of adaptive control systems. So much attention has not been paid on the control performances such as transient performance and other performances (Åström, *et al.*, 1989; Ioannou, *et al.*, 1996). On the contrary, the backstepping procedures in the last few years, have not only made it possible to analyze the stability of adaptive control systems in simpler forms, but also made it possible to discuss the transient performance of responses (such as $\mathcal{L}^{\infty}/\mathcal{L}^2$ performances) of many kinds of adaptive and nonlinear control systems (Krstić, *et al.*, 1995). Furthermore, recent researches on nonlinear $\mathcal{H}^{\infty}$ control and inverse optimality, could derive adaptive or nonlinear control systems which are optimal to certain meaningful cost functionals (Lu, *et al.*, 1997; Krstić, *et al.*, 1998; Miyasato, 1999).

Additionally, in the past study of adaptive control, there has been another problem that the stability analysis of adaptive control systems have been focused on time-invariant processes mainly; no enough discussion for the case of time-varying systems has been made. Although several approaches have been examined for time-varying processes in the study of robust adaptive control schemes (Ioannou, *et al.*, 1996), the results in those works could be applied to limited classes of time-varying systems, that is, only sufficiently small variations of time-varying parameters are accepted in those robust adaptive control schemes.

The purpose of the present paper is to provide a new class of adaptive nonlinear $\mathcal{H}^{\infty}$ control systems for processes with bounded variations of parameters, where the control performances are discussed explicitly, and the stability analysis for time-varying systems are carried out successfully. Those control schemes are derived as solutions of particular nonlinear $\mathcal{H}^{\infty}$ control problems, where unknown system parameters are regarded as exogenous disturbances to the processes, and thus, in the resulting control systems, the $\mathcal{L}^2$ gains from system parameters to generalized outputs are made less than $\gamma(>0)$ (the prescribed positive constants). Not only the general forms of the control schemes, but also two explicit descriptions of those general ones, are provided. It is shown that the proposed control strategy can be applied to any time-varying (or time-invariant) systems, and the resulting control systems are bounded for arbitrarily large but bounded variations of time-varying parameters. Also, the control schemes are shown to be sub-optimal to some $\mathcal{H}^\infty$ cost functionals (or certain differential games), when the high-frequency gains are time-invariant.

2. PROBLEM STATEMENT

A single-input single-output nonlinear system is given by

$$\frac{d}{dt}e(t) = \mathcal{L}(e(t)) + \mathcal{L}(f(e,t)) + b_0 u(t) + \Phi^T\omega_1(t), \quad (1)$$

where $e(t)$ is a control variable (an output or a tracking error, *et al.*), $u(t)$ is a control input, and $f(e,t)$ is an unknown nonlinear term. $\omega_1(t)$ is a vector composed of measurable signals, b_0 and Φ are unknown system parameters which can be time-varying. $\mathcal{L}(\cdot)$ is an unstructured element defined by

$$\mathcal{L}(v(t)) = G_0(s)v(t), \qquad (G_0 \in \mathcal{RH}^\infty). \quad (2)$$

The following assumptions are introduced.

Assumption 1 : AS-1-1) *Although b_0 can be time-varying, the sign of it remains unchanged ($b_0 > 0$ or $b_0 < 0$), and is known a priori. It is assumed that $b_0 > 0$ without loss of generality.* **AS-1-2)** *$f(e,t)$ is evaluated by*

$$f(e,t)^2 \le f_0 \cdot \phi(e) \cdot e^2, \quad \phi(0) = 0, \quad (3)$$

where f_0 is an unknown positive constant, and $\phi(e)(>0)$ is a known function of e which is continuous with respect to its argument. **AS-1-3)** The magnitude of $\omega_1(t)$ is evaluated as follows:

$$\|\omega_1(t)\| \le M_1 \sup_{t\ge\tau}|e(\tau)| + M_2, \quad (M_1, M_2 > 0). \quad (4)$$

AS-1-4) The upper bounds on the norm of parameters Φ, and θ_1, θ_2, which will be introduced later, are known. Also, the upper bound $\bar{b}_0$ and lower bound $\underline{b}_0$on the high-frequency gain b_0 are known such that

$$0 < \delta \le \underline{b}_0 \le b_0 \le \bar{b}_0 \le M < \infty, \quad (5)$$

and the upper bound $\bar{p}$ and lower bound $\underline{p}$ on the parameter p are known, too.

$$p \equiv \frac{1}{b_0}, \quad \bar{p} \equiv \frac{1}{\underline{b}_0}, \quad \underline{p} \equiv \frac{1}{\bar{b}_0}. \quad (6)$$

The control problem of this paper is to determine a control input $u(t)$ adaptively such that the overall system is stabilized for arbitrary but bounded time-varying system parameters, and additionally, the control variable $e(t)$ converges to zero asymptotically in the ideal case (stability condition), while the resulting control system becomes optimal or sub-optimal to some meaningful cost functionals (optimality condition).

3. ADAPTIVE NONLINEAR CONTROL SCHEME

In the present section, the design method of the adaptive nonlinear $\mathcal{H}^\infty$ control scheme is shown, and several main results are provided. The control input $u(t)$ is divided into the form

$$\begin{aligned} u(t) &= -\hat{p}(t)\hat{\Phi}(t)^T\omega_1(t) + v(t) \\ &\equiv -\hat{p}(t)v_0(t) + v(t), \end{aligned} \quad (7)$$

and $v(t)$ is regarded as a new input signal to be determined based on $\mathcal{H}^\infty$ control strategy. The projection type adaptive laws (Ioannou, *et al.*, 1996) are utilized for the tuning of $\hat{\Phi}(t)$ and $\hat{p}(t)$.

$$\begin{aligned} &\text{if } \|\hat{\Phi}(t)\| < M, \\ &\quad \text{or } \|\hat{\Phi}(t)\| = M \;\&\; \hat{\Phi}(t)^T G_3\omega_1(t)e(t) \le 0, \\ &\quad \dot{\hat{\Phi}}(t) = G_3\omega_1(t)e(t), \end{aligned} \quad (8)$$

$$\begin{aligned} &\text{if } \|\hat{\Phi}(t)\| = M \;\&\; \hat{\Phi}(t)^T G_3\omega_1(t)e(t) > 0, \\ &\quad \dot{\hat{\Phi}}(t) = G_3\omega_1(t)e(t) \\ &\qquad -G_3\frac{\hat{\Phi}(t)\hat{\Phi}(t)^T}{\hat{\Phi}(t)^T G_3\hat{\Phi}(t)}G_1\omega_1(t)e(t), \end{aligned} \quad (9)$$

$$\begin{aligned} &\text{if } 0 < \delta < \hat{p}(t) < M, \\ &\quad \text{or } \hat{p}(t) = \delta \;\&\; v_0(t)e(t) \ge 0, \\ &\quad \text{or } \hat{p}(t) = M \;\&\; v_0(t)e(t) \le 0, \\ &\quad \dot{\hat{p}}(t) = g_4 v_0(t)e(t), \end{aligned} \quad (10)$$

$$\begin{aligned} &\text{if } \hat{p}(t) = \delta \;\&\; v_0(t)e(t) < 0, \\ &\quad \text{or } \hat{p}(t) = M \;\&\; v_0(t)e(t) > 0, \\ &\quad \dot{\hat{p}}(t) = 0, \end{aligned} \quad (11)$$

$$(G_3 = G_3^T > 0, \quad g_4 > 0),$$

where M and δ are properly selected positive constants based on **AS-1-4**. The initial conditions for those tuning parameters are such that $\|\hat{\Phi}(0)\| \le M$, $\delta \le \hat{p}(0) \le M$. Also, $\hat{\theta}_1(t)$, $\hat{\theta}_2(t)$, $\hat{b}_0(t)$ are added such that

$$\text{if } \hat{\theta}_1(t) < M, \quad \dot{\hat{\theta}}_1(t) = g_1 e(t)^2, \quad (12)$$

$$\text{if } \hat{\theta}_1(t) = M, \quad \dot{\hat{\theta}}_1(t) = 0, \quad (13)$$

$$\text{if } \hat{\theta}_2(t) < M, \quad \dot{\hat{\theta}}_2(t) = g_2\phi(e(t)e(t)^2, \quad (14)$$

$$\text{if } \hat{\theta}_2(t) = M, \quad \dot{\hat{\theta}}_2(t) = 0, \quad (15)$$

$$\begin{aligned} &\text{if } 0 < \delta < \hat{b}_0(t) < M, \\ &\quad \text{or } \hat{b}_0(t) = \delta \;\&\; v(t)e(t) \ge 0, \\ &\quad \text{or } \hat{b}_0(t) = M \;\&\; v(t)e(t) \le 0, \\ &\quad \dot{\hat{b}}_0(t) = g_5 v(t)e(t), \end{aligned} \quad (16)$$

$$\begin{aligned} &\text{if } \hat{b}_0(t) = \delta \;\&\; v(t)e(t) < 0, \\ &\quad \text{or } \hat{b}_0(t) = M \;\&\; v(t)e(t) > 0, \\ &\quad \dot{\hat{b}}_0(t) = 0, \end{aligned} \quad (17)$$

$$(g_1, g_2, g_5 > 0),$$

where M and δ are determined similarly, and the initial conditions are that $0 \leq \hat{\theta}_1(0) \leq M$, $0 \leq \hat{\theta}_2(0) \leq M$, $\delta \leq \hat{b}_0(0) \leq M$. $\hat{\theta}_1(t)$, $\hat{\theta}_2(t)$ are current estimates of θ_1, θ_2 (positive) satisfying the following inequality.

$$\int_0^t e(\tau)\{\mathcal{L}(e(\tau)+\mathcal{L}(f(e,\tau))\}d\tau$$
$$\leq \int_0^t \{\theta_1(\tau)e(\tau)^2+\theta_2(\tau)\phi(e(\tau))e(\tau)^2\}d\tau. \quad (18)$$

Those tuning parameters $\hat{\theta}_1(t)$, $\hat{\theta}_2(t)$, $\hat{\Phi}(t)$, $\hat{p}(t)$, $\hat{b}_0(t)$ are made uniformly bounded by those projection type adaptive laws (Ioannou, *et al.* (1995)). For this adaptive system, define a positive function $V(t)$ by

$$V(t)=\frac{1}{2}e(t)^2+\frac{1}{2}\sum_{i=1}^{2}\{\hat{\theta}_i(t)-\theta_i^*\}^2/g_i$$
$$+\frac{1}{2}\{\hat{\Phi}(t)-\Phi^*\}^T G_3^{-1}\{\hat{\Phi}(t)-\Phi^*\}$$
$$+\frac{\underline{b}_0}{2}\{\hat{p}(t)-\bar{p}\}^2/g_4+\frac{1}{2}\{\hat{b}_0(t)-\underline{b}_0\}^2/g_5, \quad (19)$$

where θ_1^*, θ_2^*, Φ^* are nominal values (time-invariant) of the system parameters θ_1, θ_2, Φ (time-varying), respectively. Take the time derivative of $V(t)$

$$\dot{V}(t)\leq e(t)\{\mathcal{L}(e(t))+\mathcal{L}(f(e,t))+\Phi^T\omega_1(t)\}$$
$$-\theta_1^*e(t)^2-\theta_2^*\phi(e(t))e(t)^2-\Phi^{*T}\omega_1(t)e(t)$$
$$+(\underline{b}_0-b_0)\hat{p}(t)\hat{\Phi}(t)^T\omega_1(t)e(t)+\hat{\theta}_1(t)e(t)^2$$
$$+\hat{\theta}_2(t)\phi(e(t))e(t)^2+\{\hat{b}_0(t)+(b_0-\underline{b}_0)\}v(t)e(t), \quad (20)$$

and integrate $\dot{V}(t)$ into the following form, where (18) is also considered.

$$V(t)-V(0)\leq\int_0^t(\theta_1-\theta_1^*)e(\tau)^2d\tau$$
$$+\int_0^t(\theta_2-\theta_2^*)\phi(e(\tau))e(\tau)^2d\tau$$
$$+\int_0^t(\Phi-\Phi^*)^T\omega_1(\tau)e(\tau)d\tau$$
$$+\int_0^t(\underline{b}_0-b_0)\hat{p}(\tau)\hat{\Phi}(\tau)^T\omega_1(\tau)e(\tau)d\tau$$
$$+\int_0^t\Big[\hat{\theta}_1(\tau)e(\tau)^2+\hat{\theta}_2(\tau)\phi(e(\tau))e(\tau)^2$$
$$+\{\hat{b}_0(t)+(b_0-\underline{b}_0)\}v(\tau)e(\tau)\Big]d\tau$$
$$\equiv\int_0^t\tilde{\Theta}(\tau)^T\omega(\tau)e(\tau)d\tau$$
$$+\int_0^t\Big[\hat{\theta}_1(\tau)e(\tau)^2+\hat{\theta}_2(\tau)\phi(e(\tau))e(\tau)^2$$
$$+\{\hat{b}_0(t)+\tilde{b}_0(t)\}v(\tau)e(\tau)\Big]d\tau, \quad (21)$$

$$\tilde{\Theta}\equiv[\theta_1-\theta_1^*,\ \theta_2-\theta_2^*,\ (\Phi-\Phi^*)^T,\ \underline{b}_0-b_0]^T, \quad (22)$$
$$\omega(t)\equiv[e(t),\ \phi(e(t))e(t),\ \omega_1(t)^T,\ \hat{p}(t)\hat{\Phi}(t)^T\omega_1(t)]^T, \quad (23)$$
$$\tilde{b}_0(t)\equiv b_0-\underline{b}_0\ (\geq 0). \quad (24)$$

From that evaluation of $V(t)$, the following virtual process is introduced.

$$\dot{e} = \hat{\theta}_1 e+\hat{\theta}_2\phi(e)e+\omega^T\tilde{\Theta}+(\hat{b}_0+\tilde{b}_0)v$$
$$\equiv f(e)+g_1 d+g_2 v, \quad (25)$$
$$(f(e)=\hat{\theta}_1 e+\hat{\theta}_2\phi(e)e,\ g_1=\omega^T,\ d=\tilde{\Theta},$$
$$g_2=\hat{b}_0+\tilde{b}_0). \quad (26)$$

That process is stabilized by v via the nonlinear $\mathcal{H}^\infty$ control scheme, where $\tilde{\Theta}$ is regarded as an exogenous disturbance. For this purpose, consider the following Hamilton-Jacobi-Isaacs equation

$$\frac{\partial\tilde{V}}{\partial e}f+\frac{1}{4}\left(\frac{\|g_1\|^2}{\gamma^2}-\frac{g_2^2}{r}\right)\left(\frac{\partial\tilde{V}}{\partial e}\right)^2+he^2\leq 0, \quad (27)$$

where the solution $\tilde{V}$ is given by the next equation.

$$\tilde{V}(t)=\frac{1}{2}e(t)^2. \quad (28)$$

h and r are positive functions to be determined from the inequality (27) based on inverse optimality for the given solution $\tilde{V}(t)$ (28) and the positive constant γ. From Hamilton-Jacobi-Isaacs equation (27) and its solution (28), it follows that

$$\hat{\theta}_1+\hat{\theta}_2\phi(e)+\frac{1}{4}\left\{\frac{\|\omega\|^2}{\gamma^2}-\frac{(\hat{b}_0+\tilde{b}_0)^2}{r}\right\}+h\leq 0. \quad (29)$$

Since the unknown element $\tilde{b}_0$ is included in the above inequality (29), h and r satisfying the next relation (30) are to be obtained, which is a sufficient condition for the original inequality (29), since $\tilde{b}_0\geq 0$ (24).

$$\hat{\theta}_1+\hat{\theta}_2\phi(e)+\frac{1}{4}\left\{\frac{\|\omega\|^2}{\gamma^2}-\frac{\hat{b}_0^2}{r}\right\}+h\leq 0. \quad (30)$$

Once the positive functions h and r are obtained, then the corresponding control input is given by

$$v=-\frac{1}{2r}g_2\frac{\partial\tilde{V}}{\partial e}=-\frac{1}{2r}(\hat{b}_0+\tilde{b}_0)e. \quad (31)$$

Since the unknown element $\tilde{b}_0$ is included in the above equation, the actual input is replaced by

$$v=-\frac{1}{2r}\hat{b}_0 e. \quad (32)$$

Then, the following main theorems are given.

Theorem 1 : *The adaptive control system described above ((30), (32) and the adaptive laws) is uniformly bounded for arbitrary bounded variations of system parameters* b_0, p, θ_1, θ_2, Φ.

(Proof) By introducing state variables $z_1(t)$, $z_2(t)$ of the stable systems (state-space representation (F_1, g_1), (F_2, g_2)), the unstructured elements $\mathcal{L}(e(t))$ and $\mathcal{L}(f(e,t))$ are written as follows:

$$\dot{z}_1(t)=F_1z_1(t)+g_1e(t), \quad (33)$$
$$\dot{z}_2(t)=F_2z_2(t)+g_2f(e,t), \quad (34)$$
$$P_iF_i+F_i^TP_i=-I,\quad (P_i=P_i^T>0), \quad (35)$$
$$\|\mathcal{L}(e(t))\|^2\leq M_1\|z_1(t)\|^2+M_2e(t)^2, \quad (36)$$
$$\|\mathcal{L}(f(y,t))\|^2\leq M_3\|z_2(t)\|^2+M_4\phi(e(t))e(t)^2, \quad (37)$$
$$(M_1,\ M_2,\ M_3,\ M_4>0).$$

Adding those $z_1(t)$ and $z_2(t)$, $V_1(t)$ is defined by

$$V_1(t) = V(t) + \sum_{i=1}^{2} z_i(t)^T P_i z_i(t). \tag{38}$$

Take the time derivative of $V_1(t)$.

$$\begin{aligned}
\dot{V}_1(t) &= e(t)\{\mathcal{L}(e(t)) + \mathcal{L}(f(e,t)) + \Phi^T\omega_1(t)\} \\
&\quad -\theta_1^* e(t)^2 - \theta_2^*\phi(e(t))e(t)^2 - \Phi^{*T}\omega(t)e(t) \\
&\quad +(\underline{b}_0 - b_0)\hat{p}(t)\hat{\Phi}(t)^T\omega(t)e(t) + \hat{\theta}_1(t)e(t)^2 \\
&\quad +\hat{\theta}_2(t)\phi(e(t))e(t)^2 + \{\hat{b}_0(t) + \underline{b}_0\}v(t)e(t) \\
&\quad -\sum_{i=1}^{2}\|z_i(t)\|^2 + 2z_1(t)^T P_1 g_1 e(t) \\
&\quad +2z_2(t)^T P_2 g_2 f(e,t) \\
&\le \tilde{\Theta}(t)^T\omega(t)e(t) \\
&\quad +\hat{\theta}_1(t)e(t)^2 + \hat{\theta}_2(t)\phi(e(t))e(t)^2 \\
&\quad +\{\hat{b}_0(t) + \tilde{b}_0(t)\}v(t)e(t) - \sum_{i=1}^{2}\delta_i\|z_i(t)\|^2,
\end{aligned} \tag{39}$$

$$\tilde{\Theta} \equiv [M_1 - \theta_1^*,\ M_2 - \theta_2^*,\ (\Phi - \Phi^*)^T,\ \underline{b}_0 - b_0]^T, \tag{40}$$
$$(0 < M_1,\ M_2 < \infty).$$

From (30) and (32), $\dot{V}_1(t)$ is evaluated by

$$\begin{aligned}
\dot{V}_1(t) &\le -he(t)^2 - \frac{e(t)^2}{4}\left\{\frac{\|\omega(t)\|^2}{\gamma^2} - \frac{\hat{b}_0^2}{r}\right\} \\
&\quad +\tilde{\Theta}(t)^T\omega(t)e(t) \\
&\quad +\hat{b}_0 v(t)e(t) + \tilde{b}_0 v(t)e(t) - \sum_{i=1}^{2}\delta_i\|z_i(t)\|^2 \\
&= -\{he(t)^2 + rv(t)^2\} + r\left\{\frac{\hat{b}_0}{2r}e(t) + v(t)\right\}^2 \\
&\quad -\gamma^2\left\|\tilde{\Theta} - \frac{\omega(t)}{2\gamma^2}e(t)\right\|^2 + \gamma^2\|\tilde{\Theta}\|^2 + \tilde{b}_0 v(t)e(t) \\
&\quad -\sum_{i=1}^{2}\delta_i\|z_i(t)\|^2 \\
&\le -\{he(t)^2 + rv(t)^2\} - \gamma^2\left\|\tilde{\Theta}_1 - \frac{\bar{\omega}(t)}{2\gamma^2}e(t)\right\|^2 \\
&\quad +\gamma^2\|\tilde{\Theta}\|^2 - \frac{\hat{b}_0\tilde{b}_0}{2r}e(t)^2 \\
&\quad -\frac{1}{2}\sum_{i=1}^{2}\{\hat{\theta}_i(t) - \theta_i^*\}^2/g_i \\
&\quad -\frac{1}{2}\{\hat{\Phi}(t) - \Phi^*\}^T G_3^{-1}\{\hat{\Phi}(t) - \Phi^*\} \\
&\quad -\frac{\underline{b}_0}{2}\{\hat{p}(t) - \bar{p}\}^2/g_4 - \frac{1}{2}\{\hat{b}_0(t) - \underline{b}_0\}^2/g_5 \\
&\quad -\sum_{i=1}^{2}\delta_i\|z_i(t)\|^2 + \bar{D} \\
&\le -\delta V_1(t) + \bar{D} + \gamma^2\|\tilde{\Theta}\|^2,
\end{aligned} \tag{41}$$

where $0 < \delta,\ \delta_i,\ \bar{D} < \infty$, and the boundedness of $\hat{\theta}_1(t)$, $\hat{\theta}_2(t)$, $\hat{\Phi}(t)$, $\hat{p}(t)$, $\hat{b}_0(t)$ is considered. Hence, it is shown that the adaptive system remains bounded for any bounded variations of system parameters. □

Theorem 2 : *For that adaptive control systems, it is assumed that b_0 is time-invariant. Then, v (32) is a sub-optimal control input which minimizes the upper bound on the following cost functional.*

$$\begin{aligned}
J(t) \equiv \sup_{\tilde{\Theta}\in\mathcal{L}^2} \Bigg[&\int_0^t (he(\tau)^2 + rv(\tau)^2)d\tau \\
&+\frac{1}{2}e(t)^2 + \frac{1}{2}\sum_{i=1}^{2}\{\hat{\theta}_i(t) - \theta_i^*\}^2/g_i \\
&+\frac{1}{2}\{\hat{\Phi}(t) - \Phi^*\}^T G_3^{-1}\{\hat{\Phi}(t) - \Phi^*\} \\
&+\frac{\underline{b}_0}{2}\{\hat{p}(t) - \bar{p}\}^2/g_4 + \frac{1}{2}\{\hat{b}_0(t) - \underline{b}_0\}^2/g_5 \\
&-\gamma^2\int_0^t \|\tilde{\Theta}(\tau)\|^2 d\tau\Bigg].
\end{aligned} \tag{42}$$

Also the next inequality holds.

$$\begin{aligned}
&\int_0^t (he(\tau)^2 + rv(\tau)^2)d\tau + \frac{1}{2}e(t)^2 \\
&\quad +\frac{1}{2}\sum_{i=1}^{2}\{\hat{\theta}_i(t) - \theta_i^*\}^2/g_i \\
&\quad +\frac{1}{2}\{\hat{\Phi}(t) - \Phi^*\}^T G_3^{-1}\{\hat{\Phi}(t) - \Phi^*\} \\
&\quad +\frac{\underline{b}_0}{2}\{\hat{p}(t) - \bar{p}\}^2/g_4 + \frac{1}{2}\{\hat{b}_0(t) - \underline{b}_0\}^2/g_5 \\
&\le \frac{1}{2}e(0)^2 + \frac{1}{2}\sum_{i=1}^{2}\{\hat{\theta}_i(0) - \theta_i^*\}^2/g_i \\
&\quad +\frac{1}{2}\{\hat{\Phi}(0) - \Phi^*\}^T G_3^{-1}\{\hat{\Phi}(0) - \Phi^*\} \\
&\quad +\frac{\underline{b}_0}{2}\{\hat{p}(0) - \bar{p}\}^2/g_4 + \frac{1}{2}\{\hat{b}_0(0) - \underline{b}_0\}^2/g_5 \\
&\quad +\gamma^2\int_0^t \|\tilde{\Theta}(\tau)\|^2 d\tau.
\end{aligned} \tag{43}$$

Especially, if $\tilde{\Theta} \in \mathcal{L}^2$, then $e(t) \to 0$ as $t \to \infty$.

Theorem 3 : *For that adaptive system, it is assumed that b_0 is not time-invariant. Then, the following inequality is derived.*

$$\begin{aligned}
&\int_0^t (he(\tau)^2 + rv(\tau)^2)d\tau + \frac{1}{2}e(t)^2 \\
&\quad +\frac{1}{2}\sum_{i=1}^{2}\{\hat{\theta}_i(t) - \theta_i^*\}^2/g_i \\
&\quad +\frac{1}{2}\{\hat{\Phi}(t) - \Phi^*\}^T G_3^{-1}\{\hat{\Phi}(t) - \Phi^*\} \\
&\quad +\frac{\underline{b}_0}{2}\{\hat{p}(t) - \bar{p}\}^2/g_4 + \frac{1}{2}\{\hat{b}_0(t) - \underline{b}_0\}^2/g_5 \\
&\quad +\int_0^t \frac{\hat{b}_0(\tau)\tilde{b}_0}{2r}e(\tau)^2 d\tau \\
&\le \frac{1}{2}e(0)^2 + \frac{1}{2}\sum_{i=1}^{2}\{\hat{\theta}_i(0) - \theta_i^*\}^2/g_i \\
&\quad +\frac{1}{2}\{\hat{\Phi}(0) - \Phi^*\}^T G_3^{-1}\{\hat{\Phi}(0) - \Phi^*\} \\
&\quad +\frac{\underline{b}_0}{2}\{\hat{p}(0) - \bar{p}\}^2/g_4 + \frac{1}{2}\{\hat{b}_0(0) - \underline{b}_0\}^2/g_5 \\
&\quad +\gamma^2\int_0^t \|\tilde{\Theta}(\tau)\|^2 d\tau.
\end{aligned} \tag{44}$$

Especially, if $\tilde{\Theta} \in \mathcal{L}^2$, then, $e(t) \to 0$ as $t \to \infty$, and the adaptive system converges to the sub-optimal control system described in Theorem 2.

(Proof) From the inequality (21), the properties of the control input and the solutions of Hamilton-Jacobi-Isaacs equation, it follows that

$$
\begin{aligned}
V(t) - V(0) &\leq \int_0^t \tilde{\Theta}(\tau)^T \omega(\tau) e(\tau) d\tau \\
&+ \int_0^t \left[\hat{\theta}_1(\tau)e(\tau)^2 + \hat{\theta}_2(\tau)\phi(e(\tau))e(\tau)^2 \right. \\
&\left. + \{\hat{b}_0(t) + \tilde{b}_0(t)\} v(\tau)e(\tau)\right] d\tau \\
&\leq \int_0^t \left\{ -\frac{\|\omega(\tau)\|^2}{\gamma^2} + \frac{1}{r} \right\} \frac{e(\tau)^2}{4} d\tau - \int_0^t he(\tau)^2 d\tau \\
&+ \int_0^t e(\tau)\{\tilde{\Theta}^T\omega(\tau) + (\hat{b}_0 + \tilde{b}_0)v(\tau)\} d\tau \\
&= \int_0^t r\left\{ \frac{\hat{b}_0(\tau)e(\tau)}{2r} + v(\tau) \right\}^2 d\tau \\
&- \int_0^t \{he(\tau)^2 + rv(\tau)^2\} d\tau \\
&- \int_0^t \gamma^2 \left\| \tilde{\Theta}(\tau) - \frac{1}{2\gamma^2}\omega(\tau)e(\tau) \right\|^2 d\tau \\
&+ \gamma^2 \int_0^t \|\tilde{\Theta}(\tau)\|^2 d\tau + \int_0^t \tilde{b}_0(\tau) v(\tau) e(\tau) d\tau. \qquad (45)
\end{aligned}
$$

Then, Theorem 2 and Theorem 3 are shown. □

Remark : The sub-optimality in Theorem 2 comes from the inequalities (18), (29). If those relations hold for equality conditions, then v is a optimal control input which minimizes $J(t)$ (42).

Theorem 4 : *Assume that* $(\theta_1 - \theta_1^*)$, $(\theta_2 - \theta_2^*)$, $(\Phi - \Phi^*)$, $(b_0 - b_0^*)$, $(p - p^*) \in \mathcal{L}^2$ *and* $(b_0 - b_0^*) \to 0$ *(as* $t \to \infty$*) for certain constants* θ_1^*, θ_2^*, Φ^*, b_0^*, p^* *(*$b_0^* p^* = 1$*). Then, it follows that* $e(t) \to 0$ *(as* $t \to \infty$*).*

(Proof) $V(t)$ (≥ 0) is newly re-defined by

$$
\begin{aligned}
V(t) &= \frac{1}{2}e(t)^2 + \frac{1}{2}\sum_{i=1}^{2}\{\hat{\theta}_i(t) - \theta_i^*\}^2/g_i \\
&+ \frac{1}{2}\{\hat{\Phi}(t) - \Phi^*\}^T G_3^{-1}\{\hat{\Phi}(t) - \Phi^*\} \\
&+ \frac{b_0^*}{2}\{\hat{p}(t) - p^*\}^2/g_4 + \frac{1}{2}\{\hat{b}_0(t) - b_0^*\}^2/g_5. \qquad (46)
\end{aligned}
$$

Take the time derivative of it

$$
\begin{aligned}
\dot{V}(t) &= e(t)\{\mathcal{L}(e(t)) + \mathcal{L}(f(e,t)) + \Phi^T\omega_1(t)\} \\
&- \theta_1^* e(t)^2 - \theta_2^*\phi(e(t))e(t)^2 - \Phi^{*T}\omega(t)e(t) \\
&+ (b_0^* - b_0)\hat{p}(t)\hat{\Phi}(t)^T\omega(t)e(t) + \hat{\theta}_1(t)e(t)^2 \\
&+ \hat{\theta}_2(t)\phi(e(t))e(t)^2 + \{\hat{b}_0(t) + (b_0 - b_0^*)\}v(t)e(t), \qquad (47)
\end{aligned}
$$

and integrate $\dot{V}(t)$ into the following form, where (18) is also considered.

$$
\begin{aligned}
V(t) - V(0) &\leq \int_0^t (\theta_1 - \theta_1^*)e(\tau)^2 d\tau \\
&+ \int_0^t (\theta_2 - \theta_2^*)\phi(e(\tau))e(\tau)^2 d\tau \\
&+ \int_0^t (\Phi - \Phi^*)^T\omega_1(\tau)e(\tau) d\tau \\
&+ \int_0^t (b_0^* - b_0)\hat{p}(\tau)\hat{\Phi}(\tau)^T\omega_1(\tau)e(\tau) d\tau \\
&+ \int_0^t \left[\hat{\theta}_1(\tau)e(\tau)^2 + \hat{\theta}_2(\tau)\phi(e(\tau))e(\tau)^2 \right. \\
&\left. + \{\hat{b}_0(t) + (b_0 - b_0^*)\}v(\tau)e(\tau)\right] d\tau \\
&\equiv \int_0^t \tilde{\Theta}(\tau)^T\omega(\tau)e(\tau) d\tau \\
&+ \int_0^t \left[\hat{\theta}_1(\tau)e(\tau)^2 + \hat{\theta}_2(\tau)\phi(e(\tau))e(\tau)^2 \right. \\
&\left. + \{\hat{b}_0(t) + \tilde{b}_0(t)\}v(\tau)e(\tau)\right] d\tau, \qquad (48)
\end{aligned}
$$

$$\tilde{\Theta} \equiv [\theta_1 - \theta_1^*,\ \theta_2 - \theta_2^*,\ (\Phi - \Phi^*)^T,\ b_0^* - b_0]^T, \qquad (49)$$

$$\tilde{b}_0(t) \equiv b_0 - b_0^*. \qquad (50)$$

Contrary to the previous $\tilde{b}_0(t)$ (24), for this new $\tilde{b}_0(t)$, it does not hold that $\tilde{b}_0(t) \geq 0$. However, similarly, the next relation is given.

$$
\begin{aligned}
&\int_0^t (he(\tau)^2 + rv(\tau)^2) d\tau + \frac{1}{2}e(t)^2 \\
&+ \frac{1}{2}\sum_{i=1}^{2}\{\hat{\theta}_i(t) - \theta_i^*\}^2/g_i \\
&+ \frac{1}{2}\{\hat{\Phi}(t) - \Phi^*\}^T G_3^{-1}\{\hat{\Phi}(t) - \Phi^*\} \\
&+ \frac{b_0^*}{2}\{\hat{p}(t) - p^*\}^2/g_4 + \frac{1}{2}\{\hat{b}_0(t) - b_0^*\}^2/g_5 \\
&+ \int_0^t \frac{\hat{b}_0(\tau)\tilde{b}_0(\tau)}{2r} e(\tau)^2 d\tau \\
\leq\ & \frac{1}{2}e(0)^2 + \frac{1}{2}\sum_{i=1}^{2}\{\hat{\theta}_i(0) - \theta_i^*\}^2/g_i \\
&+ \frac{1}{2}\{\hat{\Phi}(0) - \Phi^*\}^T G_3^{-1}\{\hat{\Phi}(0) - \Phi^*\} \\
&+ \frac{b_0^*}{2}\{\hat{p}(0) - p^*\}^2/g_4 + \frac{1}{2}\{\hat{b}_0(0) - b_0^*\}^2/g_5 \\
&+ \gamma^2 \int_0^t \|\tilde{\Theta}(\tau)\|^2 d\tau. \qquad (51)
\end{aligned}
$$

Since $\tilde{b}_0(t) \to 0$ as $t \to \infty$, it follows that

$$
\begin{aligned}
he(t)^2 + \frac{\hat{b}_0(t)\tilde{b}_0(t)}{2r}e(t)^2 \geq \delta e(t)^2, \qquad (52) \\
(\forall t \geq T,\ \exists \delta > 0),
\end{aligned}
$$

for sufficiently large $T > 0$. Then, rewriting the inequality (51) with the initial time T, it is shown that $e(t) \to 0$, where the boundedness of the adaptive system is also considered. □

Up to now, the general forms of the control schemes were provided by (30), (32), Theorem 1 ∼ Theorem 4. Next, h and r are solved, and the explicit descriptions of the control schemes are given by assuming specified forms to h and r.

Solution I : From (30), r can be chosen such that

$$\frac{1}{r} = \frac{k_1 + k_2\|\omega\|^2 + k_3\{\hat{\theta}_1 + \hat{\theta}_2\phi(e)\}}{r_0}$$
$$\Leftrightarrow r = \frac{r_0}{k_1 + k_2\|\omega\|^2 + k_3\{\hat{\theta}_1 + \hat{\theta}_2\phi(e)\}}, \quad (53)$$

where k_1, k_2, k_3, $r_0(> 0)$ are design parameters. Then, the corresponding hx^2 is obtained as follows:

$$hx^2 \le \left[\hat{b}_0^2 k_1\gamma^2 + (\hat{b}_0^2 k_2\gamma^2 - r_0)\|\omega\|^2 + (\hat{b}_0^2 k_3\gamma^2 - 4r_0\gamma^2)\{\hat{\theta}_1 + \hat{\theta}_2\phi(e)\}\right] x^2/(4r_0\gamma^2). \quad (54)$$

In order that hx^2 is positive definite for any x^2 and $\hat{\theta}_1$, $\hat{\theta}_2$ (≥ 0), k_2 and k_3 should be chosen such that

$$k_2 \ge \frac{r_0}{\hat{b}_0^2\gamma^2} \left(\Leftarrow k_2 \ge \frac{r_0}{\delta^2\gamma^2}\right), \quad (55)$$
$$k_3 \ge \frac{4r_0}{\hat{b}_0^2} \left(\Leftarrow k_3 \ge \frac{4r_0}{\delta^2}\right), \quad (56)$$
$$(0 < \delta \le b_0, \hat{b}_0). \quad (57)$$

The explicit description of the control input is

$$v = -\frac{\hat{b}_0}{2r}e = -\frac{\hat{b}_0[k_1 + k_2\|\omega\|^2 + k_3\{\hat{\theta}_1 + \hat{\theta}_2\phi(e)\}]}{2r_0}e. \quad (58)$$

Solution II : Next, r and h are obtained by setting

$$h = ar, \quad (0 < a < \infty), \quad (59)$$

where a is a positive constant, which prescribes the ratio between r and h. For equality condition of (30), r is obtained as the positive solution of

$$ar^2 + G(\hat{\Theta}, \gamma, e)r - \frac{\hat{b}_0^2}{4} = 0, \quad (60)$$
$$G(\hat{\Theta}, \gamma, e) = \hat{\theta}_1 + \hat{\theta}_2\phi(e) + \frac{\|\omega\|^2}{4\gamma^2}. \quad (61)$$

Hence, r and h are

$$r = \frac{-G(\hat{\Theta}, \gamma, e) + \sqrt{G(\hat{\Theta}, \gamma, e)^2 + a\hat{b}^2}}{2a} = \frac{\hat{b}_0^2}{2\left\{\sqrt{G(\hat{\Theta}, \gamma, e)^2 + a\hat{b}_0^2} + G(\hat{\Theta}, \gamma, e)\right\}}, \quad (62)$$
$$h = \frac{-G(\hat{\Theta}, \gamma, e) + \sqrt{G(\hat{\Theta}, \gamma, e)^2 + a\hat{b}_0^2}}{2}, \quad (63)$$

and the explicit description of the control input is

$$v = -\frac{\hat{b}_0}{2r}e = -\frac{1}{\hat{b}_0}\left\{\sqrt{G(\hat{\Theta}, \gamma, e)^2 + a\hat{b}_0} + G(\hat{\Theta}, \gamma, e)\right\}x. (64)$$

For those two explicit v, the following theorem is given.

Theorem 5 *For those two solutions of v (***Solution I*** and* **Solution II***), the residual regions of e can be made arbitrarily small by proper choices of the design parameters k_1, k_2, k_3, r_0, a, γ (sufficiently large k_1, k_2, k_3, a, and sufficiently small r_0, γ).*

(Proof) By the proper choices of the design parameters k_1, k_2, k_3, r_0, a, γ, the positive function h (54), (63) can be made arbitrarily large, and $\gamma^2\|\tilde{\Theta}\|^2$ sufficiently small, while other terms remain unchanged in (45). Since it holds that

$$\frac{1}{T}\int_0^T e(t)^2 dt \le \frac{V(0) - V(T)}{Th_0} + \frac{\gamma^2\int_0^T\|\tilde{\Theta}\|^2 dt}{Th_0}$$
$$\le \frac{M_1}{Th_0} + \frac{\gamma^2 M_2}{h_0}, \quad (65)$$
$$0 < h_0 \le h, \quad 0 < M_1, M_2 < \infty, \quad (66)$$

and h_0 can be made arbitrarily large, then $\frac{1}{T}\int_0^T e(t)^2 dt$ can be made arbitrarily small. □

4. CONCLUDING REMARKS

Solution I in this manuscript can be seen as the specified form of κ-term compensation in the integrator backstepping procedures (Krstić, *et al.*, 1995). However, the difference is that those are derived as solutions of particular $\mathcal{H}^\infty$ control problems, and that the $\mathcal{L}^2$ gains from unknown system parameters to the generalized outputs are prescribed explicitly .

In this manuscript, only the case where the relative degree is 1 was considered. The general version where the relative degree is greater than 1, becomes more complex than the present one, but can be synthesized via backstepping procedures, and will be shown in the future works.

REFERENCES

Åström, K.J. and B. Wittenmark (1989). *Adaptive Control.* Addison-Wesley.

Ioannou, P.A. and J. Sun (1996). *Robust Adaptive Control.* Prentice-Hall, New Jersey.

Krstić, M., I. Kanellakopoulos and P.V. Kokotović (1995). *Nonlinear and Adaptive Control Design.* John Wiley & Suns, New York.

Krstić, M. and H. Deng (1998). *Stabilization of Nonlinear Uncertain Systems.* Springer.

Lu, W.M. and A. Packard (1997). Adaptive $\mathcal{H}^\infty$-controller for nonlinear systems. *Proceedings of 1997 American Control Conference,* 1208-1212, Albuquerque, USA.

Miyasato, Y. (1999). Redesign of adaptive control systems based on the notion of optimality. *Proceedings of 38th IEEE Conference on Decision and Control,* 3315-3320, Phoenix, USA.

ROBUST OUTPUT TRACKING FOR UNCERTAIN STRICT-FEEDBACK SYSTEMS WITH UNKNOWN VIRTUAL CONTROL COEFFICIENTS [1]

Gürdal Arslan * Tamer Başar *

* *Coordinated Science Laboratory, University of Illinois; 1308 West Main Street; Urbana, IL 61801-2307 / USA*

Abstract: In this paper, a class of single-input single-output nonlinear systems in strict-feedback form with unknown virtual control coefficients is considered. The nonlinearities in the system dynamics are taken to be unknown, and are approximated in terms of radial basis functions (using a particular neural-net architecture). The only crucial assumption made on the unknown part of the system is that the parameters representing the neural network architecture and the virtual control coefficients lie in some known compact sets. In this setup, a robust state-feedback controller is designed, under which the system output tracks a given signal arbitrarily well, and all signals in the closed-loop system remain bounded. Moreover, a relevant disturbance attenuation inequality is satisfied by the closed-loop signals. *Copyright ©2000 IFAC*

Keywords: Adaptive control, Tracking, Neural networks, Nonlinear systems, Robust estimation.

1. INTRODUCTION

Last several years have witnessed intense activity on the topic of control of uncertain strict-feedback systems. The majority of this work has focused on the case where the uncertain part of the system is linearly parametrized. For example, in (Pan and Başar, 1998), disturbance attenuating state-feedback adaptive controllers have been presented for the linearly parametrized uncertainty case, using the H^∞ identifiers developed in (Didinsky and Başar, 1995). The results of (Pan and Başar, 1998) have then been extended to the output-feedback case in (Tezcan and Başar, 1999). The more general case where the uncertainty is not linearly parametrized has also received a great deal of attention, recently. Some selective references for this more general case are (Lu and Başar, 1998; Sanner and Slotine, 1992; Polycarpou and Mears, 1998; Narendra and Mukhopadhyay, 1997; Yu and Annaswamy, 1995; Arslan and Başar, 1999a; Arslan and Başar, 1999b). In the majority of the previous work on this topic, some type of neural network approximation has been used to cope with the general unknown nonlinearities. The universal approximation property, see (Park and Sandberg, 1993; Cardaliaguet and Euvrard, 1992; Powell, 1992), has always been invoked to justify the use of neural network approximations. Indeed, in our recent papers (Arslan and Başar, 1999b; Arslan and Başar, 1999c), we have generalized the results of (Pan and Başar, 1998; Tezcan and Başar, 1999), respectively, by approximating the general nonlinearities in terms of radial basis functions. Yet another generalization, however, is to consider the case where the virtual control coefficients are also unknown, which is the subject of this paper. We should point out that a simple coordinate transformation that would lead to renaming the virtual control inputs along with their unknown coefficients as new states does not reduce this class of systems to a form to which existing results are directly applicable. Estimating

[1] This work was supported in part by the U.S. Department of Energy under Grant DOE-DEFG-02-94ER13939.

the unknown control coefficients and incorporating these estimates into the backstepping control design appear to be a more viable approach, and this is what is done in this paper. More precisely, the modified versions of the identifiers presented in (Pan and Başar, 1998) are used to estimate the unknown parameter vectors that represent the neural network approximations of the unknown nonlinearities as well as the unknown virtual control coefficients. It is assumed that the unknown parameters to be estimated lie in a known compact set, and smooth projection is used to keep the parameter estimates within some small neighborhoods of those known compact sets. The resulting controller drives the output tracking error into a small neighborhood of the origin as quickly as desired, and keeps all other signals bounded. Moreover, a disturbance attenuation inequality with respect to an equivalent disturbance term, which is composed of the external disturbances and the approximation errors, is satisfied by the closed-loop signals. In the absence approximation errors, this results in asymptotic tracking, provided that the external disturbances are square-integrable. In a nutshell, the results of this work extend those of (Pan and Başar, 1998; Arslan and Başar, 1999b) to the more general case where the virtual control coefficients are also unknown.

Our objectives are defined in more precise terms in Section 2. In Section 3, the identifier adopted in the paper is introduced. In Section 4, the controller design steps with full state and derivative information are presented. Some simulation results are included in Section 5 to demonstrate the effectiveness of the proposed controller. Finally, the paper ends with the concluding remarks of Section 6.

2. PROBLEM FORMULATION

A strict-feedback nonlinear system described by the following equations is considered:

$$\begin{aligned} \dot{x}_i &= \varphi_i^T(x_{[i]})\vartheta_i x_{i+1} + g_i(x_{[i]}) + h_i^T(x_{[i]})w_i, \\ &\quad i = 1, \dots, n-1 \qquad (1) \\ \dot{x}_n &= \varphi_n^T(x_{[n]})\vartheta_n u + g_n(x_{[n]}) + h_n^T(x_{[n]})w_n \\ y &= x_1 \end{aligned}$$

where $x := [x_1, \dots, x_n]^T =: x_{[n]}$ is the state vector, u is the scalar control input, y is the scalar output, and w_i is the q_i-dimensional uniformly bounded external disturbance. The nonlinear functions φ_i and h_i are known and sufficiently smooth, whereas the nonlinear functions g_i are unknown, but still sufficiently smooth. The parameter vectors $\vartheta_i \in \mathcal{R}^{m_i}$ are also unknown. The objective is to force the output y track a prespecified signal $y_d(t)$. The unknown nonlinear terms g_i will be approximated in terms of a set of basis functions, ϕ_i, of an appropriate neural network defined as: $\forall i \in \{1, \dots, n\}$

$$\phi_i(x_{[i]}) := \begin{bmatrix} \exp\left(-\sigma_i \left|x_{[i]} - c_i^1\right|^2\right) \\ \vdots \\ \exp\left(-\sigma_i \left|x_{[i]} - c_i^{d_i}\right|^2\right) \end{bmatrix} \qquad (2)$$

for some numbers $\sigma_i > 0$, some vectors $c_i^j \in \mathcal{R}^i$, $j \in \{1, \dots, d_i\}$, and some integers $d_i > 0$. The vectors c_i^j are called the centers of the basis functions, and $c_i^j \neq c_i^k$ for all $j, k \in \{1, \dots, d_i\}$, $j \neq k$. The optimal parameter sets Ω_i are defined as:

$$\Omega_i := \{\theta_i : \theta_i = \arg \min_{\breve{\theta}_i \in \mathcal{R}^{d_i}} \max_{x_{[i]} \in B_{R_i}} |g_i - \phi_i^T \breve{\theta}_i|\},$$

where B_{R_i} are some compact rectangular sets in $\mathcal{R}^i$. In other words, if the functions g_i were known, and one wanted to use only an approximation for them in terms of the given basis functions, the coefficients of the approximant for g_i would be the components of θ_i. For each $i \in \{1, \dots, n\}$, a $\theta_i \in \Omega_i$ is fixed, and the approximation error is defined as:

$$e_i(x_{[i]}) := g_i(x_{[i]}) - \phi_i^T(x_{[i]})\theta_i.$$

Note that the functions $e_i(x_{[i]})$ are continuous, and hence are uniformly bounded on any compact set. Finally, the following assumptions, which will be used throughout the paper, are made.

Assumption 1. $|h_i(x_{[i]})|^2 > k_h$, $\forall x_{[i]} \in \mathcal{R}^i$, $\forall i \in \{1, \dots, n\}$, and for some $k_h > 0$.

Assumption 2. For each $i \in \{1, \dots, n\}$, there exists a known sufficiently smooth, convex, radially unbounded function $P_i : \mathcal{R}^{m_i + d_i} \mapsto [0, \infty)$ such that $P_i(\xi_i) \leq 1$, where $\xi_i := [\vartheta_i^T \ \theta_i^T]^T$. It is further assumed that if $P_i(\bar{\xi}_i) \leq 1$ for any $\bar{\xi}_i \in \mathcal{R}^{m_i + d_i}$, then $\varphi_i^T(x_{[i]})\bar{\xi}_{i1} \geq 2\varsigma_i$, $\forall x_{[i]} \in \mathcal{R}^i$, where $\bar{\xi}_{i1}$ is the vector containing the first m_i components of $\bar{\xi}_i$, and ς_i is some known positive number.

Assumption 3. The reference trajectory $y_d(t)$ is n times continuously differentiable, where $y_d(t)$ and its derivatives $y_d^{(1)}(t), \dots, y_d^{(n)}(t)$ are known and uniformly bounded by some constant, say, $M_d > 0$. Also the notation $y_d^{[i]}$ is used to denote $\left[y_d, y_d^{(1)}, \dots, y_d^{(i)}\right]^T$.

Assumption 4. $\|w_i\| \leq M_w$, $\forall i \in \{1, \dots, n\}$, and for some known $M_w > 0$.

Assumption 5. The state variables $x_i(t)$ and their derivatives $\dot{x}_i(t)$ are available for both identification and control purposes.

Remark. It is possible to relax this assumption and develop a counterpart of this theory when $\dot{x}_i(t)$, $i \in \{1, \dots, n\}$, are not available. This has not been pursued here due to page limitations, but can be found in a longer version of the paper available from the authors.

3. THE IDENTIFIER

Since θ_i and ϑ_i are not known, they have to be identified online. For this purpose, the following modified version of the identifier introduced in (Pan and Başar, 1998), is adopted:

$$\dot{\hat{\xi}}_i = \Phi_i(\dot{x}_i - \Phi_i^T\hat{\xi}_i)/\left|h_i\right|^2 - \mathcal{P}_i(\hat{\xi}_i)$$
$$\mathcal{P}_i(\hat{\xi}_i) := I_{\{P_i(\hat{\xi}_i)>1\}}\frac{(P_i(\hat{\xi}_i)-1)^{n+1}}{1+\epsilon_i - P_i(\hat{\xi}_i)}\frac{\partial P_i}{\partial \xi_i}(\hat{\xi}_i)$$
$$\Phi_i := [\varphi_i^T x_{i+1} \ \ \phi_i^T]^T, \quad \hat{\xi}_i := [\hat{\vartheta}_i^T \ \ \hat{\theta}_i^T]^T \quad (3)$$

where $x_{n+1} := u$, $\hat{\xi}_i(0)$ is such that $P_i(\hat{\xi}_i(0)) \leq 1$, and $\epsilon_i > 0$ are picked such that $P_i(\hat{\xi}_i) \leq 1 + \epsilon_i$ implies $\varphi_i^T\hat{\vartheta}_i \geq \varsigma_i$, $\forall i \in \{1,...,n\}$. The projection functions $\mathcal{P}_i(.)$ are sufficiently smooth to permit the backstepping design to be carried out as will be clear later, and they are utilized in (3) to obtain the following properties: $\forall t \geq 0$

$$P_i(\hat{\xi}_i(t)) \leq 1 + \epsilon_i$$
$$\left|\hat{\theta}_i(t)\right| \leq M_{\tilde{\theta}_i}, \text{ for some } M_{\tilde{\theta}_i} > 0$$
$$\left|\hat{\vartheta}_i(t)\right| \leq M_{\tilde{\vartheta}_i}, \text{ for some } M_{\tilde{\vartheta}_i} > 0$$
$$\varphi_i^T(x_{[i]})\hat{\vartheta}_i(t) \geq \varsigma_i, \ \forall x_{[i]} \in \mathcal{R}^i$$

where the parameter errors are defined as $\tilde{\vartheta}_i(t) := \vartheta_i - \hat{\vartheta}_i(t)$ and $\tilde{\theta}_i(t) := \theta_i - \hat{\theta}_i(t)$, $\forall i \in \{1,...,n\}$. In the next section, adaptive controllers are constructed, which use the parameter estimates generated by (3).

4. CONTROLLER DESIGN

In this section, the detailed design steps of a controller that forces the output of the plant track a given reference signal are presented. The special structure of the nonlinear plant (1) allows one to employ the backstepping design methodology. At each design step, a virtual control input is picked and a desired value is prescribed for it to make a relevant error term from the previous step sufficiently small. This procedure is repeated until the last step is reached where the actual control input is designed. After completing the controller design which achieves output tracking, the section is concluded with a discussion on the disturbance attenuation properties.

Step 1: The output tracking error is defined as:

$$z_1 := x_1 - y_d.$$

It follows that:

$$\dot{z}_1 = \varphi_1^T\vartheta_1 x_2 + g_1 + h_1^T w_1 - y_d^{(1)} \quad (4)$$

where the same notation is used to denote the equivalent forms of the functions in terms of the new variable z_1. This slight abuse of notation will be exercised throughout the section. A reasonable choice for the virtual control input x_2 would be to cancel the term $y_d^{(1)}$ exactly, to lessen the effect of g_1 by using its estimate, and to dominate the term $h_1^T w_1$ by an appropriate negative feedback. Thus, the desired value of x_2 is picked as:

$$\alpha_1 := \frac{1}{\varphi_1^T\hat{\vartheta}_1}\left[y_d^{(1)} - \frac{\left|h_1\right|^2 z_1}{4\gamma_1^2} - \beta_1 z_1 - \phi_1^T\hat{\theta}_1\right]$$

where $\beta_1(z_1, y_d^{[1]}, \hat{\xi}_1) > 0$ is a sufficiently smooth design function, and $\gamma_1 > 0$ is a scalar. Since x_2 is not the actual control input, the error term z_2 is defined as: $z_2 := x_2 - \alpha_1(z_1, y_d^{[1]}, \hat{\xi}_1)$, and (4) is rewritten as:

$$\dot{z}_1 = \varphi_1^T\hat{\vartheta}_1 z_2 - \beta_1 z_1 + h_1^T v_1 - \left|h_1\right|^2 z_1/(4\gamma_1^2),$$

where $v_1 := w_1 + h_1(\varphi_1^T\tilde{\vartheta}_1 x_2 + \phi_1^T\tilde{\theta}_1 + e_1)/\left|h_1\right|^2$. Finally, the function V_1 is introduced as: $V_1 := z_1^2/2$, whose time-derivative satisfies:

$$\dot{V}_1 \leq \varphi_1^T\hat{\vartheta}_1 z_1 z_2 - \beta_1 z_1^2 + \gamma_1^2\left|v_1\right|^2. \quad (5)$$

Remark 4.1. If x_2 were the control input, z_2 would be made equal to zero by choosing $x_2 = \alpha_1$. But, since x_2 is not the actual control input, some arbitrary upper bounds $M_2 > 0$ and $M_1 > 0$ on $|z_2|$ and $|z_1|$, respectively, are prescribed. It is observed that if $|z_2(t)| \leq M_2$, $P_1(\hat{\xi}_1(t)) \leq 1 + \epsilon_1$, and $\beta_1 > k_{\beta_1}$, where $k_{\beta_1} > 0$ is a sufficiently large number, then

$$|z_1(t)| = M_1 \Rightarrow \dot{V}_1(t) < 0.$$

Step k: $(1 < k < r)$ Assume the following structure from the $(k-1)$-th step: For $i \in \{1,...,k-1\}$

$$z_{i+1} = x_{i+1} - \alpha_i(z_{[i]}, y_d^{[i]}, \hat{\xi}_{[i]})$$
$$\dot{z}_i = \varphi_i^T\hat{\vartheta}_i z_{i+1} - \beta_i z_i - \varphi_{i-1}^T\hat{\vartheta}_{i-1}z_{i-1} + h_i^T v_i$$
$$+\sum_{j=1}^{i-1} p_{ij}^T v_j - \frac{\left|h_i\right|^2}{4\gamma_i^2}z_i - \sum_{j=1}^{i-1}\frac{\left|p_{ij}\right|^2}{4\gamma_j^2}z_i.$$

Using the induction hypothesis above, we compute the derivative of z_k as

$$\dot{z}_k = \varphi_k^T \vartheta_k x_{k+1} + g_k + h_k^T w_k - \sum_{i=1}^{k} \frac{\partial \alpha_{k-1}}{\partial y_d^{(i-1)}} y_d^{(i)}$$

$$- \sum_{i=1}^{k-1} \frac{\partial \alpha_{k-1}}{\partial z_i} \left[\varphi_i^T \vartheta_i z_{i+1} - \beta_i z_i + h_i^T v_i \right.$$

$$\left. + \sum_{j=1}^{i-1} p_{ij}^T v_j - \frac{|h_i|^2}{4\gamma_i^2} z_i - \sum_{j=1}^{i-1} \frac{|p_{ij}|^2}{4\gamma_j^2} z_i \right]$$

$$- \sum_{i=1}^{k-1} \frac{\partial \alpha_{k-1}}{\partial \hat{\xi}_i^T} \Phi_i \left[\frac{h_i^T v_i}{|h_i|^2} - \mathcal{P}_i \right]$$

which can be written as:

$$\dot{z}_k = \varphi_k^T \vartheta_k x_{k+1} + g_k + h_k^T w_k + r_k + \sum_{j=1}^{k-1} p_{kj}^T v_j$$

where

$$r_k := - \sum_{i=1}^{k-1} \frac{\partial \alpha_{k-1}}{\partial z_i} \left[\varphi_i^T \vartheta_i z_{i+1} - \beta_i z_i - \frac{|h_i|^2}{4\gamma_i^2} z_i \right.$$

$$\left. - \sum_{j=1}^{i-1} \frac{|p_{ij}|^2}{4\gamma_j^2} z_i \right] - \sum_{i=1}^{k} \frac{\partial \alpha_{k-1}}{\partial y_d^{(i-1)}} y_d^{(i)}$$

$$+ \sum_{i=1}^{k-1} \frac{\partial \alpha_{k-1}}{\partial \hat{\xi}_i^T} \Phi_i \mathcal{P}_i$$

$$p_{kj} := - \frac{\partial \alpha_{k-1}}{\partial z_j} h_j - \frac{h_j \Phi_j^T}{|h_j|^2} \frac{\partial \alpha_{k-1}}{\partial \hat{\xi}_j}$$

$$- \sum_{m=j+1}^{k-1} \frac{\partial \alpha_{k-1}}{\partial z_m} p_{mj}, \quad j \in \{1, ..., k-1\}.$$

The desired value of x_{k+1} is picked as:

$$\alpha_k = \frac{1}{\varphi_k^T \hat{\vartheta}_k} \left[-r_k - \beta_k z_k - \varphi_{k-1}^T \hat{\vartheta}_{k-1} z_{k-1} \right.$$

$$\left. - \frac{|h_k|^2}{4\gamma_k^2} z_k - \sum_{j=1}^{k-1} \frac{|p_{kj}|^2}{4\gamma_j^2} z_k - \phi_k^T \hat{\theta}_k \right] \qquad (6)$$

where $\beta_k(z_{[k]}, y_d^{[k]}, \hat{\xi}_{[k]}) > 0$ is a sufficiently smooth design function, and $\gamma_k > 0$ is a scalar. Since x_{k+1} is not the actual control input, the error term z_{k+1} is defined as: $z_{k+1} := x_{k+1} - \alpha_k(z_{[k]}, y_d^{[k]}, \hat{\xi}_{[k]})$, and $\dot{z}_k$ is rewritten as:

$$\dot{z}_k = \varphi_k^T \hat{\vartheta}_k z_{k+1} - \beta_k z_k - \varphi_{k-1}^T \hat{\vartheta}_{k-1} z_{k-1} + h_k^T v_k$$

$$+ \sum_{j=1}^{k-1} p_{kj}^T v_j - \frac{|h_k|^2}{4\gamma_k^2} z_k - \sum_{j=1}^{k-1} \frac{|p_{kj}|^2}{4\gamma_j^2} z_k.$$

where $v_k := w_k + h_k(\varphi_k^T \tilde{\vartheta}_k x_{k+1} + \phi_k^T \tilde{\theta}_k + e_k) / |h_k|^2$. Now, the time-derivative of $V_k := z_k^2/2$ satisfies:

$$\dot{V}_k \leq \varphi_k^T \hat{\vartheta}_k z_k z_{k+1} - \beta_k z_k^2 - \varphi_{k-1}^T \hat{\vartheta}_{k-1} z_{k-1} z_k$$

$$+ \sum_{j=1}^{k} \gamma_j^2 |v_j|^2 \qquad (7)$$

Remark 4.k. If x_{k+1} were the control input, z_{k+1} would be made equal to zero, by choosing $x_{k+1} = \alpha_k$. But, since x_{k+1} is not the control input, an arbitrary upper bound $M_{k+1} > 0$ on $|z_{k+1}|$ is prescribed. It is observed that if $|z_i(t)| \leq M_i$, $\forall i \in \{1, ..., k-1, k+1\}$, $P_i(\hat{\xi}_i(t)) \leq 1 + \epsilon_i$, $\forall i \in \{1, ..., k\}$, and $\beta_k > k_{\beta_k}$, where k_{β_k} is a sufficiently large number, then

$$|z_k(t)| = M_k \Rightarrow \dot{V}_k(t) < 0.$$

Since all of the relevant definitions and results of Step k are consistent with the induction hypothesis, it is concluded that the induction hypothesis holds true for all $k \in \{1, ..., n-1\}$.

Step n: It is easy to see that the results of Step k holds true also for $k = n$ with $x_{n+1} := u$, where u is the actual control input. Thus, z_{n+1} can be made zero , by picking the control input as: $u = \alpha_n$, where α_n is obtained by setting $k = n$ in (6). This control input renders the time-derivative of the function $V_n := z_n^2/2$ as:

$$\dot{V}_n \leq -\beta_n z_n^2 + \sum_{j=1}^{n} \gamma_j^2 |v_j|^2 - \varphi_{n-1}^T \hat{\vartheta}_{n-1} z_{n-1} z_n.$$

Remark 4.n. It is observed that if $|z_i(t)| \leq M_i$, $\forall i \in \{1, ..., n-1\}$, $P_i(\hat{\xi}_i(t)) \leq 1 + \epsilon_i$, $\forall i \in \{1, ..., n\}$, and $\beta_n > k_{\beta_n}$, where k_{β_n} is a sufficiently large number, then

$$|z_n(t)| = M_n \Rightarrow \dot{V}_n(t) < 0.$$

This now leads to:

Theorem 1. *Consider the closed-loop system described by (1), (3), and the designed controller, under Assumptions 1,...,5. Assume further that $|z_i(0)| \leq M_i$, and $\beta_i > k_{\beta_i}$, $i \in \{1, ..., n\}$, where M_i, $i \in \{1, ..., n\}$, are arbitrary positive numbers, and k_{β_i} are as specified in Remarks 5.1, 5.2,...,5.n. Then:*

1. $\|z_i(t)\| \leq M_i$, $\|P_i(\hat{\xi}_i(t))\| \leq 1 + \epsilon_i$, $i = 1, ..., n$.
2. *Given $\bar{T} > 0$, and $\rho > 0$ such that $\rho \leq M_1$, if*

$$\beta_1 \geq \max \left\{ \frac{M_1^2 - \rho^2 + 2K\bar{T}}{2\rho^2 \bar{T}}, k_{\beta_1} \right\},$$

where

$$K := \sup\{ |\varphi_1^T \hat{\vartheta}_1 z_1 z_2 + \gamma_1^2 |v_1|^2 | :$$
$$|z_1| \leq M_1, |z_2| \leq M_2, P_1(\hat{\xi}_1) \leq 1 + \epsilon_1,$$
$$|y_d| \leq M_d, |y_d^{(1)}| \leq M_d, |w_1| \leq M_w \},$$

then $\sup_{t \geq \bar{T}} |z_1(t)| \leq \rho$.

Proof. Consider the continuous function

$$V(t) := \tfrac{1}{2} \max \left\{ \frac{z_1^2(t)}{M_1^2}, ..., \frac{z_n^2(t)}{M_n^2} \right\}.$$

Clearly, $V(0) \leq \frac{1}{2}$. Let $t^* \geq 0$ be the first time, if ever, such that $V(t^*) = \frac{1}{2}$. Obviously,

$|z_i(t)| \leq M_i, \forall i \in \{1,...,n\}$, for all $t \in [0, t^*]$. Now, let $I_1 := \{i \in \{1,...,n\} : |z_i(t^*)| = M_i\}$. From Remarks 5.1, 5.2,...,5.n, the upper-right derivative of $V(t)$ at $t = t^*$ is

$$D^+V(t)\big|_{t=t^*} = \max_{i \in I_1} \frac{\frac{d}{dt}\left(\frac{1}{2}z_i^2(t)\right)_{t=t^*}}{M_i^2} < 0$$

which proves the first part of the theorem.

To prove the second part of the theorem, we note that, from the first part of the theorem, we have $\|z_i(t)\| \leq M_i$, $P_i(\hat{\xi}_i(t)) \leq 1 + \epsilon_i$, $\forall i \in \{1,...,n\}$. Now, from (5), whenever $|z_1(t)| \geq \rho$, the following inequality clearly holds

$$\dot{V}_1 < 0. \tag{8}$$

The inequality (8) indicates that if $|z_1(\tilde{t})| \leq \rho$ for any $\tilde{t} \geq 0$, then $|z_1(t)| \leq \rho$ for all $t \geq \tilde{t}$. Now, if $\rho < |z_1(0)| \leq M_1$, assume to the contrary that $|z_1(t)| > \rho$ for all $0 \leq t \leq \bar{T}$. Then, we have the following contradiction

$$|z_1(\bar{T})|^2/2 < \rho^2/2.$$

Hence, we have $|z_1(t)| \leq \rho$ for all $t \geq \bar{T}$, completing the proof.

Next, the disturbance attenuation properties are discussed. To obtain a meaningful disturbance attenuation inequality, the following function is introduced: $W_i := |\tilde{\xi}_i|^2$, whose time-derivative along (3) is:

$$\dot{W}_i = -\frac{|\Phi_i^T\tilde{\xi}_i|^2}{|h_i|^2} + \left|w_i + \frac{h_i e_i}{|h_i|^2}\right|^2 - |v_i|^2 + 2\tilde{\xi}_i^T\mathcal{P}_i$$

We note that the last term $2\tilde{\xi}_i^T\mathcal{P}_i$ in the above equality is nonpositive. Now, let U be defined as: $U := \sum_{i=1}^n (V_i + \gamma^2 W_i)$ for some disturbance attenuation level $\gamma > 0$. It is straightforward to show that the time-derivative of U satisfies:

$$\begin{aligned}\dot{U} \leq &-\sum_{i=1}^n \beta_i z_i^2 - \gamma^2 \sum_{i=1}^n |\Phi_i^T\tilde{\xi}_i|^2/|h_i|^2 \\ &+\sum_{i=1}^n \left[((n+1-i)\gamma_i^2 - \gamma^2)|v_i|^2 \right. \\ &\left. +\gamma^2 \left|w_i + h_i e_i/|h_i|^2\right|^2\right].\end{aligned} \tag{9}$$

This now leads to the following result.

Theorem 2. *Consider the closed-loop system described by (1), (3), and the designed controller, under Assumptions 1,...,5. Assume further that the hypothesis of Theorem 1 holds. Then:*

If $\gamma_i := \gamma/\sqrt{n-i+1}$, $\forall i \in \{1,...,n\}$, for some $\gamma > 0$, then the following disturbance attenuation inequality is satisfied: $\forall t \geq 0$

$$\int_0^t \sum_{i=1}^n \left[\beta_i z_i^2 + \gamma^2|\Phi_i^T\tilde{\xi}_i|^2/|h_i|^2\right] ds + \leq \gamma^2 \int_0^t \sum_{i=1}^n \left|w_i + h_i e_i/|h_i|^2\right|^2 ds + U(0).$$

Moreover, if $e_i = 0$, and $w_i \in \mathcal{L}_2$, $\forall i \in \{1,...,n\}$, then we obtain asymptotic tracking, i.e.,

$$\lim_{t\to\infty} z_1(t) = 0.$$

Proof. The integration of (9) yields the disturbance attenuation inequality. Moreover, if $e_i = 0$, and $w_i \in \mathcal{L}_2$, $\forall i \in \{1,...,n\}$, we conclude that $z_1(t) \in \mathcal{L}_2$. Finally, since, from the first part of Theorem 1, $\dot{z}_1(t)$ is uniformly bounded, we have $\lim_{t\to\infty} z_1(t) = 0$. □

5. AN ILLUSTRATIVE EXAMPLE

A second order system of the form (1) is considered, where

$$\begin{aligned}
g_1 &= \phi_1^T[1,-1]^T + 0.01x_1^4 + 0.1\sin(x_1)\\
g_2 &= \phi_2^T[2,-2]^T + 0.1x_1^2 + 0.1x_2^2\\
\phi_1 &= [\exp(-(x_1-1)^2), \exp(-(x_1+1)^2)]^T\\
\phi_2 &= [\exp(-(x_1-1)^2 - (x_2-1)^2),\\
&\qquad \exp(-(x_1+1)^2 - (x_2+1)^2)]^T\\
\varphi_1 &= 1 + x_1^2/10,\ \varphi_2 = 1.1 + \sin(x_2)/10\\
\vartheta_1 &= \vartheta_2 = 2,\ h_1 = [1, x_1]^T,\ h_2 = [0,1]^T\\
w_1 &= w_2 = [0.5\sin(5t), 0.1\cos(20t)\\
&\qquad +0.2\sin(10t)]^T.
\end{aligned}$$

The functions g_1 and g_2 are approximated within $B_{R_1} := \{x_1 \in \mathcal{R} : |x_1| \leq 1\}$, and $B_{R_2} := \{x_{[2]} \in \mathcal{R}^2 : |x_1| \leq 1, |x_2| \leq 1.5\}$, by using ϕ_1, ϕ_2, respectively. The optimal parameters can be computed to be: $\bar{\theta}_1 = \{1.08, -1.08\}$, $\bar{\theta}_2 = \{2.48, -1.52\}$. The unknown parameters are assumed to satisfy: $(\vartheta_i - 3)^2/4 + |\theta_i|^2/50 \leq 1$, $i = 1, 2$. The reference signal is $y_d(t) = \sin(t)$. The designed controller is used with the following parameters: $\beta_1 = 5.0$, $\beta_2 = 1.0$, $\gamma_1 = \gamma_2 = 5.0$, $\epsilon_1 = \epsilon_2 = 0.2$. The simulation results are shown in Figs. 1, and 2, and as can be seen, the controller keeps all the signals bounded. Fig. 1 shows that, after an initial deviation, the output of the system always stays within a small neighborhood of the reference signal. The parameter estimates approach their true values reasonably fast, and then experience small oscillations, as shown in Fig. 2. It is plausible that this parameter "convergence" result is due to the persistent nature of the reference signal and the disturbances. In general, however, no such result can be expected.

6. CONCLUSIONS

An adaptive controller design has been presented for uncertain strict-feedback systems with unknown virtual control coefficients. The controller design procedure involves the use of the modified versions of the identifiers developed in (Pan and Başar, 1998), the backstepping design methodology, and the use of radially unbounded functions. The resulting controller achieves the output tracking of a reference signal arbitrarily well at the expense of increased control effort, as stated in Theorem 1. Also, the closed-loop signals remain bounded and satisfy a relevant disturbance attenuation inequality.

One extension of this work would be to the case where only the output is available for control and identification purposes. This would constitute an extension of the results of (Arslan and Başar, 1999c) to the case where the virtual control coefficients are also unknown, and is currently being studied.

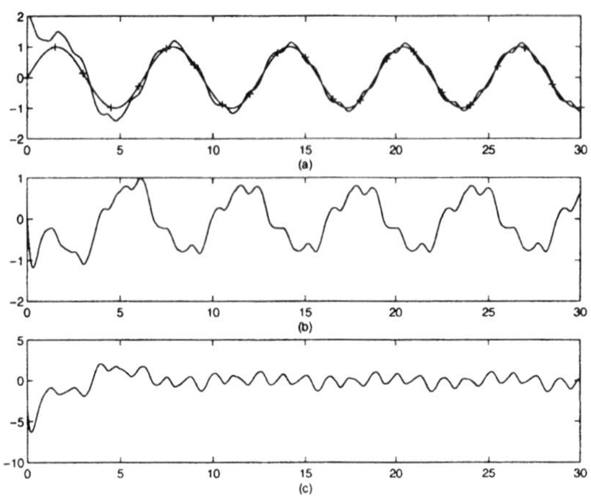

Fig. 1. (a) $y(t)$ (-) and $y_d(t)$ (+) (b) $x_2(t)$ (c) $u(t)$.

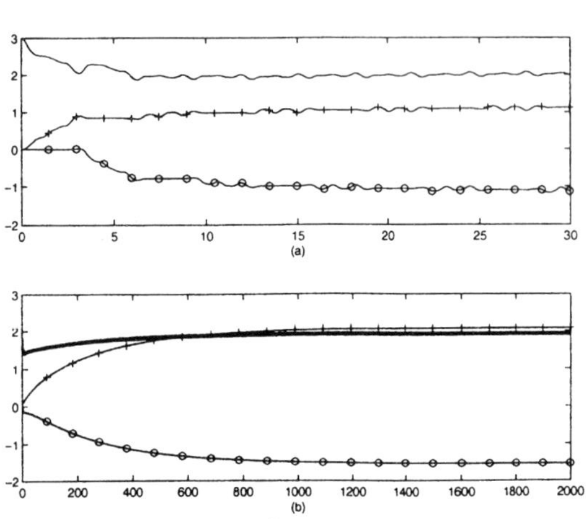

Fig. 2. (a) $\vartheta_1(t)$ (-), $\theta_{11}(t)$ (+), $\theta_{12}(t)$ (o) (b) $\vartheta_2(t)$ (-), $\theta_{21}(t)$ (+), $\theta_{22}(t)$ (o).

7. REFERENCES

Arslan, G. and T. Başar (1999a). Robust identification for output tracking in systems with unknown nonlinearities. In: *Proc. of the 1999 ACC*. pp. 1057–1061.

Arslan, G. and T. Başar (1999b). Robust output tracking for strict-feedback systems using neural-net based approximators for nonlinearities. In: *Proc. of the 1999 CDC*. pp. 2987–2992.

Arslan, G. and T. Başar (1999c). Robust output-feedback control of strict-feedback systems with unknown nonlinearities. In: *Proc. of the 1999 CDC*. pp. 4748–4753.

Cardaliaguet, P. and G. Euvrard (1992). Approximation of a function and its derivative with a neural network and. *Neural Networks* **5**, 207–220.

Didinsky, G. and T. Başar (1995). Parameter identification for uncertain plants using h^∞ methods. *Automatica* **39**, 1227–1250.

Lu, S. and T. Başar (1998). Robust nonlinear system identification using neural-network models. *IEEE Transactions on Neural Networks* **9**, 407–429.

Narendra, K.S. and S. Mukhopadhyay (1997). Adaptive control using neural networks and approximate models. *IEEE Transactions on Neural Networks* **8**, 475–485.

Pan, Z. and T. Başar (1998). Adaptive controller design for tracking and disturbance attenuation in parametric-strict-feedback nonlinear systems. *IEEE Transactions on Automatic Control* **43**, 1066–1083.

Park, J. and I.W. Sandberg (1993). Approximation and radial-basis-function networks. *Neural Computation* **5**, 305–316.

Polycarpou, M.M. and M.J. Mears (1998). Stable adaptive tracking of uncertain systems using nonlinearly parameterized on-line approximators. *International Journal of Control* **70**, 363–384.

Powell, M.J.D. (1992). The theory of radial basis function approximation in 1990. In: *Advances in Numerical Analysis* (W. Light, Ed.). Vol. 2. pp. 102–205. Clarendon Press, Oxford.

Sanner, R.M. and J-J.E. Slotine (1992). Gaussian networks for direct adaptive control. *IEEE Transactions on Neural Networks* **3**, 837–863.

Tezcan, I.E. and T. Başar (1999). Disturbance attenuating adaptive controllers for parametric strict feedback nonlinear systems with output measurements. *Journal of Dynamic Systems, Measurement & Control* **121**, 48–57.

Yu, S-H. and A.M. Annaswamy (1995). Stable neural controllers for nonlinear dynamic systems. *Automatica* **34**, 641–650.

PARTIAL CONVERGENCE OF COUPLED TIME-VARYING SYSTEMS

Keum-Shik Hong* and Kyung-Jinn Yang**

* *School of Mechanical Engineering, Pusan National University, 30 Changjeon-dong Kumjeong-ku, Pusan, 609-735, Korea. Tel: +82-51-510-2454, Fax: +82-51-514-0685, Email: kshong@hyowon.pusan.ac.kr*
** *Department of Mechanical and Intelligent Systems Engineering, Graduate College Pusan National University, Email: jinnky@hyowon.pusan.ac.kr*

Abstract: Asymptotic behavior of a partial state of coupled nonautonomous ordinary and/or partial differential equations is investigated. The system assumes the existence and uniqueness of the solution and the existence of a Lyapunov function candidate. Although ordinary Lyapunov analysis can not guarantee the asymptotic convergence of the state, the analysis in this paper assures the asymptotic convergence to zero of the partial state which remains in the time derivative of the Lyapunov function candidate. The results are applied to adaptive systems. *Copyright ©2000 IFAC*

Keywords: Time-varying system, stability, uniform ultimate boundedness, adaptive control, parabolic partial differential equation.

1. INTRODUCTION

To conclude various stabilities in the sense of Lyapunov (Sastry and Bodson, 1989), the associated theorems require that the time derivative of a Lyapunov function candidate, $\dot{V}$, should be at least ≤ 0. Therefore, if $\dot{V}$ allows positive values near an equilibrium point, no stability can be asserted. In this paper, however, motivated by the error equations from the adaptive control of time-varying finite or infinite dimensional systems, new convergence results of a partial state of coupled dynamic systems are investigated. In the model reference adaptive control of time-varying plants, the derivative of a Lyapunov function, which is introduced to derive adaptation laws, is not negative semi-definite in general.

Briefly mentioning a few related papers, adaptive tracking controls for linear time-varying systems with uncertain slowly varying parameters are given in (Middleton and Goodwin, 1988; Ravikanth and Meyn, 1999; Tsakalis and Ioannou, 1987). For nonlinear systems with time-varying parameters, adaptive controls are given in (Freeman, *et al.*, 1998; Marino and Tomei, 1993; Polycarpou and Mears, 1998). Compared to the finite dimensional case, the adaptive control of infinite dimensional systems is not well developed and has only recently been studied in (Balas, 1998; Baumeister, *et al.*, 1997; Hong and Bentsman, 1994; Kobayashi, 1988) for the time-invariant system and in (Hong, *et al.*, 1999) for the slowly time-varying system.

The main contribution of this paper is Theorem 1, which asserts the asymptotic convergence of a partial state of coupled nonautonomous dynamic systems to zero. An application of Theorem 1 is demonstrated in Section 3, but the plausible applications are not limited to adaptive systems. This paper clarifies that the converging behavior of the state error dynamics to zero with the Lyapunov redesign method is a general property and is not dependent on the type of the time-varying system considered. $\|\cdot\|$ will be adopted as a genetic norm in any Banach space.

2. MAIN RESULTS

Theorem 1: Consider coupled nonautonomous partial and ordinary differential equations as

$$\dot{e}(t) = A(t,\theta(t))e(t) + f(t,e(t),\theta(t)), \quad e(0) = e_0 \quad (1)$$

$$\dot{\theta}(t) = g(t,e(t),\theta(t)), \quad \theta(0) = \theta_0 \quad (2)$$

where $e(t) \in X$ and $\theta(t) \in R^n$. X is a Banach space and $A(t,\theta(t))$ is a family of operators on X. The following assumptions are made:

(C1) There exist unique solutions to (1)-(2). The solution of (1) is given in the variation of constants formula of the following form

$$e(t) = \Phi(t,0)e_0 + \int_0^t \Phi(t,\sigma)f(\sigma,e(\sigma),\theta(\sigma))d\sigma \quad (3)$$

where $\Phi(t,s)$, $0 \le s \le t$, is the evolution operator associated with $A(t,\theta(t))$, and $\|\Phi(t,s)\| \le M_1$, where M_1 is a positive constant, for $0 \le s \le t < \infty$.

(C2) For all $t \ge 0$

$$\|f(t,e(t),\theta(t))\| \le \alpha_0(\theta(t))\|e(t)\| + c_0 \quad (4)$$

where c_0 is a positive constant and $\alpha_0 : R^n \to R^+$ is bounded for finite value of $\theta(t)$.

(C3) There exists a functional $V(t,e(t),\theta(t)): R^+ \times X \times R^n \to R^+$ for system (1)-(2) such that

$$\alpha_1(\|e(t)\|) + \beta_1(\|\theta(t)\|) \le V(t,e(t),\theta(t)) \le \alpha_2(\|e(t)\|) + \beta_2(\|\theta(t)\|) \quad (5)$$

where for $i = 1,2$, $\alpha_i(\cdot)$ and $\beta_i(\cdot)$ are continuous, strictly increasing functions with $\alpha_i(0) = \beta_i(0) = 0$, and $\lim_{r\to\infty} \alpha_i(r) = \lim_{r\to\infty} \beta_i(r) = \infty$.

(C4) The derivative of V satisfies:

$$\dot{V}(t,e(t),\theta(t))\Big|_{(1)-(2)} \le -\alpha_3(\|e(t)\|) + \upsilon(t) \quad (6)$$

where $\alpha_3(\cdot)$ is a continuous, strictly increasing function with $\alpha_3(0) = 0$ and $\upsilon(t)$ is uniformly bounded by ρ.

Then, for every $e_0 \in X$ and $\theta_0 \in R^n$, i) $e(t)$ and $\theta(t)$ are uniformly ultimately bounded. ii) Furthermore, if $\int_0^\infty \upsilon(t)\,dt < \infty$, then $e(t) \to 0$ as $t \to \infty$.

Proof: (i) (C3) and (C4) establish that $\theta(t)$ and $e(t)$ are uniformly ultimately bounded. The proof of uniform ultimate boundedness is collected in Appendix A considering the length of the proof. Let the uniform bound be $\bar{d}$.

(ii) Now, let the solution of (1) at time t starting with the initial state $e(s)$ at initial time s be of the form:

$$e(t) = \Phi(t,s)e(s) + \int_s^t \Phi(t,\sigma)f(\sigma,e(\sigma),\theta(\sigma))d\sigma . (7)$$

Then, a two parameter family of map $S(t,s)$ on X can be defined as

$$S(t,s)e(s) := e(t,e(s),s), \quad 0 \le s \le t < \infty . \quad (8)$$

Then, by the uniqueness and continuous dependence of the solution on the triple $\{t,e(s),s\}$, the mapping $S(t,s)$ on X becomes an evolution process (Walker, 1980) such that

1) $S(\cdot,s)e(s): R^+ \to X$ is continuous (right continuous at $t = s$),

2) $S(t,\cdot)(\cdot): R^+ \times X \to X$ is continuous,

3) $S(s,s)e(s) = e(s)$,

4) $S(t,s)e(s) = S(t,r)S(r,s)e(s)$, for all $e(s) \in X$ and $0 \le s \le r \le t < \infty$.

Now, the integration of both sides of (6) from 0 to ∞ yields:

$$\int_0^\infty \alpha_3(\|e(t)\|)\,dt = \int_0^\infty \alpha_3(\|S(t,0)\,e_0\|)\,dt \le V(0) - V(\infty) + \int_0^\infty \upsilon(t)dt < \infty \quad (9)$$

where the initial time and state are chosen to be zero, respectively. Note also that $e(t) = S(t,0)e_0$ has been used in deriving the first equality.

Indeed, conclusion ii) of the theorem can be proven by using a contradiction argument. Suppose that $e(t) = S(t,0)e_0 \not\to 0$ as $t \to \infty$. Then, there exists a $\mu > 0$ and an infinite sequence $t_j \to \infty$ such that $\|S(t_j,0)e_0\| \ge \mu$. Now, however small this μ is, there exist constants $M_2 > 0$ and $\mu_0 > 0$ such that

$$M_2 \ge \sup_{\|\theta(t)\| \le \bar{d}} \alpha_0(\theta(t)) \quad (10)$$

$$\frac{\mu}{\exp(1)} - \frac{c_0}{M_2} \ge \mu_0 > 0 \quad (11)$$

where $\alpha_0(\cdot)$ and c_0 were defined in (4) and exp denotes exponential function. Note that (11) is always satisfied if $c_0 = 0$. Therefore, taking norms on both sides of (7) yields:

$$\|e(t)\| \le \|\Phi(t,s)\|\|e(s)\| + \int_s^t \|\Phi(t,\sigma)\|\left(\alpha_0(\theta(\sigma))\|e(\sigma)\| + c_0\right)d\sigma$$
$$\le M_1\|e(s)\| + M_1 M_2 \int_s^t \left(\|e(\sigma)\| + \frac{c_0}{M_2}\right)d\sigma \quad (12)$$

where M_2 is chosen to satisfy (10). Applying the Bellman-Gronwall's inequality (Sastry and Bodson, 1989) to (12) yields for all $t \ge s \ge 0$

$$\|e(t)\| \le \left(M_1\|e(s)\| + \frac{c_0}{M_2}\right)\exp\left(M_1 M_2 (t-s)\right). \quad (13)$$

Now, without loss of generality, we can assume that the sequence t_j satisfies $t_{j+1} - t_j > (M_1 M_2)^{-1}$. If we set $\Delta_j = [t_j - (M_1M_2)^{-1}, t_j]$, then $m(\Delta_j) = (M_1 M_2)^{-1} > 0$, where m is the Lebesgue measure, and the intervals Δ_j do not overlap. For $t \in \Delta_j$

$$\mu \le \|e(t_j)\| = \|S(t_j,t)e(t)\|$$
$$\le \left(M_1\|e(t)\| + \frac{c_0}{M_2}\right)\exp\left(M_1 M_2 (t_j - t)\right)$$
$$\le \left(M_1\|e(t)\| + \frac{c_0}{M_2}\right)\exp(1)$$

where the second inequality above is obtained from (13). Therefore, we have $\|e(t)\| \ge \frac{\mu_0}{M_1} > 0$ for all $t \in \Delta_j$. Hence,

$$\int_0^\infty \alpha_3\left(\|S(t,0)e_0\|\right)dt \ge \sum_{j=1}^{\infty} \int_{\Delta_j} \alpha_3\left(\|S(t,0)e_0\|\right)dt$$
$$\ge \sum_{j=1}^{\infty} \int_{\Delta_j} \alpha_3\left(\mu_0 / M_1\right)dt = \alpha_3\left(\mu_0 / M_1\right)\sum_{j=1}^{\infty} m(\Delta_j) = \infty$$

contradicting (9). Thus, we must have $e(t) \to 0$ as $t \to \infty$. □

Remark 1: Even though the existence and uniqueness of the solution has been assumed in Theorem 1, the class of systems that permits application of Theorem 1 is not empty, as will be shown in Section 3. The specific conditions for the existence and uniqueness of solution will be different, depending on the types of systems in the considered Banach spaces. Also, there is no general existence theorem available yet for general infinite dimensional systems. By defining $B(t) := A(t,\theta(t))$, the conditions (with respect to $B(t)$) for the existence of the evolution operator $\Phi(t,s)$ are found in (Pazy, 1983) or (Lasiecka, *et al*, 1997) for the hyperbolic system and in (Pazy, 1983) for the parabolic system. Furthermore, with the same conditions $\|\Phi(t,s)\| \le M\exp(\omega(t-s))$ (ω is a stability constant) for the hyperbolic case (Pazy, 1983), and $\|\Phi(t,s)\| \le M$ for the parabolic case (Pazy, 1983), are achieved.

Remark 2: In model reference adaptive controls, the state $e(t)$ represents the error dynamics between the closed loop plant and the reference model. The state $\theta(t)$ denotes the parameter estimation error vector.

The corollary next establishes the asymptotic convergence of a partial state of coupled nonautonomous ordinary differential equations.

Corollary 1: Consider coupled nonautonomous ordinary differential equations as

$$\dot{e}(t) = f(t,e(t),\theta(t)), \qquad e(0) = e_0 \qquad (14)$$
$$\dot{\theta}(t) = g(t,e(t),\theta(t)), \qquad \theta(0) = \theta_0 \qquad (15)$$

where $e(t) \in R^n$ and $\theta(t) \in R^m$. The following assumptions are made:

(D1) f and g are Lebesgue measurable in t, locally Lipschitz continuous in $e(t)$ and $\theta(t)$, and $f(t,0,0) = 0$ and $g(t,0,0) = 0$ for all $t \ge 0$.

(D2) For all $t \ge 0$

$$\|f(t,e(t),\theta(t))\| \le \alpha_0(\theta(t))\|e(t)\| + c_0 \qquad (16)$$

where c_0 is a positive constant and $\alpha_0 : R^m \to R^+$ is bounded for finite value of $\theta(t)$.

(D3) There exists a continuously differentiable function $V(t,e(t),\theta(t))$: $R^+ \times R^{n+m} \to R^+$ such that

$$\alpha_1\left(\|e(t)\|\right) + \beta_1\left(\|\theta(t)\|\right) \le V(t,e(t),\theta(t))$$
$$\le \alpha_2\left(\|e(t)\|\right) + \beta_2\left(\|\theta(t)\|\right) \quad (17)$$

where for $i = 1,2$, $\alpha_i(\cdot)$ and $\beta_i(\cdot)$ are continuous, strictly increasing functions with $\alpha_i(0) = \beta_i(0) = 0$, and $\lim_{r\to\infty} \alpha_i(r) = \lim_{r\to\infty} \beta_i(r) = \infty$.

(D4) The derivative of V satisfies

$$\dot{V}(t,e(t),\theta(t))\Big|_{(14)-(15)} \le -\alpha_3\left(\|e(t)\|\right) + \upsilon(t) \quad (18)$$

where $\alpha_3(\cdot)$ is a continuous, strictly increasing function with $\alpha_3(0) = 0$ and $\upsilon(t)$ permits positive values but uniformly bounded by ρ.

Then, for every $e_0 \in R^n$ and $\theta_0 \in R^m$, i) $e(t)$ and

$\theta(t)$ are uniformly ultimately bounded. ii) Furthermore, if $\int_0^\infty \upsilon(t)\,dt < \infty$, then $e(t) \to 0$ as $t \to \infty$.

Proof: The sequence of proof is the same as that of Theorem 1. The only difference is: System (1) is semilinear and system (14) is nonlinear.

(i) (D1) establishes the local existence and uniqueness of the solutions of (14)-(15) (Hale, 1969). (D3) and (D4) establish that $e(t)$ and $\theta(t)$ are uniformly ultimately bounded.

(ii) Let the unique solution of (14), starting with the initial state $e(s)$ at initial time s, at time t be of the form

$$e(t) = e(s) + \int_s^t f(\sigma, e(\sigma), \theta(\sigma))\,d\sigma := e(t, e(s), s). \tag{19}$$

Taking norms using (16) yields:

$$\begin{aligned} \|e(t)\| &\le \|e(s)\| + \int_s^t \left(\alpha_0(\theta(\sigma))\|e(\sigma)\| + c_0\right) d\sigma \\ &\le \|e(s)\| + M_2 \int_s^t \left(\|e(\sigma)\| + \frac{c_0}{M_2}\right) d\sigma . \end{aligned} \tag{20}$$

Applying the Bellman-Gronwall's inequality yields for all $t \ge s \ge 0$

$$\|e(t)\| \le \left(\|e(s)\| + \frac{c_0}{M_2}\right) \exp(M_2(t-s)). \tag{21}$$

Therefore, choosing $\Delta_j = [t_j - M_2^{-1}, t_j]$, $\|e(t)\| \ge \mu_0 > 0$ for all $t \in \Delta_j$.

The rest of the proof is the same as Theorem 1. □

Now, assume that (15) vanishes. Then, the following corollary is easily obtained.

Corollary 2: Consider the nonautonomous system

$$\dot e(t) = f(t, e(t)), \quad e(0) = e_0 \in R^n \tag{22}$$

where f is Lebesgue measurable in t, locally Lipschitz continuous in $e(t)$, and $f(t,0) = 0$ for all $t \ge 0$. Assume the following conditions

(D2)′ $\|f(t, e(t))\| \le \alpha_0 \|e(t)\| + c_0$, $\forall t \ge 0$

(D3)′ $\alpha_1(\|e(t)\|) \le V(t, e(t)) \le \alpha_2(\|e(t)\|)$

(D4)′ $\dot V(t, e(t))\big|_{(22)} \le -\alpha_3(\|e(t)\|) + \upsilon(t)$

where $\alpha_0 > 0$, $c_0 \ge 0$, $\alpha_i(\cdot)\,(i = 1,2,3)$ are continuous, strictly increasing functions with $\alpha_i(0) = 0$ and $\lim_{r\to\infty} \alpha_1(r) = \lim_{r\to\infty} \alpha_2(r) = \infty$, $V(\cdot,\cdot): R^+ \times R^n \to R^+$ is a continuously differentiable function, and $\upsilon(t)$ is uniformly bounded by ρ. Then, if $\int_0^\infty \upsilon(t)\,dt < \infty$, then $e(t) \to 0$ as $t \to \infty$. □

Remark 3: For the finite-dimensional case, the conditions in Corollary 1 are equivalent to the boundedness of $\dot e$. Hence $e(t) \in L_2(0,\infty)$ together with Barbalat's lemma implies that $e(t) \to 0$ as $t \to \infty$. The use of Barbalat's lemma is standard in adaptive control literature.

Remark 4: Equations (13) and (21) imply that the solutions existing for all $t \ge 0$ cannot grow faster than any exponential function.

3. AN EXAMPLE: MRAC OF A PARABOLIC PDE

In this section, the applicability of Theorem 1 is demonstrated by considering a model reference adaptive control of a nonautonomous parabolic partial differential equation (PDE). The asymptotic convergence of the state error to zero is shown. Consider the distributed parameter system described by the linear parabolic PDE with time-varying parameters as follows:

$$\begin{aligned} &\dot\xi = a(t)\xi_{xx} + b(t)\xi + u, \quad 0 < x < 1, \quad t > 0 \\ &\xi(0,t) = g_a(t), \quad \xi(1,t) = g_b(t), \quad \xi(x,0) = \xi_0(x) \end{aligned} \tag{23}$$

where x denotes the spatial variable, $u = u(x,t)$ is a control input, $g_a(t)$ and $g_b(t)$ are boundary conditions and are analytic, $\xi_0(x)$ is the initial state, and $\xi = \xi(x,t)$ is the observed distributed state. $\xi_{xx} = \partial^2\xi/\partial x^2$. $a(t)$ and $b(t)$ are unknown bounded time-varying parameters. Along with system (23), consider a reference model with the same boundary conditions as

$$\begin{aligned} &\dot\xi_m = a_m\xi_{mxx} + b_m\xi_m + r, \quad 0 < x < 1, \quad t > 0 \\ &\xi_m(0,t) = g_a(t), \quad \xi_m(1,t) = g_b(t), \quad \xi_m(x,0) = \xi_{m0}(x) \end{aligned} \tag{24}$$

where the subscript m stands for model, and $r = r(x,t)$ is the reference signal. $a_m > 0$ and $b_m < 0$ are constant coefficients. The control law is given by

$$u = (a_m - \hat a)\xi_{xx} + (b_m - \hat b)\xi + r \tag{25}$$

where $\hat a$ and $\hat b$ are adaptive estimates to be specified. Substituting (25) into (23) yields the following closed loop plant equation:

$$\dot\xi = (a_m - \tilde a)\xi_{xx} + (b_m - \tilde b)\xi + r \tag{26}$$

where $\widetilde{a}=\hat{a}-a$ and $\widetilde{b}=\hat{b}-b$ are parameter estimation errors.

Introducing the state error $e=\xi-\xi_m$, the following state error equation is obtained:

$$\dot{e}=(a_m-\widetilde{a})e_{xx}+(b_m-\widetilde{b})e-(\widetilde{a}\xi_{mxx}+\widetilde{b}\xi_m)$$
$$e(0,t)=e(1,t)=0\,,\quad e(x,0)=\xi_0(x)-\xi_{m0}(x)\,. \quad (27)$$

Now, consider a functional $V:L_2(0,1)\times R^2\to R^+$ such that

$$V(e,\widetilde{a},\widetilde{b})=\frac{1}{2}\langle e,e\rangle+\frac{1}{2\gamma_a}\widetilde{a}^2+\frac{1}{2\gamma_b}\widetilde{b}^2 \quad (28)$$

where γ_a and γ_b are positive constants; $\langle\cdot,\cdot\rangle$ is the inner product in the space $L_2(0,1)$ defined as $\langle h,g\rangle=\int_0^1 h(x)g(x)dx$ and with the induced norm $\|\cdot\|$. Differentiating (28) with respect to t along (27) yields:

$$\begin{aligned}\dot{V}(t)&=\langle e,\dot{e}\rangle+\frac{1}{\gamma_a}\widetilde{a}\dot{\widetilde{a}}+\frac{1}{\gamma_b}\widetilde{b}\dot{\widetilde{b}}\\&=a_m\langle e_x,e_{xx}\rangle+b_m\langle e,e\rangle\\&\quad+\widetilde{a}\left(-\langle e,e_{xx}\rangle-\langle e,\xi_{mxx}\rangle+\frac{1}{\gamma_a}\dot{\widetilde{a}}\right)\\&\quad+\widetilde{b}\left(-\langle e,e\rangle-\langle e,\xi_m\rangle+\frac{1}{\gamma_b}\dot{\widetilde{b}}\right).\end{aligned} \quad (29)$$

Let the differential equations of the adaptive estimators in (25) be given as

$$\dot{\hat{a}}=\gamma_a\left(\langle e,e_{xx}\rangle+\langle e,\xi_{mxx}\rangle\right)$$
$$\dot{\hat{b}}=\gamma_b\left(\langle e,e\rangle+\langle e,\xi_m\rangle\right). \quad (30)$$

Then, substituting (30) into (29) yields:

$$\dot{V}\le\left(-a_m\pi^2+b_m\right)\|e\|^2-\frac{1}{\gamma_a}\widetilde{a}\dot{a}-\frac{1}{\gamma_b}\widetilde{b}\dot{b}. \quad (31)$$

The existence and uniqueness of the solutions for coupled nonautonomous dynamic systems (27) and (30) have been established in (Hong, *et al*, 1999). Consequently, if $\widetilde{a}$ and $\widetilde{b}$ are bounded, $\dot{a}\in L_1$, and $\dot{b}\in L_1$, then $\int_0^\infty\left(\frac{1}{\gamma_a}\widetilde{a}\dot{a}+\frac{1}{\gamma_b}\widetilde{b}\dot{b}\right)dt<\infty$. Therefore, by applying Theorem 1 the tracking error $e(t)$ converges to zero as $t\to\infty$.

4. CONCLUSIONS

The asymptotic behavior of a part of the solutions of coupled time-varying dynamic systems was investigated. Since the derivative of a Lyapunov function candidate was not ≤ 0, only uniform ultimate boundedness would have been concluded. However, further investigation in this paper has shown that the partial state which remains in the derivative of the Lyapunov function candidate converges to zero asymptotically. The results were applied to the state error convergence to zero in adaptive control of time-varying plants.

REFERENCES

Balas, M.J. (1998). Stable feedback control of linear distributed parameter systems: Time and frequency domain conditions. *J. Math. Anal. Appl.*, **225**, pp. 144- 167.

Baumeister, J., W. Scondo, M.A. Demetriou, and I.G. Rosen (1997). On-line parameter estimation for infinite-dimensional dynamical systems. *SIAM J. Control Optim.*, **35**(2), pp. 678-713.

Freeman, R.A., M. Krstic, and P.V. Kokotovic (1998). Robustness of adaptive nonlinear control to bounded uncertainties. *Automatica*, **34**(10), pp. 1227-1230.

Hale, J.K. (1969). *Ordinary Differential Equations*. New York: Wiley.

Hong, K.S. and J. Bentsman (1994). Direct adaptive control of parabolic systems: Algorithm synthesis, convergence and stability analysis. *IEEE Trans. Automatic Control*, **39**(10), pp. 2018-2033.

Hong, K.S., V. Solo, and J. Bentsman (1999). Adaptive control of a time-varying parabolic system: Averaging analysis. *Proc. 14th World Congress of IFAC*, Beijing, China, Preprint Number E-2c-16-1, **E**, pp. 351-356.

Kobayashi, T. (1988). Finite dimensional adaptive control for infinite dimensional systems. *Int. J. Control*, **48**(1), pp. 289-302.

Lasiecka, I., L. Pandolfi, and R. Triggiani (1997). A singular control approach to highly damped second-order abstract equations and applications. *Appl. Math. Optim.*, **36**, pp. 67-107.

Marino, R. and P. Tomei (1993). Global adaptive output-feedback control of nonlinear systems. Part I: Linear parameterization. *IEEE Trans. Automatic Control*, **38**, pp. 17-32.

Middleton, R.H. and G.C. Goodwin (1988). Adaptive control of time varying linear plants. *IEEE Trans. Automatic Control*, **33**, pp. 150-155.

Pazy, A. (1983). *Semigroups of Linear Operators and Applications to Partial Differential Equations*, pp. 134-135 and 149-150. New York: Springer-Verlag.

Polycarpou, M.M and M.J. Mears (1998). Stable adaptive tracking of uncertain systems using nonlinearly parameterized on-line approximators. *Int. J. Control*, **70**(3), pp. 363-384.

Ravikanth, R. and S.P. Meyn (1999). Bounds on achievable performance in the identification and adaptive control of time-varying systems. *IEEE Trans. Automatic Control*, **44**(4), pp. 670-682.
Sastry, S. and M. Bodson (1989). *Adaptive Control: Stability, Convergence, and Robustness,* pp. 23 and 27. Englewood Cliffs, NJ: Prentice-Hall.
Tsakalis, K.S. and P.A. Ioannou (1987). Adaptive control of linear time-varying plants. *Automatica,* **33**, pp. 893-900.
Walker, J.A. (1980). *Dynamical Systems and Evolution Equations*, pp. 12 and 49. New York: Plenum.

APPENDIX A: UNIFORM ULTIMATE BOUNDEDNESS

Proof of Assertion i) of Theorem 1: Let the solutions of (1)-(2) be $e(\cdot):[t_0,t_1]\rightarrow X$ and $\theta(\cdot):[t_0,t_1]\rightarrow R^n$ with initial conditions $\|e(t_0)\|\le r$ and $\|\theta(t_0)\|\le r$, respectively. Let $\bar{r}\stackrel{\Delta}{=}\max\{r,R\}$, where $R\stackrel{\Delta}{=}\alpha_3^{-1}(\rho)$. Hence, $\|e(t_0)\|\le\bar{r}$, $\|\theta(t_0)\|\le\bar{r}$, and $R\le\bar{r}$. Let

$$d(\bar{r})=\max\left\{\left(\alpha_1^{-1}\circ\alpha_2\right)(\bar{r}),\left(\beta_1^{-1}\circ\beta_2\right)(\bar{r})\right\}. \quad (A1)$$

Furthermore, in view of (5)

$$\alpha_1(\bar{r})\le\alpha_2(\bar{r}) \text{ and } \beta_1(\bar{r})\le\beta_2(\bar{r}). \quad (A2)$$

Uniform Boundedness:

Now, show the uniform boundedness of $e(t)$ and $\theta(t)$ over $[t_0,t_1]$ by using a contradiction argument. Suppose there is a $t_3>t_0$ such that

$$\|e(t_3)\|>d(\bar{r}) \text{ and } \|\theta(t_3)\|>d(\bar{r}). \quad (A3)$$

Since $e(\cdot)$ and $\theta(\cdot)$ are continuous and

$$\|e(t_0)\|\le\bar{r}\le d(\bar{r})<\|e(t_3)\| \text{ and}$$
$$\|\theta(t_0)\|\le\bar{r}\le d(\bar{r})<\|\theta(t_3)\|$$

there exists a $t_2\in[t_0,t_3]$ such that $\|e(t_2)\|=\bar{r}$ and $\|\theta(t_2)\|=\bar{r}$, and for $\forall t\in[t_2,t_3]$

$$\|e(t)\|\ge\bar{r}\ge R \text{ and } \|\theta(t)\|\ge\bar{r}. \quad (A4)$$

Now using conditions (5)-(6) and (A4) yields:

$$\begin{aligned}\alpha_1\left(\|e(t_3)\|\right)+\beta_1\left(\|\theta(t_3)\|\right)&\le V\left(t_3,e(t_3),\theta(t_3)\right)\\ &\le\alpha_2(\bar{r})+\beta_2(\bar{r})+\int_{t_2}^{t_3}\{-\alpha_3(R)+\rho\}\,d\sigma\\ &=\alpha_2(\bar{r})+\beta_2(\bar{r})\end{aligned} \quad (A5)$$

where $R=\alpha_3^{-1}(\rho)$ has been used in getting the last equality. Therefore, the fact that (A5) should hold for both $e(\cdot)$ and $\theta(\cdot)$ implies $\alpha_1\left(\|e(t_3)\|\right)\le\alpha_2(\bar{r})$ and $\beta_1\left(\|\theta(t_3)\|\right)\le\beta_2(\bar{r})$.

Hence, $\|e(t_3)\|\le\left(\alpha_1^{-1}\circ\alpha_2\right)(\bar{r})\le d(\bar{r})$ and $\|\theta(t_3)\|\le\left(\beta_1^{-1}\circ\beta_2\right)(\bar{r})\le d(\bar{r})$. This now contradicts (A3). Therefore, $\|e(t)\|\le d(\bar{r})$ and $\|\theta(t)\|\le d(\bar{r})$.

Uniform Ultimate Boundedness:

The uniform boundedness above is now extended over $[t_0,\infty)$. Let

$$\bar{d}>\max\left\{\left(\alpha_1^{-1}\circ\alpha_2\right)(R),\left(\beta_1^{-1}\circ\beta_2\right)(R)\right\} \text{ and}$$
$$\bar{R}=\max\left\{\left(\alpha_2^{-1}\circ\alpha_1\right)(\bar{d}),\left(\beta_2^{-1}\circ\beta_1\right)(\bar{d})\right\}$$

so that $\bar{R}>R$ and

$$d(\bar{R})=\max\left\{\left(\alpha_1^{-1}\circ\alpha_2\right)(\bar{R}),\left(\beta_1^{-1}\circ\beta_2\right)(\bar{R})\right\}\stackrel{\Delta}{=}\bar{d}.$$

If $r\le\bar{R}$, then $\|e(t_0)\|\le\bar{R}$ and $\|\theta(t_0)\|\le\bar{R}$. Hence, in view of the uniform boundedness results above, $\|e(t)\|\le d(\bar{R})=\bar{d}$ and $\|\theta(t)\|\le d(\bar{R})=\bar{d}$ for $\forall t\in[t_0,\infty)$ so that $T(\bar{d},r)=0$. Next consider $r>\bar{R}$, and suppose that for $\forall t\in[t_0,t_1]$

$$\|e(t)\|>\bar{R} \text{ and } \|\theta(t)\|>\bar{R} \quad (A6)$$

where $t_1=t_0+T(\bar{d},r)$ and

$$T(\bar{d},r)\stackrel{\Delta}{=}\frac{\alpha_2(r)+\beta_2(r)-\alpha_1(\bar{R})-\beta_1(\bar{R})}{\alpha_3(\bar{R})-\rho}.$$

Then, using conditions (5)-(6) and (A6) yields:

$$\begin{aligned}\alpha_1\left(\|e(t_1)\|\right)+\beta_1\left(\|\theta(t_1)\|\right)&\le V\left(t_1,e(t_1),\theta(t_1)\right)\\ &\le\alpha_2(r)+\beta_2(r)+T(\bar{d},r)\{-\alpha_3(\bar{R})+\rho\}\\ &=\alpha_1(\bar{R})+\beta_1(\bar{R}).\end{aligned}$$

That is, $\|e(t_1)\|\le\bar{R}$ and $\|\theta(t_1)\|\le\bar{R}$. But, this contradicts supposition (A6). Hence, there must be a $t_2\in[t_0,t_1]$ such that $\|e(t_2)\|\le\bar{R}$ and $\|\theta(t_2)\|\le\bar{R}$. Then, as a consequence of the uniform boundedness result above, for $\forall t\ge t_2$

$$\|e(t)\|\le d(\bar{R})=\bar{d} \text{ and } \|\theta(t)\|\le d(\bar{R})=\bar{d}.$$

Hence, we have that $\|e(t)\|\le\bar{d}$ and $\|\theta(t)\|\le\bar{d}$, $\forall t\ge t_0+T(\bar{d},r)$. □

ADAPTIVE CONTROL OF STOCHASTIC STRICT-FEEDBACK SYSTEMS UNDER A RISK-SENSITIVE CRITERION [1]

Gürdal Arslan * Tamer Başar *

* *Coordinated Science Laboratory, University of Illinois; 1308 West Main Street; Urbana, IL 61801-2307 / USA*

Abstract: In this paper, a class of single-input single-output stochastic nonlinear systems in strict-feedback form with a risk sensitive cost criterion is considered. The uncertainty in the system description is assumed to be linearly parameterized. By employing the backstepping design technique, a state-feedback adaptive controller is constructed such that a meaningful risk sensitive cost remains bounded by a positive definite function of the initial conditions. Moreover, all closed-loop signals stay bounded in probability and the states of the nonlinear plant asymptotically approach origin in probability.

Keywords: Stochastic control, Adaptive control, Nonlinear systems, Risk, Uncertain dynamic systems.

1. INTRODUCTION

As a result of intensive research on the topic of nonlinear systems, a profusion and variety of results are available today on controlling a nonlinear strict-feedback system, particularly in a deterministic framework. The book (Krstić *et al.*, 1995) overviews numerous design techniques to asymptotically stabilize a noise-free strict-feedback system. The disturbance attenuation issues are studied in (Pan and Başar, 1998; Tezcan and Başar, 1999) for uncertain strict-feedback systems where the uncertainty is linearly parameterized, whereas the local optimality is introduced in (Ezal *et al.*, 1999; Ezal *et al.*, 1997). Also, the use of neural-network approximation to control such systems with more general uncertainty structures has been recently popular; see (Sanner and Slotine, 1992; Polycarpou and Mears, 1998; Narendra and Mukhopadhyay, 1997; Arslan and Başar, 1999a; Arslan and Başar, 1999b). Yet, another recent trend has been to obtain the counterparts of these results in a stochastic-framework where the external disturbances are assumed to be random. The recent book (Krstić and Deng, 1998) has several chapters on this topic, primarily on stabilization of stochastic strict feedback systems. A more relevant reference to the present work, however, is (Pan and Başar, 1999), which has presented a control design that results in closed-loop signals achieving an arbitrarily small average risk sensitive cost. In this paper, the results of (Pan and Başar, 1999) are extended to strict-feedback systems where the system model includes also additional linearly parameterized uncertainty. Thus, the main objective of this paper is to design adaptive state-feedback controllers for a class of uncertain stochastic nonlinear systems under a risk-sensitive cost criterion, so that not only the main value but also the variance of a meaningful integral cost is penalized; see (Pan and Başar, 1999) for this interpretation of the risk-sensitive cost. It should be noted that it is possible to construct identifiers, which are stochastic counterparts of those presented in (Didinsky and Başar, 1995), to estimate the unknown parameters

[1] This work was supported in part by the U.S. Department of Energy under Grant DOE-DEFG-02-94ER13939.

based on the relationship between risk-sensitive stochastic designs and stochastic games. One can then attempt to use these estimates to design an adaptive controller which makes a relevant risk sensitive cost for the closed-loop system bounded. Nevertheless, for simplicity and to avoid technical difficulties, in this paper a simpler adaptive controller is constructed such that a meaningful risk sensitive cost for the closed-loop system remains bounded by a positive definite function of the initial conditions. The controller design is based on the well-known back-stepping methodology. The proposed adaptive controller also keeps the closed-loop signals bounded in probability and makes the states of the plant asymptotically approach the origin.

It should, finally, be pointed out that this paper tackles a stochastic **nonlinear** adaptive control problem under a risk-sensitive cost criterion. Therefore, an optimum risk-sensitive controller where the cost function contains a priori fixed weights on the states and the control action seems to be quite difficult to obtain. The approach taken in this paper leads to a suboptimal yet computable finite-dimensional controller. The design procedure offers substantial flexibility in choosing the weights on the state variables. However, no cost on the control action is imposed, and as a result the proposed controller could be a high-gain controller, particularly when the performance specifications on state variables are tight. Therefore, the design parameters should be picked judiciously to trade off between a good transient response of the states and the control action.

In the sequel, the objectives of this paper is defined in more precise terms; see Section 2. Then, in Section 3, the controller design steps with full state measurements are presented. In Section 4, some simulation results are included to demonstrate the effectiveness of the proposed control design. Finally, the paper ends with the concluding remarks of Section 5.

2. PROBLEM FORMULATION

A nonlinear system described by the following Itô differential equations is considered:

$$\begin{aligned} dx_i &= [x_{i+1} + f_i(x_{[i]}) + \phi_i^T(x_{[i]})\theta_i]dt \\ &\quad + h_i^T(x_{[i]})dw, \quad i = 1, ..., n-1 \\ dx_n &= [b(x_{[n]})u + f_n(x_{[n]}) + \phi_n^T(x_{[n]})\theta_n]dt \\ &\quad + h_n^T(x_{[n]})dw \end{aligned} \tag{1}$$

where $x := [x_1, ..., x_n]^T =: x_{[n]}$ is the state vector, u is the scalar control input, $y := x_1$ is the output, $\theta_i \in \mathcal{R}^{d_i}$ are the unknown parameters, w is the $\mathcal{R}^q$-valued standard Wiener process, and the initial condition $x(0)$ is fixed. The functions f_i, ϕ_i, h_i, b, and $1/b$ are known, and sufficiently smooth. It is further assumed that $f_i(0_{i\times 1}) = 0$, $\phi_i(0_{i\times 1}) = 0_{d_i\times 1}$, $h_i(0_{i\times 1}) = 0_{q\times 1}$, $i = 1, ..., n$. The objective is to design a state-feedback controller such that the states of the plant (1) asymptotically approach origin, and the following property holds:

$$\frac{2}{\theta} \ln E \exp\left[\frac{\theta}{2}\int_0^T [y^2 + l(t,x)]dt\right] \leq \gamma(x(0), \rho - \bar{\rho}), \qquad \forall T \geq 0 \tag{2}$$

where $\theta > 0$ is a given constant, $l(t, x) \geq 0$ is some weight function, $\gamma(.,.)$ is some positive definite function, $\rho := [|\theta_1|, ..., |\theta_n|]^T$, and $\bar{\rho}$ is an arbitrary initial estimate for ρ. It is further required that the closed-loop system is globally stable in probability. Although, it is not possible to pick the weight function $l(t, x)$ arbitrarily, the design procedure will give the designer a great degree of freedom in shaping this function.

3. CONTROLLER DESIGN

In this section, the design steps of an adaptive controller that achieves asymptotic stabilization of (1) and the satisfaction of (2) are presented.

Step 1: Defining $z_1 := x_1$ yields:

$$dz_1 = (x_2 + \bar{f}_1 + \bar{\phi}_1^T\theta_1)dt + \bar{h}_1^T dw \tag{3}$$

where the functions with an over-bar symbol denote the equivalents of those without an over-bar symbol in terms of z_1. This notation will be used throughout to denote the equivalent forms of the functions in terms of the transformed variables $z_{[n]} := [z_1, ..., z_n]^T$. (3) can be rewritten as:

$$dz_1 = (x_2 + r_1 + p_{11}^T\theta_1)dt + s_1^T dw$$

where $r_1(z_1) := \bar{f}_1(z_1)$ represents the terms that could have been canceled if x_2 were the control input, $p_{11}(z_1) := \bar{\phi}_1(z_1)$ is the multiplier of the unknown parameter θ_1, and $s_1(z_1) := \bar{h}_1(z_1)$ is the multiplier of the random disturbance. Next, the quadratic function $V_1 := z_1^2/2$ is introduced. Its Itô differential is:

$$\begin{aligned} dV_1 &= z_1[(x_2 + r_1 + p_{11}^T\theta_1)dt + s_1^T dw] \\ &\quad + (|s_1|^2/2)dt. \end{aligned} \tag{4}$$

From Yang's inequality and the fact that p_{11} and s_1 are smooth and vanish at origin, it follows that:

$$z_1 p_{11}^T\theta_1 + \frac{|s_1|^2}{2} \leq z_1^2(\eta_1 + \psi_{1v}(z_1)) \tag{5}$$

where $\eta_1 := |\theta_1|^2/2$ is unknown, and $\psi_{1v}(z_1) \geq 0$ is some known smooth function. Using (5) in (4) leads to:

$$dV_1 \leq z_1[(x_2 + k_1 + z_1\eta_1)dt + \sigma_1^T dw] \quad (6)$$

where $k_1(z_1) := r_1 + z_1\psi_{1v}$, $\sigma_1(z_1) := s_1$). Now, the error term z_2 between x_2 and its desired value is defined as: $z_2 := x_2 + \alpha_1(z_1, \hat{\eta}_1)$, where

$$\begin{aligned}\alpha_1 &:= k_1 + z_1(\hat{\eta}_1 + n\theta|\sigma_1|^2/4 + \beta_1 + n),\\ \dot{\hat{\eta}}_1 &:= \delta_1(z_1) := z_1^2, \qquad \hat{\eta}_1(0) = \bar{\eta}_1,\end{aligned}$$

$\beta_1(z_1, \hat{\eta}_1) \geq 0$ is some design function, and $\bar{\eta}_1 \geq 0$ is an arbitrary initial estimate for η_1. With this, (6) can be written as:

$$\begin{aligned}dV_1 \leq\ & (z_1 z_2 - (\beta_1 + n)z_1^2 + \tilde{\eta}_1 z_1^2)dt\\ & -(n\theta z_1^2|\sigma_1|^2/4)dt + z_1\sigma_1^T dw\end{aligned}$$

where $\tilde{\eta}_1 := \eta_1 - \hat{\eta}_1$. Finally, $W_1 := V_1 + \tilde{\eta}_1^2/2$ is introduced, where its Itô differential satisfies:

$$\begin{aligned}dW_1 \leq\ & (z_1 z_2 - (\beta_1 + n)z_1^2 - n\theta z_1^2|\sigma_1|^2/4)dt\\ & +z_1\sigma_1^T dw.\end{aligned}$$

This completes the first step.

Step 2: The Itô differential of z_2 is

$$\begin{aligned}dz_2 =\ & (x_3 + \bar{f}_2 + \bar{\phi}_2^T\theta_2)dt + \bar{h}_2^T dw\\ & +\frac{\partial\alpha_1}{\partial z_1}[(z_2 - \alpha_1 + r_1 + p_{11}^T\theta_1)dt + s_1^T dw]\\ & +\frac{\partial\alpha_1}{\partial\hat{\eta}_1}\delta_1 dt + \frac{1}{2}\frac{\partial^2\alpha_1}{\partial z_1^2}|s_1|^2 dt\end{aligned}$$

which could be written as:

$$dz_2 = (x_3 + r_2 + \sum_{j=1}^{2} p_{2j}^T\theta_j)dt + s_2^T dw \quad (7)$$

where

$$\begin{aligned}r_2(z_{[2]}, \hat{\eta}_1) &= \bar{f}_2 + \frac{\partial\alpha_1}{\partial z_1}[z_2 - \alpha_1 + r_1]\\ &\quad +\frac{\partial\alpha_1}{\partial\hat{\eta}_1}\delta_1 + \frac{1}{2}\frac{\partial^2\alpha_1}{\partial z_1^2}|s_1|^2\\ p_{21}(z_{[2]}, \hat{\eta}_1) &= \frac{\partial\alpha_1}{\partial z_1}p_{11}, \qquad p_{22}(z_{[2]}, \hat{\eta}_1) = \bar{\phi}_2\\ s_2(z_{[2]}, \hat{\eta}_1) &= \bar{h}_2 + \frac{\partial\alpha_1}{\partial z_1}s_1.\end{aligned}$$

Since s_2 and p_{2j}, $j = 1, 2$, are smooth functions, and $s_2(z_{[2]} = 0_{2\times1}, \hat{\eta}_1) = 0_{q\times1}$, $p_{2j}(z_{[2]} = 0_{2\times1}, \hat{\eta}_1) = 0_{d_j\times1}$, $j = 1, 2$, it follows that:

$$|s_2|^2 + \sum_{j=1}^{2}|p_{2j}|^2 \leq 2z_2^2\psi_{2v}(z_{[2]}, \hat{\eta}_1) + 2z_1^2\psi_{2c}(z_1, \hat{\eta}_1)$$

for some smooth functions $\psi_{2v}(z_{[2]}, \hat{\eta}_1) \geq 0$ and $\psi_{2c}(z_1, \hat{\eta}_1) \geq 0$. Now, the smooth function $V_2 = W_1 + \Xi_2(z_1, \hat{\eta}_1)z_2^2/2$ is introduced, where $\Xi_2(z_1, \hat{\eta}_1) := 1/(1 + \psi_{2c}(z_1, \hat{\eta}_1))$. The Itô differential of V_2 is calculated as:

$$\begin{aligned}dV_2 =\ & dW_1 + \Xi_2 z_2[(x_3 + r_2 + \sum_{j=1}^{2} p_{2j}^T\theta_j)dt\\ & +s_2^T dw] + \frac{z_2^2}{2}\frac{\partial\Xi_2}{\partial z_1}[(z_2 - \alpha_1 + r_1 + p_{11}^T\theta_1)dt\\ & +s_1^T dw] + \frac{z_2^2}{2}\frac{\partial\Xi_2}{\partial\hat{\eta}_1}\delta_1 dt + \frac{1}{2}\Xi_2|s_2|^2 dt\\ & +z_2\left[\frac{\partial\Xi_2}{\partial z_1}s_2^T s_1 + \frac{z_2}{4}\frac{\partial\Xi_2^2}{\partial z_1^2}|s_1|^2\right]dt.\end{aligned} \quad (8)$$

Again, Yang's inequality yields:

$$\begin{aligned}&\Xi_2 z_2\sum_{j=1}^{2} p_{2j}^T\theta_j + \frac{z_2^2}{2}\frac{\partial\Xi_2}{\partial z_1}p_{11}^T\theta_1 + \frac{\Xi_2}{2}|s_2|^2 \quad (9)\\ &\leq \Xi_2 z_2^2[\eta_2 + \psi_{2v}] + \frac{z_2^2}{4\Xi_2}\left|\frac{\partial\Xi_2}{\partial z_1}p_{11}\right|^2 + \Xi_2\psi_{2c}z_1^2\end{aligned}$$

where $\eta_2 = \sum_{j=1}^{2}|\theta_j|^2/2 + |\theta_1|^2/4$ is an unknown parameter. Using (9) in (8) leads to:

$$\begin{aligned}dV_2 \leq\ & dW_1 + \Xi_2 z_2[(x_3 + k_2 + z_2\eta_2)dt + \sigma_2^T dw]\\ & +\Xi_2\psi_{2c}z_1^2 dt\end{aligned} \quad (10)$$

where

$$\begin{aligned}k_2 &:= r_2 + \frac{1}{\Xi_2}\left[\frac{z_2}{2}\frac{\partial\Xi_2}{\partial z_1}(z_2 - \alpha_1 + r_1) + \frac{z_2}{2}\frac{\partial\Xi_2}{\partial\hat{\eta}_1}\delta_1\right.\\ &\quad +\frac{\partial\Xi_2}{\partial z_1}s_2^T s_1 + \frac{z_2}{4}\frac{\partial\Xi_2^2}{\partial z_1^2}|s_1|^2\\ &\quad \left.+\frac{z_2}{4\Xi_2}\left|\frac{\partial\Xi_2}{\partial z_1}p_{11}\right|^2\right] + \psi_{2v}z_2\\ \sigma_2 &:= s_2 + \frac{z_2}{2\Xi_2}\frac{\partial\Xi_2}{\partial z_1}s_1.\end{aligned}$$

The error term z_3 between x_3 and its desired value is defined as: $z_3 := x_3 + \alpha_2(z_{[2]}, \hat{\eta}_{[2]})$, where

$$\begin{aligned}\alpha_2 &= k_2 + z_2(\hat{\eta}_2 + n\theta\Xi_2|\sigma_2|^2/4 + \beta_2/\Xi_2)\\ &\quad +[(n-2)z_2 + z_1]/\Xi_2,\\ \dot{\hat{\eta}}_2 &:= \delta_2(z_{[2]}, \hat{\eta}_1) := \Xi_2 z_2^2, \qquad \hat{\eta}_2(0) = \bar{\eta}_2,\end{aligned}$$

$\beta_2(z_{[2]}, \hat{\eta}_{[2]}) \geq 0$ is some design function, and $\bar{\eta}_2$ is an arbitrary initial estimate for η_2. (10) is written in terms of z_3 as:

$$\begin{aligned}dV_2 \leq\ & dW_1 + [\Xi_2\psi_{2c}z_1^2 - (n-2)z_2^2 - z_2 z_1]dt\\ & +\Xi_2[z_2 z_3 - (n\theta\Xi_2|\sigma_2|^2/4 + \beta_2/\Xi_2)z_2^2]dt\\ & +\Xi_2\left[z_2^2\tilde{\eta}_2 dt + z_2\sigma_2^T dw\right]\end{aligned}$$

where $\tilde{\eta}_2 := \eta_2 - \hat{\eta}_2$. Finally, $W_2 := V_2 + \tilde{\eta}_2^2/2$ is introduced, where its Itô differential satisfies:

$$dW_2 \le -z_1^2 dt + \Xi_2 z_2 z_3 dt - \sum_{i=1}^{2} (n\theta \Xi_i^2 |\sigma_i|^2/4) z_i^2 dt$$
$$+ \sum_{i=1}^{2} \Xi_i \left[-\beta_i z_i^2/\Xi_i dt + z_i \sigma_i^T dw \right]$$
$$-(n-2)|z_{[2]}|^2 - \underbrace{(1 - \Xi_2 \psi_{2c})}_{\ge 0} z_1^2 dt$$

where $\Xi_1 := 1$. This completes the second step.

Step k: $(k = 3, ..., n-1)$ Assume the following structure from the previous step: $\forall i \in \{1, ..., k-1\}$

$$z_m = x_m + \alpha_{m-1}(z_{[m-1]}, \hat{\eta}_{[m-1]}) \ni$$
$$\alpha_0 := 0, \ \alpha_{m-1}|_{z_{[m-1]}=0_{m-1\times 1}} = 0,$$
$$m = 1, ..., k$$
$$dz_i = (z_{i+1} - \alpha_i + r_i + \sum_{j=1}^{i} p_{ij}^T \theta_j) dt + s_i^T dw,$$
$$r_i = r_i(z_{[i]}, \hat{\eta}_{[i-1]}), \ \ni \ r_i|_{z_{[i]}=0_{i\times 1}} = 0,$$
$$p_{ij} = p_{ij}(z_{[i]}, \hat{\eta}_{[i-1]}), \ \ni \ p_{ij}|_{z_{[i]}=0_{i\times 1}} = 0_{d_i \times 1},$$
$$j = 1, ..., i$$
$$s_i = s_i(z_{[i]}, \hat{\eta}_{[i-1]}), \ \ni \ s_i|_{z_{[i]}=0_{i\times 1}} = 0_{q\times 1},$$
$$\dot{\hat{\eta}}_i = \delta_i(z_{[i]}, \hat{\eta}_{[i-1]}), \ \ni \ \delta_i|_{z_{[i]}=0_{i\times 1}} = 0,$$
$$W_{k-1} = \sum_{i=1}^{k-1} [\Xi_i z_i^2 + \tilde{\eta}_i^2]/2 \text{ such that}$$
$$\Xi_i(z_{[i-1]}, \hat{\eta}_{[i-1]}) > 0,$$
$$dW_{k-1} = [-z_1^2 + \Xi_{k-1} z_{k-1} z_k] dt$$
$$-(n-k+1)|z_{[k-1]}|^2 dt - \sum_{i=1}^{k-1} \beta_i z_i^2 dt$$
$$- \sum_{i=1}^{k-1} \Xi_i \left[(n\theta \Xi_i |\sigma_i|^2/4) z_i^2 dt - z_i \sigma_i^T dw \right].$$

From this, the Itô differential of z_k is obtained as:

$$dz_k = (x_{k+1} + \bar{f}_k + \bar{\phi}_k^T \theta_k) dt + \bar{h}_k^T dw$$
$$+ \sum_{i=1}^{k-1} \frac{\partial \alpha_{k-1}}{\partial z_i} [(z_{i+1} - \alpha_i + r_i + \sum_{j=1}^{i} p_{ij}^T \theta_j) dt$$
$$+ s_i^T dw] + \sum_{i=1}^{k-1} \frac{\partial \alpha_{k-1}}{\partial \hat{\eta}_i} \delta_i dt$$
$$+ \frac{1}{2} \sum_{i,j \in \{1,...,k-1\}} \frac{\partial^2 \alpha_{k-1}}{\partial z_i \partial z_j} s_i^T s_j$$

which could be written as:

$$dz_k = (x_{k+1} + r_k + \sum_{j=1}^{k} p_{kj}^T \theta_j) dt + s_k^T dw \qquad (11)$$

where

$$r_k = \bar{f}_k + \sum_{i=1}^{k-1} \frac{\partial \alpha_{k-1}}{\partial z_i} [z_{i+1} - \alpha_i + r_i]$$
$$+ \sum_{i=1}^{k-1} \frac{\partial \alpha_{k-1}}{\partial \hat{\eta}_i} \delta_i$$
$$+ \frac{1}{2} \sum_{i,j \in \{1,...,k-1\}} \frac{\partial^2 \alpha_{k-1}}{\partial z_i \partial z_j} s_i^T s_j$$
$$p_{kj} = \sum_{m=j}^{k-1} \frac{\partial \alpha_{k-1}}{\partial z_m} p_{mj}, \qquad j = 1, ..., k-1$$
$$p_{kk} = \bar{\phi}_k, \quad s_k = \bar{h}_k + \sum_{i=1}^{k-1} \frac{\partial \alpha_{k-1}}{\partial z_i} s_i.$$

Since s_k and p_{kj}, $j = 1, ..., k$, are smooth, and vanish whenever $z_{[k]} = 0_{k\times 1}$, it follows that:

$$|s_k|^2 + \sum_{j=1}^{k} |p_{kj}|^2 \le 2 z_k^2 \psi_{kv} + 2|z_{[k-1]}|^2 \psi_{kc}$$

for some smooth functions $\psi_{kv}(z_{[k]}, \hat{\eta}_{[k-1]}) \ge 0$ and $\psi_{kc}(z_{[k-1]}, \hat{\eta}_{[k-1]}) \ge 0$. The smooth function V_k is defined as $V_k = W_{k-1} + \Xi_k z_k^2/2$ where $\Xi_k(z_{[k-1]}, \hat{\eta}_{[k-1]}) := 1/(1 + \psi_{kc}(z_{[k-1]}, \hat{\eta}_{[k-1]}))$. The Itô differential of V_k is calculated as:

$$dV_k = dW_{k-1} + \Xi_k z_k [(x_{k+1} + r_k + \sum_{j=1}^{k} p_{kj}^T \theta_j) dt$$
$$+ s_k^T dw] + \frac{z_k^2}{2} \sum_{i=1}^{k-1} \frac{\partial \Xi_k}{\partial z_i} [(z_{i+1} - \alpha_i + r_i$$
$$+ \sum_{j=1}^{i} p_{ij}^T \theta_j) dt + s_i^T dw] + \frac{z_k^2}{2} \sum_{i=1}^{k-1} \frac{\partial \Xi_k}{\partial \hat{\eta}_i} \delta_i dt$$
$$+ \frac{\Xi_k}{2} |s_k|^2 dt + z_k \sum_{i=1}^{k-1} \frac{\partial \Xi_k}{\partial z_i} s_k^T s_i dt$$
$$+ \frac{z_k^2}{4} \sum_{i,j \in \{1,...,k-1\}} \frac{\partial^2 \Xi_k}{\partial z_i \partial z_j} s_i^T s_j dt. \qquad (12)$$

From Yang's inequality, it again follows that:

$$\Xi_k z_k \sum_{j=1}^{k} p_{kj}^T \theta_j + \frac{z_k^2}{2} \sum_{i=1}^{k-1} \frac{\partial \Xi_k}{\partial z_i} \sum_{j=1}^{i} p_{ij}^T \theta_j + \frac{\Xi_k}{2} |s_k|^2$$
$$\le \Xi_k z_k^2 \left[\eta_k + \frac{1}{4\Xi_k^2} \sum_{i=1}^{k-1} \sum_{j=1}^{i} \left| \frac{\partial \Xi_k}{\partial z_i} p_{ij} \right|^2 + \psi_{kv} \right]$$
$$+ \Xi_k \psi_{kc} |z_{[k-1]}|^2 \qquad (13)$$

where $\eta_k = \sum_{j=1}^{k} |\theta_j|^2/2 + \sum_{i=1}^{k-1} \sum_{j=1}^{i} |\theta_j|^2/4$ is an unknown parameter. Using (13) in (12) yields:

$$dV_k \le dW_{k-1} + \Xi_k \psi_{kc} |z_{[k-1]}|^2 dt \qquad (14)$$
$$+ \Xi_k z_k [(x_{k+1} + k_k + z_k \eta_k) dt + \sigma_k^T dw]$$

where

$$k_k := r_k + \frac{1}{\Xi_k}\left[\frac{z_k}{2}\sum_{i=1}^{k-1}\frac{\partial \Xi_k}{\partial z_i}(z_{i+1} - \alpha_i + r_i)\right.$$
$$+\frac{z_k}{2}\sum_{i=1}^{k-1}\frac{\partial \Xi_k}{\partial \hat{\eta}_i}\delta_i + \sum_{i=1}^{k-1}\frac{\partial \Xi_k}{\partial z_i}s_k^T s_i$$
$$+\frac{z_k}{4}\sum_{i,j\in\{1,...,k-1\}}\frac{\partial^2 \Xi_k}{\partial z_i \partial z_j}s_i^T s_j$$
$$\left.+\frac{z_k}{4\Xi_k}\sum_{i=1}^{k-1}\sum_{j=1}^{i}|\frac{\partial \Xi_k}{\partial z_i}p_{ij}|^2\right] + \psi_{kv}z_k$$
$$\sigma_k := s_k + \frac{z_k}{2\Xi_k}\sum_{i=1}^{k-1}\frac{\partial \Xi_k}{\partial z_i}s_i.$$

The error term z_{k+1} is defined as: $z_{k+1} := x_{k+1} + \alpha_k(z_{[k]}, \hat{\eta}_{[k]})$, where

$$\alpha_k = k_k + z_k(\hat{\eta}_k + n\theta\Xi_k|\sigma_k|^2/4 + \beta_k/\Xi_k)$$
$$+[(n-k)z_k + \Xi_{k-1}z_{k-1}]/\Xi_k, \quad (15)$$
$$\dot{\hat{\eta}}_k := \delta_k(z_{[k]}, \hat{\eta}_{[k-1]}) := \Xi_k z_k^2, \quad \hat{\eta}_k(0) = \bar{\eta}_k,$$

$\beta_k(z_{[k]}, \hat{\eta}_{[k]}) \geq 0$ is some design function, and $\bar{\eta}_k$ is an arbitrary initial estimate for η_k. (14) can be written in terms of z_{k+1} as:

$$dV_k \leq dW_{k-1} + \Xi_k[z_k z_{k+1} - (n\theta\Xi_k|\sigma_k|^2/4)z_k^2]dt$$
$$+\Xi_k\left[[-(\beta_k/\Xi_k)z_k^2 + z_k^2\tilde{\eta}_k]dt + z_k\sigma_k^T dw\right]$$
$$+[\Xi_k\psi_{kc}|z_{[k-1]}|^2 - (n-k)z_k^2]dt$$
$$-[\Xi_{k-1}z_k z_{k-1}]dt$$

where $\tilde{\eta}_k := \eta_k - \hat{\eta}_k$. Finally, $W_k := V_k + \tilde{\eta}_k^2/2$ is introduced, where its Itô differential satisfies:

$$dW_k \leq -z_1^2 dt + \Xi_k z_k z_{k+1}dt - (n-k)|z_{[k]}|^2 dt$$
$$-\sum_{i=1}^{k}(n\theta\Xi_i^2|\sigma_i|^2/4 + \beta_i)z_i^2 dt$$
$$+\sum_{i=1}^{k}\Xi_i z_i\sigma_i^T dw - \underbrace{(1 - \Xi_k\psi_{kc})}_{\geq 0}|z_{[k-1]}|^2$$

Since all of the relevant definitions and results of Step k are consistent with the induction hypothesis, it is concluded that the induction hypothesis holds true for all $k \in \{1, ..., n-1\}$.

Step n: It is easy to see that the results of Step k holds true also for $k = n$ if x_{n+1} is defined as $x_{n+1} := \bar{b}u$, where u is the actual control input. Thus, z_{n+1} can be made zero , by picking the control input as:

$$u = -\alpha_n(z_{[n]}, \hat{\eta}_{[n]})/\bar{b}(z_{[n]}, \hat{\eta}_{[n-1]}) \quad (16)$$

where α_n is obtained by setting $k = n$ in (15). The control input (16) renders the Itô differential of the smooth function $W = \sum_{i=1}^{n}[\Xi_i z_i^2 + \tilde{\eta}_i^2]/2$, which is positive definite in $(z_{[n]}, \tilde{\eta}_{[n]})$, as:

$$dW \leq -z_1^2 dt - \sum_{i=1}^{n}(n\theta\Xi_i^2|\sigma_i|^2/4)dt \quad (17)$$
$$+\sum_{i=1}^{n}\Xi_i\left[-(\beta_i/\Xi_i)z_i^2 dt + z_i\sigma_i^T dw\right],$$

which completes the last step. This now leads to:

Theorem 1. *Consider the nonlinear system (1). If the design functions satisfy: $\beta_i \geq k_\beta$ for some $k_\beta > 0\ \forall i \in \{1, ..., n\}$, then for any risk-sensitivity parameter $\theta > 0$, the controller (16) achieves:*

1)

$$\frac{2}{\theta}\ln E\exp\frac{\theta}{2}\left[W(T) + \int_0^T[z_1^2 + \sum_{i=1}^{n}\beta_i z_i^2]dt\right]$$
$$\leq W(0),\ \forall T \geq 0.$$

2) The equilibrium point $[z_1, ..., z_n, \tilde{\eta}_1, ..., \tilde{\eta}_n]^T = 0_{2n\times 1}$ of the closed-loop system is globally stable in probability, i.e., $\forall\epsilon > 0$ there exits a class $\mathcal{K}$ function $\bar{\gamma}(.)$ such that $\forall t \geq 0$

$$\mathcal{P}\left\{\left|\begin{array}{c} z_{[n]}(t) \\ \tilde{\eta}_{[n]}(t)\end{array}\right| \leq \bar{\gamma}\left(\left|\begin{array}{c} z_{[n]}(0) \\ \tilde{\eta}_{[n]}(0)\end{array}\right|\right)\right\} \geq 1 - \epsilon.$$

3) $\mathcal{P}\left\{\lim_{t\to\infty}|z_{[n]}(t)| = 0\right\} = 1$.

Proof. From (17), it follows that:

$$W(T) + \int_0^T[z_1^2 + \sum_{i=1}^{n}\beta_i z_i^2]dt$$
$$\leq W(0) + \int_0^T\sigma^T dw - \frac{\theta}{4}\int_0^T|\sigma|^2 dt, \quad \forall T \geq 0$$

where $\sigma = \sum_{i=1}^{n}\Xi_i z_i\sigma_i$. This implies:

$$\frac{2}{\theta}\ln E\exp\frac{\theta}{2}\left[W(T) + \int_0^T[z_1^2 + \sum_{i=1}^{n}\beta_i z_i^2]dt\right]$$
$$\leq W(0) + \frac{2}{\theta}\ln E\exp\frac{\theta}{2}\int_0^T[\sigma^T dw - \frac{\theta}{4}|\sigma|^2 dt],$$

for all $T \geq 0$. Theorem 11 of (Pan and Başar, 1999) shows that the last term on the right-hand-side of the above inequality is nonpositive, which proves the first part. The other parts follow from Theorem 3.2 of (Krstić and Deng, 1998).

4. AN ILLUSTRATIVE EXAMPLE

To verify the results of Section 3, a second order system of the form (1) is considered, where:

$$f_1 = f_2 = 0, \quad \phi_1 = x_1^2, \quad \phi_2 = x_1 x_2$$
$$\theta_1 = 2, \quad \theta_2 = -2, \quad b = 10$$
$$h_1 = [2x_1, 0]^T, \ h_2 = [3x_2, 3x_2(1 + \cos(x_1))]^T.$$

The design functions are picked as: $\beta_1 = 0.5$, and $\beta_2 = 0.001$. To illustrate the effect of the risk-sensitivity parameter, the simulation is performed for two different values of θ: $\theta = 0$ and $\theta = 1.5$. Figures 1, and 2 shows the sample paths of the states and the control action under the same random disturbance for $\theta = 0$ and $\theta = 1.5$, respectively. Clearly, in both cases, the controller asymptotically stabilizes the system. In the risk-sensitive case, i.e., $\theta = 1.5$, the controller becomes a high gain controller and the state variables experience less fluctuations than the ones in the risk-neutral case, i.e., $\theta = 0$. This illustrates the trade-off between the transient performance of the system and the available control action via the risk-sensitivity parameter θ.

5. CONCLUSIONS

In this paper, an adaptive controller design has been presented for uncertain stochastic strict-feedback systems, where the uncertainty is linearly parameterized. The controller design involves the backstepping design methodology, and the use of radially unbounded functions. The resulting controller keeps a meaningful risk-sensitive cost bounded by some positive definite function of

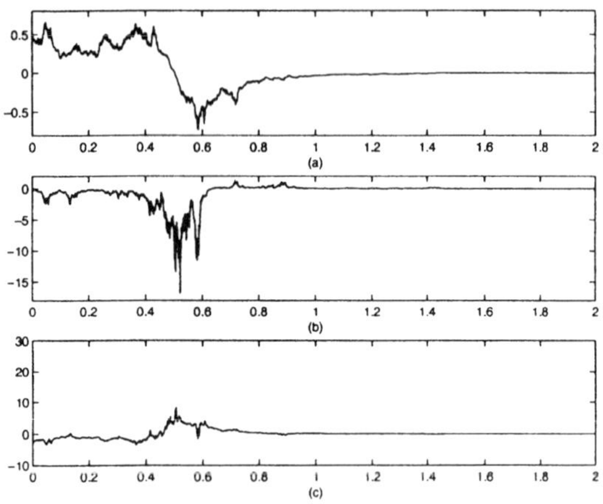

Fig. 1. ($\theta = 0$): (a) $y(t)$ (b) $x_2(t)$ (c) $u(t)$.

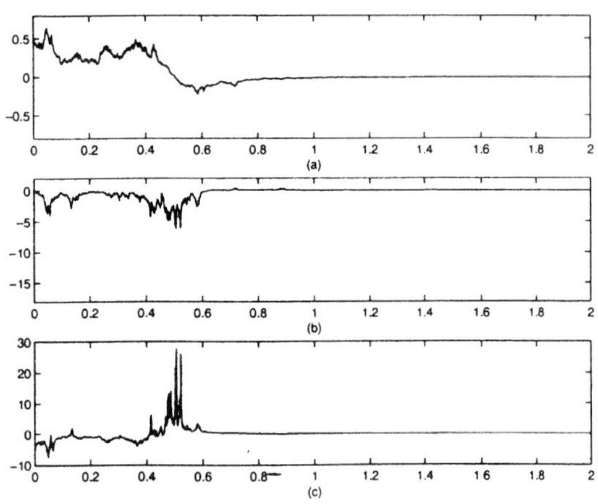

Fig. 2. ($\theta = 1.5$): (a) $y(t)$ (b) $x_2(t)$ (c) $u(t)$.

the initial conditions. Also, the closed-loop signals remain bounded in probability and the output asymptotically converges to zero.

6. REFERENCES

Arslan, G. and T. Başar (1999a). Robust identification for output tracking in systems with unknown nonlinearities. In: *Proc. of the 1999 ACC.* pp. 1057–1061.

Arslan, G. and T. Başar (1999b). Robust output tracking for strict-feedback systems using neural-net based approximators for nonlinearities. In: *Proc. of the 1999 CDC.* pp. 2987–2992.

Didinsky, G. and T. Başar (1995). Parameter identification for uncertain plants using h^∞ methods. *Automatica* **39**, 1227–1250.

Ezal, K., A.R. Teel, P. Kokotovič and T. Başar (1999). Disturbance attenuating output-feedback control of nonlinear systems with local optimality. In: *Proc. of the 1999 ACC.*

Ezal, K., Z. Pan and P. Kokotovič (1997). Locally optimal backstepping design. In: *Proc. of the 1997 CDC.* pp. 1767–1773.

Krstič, M. and H. Deng (1998). *Stabilization of Nonlinear Uncertain Systems.* Springer-Verlag. Berlin.

Krstič, M., I. Kanellakopoulos and P. Kokotovič (1995). *Nonlinear and Adaptive Control Design.* Wiley. New York.

Narendra, K.S. and S. Mukhopadhyay (1997). Adaptive control using neural networks and approximate models. *IEEE Transactions on Neural Networks* **8**, 475–485.

Pan, Z. and T. Başar (1998). Adaptive controller design for tracking and disturbance attenuation in parametric-strict-feedback nonlinear systems. *IEEE Transactions on Automatic Control* **43**, 1066–1083.

Pan, Z. and T. Başar (1999). Backstepping controller design for nonlinear stochastic systems under a risk-sensitive cost criterion. *SIAM Journal on Control & Optimization* **37**, 957–995.

Polycarpou, M.M. and M.J. Mears (1998). Stable adaptive tracking of uncertain systems using nonlinearly parameterized on-line approximators. *International Journal of Control* **70**, 363–384.

Sanner, R.M. and J-J.E. Slotine (1992). Gaussian networks for direct adaptive control. *IEEE Transactions on Neural Networks* **3**, 837–863.

Tezcan, I.E. and T. Başar (1999). Disturbance attenuating adaptive controllers for parametric strict feedback nonlinear systems with output measurements. *Journal of Dynamic Systems, Measurement & Control* **121**, 48–57.